Compendium of Plant Genomes

Series Editor

Chittaranjan Kole, Raja Ramanna Fellow, Government of India, ICAR-National Research Center on Plant Biotechnology, Pusa, New Delhi, India

Whole-genome sequencing is at the cutting edge of life sciences in the new millennium. Since the first genome sequencing of the model plant Arabidopsis thaliana in 2000, whole genomes of about 100 plant species have been sequenced and genome sequences of several other plants are in the pipeline. Research publications on these genome initiatives are scattered on dedicated web sites and in journals with all too brief descriptions. The individual volumes elucidate the background history of the national and international genome initiatives; public and private partners involved; strategies and genomic resources and tools utilized; enumeration on the sequences and their assembly; repetitive sequences; gene annotation and genome duplication. In addition, synteny with other sequences, comparison of gene families and most importantly potential of the genome sequence information for gene pool characterization and genetic improvement of crop plants are described.

Interested in editing a volume on a crop or model plant?
Please contact Prof. C. Kole, Series Editor, at ckoleorg@gmail.com

More information about this series at http://www.springer.com/series/11805

Schuyler S. Korban
Editor

The Apple Genome

Springer

Editor
Schuyler S. Korban
Department of Natural Resources
and Environmental Sciences
University of Illinois
at Urbana–Champaign
Urbana, IL, USA

ISSN 2199-4781 ISSN 2199-479X (electronic)
Compendium of Plant Genomes
ISBN 978-3-030-74681-0 ISBN 978-3-030-74682-7 (eBook)
https://doi.org/10.1007/978-3-030-74682-7

This Springer imprint is published by the registered company Springer Nature Switzerland AG
The registered company address is: Gewerbestrasse 11, 6330 Cham, Switzerland

This book series is dedicated to my wife Phullara and our children Sourav and Devleena

Chittaranjan Kole

Preface to the Series

Genome sequencing has emerged as the leading discipline in the plant sciences coinciding with the start of the new century. For much of the twentieth century, plant geneticists were only successful in delineating putative chromosomal location, function, and changes in genes indirectly through the use of a number of "markers" physically linked to them. These included visible or morphological, cytological, protein, and molecular or DNA markers. Among them, the first DNA marker, the RFLPs, introduced a revolutionary change in plant genetics and breeding in the mid-1980s, mainly because of their infinite number and thus potential to cover maximum chromosomal regions, phenotypic neutrality, absence of epistasis, and codominant nature. An array of other hybridization-based markers, PCR-based markers, and markers based on both facilitated construction of genetic linkage maps, mapping of genes controlling simply inherited traits, and even gene clusters (QTLs) controlling polygenic traits in a large number of model and crop plants. During this period, a number of new mapping populations beyond F2 were utilized and a number of computer programs were developed for map construction, mapping of genes, and for mapping of polygenic clusters or QTLs. Molecular markers were also used in the studies of evolution and phylogenetic relationship, genetic diversity, DNA fingerprinting, and map-based cloning. Markers tightly linked to the genes were used in crop improvement employing the so-called marker-assisted selection. These strategies of molecular genetic mapping and molecular breeding made a spectacular impact during the last one and a half decades of the twentieth century. But still they remained "indirect" approaches for elucidation and utilization of plant genomes since much of the chromosomes remained unknown and the complete chemical depiction of them was yet to be unraveled.

Physical mapping of genomes was the obvious consequence that facilitated the development of the "genomic resources" including BAC and YAC libraries to develop physical maps in some plant genomes. Subsequently, integrated genetic–physical maps were also developed in many plants. This led to the concept of structural genomics. Later on, emphasis was laid on EST and transcriptome analysis to decipher the function of the active gene sequences leading to another concept defined as functional genomics. The advent of techniques of bacteriophage gene and DNA sequencing in the 1970s was extended to facilitate sequencing of these genomic resources in the last decade of the twentieth century.

As expected, sequencing of chromosomal regions would have led to too much data to store, characterize, and utilize with the-then available computer software could handle. But the development of information technology made the life of biologists easier by leading to a swift and sweet marriage of biology and informatics, and a new subject was born—bioinformatics.

Thus, the evolution of the concepts, strategies, and tools of sequencing and bioinformatics reinforced the subject of genomics—structural and functional. Today, genome sequencing has traveled much beyond biology and involves biophysics, biochemistry, and bioinformatics!

Thanks to the efforts of both public and private agencies, genome sequencing strategies are evolving very fast, leading to cheaper, quicker, and automated techniques right from clone-by-clone and whole-genome shotgun approaches to a succession of second-generation sequencing methods. The development of software of different generations facilitated this genome sequencing. At the same time, newer concepts and strategies were emerging to handle sequencing of the complex genomes, particularly the polyploids.

It became a reality to chemically—and so directly—define plant genomes, popularly called whole-genome sequencing or simply genome sequencing.

The history of plant genome sequencing will always cite the sequencing of the genome of the model plant Arabidopsis thaliana in 2000 that was followed by sequencing the genome of the crop and model plant rice in 2002. Since then, the number of sequenced genomes of higher plants has been increasing exponentially, mainly due to the development of cheaper and quicker genomic techniques and, most importantly, the development of collaborative platforms such as national and international consortia involving partners from public and/or private agencies.

As I write this preface for the first volume of the new series "Compendium of Plant Genomes," a net search tells me that complete or nearly complete whole-genome sequencing of 45 crop plants, eight crop and model plants, eight model plants, 15 crop progenitors and relatives, and three basal plants is accomplished, the majority of which are in the public domain. This means that we nowadays know many of our model and crop plants chemically, i.e., directly, and we may depict them and utilize them precisely better than ever. Genome sequencing has covered all groups of crop plants. Hence, information on the precise depiction of plant genomes and the scope of their utilization are growing rapidly every day. However, the information is scattered in research articles and review papers in journals and dedicated Web pages of the consortia and databases. There is no compilation of plant genomes and the opportunity of using the information in sequence-assisted breeding or further genomic studies. This is the underlying rationale for starting this book series, with each volume dedicated to a particular plant.

Plant genome science has emerged as an important subject in academia, and the present compendium of plant genomes will be highly useful to both students and teaching faculties. Most importantly, research scientists involved in genomics research will have access to systematic deliberations on the plant genomes of their interest. Elucidation of plant genomes is of interest not only for the geneticists and breeders, but also for practitioners of an array of plant science disciplines, such as taxonomy, evolution, cytology, physiology, pathology, entomology, nematology, crop production, biochemistry, and obviously bioinformatics. It must be mentioned that information regarding each plant genome is ever-growing. The contents of the volumes of this compendium are, therefore, focusing on the basic aspects of the genomes and their utility. They include information on the academic and/or economic importance of the plants, description of their genomes from a molecular genetic and cytogenetic point of view, and the genomic resources developed. Detailed deliberations focus on the background history of the national and international genome initiatives, public and private partners involved, strategies and genomic resources and tools utilized, enumeration on the sequences and their assembly, repetitive sequences, gene annotation, and genome duplication. In addition, synteny with other sequences, comparison of gene families, and, most importantly, the potential of the genome sequence information for gene pool characterization through genotyping by sequencing (GBS) and genetic improvement of crop plants have been described. As expected, there is a lot of variation of these topics in the volumes based on the information available on the crop, model, or reference plants.

I must confess that as the series editor, it has been a daunting task for me to work on such a huge and broad knowledge base that spans so many diverse plant species. However, pioneering scientists with lifetime experience and expertise on the particular crops did excellent jobs editing the respective volumes. I myself have been a small science worker on plant genomes since the mid-1980s and that provided me the opportunity to personally know several stalwarts of plant genomics from all over the globe. Most, if not all, of the volume editors are my longtime friends and colleagues. It has been highly comfortable and enriching for me to work with them on this book series. To be honest, while working on this series I have been and will remain a student first, a science worker second, and a series editor last. And I must express my gratitude to the volume editors and the chapter authors for providing me the opportunity to work with them on this compendium.

I also wish to mention here my thanks and gratitude to the Springer staff, particularly Dr. Christina Eckey and Dr. Jutta Lindenborn for the earlier set of volumes and presently Ing. Zuzana Bernhart for all their timely help and support.

I always had to set aside additional hours to edit books beside my professional and personal commitments—hours I could and should have given to my wife, Phullara, and our kids, Sourav and Devleena. I must mention that they not only allowed me the freedom to take away those hours from them but also offered their support in the editing job itself. I am really not sure whether my dedication of this compendium to them will suffice to do justice to their sacrifices for the interest of science and the science community.

New Delhi, India Chittaranjan Kole

This volume is dedicated to my wonderful family; my supportive and lovely wife Tamra; and our much loved and bright sons Christian, Charles, and Colin

Schuyler S. Korban

Preface

The cultivated apple (*Malus* × *domestica* Borkh.) is one of the most important tree fruit crops of temperate regions of the world. It is widely cultivated and grown in North America, Europe, and Asia. The apple fruit is a highly desirable fruit due to its flavor, sugar and acid content, metabolites, aroma, as well as its overall texture and palatability. Furthermore, it is a rich source of important nutrients, including antioxidants, vitamins, and dietary fiber. As a result, consumption of apple fruits elicits various health benefits. The leading apple-producing countries include China, USA, Turkey, Italy, Russia, and other countries in Eastern Europe, among others.

The apple belongs to the genus *Malus*, subfamily Maloideae, and family Rosaceae. There are between 25 and 30 *Malus* species. Although there are over 10,000 documented cultivars of apple, commercial cultivation of apple is limited to relatively fewer numbers of cultivars. Thus, the limited gene pool coupled with difficulties associated with traditional tree breeding and genetic improvement efforts have necessitated the development of modern tools of marker-assisted breeding, as well as of other genomic resources and biotechnology tools to overcome these constraints and accelerate genetic improvement efforts. Over the past 25 years, various research efforts have been undertaken to expand and utilize these resources, including various biological and molecular markers, mapping, expressed sequence tags (ESTs), bacterial artificial chromosome (BAC) libraries, genetic transformation, genome sequencing, RNA-Seq, genome-wide association, and comparative genomics, among others, to better investigate and understand the structural, functional, evolutionary pathway of the apple genome, and its domestication.

This volume on *The Apple Genome* covers information on the economics; botany, taxonomy, and origin; germplasm resources; cytogenetics and nuclear DNA; genetic improvement efforts of scion cultivars; genetic and genomic improvement efforts of rootstocks; genetic and physical mapping; genomic resources; genome and epigenome; regulatory sequences; utility of whole genome sequencing and gene editing in trait dissection; flowering and juvenility; cold hardiness and dormancy; fruit color development; fruit acidity and sugar content; metabolomics; biology and genomics of microbiome; apple domestication; and opportunities and challenges for genetic improvement of the apple.

The apple tree is cross-pollinated, self-incompatible, and with a long juvenility period of 5 to 6 years. The apple has a long history of having undergone interspecific hybridization from its center origin in the Shian Tan region in China along its path along the Silk Route, as it moved to Europe and then to North America, and to other parts of the world. Most early commercially grown cultivars have been selected as chance seedlings. However, apple breeding programs, particularly in North America, Europe, Japan, New Zealand, and China, among others, have contributed to the release of cultivars with desirable fruit quality traits, as well as with other desirable horticultural traits that have become important cultivars, grown either in traditional commercial orchards or in new organic orchard farms. Moreover, commercial apple rootstocks have primarily been developed from breeding programs in the United Kingdom, North America, and various European countries, such as Germany, Poland, as well as in Russia, among others. As with other tree fruit breeding programs, classical apple breeding is a long-term and expensive effort. However, recent significant advances in apple genomics have ushered a new era for pursuing major apple genetic improvement initiatives and efforts.

In recent years, modern genetic and genomic tools have contributed to the development of significant valuable resources, including those myriad molecular markers, genetic and physical mapping, genetic transformation, structural and functional genomics, genome sequencing, gene microarrays and chips, metabolomes, genome-wide association studies, comparative genomic studies, as well as gene editing. These tools and resources have provided new knowledge of the apple genome and those genes and quantitative trait loci that control various vegetative and reproductive traits, as well as those controlling various biotic and abiotic stress traits, including those economic traits associated with propagation, response to controlled fruit storage conditions, as well as those involved in the complex microbiome system of the apple.

All collective genetic tools, as well as genomic, transcriptomic, metabolomics, and biome resources, are providing unprecedented opportunities for exploring and unraveling genetic variation, evolution, and domestication of apple, as well as in pursuing more robust and integrated approaches for genetic improvement and enhancement of apple cultivars, as well as of apple rootstocks. This wealth of new resources will have a major impact on our knowledge of the apple genome, along with a better understanding of biological networks and pathways that are of particular importance, and are unique to the apple. In turn, these resources and knowledge will have significant impacts on efforts for genetic improvement of apple and will have significant impact on the apple industry.

The Apple Genome book covers our current knowledge of world economics of the apple industry; botanical and taxonomic classifications; apple genetic diversity and conservation; ploidy and cytogenetics; genetic studies and genetic improvement efforts of apple cultivars; apple rootstock genetic and genomics efforts; genetic and physical linkage maps; whole genome sequencing strategies and exploration of the epigenome; analysis of regulatory sequences; gene editing efforts; genetic and biotechnology analysis of flowering and juvenility; genetic and genomic understanding of dormancy and cold hardiness; genetics and genomic analysis of fruit color development; genomic analysis of fruit acid and sugar contents; metabolomic analysis of fruit quality; fundamental analysis of the apple microbiome; origin of the apple through the domestication process; and potential future prospects for omics in genetic improvement efforts of apples.

All 18 chapters included in this volume provide a broad and a wealth of information and comprehensive overview of the status of early and ongoing efforts to delineate and better understand the genetics, breeding, genomics, and other advances in omics for benefit for the apple. This volume offers knowledge, research efforts, research findings, and delves into opportunities and challenges, as well as approaches and technologies that will support future research and discovery efforts that will not only be of benefit for genetic improvement efforts, but also to the apple industry as a whole.

Urbana, IL, USA Schuyler S. Korban

Contents

Contributors

Ahmed Abdelfattah Department of Ecology, Environment and Plant Sciences, Stockholm University, Stockholm, Sweden;
Institute of Environmental Biotechnology, Graz University of Technology, Graz, Austria

Timothy S. Artlip USDA-ARS-Appalachian Fruit Research Station, Kearneysville, WV, USA

Sandrine Balzergue Univ Angers, Institut Agro, INRAE, IRHS, SFR QUASAV, Angers, France

Luca Bianco Research and Innovation Center, Fondazione Edmund Mach, San Michele all'Adige, Italy

Paula Bramel Global Crop Diversity Trust, Bonn, Germany

Stefano Brizzolara Institute of Life Sciences, Scuola Superiore Sant'Anna, Pisa, Italy

Jean-Marc Celton Univ Angers, Institut Agro, INRAE, IRHS, SFR QUASAV, Angers, France

C. Thomas Chao USDA-ARS Plant Genetic Resources Unit, New York State Agricultural Experiment Station, Geneva, NY, USA

Xuesen Chen State Key Laboratory of Crop Biology, College of Horticulture Science and Engineering, Shangdong Agricultural University, Shangdong, China

Amandine Cornille Université Paris-Saclay, INRAE, CNRS, AgroParisTech, GQE—Le Moulon, Gif-sur-Yvette, France

Samir Droby Department of Postharvest Science, Agricultural Research Organization (ARO), Rishon LeZion, Israel

Charles-Eric Durel Univ Angers, Institut Agro, INRAE, IRHS SFR QUASAV, Angers, France

Kate M. Evans Department of Horticulture, Tree Fruit Research and Extension Center, Washington State University, Wenatchee, WA, USA

Gennaro Fazio Plant Genetic Resources Unit, USDA ARS, Geneva, NY, USA

Shiri Freilich Institute of Plant Sciences, Newe Ya'ar Research Center, the Agricultural Research Organization, Ramat Yishay, Israel

Ben Gutierrez USDA-ARS Plant Genetic Resources Unit, New York State Agricultural Experiment Station, Geneva, NY, USA

Yuepeng Han CAS Key Laboratory of Plant Germplasm Enhancement and Specialty Agriculture, Wuhan Botanical Garden, the Innovative Academy of Seed Design, Chinese Academy of Sciences, Wuhan, China;
Center of Economic Botany, Core Botanical Gardens, Chinese Academy of Sciences, Wuhan, China;
Sino-African Joint Research Center, Chinese Academy of Sciences, Wuhan, China

Yu-Jin Hao National Key Laboratory of Crop Biology, Shangdong Agricultural University, Shangdong, China

Awais Khan Plant Pathology and Plant-Microbe Biology Section, Cornell University, Geneva, NY, USA

Schuyler S. Korban Department of Natural Resources and Environmental Sciences, University of Illinois at Urbana-Champaign, Urbana, IL, USA

Sarah A. Kostick Department of Horticulture, Tree Fruit Research and Extension Center, Washington State University, Wenatchee, WA, USA

Nobuhiro Kotoda Faculty of Agriculture, Saga University, Saga, Japan

Gareth Linsmith Research and Innovation Center, Fondazione Edmund Mach, San Michele all'Adige, Italy;
Earlham Institute, Norwich Research Park, Norwich, UK;
Center for Plant Systems Biology, VIB, Gent, Belgium;
Department of Plant Biotechnology and Bioinformatics, Ghent University, Ghent, Belgium

Jianyang Liu School of Plant and Environmental Sciences, Alson H. Smith Jr. Agricultural Research and Extension Center, Virginia Tech, Winchester, VA, USA

Agnieszka Marasek-Ciołakowska Department of Applied Biology, The National Institute of Horticultural Research, Skierniewice, Poland

Mark Mazzola Department of Plant Pathology, Stellenbosch University, Matieland, South Africa

Chikako Nishitani NARO Institute of Fruit Tree and Tea Science, Tsukuba, Ibaraki, Japan

Desmond O'Rourke Belrose Inc., Pullman, WA, USA

Keishi Osakabe Faculty of Bioscience and Bioindustry, Fujii Memorial Institute of Medical Sciences, Tokushima University, Tokushima, Japan

Yuriko Osakabe School of Life Science and Technology, Tokyo Institute of Technology, Midori-ku, Yokohama, Japan

Małgorzata Podwyszyńska Department of Applied Biology, The National Institute of Horticultural Research, Skierniewice, Poland

Sherif M. Sherif School of Plant and Environmental Sciences, Alson H. Smith Jr. Agricultural Research and Extension Center, Virginia Tech, Winchester, VA, USA

Jugpreet Singh Plant Pathology and Plant-Microbe Biology Section, Cornell University, Geneva, NY, USA

Soon Li Teh Department of Horticulture, Tree Fruit Research and Extension Center, Washington State University, Wenatchee, WA, USA

Leonardo Tenori Interuniversity Consortium for Magnetic Resonance of Metalloproteins (CIRMMP), Sesto Fiorentino, Florence, Italy

Michela Troggio Research and Innovation Center, Fondazione Edmund Mach, San Michele all'Adige, Italy

Gayle M. Volk USDA-ARS National Laboratory for Genetic Resources Preservation, Fort Collins, CO, USA

Nan Wang State Key Laboratory of Crop Biology, College of Horticulture Science and Engineering, Shangdong Agricultural University, Shangdong, China

Xiao-Fei Wang National Key Laboratory of Crop Biology, Shangdong Agricultural University, Shangdong, China

Susan R. Whitehead Department of Biological Sciences, Virginia Tech, Blacksburg, VA, USA

Michael E. Wisniewski Department of Biological Sciences, Virginia Tech, Blacksburg, VA, USA

Bo Zhang CAS Key Laboratory of Plant Germplasm Enhancement and Specialty Agriculture, Wuhan Botanical Garden, the Innovative Academy of Seed Design, Chinese Academy of Sciences, Wuhan, China

Abbreviations

(CBFs)	C-repeat binding transcription factor(s)
4CL	4-Coumarate: Coenzyme A ligase
ABA	Abscisic Acid
ACO1	1-aminocyclopropane-1-carboxylic acid (ACC) oxidase 1
ACP	Anaerobic compensation point
ACS1	1-aminocyclopropane-1-carboxylic acid (ACC) synthase 1
AFLP(s)	Amplified fragment length polymorphism(s)
ALMT	Aluminum-activated malate transporter-like
ANS	Anthocyanidin synthase
AP2/ERF	APETALA2/ethylene response factor
ARD	Apple replant disease
ASVs	Amplicon sequence variants
BAC	Bacterial artificial chromosome
bHLH	Basic helix-loop-helix
BTB	Bric-à-brac, tramtrack and broad complex
CA	Controlled atmosphere
CAP(s)	Cleaved amplified polymorphic sequence(s)
CF	Chlorophyll fluorescence
CHI	Chalcone isomerase
CHS	Chalcone synthase
CK	Cytokinin
CNVR(s)	Copy number variation(s)
CR(s)	Chilling requirement(s)
CRISPR/cas9	Clusters of regularly interspaced short palindromic repeats/ cas9-associated protein
CRT	C-repeat
CU(s)	Chilling unit(s)
DA	Deacclimation
DAFB	Days after full bloom
DAM	Dormancy-associated MADS
DCA	Dynamic controlled atmosphere
DEGs	Differentially expressed genes
DFR	Dihydroflavonol-4-reductase
DMRs	Differentially methylated regions
Dof	DNA-binding one zinc finger

DRE	Dehydration Response Element
DREB1	DRE Binding1
DSB(s)	DNA double-strand break(s)
ERF(s)	Ethylene response factor(s)
ESTs	Expressed sequence tags
F3H	Flavonoid 3-hydroxylase
FBS	Fruit bearing shoots
FISH	Fluorescence in situ hybridization
FLP	Flippase gene
FRT	Flippase recognition target site
G × E	Genotype × Environment
GA	Gibberellins/gibberellic acid
GABA	Gamma-aminobutyric acid
GBS	Genotyping-by-sequencing
GC	Gas chromatography
gRNA	Guide RNA
GS	Genomic selection
GT	Gene targeting
GWAS	Genome-wide association study(ies)
HDR	Homology-directed repair
HR	Heat requirement
ILOS	Initial low O_2 stress
JA	Jasmonic acid
LBD	Lateral organ boundaries domain
LC	Liquid chromatography
LD	Linkage disequilibrium
LG(s)	Linkage group(s)
LTR	Long terminal repeat
MAB	Marker-assisted breeding
MAPS	Marker-assisted parent selection
MAS	Marker-assisted selection
MASS	Marker-assisted seedling selection
mQTLs	Metabolite quantitative trait loci
MS	Mass spectroscopy
MYB	MYB (my elob lastosis)
NAC	NAM/ATAF1,2/CUC2
NBS	Nucleotide-binding site
NGS	Next-generation sequencing
NHEJ	Non-homologous end joining
NMR	Nuclear magnetic resonance
OTUs	Operational taxonomic units
PA	Proanthocyanidin
PAHs	Polycyclic aromatic hydrocarbons
PAL	Phenylalanine ammonia lyase
PAMs	Protospacer adjacent motifs
PBDEs	Polybrominated diphenyl ethers

PCB	Polychlorinated biphenyls
PD	Plasmodesmata
PEG	Polyethylene glycol
QqQ	Triple quadrupoles
QTL(s)	Quantitative trait locus/loci
RA	Reacclimation
RAPD(s)	Random amplified fragment length polymorphism(s)
RNAi	RNA interference
RNP	Ribonucleoprotein
SCAR(s)	Sequence characterized amplified region(s)
SD(s)	Short day(s)
SNP(s)	Single nucleotide polymorphism(s)
SS	Succulent shoots
SSC	Soluble solids content
SSR(s)	Simple sequence repeat(s)
STS(s)	Sequence tag site(s)
TA	Titratable acidity
TALEN	Transcription activator-like effector nuclease
TCP	Teosinte branched/cycloidea/proliferating cell factor
TE(s)	Transposable element(s)
TF(s)	Transcription(al) factor(s)
UFGT	UDP-Glc:flavonoid-3-O-glucosyltransferase
ULO	Ultra low oxygen
VOC(s)	Volatile organic compound(s)
WGD(s)	Whole genome duplication(s)
WGS(s)	Whole genome sequence(s)
WUE	Water use efficiency
ZFN	Zinc finger nuclease

1 Economic Importance of the World Apple Industry

Desmond O'Rourke

Abstract

The apple industry is a global economic powerhouse whose tentacles reach into every aspect of the world economy. Therefore, it is important to explore these inherent links and assess objective measures of the industry's relative economic importance. Among major fruits, the apple is second in volume to the banana. However, it far outdistances the banana in total revenue and revenue per hectare. While major apple exporting countries have remained relatively stable over time, imports have grown more rapidly in many developing countries than in traditional markets. Understanding the apple system is complicated due to the fact that vestiges of the industry's agrarian past and legacy distribution systems co-exist with the most progressive segments of the industry. Thus, it is critical to consider those major factors that underlie the industry's evolution, including land, capital, labor, water, and purchased inputs. The latter have become crucial to advances in the industry's productivity. Governments around the world have played changing roles in the industry, as arbiters of grades and standards, sources of capital, providers of emergency relief, and, for many years, central planners for the entire national industry in communist countries. The apple product is also changing from relatively stable open access commodities to exclusive club cultivars with unique names, brands, and labels. Fresh apple marketers now employ marketing services long used by major consumer packaged goods manufacturers. Extrinsic factors, such as the environmental credentials of the producer or marketer, have become as important as intrinsic factors such as fruit size and color. The firms, districts, and countries that can most rapidly adapt to future changes will become increasingly important to the industry. Those who fail to adapt will continue to face shrinking opportunities.

1.1 Evolution of the World Commercial Apple Industry

The apple is one of the world's oldest permanent crops. For millennia, small farms have reserved plots of land for apple trees intended solely for household use in either fresh or preserved form. Over time, commercial orchards have been established near towns so that their products could be sold in local markets. Moreover, the choice of cultivars for planting has been based on tradition as well as on trial and error. Only primitive methods have been available for storing apples much beyond harvest. Sorting, packing,

D. O'Rourke (✉)
Belrose Inc., Pullman, WA 99163, USA
e-mail: belrose@pullman.com

S. S. Korban (ed.), *The Apple Genome*, Compendium of Plant Genomes,
https://doi.org/10.1007/978-3-030-74682-7_1

and marketing have been left to the individual farmer, and to his or her family. At the time, transactions have been primarily direct between the producer and the final consumer.

From that humble beginning, the apple industry has become a global economic powerhouse whose tentacles reach into many aspects of the wider world economy. This chapter briefly surveys how the world apple industry is integrated into the world economy and attempts to provide some objective measures of the economic importance of this industry.

Ferree and Warrington (2013) noted that "Remains of apples have been reported in historic sites dating to 6500 B.C. Long-distance trade routes between the Mediterranean area and various areas of Asia developed as early as 3500 B.C. and fostered the spread of both fresh and processed apples." With the industrial revolution, advances in transportation and storage technology allowed fresh apples to be sold commercially over longer time periods, and over greater distances. By the beginning of the twentieth century, fresh apples could be shipped from distant colonies to the mother country. For example, Australia, New Zealand, and South Africa shipped off-season supplies of fresh apples to Great Britain. Most long-distance supplies were initially sold at auction markets that had developed near either major ports or railroad terminals in the receiving country. Wholesalers, brokers, and jobbers bought at auction markets and redistributed apples to homes, retail stores, and restaurants.

As the volume of commercial apples increased, and the gap in time and distance between producers and consumers became wider, it was necessary to introduce grades and standards to facilitate transactions. Suppliers were forced to become more discriminating in segregating products of different quality to meet the demands of their distant customers. As different suppliers had established different standards, for much of the twentieth century, it was agreed to by both buyers and sellers that only a government agency could be trusted to develop, test, and guarantee consistent grades and standards. Thus, government agencies became an integral part of both domestic and international fruit marketing systems by establishing minimum standards for products allowed for either trade or importation. The Great Depression of the 1930s brought many apple producers close to bankruptcy and triggered various government initiatives to support domestic producers via domestic subsidies, tariffs, and quotas to reduce imports. Following World War II, many countries in both Eastern Europe and Asia adopted centrally planned economic systems. Government intervention became omnipresent in the collectivization of orchard ownership, price setting, product distribution, and control of imports. The system gradually fell apart following the collapse of central planning in the 1990s.

Following the end of World War II, along with the expansion of self-service supermarkets, retail grocery chains became increasingly important buyers of most agricultural production, and the role of auction markets in apple sales was greatly reduced. Large retailers increasingly bought directly from large suppliers around the world and began to apply their own standards for purchasing apples. Government grades and standards were deemed as minimum acceptable standards in domestic commerce, but governments retained important roles in international trade.

Under pressure from various activist groups, many retailers began to require suppliers to meet standards developed by private agencies, such as GlobalGap and the Safe Quality Food (SQF) that set guidelines for safe chemical use, environmental care, worker treatment, and wildlife protection, among others. To buttress their reputation for social responsibility, individual retailers often added their own more stringent standards.

Under pressure from large retailers, the apple industry in developed countries has evolved into large integrated grower–packer–marketer organizations. This development would have been predicted by Ronald Coase's theory of the firm (Coase 1937). He has argued that firms with multiple functions are a response to the high cost of using markets. Integrated grower–packer–marketers avoid costs of numerous transactions between growers and packers, as well as those between packers and marketers, therein providing the integrated firm with greater control over their entire system. Yet, separate markets for growers, packers, and marketers persist alongside integrated formats.

Understanding the economics of the world apple industry is complex due to many vestiges of its agrarian past along with its legacy distribution systems that currently co-exist with the most advanced sectors of the industry. There are small direct marketers and large integrated grower–packer–marketers. There are massive retailers and dwindling auction markets. There are government grades and standards along with more stringent private sector standards. There are firms that are constantly seeking innovation as well as those that are content with the status quo. As market competition has increased, traditional agrarian practices have been increasingly replaced by practices based on the latest science and technology, although agrarian practices also remain. For these reasons, it can be risky to generalize about many aspects of the world apple industry.

This chapter will focus primarily on the most dynamic and innovative segment of the world apple industry. This progressive segment is gradually replacing the more traditional sectors in the industry. It increasingly applies science and technology to every aspect of its production, packing, and marketing activities.

1.1.1 Economic Factors Impacting Individual Firms

A number of economic factors of production have been critical to the success of individual apple firms. These include land, capital, labor, water, and purchased inputs. Each of these factors is discussed in detail in the following sections.

1.1.2 Land, a Building Block for an Industry

As noted above, land for apple orchards was often a corner lot on a small farm that produced apples for household use. Commercial orchards initially used land adjacent to markets in towns and cities. Advances in storage and transportation allowed apple orchards to be established at longer distances from markets. Commercial orchards flourished in localities possessing comparative advantages in apple production. However, most orchards remained confined to traditional growing areas in rural settings.

The development of new apple cultivars with higher levels of tolerance for different environmental conditions, increased availability of irrigation systems, and improvements in transportation and communication have allowed for larger apple firms to extend production to many different localities. Today, production units under the same ownership are located in different states, countries, or even continents. Firms have wide discretion in where they will locate their apple production.

Another significant development in land use in apple production has been the increased density of plantings. Early experiments on increased density plantings have been conducted in Belgium and in the Netherlands due to scarcity and cost of land. It has been observed that higher numbers of small-sized trees, using dwarfing rootstocks along with posts and wires for support, could be planted per hectare. Moreover, other benefits became rapidly apparent, including earlier cropping and higher yields of fruit, as well as ease of harvesting. Furthermore, average costs of production dropped as fixed costs could be spread over a larger volume of production. The International Dwarf Fruit Tree Association, now known as the International Fruit Tree Association, used conferences and study tours to spread the message around the world. Producers in different countries learned from each other's experiences. Research studies to identify best-adapted and optimum rootstocks, tree spacing, and tree architecture for intensive plantings remain in progress.

Wherein traditional orchards often have had only 300 trees per hectare, most new plantings now have 3,000 or more trees per hectare. This has had two major effects. Production has increasingly moved to land that is most productive under intensive planting systems.

Furthermore, fewer hectares are now required to generate the same volume of production.

1.1.3 Capital, the Engine of Progressiveness

Access to capital has been critical to loosening of geographical limitations on apple firms, and to the application of progressive techniques and systems in the apple industry. Capital enabled apple firms to acquire land, access outside services, apply new technologies, and expand operations at every level to gain economies of scale. Access to capital has not been uniform across the apple world. Many firms in many countries remain handicapped by their inability to access adequate capital.

Capital can be gained and/or sought from a number of sources, including retained earnings, commercial lenders, investment companies, and governments. The first source is internal capital, generated from retained earnings. In general, the best-performing apple firms have been able to generate earnings over and above those needed to compensate for capital, labor, and entrepreneurship invested in the business. These retained earnings allow the producer to either expand operations or further invest in enhancing the business. Integrated grower–packer–shippers have a clear advantage over producer firms in generating surplus earnings as their revenue from packing and marketing tends to be more stable than that from orchard operations, and so they tend to be among the best-performing firms. Middle performing firms often operate in an environment where the firm can be kept afloat, but there is no surplus for expansion. In contrast, the worst-performing firms frequently use up their available capital for operator living expenses and other survival needs. A temporary downturn in prices can put them out of business.

A second major source of capital is from commercial lenders. As these lenders require collateral, they are more likely to lend to the best-performing firms, and least likely to lend to struggling firms. They also tend to be inconsistent suppliers of capital, favoring lending to the apple industry when apple prices are high, and abandoning the industry when apple prices are low, and when external financing is most needed.

The third source of capital has been long-term investments by insurance companies, pension systems, and other organizations that consider permanent crops as an important part of a diversified portfolio. However, because of their scale, these entities tend to invest only in large operations.

A final important source of capital has been various levels of government, including local, state, national, and international entities. International bodies, such as the International Monetary Fund or the World Bank, have funded orchard development in selected countries as part of efforts to stimulate emerging or transition economies. Many individual countries have provided capital to growers for replanting schemes to help modernize orchards. Within the European Union, the costs of such schemes have often been shared between European, national, and district governments. However, the availability of capital from governments tends to be quite erratic, and subject to disruption from overall budget pressures. For example, many apple-producing countries in Eastern Europe that joined the European Union in 2004 hoped to receive large injections of capital from EU funds. However, budget constraints and the Great Recession of 2008 greatly limited available aid.

1.1.4 Labor, the Great Vulnerability

While the total pool of capital for which individual apple firms can compete is very large, the pool of labor for which they compete is limited, and under constant threat. The demand for orchard labor tends to peak during early-season pruning and during harvest. A century ago, in rural areas, families were large, and there was much underemployed labor that could be utilized at these peak times. However, declines in family size as well as the greater opportunities for outmigration from rural areas since World War II have reduced that surplus labor.

In many major apple-producing countries, seasonal labor needs in orchards have been met by migrant workers who have moved temporarily across state or international borders, and then returned back to their home base when their labor is no longer needed. However, this stream of migrants has dwindled, and tighter immigration controls have made voluntary movement more difficult. Potential farm workers increasingly opt for full-time employment in industries that provide health and unemployment insurance, along with other fringe benefits.

More countries have reverted to organized labor schemes to bring in teams of orchard workers for limited periods, and for predetermined activities. Participating apple firms must meet minimum standards for transporting, housing, and paying these workers. Although these schemes have continued to expand, spot shortages of orchard workers have become more widespread, and now threaten the viability of many orchards.

For many years, the apple industry has been exploring the use of mechanical harvesting to substitute for labor at harvest time. This is essentially a replacement of capital for labor. While successful machine harvesting has remained elusive, labor productivity has been improved by the use of innovations, such as harvest platforms. Breakthroughs in machine harvesting could occur at any time. However, as noted previously, many apple firms do not have access to the capital needed to afford automated harvest technology. However, without harvest labor or effective mechanical replacements, the future of many apple orchards will remain in jeopardy.

1.1.5 Water, the Life Blood of the Industry

Most traditional apple orchards depended on natural rainfall for their water needs. However, the timing and amount of water available were rarely matched to the needs of the maturing fruit. Water could also be provided at the local level from free-flowing streams and wells. After rivers were dammed to funnel water to public irrigation schemes, apple production became feasible in desert areas. Water would be provided at appropriate times and in precise amounts needed to optimize fruit development.

Rill (shallow channel) irrigation relies on gravity to deliver water as needed. Overhead sprinklers can be used if power is available. As water supplies have become more limited and expensive, many growers have turned to drip irrigation, which can deliver precise amounts of water to each individual tree. Each irrigation method has strengths and weaknesses in terms of delivering optimal moisture for fruit production. For example, overhead irrigation is often frequently used for frost control and/or for fertigation (delivering fertilizers or chemicals). Studies are still ongoing for optimizing water delivery. In recent years, drip irrigation has become a common practice in modern orchards in Europe, New Zealand, Australia, Israel, and the USA, among others.

In general, under irrigated systems, yields are higher and fruit quality is more consistent than under natural rainfall conditions. Even in rainy climates, many progressive growers have introduced supplementary irrigation to make up for water shortfalls. However, this cannot easily compensate for periods of excessive rainfall.

Public irrigation schemes provide distinct advantages to participating producers. However, only in rare cases do producers pay the full cost of these systems. Competitors deem this as an unfair subsidy. In addition, damming of rivers has become controversial due to harmful effects on fisheries, wildlife, and the environment. New public irrigation schemes have become exceedingly rare in developed countries.

1.1.6 Purchased Inputs, a Key to Progress

While land, capital, labor, and water are essential ingredients for any apple enterprise, access to and exploitation of purchased inputs have allowed apple firms to set and achieve unprecedented improvements in productivity. In the past, most purchased inputs available to apple firms have required little intellectual effort on the part of the operator. For example, fuel to power a tractor, fertilizer to improve the soil, or broad-spectrum chemicals to control major pests and diseases could be used following simple manufacturer instructions.

However, in the last two decades, the nature, diversity, and quality of purchased inputs have exploded. Scientific advances, both within agriculture and in many fields often deemed unrelated to agriculture, have resulted in technologies that can be utilized in the apple industry. However, for such technological advances to be effectively adopted, they often require the services of expert employees, and/or of substantial intellectual effort.

Among the categories of purchased inputs are the following:

- Engineered products, such as tree support systems, sprinkler systems, wind machines, drip irrigation systems, and protective canopies.
- Plant materials, including rootstocks, scionwood, grafted trees, grass seed, and windbreaks, among others.
- Chemical materials that promote tree and fruit growth as well as control pathogens and pests.
- Biological materials such as beneficial insects and honeybees.
- Transportation equipment such as tractors, straddle trucks, and refrigerated containers.
- Electronic devices, such as computers, cameras, smartphones, satellite systems, and unmanned aerial vehicles (drones), often interconnected to monitor, interpret, and manage activities in the orchard, storage facilities, packing houses, or processing facilities.
- A multitude of other inputs for measuring, monitoring, and control devices in which genetic engineering, electronic, or other technologies are already embedded.

These purchased inputs have helped growers, packers, and processors better understand and manage the beneficial effects of soil, water, light, and temperature, as well as minimize the adverse effects of pests, diseases, and other deterrents of successful fruit production, marketing, and distribution.

All these purchased inputs extend the economic footprint of the apple industry. Collectively, they are both a cost to the industry and a tool for improving productivity and quality.

1.1.7 Role of Storage in Apple Marketing

Advances in storage techniques have transformed apple markets by allowing quality products to become available for many months after harvest. Temperature control has long been recognized as prolonging the shelf-life of apple fruits. China has used caves, with their natural low temperatures, to prolong storage life. For a century, packing houses elsewhere have used refrigerated storage, powered by electricity to store apples at low temperatures. The next major advance in storage techniques is the development of controlled atmosphere (CA) storage, wherein the oxygen concentration of air (21%) is reduced from 21% down to 1–3% by flushing the storage room with nitrogen gas, while carbon dioxide levels are maintained at 3% or even lower, and moisture levels are manipulated in sealed chambers to slow fruit respiration. For some cultivars, such as 'Red Delicious', standard controlled atmosphere conditions have kept fruit saleable for the entire marketing season. However, modified storage protocols have been gradually developed to extend the shelf-life of other mainstream apple cultivars. By the 1980s, CA storage has become a major tool for packers and marketers to prolong the shelf-life of apples around the world.

Another major technological advance in fruit storage has come with the introduction of SmartFresh. The active ingredient in SmartFresh is 1-methylcyclopropene (1-MCP), which is injected into sealed storage rooms. This technology is undergoing some modifications to improve its effectiveness for current and newly developed apple cultivars.

1.2 The Apple Product is Changing

The revenue side of the apple industry is also diverse and changing. Many apple-producing countries still rely heavily on traditional apple cultivars to supply both fresh and processing markets. Processors, retailers, and consumers have been content to buy familiar products.

However, the main criticism of familiar products is that they provide few opportunities for both enhancing prices and increasing firm revenues. In contrast, since the 1980s, progressive operators have been able to gain price premiums for a succession of new apple cultivars, such as 'Fuji' and 'Gala', or for socially desirable production methods, such as integrated, sustainable, or organic production.

When production increases of public cultivars such as 'Fuji' and 'Gala' eroded price premiums, progressive firms turned increasingly to club cultivars. Although some breeders still sell production and marketing rights for their new cultivars to all comers, most now seek to gain long-term control through the club system. The underlying principle of clubs was "managed scarcity", keeping the supply of the new cultivars below potential demand in order to maintain price premiums. In an effort to maintain quality, club sponsors attempted to limit plantings to environments deemed most favorable for a particular new cultivar. Producers had to compete to gain entry into these clubs, thus paying substantial licensing and marketing fees to retain membership. In turn, these clubs were expected to provide marketing and promotional services to ensure a price premium high enough to more than offset membership costs.

In the early years of club development, many producing areas and individual producers were excluded from club membership. In response, state or regional breeding programs were encouraged to hasten the development of new apple cultivars suitable for even disadvantaged localities. This greatly expanded opportunities for producers to participate in either one or more clubs.

1.2.1 How Many Cultivar Clubs are Too Many?

Early clubs were welcomed by both retailers and consumers. In response, many additional clubs were set up to promote new cultivars. Consumers welcomed novel tastes, and retailers were able to increase gross profit margins per unit sold.

However, most retailers have only finite shelf space that they can allocate to apples, and consumers in developing countries have not been willing to increase their overall consumption of fresh apples. The net effect of the expansion of such clubs has been the gradual displacement of mainstream cultivars. The shelf space for cultivars such as 'Red Delicious', 'Golden Delicious', 'Braeburn', 'Jonagold', and 'Cameo' has already eroded. 'Gala' remains particularly vulnerable to displacement.

As more marketers attempt to sell larger volumes of their club cultivars to the same retailers, their collective ability to manage controlled scarcity will be eroded. Inevitably, this will lower the prices of club cultivars. It is still too early to determine which clubs or how many clubs will be able to maintain an acceptable premium to sustain the format.

1.2.2 Nurseries Face Challenges

Traditionally, apple growers propagated their own trees and made replacements very slowly over time. However, as the number of commercial growers expanded, specialized nurseries developed to provide uniform plant materials.

After World War II, the average life span of an apple tree was about 50 years. This meant that only about 2% of trees were replanted each year. Most replants were of the same mainstream cultivars. Over time, nurseries competed in providing improved strains of mainstream cultivars that promised to provide higher yields, compact tree growth habits, enhanced disease resistance, superior color, or longer storage life.

The introduction of new cultivars, such as 'Granny Smith', 'Fuji', 'Gala', and 'Braeburn' in the 1970s and 1980s, has prompted nurseries to become more adept at predicting the future demand for different cultivars and numerous strains. The proliferation of club cultivars has raised new concerns. As most club cultivars are non-public, nurseries have to compete for the rights to propagate new cultivar releases. In many cases, selected nurseries have become key partners with the initial sponsors of club cultivars and have had to place some of their capital at risk in the roll-out and management of club cultivars and/or of new cultivar releases. As with marketing operations, nurseries have navigated to search the apple world for access to new cultivars that offer the best potential for future growth.

Nurseries have to keep abreast of the latest developments in apple breeding programs and new genetic tools/technologies, both for the release of new apple cultivars, and/or for testing of the true genetic identity of these new cultivars or strain releases to deal with any claim disputes.

1.2.3 Painting the 'Lady'

Promotional campaigns have been used for decades by the apple industry. However, most promotions were aimed at either wholesalers or retailers, and not at the consumer. For example, many traditional labels sought to distinguish themselves with depictions of children, rabbits, swans, fish, wildlife, and other images with little direct connections with the fruit.

Modern marketers are increasingly trying to differentiate club cultivars they sponsor by using promotional tools that packaged goods manufacturers have been using for decades. These include distinctive image-building, creating the cultivar's "story", branding, packaging, and promotion that reinforce a memorable marketing message.

The most recognizable pioneer of this approach was the global 'Pink Lady' cultivar program, managed in Australia and based on the 'Cripps Pink' cultivar. The images, logos, packaging, public relations, and promotion efforts were coordinated and based on the pinkish coloration of this apple, and they were geared to appeal to young adult upwardly mobile females. The 'Jazz' apple from New Zealand drew from the slightly daring and exciting image of the musical jazz scene.

More recent new cultivars, such as 'Ambrosia' from Canada and 'Kanzi' from Belgium, have capitalized on favorable cultural associations. 'Ambrosia' is the "nectar of the gods", while 'Kanzi' means "treasure" in Swahili.

While intrinsic properties of new cultivars, such as color, taste, and texture, are important, it is those extrinsic (created) properties, such as a product's association with status, social responsibility, affluence, or indulgence that will become increasingly important in the battle for the apple consumer's attention. Thus, marketing services that have long been used in the packaged goods industry will become a vital element in the front lines of the apple production industry. However, just as in the production sector, the level of sophistication and progressiveness in apple marketing will vary across the board. Those that can use marketing services most effectively will have a distinct advantage in competition. The organizations with the strongest marketing arms will be the biggest winners.

The success of 'Honeycrisp' has proved to be an exception. Bred under Minnesota conditions, and sold as an open-access cultivar, it has presented many production challenges in other microclimates. However, 'Honeycrsip' has become popular with North American consumers, and by 2017, it is the sixth-largest apple in volume produced in both the USA and Canada. It has not had similar success in Europe where it must compete with a different mix of popular apple cultivars.

Another development driven by changes in consumer tastes and preferences has been the growth in popularity of apples produced using organic methods. In the USA, organic apples continue to sell at retail and at a premium of over 30%. Due to relatively fixed marketing costs, the premium at the shipping point often approaches 100%. It appears that organic consumers erroneously assume that organic apples are chemical-free. Retailers are happy to indulge such consumers, and prices are likely to remain relatively high as organic production is not economic except in desert climates with low pest pressures. Producers in such areas will continue to expand organic production until the price advantage of organic over conventional production is eroded.

1.3 A Global Industry

The apple industry has become a truly global industry. While the industry in many countries has been initially based on plant materials developed in Europe and then has spread over to distant colonies, there is little direct contact among operating industries in different continents. However, the success of Southern Hemisphere suppliers in Northern Hemisphere markets with 'Granny Smith', 'Fuji', 'Gala', and 'Braeburn' has led many Northern Hemisphere suppliers to plant the same cultivars and to attempt to learn from experiences of their southern rivals.

In turn, due to cost pressures, producers, packers, and marketers in both Europe and North America took the lead in introducing new production practices, such as high-density plantings, as well as innovations in packing, such as computer-controlled sorting systems, and cold storage, such as CA storage. Suppliers in Asia and in the Southern Hemisphere came to observe and learn. Many different forms of contact increased, from private visits to organized tours, such as those conducted by the International Fruit Tree Association, to international nursery associations, and to young people interning in other countries. With instant communication, innovations in any country could now rapidly be transmitted around the world in print and online.

Much of the apple production in Southern Hemisphere countries, such as Chile and New Zealand, is geared primarily towards export markets. However, production has now vastly grown among major Northern Hemisphere producers, including France, Italy, Poland, the USA, and China thereby rendering continued expansion of exports crucial to the viability of their apple industries. In recent years, about 40% of total apple production in France, Italy, and Poland is exported. However, to meet retailer demands for year-round supplies, marketers of established cultivars, such as 'Gala' and 'Fuji', have imported off-season supplies from other hemispheres. Many apple club organizations have set up permanent global arrangements, wherein planting, production, and marketing quotas are preset for each country, while packing, storage, and marketing rights are allocated for different territories and for different seasons. This allows for supplementary supplies to be imported from other franchised territories. As in other elements of the apple industry, many operators continue to think only in local terms, while the more progressive firms are increasingly thinking globally.

1.4 Quantitative Estimates of the Economics of the World Apple Industry

Quantitative estimates of the value of the world apple industry should be handled with caution for several reasons. Often data become available several years after the fact, so it is difficult to gauge current readings of the industry's worth. For the sake of comparability, most global data are presented in terms of US dollars. However, changes in exchange rates between the US dollar and national currencies can distort trends in the worth of specific national industries, and periods of rapid inflation can make comparisons between time periods more precarious.

The following four charts present gross data for the world apple industry provided by the United Nations FAOSTAT website. These figure compare data for 2007 with those for the latest period covered, 2017, for the world's three largest fruits, including apples, bananas, and oranges. In 2007, the area harvested for apples is slightly less than that for bananas, and more than 22% greater than that for oranges (Fig. 1.1). By 2017, the banana area has increased by 12%, while that for the apple area has increased by only 2%, and the orange area has decreased by over 2%.

World production of apples was very similar to that of oranges in 2007, but 30% below that for bananas (Fig. 1.2). Between 2007 and 2017,

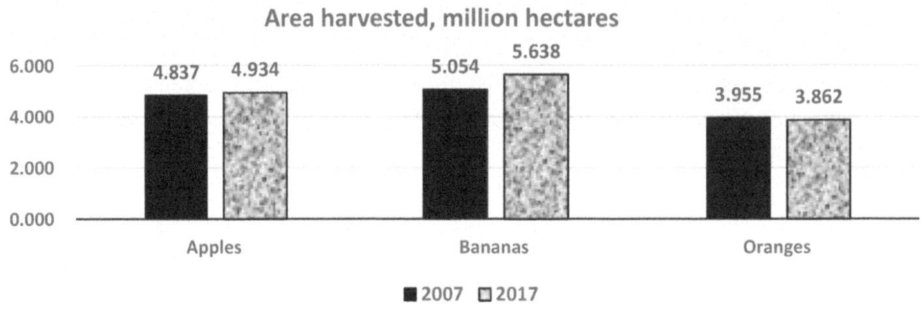

Fig. 1.1 World area harvested of apples, bananas, and oranges between 2007 and 2017 (million hectares). *Source* United Nations, Food and Agriculture Organization, FAOSTAT database (2019). Accessed September 10, 2019

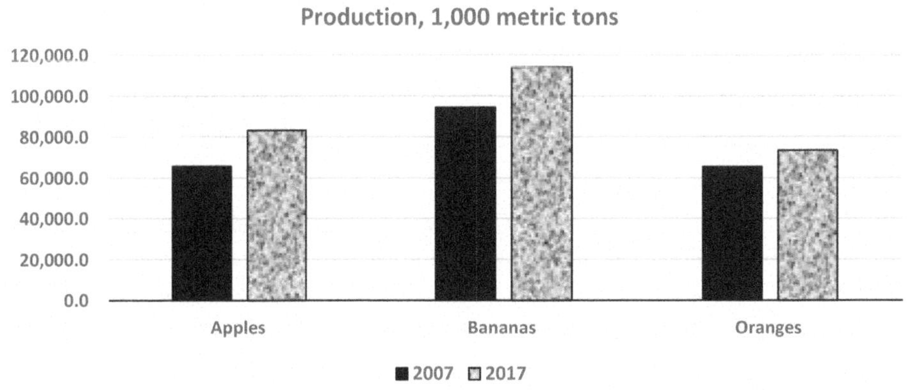

Fig. 1.2 World production of apples, bananas, and oranges in 2007 and 2017 (1,000 metric tons). *Source* United Nations, Food and Agriculture Organization, FAOSTAT database (2019). Accessed September 10, 2019

banana production grew by 20.7%, and apple production by 26.6%; whereas, production of oranges grew by only 12.2%.

Average yields of apple, banana, and orange plantings offer misleading comparisons, as the value per ton varies widely. To take into account, these different unit values, FAOSTAT also reports on the world gross production value for each fruit. These data are reported in constant 2004–2006 US dollars. In 2006, the world gross production value of apples is almost $27 billion, while that for banana production is 6.7% less and that for orange is more than 50% less (Fig. 1.3). In 2016, the latest year for which data are available, the world value of apple production is 41.2% higher than that in 2006, that of bananas is 27.8% higher, but that of oranges is only 11.9% higher.

A more accurate measure of the comparative value of a bearing hectare of apples compared to a hectare of either bananas or oranges is provided (Fig. 1.4). In 2004, a one-hectare of apples has yielded $5,288. In contrast, a hectare of bananas has yielded 6.4% less, and a hectare of oranges has yielded 37.1% less. Between 2006 and 2016, this gap has widened even further as the apple value has grown by 38.3%, compared to growths of 19.9% for bananas and only 7.5% for oranges. Apples continue to outperform the two other major fruits in economic value.

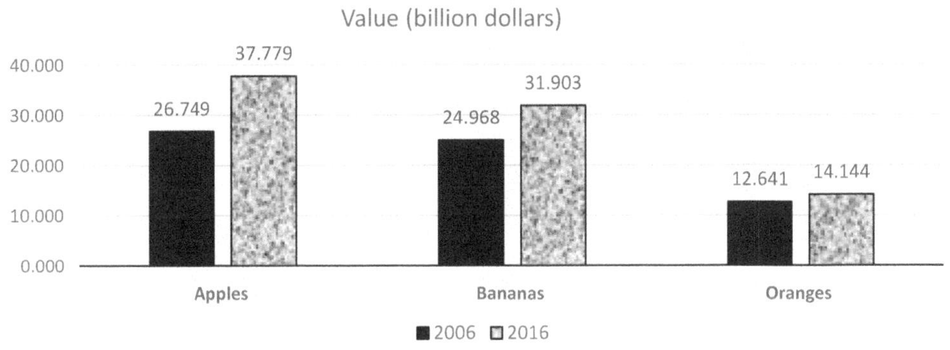

Fig. 1.3 World gross production value of apples, bananas, and oranges in 2006 and 2016 (billion dollars—at constant 2004–2006 dollars). *Source* United Nations, Food and Agriculture Organization, FAOSTAT database (2019). Accessed September 10, 2019

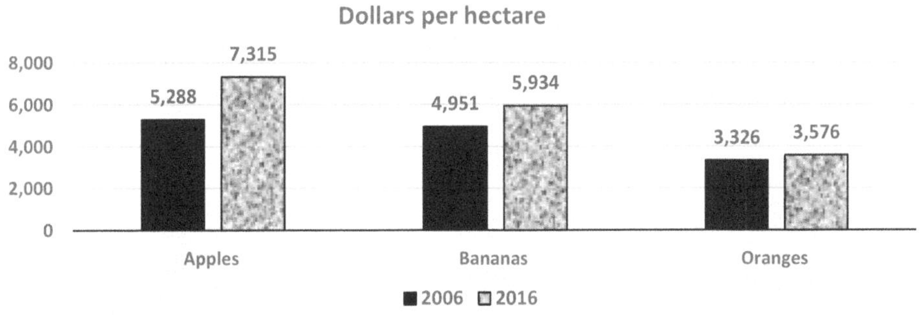

Fig. 1.4 World gross value per hectare harvested of apples, bananas, and oranges in 2006 and 2016 (Constant 2004–2006 dollars). *Source* United Nations, Food and Agriculture Organization, FAOSTAT database (2019). Accessed September 10, 2019

Another indicator of global economic value is the percent of each major fruit exported. Using FAOSTAT data, percentages for apples, bananas, and oranges for 2006 and 2016 are presented in Fig. 1.5. It is observed that bananas have recorded the highest percentages of world production exported in both years, far ahead than those of both apples and oranges. However, oranges have recorded a 1 percentage point gain in the share exported between 2006 and 2016.

Similarly, FAOSTAT data have been used to compare average values of exports (Fig. 1.6). It is observed that the average value per metric ton of apples exported in 2006 was 21.1% greater than that for oranges and 83.4% greater than that for bananas (Fig. 1.6). Moreover, between 2006 and 2016, the average value of apples exported rose by 28.7%, while that of bananas by 52.8%, and of oranges by 31.9%, thus maintaining the apple's lead in average value exported.

In order to evaluate and compare apple exports and imports in different countries, data are extracted from FAOSTAT (for 2008) and from the United Nations Comtrade online database (for 2018). It is revealed that the relative importance of apple exports and imports to different countries is likely to continue to change (Table 1.1).

In particular, Table 1.1 shows how total world exports, and those of countries that were among the top 10 fresh apple exporters in 2008, fared by 2018, the most recent year for which data were available. It was found that the total world fresh apple exports increased modestly in volume over

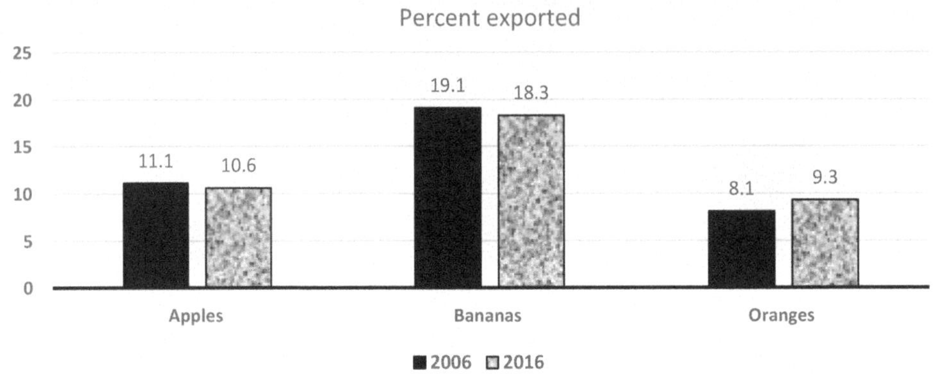

Fig. 1.5 Export share of world production of apples, bananas, and oranges in 2006 and 2016 (Percent). *Source* United Nations, Food and Agriculture Organization, FAOSTAT database (2019). Accessed September 10, 2019

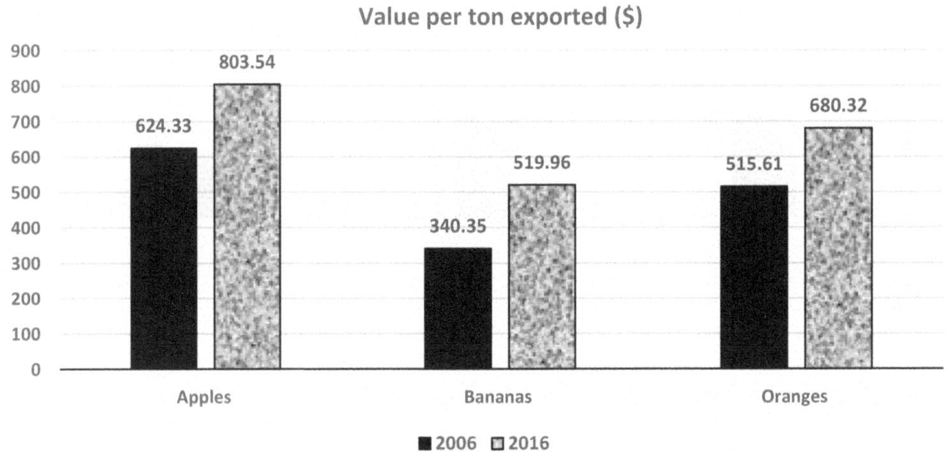

Fig. 1.6 Average value of apples, bananas, and oranges exported in 2006 and 2016($ per metric ton). *Source* United Nations, Food and Agriculture Organization, FAOSTAT database (2019). Accessed September 10, 2019

the decade, while average price grew less than inflation. All, but for one of the countries, ranked among the top 10 exporters of fresh apples in 2008 were also in the top 10 in 2018. Belgium lost its place to Moldova. China retained its number one ranking in both years. Poland jumped four places in the rankings, and New Zealand jumped two places. In contrast, Chile and France fell by two places, and the Netherlands and Belgium dropped by four places. The role of the Netherlands and Belgium in trans-shipments dropped sharply in the 2008–2018 decade.

Interestingly, there were equally dramatic changes in volume and average price of fresh apple exports by major exporting countries over the decade. The volume of exports from Poland, New Zealand, and the USA rose by over 30%, while those of the Netherlands and Belgium fell by more than half. Average prices were more than 50% higher for China and New Zealand, and more than 20% higher for Chile, the Netherlands, and South Africa. Four countries,

Table 1.1 World and top 10 apple exporters in 2008 (Rank and Volume in 1,000 metric tons in 2008 and 2018)

Country	Rank (2008) (#)	Volume (2008) (1,000 mt)	Rank (2018) (#)	Volume (2018) (1,000 mt)	Volume change (2008–2018) (%)	Average price change (2008–2018) (%)
World	–	*7,807.9*	–	*8,032.4*	*+ 2.9*	*+ 12.5*
China	1	1,188.1	1	1,118.5	–5.9	+ 90.5
Chile	2	766.3	4	779.0	+ 1.7	+ 24.0
USA	3	712.5	2	928.8	+ 30.4	+ 4.0
France	4	684.1	6	611.5	−10.6	−17.4
Italy	5	683.4	5	686.4	+ 0.4	−1.3
Netherlands	6	391.8	10	194.2	–50.4	+ 22.4
Poland	7	371.0	3	794.3	+ 114.1	–4.5
South Africa	8	358.1	7	448.6	+ 25.3	+ 26.6
Belgium	9	261.9	13	122.0	–53.4	−15.8
New Zealand	10	260.8	8	369.4	+ 41.6	+ 56.7

Sources 2008, UN, FAO online database. 2018, UN, Comtrade, online database. Accessed September 10, 2019

including France, Italy, Poland, and Belgium, suffered price declines. Price differences between exporters remained wide in both seasons. For example, average US export prices were more than twice those of Poland in both 2008 and 2018. While there was a steady trend in trade patterns, disruptions continued to arise from changes in apple demand and supply, as well as from abrupt policy changes, such as the embargo placed by the Russian Federation in 2014 on imports of fruits from the European Union, USA, and Australia.

The total world fresh apple imports along with both volume and rank of the top 10 apple importing countries in 2008, as well as their status in 2018 are presented in Table 1.2. Preliminary estimates suggest that the world volume of apple imports has declined between 2008 and 2018, while the average price has risen slightly faster than inflation. The results for major importing countries are mixed. Only 4 of the top 10 importers in 2008 have imported more fresh apples in 2018, while 6 have imported less. However, unlike the top 10 exporters, 3 of the top 10 importers in 2008, including Belgium, the USA, and the United Arab Emirates, are no longer in the top 10 in 2018. In addition, the Netherlands has dropped from fourth to sixth place, and Spain has dropped from fifth to tenth place. Interestingly, Belgium and the Netherlands have been major transit points for fresh fruit destined for other markets in continental Europe. Therefore, their declining imports reflect how large retailers in continental Europe have sourced more of their fresh produce directly and have taken their market share from traditional importers and wholesalers in Belgium and the Netherlands.

New entrants among the top 10 importers in 2018 (not shown in Table 1.2) are India, Belarus, and Saudi Arabia. Belarus' rise is linked to diversion of trade flows due to the 2014 Russian embargo on imports of produce (including fresh apples) from member countries of the European Union. By 2018, India has become the world's fourth-largest importer of fresh apples. Other Asian countries, including Hong Kong, the Philippines, Indonesia, and Thailand, have become increasingly important importers of fresh apples as Europe's influence on apple imports has waned.

Similar trade patterns to those of the last decade are likely to persist in the next decade. Major fresh apple exporting countries are likely

Table 1.2 World and top 10 apple importers in 2008 (Rank and Volume in 1,000 metric tons in 2008 and 2018)

Country	Rank (2008) (#)	Volume (2008) (1,000 mt)	Rank (2018) (#)	Volume (2018) (1,000 mt)	Volume change (2008–2018) (%)	Average price change (2008–2018) (%)
World	–	*7,372.1*	–	*6,777.8*	*−8.1*	*+ 19.8*
Russian Federation	1	1,062.9	1	843.5	−20.4	+ 25.3
Germany	2	613.3	2	653.6	+ 6.6	−0.3
UK	3	481.8	3	391.8	−18.7	+ 1.0
Netherlands	4	396.4	6	238.1	−39.9	+ 2.4
Spain	5	227.9	10	185.7	−18.5	+ 13.2
Mexico	6	188.4	5	278.9	+ 48.0	+ 9.4
Belgium	7	177.7	15	146.3	−17.7	−3.0
Canada	8	166.2	7	219.2	+ 31.9	−14.4
USA	9	165.3	18	131.0	−20.8	+ 64.3
United Arab Emir	10	157.8	12	169.9	+ 7.7	+ 11.0

Sources For 2008, UN, FAO online database. For 2018, UN, Comtrade, online database. Accessed September 10, 2019

to continue to dominate fresh apple exports. However, the volume and value of fresh apple imports are heavily dependent on changes in per capita income in each importing country. That, in turn, can be influenced in each country by global demand for its principal exports, attractiveness for foreign investment, and astuteness of economic policies and trade alliances engineered by the government.

1.5 Apple Industry Economic Connections

The apple industry has multiple, and ever expanding, economic linkages with other sectors of the economy. Attempts at the measurement of these connections have tended to be ad hoc in nature and not easily expanded to a global basis. These connections differ for the production (orchard) segment of the industry that generates the raw material; for packing and processing segments that prepare apple products for market; and for the marketing segment that ensures delivery of apple products to retailers, and to consumers.

1.5.1 Production Segment Connections

Occasional studies of production costs for apples highlight those major factors involved in the establishment and production of different cultivars of apples. Most of these studies utilize data from representative growers for a single cultivar, thus they are not statistically valid indicators of the economic experience of all apple producers in an entire growing district or region. In addition, different analysts include different production materials and activities; therefore, it remains difficult to compare unit production costs between countries and/or growing districts.

Major expenses for establishing an orchard block include the following: acquiring and preparing the needed land, purchasing rootstocks and/or grafted trees, setting up trellising and irrigation systems appropriate to the site and to the cultivar, planting ground cover (if necessary or desired), acquiring machinery and equipment required for establishment activities, and paying interest on the capital invested in the enterprise. In each successive year after planting, additional production inputs, such as chemicals and pesticides, fuel and oil, wind machines, protective covers and orchard labor, are required. Furthermore, as output increases, more harvest labor and equipment, including fruit storage facilities, are required, as well as more housing and transportation must be provided for this labor.

Orchard owners and managers interact regularly with suppliers of diverse inputs, including nurseries, sellers of fertilizers and chemicals, engineering companies, computing, electronic and communication companies, and banks or other financial intermediaries. Increasingly, to optimize the use of inputs or quality of outputs, producers draw on the expertise of both public and private consultants. They must assess the relative costs of each input and their relative contribution to gross and net revenues. In economic terms, they must continually weigh the marginal cost of each additional input against its marginal productivity. As production systems become increasingly more complex, producers must make more and more inter-related decisions about marginal productivity. While all producers seek to maximize the gap between total revenue and total costs, individual producers vary in how they reach this goal, depending on their choice of necessary inputs and the productivity of their operations.

1.5.2 Packing and Processing Connections

Quantitative studies of the economics of apple packing and processing are even rarer than those available for orchard operations. For both cases, the raw product is received in bulk form, and then converted into either individual packs or containers to meet the demands of wholesalers, retailers, and consumers.

Most packing and processing plants require buildings, machinery, and equipment (whether fixed lines or mobile equipment, such as forklift trucks) to move the raw product to different work stations within the plant. They require storage for raw and packed products, labor for sorting and packing, and packing materials. The raw product goes through a series of steps before becoming a marketable product. These include dumping onto a packing line, sorting, packing, combining into pallet units, and pre-shipment storage.

Traditionally, about one-third of the cost of packing apples for the fresh market has been in fixed plant overhead, one-third in labor, and one-third in packing and other materials. However, more progressive firms are in the process of continually replacing labor with capital. For example, many, once mechanical operations have now been automated, sorting by humans is supplemented by electronic sorters, and packing operations are often machine-aided.

Traditionally, outputs of packing houses and processors are standard cartons of bulk products. However, the apple industry increasingly imitates manufacturers of packaged consumer goods in using outer containers of many different weights and dimensions to meet the requirements of major retailers, and in using inner packages, such bags, pouches, tubes, kiddie packs and other special formats, rather than the traditional bulk, loose product. Different large retailers often have conflicting specifications for container size and shape, packaging materials, labeling, branding, and promotional claims. Packing houses must work closely with their major marketers in meeting the exact needs of their retail customers.

These developments have presented packing houses and processors with conflicting business challenges. Many continue to seek out economies of scale by building larger plants, installing faster machinery, and increasing plant throughput to lower fixed costs per unit of output. As long as the plant is turning out a standard product, such a strategy is deemed effective. However, as packing houses and processors have responded to customer demands for more diverse products and pack types, they have had to incur additional costs for selective sorting, packaging materials, and packaging machines for different items. As these specialty costs have begun to offset economies of scale, many larger operators have increasingly turned to use high-powered electronics to efficiently handle the diversity of items. Some larger operators have begun to dedicate separate operations, or separate lines, to individual fresh apple products. For example, some large operators dedicate a separate plant to handling organic apples. Such trade-offs between meeting retailer needs and securing economies of scale are likely to become more common.

Apple processing is now dominated by a small number of countries, and by a few individual firms that specialize in the production of apple juice concentrate. The concentrate industry has several unique characteristics. It continues to be based on a processing technology developed in the 1970s. In general, concentrate plants are only economically feasible if they can acquire raw apples below the full cost of their production, usually apples that cannot meet fresh market standards.

Concentrate processing is now dominated by a few countries, notably China, Poland, Chile, and Argentina, while import markets are dominated by the USA and western Europe. Returns to concentrate manufacturers vary widely as prices respond to changing availability of concentrate supplies. In contrast, most users of concentrate products such as single-strength apple juice, juice blends, or other fruit-based products can pass on costs to their customers while maintaining a satisfactory profit. Many are subsidiaries of major international consumer product companies that have sophisticated product development, marketing, and promotion programs. They will continue to utilize apple juice concentrate as long as it can be used for profitable consumer products.

1.5.3 The Evolution of Price Setting in Fresh Apples

Until recently, the spot (or cash) market in fresh apples dominated dealings between marketers and their wholesale and retail customers. Most of this business was conducted by long-distance telephone or telefax. Deals were usually price-driven. Wholesale or retail buyers sought quotes from potential suppliers for a desired cultivar and grade at the best possible price. Marketers would bid against each other for the most desirable clients. As the internet became more widely accessible, some major retailers attempted to use online auctions to achieve the same ends.

When marketers were small and numerous, they welcomed generic promotions of apples. For the second half of the twentieth century, apple promotion was dominated by grower-funded organizations. Some were national in scope, such as ENZA Limited in New Zealand or Capespan in South Africa. Other bodies, such as the Washington Apple Commission in the USA or BC Tree Fruits in Canada, represented the state or provincial growers. Many generic promotion organizations for fresh market apples either disappeared or shrunk in influence as marketers became fewer in number, yet larger in size.

Food retailers have also become larger in size through mergers and consolidations. As the number of retail outlets controlled by retail chains has grown and outlets become geographically dispersed, large retailers incurred heavy penalties if they were out-of-stock on any item. Increasingly, retailers have placed the burden on large marketers for ensuring the availability of desired items for all 12 months of a year. Retailers have also sought assistance from marketers to ensure that stocks move rapidly off retail shelves. Therefore, in addition to ensuring that desired apples are continually being replenished, apple marketers have become more heavily involved in merchandising and promotion activities to stimulate continuous sales at the retail level.

Marketers have become obligated to take on new roles, such as experimenting with new packaging designs, as well as conducting in-store sampling and organized promotional contests for their retail customers. They have also increasingly attempted to influence consumers through social media, such as Facebook, Twitter, or YouTube, as well as through firm blogs, contests, and/or co-opting food influencers to promote their products.

Large marketers have also played a major role in acquiring exclusive rights to club (or managed) apple cultivars. Almost every major apple marketer now has exclusive national or regional rights to at least one club cultivar. Their goals for new cultivars have varied from providing selected retailers with a novelty apple available for only a few weeks, to establishing a cultivar that would challenge mainstream cultivars such as 'Delicious' or 'Gala' throughout the year.

To date, the benefits of new club cultivars have generally outweighed costs. Consumers have been willing to pay premium prices for desirable new features. Retailers have enjoyed higher margins on club apples. Marketers have been able to earn higher prices than mainstream cultivars, and they have been able to pass back substantially higher returns to affiliated growers. The biggest losers to date have been producers that have not, for various reasons, been able to gain access to club cultivars.

However, the volume of club cultivars is expected to grow very rapidly in the next decade. If club apples are to maintain their price premiums, more consumers who appreciate novel tastes are needed. Thereby, retailers will have to devote more shelf space to new cultivars, and marketers will have to continue to gain a competitive edge with retailers by supplying a steady stream of new cultivars. As the volume of club cultivars expands, their price premiums are likely to come under pressure. In addition, they are more likely to cannibalize the market share held by traditional mainstream cultivars. Apple markets could remain in flux for many years.

1.6 The Apple's Place in Society and in Business

The apple is likely to play a prominent role in society and in business for many years to come.

1.6.1 The Apple's Place in Society

The production of apples has adapted to many environments around the world. The apple also responds well to long-term cold storage and can withstand the rigors of transportation over long distances. As a result, it has become one of the leading fruits consumed in many societies. In most developed countries, the apple is the first or second most widely consumed fresh fruit. Consumption has also grown rapidly in many developing countries as a rising middle class seeks to diversify from local tropical or semitropical fruits.

Thus, the world apple industry must deal with two offsetting trends. In developed countries, demand for apples continues to be under threat, while in developing countries, demand for apples has been on a strong upward trend. As populations of developed countries have become older and more affluent, their demands for alternative exotic fruits have expanded, and thereby eroding the demands for traditional fruits such as apples. In developing countries, an opposite phenomenon is observed, wherein apples are deemed as a good alternative, and increasingly desirable, fruit.

Apple producers, packers, and marketers in the developed world are dealing with reduced demand in their domestic markets. If they wish to expand, they will have to increase their exports to the rest of the world. However, the production of apples has also grown very rapidly in lower-income countries, including China, Turkey, Chile, South Africa, and Poland, and these countries are also seeking to expand sales to the rest of the world. The outcome of this clash between two major opposing forces cannot be predicted with certainty. However, it will continue to rearrange the role of apple consumption in many countries.

1.6.2 The Apple's Place in Business

The apple industry supports many forms of transportation, including truck, rail, and ship, as well as many fixed resources, including ports and wholesale markets, in moving its products around the world. In addition, apples play a special role in production departments of most retail food stores, as they are available year-round. In contrast to oranges or bananas, they provide consumers with many options in terms of cultivar, size, color, and taste, among others. They also suffer from very little shrink on retail shelves compared to competing fruits. Finally, they provide consistently large gross profit margins for retailers.

For many retailers, the quality and selection of their produce departments have become major competitive factors in wooing shoppers. However, their appeal is based on two competing trends. One is the desire of retailers and consumers for access to continual supplies of standard produce items. The other is their desire for new, exciting, and innovative fruit offerings. The upshot of these competing desires is that apples will remain a major element in retail production departments. However, the shelf space devoted to apples will face continuous threats from more novel fruits. In turn, the shelf space devoted to established apple cultivars will be subjected to constant pressure from newer apple cultivars. This suggests that the apple industry will have to continue to devote considerable resources to new product innovation with the full awareness that such innovation may displace mainstream cultivars from retail shelves. The apple industry will face a challenge in balancing familiarity with innovation while retaining adequate return on investment.

1.6.3 The Role of Continuing Innovation

Leading firms and producing districts in the world apple industry have become committed to finding and exploiting innovations, whether developed within the apple industry, or in the rest of the economy. Pressures from consumer activists and government regulators, demands of major retailers, expansion of alternative fruits, and competition among firms for a share of the apple market, are likely to stimulate continuing innovation among leading apple firms.

Those apple firms, whether growers, packers, processors, or marketers, that fail to keep up with the leading firms and producing districts, will find themselves squeezed out of the industry. The world apple industry can expect dynamic change to continue indefinitely.

1.7 Concluding Remarks and Future Prospects

The economics of the world apple industry will continue to change rapidly in response to changing pressures from the food distribution system, and the applicability of emerging new technologies. Those firms, districts, and countries that best discern future trends, and have the personnel and capital to implement the needed changes, will become an increasingly important share of the world apple industry.

References

Coase R (1937) The nature of the firm. Economica 4 (16):386–405

Ferree DC, Warrington IJ (2003) Apples, botany, production and uses. CABI Publish

United Nations, Food and Agriculture Organization, FAOSTAT database. Accessed 10 Sep 2019

2 Botany, Taxonomy, and Origins of the Apple

Gayle M. Volk, Amandine Cornille, Charles-Eric Durel, and Ben Gutierrez

Abstract

Apples are important from both a cultural and nutritional standpoint. The fruit produced from the cultivated apple, *Malus* × *domestica* Borkh., is grown in temperate regions throughout the world. Ornamental crabapple trees represent other *Malus* species and are generally valued for their landscape properties rather than their fruits. Genetic analyses of apple cultivars and species have revealed domestication pathways, pedigree relationships, as well as data that guide genebank collection management practices. Molecular information is also critical for breeding programs that are using new techniques to identify novel genetic combinations with enhanced biotic and abiotic stress resistance as well as desirable fruit quality and production traits. This chapter includes information about wild apple species, origins of cultivated apples, genetic assessments, as well as some basic information about tree phenology, architecture, and propagation methods.

2.1 Introduction

Malus spp., belonging to the family Rosaceae, includes approximately 61 apple wild species and hybrids, traditionally taxonomically organized based on key morphological distinctions (Qian et al. 2006), and a single cultivated species, *Malus* × *domestica* Borkh. Worldwide, apples are one of the most economically important fruit crops (Bramel and Volk 2019). Fruit from *M.* × *domestica* is harvested for fresh consumption, cider, and processing uses, while other species of *Malus* are cultivated as either rootstocks or ornamentals. Wild *Malus* species are primarily found in temperate climates throughout much of the northern hemisphere, with China as a primary center of origin (Juniper and Mabberley 2006). *Malus* × *domestica* is proposed to have originated primarily from its progenitor, *M. sieversii*, which is native to Central Asia and Western China (Harris et al. 2002; Cornille et al. 2015, 2019).

G. M. Volk (✉)
USDA-ARS National Laboratory for Genetic Resources Preservation, 1111 S. Mason St., Fort Collins, CO 80521, USA
e-mail: Gayle.Volk@usda.gov

A. Cornille
Université Paris-Saclay, INRAE, CNRS, AgroParisTech, GQE—Le Moulon, 91190 Gif-sur-Yvette, France

C.-E. Durel
Univ Angers, Institut Agro, INRAE, IRHS SFR QUASAV, 49000 Angers, France

B. Gutierrez
USDA-ARS Plant Genetic Resources Unit, 630 West North Street, Geneva, NY 14456, USA

S. S. Korban (ed.), *The Apple Genome*, Compendium of Plant Genomes,
https://doi.org/10.1007/978-3-030-74682-7_2

2.2 Wild *Malus* Species and Hybrids

Most *Malus* species have a haploid genome with 17 chromosomes originating from genome duplication (Daccord et al. 2017). However, there is a range in genome size and ploidy among and within species (Höfer and Meister 2010; Chagné et al. 2015). Most dessert apple cultivars are either diploid or triploid, while wild species of *Malus* may be diploid, triploid, or tetraploid. *Malus* genera have been subdivided into eight sections, likely corresponding to their geographic origins (Fig. 2.1, Table 2.1).

Section *Malus* includes the cultivated apple, *M.* × *domestica*, along with its primary crop relatives including *M. sieversii* (Ledeb.) M. Roem., *M. orientalis* Uglitzk., and *M. sylvestris* Mill., as well as a number of other related species (Table 2.1). These three species, as well as *M. prunifolia* (Willd.) Borkh., share regions of nuclear and chloroplast sequences with *M.* × *domestica* (Fig. 2.2; Velasco et al. 2010; Nikiforova et al. 2013; Volk et al. 2015). Moreover, microsatellite, or simple sequence repeats (SSRs), markers have revealed that *M.* × *domestica* and *M. sieversii* form distinct genetic groups, with *M.* × *domestica* being a panmictic group (Cornille et al. 2012). Assessments of the wild apple genetic diversity using microsatellite markers have compared diversity among and within sampling locations in Kazakhstan (*M. sieversii*), the Caucasus region (*M. orientalis*), and Europe (*M. sylvestris*). It is found that the three wild species form distinct genetic groups in Eurasia but are only weakly genetically differentiated from each other (Richards et al. 2009; Cornille et al. 2013, 2015; Volk et al. 2008). Furthermore, data suggest that there is substantial gene flow among populations and species. These findings support the existence of wild *Malus* populations that are highly outcrossing (Richards et al. 2009) and that range contractions may have occurred during glacial periods in Europe (Cornille et al. 2013).

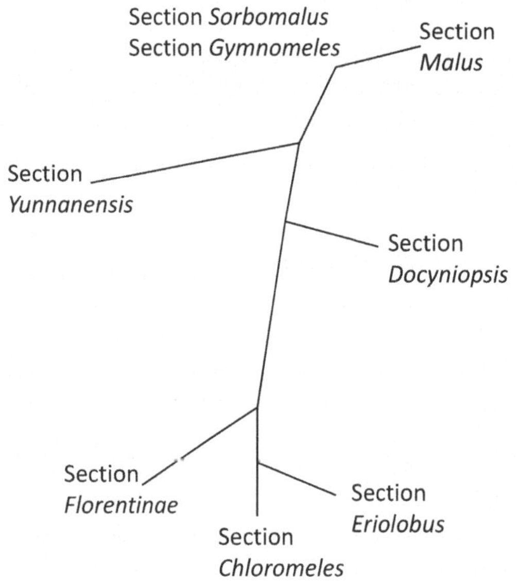

Fig. 2.1 Relationships among sections of *Malus* based on chloroplast sequence data of *Malus* as adapted from Volk et al. (2015) and Robinson et al. (2001)

Sections *Sorbomalus*, *Gymnomeles*, and *Yunnanensis* are comprised of wild species native to China and the Far East (Fig. 2.1 and Table 2.1, USDA 2020; Zhi-Qin 1999; Volk et al. 2015; Yu 1988). An exception is *M. fusca* (Raf.) C.K. Schneid. (Section *Sorbomalus*), which is native to the western coast of North America, and it is suggested to have arrived in North America across the Bering Strait (Qian et al. 2006; Robinson et al. 2001; Routson et al. 2012).

Section *Docyniopsis* includes the only tropical apple species, *M. doumeri* (Bois) A. Chev., that grows in the wild in Southern China, and as far south as Vietnam. Moreover, *M. leiocalyca* S.Z. Huang and *M. tschonoskii* (Maxim.) C.K. Schneid. have also been assigned to this section (Robinson et al. 2001).

Each of sections *Florentinae* (*M. florentina* (Zucc.) C.K. Schneid., native to Southern Europe) and *Eriolobus* (*M. trilobata* (Poir.) C.K. Schneid., native to the Middle East and Southern Europe) has a single species. These are deemed as likely European relict species (Qian et al. 2008) that share common ancestries with North American *Malus* species in Section *Chloromeles* (Fig. 2.1, Forte et al. 2002). Ancestors of the species within Sections *Florentinae*, *Eriolobus*, and *Chloromeles* may have inhabited the northern landmass in the Cretaceous Period when current North America and Europe have been in

Table 2.1 Sections of *Malus**

Malus* Mill. Sect. *Malus
Malus chitralensis Vassilcz
Malus crescimannoi Raimondo
Malus × *domestica* (Suckow) Borkh
Malus muliensis T.C. Ku
Malus orientalis Uglitzk
Malus prunifolia (Willd.) Borkh
Malus sieversii (Ledeb.) M. Roem
Malus spectabilis (Aiton) Borkh
Malus sylvestris (L.) Mill
Malus zhaojiaoensis N.G. Jiang
***Malus* Mill. Sect. Sorbomalus**
Malus fusca (Raf.) C.K. Schneid
Malus kansuensis (Batalin) C.K. Schneid
Malus komarovii (Sarg.) Rehder
Malus sargentii Rehder
Malus toringo (Siebold) de Vriese
Malus toringoides (Rehder) Hughes
Malus transitoria (Batalin) C.K. Schneid
Malus zumi (Matsum.) Rehder
***Malus* Mill. Sect. Gymnomeles**
Malus baccata (L.) Borkh
Malus halliana Koehne
Malus hupehensis (Pamp.) Rehder
Malus mandshurica (Maxim.) Kom. Ex Skvortsov
Malus sikkimensis (Wenz.) Koehne ex C.K. Schneid
Malus spontanea (Makino) Makino
***Malus* Mill. Sect. Yunnanensis**
Malus honanensis (Rehder)
Malus ombrophila Hand.-Mazz
Malus prattii (Hemsl.) C.K. Schneid
Malus yunnanensis C.K. Schneid
***Malus* Mill. Sect. Docyniopsis**
Malus doumeri (Bois) A. Chev
Malus leiocalyca S.Z. Huang
Malus tschonoskii (Maxim.) C.K. Schneid
***Malus* Mill Sect. Florentinae**
Malus florentina (Zucc.) C.K. Schneid
***Malus* Mill Sect. Eriolobus**
Malus trilobata (Poir.) C.K. Schneid
Malus* Mill. Sect. *Malus
***Malus* Mill. Sect. Chloromeles**
Malus angustifolia (Aiton) Michx
Malus coronaria (L.) Mill
Malus ioensis (Alph. Wood) Britton

*This list is based on information collected from USDA (2020), Robinson et al. (2001), and Qian et al. (2006, 2008)

close proximity, and have then diverged when these continents have separated (Forte et al. 2002).

Section *Chloromeles* consists of *M. coronaria* (L.) Mill., *M. angustifolia* (Aiton) Michx., and *M. ioensis* (Alph. Wood) Britton, these are found in eastern North America. Moreover, these appear to be the most distant species from *M.* × *domestica* and from Sections *Malus* and *Sorbomalus* based on assessments of chloroplast sequence data (Lo and Donoghue 2012; Robinson et al. 2001; Volk et al. 2015).

Most research on *Malus* focuses heavily on the cultivated, consumed apple. While dessert apples have sweet fruit, crabapples represent their counterparts, wherein fruit is bitter, acidic, and in terms of human edibility, unappealing. The term crabapple could be used to describe nearly all species other than *M.* × *domestica* species. Generally, crabapples bear smaller fruit, although fruit size varies significantly (Cornille et al. 2014). In general, crabapple fruit size ranges from 10 to 50 mm (USDA 2020). Although poor in fruit quality, crabapples have a significant economic impact, particularly as ornamentals (Fiala 1994). Additionally, they are planted in commercial orchards as pollinizers. Many ornamental crabapples are either specific selections from *Malus* wild species or have developed as hybrids between two *Malus* species, as *Malus* species are interfertile/cross-compatible (Table 2.2).

2.3 Cultivar Origins

Desirable apple trees have been originally identified in their native settings. The first drivers of the domestication of apple are most probably

Fig. 2.2 Relationships among *Malus* × *domestica* and primary crop wild relatives based on chloroplast sequence data (adapted from Volk et al. 2015)

fruit size and organoleptic traits (Yao et al. 2015; Duan et al. 2017). *Malus* × *domestica* is proposed to have arisen from the progenitor species *M. sieversii* (native to Western China, Kazakhstan, and other Central Asian countries). Wild *M. sieversii* trees exhibit a wide range of fruit diversity (Fig. 2.3), and these may have served as original sources for domestication events in China and Russia, as well as for western cultivars. Then, during its journey along the Silk Routes through the Caucasus region (native to *M. orientalis*) and Europe (native to *M. sylvestris*), *M. sieversii* underwent several hybridization events with local wild apples (Fig. 2.4; Cornille et al. 2019; Duan et al. 2017; Gao et al. 2015; Luby 2003; Peace et al. 2019). When Greeks and Romans brought apples into Europe, about 1,500 years ago, *M. sieversii* was introgressed to *M. sylvestris*, the European crab apple (Cornille et al. 2012, 2014, 2019).

During its journey along the Silk Routes, desirable *M.* × *domestica* trees were vegetatively propagated locally to preserve desirable genetic combinations. Traits that were especially favored included lower acidity and higher sugar content, thus resulting in numerous cultivars with sweet or sub-acid fruits (Duan et al. 2017; Ma et al. 2018). Domestication, and recent breeding efforts, also changed fruit metabolic compositions and lowered levels of phenolic compounds influencing astringency, bitterness, and color (Lea and Arnold 1978; Sanoner et al. 1999; Khan

Table 2.2 Taxonomy of *Malus* interspecific hybrids and progenitor species as listed in GRIN-Global (USDA 2020)

Taxon	Proposed parentage
M. × *adstringens*	*M. baccata* × *M. pumila*
M. × *arnoldiana*	*M. baccata* × *M. floribunda*
M. × *asiatica*	*M. sieversii* × *M. baccata*
M. × *astracanica*	*M. prunifolia* × *M. pumila*
M. × *atrosanguinea*	*M. halliana* × *M. toringo*
M. × *dawsoniana*	*M.* × *domestica* × *M. fusca*
M. × *gloriosa*	*M.* × *scheideckeri* × *M. pumila* 'Niedzwetzkyana'
M. × *hartwigii*	*M. baccata* × *M. halliana*
M. × *magdeburgensis*	*M. pumila* × *M. spectabilis*
M. × *micromalus*	*M. spectabilis* × *M. baccata*
M. × *moerlandsii*	*M.* × *purpurea* × *M. toringo*
M. × *platycarpa*	*M.* × *domestica* × *M. coronaria*
M. × *purpurea*	*M.* × *atrosanguinea* × *M. pumila* 'Niedzwetzkyana'
M. × *robusta*	*M. baccata* × *M. prunifolia*
M. × *scheideckeri*	*M. floribunda* × *M. prunifolia*
M. × *soulardii*	*M. ioensis* × *M. pumila*
M. × *sublobata*	*M. prunifolia* × *M. toringo*
M. × *xiaojinensis*	*M. toringoides* × *M. kansuensis*

et al. 2014), particularly as fruits were increasingly used for fresh and dessert consumption, in contrast to making alcoholic ciders. In Europe, Russia, and China, particular offsprings were named as cultivars and propagated.

Some apple cultivars are likely to be from the Roman period, including 'Annurca' and 'Decio'. Moreover, there are many examples of Old-World apple cultivars, such as 'Winter Pearmain' (the year 1200), 'Calville Blanc d'Hiver' (1598), 'Court Pendu Plat' (1613), 'Blenheim Orange' (1740), 'Antonovka' (1826), and 'Yellow Transparent' (1850) (Fig. 2.5).

Apple seeds and trees moved along as Europeans migrated to North America. Apple seeds were planted throughout North America, and in some cases, seedling trees were named, propagated, and thereafter referred to as cultivars. The plethora of seedling apple trees and hundreds of cultivars that were named led to the designation of North America as a likely secondary center of origin for apple cultivars. Early North American apple cultivars included 'Roxbury Russet' (1600), 'Rhode Island Greening' (1650), and 'Esopus Spitzenburg' (1735). Interestingly, molecular marker-based pedigree inferences identified pedigrees of cultivars derived from North American seedling trees (Muranty et al. 2020). For example, 'Esopus Spitzenberg' was found to be directly derived from the French cultivar 'Reinette Franche' (1510). By the mid-1800s, the most popular apple cultivar in the USA was 'Ben Davis', a cultivar identified from a North American seedling in about 1800 (Fig. 2.6).

Centuries of domestication, selection, and propagation have resulted in a wide range of genetic diversity among apple cultivars. Some have bitter, phenolic traits desirable for hard cider production. Other cultivars have a wide range of fruit shapes and skin color, including stellar shape (e.g., 'Api Etoilée'), ribbed apple (e.g., 'Calville Blanc d'Hiver'), fully russeted fruit (e.g., 'Renetta Grigia di Torriana'), or burgundy-colored apple mutants (e.g., 'Bravo'™). Furthermore, new cultivars have enhanced biotic and/or abiotic stress resistance, as well as tree architectures that lend themselves to modern horticultural practices, including automated harvesting. Modern cultivars may also have an extremely long fruit shelf-life, particularly when stored under controlled atmosphere conditions (Stanger et al. 2018). In addition, modern cultivars and production practices result in high crop loads (i.e., number of fruits per tree), thus necessitating fruit thinning to avoid biennial bearing. Another innovation for breeding programs is the identification and use of wild apple species carrying desirable genes and alleles for disease resistance. For example, *M. floribunda* '821' is a source for *Vf* (*Rvi6*) apple scab

Fig. 2.3 Diversity of fruits collected from *Malus sieversii* trees growing at the USDA National Plant Germplasm System Apple Collection (Peggy Greb, USDA Image Gallery)

resistance allele for several commercial cultivars (Janick 2002).

Today, there are many well-known apple cultivars that are sold as specialty cultivars around the world, including those from Europe ('Cox's Orange Pippin'), Russia ('Antonovka', 1826), Japan ('Fuji', 1930s), Australia ('Granny Smith', 1868), New Zealand ('Braeburn', 1952; 'Gala', 1930s), and the USA '(Golden Delicious', 1914; 'Red Delicious', 1880; 'Honeycrisp', 1991) (Fig. 2.7). Surprisingly, many of these well-known apple cultivars have been identified as chance seedlings or bred decades or even hundreds of years ago. Breeding programs continue to select for improved disease resistance and stress tolerance, fruit quality, and storage characteristics (Laurens et al. 2010).

Sometimes apple trees spontaneously mutate, and these mutations result in the observable change(s) that can be maintained following clonal propagation. These "sports" of a cultivar are nearly identical genetically to the original cultivar and are usually not detected using most genetic techniques. Sport variation ranges from mild to extreme. For example, 'Red Delicious' has a series of sports with various fruit red color pigmentation/or patterns, and 'Sargeant Russet Golden' is a russeted sport mutant of 'Golden Delicious' (Fig. 2.8; Gross et al. 2012). Another form of observed mutations, chimeras, is most often visible as sectional variations in apple fruit coloration. However, such chimeras are generally not maintained during clonal propagation.

2.4 Genetic Characterization

Genetic characterization of apple cultivars has long been grounded in the so-called pomological descriptions, based on accurate phenotypic observations of various fruit and tree traits (Juniper and Mabberley 2006). In his 'Dictionnaire de Pomologie', A. Leroy cites the first classifications performed by Greek writers such as Théophraste who subdivided Greek apple cultivars into six types according to their phenology or fruit taste (Leroy 1873). In modern times, biochemical analyses have been used to characterize sets of apple cultivars in addition to morphological traits. However, such biochemical analyses are more geared for purposes of description rather than classification.

Genetic characterization of apple cultivars began with isozyme analyses in the 1980s and 1990s (Weeden and Lamb 1985; Menendez et al. 1986; Korban and Bournival 1987; Battle et al. 1995). Subsequently, analyses of microsatellite/SSR markers have led to the collection and species diversity assessments, as well as for use in genetic fingerprinting for cultivar identity analyses. These markers can also be used to determine cultivar/genotype trueness-to-type, as well as identify duplications and sport-families within collections. Assessments have been performed for national apple genebank collections in France, the UK, the USA, and elsewhere (Lassois et al. 2016; Hokanson et al. 2001; Gross

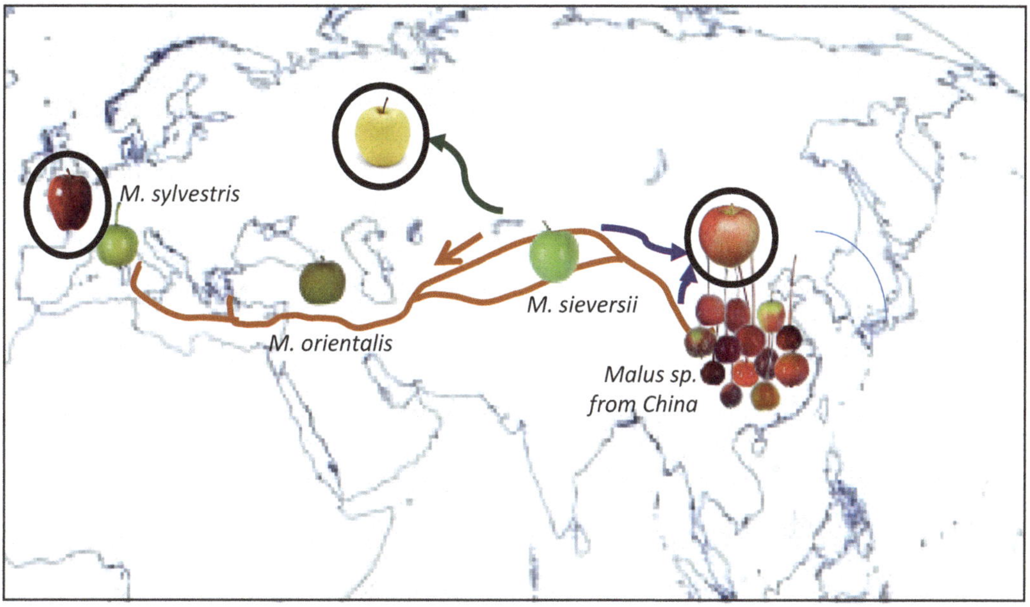

Fig. 2.4 Proposed routes of *Malus* domestication, with circled fruit denoting likely domestication lineages. Orange-colored arrows denote domestication routes of apples of European ancestry, while Black-colored arrows denote domestication route of apples in Russia, and Purple-colored arrows denote domestication of apples in East Asia

Fig. 2.5 Watercolor paintings of apple fruits from Old World apple cultivars. **a** 'Blenheim Orange', United Kingdom, 1740; and **b** 'Yellow Transparent', Russia, 1850 (National Agricultural Library 2020)

Fig. 2.6 Fruits of historic North American apple cultivars. **a** 'Roxbury Russet' (1600 AD); **b** 'Rhode Island Greening' (1650 AD); **c** 'Esopus Spitzenburg' (1735 AD); and **d** 'Ben Davis' (1800 AD) (USDA 2020)

et al. 2012; Ordidge et al. 2018). Microsatellite assessments of collections have also facilitated international collection comparisons (Evans et al. 2011; Urrestarazu et al. 2016).

Genebank collections serve as reference sets of known cultivars to aid in the identification of historic fruit trees on public and private lands (Routson et al. 2009; Lassois et al. 2016; Magby et al. 2019). Recently, single nucleotide polymorphism (SNP) arrays and genomic sequencing have rapidly advanced our capacity to determine pedigree relationships, pursue marker-based selection breeding programs, and target specific cultivars/genotypes with alleles of particular interest to breeding programs (Cornille et al. 2019; Peace et al. 2019; Ordidge et al. 2018; Muranty et al. 2020; Howard et al. 2017; Baumgartner et al. 2016). The most frequently cited apple SNP arrays are the International RosBREED SNP Consortium (IRSC) Illumina Infinium 8K SNP array (Chagné et al. 2012), the Illumina Infinium 20K SNP array (Bianco et al. 2014), and the Affymetrix Axiom 487K SNP array (Bianco et al. 2016).

2.5 Tree Phenology and Architecture

Self-rooted apple trees grow to heights ranging between 2 and 20 m and have juvenile periods ranging between 4 and 12 years (Fischer 1994). Juvenility is a significant challenge for breeding and genetic research in apples. An early bearing can be induced by grafting seedling trees onto dwarfing rootstocks (Fazio et al. 2014). Transgenic approaches can drastically shorten generation cycles to 1 year or less (Flachowsky et al. 2011).

In general, inflorescences develop on first-year fruiting shoots (i.e., 2-year-old wood), brindles (medium shoots), or spurs (compact

Fig. 2.7 A fruit display of internationally known cultivars of 'Golden Delicious', 'Gala', 'Granny Smith', and 'Red Delicious' apples (Scott Bauder, USDA Image Gallery)

shoots). Bourse shoots (short vegetative shoots) develop proximally from the inflorescence meristem and terminate in vegetative or inflorescence buds (Costes et al. 2006). Floral bud development is initiated after full-bloom when terminal meristems either remain vegetative or commit to flowering (Foster et al. 2003). The development of floral primordia is influenced by genetic, physiological, and environmental factors (Koutinas et al. 2010). Flowers develop from inflorescences (consisting of 5–7 flowers) with a distinctive king flower. Flowers are complete, typically arranged with five petals, ranging from white to red in color. *Malus* trees are either deciduous or semi-deciduous with an alternate leaf arrangement. Leaves are typically serrulate elliptical to ovate. Fruit is a pome ranging from 3 to 12 cm in diameter, usually with five carpels, a red, yellow, or green exocarp, and typically a white mesocarp. The persistent calyx may be open or closed, pedicle length is typically 1.5 cm but ranges from 0.5 to 4.0 cm (USDA 2020).

Growth habits of apple trees can be classified into distinct categories based on the positioning of fruits and branches, with columnar and weeping representing the two extremes. High-

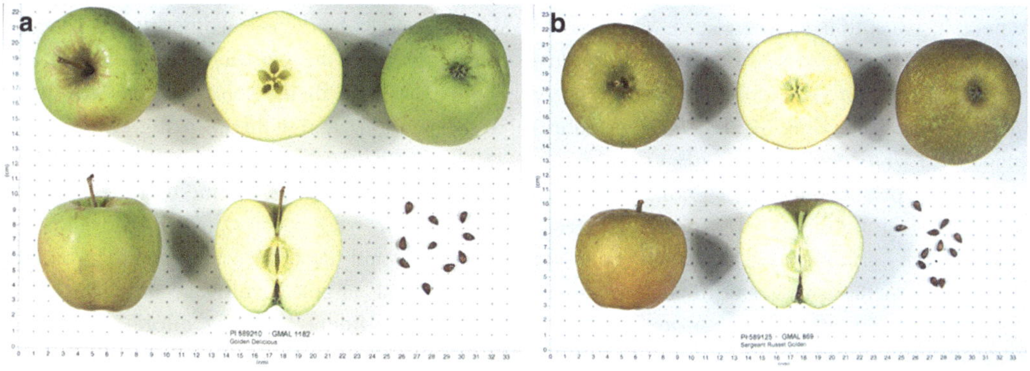

Fig. 2.8 Fruits of **a** 'Golden Delicious' and **b** 'Sargeant Russet Golden', a russetted sport of 'Golden Delicious' (USDA 2020)

density planting favors more compact columnar and spur types to maximize production and sustainability. The genetic characteristics that contribute to distinct architectural traits have been identified (Kenis and Keuleman 2007; Segura et al. 2009; Okada et al. 2020).

2.6 Propagation

Wild *Malus* trees propagate by seeds that are often spread by animals. *Malus* species, including *M.* × *domestica*, are self-incompatible (Igic et al. 2008). They exhibit gametophytic self-incompatibility, which is determined by the multigenic *S*-locus with two major components (Kubo et al. 2010; Pratas et al. 2018). One of these two components is a pistil specificity *S*-RNase that inhibits pollen tube growth from self-pollen, and the second consists of *S*-pollen specificity F-box proteins that recognize and degrade non-self *S*-RNases (Kubo et al. 2010; Pratas et al. 2018).

Due to self-incompatibility, apple trees must be pollinated by a different cultivar/genotype as a pollen source. Therefore, seeds collected from these fruits are hybrids, and if germinated, they grow into seedlings that are not true-to-type to their maternal parent as they are highly heterozygous. A notable exception is when seeds are derived asexually from maternal tissues and are true-to-type through apomixis. Apomixis is observed in *Malus* species, including *M. hupehensis* (Pamp.) Rehder and *M. toringo* (Siebold) de Vriese.

Cultivar-specific allelic combinations are maintained through asexual propagation techniques, with grafting as the most common propagation method. Either dormant or summer buds are budded/grafted onto rootstocks, thereby resulting in a two-genotype compound tree system, consisting of a combination of a rootstock (below-ground) and a scion cultivar (above-ground) (Fig. 2.9). This clonal propagation method dates back approximately 4000 years ago (Zohary 2000), and it has been used for centuries to maintain identities of unique apple selections. Remarkably, 'Winter Pearmain' fruit consumed today has the same genetic profile as that of fruit grown in 1200 AD. In some cases, somatic mutations may accumulate as a result of asexual propagation methods, but the rate(s) of incidence of such mutations has not been well documented.

Until the mid-1800s, rootstocks were derived from seeds, and germinating seedlings were grown simply for use in the vegetative propagation of scion cultivars. Modern apple clonal rootstocks have transformed apple production worldwide, as these have been selected for particular desirable traits that they confer to scions, including disease resistance and dwarfing, fruit quality, tree habit, nutrient assimilation, and tolerance to biotic and abiotic stress (Marini and Fazio 2018; also see Chap. 6 in this volume). Modern rootstocks are vegetatively propagated using mound layering or in vitro tissue culture systems (Fig. 2.9, Teixeira da Silva et al. 2019).

2.7 Conclusions

Malus species have critical importance in world agriculture, yet the genetic relationships among wild species, their contributions to apple domestication, and cultivar pedigrees are currently under investigation and will be elucidated. Genomic approaches are being used to assess the diversity of wild species, as well as those of important and desirable cultivars. These wild species and targeted cultivars are critical for current and future breeding efforts to enhance nutritional content, productivity, resistance to various biotic and abiotic stress, as well as sustainably of apple produced for future generations.

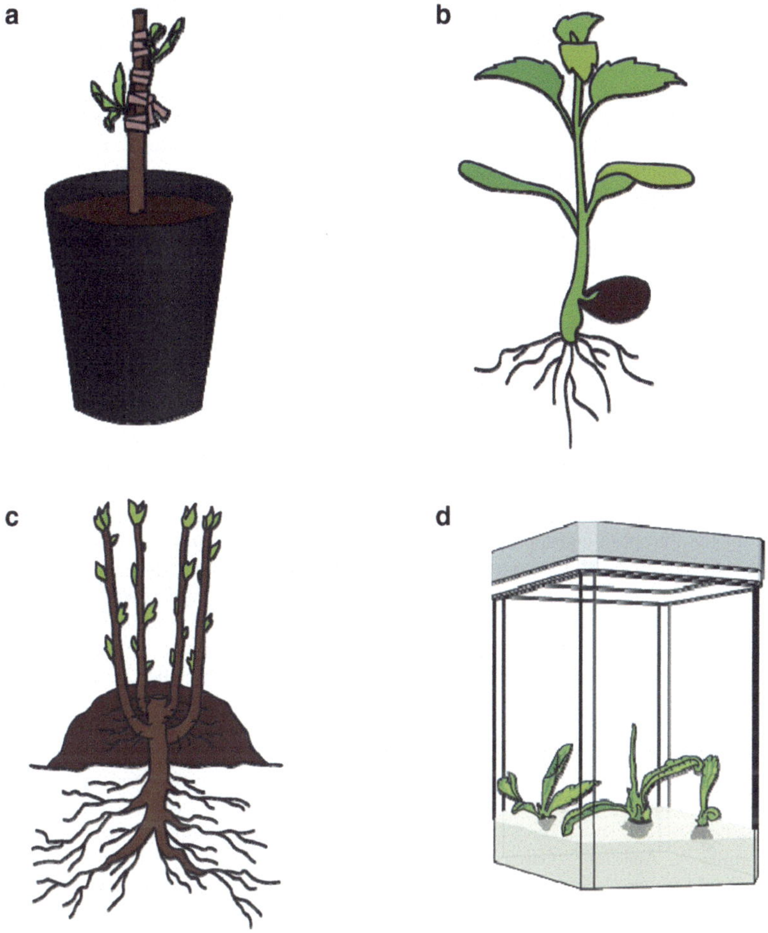

Fig. 2.9 Illustrations of *Malus* propagation methods. **a** Grafting; **b** seedling (derived sexually or by apomixis); **c** mound layering; and **d** in vitro tissue culture

Acknowledgements We thank Katheryn Chen for providing the illustrations in Fig. 2.9. USDA is an equal opportunity provider, employer, and lender.

References

Battle I, Alston FH, Evans KM (1995) The use of the isoenzymic marker gene *Got-1* in the recognition of incompatibility *S* alleles in apple. Theor Appl Genet 90:303–306

Baumgartner IO, Kellerhals M, Costa F, Dondini L, Pagliarani G, Gregori R, Tartarini S, Leumann L, Laurens F, Patocchi A (2016) Development of SNP-based assays for disease resistance and fruit quality traits in apple (*Malus* × *domestica* Borkh.) and validation in breeding pilot studies. Tree Genet Genomes 12:35. https://doi.org/10.1007/s11295-016-0994-y

Bianco L, Cestaro A, Sargent DJ, Banchi E, Derdak S, Di Guardo M, Salvi S, Jansen J, Viola R, Gut I, Laurens F, Chagné D, Velasco R, van de Weg E, Troggio M (2014) Development and validation of a 20K single nucleotide polymorphism (SNP) whole genome genotyping array for apple (*Malus* × *domestica* Borkh). PLoS ONE 9(10). https://doi.org/10.1371/journal.pone.0110377

Bianco L, Cestaro A, Linsmith G, Muranty H, Denancé C, Théron A, Poncet C, Micheletti D, Kerschbamer E, Di Pierro EA, Larger S, Pindo M, Van de Weg E, Davassi A, Laurens F, Velasco R, Durel C-E, Troggio M (2016) Development and validation of the Axiom Apple 480K SNP genotyping array. Plant J 86:62–74. https://doi.org/10.1111/tpj.13145

Bramel PJ, Volk G (2019) A global strategy for the conservation and use of apple genetic resources. Global Crop Diversity Trust. Bonn, Germany. https://doi.org/10.13140/RG.2.2.34072.34562

Chagné D, Crowhurst RN, Troggio M, Davey MW, Gilmore B, Lawley C, Vanderzande S, Hellens RP, Kumar S, Cestaro A, Velasco R, Main D, Rees JD, Iezzoni A, Mockler T, Wilhelm L, van de Weg E, Gardiner SE, Bassil N, Peace C (2012) Genome-wide

SNP detection, validation, and development of an 9K SNP array for apple. PLoS ONE 7(2). https://doi.org/10.1371/journal.pone.0031745

Chagné D, Kirk C, Whitworth C, Erasmuson S, Bicknell R, Sargent DJ, Kumar S, Troggio M (2015) Polyploid and aneuploid detection in apple using a single nucleotide polymorphism array. Tree Genet Genomes 11:94. https://doi.org/10.1007/s11295-015-0920-8

Cornille A, Gladieux P, Smulders MJM, Roldán-Ruiz I, Laurens F, Le Cam B, Nersesyan A, Clavel J, Olonova M, Feugey L, Gabrielyan I, Zhang X-G, Tenaillon MI, Giraud T (2012) New insight into the history of domesticated apple: secondary contribution of the European wild apple to the genome of cultivated varieties. PLoS Genet 8(5): e1002703. https://doi.org/10.1371/journal.pgen.1002703

Cornille A, Giraud T, Bellard C, Tellier A, Le Cam B, Smulders MJM, Kleinschmit J, Roldan-Ruiz I, Gladieux P (2013) Postglacial recolonization history of the European crabapple (*Malus sylvestris* Mill.), a wild contributor to the domesticated apple. Mol Ecol 22:2249–2263

Cornille A, Giraud T, Smulders MJM, Roldán-Ruiz I, Gladieux P (2014) The domestication and evolutionary ecology of apples. Trends Genet 30:57–64

Cornille A, Feurtey A, Gélin U, Ropars J, Misvanderbrugge K, Gladieux P, Giraud T (2015) Anthropogenic and natural drivers of gene flow in a temperate wild fruit tree: a basis for conservation and breeding programs in apples. Evol Appl 8:373–384

Cornille A, Antolín F, Garcia E, Vernesi C, Fietta A, Brinkkemper O, Kirleis W, Schlumbaum A, Roldán-Ruiz I (2019) A multifaceted overview of apple tree domestication. Trends Plant Sci 24:770–782. https://doi.org/10.1016/j.tplants.2019.05.007

Costes E, Lauri P, Regnard J (2006) Analyzing fruit tree architecture: implications for tree management and fruit production. In: Janick J (ed) Horticultural reviews. John Wiley & Sons, pp 1–62

Daccord N, Celton J-M, Linsmith G, Becker C, Choisne N, Schijlen E, van de Geest H, Bianco L, Micheletti D, Velasco R, Di Pierro EA, Gouzy J, Rees DJG, Guérif P, Muranty H, Durel C-E, Laurens F, Lespinasse Y, Gaillard S, Aubourg S, Quesneville H, Weigel D, van de Weg E, Troggio M, Bucher E (2017) High-quality *de novo* assembly of the apple genome and methylome dynamics of early fruit development. Nat Genet 49:1099–1106. https://doi.org/10.1038/ng.3886

Duan N, Bai Y, Sun H, Wang N, Ma Y, Li M, Wang X, Jiao C, Legall N, Mao L, Wan S, Wang K, He T, Feng S, Zhang Z, Mao Z, Shen X, Chen X, Jiang Y, Wu S, Yin C, Ge S, Yang L, Jiang S, Xu H, Liu J, Wang D, Qu C, Wang Y, Zuo W, Xiang L, Liu C, Zhang D, Gao Y, Xu Y, Xu K, Chao T, Fazio G, Shu H, Zhong GY, Cheng L, Fei Z, Chen X (2017) Genome re-sequencing reveals the history of apple and supports a two-stage model for fruit enlargement. Nat Commun 8:249. https://doi.org/10.1038/s41467-017-00336-7

Evans K, Patocchi A, Rezzonico F, Mathis F, Durel C, Fernández-Fernández F, Boudichevskaia A, Dunemann F, Stankiewicz-Kosyl M, Gianfranceschi L, Komjanc M, Lateur M, Madduri M, Noordijk Y, van de Weg WE (2011) Genotyping of pedigreed apple breeding material with a genome-covering set of SSRs: trueness-to-type of cultivars and their parentages. Mol Breed 28:535–547

Fazio G, Wan Y, Kviklys D, Romero L, Adams R, Strickland D, Robinson T (2014) *Dw2*, a new dwarfing locus in apple rootstocks and its relationship to induction of early bearing in apple scions. J Am Soc Hortic Sci 139:87–98. https://doi.org/10.21273/JASHS.139.2.87

Fiala JL (1994) Flowering crabapples. The genus Malus. Timber Press, Portland, OR. 273p

Fischer C (1994) Shortening of the juvenile period in apple breeding. In: Schmidt H, Kellerhals M (eds) Progress in temperate fruit breeding: proceedings of the eucarpia fruit breeding section meeting held at Wädenswil/Einsiedeln, Switzerland from August 30 to September 3, 1993. Springer Netherlands, Dordrecht, pp 161–164. https://doi.org/10.1007/978-94-011-0467-8_32

Flachowsky H, Le Roux P-M, Peil A, Patocchi A, Richter K, Hanke M-V (2011) Application of a high-speed breeding technology to apple (*Malus* x *domestica*) based on transgenic early flowering plants and marker-assisted selection. New Phytol 192:364–377. https://doi.org/10.1111/j.1469-8137.2011.03813.x

Forte AV, Ignatov AN, Ponomarenko VV, Dorokhov DB, Savelyev NI (2002) Phylogeny of *Malus* (apple tree) species inferred from the morphological traits and molecular DNA analysis. Russ J Genet 38:1150–1160

Foster T, Johnston R, Seleznyova A (2003) A morphological and quantitative characterization of early floral development in apple (*Malus* × *domestica* Borkh.). Ann Bot 92:199–206. https://doi.org/10.1093/aob/mcg120

Gao Y, Liu F, Wang K, Wang D, Gong X, Liu L, Richards CM, Henk AD, Volk GM (2015) Genetic diversity of *Malus* cultivars and wild relatives in the Chinese National Repository of Apple Germplasm Resources. Tree Genet Genomes 11:106. https://doi.org/10.1007/s11295-015-0913-7

Gross B, Volk GM, Richards CM, Forsline P, Fazio G, Chao C-T (2012) Identification of "duplicate" accessions within the USDA-ARS National Plant Germplasm System *Malus* collection. J Am Soc Hort Sci 137:333–342

Harris SA, Robinson JP, Juniper BE (2002) Genetic clues to the origin of the apple. Trends Genet 18:426–430

Höfer M, Meister A (2010) Genome size variation in *Malus* species. J Bot 2010. https://doi.org/10.1155/2010/480873

Hokanson SC, Lamboy WF, Szewc-McFadden AK, McFerson JR (2001) Microsatellite (SSR) variation in a collection of *Malus* (apple) species and hybrids. Euphytica 118:281–294

Howard NP, van de Weg E, Bedford DS, Peace CP, Vanderzande S, Clark MD, Teh SL, Cai L,

Igic B, Lande R, Kohn JR (2008) Loss of self-incompatibility and its evolutionary consequences. Int J Plant Sci 169:93–104. https://doi.org/10.1086/523362

Janick J (2002) History of the PRI apple breeding program. Acta Hortic 55–60. https://doi.org/10.17660/ActaHortic.2002.595.7

Juniper BE, Mabberley DJ (2006) The Story of the Apple, Timber Press, Inc.

Kenis K, Keuleman J (2007) Study of tree architecture of apple (*Malus* × *domestica* Borkh.) by QTL analysis of growth traits. Mol Breed 19:193–208. https://doi.org/10.1007/s11032-006-9022-5

Khan SA, Tikunov Y, Chibon PY, Maliepaard C, Beekwilder J, Jacobsen E, Schouten HJ (2014) Metabolic diversity in apple germplasm. Plant Breed 133:281–290. https://doi.org/10.1111/pbr.12134

Korban SS, Bournival BL (1987) Catalase, esterase and peroxidase enzymes in seeds and leaves of *Malus domestica* Borkh. Scient Hortic 32:213–219

Koutinas N, Pepelyankov G, Lichev V (2010) Flower induction and flower bud development in apple and sweet cherry. Biotech Biotechnol Equip 24:1549–1558. https://doi.org/10.2478/V10133-010-0003-9

Kubo K, Entani T, Takara A, Wang N, Fields AM, Hua Z, Toyoda M, Kawashima S, Ando T, Isogai A, Kao T, Takayama S (2010) Collaborative non-self recognition system in S-RNase–based self-incompatibility. Science 330:796–799. https://doi.org/10.1126/science.1195243

Lassois L, Denancé C, Ravon E, Guyader A, Guisnel R, Hibrand-Saint-Oyant L, Poncet C, Lasserre-Zuber P, Feugey L, Durel CE (2016) Genetic diversity, population structure, parentage analysis and construction of core collections in the French apple germplasm based on SSR markers. Plant Mol Biol Rep 34:827–844. https://doi.org/10.1007/s11105-015-0966-7

Laurens F, Aranzana MJ, Arus P, Bassi D, Bonany J, Corelli L, Durel CE, Mes J, Pascal T, Patocchi A, Peil A, Quilot B, Salvi S, Tartarini S, Troggio M, Vecchietti A, Velasco R, van de Weg E (2010) Review of fruit genetics and breeding programmes and a new European initiative to increase fruit breeding efficiency. Acta Hortic 929:95–102

Lea AGH, Arnold GM (1978) The phenolics of ciders: bitterness and astringency. J Sci Food Agric 29:478–483

Leroy A (1873) Dictionnaire de pomologie: contenant l'histoire, la description, la figure des fruits anciens et des fruits modernes les plus généralement connus et cultivés. Imprimerie Lachèse, Belleuvre et Dolbeau, Paris

Lo EYY, Donoghue MJ (2012) Expanded phylogenetic and dating analyses of the apples and their relatives (Pyreae, Rosaceae). Mol Phylogenet Evol 63:230–243

Luby JJ (2003) Taxonomic classification and brief history. In: Ferree DC, Warrington IJ (eds) Apples: Botany, Production and Uses. CAB Intl, Wallingford, UK. p 1–14

Ma B, Yuan Y, Gao M, Li C, Ogutu C, Li M, Ma F (2018) Determination of predominant organic acid components in *Malus* species: correlation with apple domestication. Metabolites 8:74. https://doi.org/10.3390/metabo8040074

Magby J, Volk G, Henk A, Miller S (2019) Identification of historic homestead and orchard apple cultivars in Wyoming. HortSci 54:8–16

Marini RP, Fazio G (2018) Apple rootstocks. In: History, Physiology, Management, and Breeding. Hortic Rev 45:197–312

Menendez RA, Larsen FE, Fritts R (1986) Fingerprinting apple cultivars by electrophoretic isozyme banding patterns. J Environ Hort 4:101–107

Muranty H, Denancé C, Feugey L, Crépin JL, Barbier Y, Tartarini S, Ordidge M, Troggio M, Lateur M, Nybom H, Paprstein F, Laurens F, Durel CE (2020) Using whole-genome SNP data to reconstruct a large multi-generation pedigree in apple germplasm. BMC Plant Biol 20:2. https://doi.org/10.1186/s12870-019-2171-6

National Agricultural Library (2020) USDA Pomological Watercolor Collection. https://usdawatercolors.nal.usda.gov/pom/about.xhtml Accessed 28 Aug 2020

Nikiforova SV, Cavalieri D, Velasco R, Goremykin V (2013) Phylogenetic analysis of 47 chloroplast genomes clarifies the contribution of wild species to the domesticated apple maternal line. Mol Biol Evol 30:1751–1760

Okada K, Wada M, Takebayashi Y, Kojima M, Sakakibara H, Nakayasu M, Mizutani M, Nakajima M, Moriya S, Shimizu T, Abe K (2020) Columnar growth phenotype in apple results from gibberellin deficiency by ectopic expression of a dioxygenase gene. Tree Physiol 40:1205–1216. https://doi.org/10.1093/treephys/tpaa049

Ordidge M, Kirdwichai P, Baksh MF, Venison EP, Gibbings JG, Dunwell JM (2018) Genetic analysis of a major international collection of cultivated apple varieties reveals previously unknown historic heteroploid and inbred relationships. PLoS ONE 13. https://doi.org/10.1371/journal.pone.0202405

Peace DP, Bianco L, Troggio M, Van de Weg E, Howard NP, Cornille A, Durel CE, Myles S, Migicovsky Z, Schaffer R, Costes E, Fazio G, Yamane H, van Nocker S, Gottschalk C, Costa F, Chagné D, Zhang X, Patocchi A, Gardiner S, Hardner C, Kumar S, Laurens F, Bucher E, Main D, Jung S, Vanderzande S (2019) Apple whole genome sequences: recent advances and new prospects. Hortic Res 6:59. https://doi.org/10.1038/s41438-019-0141-7

Pratas MI, Aguiar B, Vieira J, Nunes V, Teixeira V, Fonseca NA, Iezzoni A, van Nocker S, Vieira CP (2018) Inferences on specificity recognition at the *Malus* × *domestica* gametophytic self-incompatibility system. Sci Rep 8:1–17. https://doi.org/10.1038/s41598-018-19820-1

Qian G-Z, Liu L-F, Tang G-G (2006) A new section in *Malus* (Rosaceae) from China. Ann Bot Fenn 43:68–73

Qian G-Z, Liu L-F, Hong D-Y, Tang G-G (2008) Taxonomic study of Malus section Florentinae (Rosaceae). Bot J Linn Soc 158:223–227

Richards CM, Volk GM, Reeves PA, Reilley AA, Henk AD (2009) Selection of stratified core sets representing wild apple (*Malus sieversii*). J Am Soc Hort Sci 134:228–235

Robinson JP, Harris SA, Juniper BE (2001) Taxonomy of the genus *Malus* Mill. (Rosaceae) with emphasis on the cultivated apple *Malus Domestica* Borkh. . Plant Syst Evol 226:35–58

Routson KJ, Reilley AA, Henk AD, Volk GM (2009) Identification of historic apple trees in the Southwestern United States and implications for conservation. HortSci 44:589–594

Routson KJ, Volk GM, Richards CM, Smith SE, Nabhan GP, de Echeverria VW (2012) Genetic variation and distribution of pacific crabapple. J Am Soc Hort Sci 137:325–332

Sanoner P, Guyot S, Marnet N, Molle D, Drilleau JF (1999) Polyphenol profiles of French cider apple varieties (*Malus domestica* sp.). Agric Food Chem 47:4847–4853

Segura V, Cilas C, Costes E (2009) Dissecting apple tree architecture into genetic, ontogenetic and environmental effects: mixed linear modelling of repeated spatial and temporal measures. New Phytol 178:302–314. https://doi.org/10.1111/j.1469-8137.2007.02374.x

Stanger MC, Steffens CA, Soethe C, Moreira MA, do Amarante CVT, Both V, Brackmann A, (2018) Phenolic compounds content and antioxidant activity of 'Galaxy' apples stored in dynamic controlled atmosphere and ultralow oxygen conditions. Postharv Biol Technol 144:70–76

Teixeira da Silva JA, Gulyás A, Magyar-Tábori K, Wang M-R, Wang Q-C, Dobránszki J (2019) In vitro tissue culture of apple and other *Malus* species: recent advances and applications. Planta 249:975–1006. https://doi.org/10.1007/s00425-019-03100-x

Urrestarazu J, Denancé C, Ravon E, Guyader A, Guisnel R, Feugey L, Poncet C, Lateur M, Houben P, Ordidge M, Fernandez-Fernandez F, Evans KM, Paprstein F, Sedlak J, Nybom H, Garkava-Gustavsson L, Miranda C, Gassmann J, Kellerhals M, Suprun I, Pikunova AV, Krasova NG, Torutaeva E, Dondini L, Tartarini S, Laurens F, Durel CE (2016) Analysis of the genetic diversity and structure across a wide range of germplasm reveals prominent gene flow in apple at the European level. BMC Plant Biol 16:130. https://doi.org/10.1186/s12870-016-0818-0

USDA (2020) National genetic resources program. Germplasm Resources Information Network (GRIN). USDA ARS Natl Germplasm Resour Lab, Beltsville, MD. https://www.ars-grin.gov/npgs/index.html. Accessed 28 Aug 2020

Velasco R, Zharkikh A, Affourtit J, Dhingra A, Cestaro A, Kalyanaraman A, Fontana P, Bhatnagar SK, Troggio M, Pruss D, Dhingra A, Cestaro A, Kalyanaraman A, Fontana P, Bhatnagar SK, Troggio M, Pruss D, Salvi S, Pindo M, Baldi P, Castelletti S, Cavaiuolo M, Coppola G, Costa F, Cova V, Ri AD, Goremykin V, Komjanc M, Longhi S, Magnago P, Malacarne G, Malnoy M, Micheletti D, Moretto M, Perazzolli M, Si-Ammour A, Vezzulli S, Zini E, Eldredge G, Fitzgerald LM, Gutin N, Lanchbury G, Macalma T, Mitchell JT, Reid J, Wardell B, Kodira C, Chen Z, Desany B, Niazi F, Palmer M, Koepke T, Jiwan D, Schaeffer S, Krishnan V, Wu C, Chu VT, King ST, Vick J, Tao Q, Mraz A, Stormo A, Stormo K, Bogden R, Ederle D, Stella A, Vecchietti A, Kater MM, Masiero S, Lasserre P, Lespinasse Y, Allan AC, Bus V, Chagné D, Crowhurst RN, Gleave AP, Lavezzo E, Fawcett JA, Proost S, Rouzé P, Sterck L, Toppo S, Lazzari B, Hellens RP, Durel C-E, Gutin A, Bumgarner R, Gardiner SE, Skolnick M, Egholm M, Van de Peer Y, Salamini F, Viola R (2010) The genome of the domesticated apple (*Malus* × *domestica* Borkh.). Nat Genet 42:833–839. https://doi.org/10.1038/ng.654

Volk GM, Richards CM, Reilley AA, Henk AD, Reeves PA, Forsline PL, Aldwinckle HS (2008) Genetic diversity of wild *Malus orientalis* from Turkey and southern Russia. J Am Soc Hort Sci 133:383–389

Volk GM, Henk AD, Baldo A, Fazio G, Chao CT, Richards CM (2015) Chloroplast heterogeneity and historical admixture within the genus *Malus*. Am J Bot 102(7):1198–1208

Weeden NF, Lamb RC (1985) Identification of apple cultivars by isozyme phenotypes. J Am Soc Hort Sci 110:509–515

Yao JL, Xu J, Cornille A, Tomes S, Karunairetnam S, Luo Z, Bassett H, Whitworth C, Rees-George J, Ranatunga C, Snirc A, Crowhurst R, de Silva N, Warren B, Deng C, Kumar S, Chagné D, Bus VG, Volz RK, Rikkerink EH, Gardiner SE, Giraud T, MacDiarmid R, Gleave AP (2015) A microRNA allele that emerged prior to apple domestication may underlie fruit size evolution. Plant J 84:417–427

Yu TC (1988) Taxonomy of fruit trees in China. Pome Fruits. 1. *Malus*. Translated from Kuo C-K, Heueh FL, Ghang TP, Lei JX. F'in Xuo Lei. 1979. Pei-ching: Hung yeh eh-u. pa she. pp 87–122

Zhi-Qin Z (1999) The apple genetic resources in China: The wild species and their distributions, informative characteristics and utilization. Genet Resour Crop Evol 46:599–609

Zohary D, Hopf M (2000) Domestication of plants in the Old World, 3rd edn. Oxford University Press, New York, 316 p

Apple Genetic Resources: Diversity and Conservation

3

Gayle M. Volk and Paula Bramel

Abstract

Diverse apple cultivars and wild *Malus* species offer necessary allelic diversity to breeding and research programs. Wild apple species are found in native landscapes of countries throughout the Northern Hemisphere; however, some of these species are at risk of being lost, particularly if trees are unable to thrive under changing climatic conditions, disease/pathogen outbreaks, and human encroachment. Apple genebanks have been established to ensure that user communities have current and future access to these genetic resources. Herein, we describe necessary components of apple genebanks, including collection maintenance forms (trees in orchards, in vitro cultures, low-temperature medium-term storage, and cryopreservation), acquisition of genebank accessions, documentation and data management, distribution of collection materials, as well as characterization and evaluation (phenotypic evaluations and genetic tools for collection management). We conclude with a summary of international apple genebank holdings and provide some examples of how collaborations among genebanks would benefit the global community.

G. M. Volk (✉)
USDA-ARS National Laboratory for Genetic Resources Preservation, 1111 S. Mason St., Fort Collins, CO 80521, USA
e-mail: Gayle.Volk@usda.gov

P. Bramel
Global Crop Diversity Trust, Platz der Vereinten Nationen 7, 53113 Bonn, Germany

3.1 Introduction

The *Malus* gene pool includes cultivars grown for their fruit, which is used for food, vinegar, and cider. These cultivars, many of which have been grown for millennia, are classified as *M.* × *domestica*. *Malus* × *domestica* cultivars range from those having small-sized and high-tannin content fruit that store poorly to those having large, crunchy, and sweet fruit that can be shipped and stored for many months. The widespread availability of internationally recognized commercial cultivars and the long lifespans of trees ensure that they will likely be available for generations to come. In contrast, many historic and local cultivars are regionally grown or exist as remnants in local orchards, field boundaries, or parks. These may be at risk of loss, particularly if such trees are unable to thrive under changing climatic conditions and/or disease/pathogen outbreaks.

The *Malus* gene pool is much broader than the domesticated cultivars and includes wild species found in countries throughout the northern hemisphere. In fact, some of these wild *Malus* species and hybrids have been selected and grown as ornamental crabapple trees, valued for

S. S. Korban (ed.), *The Apple Genome*, Compendium of Plant Genomes,
https://doi.org/10.1007/978-3-030-74682-7_3

their bloom, shape, and bird habitats for the landscape industry. The value of diverse apple cultivars and wild species has long been recognized, and efforts have been made to conserve these genetic resources in national and local ex situ collections in field genebanks. These collections serve as sources of diverse germplasm for researchers, breeders, nurseries, and sometimes even the general public. This chapter focuses primarily on the conservation of apple cultivars as well as *Malus* species diversity within an ex situ genebank framework.

3.2 *Malus* Species Distribution and Diversity

Wild apple species are often a key component of apple genebank collections, providing access to novel genetic resources that may not currently be available from their native countries. Wild *Malus* species offer diversity necessary for future breeding programs that is not available within the crop (Brozynska et al. 2015). Wild species may provide novel alleles for disease resistance, as well as desirable plant architecture, fruit quality, rootstocks, and increased yield (Peace and Norelli 2009; Fazio et al. 2009, 2014; Brown 2012; Duan et al. 2017; Norelli et al. 2017). Although most *Malus* species are interfertile, undesirable fruit characteristics, non-uniform ripening times, and other traits that are not amenable to commercial apple production have mostly deterred the use of wild species in breeding programs, except for a few instances, such as the breeding for apple scab resistance (with *M. floribunda*). With technological advances, however, it is becoming possible to transfer desirable alleles from wild *Malus* species into the cultivated apple faster and without deleterious effects on fruit quality, quantity, and production traits (Kumar et al. 2010). New technologies, such as marker-assisted selection (MAS), genomic selection (GS), genetic engineering, genome-wide association mapping, high-throughput genotyping, and/or rapid-cycling have made it more likely to use a wider range of *Malus* wild species in breeding programs (Velasco et al. 2010; Flachowsky et al. 2011; Chagné et al. 2012; Kumar et al. 2012, 2013; Troggio et al. 2012; Evans 2013; Bianco et al. 2014, 2016; Broggini et al. 2014; Ru et al. 2015).

Traditional breeding programs are more likely to use closely related crop wild relatives to select for genes that may enhance biotic or abiotic stress resistance (Fig. 3.1; Bassett et al. 2011; Bus et al. 2005; Harshman et al. 2017; Norelli et al. 2017). The wild species *M. sieversii* (Central Asia and Western China), *M. orientalis* (Caucasus Region), *M. sylvestris* (Europe), and *M. prunifolia* (China) are closely related to and are likely progenitors of *M.* × *domestica*, as demonstrated by both chloroplast sequence and microsatellite/simple sequence repeat (SSR) data (Nikiforova et al. 2013; Lo and Donoghue 2012; Volk et al. 2015; Duan et al. 2017). The genetic diversity of *M. orientalis, M. sylvestris,* and *M. sieversii* has been compared to that of *M.* × *domestica* using 19 SSRs (Gross et al. 2014). It is found that there are similar levels of expected heterozygosity (He 0.780–0.848) and observed homozygosity (Ho 0.728–0.779) among representatives of sampled species. Allelic richness ranges from 9.01 for *M.* × *domestica* to 12.31 for *M. orientalis*, based on a sample size of 40 accessions (Gross et al. 2014). Furthermore, genome sequence data have revealed that the nucleotide diversity of *M.* × *domestica* is lower than that of both *M. sieversii* (from Kazakhstan) and *M. sylvestris;* however, *M. sieversii* from the Tien Shan Mountains (China) has a lower nucleotide diversity than that of either *M.* × *domestica* or *M. sieversii* of Kazakhstan (Duan et al. 2017).

Assessments of apple cultivars based on microsatellite/SSR data have revealed that the genetic diversity available in apple cultivars has not significantly changed between the 1600s to the present, as measured by expected and observed heterozygosity (Gross et al. 2014). However, there is an increased level of diversity in cultivars developed after the 1950s, and this could be due to the introgression of genes from the wild relatives for disease resistance (Gross et al. 2014). SSR markers have been used to identify old dessert, old cider, and modern

Fig. 3.1 Fruit images of various *Malus* species. **a** *M. sieversii*; **b** *M. orientalis*; **c** *M. sylvestris*; **d** *M. prunifolia*; **e** USDA-ARS National Plant Germplasm (NPGS) apple collection; and **f** selected *M. sieversii* fruit samples from the NPGS apple collection demonstrating phenotypic diversity within *Malus*. Images provided by USDA-Plant Genetic Resources Unit (**a**, **b**, **c**, **d**) and G. Volk (**e**, **f**)

cultivars present in the highly heterogeneous French apple genebank collection (Lassois et al. 2016). Additional genomic assessments comparing cider and dessert apple cultivars have revealed underlying genetic differences between these two groups of *M.* × *domestica* and have identified quantitative trait loci (QTLs) associated with key metabolic pathways, such as those involved in the formation of polyphenolic compounds (Leforestier et al. 2015). Furthermore, diversity assessed using 16 SSR markers has identified sets of cultivars that could be generally differentiated based on their origin, including northern and eastern Europe, western Europe, and southern Europe (Urrestarazu et al. 2016). This has suggested the presence of an underlying genetic complex involved in the adaptation of cultivars to these different regions. With this knowledge, strategies can be developed to conserve and protect wild *Malus* species in their native habitats and in ex situ genebanks to prevent loses due to environmental, biotic, and human pressures.

3.3 Apple Collections

Apple (*Malus*) collections have been established by either federal or national governments, universities, private companies, arboreta, non-profit organizations, and individuals. These collections are usually long-term investments committed to conserve cultivars and/or wild species for research, breeding, conservation, distribution, and public interest. In some cases, these collections may also be considered genebank collections, particularly if there is an emphasis on both conservation and distribution of materials.

In the present context, apple genebanks are considered to be collections of apple genetic resources that are conserved for the long-term and that are made available to a customer base. These collections may take on many different life forms, existing as trees in the field, as in vitro cultures, or as seeds. Key aspects of apple genebanks include collection maintenance and security back-ups, acquisition, distribution, database documentation, and associated phenotypic/genotypic information.

3.3.1 Collection Composition

Apple collections may include a number of different types of cultivars and species materials. Some collections are designed to bring together a set of highly desirable commercially grown fruit cultivars. These may serve as a set of cultivars to assess how cultivars perform under specific field conditions for possible inclusion in production orchards. Other collections may be comprised of landraces of either local or historical importance. In some cases, the fruit of these landrace or historical cultivars may not have the same quality attributes of major commercial cultivars, but they could possess traits that render them adaptive to local conditions or biotic/abiotic stress pressure conditions. Arboreta and botanic gardens often maintain apple tree collections; many of which may be ornamental flowering crabapples valued for their showy flowers along with small colorful fruit that may persist on the tree long into the winter months. Furthermore, apple collections may include cultivars that are novelties. These may have originated from far-away locations, have local importance, and exhibit either interesting or unusual desirable traits.

Apple collections may also focus on botanical variations within *Malus* species. In this instance, planted materials may represent wild apple species to demonstrate the wide range of diversity of traits available among apple species and to provide species exemplars for taxonomic and phenotypic evaluations, or for use in breeding programs. Finally, apple collections may be comprised of large plantings of a single wild species for use in assessing phenotypic variation and genetic diversity within a selected species. In addition, these species orchards may be used as seed sources for wild apple revegetation projects (Schnitzler et al. 2014)

Financial and physical land constraints may place limits on how extensive a specific orchard collection can be, thus forcing institutions and collection managers to ensure that their collection focus meets the needs of the intended user community. Some genebanks have used phenotypic and genotypic data to identify smaller sets of individuals that may be representative of a larger collection (Hokanson et al. 1998; Gross et al. 2013; Liang et al. 2015; Urrestarazu et al. 2016; Lassois et al. 2016). In some cases, core sets have been designed to be smaller subsets of the entire collection, and sometimes these are even a smaller set of a pre-defined part of a collection (i.e., a specific species or a collection region) (Hokanson et al. 1998; Gross et al. 2013). Core collections have been identified, using both morphological and SSR data, for *M. sieversii* and *M. orientalis* collections in the USDA-ARS National Plant Germplasm System (NPGS) (Volk et al. 2005, 2009; Richards et al. 2009). Core subsets for cider and dessert apples have also been established for French apple collections (Lassois et al. 2016; LeForestier et al. 2016).

3.3.2 Maintenance Forms

The majority of the world's apple genebanks is maintained as field collections, with either one or several trees of each cultivar or species representative grown in either a single or in duplicated orchards. These field collections may be supplemented with in vitro collections, cryopreserved back-ups, and seeds in long-term storage.

3.3.2.1 Field Collections

Most apple collections are grown in the field as cultivars/species (commercial, historic/landrace, and ornamental, among others) and are grafted onto rootstocks. Specific apple cultivars/species are clonally propagated and maintained by grafting/budding, as they are not true-to-type when seeds from a specific cultivar/species are planted. Apple trees are self-incompatible, with bees delivering pollen from neighboring trees of different genetic backgrounds. Seeds resulting from outcrossed cultivars are hybrids, and have combinations of genes from both parents; moreover, recombination events usually result in seeds likely carrying undesirable combinations of traits.

Traditionally, rootstocks are propagated from seeds collected from open-pollinated apple trees, grown into young seedlings (1–2-year-old), and used for grafting/budding of fruiting (scion) cultivars (in the form of either scionwood or budwood). In the 1800s, European apple producers recognized the value of breeding apple rootstocks that yield lines or clones that provide fruiting cultivars with enhanced disease resistance, preferred tree architecture, dwarf habit, and precocity. In particular, the choice of a rootstock helps in the control of tree architecture, as well as of disease/pest tolerance/resistance of a grafted tree (Marini and Fazio 2018; see Chap. 6 in this volume).

Apple field collections are time-consuming and costly, requiring field maintenance including mowing, pruning, protection from animal grazing (voles, deer), and at least a minimum number of spray applications to control insects and diseases. Year-round field maintenance is expensive and labor-intensive, particularly as orchards age, and require re-propagation. Collections also pose significant challenges with respect to the maintenance of many cultivars/species of differing phenologies, thereby rendering their maintenance quite different from production orchards of a few cultivars. Diverse cultivars/species will have different bloom and harvest phenologies, as well as varying levels of pest/pathogen vulnerabilities, thus requiring different spray regimes.

The emphasis on a field genebank or an orchard for apples renders these genebank collections quite different from other genebank collections that are primarily dedicated for seed storage, such as those for grains, legumes, and vegetables. As a field collection, apple genebanks offer several advantages. For one, field materials can be readily distributed as either dormant budwoods or scionwoods that are then used for clonal propagation via budding or grafting, respectively (Fig. 3.2a, b). In addition, data can be collected from orchard trees throughout the year, and controlled crosses or hybridizations can be made during the bloom season. Furthermore, fruit is available at harvest for testing and/or for evaluation. Although field collections are convenient to the user community, they are also vulnerable to pests, diseases, and environmental threats.

A key consideration for field collections is to assess whether or not the collection is conserved in a manner that renders it less vulnerable to environmental and biotic threats. In some cases, a genebank accession is maintained as multiple clonal trees within a primary planting or at a secondary site. In other cases, original seeds from germplasm collection trips are maintained in cold storage at single or multiple locations, thereby rendering collections less vulnerable to loss. Furthermore, in vitro cultures may be established to complement field collections, as well as to serve as sources of materials that may be free of endophytes or diseases.

Finally, cryopreservation technologies can be used to preserve either dormant budwood or in vitro shoot-tip cultures of apple cultivars, thus ensuring that specific allelic combinations are conserved for the long term at an offsite location. Another key consideration is whether or not the appropriate conservation target is conserved; i.e., specific cultivars can be maintained as clonal trees in the field, as in vitro shoot cultures (Fig. 3.2c), or as cryopreserved dormant buds or shoot-tips (Fig. 3.2d, e). Wild populations may best be conserved as seed collections, possibly augmented by orchard plantings for use in phenotyping, genetic assessments, and breeding.

3.3.2.2 In Vitro Collections

In some cases, apple collections may be maintained, in part, as proliferating in vitro shoot cultures, or in a greenhouse. Often, the expense of in vitro collections limits the use of this maintenance form. However, such in vitro shoot cultures can be particularly useful for large-scale propagation of clonal rootstocks and for disease-free propagation materials (Fig. 3.2c–f, Webster and Jones 1989).

3.3.2.3 Cold Storage

Within genebanks, seeds of accessions of wild *Malus* species may be stored in either a freezer or under cryogenic conditions to extend their longevity (Fig. 3.2g; Kushnarenko et al. 2010; Michalak et al. 2015). Orthodox apple seeds are equilibrated to optimal moisture content (12% moisture content at the storage temperature, obtained by using methods described by the Center for Plant Conservation 2021) prior to storage at -18 °C freezer or under liquid nitrogen conditions, as they can often survive for decades or even for centuries (Walters et al. 2004, 2005). Seed conservation may be particularly critical to conserve *Malus* species when native populations may no longer be available for acquisition by genebanks. Long-term seed storage effectively conserves seedlots of wild *Malus* species populations.

3.3.2.4 Liquid Nitrogen Storage

Some genebanks have a long-term conservation component that takes advantage of cryopreservation techniques. Cryopreservation, involving storage of propagules in either a liquid or vapor phase of liquid nitrogen, facilitates storage of plant genetic resources in secure secondary sites at relatively low annual costs (compared with field collection maintenance), once accessions are in cryopreserved form. For apples, cryopreservation methods have been developed to conserve either dormant buds or shoot-tips (Wang et al. 2018). Apple cryopreservation technologies have been first implemented at the USDA NPGS-National Laboratory for Genetic Resources Preservation as dormant buds (Forsline et al. 1998; Towill et al. 2004). The general protocol involves the collection of dormant scionwood (budwood) in mid-winter, desiccation to about 30% moisture content, and then slow-cooling to −30 to −35 °C prior to placement in liquid nitrogen vapor for long-term storage (Fig. 3.2a, Forsline et al.1998; Towill et al. 2004; Höfer 2015). Viability is then measured by warming, rehydrating, and grafting of cryopreserved vegetative buds onto rootstocks (Fig. 3.2b, e). *Malus* species vary in their response to cryopreservation, but many apple cultivars and wild relatives can be cryopreserved with high levels of recovery (Towill et al. 2004; Volk et al. 2017).

Shoot-tip cryopreservation methods have also been developed for apple. These methods may be particularly applicable for apple cultivars that are either not amenable for dormant bud cryopreservation techniques or when apple collections are grown in regions that do not acquire adequate levels of winter dormancy for successful dormant bud cryo-processing. Shoot-tip cryopreservation methods utilize 1 mm excised shoot-tips, usually collected from in vitro stock plants or proliferating shoot cultures, and take advantage of

Fig. 3.2 *Malus* propagules for preservation. **a** Nodal sections of dormant buds. **b** Grafted dormant bud exhibiting regrowth after cryopreservation. **c** Proliferating shoot cultures. **d** Shoot-tips that have regrown into plants after cryopreservation. **e** Cryopresreved dormant bud 2 weeks after grafting. **f** Shoot-tip, 3 weeks after cryopreservation (scale bar = 2 mm). **g** Seeds. Images provided by G. Volk (**a**, **c**, **g**), R. Bonnart (**b**, **e**), and J.C. Bettoni (**d**, **f**)

vitrification, as well as encapsulation-dehydration procedures (Fig. 3.2c, d and f, Wang et al. 2018).

Cryopreservation provides secure back-ups of apple collections, and it is usually implemented within the context of genebanks, wherein specialized equipment and trained staff are available to process and maintain cryo-collections. The largest genebank collection of apple dormant buds is located at the USDA-ARS-NLGRP, wherein 2,155 apple accessions have been cryopreserved as dormant buds (Jenderek and Reed 2017).

3.3.3 Acquisition of Genebank Accessions

Apple collections acquire their materials from a number of different sources. Many collections have originated as sets of cultivars assimilated by apple breeders. When the value of these collections is recognized, these may then be converted into genebank collections. New acquisitions of cultivars may be received through donations, requests, or exchanges with either individuals or other collection sites. These acquisitions may be straightforward if both collections are in the same region. However, if cultivars are imported from other countries, they may have to go through an intensive quarantine, disease, and clean-up processes to ensure that imported plant materials do not introduce diseases into the recipient country.

Seeds are often easier to import through quarantine than vegetative propagules because many diseases are not spread through seeds. As noted previously, seeds do not reproduce true-to-type phenotypes as clonal vegetative tissues, but they do capture species diversity. As a result, most plant collection explorations capturing wild species diversity focus on the collection and import of seeds.

Accurate acquisition documents of either cultivars (source, history, and pedigree) or wild-collected materials increase the value of materials of plant collections, as poorly documented materials are usually less requested than those with reliable provenance data. Careful consideration should be taken into the collection strategy pursued when seeds are collected from natural populations of apple species. Standard documentation information for wild-collected seeds includes collection date, geographic coordinates of the source tree, image of the source tree, phenotypic data (fruit size, shape, and photographic image) of the source fruit, tree circumference, and surrounding vegetation. In some cases, seeds are pooled from trees in each population, but it may be preferable to keep seeds from each source tree as separate seedlots. By keeping source tree seedlots separate, subsequent phenotypic and genotypic analyses can assess variability within source trees, as well as within and among populations.

3.3.4 Documentation and Data Management

Accurate, detailed passport, and orchard locator information are necessary for apple collection documentation. Source history, cultivar, pedigree, planting date, and rootstock type are required to track age, condition, and vulnerability of trees. Additional information such as phenotypic traits, images, and genotype/genetic data of the collection materials can be particularly useful. Phenotypic data aid users and repository managers in understanding the susceptibility (or resistance) of the specific collection materials to diseases and pests. If available, it also provides valuable information about flowering time (under environmental conditions of the collection), fruit maturity date, fruit quality traits, and possibly yield (depending upon pruning practices). Photographic images are useful for visual confirmation of fruit appearance, particularly when fruit data are needed during non-fruiting seasons for comparative purposes (determining if a cultivar designation is accurate or if the fruit appearance of one tree matches that of another tree of the same name or of unknown identity). Genotypic data are critical for molecular comparisons, particularly of those related to identity and genetic relatedness among accessions in a collection or among different collections. These data have also been useful to determine if scions are "true-to-type" instead of being suckers grown from rootstocks in poorly maintained trees.

Documentation methods vary depending on the use and purpose of a collection. In some cases, spreadsheets or personal databases render information readily accessible in an acceptable format. Collections that are available for distribution may have traditionally used printed catalogs, which have been mostly replaced by searchable online websites and databases. Genebank collections generally have larger-scale databases allowing for these data to be available for downloads, and for tracking distribution and inventory records. Ideally, inventory databases are interoperable and provide access to availability, passport, phenotypic, and genotypic data from many sources.

There are many genebank databases with information that are publicly available. GRIN-Global is the database for the NPGS (U.S. Department of Agriculture 2021). An example of regional databases is EURISCO, a European database for sharing information about materials in the European Cooperative Programme for Plant Genetic Resources (ECPGR) (Van Hintum and Knüpffer 2010). Genesys is a database that shares passport information (cultivar name, species name, holding institute, biological status, country and geographic coordinates of origin, type of germplasm storage, and availability for distribution) from collections around the world, which have provided such data (Genesys 2021). The Global Biodiversity Information Facility (GBIF) maintains a global database of information on materials collected and maintained as either live-plant materials or as herbarium collections (GBIF 2017). This database includes basic information such as genus, species, georeference of collecting site, and data source. FAO's WIEWS database serves as a repository of metadata on holdings of over 600 genebanks in 82 countries.

3.3.5 Distribution of Genebank Materials

Genebanks often distribute germplasm to their user community upon request. Dormant budwood may be collected from trees, and then distributed during the winter months. Alternatively, actively growing shoots, leaves, pollen, and even fruit can be distributed. In some cases, grafted trees may be available. Propagule distributions are most easily handled within countries or within phytosanitary regions (sets of European countries).

The movement of apple germplasm across international borders is regulated by national phytosanitary regulations. For apple seeds, this often involves obtaining phytosanitary certificates and having inspections performed at the point of entry into a country. Clonal propagules, living tissues, or trees often undergo extensive disease screening in quarantine programs to ensure that no new diseases are introduced into a recipient country or region.

3.3.6 Phenotypic Characterization and Evaluation

Ideally, genebanks provide well-described, true-to-type cultivars (or species) to their users. However, characterization and evaluation of collections to ensure trueness-to-type are time-consuming and expensive. National and regional efforts have sought to standardize phenotypic data, with the goal of ensuring that collected data are useful to breeding and research programs. One of the major challenges is to account for environmental variation in phenotypic responses as collections are often kept only at a single location, with cultivars represented by either a single or a few trees in orchard blocks. Sets of defined descriptors have been published by projects (such as RosBREED), genebanks (such as USDA and China), and from the international community (such as the International Board for Plant Genetic Resources) (Evans et al. 2012; Gianfranceschi and Soglio 2004; U.S. Department of Agriculture 2021; Watkins and Smith 1982; Wang et al. 2005).

3.3.7 Genetic Tools for Collection Management

Genetic marker and sequence information collected for genebank inventories facilitate comparisons of the genetic composition between collections. This information has been used to confirm identities, identify unknowns, confirm sets of duplicates or sport-families, pedigree relationships and compare collections grown at distant locations (Evans et al. 2011; van Treuren et al. 2010). Due to high levels of allelic diversity among apple cultivars and clonal materials of wild species, relatively few markers have successfully differentiated among these materials (aside from sports of cultivars) (Foroni et al. 2012; Moriya et al. 2011). A recent large collaborative research project has revealed both within-collection and among-collection genetic variations using microsatellite markers and includes apple collections available in the following countries: France, Czech Republic, Spain, Switzerland, UK, Italy, Belgium, and Russia (Urrestarazu et al. 2016). Availability of single nucleotide polymorphism (SNP) marker platforms has facilitated the use of markers in breeding programs, pedigree identifications, as well as for genebank collection comparisons (Howard et al. 2017; Peace et al. 2019). An Illumina 8K SNP array, an Illumina 20K SNP array, and an Affymetrix Apple480K array are available (Chagné et al. 2012; Bianco et al. 2014, 2016; Vanderzande et al. 2019). Ongoing coordinated national and regional efforts among genebanks to compare collections at the phenotypic and genetic levels, as well as efforts to confirm cultivar/species/accession identities will enable greater use of genebank collections in the future (Kellerhals e al. 2012; Urrestarazu et al. 2016; Liang et al. 2015; Lassois et al. 2016).

3.4 Worldwide Apple Conservation

Apple genebanks and collections have sought out both common and rare apple cultivars in attempts to ensure the long-term conservation of cultivars/germplasm. These collections are often established for different purposes, such as use in breeding programs, long-term conservation, distribution of budwood or grafted trees, and for public interest, among others. Although collections are often prized for their unusual holdings, the wide range in phenotypic diversity as well as flowering and fruiting timeframes may render collection management difficult, particularly for controlling diseases and pests. It is expensive to maintain apple collections for the long term, especially if the goal is to ensure that none of these cultivars is lost.

There are dozens of apple genebanks around the world and information gathered from collection managers has been used to develop a global strategy for the conservation and use of genetic resources (Bramel and Volk 2019; Volk and Bramel 2017). Responses from 35 genebanks in a 2014 survey have revealed the existence of a total of 33,588 *Malus* accessions that are held ex situ in the form of cultivars, wild species, and seeds. Nearly all of the accessions in these collections are grown as grafted trees in the field. Additional information about the collection responses is available in Bramel and Volk (2019). Sharing of information about accession sources, conditions, morphological evaluations, and genetic characterizations of all these materials will facilitate future efforts to align national or local conservation efforts and to ensure the global security of apple genebanks.

Due to challenges in germplasm transport across international borders, it is important that both genebank managers and user communities are fully aware of which genebank accessions match those held by other collections, as well as which accessions are indeed unique, and are thus particularly vulnerable to lose. Genetic and phenotypic assessments may identify local landrace cultivars that are of potential interest to regional or international programs (Bramel and Volk 2019).

Wild apple species are native to many temperate countries in the northern hemisphere (Volk and Bramel 2017). These genetic resources within national borders could be conserved as either ex situ materials in genebanks or in situ as protected areas in the landscape. It benefits the international community when countries choose to protect their wild genetic resources, even if, due to political reasons, they are unable to distribute and share them with other countries at this time. This complimentary ex situ and in situ approach toward genebanking helps to ensure the long-term survival of key apple cultivars, landraces, and wild species for use in the future (Bramel and Volk 2019).

References

Bianco L, Cestaro A, Sargent DJ, Banchi E, Derdak S, Di Guardo M, Salvi S, Jansen J, Viola R, Gut I, Laurens F, Chagné D, Velasco R, van de Weg E, Troggio M (2014) Development and validation of a 20K single nucleotide polymorphism (SNP) whole genome genotyping array for apple (*Malus* × *domestica* Borkh.). PLoS One 9:e110377

Bianco L, Cestaro A, Linsmith G, Murranty H, Denancé C, Théron A, Poncet C, Micheletti D, Kerschbamer E, Di Pierro EA, Larger S, Pindo M, Van de Weg E, Davassi A, Laurens F, Velasco R, Durel C-E, Troggio M (2016) Development and validation of the Axiom Apple 480K SNP genotyping array. Plant J 86:62–74

Bramel PJ, Volk G (2019) A global strategy for the conservation and use of apple genetic resources. Global Crop Diversity Trust. Bonn, Germany. https://doi.org/10.13140/RG.2.2.34072.34562

Broggini GAL, Wöhner T, Fahrentrapp J, Kost TD, Flachowsky H, Peil A, Hanke M-V, Richter K, Patocchi A, Gessler C (2014) Engineering fire blight resistance into the apple cultivar 'Gala' using the *FB_MR5* CC-NBS-LRR resistance gene of *Malus* × *robusta* 5. Plant Biotechnol J 12:728–733

Brown S (2012) Apple. In: Byrne DH (ed) Badenes ML. Fruit breeding, Springer Publ, pp 329–367

Brozynska M, Furtado A, Henry RJ (2015) Genomics of crop wild relatives: expanding the gene pool for crop improvement. Plant Biotechnol J 14:1070–1085. https://doi.org/10.1111/pbi.12454

Bus VGM, Laurens FND, van de Weg WE, Rusholme RL, Rikkerink EHA, Gardiner SE, Bassett HCM, Kodde LP, Plummer KM (2005) The *Vh8* locus of a new gene-for-gene interaction between *Venturia inaequalis* and the wild apple *Malus sieversii* is closely linked to the Vh2 locus in *Malus pumila* R12740–7A. New Phytol 166:1035–1049

Center for Plant Conservation (2021) Cleaning, processing, drying, and storing orthodox seeds in CPC best plant conservation practices to support species survival in the wild. https://academy.saveplants.org/node/67. Accessed 13 Jan 2021

Chagné D, Crowhurst RN, Troggio M, Davey MW, Gilmore B, Lawley C, Vanderzande S, Hellens RP, Kumar S, Cestaro A, Velasco R, Main D, Rees JD, Iezzoni A, Mockler T, Wilhelm L, Van de Weg E, Gardiner SE, Bassil N, Peace C (2012) Genome-wide SNP detection, validation, and development of an 8K SNP array for apple. PLoS ONE 7(2). https://doi.org/10.1371/journal.pone.0031745

Duan N, Bai Y, Sun H, Wang N, Ma Y, Li M, Wang X, Jiao C, Legall N, Mao L, Wan S, Wang K, He T, Feng S, Zhang Z, Mao Z, Shen X, Chen X, Jiang Y, Wu S, Yin C, Ge S, Yang L, Jiang S, Xu H, Liu J, Wang D, Qu C, Wang Y, Zuo W, Xiang L, Liu C, Zhang D, Gao Y, Xu Y, Xu K, Chao T, Fazio G, Shu H, Zhong G-Y, Cheng L, Fei Z, Chen X (2017) Genome re-sequencing reveals the history of apple and supports a two-stage model for fruit enlargement. Nat Commun 8:249. https://doi.org/10.1038/s41467-017-00336-7

Evans K (2013) Apple breeding in the Pacific Northwest. Acta Hortic 976:75–78

Evans KM, Patocchi A, Rezzonico F, Mathis F, Durel CE, Fernández- Fernández F, Boudichevskaia A, Dunemann F, Stankiewicz-Kosyl M, Gianfranceschi L, Komjanc M, Lateur M, Madduri M, Noordijk Y, van de Weg E (2011) Genotyping of pedigreed apple breeding material with a genome-covering set of SSRs: trueness-to-type of cultivars and their parentages. Mol Breed 28:535–547

Evans K, Guan Y, Luby J, Clark M, Schmitz C, Brown S, Orcheski B, Peace C, van de Weg E, Iezzoni A (2012) Large-scale standardized phenotyping of apple in RosBREED. Acta Hortic 945:223–238

Fazio G, Aldwinckle HS, Volk GM, Richards CM, Janisiewicz W, Forsline PL (2009) Progress in evaluating *Malus sieversii* for disease resistance and horticultural traits. Acta Hortic 814:59–66

Fazio G, Chao CT, Forsline P, Richards C, Volk G (2014) Tree and root architecture of *Malus sieversii* seedlings for rootstock breeding. Acta Hortic 1058:585–594

Flachowsky H, Le Roux PM, Peil A, Patocchi A, Richter K, Hanke M-V (2011) Application of a high-speed breeding technology to apple (*Malus* × *domestica*) based on transgenic early flowering plants and marker-assisted selection. New Phytol 192:364–377

Foroni I, Baptista C, Monteiro L, Lopes MS, Mendonça D, Melo M, Carvalho C, Monjardino P, Lopes DJH, Machado A (2012) The use of microsatellites to analyze relationships and to decipher homonyms and synonyms in Azorean apples (*Malus* × *domestica* Borkh.). Plant Syst Evol 298:1297–1313

Forsline PL, Towill LE, Waddell JW, Stushnoff C, Lamboy WF, McFerson JR (1998) Recovery and longevity of cryopreserved dormant apple buds. J Am Soc Hort Sci 123:365–370

GBIF (Global Biodiversity Information Facility) (2017) Explore occurrences. https://www.gbif.org/occurrence

Genesys (2021) A gateway to genetic resources. https://www.genesys-pgr.org. Accessed 8 Jan 2021

Gianfranceschi L, Soglio V (2004) The European Project HiDRAS: innovative multidisciplinary approaches to breeding high quality disease resistant apples. Acta Hortic 663:327–330

Gross BL, Volk GM, Richards CM, Reeves PA, Henk AD, Forsline PL, Szewc-McFadden A, Fazio G, Chao CT (2013) Diversity captured in the USDA-ARS-national plant germplasm system apple core collection. J Am Soc Hort. Sci 138:375–381

Gross B, Henk A, Richards C, Fazio G, Volk G (2014) Genetic diversity in *Malus* × *domestica* (Rosaceae) through time in response to domestication. Am J Bot 101(10):1770–1779

Harshman JM, Evans KM, Allen H, Potts R, Flamenco J, Aldwinckle HS, Wisniewski ME, Norelli JL (2017) Fire Blight Resistance in Wild Accessions of *Malus Sieversii*. 101:1738–1745

Höfer M (2015) Cryopreservation of winter-dormant apple buds: establishment of a duplicate collection of *Malus* germplasm. Plant Cell Tiss Org Cult 121:647–656

Hokanson SC, Szewc-McFadden AK, Lamboy WF, McFerson JR (1998) Microsatellite (SSR) markers reveal genetic identities, genetic diversity and relationships in a *Malus* × *domestica* Borkh. core subset collection. Theor Appl Genet 97:671–683

Howard NP, van de Weg E, Bedford DS, Peace CP, Vanderzande S, Clark MD, Teh SL, Cai L, Luby JJ (2017) Elucidation of the 'Honeycrisp' pedigree through haplotype analysis with a multi-family integrated SNP linkage map and a large apple (*Malus* × *domestica*) pedigree-connected SNP data set. Hortic Res 4:17003. https://doi.org/10.1038/hortres.2017.3

Jenderek MJ, Reed BM (2017) Cryopreserved storage of clonal germplasm in the USDA National Plant

Germplasm System. Vitro Cell Dev Biol-Plant 53:299–308

Kellerhals M, Szalatnay D, Hunziker K, Duffy B, Nybom H, Ahmadi-Afzadi M, Höfer M, Richter K, Lateur M (2012) European pome fruit genetic resources evaluated for disease resistance. Trees 26:179–189

Kumar S, Volz RK, Alspach PA, Bus VGM (2010) Development of a recurrent apple breeding programme in New Zealand: a synthesis of results, and a proposed revised breeding strategy. Euphytica 173:207–222

Kumar S, Bink MCAM, Volz RK, Bus VGM, Chagné D (2012) Towards genomic selection in apple (*Malus* × *domestica* Borkh.) breeding programmes: prospects, challenges and strategies. Tree Genet Genomes 8:1–14

Kumar S, Garrick DJ, Bink MCAM, Whitworth C, Chagné D, Volz RK (2013) Novel genomic approaches unravel genetic architecture of complex traits in apple. BMC Genomics 14:393

Kushnarenko S, Salnikov E, Nurtazin M, Mukhitdinova Z, Rakhimbaev I, Reed BM (2010) Characterization and cryopreservation of *Malus sieversii* seeds. Asian Aust J Plant Sci Biotechnol 4:5–9

Lassois L, Denancé C, Ravon E, Guyader A, Guisnel R, Hibrand-Saint-Oyant L, Poncet C, Lasserre-Zuber P, Feugey L, Durel C-E (2016) Genetic diversity, population structure, parentage analysis, and construction of core collections in the French apple germplasm based on SSR markers. Plant Mol Biol Rep 34:827–844

Leforestier D, Ravon E, Muranty H, Cornille A, Lemaire C, Giraud T, Durel C-E, Branca A (2015) Genomic basis of the differences between cider and dessert apple varieties. Evol Appl 8:650–661

Liang W, Dondini L, De Franceschi P, Paris R, Sansavini S, Tartarini S (2015) Genetic diversity, population structure and construction of a core collection of apple cultivars from Italian germplasm. Plant Mol Biol Rep 33:458–473. https://doi.org/10.1007/s11105-014-0754-9

Lo EYY, Donoghue MJ (2012) Expanded phylogenetic and dating analyses of the apples and their relatives (Pyreae, Rosaceae). Mol Phylgenet Evol 63:230–243

Marini RP, Fazio G (2018) Apple rootstocks: History, physiology, management, and breeding. Hortic Rev 45:197–312

Michalak M, Plitta-Michalak BP, Chmielarz P (2015) Desiccation tolerance and cryopreservation of wild apple (*Malus sylvestris*) seeds. Seed Sci Technol 43:480–491

Moriya S, Iwanami H, Okada K, Yamamoto T, Abe K (2011) A practical method for apple cultivar identification and parent-offspring analysis using simple sequence repeat markers. Euphytica 177:135–150

Nikiforova SV, Cavalieri D, Velasco R, Goremykin V (2013) Phylogenetic analysis of 47 chloroplast genomes clarifies the contribution of wild species to the domesticated apple maternal line. Mol Breed Evol 30:1750–1760

Norelli JL, Wisniewski M, Fazio G, Burchard E, Gutierrez B, Levin E, Droby S (2017) Genotyping-by-sequencing markers facilitate the identification of quantitative trait loci controlling resistance to *Penicillium expansum* in *Malus sieversii*. PLoS ONE 12(3). https://doi.org/10.1371/journal.pone.0172949

Peace C, Norelli J (2009) Genomics approaches to crop improvement in the Rosaceae. In: Folta KM, Gardiner SE (eds) Genetics and genomics of Rosaceae. Springer, New York, pp 19–53

Peace CP, Bianco L, Troggio M, van de Weg E, Howard NP, Cornille A, Durel C-E, Myles S, Migicovsky Z, Schaffer RJ, Costes E, Fazio G, Yamane H, van Nocker S, Gottschalk C, Costa F, Chagné ZX, Patocchi A, Gardiner SE, Hardner C, Kumar S, Laurens F, Bucher E, Main D, Jung S, Vandersande S (2019) Apple whole genome sequences: recent advances and new prospects. Hortic Res 6:59. https://doi.org/10.1038/s41438-019-0141-7

Richards CM, Volk GM, Reeves PA, Reilley AA, Henk AD (2009) Selection of stratified core sets representing wild apple (*Malus sieversii*). J Am Soc Hort Sci 134:228–235

Ru S, Main D, Evans K, Peace C (2015) Current applications, challenges, and perspectives of marker-assisted seedling selection in Rosaceae tree fruit breeding. Tree Genet Genome 11:8. https://doi.org/10.1007/s11295-015-0834-5

Schnitzler A, Arnold C, Cornille A, Bachmann O, Schnitzler C (2014) Wild European apple (*Malus sylvestris* (L.) Mill.) population dynamics: Insight from genetics and ecology in the Rhine valley. Priorities for a future conservation programme. PLoS One 9(5):e96596. https://doi.org/10.1371/journal.pone.0096596

Towill LE, Forsline PL, Walters C, Waddell JW, Laufmann J (2004) Cryopreservation of *Malus* germplasm using a winter vegetative bud method: Results from 1915 accessions. CryoLetters 25:323–334

Troggio M, Gleave A, Salvi S, Chagné D, Cestaro A, Kumar S, Crowhurst RN, Gardiner SE (2012) Apple, from genome to breeding. Tree Genet Genomes 8:509–529

U.S. Department of Agriculture (2021) Germplasm Resources Information Network (GRIN-Global). 8 Jan 2021. https://npgsweb.ars-grin.gov/gringlobal/search.aspx?

Urrestarazu J, Denancé C, Ravon E, Guyader A, Guisnel R, Feugey L, Poncet C, Lateur M, Houben P, Ordidge M, Fernandez-Fernandez F, Evans KM, Paprstein F, Sedlak J, Nybom H, Garkava-Gustavsson L, Miranda C, Gassmann J, Kellerhals M, Suprun I, Pikunova AV, Krasova NG, Torutaeva E, Dondini L, Tartarini S, Laurens F, Durel C-E (2016) Analysis of the genetic diversity and structure across a wide range of germplasm reveals prominent gene flow in apple at the European level. BMC Plant Biol 16:130. https://doi.org/10.1186/212870-016-0818-0

Van Hintum T, Knüpffer H (2010) Current taxonomic composition of European genebank material documented in EURISCO. Plant Genet Resour 8 (2):182–188

Van Treuren R, Kemp H, Ernsting G, Jongejans B, Houtman H, Visser L (2010) Microsatellite genotyping of apple (*Malus* x *domestica* Borkh.) genetic resources in the Netherlands: application in collection management and variety identification. Genet Resour Crop Evol 57:853–865

Vanderzande S., Howard NP, Cai L, Da Silva Linge C, Antanaviciute L, Bink MCAM, Kruisselbrink JW, Bassil N, Gasic K, Iezzoni A, Van de Weg E, Peace C (2019) High-quality, genome-wide SNP genotypic data for pedigreed germplasm of the diploid outbreeding species apple, peach, and sweet cherry through a common workflow. PLoS One 14(6):e0210928. https://doi.org/10.1371/journal.pone.0210928

Velasco R, Zharkikh A, Affourtit J, Dhingra A, Cestaro A, Kalyanaraman A, Fontana P, Bhatnagar SK, Troggio M, Pruss D, Dhingra A, Cestaro A, Kalyanaraman A, Fontana P, Bhatnagar SK, Troggio M, Pruss D, Salvi S, Pindo M, Baldi P, Castelletti S, Cavaiuolo M, Coppola G, Costa F, Cova V, Ri AD, Goremykin V, Komjanc M, Longhi S, Magnago P, Malacarne G, Malnoy M, Micheletti D, Moretto M, Perazzolli M, Si-Ammour A, Vezzulli S, Zini E, Eldredge G, Fitzgerald LM, Gutin N, Lanchbury G, Macalma T, Mitchell JT, Reid J, Wardell B, Kodira C, Chen Z, Desany B, Niazi F, Palmer M, Koepke T, Jiwan D, Schaeffer S, Krishnan V, Wu C, Chu VT, King ST, Vick J, Tao Q, Mraz A, Stormo A, Stormo K, Bogden R, Ederle D, Stella A, Vecchietti A, Kater MM, Masiero S, Lasserre P, Lespinasse Y, Allan AC, Bus V, Chagné D, Crowhurst RN, Gleave AP, Lavezzo E, Fawcett JA, Proost S, Rouzé P, Sterck L, Toppo S, Lazzari B, Hellens RP, Durel C-E, Gutin A, Bumgarner R, Gardiner SE, Skolnick M, Egholm M, Van de Peer Y, Salamini F, Viola R (2010) The genome of the domesticated apple (*Malus* × *domestica* Borkh.). Nat Genet 42:883–889

Volk GM, Bramel P (2017) A strategy to conserve worldwide apple genetic resources: survey results. Acta Hortic 1172:99–105

Volk GM, Richards CM, Reilley AA, Henk AD, Forsline PL, Aldwinckle HS (2005) *Ex situ* conservation of vegetatively-propagated species: development of a seed-based core collection for *Malus sieversii*. J Am Soc Hort Sci 130:203–210

Volk GM, Richards CM, Henk A, Reilley AA, Reeves PA, Forsline PL, Aldwinkle HS (2009) Capturing the diversity of wild *Malus orientalis* from Georgia, Armenia, Russia and Turkey. J Am Soc Hort Sci 134:453–459

Volk GM, Henk AD, Baldo A, Fazio G, Chao CT, Richards CM (2015) Chloroplast heterogeneity and historical admixture within the genus *Malus*. Am J Bot 102(7):1198–1208

Volk GM, Jenderek M, Chao CT (2017) Prioritization of *Malus* accessions for collection cryopreservation at the USDA-ARS National Center for Genetic Resources Preservation. Acta Hortic 1172:267–272

Walters C, Wheeler L, Stanwood PC (2004) Longevity of cryogenically stored seeds. Cryobiology 48(3):229–244

Walters C, Wheeler LM, Grotenhuis JM (2005) Longevity of seeds stored in a genebank: species characteristics. Seed Sci Res 15(1):1–20

Wang K, Liu F, Cao Y (2005) Descriptors and data standard for apple (*Malus* spp. Mill). China Agriculture Press, , Beijing, China

Wang M-R, Chen L, Teixeira da Silva JA, Volk GM, Wang Q-C (2018) Cryobiotechnology of apple (*Malus* spp.): Development, progress and future prospects. Plant Cell Rep 37:689–709

Watkins R, Smith RA (1982) Apple descriptors. IBPGR Secretariat, Rome, p 49

Webster CA, Jones OP (1989) Micropropagation of the apple rootstock M.9: Effect of sustained subculture on apparent rejuvenation in vitro. J Hort Sci 64(4):421–428

4 Ploidy, Genome Size, and Cytogenetics of Apple

Małgorzata Podwyszyńska and Agnieszka Marasek-Ciołakowska

Abstract

Apple (*Malus*) is one of the most important fruit crops. The cultivated apple, *M.* × *domestica*, can be either diploid or triploid, whereby the latter accounts for 10% of all currently cultivated apples. In this chapter, we will explore the hypothesis on the origin of the apple genus and its haploid chromosome number of $x = 17$. Furthermore, special attention will be paid to genome size which is commonly evaluated using flow cytometry. Genome size is a multipurpose parameter used in many fields of plant science regarding species evolution and taxonomy, cell biology, cell cycle, endoreplication, and organ development as well as in breeding efforts. Based on data collected for 120 *Malus* taxa in The Plant DNA C-values Database for species and hybrid species with the same number of chromosomes, it is found that the rate of genome size variation is relatively low. The reported nuclear genome sizes for *Malus* accessions, including species, hybrids, and cultivars of *M.* × *domestica*, range from 1.50 to 1.78 pg for diploids, 2.27–2.58 pg for triploids, and 3.13–3.37 pg for tetraploids. Differences between the smallest and the largest nuclear DNA contents are as follows: 0.45 pg (36%), 0.45 pg (21%), and 0.34 pg (12%) for diploids, triploids, and tetraploids, respectively. Furthermore, the roles of mitotic and meiotic polyploidization of apple involved in new assortment development as well as the influence of genome size/ploidy level on phenotypic traits (nucleotypic effects) will be discussed. It is observed that ploidy level significantly influences resistance to biotic and abiotic stresses, as well as secondary metabolite production, among other traits. Much attention will be devoted to genome downsizing of polyploids, as well as to methods of ploidy level evaluation, including chromosome count, flow cytometry, and molecular markers. In addition, this chapter presents the findings of cytogenetic studies of the *Malus* genus. As apple chromosomes are small, numerous, and difficult to analyze with classical methods, the use of new molecular cytogenetic techniques, such as fluorescence in situ hybridization (FISH), has enhanced investigations of genome composition in apple. Moreover, further progress has been achieved by exploiting bacterial artificial chromosomes (BACs) along with fluorescence in situ hybridization, BAC-FISH, thereby enabling the determination of physical locations of useful gene loci in the *Malus* genome.

M. Podwyszyńska (✉) · A. Marasek-Ciołakowska
Department of Applied Biology, The National Institute of Horticultural Research, Konstytucji 3 Maja 1/3, 96-100 Skierniewice, Poland
e-mail: malgorzata.podwyszynska@inhort.pl

S. S. Korban (ed.), *The Apple Genome*, Compendium of Plant Genomes,
https://doi.org/10.1007/978-3-030-74682-7_4

4.1 Introduction

Plant genomes have higher tendencies to evolve compared with those of other eukaryotic genomes, as well as demonstrate greater diversity within a genus than that reported in other eukaryotes (Murat et al. 2012). In the plant kingdom, angiosperms are particularly characterized by wide variations in genome size, from the smallest genome of 63 Mb (1C = 0.065 pg) for *Gensilea margaretae* (Greilhuber et al. 2006) to the largest genome of 150 Gb (1C = 152.23 pg) for *Paris japonica* (Pellicer et al. 2010). Whole-genome duplication (WGD) is one of the most important factors in the evolution of plant species (Panchy et al. 2016). It is suggested that this process has been going on for more than 200 million years. The history of apple genome evolution dates back to the first WGDs that must have occurred between 50 and 70 million years ago (MYA) (Proost et al. 2011). The emergence of the *Malus* genus is associated with hybridization, polyploidization, and diploidization.

This chapter covers issues related to important features for apple research, such as genome size, ploidy level, and karyotype characteristics. First, attention will focus on the general characteristics of the genome, such as the basic number of chromosomes, as well as the presentation of older and more recent hypotheses related to the origin of the relatively high base number of 17 chromosomes in *Malus*. Next, topics related to different ploidy levels and related genome sizes will be discussed, as well as their impacts on adaptability and geographical distribution of individual apple species. Furthermore, genome size and ploidy level will be covered within the context of developing new cultivars with desirable traits, such as resistance to biotic and abiotic stresses, improved fruit quality, as well as other high-value traits. Knowledge of a karyotype and of chromosome characteristics, both at the physical and molecular levels, is highly important for apple breeding, particularly when encountering a changing climate and a growing human population. The use of various technologies, from the oldest traditional methods, such as chromosome counting, through the application of flow cytometry alone or in combination with molecular markers, and more recently of high-throughput genome-wide genotyping methods for detecting polyploids and aneuploids with relatively high precision are also discussed.

Cytogenetic studies of the *Malus* genus have a long history. The first chromosome analysis of this genus, dating back to the early twentieth century, is generally limited to chromosome counts and meiotic analysis of chromosome pairing (Rybin 1926; Crane and Lawrence 1930; Moffett 1931; Roscoe 1933; Einset and Imhofe 1947). Since then, chromosome numbers have been evaluated for most primary and hybrid *Malus* species along with rapid developments of biotechnology-based research methodologies. Furthermore, progress has also been made in both chromosome characterization and identification.

4.2 Basic Chromosome Numbers of the Maloideae Subfamily and Origin of the Apple Genome

The domesticated apple, *M.* × *domestica* Borkh., is a member of the family Rosaceae and the subfamily Maloideae. The Maloideae consists of approximately 1,000 species in 30 genera and has a haploid basic chromosome number of $x = 17$; whereas, other subfamilies of the family Rosaceae, including peach, raspberry, and strawberry have $x = 7$, 8, and 9, respectively (Goldblatt 1976; Folta and Gardiner 2009; Giovannoni 2010). The basic chromosome numbers in Rosaceae have been established by the 1920s (Evans and Campell 2002). Since then, many hypotheses have been proposed to explain the origins of the Maloideae and their high basic chromosome numbers (Fig. 4.1). The first hypothesis has suggested that the Maloideae are pentaploids that have evolved from progenitors with seven chromosomes, followed by loss of one of the chromosomes and then halving of the chromosome complement (Nebel 1929). Darlington and Moffett (1930) have proposed an autopoliploidy origin of the Maloideae based on observations of the formation of multivalents during meiosis. While others have suggested an allopolyploid origin involving an ancestral Amygdalloideae (x = 7) or Spiroideae (x = 9) (Sax 1933; Stebbins 1950). Allopolyploidy in apples has been supported by flavonoid chemistry (Challice 1974), isozyme studies (Weeden and Lamb 1987), and morphological traits (Phipps et al. 1991). In a series of published articles, a wide hybridization hypothesis has been postulated wherein the Maloideae must have resulted from a cross between $x = 8$ and $x = 9$ progenitors of two other subfamilies (Stebbins 1950; Phipps et al. 1991). However, Evans and Campbell (2002) have provided evidence, based on molecular phylogenetic analysis, which contradicts the wide-hybridization hypothesis, but supports the autopolyploid origin of the Maloideae from the ancestor of *Gillenia*, a genus with a base chromosome number of 9. Under this hypothesis, the base chromosome number of $x = 17$ of the Maloideae arose via aneuploidy from $x = 18$ (Evans and Campbell 2002). Similarly, based on the analysis of the apple genome, Giovannoni (2010) has suggested that the base chromosome number of $x = 17$ in the cultivated apple is the result of a WGD event in an ancestral genome ($x = 9$), followed by loss of a single chromosome. Considine et al. (2012) have proposed an alternative three-step model for the formation of the *Malus* genome as follows: an auto-aneuploidization of two sister taxa ($x = 9$, $2n = 18$) to monosomes ($x = 9$, $2n = 17$), followed by WGD in both ova and spermatozoa involving non-reductive first division restitution (FDR) to tetraploids ($x = 9$, $4n = 34$), and followed by a diploidization to the extant diploid state ($x = 17$, $2n = 34$). In contrast, it has been previously deduced that *Malus* ($x = 17$) is derived from an autopolyploidization of two sister taxa ($x = 9$, $2n = 18$), followed by diploidization, and then an aneuploidization to $x = 17$ (Velasco et al. 2010).

4.3 Ploidy Levels of *Malus* Species

Most *Malus* species are diploid with $2n = 2x = 34$, but some are triploids with $2n = 3x = 51$, and others are tetraploids with $2n = 4x = 68$ (Schuster and Büttner 1995; Höfer and Meister 2010; Tatum et al. 2005; Korban et al. 2009). Only triploid genotypes are recorded in species *M. toringoides* and *M. hupehensis* (Schuster and Büttner 1995), while some species; e.g., *M. angustifolia, M. glaucescens, M. lancifolia,* and *M. platycarpa* are only tetraploid (Schuster and Büttner 1995; Höfer and Meister 2010). Furthermore, in some other species, different ploidy levels have been recorded; e.g., *M. halliana, M. ioensis*, and *M. spectabilis* (2x and 3x); *M. coronaria* (3x and 4x); *M. sieboldii* (2x, 3x, and 4x); as well as *M. sargentii* and *M. sikkimensis* (3x, 4x, and 5x). A listing of chromosome numbers and nuclear DNA contents (pg/2C) of selected *Malus* species in subsection Pumilae are presented in Table 4.1. Interestingly, it has been widely reported that polyploidy confers ecological adaptability (Soltis et al. 2004; Ramanna and Jacobsen 2003). In many cases, neopolyploids are formed via genomic duplication, followed by diploidization, leading to other forms of high numbers of chromosomes (Soltis et al. 2004; Feldman and Levy 2012).

According to Schuster and Büttner (1995), the presence of accessions of different ploidy levels within the same species might indicate their hybrid character; e.g., triploids may have arisen from either interploid crosses between diploids and tetraploids or by a cross with unreduced gametes produced by one of two diploid parents, or by apomixis of triploid species. Triploid genotypes are generally self-sterile, although sometimes they can produce viable reduced ($n = 17$) or unreduced (n = 34 or 51) gametes (He et al. 2018). Recently, it has been reported that the ploidy of progenies of the triploid apple 'Jonagold', derived by open-pollination, along

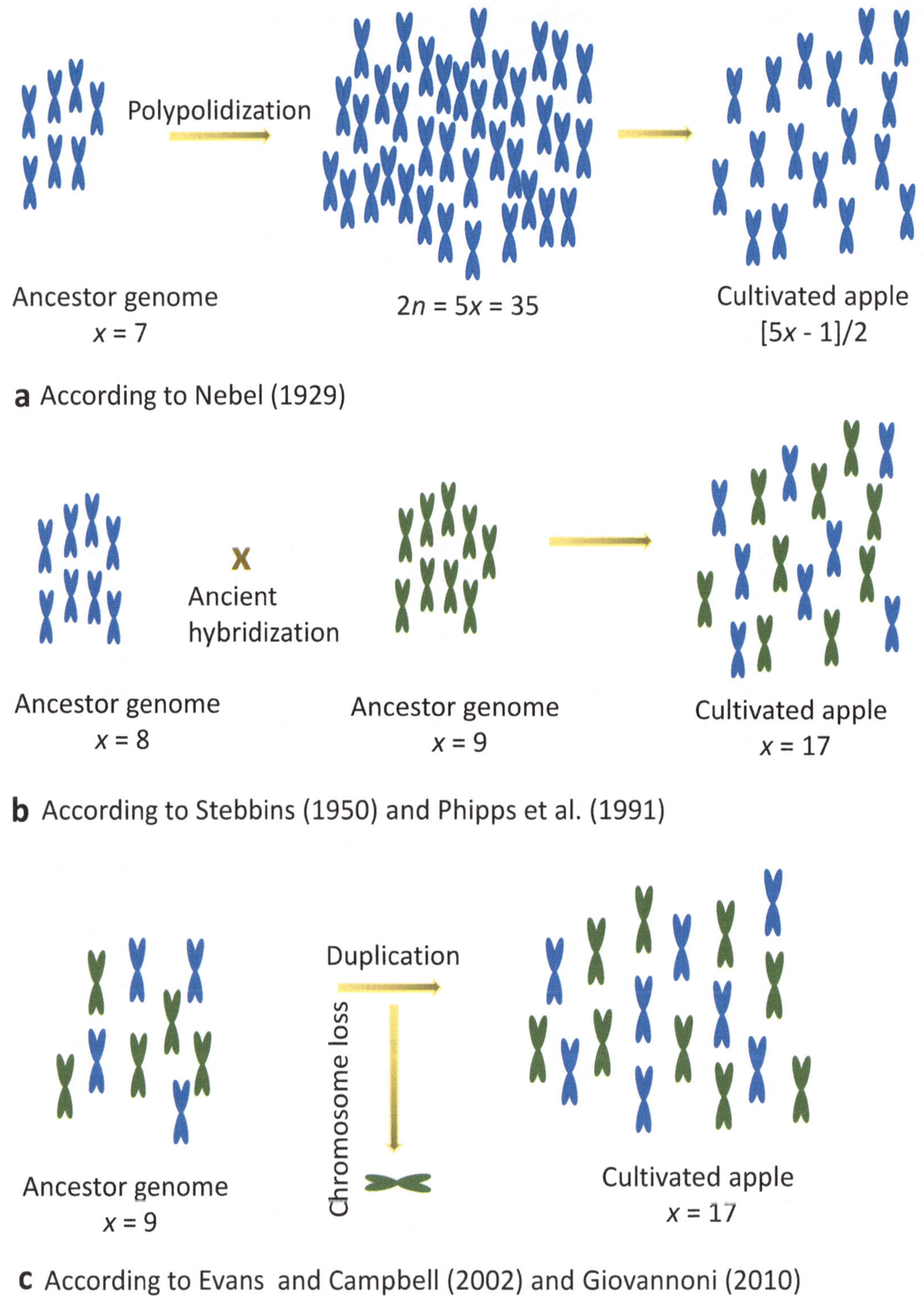

Fig. 4.1 The major proposed hypotheses pertaining to the origin of the basic chromosome number in the Maloideae subfamily. **a** Nebel (1929) hypothesis—Maloideae ($x = 17$) are pentaploids that must have evolved from progenitors with seven chromosomes followed by loss of one of chromosomes, and halving of the chromosome complement. **b** Stebbins (1950) and Phipps et al. (1991) hypothesis—a wide-hybridization hypothesis wherein the Maloideae must have resulted from a cross between $x = 8$ and $x = 9$ progenitors. **c** Evans and Campbell (2002) and Giovannoni (2010) hypothesis—the base chromosome number of $x = 17$ is the result of a whole-genome duplication (WGD) event in an ancestral genome ($x = 9$), followed by loss of a single chromosome

Table 4.1 Chromosome number, nuclear DNA content (pg/2C), and distribution of selected *Malus* species in subsection Pumilae (adapted from Höfer and Meister 2010, and Hancock et al. 2008)

Species	Chromosome number (2*n*)	DNA content (pg/2C)	Distribution
M. asiatica Nakai	34	1.500	North and North East China, Korea
M. baccata (L.) Borkh	34	1.440	North and North East China
M. × *domestica* Borkh	34, 51	1.414 (2*x*), 2.270 (3*x*)	Worldwide
M. floribunda Siebold	34	1.514	Japan
M. micromalus Makino	34	1.528	South East China, Korea
M. orientalis Uglitzk	34	1.502	Caucasia
M. prunifolia (Willd.) Borkh	34	1.500	China
M. × *purpurea* (A. Barbier) Rehd	34	1.484	Asia
M. × *sublobata* Rehd	34	1.497	Japan
M. sieversii (Ledeb.) M. Roem	34	1.501	North West China
M. spectabilis (Aiton) Borkh	34, 51	1.502 (2*x*), 2.206 (3*x*)	China
M. sylvestris (L.) Mill	34	1.484	Europe

with observed variations in mitotic chromosome numbers are predominantly aneuploids and diploids, along with some polyploid seedlings, either triploid or tetraploid.

Interspecies variability of genome size and its related ploidy level is considered within the context of adaptability to environmental conditions and to geographical distribution for many plant genera (Bennet 1987; Ramsey and Ramsey 2014). Polyploids of the plant families Poaceae and Rosaceae, for example, have been reported to be more frequent in floras of cooler climates, and even arctic regions (Brochmann et al. 2004; Ramsey et al. 2008; Green et al. 2013). It has been proposed that a higher frequency of occurrence of polyploids of various plant genera, including that in the apple genome, is correlated with specific climate zones, environments, and geographical locations (Tatum et al. 2005; Korban et al. 2009; Höfer and Meister 2010). Tatum et al. (2005) have indicated that the triploid species *M. toringo* and *M. toringoides* as well as that of the highest genome size of the diploid *M. transitoria* are distributed in colder climatic zones (II and III), and these have originated mainly from Northeastern Asia, except for *M. sikkimensis*, which has originated from heights of the Himalayan mountains with environments close to climatic zones II and III. In turn, Höfer and Meister (2010) have noted that six primary diploid species, including *M. tschonoskii*, *M. yunnanensis*, *M. ombrophila*, *M. kansuensis*, *M. honanensis*, and *M. prattii*, possessing genome sizes of <5% than the 1C median, have originated from Central Asia, the primary original habitat of *Malus*. Whereas, two diploid species (*M. trilobata* and *M. florentina*) exhibiting genome sizes of >5% than that of the 1C DNA median along with several triploid and tetraploid species, including *M.* × *heterophylla*, *M.* × *platycarpa*, *M. glaucescens*, *M. angustifolia*, *M. lancifolia*, and *M. coronaria,* all belonging to the section *Chloromeles*, are widespread in Eastern and Northern America. Moreover, those diploid species of the highest genome sizes, *M. trilobata* and *M. florentina*, are the only members of the section *Eriolobus* and section *Sorbomalus*, respectively, both belonging to series Florentinae, that are present in Southeastern Europe. Finally, Höher and Meister (2010) have concluded that species of higher genome sizes can exist far away from the center of origin for *Malus,* and they tend to cluster in separate sections or series. This may indicate that possessing a higher genome size contributes to a genetic background that is both better adapted to new environments and has a higher invasive ability to new habitats (Stebbins 1985).

4.3.1 Methods for Ploidy Level Determinations

Earlier, ploidy levels have been estimated only by using chromosome counts (Rybin 1926; Crane and Lawrence 1930; Einset and Imhofe 1947). With the development of modern methods, including flow cytometry and molecular markers, these methods are now widely used in apple breeding to assess ploidy for various purposes. At present, the ploidy level for apple is most often determined based on nuclear genome estimation using flow cytometry (see Sect. 4.4).

Ploidy levels can also be determined based on analysis of allelic frequency using simple sequence repeat (SSR) markers (Pereira-Lorenzo et al. 2007; Bisognin et al. 2009; Considine et al. 2012). SSRs are co-dominant, highly polymorphic, and reproducible (Liebhard et al. 2002). These markers are routinely used for genetic variability and for identity assessments, but there are several examples of using these markers for ploidy analysis. Pereira-Lorenzo et al. (2007) have detected 39 triploids of 140 apple accessions, mainly local Spanish cultivars, using SSR analysis conducted with 10 primer pairs homologous to SSR loci generating numerous polymorphic products, and these are then confirmed by flow cytometric analysis.

The potential of SSR markers in determining ploidy status (euploidy, aneuploidy, and karyotype analysis) with high accuracy has been fully exploited (Considine et al. 2012). While evaluating 27,542 viable F1 apple seedlings of intra- and inter-specific crosses between diploid genotypes using SSR markers, a range of non-diploid cytotypes, including 0.199% triploids, 0.0521% tetraploids, and 0.778% aneuploids were detected (Considine et al. 2012). Thus, the frequency of recovery of polyploid apple seedlings from crosses between two diploids was approximately 1%, mostly aneuploids. Furthermore, aneuploidy was more frequent among interspecific progenies, wherein only a single cytotype with one chromosome loss ($2n - 1$) and eight cytotypes with chromosomal gains were detected. Considine et al. (2012) also characterized the karyotypes of these recovered polyploids, including aneuploids. Based on the 17 linkage groups (LGs) corresponding to the 17 haploid chromosome numbers of apple (Velasco et al. 2010; Han et al. 2007) and related SSR markers, Considine et al. (2012) found that without exception, all tetraploid seedlings were derived from $2n$ egg cells fertilized with $2n$ sperm cells, while all triploids were derived from $2n$ egg cells fertilized with $1n$ sperm cells, and all aneuploids were derived from $1n$ egg cells fertilized with aneuploid sperm cells. It was confirmed that only egg cells were responsible for the development of euploids, while sperm cells were responsible for the generation of either euploids or various aneuploids. It was suggested that diploid apple exhibited unique gametic combinational patterns. In addition, SSR analysis revealed that non-reduced gametes were genetically heterozygous, thus indicating that first-division restitution (FDR) was exclusive for apple polyploidization. Non-reduced gametes derived by FDR had a higher heterozygosity along with more complex epistatic combinations from their parents than those derived by second-division restitution (SDR) (Selmecki et al. 2009). Thus, genetic differences between FDR and SDR might influence the fate of neopolyploids, either in speciation or in evolution (Comai 2005).

Recently, Chagné et al. (2015) proposed the use of single nucleotide polymorphism (SNP) analysis along with high-throughput genome-wide genotyping for detecting polyploids and aneuploids with relatively high precision. They evaluated the ploidy status of an apple germplasm collection of 663 accessions using an Illumina Infinium® II FruitBreedomics

20K apple SNP array (Bianco et al. 2014). SNP analysis for the detection of multiple sets of chromosomes relied on B allele frequency (BAF) measurements, wherein the signal ratio of the B allele was compared to that of the A allele for each SNP (Chagné et al. 2015). It was found that for homozygous diploid genotypes AA and BB, BAF was close to 0 and 1, respectively, while for heterozygous genotypes AB, BAF was 0.5. Moreover, for heterozygous triploid genotypes AAB and ABB, the BAF signal ratios were 0.33 and 0.66, respectively. Analogously, BAF values of SNPs analysis for tetraploid genotypes AAAB, AABB, and ABBB were close to 0.25, 0.5, and 0.75, respectively. Using this analysis, Chagné et al. (2015) reported that of 663 genotypes assessed, 643 (97%) were diploids, 7 genotypes were triploids, and 2 genotypes were tetraploids. They also detected several aneuploids and determined their karyotypes with a high level of accuracy. This reported high reliability of SNP analysis of ploidy status of these *Malus* accessions has been further confirmed by flow cytometry measurements. Chagné et al. (2015) recommended the use of such SNP- and BAF-based methods for genotyping efforts in breeding programs.

4.4 Genome Size Estimation as a Multi-use Parameter for Apple Studies

The nuclear genome size is a specific and constant parameter for each individual organism's DNA content, referred to as the C-value. In particular, it refers to the DNA content in a haploid gametic nucleus of an organism, irrespective of the ploidy level of the taxon (Doležel and Bartoš 2005; Bai et al. 2012; Greilhuber et al. 2005). In somatic cells of sporophytes, nuclei containing two chromosome sets at the G0/G1 cell cycle have 2C DNA contents. Within species and genera, the C-value is usually similar, but it is not the same across all. Moreover, it is closely related to the chromosome number. Therefore, the estimation of the nuclear DNA content is used by cytogeneticists, botanists, systematists, biotechnologists, and plant breeders to determine the relative genome size, ploidy level, endoreplication rate, and proportions of nuclei at specific phases of the cell cycle (Sliwinska 2018).

The relative nuclear DNA size is commonly estimated using a flow cytometry method (FCM) utilizing fluorescent dyes bound to DNA. The most commonly used fluorochrome dyes are 4′,6-diamidino-2-phenylindole (DAPI), binding to AT pairs of a DNA, and propidium iodide (PI) intercalating between strands of double-stranded nucleic acid structures. DAPI is more often utilized for ploidy estimation, while PI is used for more accurate evaluation of relative genome size, which is calculated in relation to an internal standard; i.e., a plant genotype of known nuclear DNA content (Johnston et al. 1999). Over 10 years ago, the genome size of apple cv. 'Golden Delicious' has been estimated to be 742 Mb for 17 chromosomes using whole-genome sequencing (WGS) (Velasco et al. 2010). For taxonomical, physiological, and breeding purposes, the nuclear DNA content; i.e., the relative nuclear genome size, is often measured using flow cytometry, and expressed in picograms (pg) (Bai et al. 2012). Cavalier-Smith (1985) has estimated that 1 pg of DNA contains from 0.965×10^9 bp to 0.980×10^9 bp. Nowadays, it is widely accepted that 1 pg of DNA = 0.978×10^9 bp (Doležel et al. 2003).

To ensure reliable cytometric measurements of genome size, the following principles should be followed: (i) at least three samples from three individual plants should be measured; (ii) nuclei must be intact, and in sufficient quantities; and (iii) DNA staining should be stoichiometric for either the sampled genotype or the internal standard genome for which the nuclear DNA content must be known (Doležel and Bartoš 2005). A well-selected internal standard for FCM must be already characterized for both 2C and 4C nuclear DNA contents, of different genotypes (with no overlapping picks with the sampled genotype), but close to the genome size of the sampled genotype (Johnston et al. 1999). Formerly, a chicken erythrocyte (2.33 pg DNA/2C) has been used as an internal standard for apple genome size determination using FCM analysis (Dickson et al. 1992). However, plant internal standards have been recommended for use in FCM analysis of plant tissues/genotypes (Johnston et al. 1999).

For apple, the following internal standards of well-estimated genome sizes have been used: *Zea mays* (5.43 pg DNA/2C) (Korban et al. 2009; Podwyszyńska et al. 2016), *Raphanus sativus* (1.1 pg DNA/2C) (Höfer and Meister 2010), and an F1 hybrid of *Petunia hybrida,* identified as *P* × Pc6 (2.85 DNA pg/2C) (Jedrzejczyk and Sliwinska 2010). Beginning in 1976, C-value data have been reported for numerous plant species and cultivars (Bennet and Smith 1976). Since 1997, DNA C-values have been deposited in The Plant DNA C-values Database, which currently has C-values for 12,273 species, of which 10,770 are angiosperms (Release 7.1, April 2019. Leitch IJ, Johnston E, Pellicer J, Hidalgo O, Bennett MD, https://cvalues.science.kew.org/). For *Malus*, 2C values are available for 120 taxa, including 38 species (Marie and Brown 1993; Tatum et al. 2005; Korban et al. 2009; Höfer and Meister 2010; Jedrzejczyk and Sliwinska 2010; Bou Dagher-Kharrat et al. 2013; Pustahija et al. 2013).

Based on data collected for 120 apple taxa in The Plant DNA C-values, the rate of genome size variation is relatively low among *Malus* species and hybrid species with the same number of chromosomes, as suggested by Dickson et al. (1992) and then confirmed by Korban et al. (2009). Differences among taxa having the smallest and those with the largest nuclear DNA contents are as follows: 0.45 pg (36%), 0.45 pg (21%), and 0.34 pg (12%) for diploids, triploids, and tetraploids, respectively. Therefore, the DNA nuclear content parameter is used as well for ploidy level assessment. In contrast, nuclear genome sizes of some other plant genera, such as those for lily and tulip are highly diverse (van Tuyl and Boon 1997; Zonneveld 2009). The 2C values within somatic cells of tulip are species-specific, and range from 32 up to 69 pg for diploids. Furthermore, the largest genome of tulip has approximately 3.4×10^{10} more base pairs and twice the size of chromosomes than that of the smallest genome.

Genome size and ploidy level of apple were first assessed using flow cytometry by Arumuganathan and Earle (1991). They reported on nuclear DNA contents for three cultivars of *M.* × *domestica,* ranging from 1.54 to 1.65 pg. Later on, Dickson et al. (1992) estimated genome sizes of 25 *Malus* species and apple cultivars. The DNA 2C values for 17 diploid *Malus* species, from five sections, ranged from 1.21 to 1.67 pg, while within *M.* × *domestica* cultivars, these ranged from 1.50 to 1.73 pg, and for triploid 'Jonagold', it was 2.52 pg. Genome sizes evaluated by Dickson et al. (1992) were similar to those reported by Tatum et al. (2005), Korban et al. (2009), and Podwyszyńska et al. (2016). These studies reported that nuclear genome sizes for *Malus* accessions (species, hybrids, and cultivars of *M.* × *domestica*) ranged from 1.50 to 1.78 pg for diploids, 2.27–2.58 pg for triploids, and 3.13–3.37 pg for tetraploids. Comparisons of genome size values of the same apple genotypes evaluated in various laboratories revealed that only two of six values, for cultivars 'Cortland' and 'Idared', exceeded the margins of error (Table 4.2). Whereas, Höfer and Meister (2010) estimated genome sizes of 256 *Malus* accessions and reported lower genome sizes, on average 1.484 pg (1.245–1.635 pg), 2.198 pg (2.121–2.428 pg), and 3.007 pg (2.851–3.193 pg) for diploids, triploids, and tetraploids, respectively. Moreover, they identified two pentaploids, *M. sikkimensis* and *M. sargentii,* with genome sizes of 3.540 and 3.874 pg, respectively (Höfer and Meister 2010). Interestingly, the largest genome of 4.286 pg was estimated for another taxon of *M. sargentii*, and this was deemed to be a polyploid, between 5 × and 6*x* (Höfer and Meister 2010).

It is important to point out that the above differences in *Malus* genome size estimations among different studies must have likely resulted from the use of different FCM procedures (buffers, staining time, temperature, or internal standards), as well as the use of different types of instruments (Johnston et al. 1999; Doležel et al.

Table 4.2 Comparisons of nuclear DNA contents (means ± SD) of *M.* × *domestica* cultivars analyzed by Dickson et al. 1992 (D), Tatum et al. (2005) (T), Korban et al. (2009) (K), and Podwyszyńska et al. (2016) (P) (adapted from Podwyszyńska et al. 2016)

Cultivar	2C DNA (pg)
'Cortland'	1.57 ± 0.06^{T}, 1.70 ± 0.02^{P}
'Gala'	1.57 ± 0.14^{T}, 1.66 ± 0.03^{P}
'Golden Delicious'	1.62 ± 0.02^{T}, 1.64 ± 0.06^{P}
'Idared'	1.59 ± 0.02^{T}, 1.68 ± 0.03^{P}
'Redspur Delicious'	1.66 ± 0.01^{K}, 1.69 ± 0.04^{P}
'Jonagold'	2.48 ± 0.02^{T}, 2.51 ± 0.2^{D}, 2.48 ± 0.05^{P}

2003). For example, 2C values of apple genotypes determined by Tatum et al. (2005), Korban et al. (2009), and Podwyszyńska et al. (2016) are more similar, presumably due to the fact these three studies have all used maize (2C = 5.43 pg) as an internal standard along with the same DNA staining time (1 h). Whereas, genome sizes obtained by Höfer and Meister (2010) are slightly smaller, and this is probably due to the fact that they have used radish (1.1 pg/2C) as an internal standard along with a very short time for DNA staining (1 min). Nevertheless, although genome size estimations for the same species can slightly differ among different laboratories, trends in genome size gradations among *Malus* species are found to be similar. For example, one of the smallest size genomes is that reported for *M. yunnanensis,* amounting to 1.32, 1.39, and 1.44 pg, as estimated by Höfer and Meister (2010), Dickson et al. (1992), and Korban et al. (2009), respectively. Accordingly, one of the largest diploid apple genomes is that reported for *M. florentina*, 1.65 and 1.63 pg, as estimated by Höffer and Meister (2010) and Dickson et al. (1992), respectively.

Thus, genome size evaluation using FCM is a multipurpose parameter commonly used in various fields of plant science dealing with species evolution and taxonomy, cell biology, cell cycle, endoreplication, and organ development, as well as in breeding programs.

4.4.1 Hybrid/Apomictic Status of Seedlings from Intergeneric, Interspecific, and Interploid Crosses

The nuclear DNA content measured using only FCM or in combination with molecular markers is a very convenient tool for determining hybrid/apomictic status of seedlings resulting from intergeneric, intra- and interspecific, as well as interploid crosses. As parental genomes differ significantly in size, their hybrids are expected to possess intermediate nuclear DNA content (Kamiński et al. 2016). The estimation of the genome size parameter is also important for determining ploidy levels of parental genotypes, as well as those of resulting progeny. For example, when using either triploids or tetraploids in crosses with other genotypes, irrespective of their ploidy levels, it is to be expected that a number of aneuploids will be recovered in their progeny (Bisognin et al. 2009; He et al. 2018). Likewise, when hybridizing diploids with tetraploids, it is likely that triploids will be recovered in this progeny (Bisognin et al. 2009; He et al. 2018).

Bisognin et al. (2009) aimed to obtain hybrids for breeding apple rootstocks resistant to apple proliferation disease (caused by a phytoplasma) by crossing the tetraploid *M. sieboldii* (resistant to the pathogen) with various common diploid apple rootstocks, serving as donors of agronomically valuable traits. Using SSR markers combined with genome size determinations, undesirable apomictic seedlings could be readily distinguished from sexually derived interspecific hybrid seedlings of intermediate genome size compared to those of their parental genomes. A similar strategy was pursued to detect intergeneric hybrid seedlings resulting from a cross between Japanese pear (*Pyrus pyrifolia* 'Nakai') and apple. Hybrid seedlings exhibited nuclear DNA contents of the intermediate size of those of the two parents (Gonai et al. 2006). In turn, Inoue et al. (2002) obtained apomictic seedlings of Japanese pear, whose development was induced using irradiated apple pollen, and these apomictic pear seedlings were identified and selected based on cytometric measurements (genome size) and random-amplified polymorphic DNA (RAPD) molecular marker analysis, specific to the genetic profile of the Japanese pear. Kron and Husband (2009) assessed the potential for hybridization and gene flow between the diploid domesticated apple, *M.* × *domestica*, and the tetraploid crabapple *M. coronaria* using both FCM and isoenzyme markers. They found that open-pollinated crabapple seeds consisted of 57% tetraploids (sexual or apomictic), 7.7% apomictic diploids, 7.4% hexaploids or octoploids (sexual), and 27.5% triploid or pentaploid hybrids. Hence, they suggested that there was a significant potential for gene flow from the domestic apple into native crabapple populations. Furthermore, cytometric genome size determination was most frequently used for the detection of artificially induced tetraploids of numerous plant species (Dooghe et al. 2011), including those of apple (Podwyszyńska et al. 2017). In addition, genome size analysis was also useful in detecting apple somaclonal variants among in vitro-propagated plants differing in ploidy levels from donor genotypes (Druart 2000; Noormohammadi et al. 2015).

4.4.2 Nucleotypic Effects of Genome Size

Nucleotypic effects are phenotypic changes related to the total nuclear DNA amount per cell. These effects are commonly observed among and within species for certain cell types, and in general, it has been established that there is a positive correlation between genome size and cell size (Levin 1983, Dooghe et al. 2011; Anssour et al. 2009; Snodgrass et al. 2017). In apple, nucleotypic effects are significant both within genus and within species. Cell size, stomata length, and pollen grain diameter are positively correlated with ploidy level, as well as with genome size (Lespinasse and Noiton 1986; Blanke et al. 1994; Korban et al. 2009; Podwyszyńska et al. 2016, 2020a; Xue et al. 2017). It has been reported that lengths of stomata are the shortest in diploid species, ranging from 19.47 μm in *M. yunnanesis* to 22.17 in *M. kirghisorum,* and longer in triploids, ranging from 24.50 in *M. hupehensis* to 27.07 in *M.* × *domestica* (Korban et al. 2009). It is worth noting that the shortest and the longest stomata observed for diploids are correlated with the smallest and largest genome sizes, respectively. Furthermore, high correlations between ploidy levels and stomata lengths for haploid, diploid, triploid, and tetraploid apple genotypes, amounting to 34.4, 46.7, 50.2, and 66.6 μm, respectively, have also been reported by Lespinasse and Noiton (1986). Similarly, both lengths of stomata and diameters of pollen grains are, on average, significantly higher in triploid cultivars than in diploid cultivars by 14% and 10%, respectively (Table 4.3, Figs. 4.2 and 4.3, Podwyszyńska et al. 2016). Moreover, sizes of leaf, flower, and fruit tissues are, on average, larger in triploids than in diploids by 36%, 14%, and 25%, respectively (Table 4.3). However, the sizes of plant organs should not be readily used as markers of ploidy levels. As there are many diploid cultivars that produce large leaves, flowers, and fruits. For example, the diploid Japanese

Table 4.3 Comparisons of nuclear DNA content and several phenotypic traits of diploid and triploid apple cultivars (Podwyszyńska et al. 2016)

Trait	Ploidy level				p
	2x		3x		
	Mean	Min-Max	Mean	Min-Max	
Flowering time	8–9 May	7–13 May	8 May	7–11 May	0.29
Flower diameter (mm)	46.1 b	37.8–60.7	52.4 a	42.5–58.7	0.00
Fruit weight (g)	162.7 b	87.3–257.57	204.7 a	130–242.0	0.00
Leaf area (cm^2)	36.0 b	19.1–57.4	49.1 a	32.9–70.7	0.00
Stomata length (μm)	28.4 b	26.3–29.7	32.1 a	27.9–35.1	0.00
Pollen grain diameter (μm)	30.7 b	26.4–35.7	33.7 a	31.37–35.3	0.05
Pollen viability (%)	75.6 a	48.6–86.1	22.8 b	4.0–49.5	0.00
Nuclear DNA content (pg)	1.70 b	1.58–1.78	2.51 a	2.42–2.60	0.00

Means within rows followed by the same letter are not significantly different at $p = 0.05$ using Tukey's test

cultivar 'Sekai Ichi' and the diploid Polish cultivar 'Ligostar' are characterized by their production of very large fruits (Harada et al. 2005; Podwyszyńska et al. 2016). In fact, the fruit of the diploid 'Sekai Ichi' is larger than that of the triploid 'Mutsu' (also known as 'Crispin'); however, 'Mutsu' has a larger fruit than that of the diploid 'Fuji' (Harada et al. 2005). Interestingly, it is reported that the pericarp cell size of both diploid cultivars, 'Sekai Ichi' and 'Fuji', is similar, but it is smaller than that of the triploid 'Mutsu', but the cell number in the pericarp of 'Sekai Ichi' is much higher than that of the other two cultivars (Harada et al. 2005). Apple fruit size is influenced by a combination of factors, including active cell division and increased rate of cell growth (Harada et al. 2005). Therefore, fruit size is only partially correlated with ploidy level.

Although there is a clear tendency for enhanced cell and organ growth with higher ploidy, the highest correlation is found between size of stomata and ploidy level (Lespinasse and Noiton 1986; Korban et al. 2009; Podwyszyńska et al. 2016). Thus, the morphological trait of stomata size has long been recognized as a reliable marker for ploidy level for many plant species, including apple. This marker along with FCM data has been used to determine the structure of apple mixoploids, ploidy chimeras (Podwyszyńska et al. 2016). Ploidy chimeras of 'McIntosh' and 'Jonathan', possessing diploid and tetraploids genomes (4x:2x = 3:1), exhibit the diploid characteristic of small size stomata along with specific traits for tetraploids, including poor pollen germination, large leaves, and higher chlorophyll content. These collective features indicate that these chimeral types must have 2x-4x-4x (or 2-4-4) in L1, L2, and L3 histogenic layers, respectively. Such a description of chimera is based on our knowledge of shoot meristem development, as this meristematic region usually develops from three apical histogenic layers, including L1, L2, and L3, in diploids, wherein the L1 gives rise to an epidermis with stomata, the L2 forms subepidermal tissues, including gametes, while L3 forms the corpus with vascular tissues and roots (Geier 2012).

Furthermore, genome size, and more specifically ploidy level, significantly influences various other phenotypic traits. In addition to organ size (gigas effect), ploidy level influences

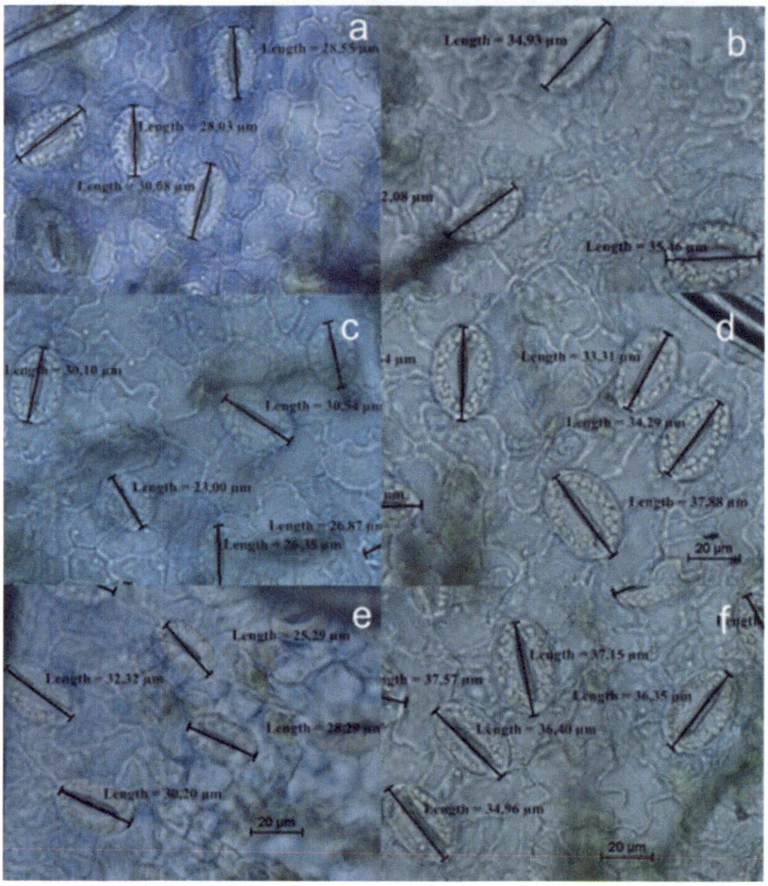

Fig. 4.2 Stomata of several apple cultivars. **a**, **c**, **e** Diploids 'Alwa', 'Ligol', and 'Ligolina', respectively. **b**, **d**, **f** Triploids 'Boscoop', 'Bursztówka Polska', and 'Witos'. Bars represent 20 µm (Podwyszyńska et al. 2016)

resistance to biotic and abiotic stresses, as well as secondary metabolite production, among many other traits (Ramanna and Jacobsen 2003; Dooghe et al. 2011; Soltis et al. 2014). Often, observed changes in polyploids are of high agronomic value. Therefore, generating triploids and tetraploids has been met with some interest by plant breeders.

4.4.3 Developmental Processes

In apple, the nuclear DNA content is very similar in all somatic tissues and organs, including fruits (Harada et al. 2005). However, there is an exception to this observation as reported for apple cv. 'Grand Gala', which is a spontaneous mutant of 'Gala'. Compared to 'Gala', 'Grand Gala' has larger-sized buds, leaves, pedicle diameters, and fruits (>38%), but reduced tree growth habit, thus indicating that it is a tetraploid. Surprisingly, flow cytometric analysis of 'Grant Gala' leaves has confirmed that it is a homogenous diploid, while FCM analysis of pericarp cells has indicated that it has a larger nuclear DNA content of 4C, reaching 57% and 77% for all cells analyzed at early and at late stages of fruit development, respectively. In addition, 8C DNA amounts of 5–8% are also observed, particularly at the early stages of fruit growth. These observed higher DNA contents are related to larger pericarp cells, by 15%, at the final stages of fruit growth. Moreover, the detection of high proportions of 4C nuclear DNA content has suggested an arrest of the G2 phase of the cell cycle, thereby resulting in endoreplication, a process of nuclear DNA replication in the absence of mitosis, thus leading to increased

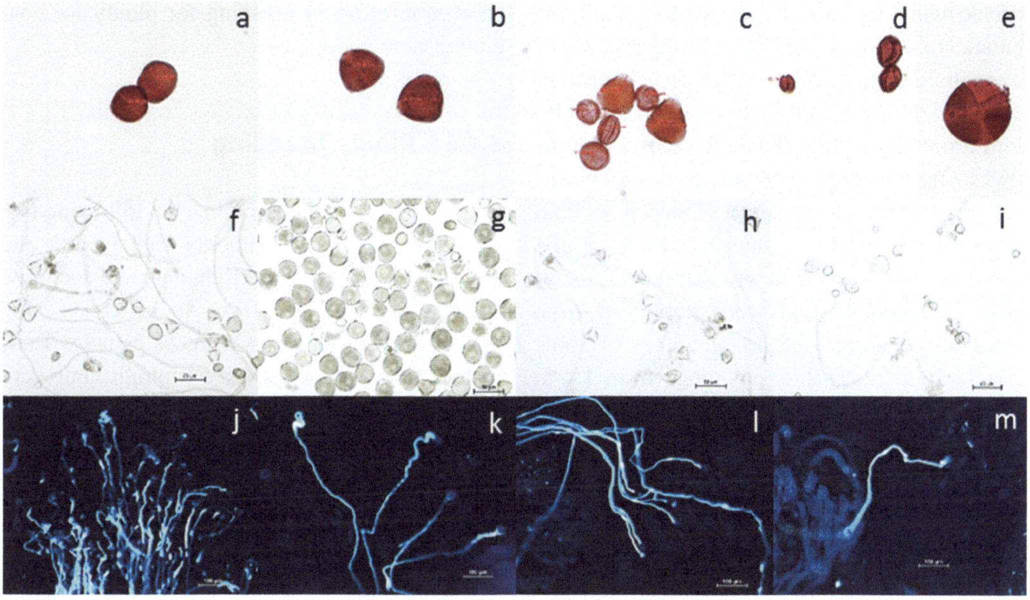

Fig. 4.3 Pollen grains of apple cultivars of different ploidy, including a diploid 'Gold Millenium' (**a**, **f**, **j**), a triploid 'Witos' (**b**, **g**, **k**), a diploid 'McIntosh' (**c**, **h**, **l**), and a mixoploid 'McIntosh' (2-4-4) (d, e, i, m). Pollen grains of these particular cultivars subjected to different treatments: stained with acetocarmine (**a, b**, **c**, **d**, **e**), bars represent 10 μm; germinating in a 20% sucrose solution (**f**, **g**, **h**, **i**), bars represent 20 μm; and germinating on stigma surfaces (**j**, **k**, **l**, **m**), bars represent 100 μm (Taken from Podwyszyńska et al. 2016)

genome size. Endoreplication, also referred to as endoreduplication, endopolyploidy, or polysomy is a developmental mechanism responsible for differentiation and morphogenesis (Barow and Meister 2003; De Veylder et al 2011), and it is common to occur in different tissues or organs; e.g., all organs of sugarbeet (Lukaszewska and Sliwinska 2007). Endoreplication has also been observed in the fruit of tomato whereby C-values can reach up to 250 C (Bertin et al. 2003). Maladi and Hirst (2010) have suggested that this phenomenon is present in apple cv. 'Grand Gala', and this is likely to be related to alterations in cell cycle gene expression, as partially confirmed using reverse transcription (RT)-polymerase chain reaction (PCR) analysis.

4.4.4 Downsizing of Polyploid Genomes

Genome size analysis of polyploids derived from the same species, such as those of *M. sargentii*, *M. sieboldii*, *M. spectabilis*, *M. hupehensis*, and *M. sikkimensis*, has revealed that mean DNA amount per basic genome (1C, calculated by dividing the 2C value by ploidy number) decreases along with increasing ploidy level (Höfer and Meister 2010). This phenomenon, referred to as genome downsizing, is widespread in angiosperms, and it has been reported for several allo- and autopolyploids of various plant species (Leitch and Bennett 2004; Eilam et al. 2010; Poggio et al. 2014).

Elimination of specific DNA sequences during polyploid formation, including the elimination of 18S-26S and/or 5S rDNA gene copies, has been recorded in polyploids of various species, such as *Triticum*, *Nicotiana*, *Festuca*, and *Brassica* (Wendel 2000). Moreover, evidence of selective DNA sequence loss in polyploids has also been documented based on analysis of 14,634 repetitive DNA sequences in genomes of the progenitor species of *Nicotiana tabacum*, *N. sylvestris*, and *N. tomentosiformis* (Renny-Byfield et al. 2011). It has been found that retroelements are mainly eliminated from the genome of *N. tabacum* when compared to its progenitors. Presumably, a similar mechanism of genome size downsizing may exist during polyploid formation within *Malus*. Recently, genome downsizing has been observed in newly generated autotetraploids of apple (Table 4.4; Podwyszyńska, original study). Interestingly, differences between the theoretical magnitude of about 2.90–2.96 pg and values obtained from cytometric measurements for genomes of apple neotetraploids (2.80–2.86 pg) range from 1.4 to 4.4%; these measurements have been carried out using *Raphanus sativus* as an internal standard (2C DNA = 1.11 pg) (Fig. 4.4). This observed phenomenon in diploids, allopolyploids, and autopolyploids of Mediterranean triticale has been widely discussed by Eilam et al. (2010). It is suggested that genome downsizing immediately following the formation of these polyploids provides the physical basis for their cytological diploidization; i.e., diploid-like meiotic behavior. Moreover, the mechanism for this phenomenon is probably necessary during the process of genome stabilization, and it is a significant source of genetic variation, either for the long term process of species evolution or during a short term effort in breeding for ploidy for generating new genotypes.

4.4.5 Ploidy Breeding

In order to develop polyploids, the following two strategies have been commonly used, meiotic and mitotic polyploidization (Ramanna and Jacobsen 2003; Dooghe et al. 2011). In general, sexual meiotic polyploidization with $2n$ gametes leads to the generation of mostly triploids, but very seldom to tetraploids, that can be divided into two main types, allopolyploids (interspecies polyploids) and autopolyploids (intraspecies polyploids). Whereas, in mitotic polyploidization, chromosome doubling occurs in somatic cells often leading to the formation of autopolyploids, usually either autotetraploids or autohexaploids when chromosome doubling is induced in either diploids or triploids, respectively. Mitotic polyploidization is also commonly used to overcome either interspecific or intergeneric hybrid sterility whereby chromosome doubling is induced in somatic tissues of such a hybrid, using colchicine, to produce allopolyploids that might be fertile.

Table 4.4 Estimation of nuclear DNA contents of newly generated apple autotetraploids, following treatment with either colchicine or amiprophos methyl (APM), relative to their diploid counterparts

Cultivar	Nuclear DNA content (pg)		
	Diploid	Tetraploid clone[a]	
'Free Redstar'	1.45 ± 0.040 b	1-S	2.85 ± 0.013 a
		5-S	2.82 ± 0.050 a
'Gala Must'	1.44 ± 0.031 b	2-S	2.83 ± 0.011 a
		4-S	2.83 ± 0.034 a
'Redchief'	1.45 ± 0.026 b	8-S	2.86 ± 0.043 a
		20-S	2.80 ± 0.039 a
'Pinova'	1.45 ± 0.039 b	3-S	2.83 ± 0,035 a
		6-L	2.80 ± 0.045 a
'Sander'	1.48 ± 0.026 b	3-L	2.87 ± 0.041 a
		6-S	2.83 ± 0.038 a

[a]S and L refer to origins of the tetraploid from either shoot (S) or leaf (L) explants treated with antimitotic agents. Statistical analysis was conducted separately for each cultivar. Mean values with the same letter are not significantly different at $p = 0.05$, according to Duncan's multiple range test

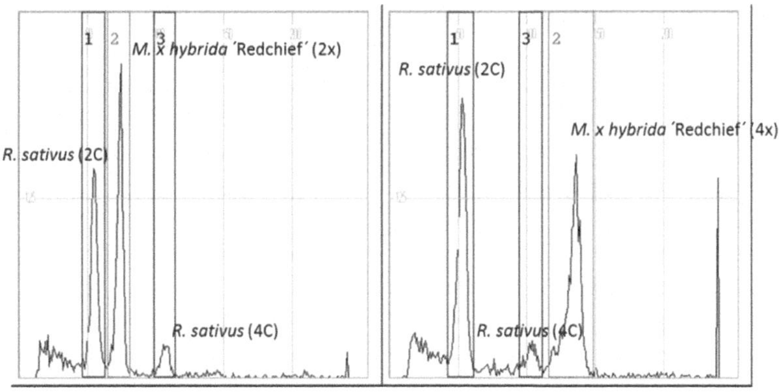

Fig. 4.4 Sample histograms of apple cytometric analysis conducted to assess nuclear DNA contents of a diploid cultivar and its corresponding induced autotetraploid. On the left, 'Redchief' and its newly generated tetraploid (on the right). *Raphanus sativus* (2C = 1.11 pg DNA) was used as an internal standard

Autopolyploidy has been reported in many species, including apple, which is induced using either in vivo or more often in vitro techniques (Sedov et al. 2017a, b, 2018; Sedysheva and Gorbacheva 2013; Podwyszyńska et al. 2017). Both allopoliploidy and autopolyploidy considerably alter both plant phenotype and genotype (Osborn et al. 2003; Chen 2007; Anssour et al. 2009; Parisod et al. 2010; Soltis et al. 2014). The most common characteristic associated with polyploidy is an enlargement of organs (referred to as a 'gigas' effect). Moreover, phenotypic variations are due to genomic changes, not only via doubling of genome size but also via alterations of DNA structure and methylation patterns, thereby resulting in modifications of cytological, biochemical, physiological, and developmental features of a plant (Wendel et al. 2016). Often, polyploidy yields either new or transgressive traits affording these new genotypes better adaptation to less favorable growth conditions compared with their diploid progenitor(s).

4.4.5.1 Triploids

Apple triploids are in demand by consumers and producers due to their large attractive fruits and other valuable traits (Janick et al. 1996; Sedov et al. 2014, 2017a, b). Triploids account, on average, for about 10% of all currently grown cultivars, but there are differences among local populations (Janick et al. 1996). For example, among Polish apple cultivars alone, the frequency of triploids accounts for 10% of all grown cultivars (Podwyszyńska et al. 2016). Interestingly, out of five known triploids, 'Witos' is a confirmed seedling of diploid parents, 'Rarytas Śląski' is a seedling of the triploid 'Blenheim Orange', 'Pagacz' is a mutant of the triploid 'Jonagold', and the remaining two triploid cultivars are most likely derived from open pollination. Pereira-Lorenzo et al. (2007) have reported that within local Spanish cultivars, 28% are triploids. Most, if not all, of the popular triploid cultivars are derived from meiotic natural polyploidization resulting from fertilization of unreduced $2n$ gametes. These include old cultivars such as 'Gravenstein', 'Belle de Boskoop', 'Baldwin', 'Rhode Island Greening', 'Blenheim Orange', 'Stayman Winesap', 'Tompkins County King', and various members of the 'Reineta' family, such as 'Reineta Blanca' and 'Reineta Gris', as well as newer cultivars, such as 'Jonagold' and 'Mutsu' (Janick et al. 1996). Due to their vigorous growth and higher fruit productivity, compared to diploid cultivars, triploids have been predominantly cultivated in the early twentieth century (Crane and Lawrence 1930).

As reported by Russian researchers, long-term breeding studies launched in the 1970s have resulted in the development of a number of triploid apple cultivars, including genotypes derived mostly via interploid crosses ($2x \times 4x$) along with a few derived from meiotic polyploidization (Sedov et al. 2014, 2017a, b, 2018). Most of these triploid cultivars are characterized with immunity to apple scab disease, more regular fruiting behavior, high-quality fruit, and higher autogamy. Some of these cultivars; e.g., 'Sinap Orlovskiy' and 'Rozhdestvenskoye', are currently grown in orchards in both Russia and Belarus. One of the most valued cultivars, 'Ministr Kiselev', has been included in the Russian State Registry in 2017.

4.4.5.2 Tetraploids

In apple, tetraploids have been generated for breeding purposes, mainly for use in sexual hybridizations with diploids in order to obtain triploid genotypes, as these are deemed optimal for commercial markets (Sedov and Makarkina 2008).

One of the best-known old tetraploid cultivars is 'Hibernal', which originated in Russia from about 1870–1880, and it is still present in some fruit collections (Vaarama 1948). The genetic origin of 'Hibernal' has not been documented, but it is likely derived from sexual polyploidization. Due to its high resistance to frost, vigorous growth, and good productivity, 'Hibernal' has been widely cultivated in the early twentieth century in northern regions of Finland, Canada, and the USA, where it has been imported from Russia. 'Hibernal' possesses typical traits for tetraploids, including thick and short branches, large dark-green leaves, and relatively large fruit with thick skin.

In the 1970s, several either homogenous or chimeral apple tetraploids of 'Antonovka Ploskaya' (2-4-4-4), 'Papirovka' (2-4-4-4), 'McIntosh' (4x), and 'Wealthy' (2-4-4-4) have been generated via colchicine treatments, used for inter- and intraploid crosses, and yielding several desirable triploid cultivars (Sedov et al. 2014, 2017a, b, 2018). These studies have revealed that $2x \times 4x$ crosses resulted in a higher number of triploids compared to reciprocal crosses ($4x$ $2x$). This is attributed to the fact that maternal tetraploids possess a relatively high capability for autogamy. Therefore, to obtain higher numbers of triploids using $4x \times 2x$ crosses, flowers of the maternal tetraploid should be emasculated (Sedov 2014). Furthermore, it has been reported that tetraploids have been also produced following hybridizations between $3x$ and $2x$, wherein both genotypes yield unreduced gametes (Schuster and Büttner 1995).

Although tetraploid cultivars are currently not cultivated commercially, they are maintained in genebanks and germplasm resources for biodiversity studies, and for utilization in breeding programs. In the last 20 years, several new apple autotetraploids have been developed using in vitro techniques (Ou et al. 2008; Jia 2009; Xue et al. 2015). These tetraploids have been induced from leaf explants, while induction of shoot explants has not yielded any tetraploids (Jia 2009). For chromosome doubling, antimitotic agents, such as colchicine and oryzalin, have been most commonly used (Ou et al. 2008; Jia 2009; Xue et al. 2015; Hias et al. 2017). However, as recent concerns over the carcinogenic activity of colchicine, other less toxic antimitotics, such as amiprophos-methyl (APM), have been successfully used for chromosome doubling under in vitro conditions (Podwyszyńska et al. 2017). Thus, treating either shoots or leaf explants with APM in in vitro cultures has resulted in the recovery of numerous mitotic tetraploids from six apple cultivars (Podwyszyńska et al. 2017).

During the last decade, phenotypes and genotypes of autotetraploid have been evaluated by several groups worldwide (Jia 2009; Zhang et al. 2015; Xue et al. 2015, 2017; Chen et al. 2017; Hias et al. 2017; Podwyszyńska et al. 2020, 2021b). Newly generated tetraploids are notably characterized by their weak growth (Xue et al. 2017; Podwyszyńska et al. 2020, 2021a, b). Moreover, these tetraploids have shorter shoots, smaller dark-green leaves with altered shapes (more roundish), but have much higher chlorophyll contents (Fig. 4.5). Analysis of 6–12-month old own-rooted tetraploid plants by measuring gas exchange and chlorophyll fluorescence (maximum PSII quantum yield, Fv/Fm) have revealed that these tetraploids have either comparable or lower photosynthetic activities, but higher transpiration rates and stomatal conductivities than their diploid counterparts (Podwyszyńska et al. 2021a, b). In addition, some tetraploid clones, depending on the cultivar, have demonstrated propensities for premature dormancy, as manifested with growth inhibition within a period of 2 months following transfer from in vitro to ex vitro conditions. However, when these tetraploids are subjected to a 12-week low temperature (5 °C) treatment, dormancy is broken, and they resume growth. This observed transient growth inhibition of newly-developed tetraploids is likely attributed to epigenetic changes resulting from alterations in DNA methylation due to chromosome doubling (Podwyszyńska et al. 2021a, b). In fact, higher DNA methylation levels are detected in tetraploid clones compared to their diploid counterparts using methylation-sensitive amplification polymorphism (MSAP) analysis. Furthermore, such tetraploid clones undergoing premature dormancy are also characterized by a high DNA methylation rate (22.95%), which is ~ threefold higher than that detected in diploids (8.3%). Therefore, it is proposed that the process of entering into and then subsequent recovery from dormancy is related to alterations in DNA methylation rates, induced by low temperature, and it is likely to be involved in genome stabilization of newly developed tetraploids.

Xue et al. (2017) have reported on increased vigor, larger organs, and enhanced physiological parameters in mitotic neotetraploids of apple cv. 'Hanfu' after reaching maturity (reproductive phase). It is observed that tetraploid 'Hanfu' lines have flowered in the fifth year; i.e., 2 years later than diploid 'Hanfu'. Furthermore, flowering of tetraploids is delayed by 5–7 days within the season. Moreover, all physiological parameters measured, such as net photosynthetic rate, transpiration rate, stomata conductance, and fluorescence performance are found to be higher in tetraploids than in diploids. However, the dwarf growth habit trait of tetraploids is maintained. Ma et al. (2016) have investigated the molecular mechanism of dwarfism in autopolyploid apple trees. It is reported that following genome doubling, differential expression of genes involved in both auxin and brassinosteroid biosynthesis pathways likely leads to reduced levels of these two growth regulators, thereby contributing to dwarfism (Ma et al. 2016).

Apple tetraploids are reported to have enhanced productivity and fruit quality along

Fig. 4.5 Plants, leaves, and stomata of 6-month-old diploid (left) and tetraploid (right) plants of 'Gala Must' apple

with increased resistance to various biotic and abiotic stresses (Sedyscheva and Gorbacheva 2013; Jia 2009; Xue et al. 2015, 2017; Chen et al. 2017; Podwyszyńska et al. 2020, 2021a, b). Sedov and Makarkina (2008) have reported that fruits of some selected apple tetraploids have higher contents of total sugars, titratable acids, and ascorbic acids compared to those detected in their diploid counterparts. Furthermore, some of these traits, such as high ascorbic acid content, are transmitted to hybrid triploid progeny following sexual hybridizations between tetraploid and diploid genotypes.

There are several reports suggesting that tetraploids of various crop species are more resistant to diseases than their diploid progenitors due to duplications of resistance genes and/or alterations of their molecular functions as a consequence of whole-genome duplication (Oswald and Nuismer 2011; Panchy et al. 2016). Chen et al. (2017) have obtained autotetraploids of 'Hanfu' and 'Gala' demonstrating significantly lower levels of susceptibility to various serious diseases, such as leaf blight and anthracnose, caused by *Alternaria alternate* and *Colletotrichum gloeosporioides*, respectively. Using real-time quantitative PCR analysis, they have investigated the molecular mechanism of enhanced resistance to infection by these two pathogens and have found that several disease resistance-related genes, including *PR1*, *WRKY29*, *CDPK*, and *MPK4*, are significantly up-regulated in autotetraploids (Chen et al. 2017). Recently, several individual tetraploid clones of cvs. 'Redchief' and 'Gala Must' with higher levels of resistance to the bacterial disease fire blight caused by *Erwinia amylovora* and tetraploid clones of 'Free Redstar' with higher levels of resistance to apple scab disease caused by *Venturia inaequalis* have been selected based on in vitro bioassays and greenhouse tests, respectively (Podwyszyńska et al. 2020, 2021a, b). In addition, it is revealed that sister tetraploid clones derived from the same diploid apple cultivar have exhibited differences in susceptibility to these diseases (Podwyszyńska et al. 2020, 2021a, b). Molecular analysis has confirmed that differences within tetraploids of a given cultivar are associated with genetic and epigenetic changes as indicated by variations in DNA structure and levels of methylation. Mean level of genetic differentiation among tetraploids analyzed in the above three apple cultivars using amplified fragment length polymorphism (AFLP) markers is 2.4% compared with their diploid counterparts.

It has been reported that autotetraploidy enhances drought stress tolerance in two apple cultivars, 'Hanfu' and 'Gala' (Zhang et al. 2015). Under drought stress, autotetraploids had higher relative water contents (RWC) and higher chlorophyll fluorescence, but lower levels of malondialdehyde (MDA) and proline compared with their counterpart diploids. It was observed that key aquaporins genes, including *MdPIP1;1* and *MdTIP1;1,* were induced in leaves in response to polyethylene glycol (PEG 6000) treatment. Furthermore, expression of these two genes induced under drought stress treatment was lower in autotetraploids than in diploids. In another study, tetraploids of the apple rootstock *M. zumi* with increased resistance to both salt and drought, compared to its diploid counterpart, were also reported (Jia 2009).

4.5 Cytogenetic Analyses

4.5.1 Identification of *Malus* Chromosomes

As size of chromosomes of apple species (*Malus* sp.) is very small and they are relatively high in number, this makes it rather difficult to conduct cytogenetic analysis. According to Bouveir et al. (2000), lengths of apple chromosomes range from 1.5–3.5 μm. However, lengths of apple chromosomes are reported to be slightly shorter, 1.0–2.7 μm by Zhu and Gardiner (1995) and 0.5–1.0 μm by Lespinasse et al. (1976). These observed differences in lengths of chromosomes may be due to different genotypes analyzed, measurement method, and slide preparation, mainly by squashing, that may increase total chromosome length and alter arm indices (Bosemark and Bormotov 1971). Furthermore, this may also depend on the position of a chromosome relative to the center of a squash. Nevertheless, there are no reported wide variations in chromosome morphology. According to Zhu and Gardiner (1995), based on the position of the centromere, all apple chromosomes are submetacentric, using terminology suggested by Levan et al. (1964), and there are no secondary constrictions. Bouveir et al. (2000) have classified chromosomes in haploid *M.* × *domestica* (cultivated apple) as metacentric (6 chromosomes) and submetacentric (11 chromosomes), with the longest chromosome bearing a satellite. It has been observed that a small satellite is present on either one or both homologues of chromosome 3 in the diploid apple rootstock *M. sieboldii* clone A106 (Zhu et al. 1995). This observed low frequency of satellites detected by conventional staining may be due to slide preparation and to the small size of apple chromosomes.

Identification of apple chromosomes has been improved by using techniques enabling longitudinal differentiation of chromosomes, such as silver staining (Zhu and Gardiner 1995) and chromosome C-banding of heterochromatin sections (Zhu et al. 1995). The cytological technique of silver (Ag) staining, capitalizing on the preferential binding of silver ions onto some of the nucleolus-linked proteins, has been used for detecting nucleolar organizing regions (Ag-NORs) in the apple rootstock MM106 (Zhu and Gardiner 1995). NORs are found to be present at telomeric regions of the short arms of chromosome pairs 6 and 14 (classified based on decreasing chromosome length). Zhu et al. (1995) have used, for the first time, C-banding techniques for the apple genome to identify pairs of homologous chromosomes in the diploid apple rootstock A106 (*M. sieboldii*). The C-banding technique has revealed the presence of C-bands on telomeric regions of the short arms of chromosomes 2, 4, 6, 8, 14, and 16. Although C-banding has enabled the identification of only 6 of the 17 chromosome pairs of apple, it has been useful to distinguish between chromosomes 4 and 5 that are similar in size and arm ratios (Zhu et al. 1995).

4.5.2 Fluorescence in Situ Hybridization for Physical Mapping of Chromosomes

Fluorescence in situ hybridization (FISH) is yet another technique that produces molecular cytogenetic markers, allowing for chromosomal identification and studies of phylogenetic relationships among species (Fukui et al. 1994). This technique is used to locate the physical position of either repeated or unique DNA sequences on chromosomes. Thus far, there has been limited work using FISH for investigating the small chromosomes of *Malus* species. Zhu and Gardiner (1995) have described the use of FISH using an 18S plus a spacer rDNA probe and silver staining of NORs in a diploid apple rootstock. In this study, rDNA genes have been localized at terminal positions of the short arms of chromosomes 6 and 14. Furthermore, two ribosomal DNA genes, 5S rDNA and 18S-25S rDNA, have been used in analyzing apple cv. 'Pinova', $2n = 2x = 34$ (Schuster et al. (1997). The simultaneous use of these probes has revealed the presence of 10 ribosomal RNA loci at different chromosomal locations; thus, offering reliable chromosome landmarks for identifying five of the 17 chromosome pairs of apple. In addition, 18S-25S rDNA loci have been identified along four chromosome pairs in telomeric regions of the short arms, while 5S rDNA loci are detected in the proximal region of the short arm of a small metacentric chromosome pair. These findings are consistent with more recent observations revealing the presence of eight telomeric 18S-5.8S-25S sites and two centromeric 5S rDNA sites, located on different chromosomes in open-pollinated seedlings of apple cv. 'Sensyu' (Yamamoto et al. 2016).

Three genes from the cultivated apple, *M.* × *domestica,* related to fruit ripening encoding endopolygalacturonase (EPG), 1-aminocyclopropane-1-carboxylic acid (ACC) oxidase, and ACC synthase were used as probes for in situ hybridization of chromosomes of the diploid apple rootstock *M. sieboldii* A106 to locate the physical positions of corresponding genes on an apple karyotype (Gardiner et al. 1994; Zhu et al. 1995). FISH revealed that genes homologous to EPG, ACC oxidase, and ACC synthase were located in different positions along 10 pairs of chromosomes. Further progress of the *Malus* genome has been made by using bacterial artificial chromosome (BAC) with FISH (BAC-FISH) to determine physical locations of useful gene loci. BAC-FISH analysis was used to assess the linkage of a monomorphic gene, *MdFBX11,* to the *S* locus region in apple (Minamikawa et al. 2010). FISH revealed that *MdFBX11* was located near the *S* locus in a subtelomeric region of a chromosome. In addition, a BAC clone, containing repetitive sequences, was detected at centromeric regions of each of the 17 chromosomes.

4.5.3 Meiotic Analysis of Chromosome Pairing

Microsporogenesis has been extensively analyzed both in diploid and triploid genotypes (Roscoe 1933; Koul et al. 1984; Singh and Wafai 1984; Singh and Wafai 1984; Dar et al. 2015; Dar et al. 2018). In general, a normal chromosome behavior, with 17 bivalents in meiotic metaphase 1, is observed in diploid apple cultivars (Zhu et al. 1995; Dar et al. 2018). However, Dar et al. (2015) have observed meiotic irregularities, such as univalents, non-disjunctions, as well as disturbed anaphase and telophase during male-meiotic analysis in six commercial apple cultivars ($2n$ = 34). The frequency of meiotic aberrations ranged from 5.06% (‘Raj Ambri’) to 7.07% (‘Golden Delicious’). In an earlier study of meiosis in the diploid ‘Kulu Kenya’, 17 bivalents are observed accompanied by five supernumerary or B chromosomes that might have arisen by fragmentation of chromosomes (Chaudhary and Mehra 1975).

Overall, it has been reported that meiotic division in diploid and triploid apple cultivars varies considerably (Koul et al. 1984). In triploid cultivars, multivalent associations are observed in meiotic metaphase 1, and segregation of chromosomes is highly irregular. Moreover, variable numbers of chromosomes are found to be directed to either pole, and some laggards are not included in daughter nuclei (Singh and Wafai 1984). In fact, analysis of chromosome behavior during meiosis has been used to unravel the mode of origin of triploid apples. Singh and Wafai (1984) have analyzed chromosome pairing in diploid and triploid cytotypes of *M. pumila* ‘Hazatbali’ and have observed the formation of 17 bivalents in diploids and a high frequency of trivalents in triploids, respectively. This latter finding indicates that the autotriploid origin of these triploids is likely derived from fertilization of n and $2n$ gametes.

In a study of meiosis in tetraploids, it is observed that the numbers of microsporogenesis disorders of tetraploids, such as lagging chromosomes, over-abundance of spindles, micronuclei, and polyads, account for 11.7–61.1% (Sedysheva and Gorbacheva 2013). However, normal diploid pollen is also present, thus available for use in polyploidy breeding efforts (Sedysheva and Gorbacheva 2013).

4.6 Concluding Remarks and Future Prospects

Knowledge of ploidy level, genome size, and karyotype characteristics is key for pursuing studies conducted by taxonomists and breeders. Evaluation of the hybrid status of seedlings derived from intergeneric and interploid hybridizations, either via interspecific or intraspecific crosses, is essential in breeding to eliminate apomictic offspring. Classical cytogenetic approaches in studies of the apple genome are well documented. However, there is relatively little work in applying in situ hybridization, FISH and GISH (genomic in situ hybridization), and chromosome painting mapping with BACs for studying genome evolution and hybrid verification in the genus *Malus*.

Phenotypic changes observed in newly developed apple polyploids, either allo- or autopolyploids, are not fully understood. Apple ploidy breeding is somewhat underestimated, although it is an important source of genetic variation. There is little information available on phenotypic, cytogenetic, and genetic changes resulting from polyploidization, particularly those relevant to resistance against biotic and abiotic stresses or to secondary metabolite contents. What is the molecular mechanism of these changes? In apple, mechanisms of genome downsizing in polyploids as well as geographical distribution of species based on genome size variations are poorly understood, in contrast to the available knowledge of other crops. To understand these processes, it is necessary to collectively exploit various modern methods in the field of cytology, cytometry, and molecular biology. Comprehensive evaluation of tested germplasm or breeding material will allow for significant progress in studying the origin, ecology, and breeding of apples.

References

Anssour S, Krugel T, Sharbel TF, Saluz HP, Bonaventure G, Baldwin IT (2009) Phenotypic, genetic, and genomic consequences of natural and synthetic polyploidization of *Nicotiana attenuata* and *Nicotiana obtusifolia*. Ann Bot 103:1207–1217

Arumuganathan K, Earle ED (1991) Nuclear DNA content of some important plant species. Plant Mol Biol Rep 9:208–218

Bai C, Alverson WS, Follansbee A, Waller DM (2012) New reports of nuclear DNA content for 407 vascular plant taxa from the United States. Ann Bot 110 (8):1623–1629

Barow M, Meister A (2003) Endopolyploidy in seed plants is differently correlated to systematics, organ, life strategy and genome size. Plant Cell Environ 26 (4):571–584

Bennett MD (1987) Variation in genomic form in plants and its ecological implications. New Phytol 106:177–200

Bennett MD, Smith JB (1976) Nuclear DNA amounts in angiosperms. Phil Trans R Soc Lond B 274(933):227–274

Bertin N, Borel C, Brunel B, Cheniclet C, Causse M (2003) Do genetic make-up and growth manipulation affect tomato fruit size by cell number, or cell size and DNA endoreduplication? Ann Bot 92(3):415–424

Bianco L, Cestaro A, Sargent DJ, Banchi E, Derdak S, Di Guardo M, Salvi S, Jansen J, Viola R, Gut I, Laurens F, Chagné D, Velasco R, van de Weg E, Troggio M (2014) Development and validation of a 20K single nucleotide polymorphism (SNP) whole genome genotyping array for apple (*Malus* × *domestica* Borkh). PloS One 9(10):e110377

Bisognin C, Seemüller E, Citterio S, Velasco R, Grando MS, Jarausch W (2009) Use of SSR markers to assess sexual vs. apomictic origin and ploidy level of breeding progeny derived from crosses of apple proliferation-resistant *Malus sieboldii* and its hybrids with *Malus* × *domestica* cultivars. Plant Breed 128:507–513

Blanke MM, Höfer M, Pring RJ (1994) Stomata and structure of tetraploid apple leaves cultured *in vitro*. Ann Bot 73:651–654

Bosemark NO, Bormotov VE (1971) Chromosome morphology in a homozygous line of sugar beet. Hereditas 69(2):205–211

Bou Dagher-Kharrat M, Abdel-Samad N, Douaihy B, Bourge M, Fridlender A, Siljak-Yakovlev S, Brown S (2013) Nuclear DNA C-values for biodiversity screening: case of the Lebanese flora. Plant Biosyst 147 (4):1228–1237

Bouvier L, Lesinasse Y, Schuster M (2000) Karyotype analysis of an haploid plant of apple (*Malus domestica*). Acta Hortic 538:321–324

Brochmann C, Brysting AK, Alsos IG, Borgen L, Grundt HH, Scheen AC, Elven R (2004) Polyploidy in arctic plants. Biol J Linn Soc 82:521–536

Cavalier-Smith T (1985) Cell volume and the evolution of eukaryote genome size. In: Cavalier-Smith T (ed) The Evolution of Genome Size. Wiley, Chichester, pp 105–184

Chagné D, Kirk C, Whitworth C, Erasmuson S, Bicknell R, Sargent D, Kumar S, Troggio M (2015) Polyploid and aneuploid detection in apple using a single nucleotide polymorphism array. Tree Genet Genomes 11(5):94. https://doi.org/10.1007/s11295-015-0920-8

Challice JS (1974) Rosaceae chemotaxonomy and the origins of the Pomoideae. Bot J Linn Soc 69:239–259

Chaudhary MS, Mehra PN (1975) Accessory chromosomes in apples. Cytologia 40(1):7–11

Chen ZJ (2007) Genetic and epigenetic mechanisms for gene expression and phenotypic variation in plant polyploids. Annu Rev Plant Biol 58:377–406

Chen M, Wang F, Zhang Z, Fu J, Ma Y (2017) Characterization of fungi resistance in two autotetraploid apple cultivars. Sci Hortic 220:27–35

Comai L (2005) The advantages and disadvantages of being polyploid. Nat Rev Genet 6(11):836–846

Considine MJ, Wan Y, D'Antuono MF, Zhou Q, Han M, Gao H, Wang M (2012) Molecular genetic features of polyploidization and aneuploidization reveal unique patterns for genome duplication in diploid *Malus*. PloS One 7(1):e29449

Crane MB, Lawrence WJC (1930) Fertility and vigour of apples in relation to chromosome number. J Genet 22:153–163

Dar JA, Wani AA, Dhar MK (2015) Morphological, biochemical and male-meiotic characterization of apple (*Malus* × *domestica* Borkh.) germplasm of Kashmir Valley. Chrom Bot 10:39–49

Dar JA, Wani AA, Dhar MK (2018) Male meiotic behavior of apple germplasm in Kashmir: towards understanding the evolution of apple. Cytologia 83:87–91

Darlington CD, Moffett AA (1930) Primary and secondary chromosome balance in *Pyrus*. J Genet 22:129–151

De Veylder L, Larkin JC, Schnittger A (2011) Molecular control and function of endoreplication in development and physiology. Trends Plant Sci 16(11):624–634

Dhooghe E, Van Laere K, Eeckhaut T, Leus L, Van Huylenbroeck J (2011) Mitotic chromosome doubling of plant tissues *in vitro*. Plant Cell Tiss Org Cul 104 (3):359–373

Dickson EE, Arumuganathan K, Kresovich S, Doyle JJ (1992) Nuclear DNA content variation within the Rosaceae. Am J Bot 79(9):1081–1086

Doležel J, Bartoš J (2005) Plant DNA flow cytometry and estimation of nuclear genome size. Ann Bot 95(1):99–110

Doležel J, Bartos J, Voglmayr H, Greilhuber J (2003) Nuclear DNA content and genome size of trout and human. Cytometry A 51(2):127–128

Druart P (2000) Aneuploïds and variants of apple (*Malus domestica* Borkh.) through *in vitro* culture techniques. Acta Hortic 520:301–307

Eilam T, Anikster Y, Millet E, Manisterski J, Feldman M (2010) Genome size in diploids, allopolyploids, and autopolyploids of Mediterranean Triticeae. J Bot 2010:1–12

Einset J, Imhofe B (1947) Chromosome numbers of apple varieties and sports. Proc Am Soc Hortic Sci 50:45–50

Evans RC, Campell CS (2002) The origin of the apple subfamily (Maloideae; Rosaceae) is clarified by DNA sequence data from duplicated GBSSI genes. Am J Bot 89:1478–1484

Feldman M, Levy AA (2012) Genome evolution due to allopolyploidization in wheat. Genetics 192(3):763–774

Folta KM, Gardiner SE (2009) Genetics and genomics of Rosaceae. Springer, New York, p 633

Fukui K, Ohmido N, Khush GS (1994) Variability in rDNA loci in genus *Oryza* detected through fluorescence *in situ* hybridization. Theor Appl Genet 87:893–899

Gardiner SE, Zhu JM, Whitehead HCM, Madie C (1994) The New Zealand apple genome mapping project. Euphytica 77:77–81

Geier T (2012) Chimeras: properties and dissociation in vegetatively propagated plants. In: Shu QY, Forster BP, Nakagawa H (eds) Plant Mutation Breeding and Biotechnology. CABI Publishing, Wallingford, pp 191–201

Giovannoni J (2010) Harvesting the apple genome. Nat Genet 10:822–823

Goldblatt P (1976) Cytotaxonomic studies on the tribe Quillajeae (Rosaceae). Ann MO Bot Gard 63:37–60

Gonai T, Manabe T, Inoue E, Hayashi M, Yamamoto T, Hayashi T, Yamamoto T, Hayashi T, Sakuma F, Kasumia M (2006) Overcoming hybrid lethality in a cross between Japanese pear and apple using gamma irradiation and confirmation of hybrid status using flow cytometry and SSR markers. Scient Hortic 109 (1):43–47

Green AF, Ramsey TS, Ramsey J (2013) Polyploidy and invasion of English ivy (*Hedera* spp., Araliaceae) in North American forests. Biol Invasions 15:2219–2241

Greilhuber J, Doležel J, Lysak MA, Bennett MD (2005) The origin, evolution and proposed stabilization of the terms 'genome size' and 'C-value' to describe nuclear DNA contents. Ann Bot 95(1):255–260

Greilhuber J, Borsch T, Müller K, Worberg A, Porembski S, Barthlott W (2006) Smallest angiosperm genomes found in Lentibulariaceae, with chromosomes of bacterial size. Plant Biol 8(06):770–777

Han Y, Gasic K, Marron B, Beever JE, Korban SS (2007) A BAC-based physical map of the apple genome. Genomics 89(5):630–637

Hancock JF, Luby JJ, Brown SK, Lobus GA (2008) Apples. In: Hancock JF (ed) Temperate fruit crop breeding: germplasm to genomics. Springer, pp 1–38

Harada T, Kurahashi W, Yanai M, Wakasa Y, Satoh T (2005) Involvement of cell proliferation and cell enlargement in increasing the fruit size of *Malus* species. Scient Hortic 105(4):447–456

He P, LiL CL, Wang H, Chang Y (2018) Variation in ploidy level and morphological traits in the progeny of the triploid apple variety Jonagold. Czech J Genet Plant Breed 54(3):135–142

Hias N, Leus L, Davey MW, Vanderzande S, Van Huylenbroeck J, Keulemans J (2017) Effect of polyploidization on morphology in two apple (*Malus* × *domestica*) genotypes. Hort Sci 44(2):55–63

Höfer M, Meister A (2010) Genome size variation in *Malus* species. J Bot 2010:1–9

Inoue E, Sakuma F, Kasumi M, Tsukihashi HH, T, (2002) Production of apomict-like seedling through intergeneric cross between Japanese pear and apple by means of pollen irradiation. Acta Hortic 587:211–215

Janick J, Cummins JN, Brown SK, Hemmat M (1996) Apples. In: Janick J, Moore JN (eds) Fruit breeding: tree and tropical fruits 1. John Willey and Sons, NY, USA, pp 1–77

Jedrzejczyk I, Sliwinska E (2010) Leaves and seeds as materials for flow cytometric estimation of the genome size of 11 Rosaceae woody species containing DNA-staining inhibitors. J Bot

Jia S (2009) *In vitro* induction and identification of tetraploid in *Malus zumi* and its evaluation of stress-tolerance. Master's Thesis, Agricultural University of Hebei, China. https://www.dissertationtopic.net/doc/517919

Johnston JS, Bennett MD, Rayburn AL, Galbraith DW, Price HJ (1999) Reference standards for determination of DNA content of plant nuclei. Am J Bot 86(5):609–613

Kamiński P, Podwyszyńska M, Starzycki M, Starzycka-Korbas E (2016) Interspecific hybridisation of cytoplasmic male-sterile rapeseed with Ogura cytoplasm and *Brassica rapa* var. *pekinensis* as a method to obtain male-sterile Chinese cabbage inbred lines. Euphytica 208(3):519–534

Korban SS, Wannarat W, Rayburn CM, Tatum TC, Raybur AL (2009) Genome size and nucleotypic variation in *Malus* germplasm. Genome 52:149–155

Koul AK, Singh R, Wafai B (1984) Assessment of the germlasm of apple (*Malus pumila* Mill.) grown in Kashmir II. Cytology of Rikhal saharanpuri, Acc. 1877 and Kharoo. Cytologia 49:313–323

Kron P, Husband BC (2009) Hybridization and the reproductive pathways mediating gene flow between native *Malus coronaria* and domestic apple. M. Domestica. Botany 87(9):864–874

Leitch IJ, Bennett MD (2004) Genome downsizing in polyploid plants. Biol J Linn Soc 82(4):651–663

Lespinasse Y, Noiton D (1986) Contribution to the study of a haploid apple (*Malus pumila*). Descriptive study and comparison with clones of different levels of ploidy. 1. Vegetative characters: internodes, leaves and stomata. Agronomie 6:659–664

Lespinasse Y, Alston FH, Watkins R (1976) Cytological techniques for use in apples breeding. Ann Appl Biol 82:349–353

Levan A, Fredga K, Sandberg AA (1964) Nomenclature for centromeric position on chromosomes. Hereditas 52(2):201–220

Levin DA (1983) Polyploidy and novelty in flowering plants. Am Nat 122:1–25

Liebhard R, Gianfranceschi L, Koller B, Ryder CD, Tarchini R, Van de Weg E, Gessler C (2002) Development and characterisation of 140 new microsatellites in apple (*Malus* × *domestica* Borkh.). Mol Breed 10(4):217–241

Lukaszewska E, Sliwinska E (2007) Most organs of sugar-beet (*Beta vulgaris* L.) plants at the vegetative and reproductive stages of development are polysomatic. Sexual Plant Rep 20(2):99–107

Ma Y, Xue H, Zhang L, Zhang F, Ou C, Wang F, Zhang Z (2016) Involvement of auxin and brassinosteroid in dwarfism of autotetraploid apple (*Malus* × *domestica*). Sci Rep 6:26719

Malladi A, Hirst PM (2010) Increase in fruit size of a spontaneous mutant of 'Gala' apple (*Malus* × *domestica* Borkh.) is facilitated by altered cell production and enhanced cell size. J Exp Bot 61(11):3003–3013

Marie D, Brown SC (1993) A cytometric exercise in plant DNA histograms, with 2C values for 70 species. Biol Cell 78(1–2):41–51

Minamikawa M, Kakui H, Wang S, Kotoda N, Kikuchi S, Koba T, Sassa H (2010) Apple S locus region represents a large cluster of related, polymorphic and pollen-specific F-box genes Plant Mol Biol 74:143–154

Moffett AA (1931) The chromosome constitution of the Pomoideae. Proc Roy Soc Lond Series B 108 (758):423–446

Murat F, Peer YVD, Salse J (2012) Decoding plant and animal genome plasticity from differential paleo-evolutionary patterns and processes. Genome Biol Evol 4(9):917–928

Nebel BR (1929) Zur cytology van *Malus* und *Vitis*. Die Gardenbauwissenschaft 1:549–592

Noormohammadi Z, Fazeli S, Sheidai M, Farahani F (2015) Molecular and genome size analyses of somaclonal variation in apple rootstocks Malling 7 and Malling 9. Acta Biol Szeged 59(2):139–149

Osborn TC, Pires JC, Birchler JA, Auger DL, Chen ZJ, Lee HS, Comai L, Madlung A, Doerge RW, Colot V, Martienssen RA (2003) Understanding mechanisms of novel gene expression in polyploids. Trends Genet 19 (3):141–147

Oswald BP, Nuismer SL (2011) A unified model of autopolyploid establishment and evolution. Am Natur 178(6):687–700

Ou CQ, Li LG, He P, Zhang ZH (2008) *In vitro* adventitious shoot regeneration and induction of tetraploid from leaves of Hanfu apple. J Fruit Sci 25:293–297

Panchy N, Lehti-Shiu M, Shiu SH (2016) Evolution of gene duplication in plants. Plant Physiol 171:2294–2316

Parisod C, Holderegger R, Brochmann C (2010) Evolutionary consequences of autopolyploidy. New Phytol 186:5–17

Pellicer J, Fay MF, Leitch IJ (2010) The largest eukaryotic genome of them all? Bot J Linn Soc 164(1):10–15

Pereira-Lorenzo S, Ramos-Cabrer AM, Díaz-Hernández MB (2007) Evaluation of genetic identity and variation of local apple cultivars (*Malus* × *domestica* Borkh.) from Spain using microsatellite markers. Genet Resour Crop Ev 54:405–420

Phipps JB, Robertson KR, Rohrer JR (1991) Origins and evolution of subfam. Maloideae (Rosaceae). Syst Bot 16:303–332

Podwyszyńska M, Kruczyńska D, Machlańska A, Dyki B, Sowik I (2016) Nuclear DNA content and ploidy level of apple cultivars including Polish ones in relation to some morphological traits. Acta Biol Cracov Bot 58 (1):81–93

Podwyszyńska M, Sowik I, Machlańska A, Kruczyńska D, Dyki B (2017) *In vitro* tetraploid induction of *Malus* × *domestica* Borkh. using leaf or shoot explants. Sci Hortic 226:379–388

Podwyszyńska M, Sowik I, Puławska J (2020) *In vitro* testing of apple tetraploids for resistance to fire blight. Acta Hortic 1282:343–350

Podwyszyńska M, Markiewicz M, Broniarek-Niemiec A, Matysiak B, Marasek-Ciolakowska A (2021a) Apple autotetraploids with enhanced resistance to apple scab (*Venturia inaequalis*) due to genome duplication-phenotypic and genetic evaluation. Int J Mol Sci 22 (2):527

Podwyszyńska M, Markiewicz M, Klamkowski K, Broniarek A, Wojtania A (2021b) The genetic background of the phenotypic variability observed in apple autotetraploids. Acta Hortic 1307:177–186

Poggio L, Realini MF, Fourastié MF, García AM, González GE (2014) Genome downsizing and karyotype constancy in diploid and polyploid congeners: a model of genome size variation. AoB Plants: 6plus029

Proost S, Pattyn P, Gerats T, Van de Peer Y (2011) Journey through the past: 150 million years of plant genome evolution. Plant J 66(1):58–65

Pustahija F, Brown SC, Bogunić F, Bašić N, Muratović E, Ollie S (2013) Small genomes dominate in plants growing on serpentine soils in West Balkans, an exhaustive study of 8 habitats covering 308 taxa. Plant Soil 373(1–2):427–453

Ramanna MS, Jacobsen E (2003) Relevance of sexual polyploidization for crop improvement – a review. Euphytica 133:3–8

Ramsey J, Ramsey TS (2014) Ecological studies of polyploidy in the 100 years following its discovery. Philo Trans Roy Soc b: Biol Sci 369(1648):20130352

Ramsey J, Robertson A, Husband B (2008) Rapid adaptive divergence in new world *Achillea*, an autopolyploid complex of ecological races. Evolution 62:639–653

Renny-Byfield S, Chester M, Kovařík A, Le Comber SC, Grandbastien M-A, Deloger M, Nichols RA, Macas J, Novák P, Chase MW, Leitch AR (2011) Next generation sequencing reveals genome downsizing in allotetraploid *Nicotiana tabacum*, predominantly through the elimination of paternally derived repetitive DNAs. Mol Biol Evol 28(10):2843–2854

Roscoe MV (1933) The chromosomal constitution of certain cultivated apple varieties. J Gen 28:157–167

Rybin VA (1926) Cytological investigations of the genus *Malus* (preliminary account). Bull Appl Bot Plant Breed (leningrad) 16:187–200

Sax K (1933) The origin of Pomoideae. J Am Soc Hortic Sci 30:147–150

Schuster M, Büttner R (1995) Chromosome numbers in the *Malus* wild species collection of the genebank Dresden-Pillnitz. Genet Resour Crop Ev 42(4):353–361

Schuster M, Fuchs J, Schubert I (1997) Cytogenetics in fruit breeding-localization of ribosomal RNA genes on chromosomes of apple. Theor Appl Genet 94:322–324

Sedov EN, Makarkina MA (2008) Biochemical composition of fruit of apple cultivar clones and tetraploid forms. Russ Agric Sci 34(2):71–73

Sedov EN, Sedysheva GA, Serova ZM, Gorbacheva NG, Melnik SA (2014) Breeding assessment of heteroploid crosses in the development of triploid apple varieties. Russ J Genet Appl Res 4(1):52–59

Sedov EN, Sedysheva GA, Krasova NG, Serova ZM, Yanchuk TV (2017a) Advantages and prospects of new triploid apple varieties for production. Hortic Viticul 2:24–30. https://vstisp.org/.../24-30-2-2017.pdf

Sedov EN, Sedysheva GA, Makarkina MA, Serova ZM (2017b) Development of triploid apple cultivars as a priority in selection. Russ J Genet Appl Res 7(7):773–780

Sedov EN, Sedysheva GA, Serova ZM, Yanchuk TV (2018) Intervalent crosses: The main way of creating triploid apple cultivars. Russ Agric Sci 44(4):306–309

Sedysheva GA, Gorbacheva NG (2013) Estimation of new tetraploid apple forms as donors of diploid gametes for selection on a polyploidy level. Univ J Plant Sci 1(2):49–54

Selmecki AM, Dulmage K, Cowen LE, Anderson JB, Berman J (2009) Acquisition of aneuploidy provides increased fitness during the evolution of antifungal drug resistance. PLoS Genet 5(10):e1000705

Singh R, Wafai BA (1984) Intravartetal polyploidy in the apple (*Malus pumila* Mill.) cultivar Hazratbali. Euphytica 33(1):209–214

Sliwinska E (2018) Flow cytometry–a modern method for exploring genome size and nuclear DNA synthesis in horticultural and medicinal plant species. Folia Hortic 30(1):103–128

Snodgrass SJ, Jareczek J, Wendel JF (2017) An examination of nucleotypic effects in diploid and polyploid cotton. AoB Plants 9(1):plw082

Soltis DE, Soltis PS, Tate JA (2004) Advances in the study of polyploidy since plant speciation. New Phytol 161(1):173–191

Soltis PS, Liu X, Marchant DB, Visger CJ, Soltis DE (2014) Polyploidy and novelty: Gottlieb's legacy. Philos Trans Roy Soc B 369(1648):20130351

Stebbins GL (1950) Variation and Evolution in Plants. Columbia University Press, New York, USA, p 643

Stebbins GL (1985) Polyploidy, hybridization, and the invasion of new habitats. Ann MO Bot Gard 72:824–832

Tatum TC, Stepanovic SS, Biradar DP, Rayburn AL, Korban SS (2005) Variation in nuclear DNA content in *Malus* species and cultivated apples. Genome 48:924–930

Vaarama A (1948) Meiosis and polyploid characters in the tetraploid apple variety Hibernal. Hereditas 34(1–2):147–160

Van Tuyl JM, Boon E (1997) Variation in DNA-content in the genus *Lilium*. Acta Hortic 430:829–836

Velasco R, Zharkikh A, Affourtit J, Dhingra A, Cestaro A, Kalyanaraman A, et al. (2010) The genome of the domesticated apple (*Malus* × *domestica* Borkh.). Nat Genet 42(10):833–839

Weeden N, Lamb R (1987) Genetics and linkage analysis of 19 isozyme loci in apple. J Am Soc Hortic Sci 112:865–872

Wendel JF (2000) Genome evolution in polyploids. In: Doyle Brandon JJ, Gaut S (eds) Plant Molecular Evolution. Springer, Dordrecht, pp 225–249

Wendel JF, Jackson SA, Meyers BC, Wing RA (2016) Evolution of plant genome architecture. Genome Biol 17:37. https://doi.org/10.1186/s13059-016-0908

Xue H, Zhang F, Zhang ZH, Fu JF, Wang F, Zhang B, Ma Y (2015) Differences in salt tolerance between diploid and autotetraploid apple seedlings exposed to salt stress. Sci Hortic 190:24–30

Xue H, Zhang B, Tian JR, Chen MM, Zhang YY, Zhang ZH, Ma Y (2017) Comparison of the morphology, growth and development of diploid and autotetraploid 'Hanfu' apple trees. Sci Hortic 225:277–285

Yamamoto M, Shigeki M, Yamamoto T (2016) Detection of ribosomal RNA genes in apple (*Malus* ×*domestica*) using fluorescence in situ hybridization. Chromosome Sci 19:33–36

Zhang F, Xue H, Lu X, Zhang B, Wang F, MaY ZZ (2015) Autotetraploidization enhances drought stress tolerance in two apple cultivars. Trees 29:1773–1780

Zhu J, Gardiner S (1995) Chromosomal localization of ribosomal DNA sequences on an apple rootstock using a digoxygenin detection system. Cell Res 5:1–7

Zhu J, Gardiner S, Lay-Yee M (1995) Physical mapping of three fruit ripening genes: Endopolygalacturonase, ACC oxidase and ACC synthase from apple (*Malus* × *domestica*) in an apple rootstock A106 (*Malus sieboldii*). Cell Res 5:243–253

Zonneveld BJM (2009) The systematic value of nuclear genome size for "all" species of *Tulipa* L. (Liliaceae). Plant Syst Evol 281:217–245

Genetics and Breeding of Apple Scions

5

Soon Li Teh, Sarah A. Kostick, and Kate M. Evans

Abstract

The development of new improved apple scion cultivars through breeding continues to be an important approach to ensure sustainable apple production and increase consumer satisfaction. Apple scion breeding programs are primarily focused on improving fruit quality, fruit storability, and resistance to diseases and pests. Recent genetic and genomic advances have resulted in practical applications (e.g., DNA-informed breeding) in apple scion breeding programs. However, phenotyping continues to be a bottleneck, thus limiting efficient genetic improvement of apple scions. This chapter provides an updated review of: (1) current apple scion breeding efforts; (2) breeding for improved fruit quality; (3) breeding for resistance to diseases (e.g., apple scab, fire blight, and powdery mildew) and pests; and (4) recent advances and future breeding directions with emphasis on genotyping, DNA-informed breeding, genomic selection, and non-destructive phenotyping in apple. Throughout this chapter, perspectives on phenotyping and genomic advances are provided.

S. L. Teh and S. A. Kostick contributed equally to this chapter.

S. L. Teh · S. A. Kostick · K. M. Evans (✉)
Department of Horticulture, Tree Fruit Research and Extension Center, Washington State University, Wenatchee, WA 98801, USA
e-mail: kate_evans@wsu.edu

5.1 Introduction and Current Breeding Efforts

Apples have a long history of human consumption and are cultivated throughout the temperate world. In 2017, world apple production exceeded 83 million metric tonnes with China, USA, Turkey, Poland, India, Iran, Italy, Chile, France, and Russia being the top 10 producers (FAOSTAT 2017). Although there are thousands of cultivars throughout the world, a small number (e.g., 'Delicious', 'Golden Delicious', 'Granny Smith', 'Fuji', and 'Gala') dominate commercial production (Way et al. 1991; Hancock et al. 2008). In recent years, newer cultivars such as 'Cripps Pink' (marketed as 'Pink Lady'®) or 'Honeycrisp' (marketed as 'Honeycrunch'® in Europe) have gained popularity (Hancock et al. 2008). The cultivated apple (*Malus* × *domestica* Borkh.; also referred to as *M. pumila* Mill.) is an interspecific hybrid of a complex origin, primarily derived from *M. sieversii*, and it is a member of the genus *Malus* in the Rosaceae family, which includes many economically important fruit crops (Hancock et al. 2008; Velasco et al. 2010). Although the majority of species in Rosaceae is characterized by haploid chromosome numbers of $x = 8$ or 9, most apple species are $2n = 2x = 34$ (Hancock et al. 2008). Apple is a highly heterozygous autopolyploid that

S. S. Korban (ed.), *The Apple Genome*, Compendium of Plant Genomes,
https://doi.org/10.1007/978-3-030-74682-7_5

behaves like a diploid (Brown 2012; Daccord et al. 2017; Velasco et al. 2010). Apples typically demonstrate gametophytic self-incompatibility and are thus out-crossers.

Due to its economic and cultural importance, apple has a long and rich history of genetic improvement through breeding that is detailed throughout popular articles and scientific literature. Several book chapters and literature reviews have discussed the history of apple improvement, breeding, and cultivar releases (e.g., Brown and Maloney 2003; Brown 2012; Gardiner et al. 2007; Hancock et al. 2008; Laurens 1999). Although the focus of this chapter is on apple scion breeding, it is important to keep in mind that an apple tree is a composite plant comprised of a rootstock and a fruiting scion. Multiple public and private apple scion breeding programs exist throughout the world (reviewed in Brown and Maloney 2003). In the USA, there are currently three active public apple scion breeding programs, including Cornell University, University of Minnesota, and Washington State University along with a retired program, the cooperative PRI (Purdue University, Rutgers University, and the University of Illinois) program. Each of these breeding programs has released cultivars that have contributed to the US apple industry with examples including 'Cortland', 'Jonagold', 'NY-1' ('SnapDragon'®), and 'NY-2' ('RubyFrost'®) from Cornell University, 'Honeycrisp', 'Minneiska' ('SweeTango'®), 'MN55' ('Rave'®, 'First Kiss'®) from the University of Minnesota, 'WA 38' ('Cosmic Crisp'®) from Washington State University, and multiple scab resistant cultivars (e.g., 'Enterprise', 'GoldRush', 'WineCrisp', and 'Juliet') from the PRI program (e.g., Brown and Terry 1997; Evans et al. 2012; Garris 2013; Janick, 2002; Luby 1991).

Apple breeding is a resource-intensive process that can take over 20 years from initial cross to cultivar release. Gametophytic self-incompatibility, long generation times, and high heterozygosity complicate the breeding process rendering genetic studies challenging. As a result, significant research efforts via interdisciplinary projects, such as HiDRAS (https://sites.unimi.it/camelot/hidras/; Gianfranceschi and Soglio 2004), FruitBreedomics (https://www.fruitbreedomics.com/; Laurens et al. 2010), RosBREED and RosBREED2 (https://www.rosbreed.org/home; Iezzoni et al. 2010, 2017), have focused on developing and evaluating tools that can be used to increase efficiency, accuracy, and/or speed of the apple breeding process. In recent decades, apple breeding programs have incorporated DNA information into breeding decisions (i.e., DNA-informed breeding). DNA-informed breeding offers several advantages, including potential cost savings, more efficient backcrossing, early generation selection, more efficient pyramiding of desirable alleles, and ultimately a more accurate choice (Collard and Mackill 2008; Kellerhals et al. 2009; Peace 2017; Wannemuehler et al. 2019; Xu and Crouch 2008). Currently, several locus-specific, trait performance-predictive assays (i.e., DNA tests) are available for different traits of apple (Evans and Peace 2017). However, realization of the full potential of DNA-informed breeding has been slow due to several challenges including lack of predictive DNA tests relevant to breeding germplasm and access to DNA-based diagnostic services, as summarized by Evans and Peace (2017), Peace (2017), and Ru et al. (2015).

While DNA-informed breeding can increase the efficiency and accuracy of selection in a breeding program, it does not necessarily reduce the length of the breeding process. Shortening the apple's long juvenility period (5–10 years) would greatly accelerate and increase the efficiency of the breeding cycle. A technique that has been used for accelerated trait introgression is fast-track or rapid-cycle breeding (technique and applications described by Flachowsky et al. 2007, 2009, 2011; Le Roux et al. 2010; Schlathölter et al. 2018; Weigl et al. 2015). This technique, which uses transgenic intermediate generations overexpressing an early flowering gene (*BpMADS4*) from silver birch, shortens the juvenile phase (Flachowsky et al. 2009; Weigl et al. 2015). After several generations, non-transgenic individuals carrying desirable trait levels are selected (Weigl et al. 2015). Although the final selection is non-transgenic, industry and consumer acceptance is not guaranteed.

Apple scion breeding programs are primarily focused on increasing marketability via improved appearance (e.g., color, size, and shape) and eating quality, a combination of taste (e.g., sweetness and acidity), flavor, and texture (e.g., firmness, crispness, and juiciness) traits. Fruit storability is also an important breeding target as most apples are stored for extended periods of time. In addition, combining fruit quality with resistance to several serious diseases (e.g., apple scab, fire blight, and powdery mildew) is of particular importance. Depending on the breeding program, other targets might include winter hardiness, abiotic stress tolerance, regular cropping, and high yield. Although each program has its specific breeding targets and methodologies, most follow a similar breeding scheme. In this chapter, the Washington State University apple breeding program (WABP) scheme is described as an example. The WABP scheme consists of four stages (described by Evans 2013 and Harshman 2015): seedling production, Phase 1, Phase 2, and Phase 3. Seedling production involves the generation of segregating seedling populations via controlled hybridizations. Depending on targets for a given cross, marker-assisted parent selection (MAPS) and/or marker-assisted seedling selection (MASS) might be used during this stage. Evans and Peace (2017) provide a detailed description of applications, benefits, and challenges associated with DNA-informed apple breeding; i.e., marker-assisted breeding (MAB), as well as how it is applied in WABP. In Phase 1, fruit from seedlings is evaluated for the first time to enable seedling selection. Phase 2 involves the collection of data wherein seedlings selected in Phase 1 are evaluated in small, replicated, multi-site trials. Finally, selections advanced from Phase 2 enter Phase 3, which focuses on commercial viability. This chapter provides an updated review of breeding for disease and pest resistance combined with fruit quality, along with perspectives on phenotyping and genomic advances.

5.2 Fruit Quality and Resistance to Diseases and Pests

5.2.1 Fruit Quality Traits

5.2.1.1 Fruit Quality: Breeding and Phenotyping Challenges

Overall fruit quality is the most important target for every apple scion breeding program as the ultimate goal is to release a cultivar that appeals to consumers so that they will purchase, enjoy eating, and then repeat purchasing. However, breeding for fruit quality is highly complex. Selecting individuals based on one or two quality traits (e.g., crispness and sweetness) can be easily accomplished, but to develop a new cultivar with high overall fruit quality, breeders need to select individuals that are balanced for multiple traits, including appearance, texture, taste, aroma, and storability (reviewed by Iwanami 2011). Many quality traits (e.g., appearance, crispness, firmness, juiciness, sweetness, tartness, and flavor) are quantitatively controlled and are highly influenced by environmental factors and management practices (Brown 2012; Iwanami 2011).

Overall, standardizing fruit sampling is critical for routine phenotypic selection; moreover, adjusting crop load and determining maturity are also key to the collection of reliable data. Additionally, there are still very few direct measures available for traits of interest beyond intensive sensory evaluation; e.g., sweetness in apple is routinely only estimated by soluble solids content (SSC) measurement using a refractometer. Furthermore, challenges with fruit sampling for phenotyping also impact the ability to elucidate genetic control of fruit quality traits. To date, marker–trait association studies of fruit quality traits have relied on a myriad of instrumental and/or sensory methodologies (Table 5.1). Moreover, lack of phenotypic standardization, large within-sample variations, and differences in genetic materials used are factors that impede direct comparisons among different studies.

Table 5.1. Fruit quality QTLs reported for apple (*Malus*)

Category[a]	Trait QTL (chromosome)	Population(s)	Reference(s)
A, EQ	Skin color (9) SSC (8) Fresh green apple (9, 12) Crispness (5, 13) Juiciness (13) Mealiness (5, 10) Skin thickness (10)	85 cultivars	Amyotte et al. (2017)
EQ	Mechanical (all except 7, 9, 13, 14) Fibrousness (7, 10) Firmness (10, 11) Crunchiness (10) Graininess (1, 7, 10) Mealiness (1, 2, 4, 7, 12) Meltiness (5,6, 7, 8) Juiciness (1, 11)	X3259 × X3263	Ben Sadok et al. (2015)
A, S	Bitter pit—*Bp-2* (16)	Redchief × X6688	Buti et al. (2015)
A	Fruit shape index (11)	Jonathan × Golden Delicious	Cao et al. (2015)
EQ, S	Ethylene (3, 13, 15) Estragole (17) Propanal Butanal (16) 1-butanol (3) Alcohols and esters (8) Farnesene (5)	Golden Delicious × Scarlet	Cappellin et al. (2015a)
EQ, S	Acetate esters (2, 15) Esters (2, 4, 5, 15) Ethylene (2, 14) Furanes (4, 5, 13) Phenyls (3) Sesquiterpenes (5)	Fuji × Delearly 124 accessions	Cappellin et al. (2015b)
A	Skin overcolor (9)	4 full-sib families	Chagné et al. (2016)
Other[b]	Triterpenes (3, 5, 9, 17)	Royal Gala × Granny Smith	Christeller et al. (2019)
EQ, S	Ethylene production—*Md-ACO1* (10) Ethylene production—*Md-ACS1* (15)	Prima × Fiesta Fuji × Mondial Gala Fuji × Braeburn	Costa et al. (2005)
S	Expansin (softening)—*Md-Exp7* (1)	Prima × Fiesta 31 cultivars	Costa et al. (2008)
EQ, S	Esters (2) Ethylene (15)	Fiesta × Discovery	Costa et al. (2013)
EQ	Mechanical and acoustic traits: PBA-QTL (2, 3, 10, 11, 14, 16) GWAS (5, 10)	6 full-sib families 233 accessions	Di Guardo et al. (2017)
EQ	Alcohols (2, 3)Esters (2, 3, 9)	Discovery × Prima	Dunemann et al. (2009)
EQ	VOC profiles (2, 10)[d]	162 accessions	Farneti et al. (2017)
EQ	Fructose (1, 3, 15) Glucose (1, 2, 3, 15, 16) Sucrose (1, 3, 4, 9, 12) Sorbitol (1, 3, 5, 9, 11, 13, 15) SSC (2, 3, 12, 13, 15)	15 families 41 accessions	Guan et al. (2015)

(continued)

Table 5.1. (continued)

Category[a]	Trait QTL (chromosome)	Population(s)	Reference(s)
S	Soft scald (2, 16) Soggy breakdown (2, 16)	4 full-sib families	Howard et al. (2018)
A, EQ	Fruit diameter (10) Stiffness (16) SMSR (10) SMSG (10) BrixR (10) BrixG (10) Acidity (16)	Telamon × Braeburn	Kenis et al. (2008)
EQ	Quercetin conjugates (1, 13) Skin phenolics (16) Flesh phenolics (16)	Prima × Fiesta	Khan et al. (2012)
EQ	Compression (1, 16) Maximum force (16) Wedge fracture (1, 6, 8, 15) Specific gravity (6) Compression stiffness modulus (6)	Prima × Fiesta	King et al. (2001)
A, EQ, S	Fruit firmness (10) Weighted cortical intensity (9) Internal browning (9) Titratable acidity (8) Fruit splitting (16) Bitter pit (16)	7 full-sib families	Kumar et al. (2013)
EQ	Alcohols (2, 15) Terpenes (2, 12) Acetate esters (4, 8) Ethyl esters (17) Other esters (1, 15, 17)	230 accessions	Kumar et al. (2015c)
EQ	Acetate esters (2) Sucrose content (1) % sucrose (1) % fructose (1)	145 Danish heritage apple cultivars	Larsen et al. (2019)
A, EQ	Flesh firmness (3, 6, 12) Fruit weight (6, 16) Size (8, 17) Acidity (8, 16)	Fiesta × Discovery	Liebhard et al. (2003)
EQ	Polygalacturonase (texture)—*Md-PG1* (10)	Fuji × Delearly Fuji × Cripps Pink	Longhi et al. (2012)
EQ	Polygalacturonase (texture)—*Md-PG1* (10)	77 cultivars Fuji × Delearly	Longhi et al. (2013)
EQ	Fructose (3) Glucose (3, 4) Sucrose (3) Sorbitol (3) Malic acid (8, 16)	Jiguan × Wangshanhong	Ma et al. (2016)
EQ	Fruit firmness (1, 10)	Prima × Fiesta	Maliepaard et al. (2001)
S	Soft scald (2, 3)	11 W-12-11 × SPA440 Ambrosia × Honeycrisp	McClure et al. (2016)
A, EQ	Fruit skin color (9) Change in firmness (10)	172 accessions	McClure et al. (2018)

(continued)

Table 5.1. (continued)

Category[a]	Trait QTL (chromosome)	Population(s)	Reference(s)
Other[c]	Catechin Epicatechin Procyanidin B1 Procyanidin B2 Procyanidin C1 (16) Quercitrin (1) Chlorogenic acid (5, 15) 4-O-caffeoylquinic acid (3, 14) Cyanidin-3-galactoside (9)	136 accessions (in 2014) 85 accessions (in 2016)	McClure et al. (2019)
A, EQ	Fruit flesh firmness (3) Fruit overcolor (9) Overcolor intensity (9)	689 accessions (data mining from USDA-GRIN)	Migicovsky et al. (2016)
EQ	Mealiness: QTL (10), GWAS (2, 9, 10)	Orin × Akane 82 accessions	Moriya et al. (2017)
A	Circumference (3, 5) Diameter (3) Length (3, 5) Weight (3, 5)	Co-op 17 × Co-op 16	Potts et al. (2014)
EQ	2-methylbutyl acetate (2)	Royal Gala × Granny Smith	Rowan et al. (2009)
EQ	Sensorial acidity: *Ma* (16), *Ma3* (8), *Ma4* (6), *Ma5* (1) Sensorial sweetness (8, 15, 16)	3 full-sib families	Rymenants et al. (2020)
S	α-farnesene (5, 10, 12, 15)	Royal Gala × Granny Smith	Souleyre et al. (2019)
EQ	Malic acid (8) Citric acid (8, 15) Acetic acid (7) Total acid (8) Fructose (1) Sucrose (1) Fruit weight (3, 5) Fruit firmness (11)	Jonathan × Golden Delicious	Sun et al. (2015)
EQ	Titratable acidity:Malic acid—*Ma* (16), *Ma3* (8)	16 full-sib families	Verma et al. (2019)
EQ	Lipoxygenases (2, 4, 5, 6, 7, 9, 11, 12, 13, 16) Esters (2, 9, 12) Hexanals (7, 12)	Discovery × Prima	Vogt et al. (2013)
EQ	Malic acid—*Ma* locus: Titratable acidity (16) pH (16)	Royal Gala × PI 613,971 Royal Gala × PI 613,988	Xu et al. (2012)

[a]Category: Appearance (A), Eating Quality (EQ), Storability (S)
[b]Triterpenes belong to a class of phytocompounds associated with perceived health-promoting benefits
[c]Polyphenols
[d]VOC: Volatile organic compound
SSC: Soluble solids content
SMSR: Mean firmness of red/sun side
SMSG: Mean firmness of green/shaded side
BrixR: Mean SSC of red/sun side
BrixG: Mean SSC of green/shaded side
PBA-QTL: Pedigree-based analysis-quantitative trait locus
GWAS: Genomewide association study

5.2.1.2 Appearance

Fruit appearance traits, including size, color, presence/absence of blemishes, and/or russet influence consumer preference; thus, they are important breeding targets, and are reviewed by Iwanami (2011). For selection in apple scion breeding programs, fruit appearance traits are often rated subjectively. For example, in the WABP, personnel subjectively assess five fruit samples for each selection using ordinal rating scales for appearance traits including size (tiny to very large), shape (flat to cylindrical), color hue (green to red/purple), percent red color coverage (<20% to >80%), red color pattern (blush to stripe), ground color (green to yellow), russet (severe to no russet), and appearance of lenticels (large and rough to none).

Depending on the appearance trait, quantitative trait loci (QTL) mapping studies have relied on phenotypic data from visual ratings (e.g., Amyotte et al. 2017; Chagné et al. 2016) and/or instrumental measurements (e.g., Chagné et al. 2016; Kenis et al. 2008). QTLs associated with various appearance traits, including red overcolor (Amyotte et al. 2017; Chagné et al. 2016), fruit diameter (Kenis et al. 2008), and fruit weight (Liebhard et al. 2003) have been mapped. Predictive DNA tests have been developed for skin overcolor that can also be used in MAB (summarized by Evans and Peace 2017).

5.2.1.3 Eating quality

Texture

Harker et al. (2006) reported that although significant variability exists for consumer preference, this is typically a function of the interaction between fruit texture and flavor. As a result, fruit texture, such as firmness and crispness, are often the most highly valued traits by both consumers (Harker et al. 2008) and producers (Yue et al. 2013), and therefore serve as major targets for breeders. Improvement of apple fruit texture is challenging as texture is complex and strongly influenced by fruit maturity and environmental conditions (Brown 2012; Iwanami 2011). Iwanami (2011) discusses the impact of ripening on fruit texture, particularly on firmness.

Breeders broadly describe apple fruit texture components as either mechanical (e.g., firmness) or acoustic (e.g., crispness and crunchiness) traits (Laurens et al. 2018). Mechanical texture components, specifically firmness, are measured instrumentally via puncture tests with a variety of mechanical penetrometers. Mohr™ Digi-Test (MDT-1; Mohr and Associates, Richland, WA) is a computerized penetrometer commonly used to estimate firmness and texture-related traits. MDT-1 output data have been shown to correlate well with sensory crispness and firmness (Evans et al. 2010; Teh et al. 2020a). The MDT-1 is currently used in the WABP for instrumental estimates of firmness and crispness, and it has been used for phenotyping in various projects, such as RosBREED (Schmitz et al. 2013).

Dissecting genetic control of fruit texture traits requires the availability of higher resolution phenotypic data than for breeding selection. Those QTL mapping studies focused on detecting and characterizing loci associated with various texture traits have relied on sensory evaluation and/or various instrumental measures (Table 5.1). For example, Amyotte et al. (2017) have measured flesh firmness using a penetrometer, as well as using a trained sensory panel to evaluate multiple sensory texture traits (i.e., chewiness, crispness, juiciness, mealiness, rate of flesh melt, and skin thickness). Costa et al. (2005) have measured firmness with a penetrometer, while Longhi et al. (2012, 2013) and Di Guardo et al. (2017) have obtained mechanical/acoustic profiles comprised of 12 acoustic and four mechanical texture traits (Costa et al. 2011). These multiple approaches used to phenotype components of fruit texture in apples demonstrate the complexities of fruit texture traits, thus hampering direct comparisons among different studies. QTLs associated with various fruit texture traits have been identified throughout the apple genome (Table 5.1), further demonstrating the complexity of fruit texture traits. Despite a large number of QTLs identified, only a few predictive DNA tests for texture trait loci have been developed that breeders can readily use in MAB, as summarized by Evans and Peace (2017).

Acidity and Sweetness

Sweetness and acidity, and the balance between them, are pivotal in consumer perception of desirable eating quality in apple (Iwanami 2011). Fructose, sucrose, glucose, and sorbitol are the primary sugars in apples (Fuleki et al. 1994), and these are strongly influenced by fruit maturity, crop load, canopy management, environmental conditions, and fruit storage (Fuleki et al. 1994; Iwanami 2011). Many apple breeding programs rely on sensory evaluation and/or measurement of SSC using a refractometer to phenotype sweetness (e.g., Evans 2013). While most mapping studies have reported on QTLs for SSC, Guan et al. (2015) and Ma et al. (2016) have identified QTLs for individual sugars (Table 5.1). Few DNA tests for either SSC or individual sugars have been reported (Evans and Peace 2017).

In a breeding program, selection for acidity is typically estimated using both sensory evaluation and titration (titratable acidity) of apple juice (e.g., Evans 2013). Most QTL studies have relied solely on titratable acidity as a phenotypic trait to map loci associated with acidity (Kenis et al. 2008; Kumar et al. 2013; Liebhard et al. 2003; Sun et al. 2015; Verma et al. 2019; Xu et al. 2012). More recently, loci associated with acidity and sweetness has been mapped using sensory evaluation data (Rymenants et al. 2020). Malic acid is the predominant acid in apple fruits, and a major-effect QTL at the *Ma* locus on chromosome 16 segregates in a Mendelian pattern (examples in Table 5.1). Additionally, several studies have detected QTLs on chromosome 8 (examples in Table 5.1). Predictive DNA tests for the *Ma* locus are available for use in MAB, as summarized by Evans and Peace (2017).

Volatile Compounds

In addition to sensory perception of sweetness and acidity, flavor is strongly influenced by olfactory sensations from volatile compounds (El Hadi et al. 2013; Iwanami 2011). Over 300 volatile compounds have been identified in apples with the majority being classified as alcohols, aldehydes, or esters (Dixon and Hewett 2000). In addition to some level of genetic control, fruit volatile composition is also influenced by environmental conditions during fruit maturity, ripening, postharvest handling, and storage conditions, as reviewed by El Hadi et al. (2013). The strong influence of fruit maturity and ripening, as well as of environmental variance on fruit volatile composition hamper standardization of fruit collection for phenotyping of volatile compounds in a breeding program or in QTL mapping studies (Song and Forney 2008). It is advisable to conduct volatile compound phenotyping across multiple environments to assess stability (Costa et al. 2013; Kumar et al. 2015c). In addition to standardization of fruit collection at harvest, quantitative analysis of aroma compounds can be limited by concentrations of volatiles below instrumental detection limits (Song and Forney 2008).

QTLs associated with volatile compounds have been identified throughout the apple genome (Table 5.1). For example, Dunemann et al. (2009) have identified 50 putative QTLs for 27 volatile compounds on 12 of 17 chromosomes with most loci being clustered on chromosomes 2, 3, and 9 (Table 5.1). The large number and variability of QTLs identified to demonstrate the complexity of flavor. So far, no predictive DNA tests for volatile compounds are available for use in MAB. Moreover, it remains to be determined which specific combinations of apple volatile compounds would potentially result in improved flavor (Iwanami 2011). It may be more beneficial to focus on identifying volatile compounds and QTLs associated with "off-flavors" that could be used by apple scion breeders for culling undesirable selections in MAB.

5.2.1.4 Storability and Physiological Disorders

Storability, the ability to maintain fruit quality throughout cold storage, is another important breeding target that is greatly impacted by fruit maturity. Watkins (2017) has provided an overview of postharvest handling/storage and factors that influence the storability of apples.

As texture deterioration in storage is usually a result of fruit softening (i.e., reduction in firmness), instrumental texture measurements after different storage durations can be used to track changes in texture. Iwanami (2011) has provided a detailed description of those different factors (e.g., ethylene, flesh tissue structure) that impact fruit softening. Multiple predictive DNA tests for QTLs for several key enzymes associated with ethylene production and fruit softening, such as those for 1-aminocyclopropane-1-carboxylic acid (ACC) synthase 1 (ACS1) and ACC oxidase 1 (ACO1), have been developed (Table 5.1), as summarized by Evans and Peace (2017).

In addition, fruit flavor components, including aroma, sweetness, and acidity change during storage. Fuleki et al. (1994) have demonstrated that sucrose levels tend to decrease while fructose and glucose levels increase during storage. Furthermore, acidity decreases throughout storage, which can result in significant changes in flavor, particularly in high-acid cultivars (Iwanami 2011). Following extended periods of cold storage, apples may develop a flat flavor or lose flavor altogether. Such loss of flavor is mostly due to significant reductions in acid content (Iwanami 2011). Verma et al. (2019) have detected the same QTLs for malic acid, *Ma* and *Ma3*, associated with titratable acidity at harvest, as well as after 10 and 20 weeks of standard atmosphere cold storage. As described by Iwanami (2011), low temperatures tend to reduce volatile production and concentration. However, specific changes in flavor due to changes in volatile compounds are likely to depend on specific volatiles combined with other factors (e.g., acidity).

Depending on the cultivar/selection, apples are prone to physiological disorders that can result in reduced storability and poor pack out. Watkins (2017) provides an overview of major physiological disorders (e.g., water core, senescent breakdown, bitter pit, superficial scald, soft scald, and soggy breakdown) of apples. In breeding programs, the presence/absence of physiological disorders is typically rated following cold storage of fruit. Most QTL studies on physiological disorders have relied on phenotypic data of incidence and/or visual severity ratings of fruit after storage (e.g., Buti et al. 2015; Howard et al. 2018; Kumar et al. 2013; Table 5.1). QTLs associated with bitter pit, internal browning, soft scald, and soggy breakdown have been identified (Table 5.1), and a predictive DNA test reported for bitter pit susceptibility (summarized by Evans and Peace 2017).

5.2.2 Disease and Pest Resistance Traits

Apple production systems worldwide are impacted by several widespread diseases and pests. Typically, apple producers rely on cultural practices and chemical applications that are often costly, unsustainable, and/or only partially effective, to manage possible outbreaks. Developing cultivars with genetic resistance to important diseases and pests would enable more sustainable apple production systems. Specifically, resistance to major diseases, such as apple scab, fire blight, and powdery mildew, are targeted in many apple breeding programs. Recently, Bus et al. (2019) have provided a thorough review of breeding tree fruit cultivars with durable disease resistance. A recent review by Luo et al. (2020) covers the prospects of achieving durable disease resistance in combination with high fruit quality in the apple as well as associated approaches and challenges. This section will primarily focus on current prospects and challenges of breeding for resistance to apple scab, fire blight, and powdery mildew.

5.2.2.1 Resistance Breeding and Phenotyping Challenges

Breeding for disease and pest resistance in apple is complicated due to several factors, such as phenotyping challenges, types of resistance, and trait heritability. Reliable disease phenotyping requires standardized assessments and environments conducive to the causal pathogen(s). Incidence and severity of infections are strongly influenced by host (e.g., vigor, growth status, age, and resistance/susceptibility genes), pathogen (e.g., effector genes and inoculum load), and environmental (e.g., temperature, humidity, and precipitation) factors. Variability in resistance/susceptibility phenotyping methodology may lead to variable/inconsistent results, necessitating evaluations of biological replicates with standardized phenotyping protocols. In a perennial system such as apple, these complexities can result in costly and time-consuming experimental endeavors (Kellerhals et al. 2017; van der Zwet et al. 2012).

Often, single-gene sources of resistance (Mendelian/qualitative) are not durable in the long term, especially in perennial production systems like apple. In many cases, new races of a pathogen might either develop or be identified that are capable of overcoming a specific host resistance gene or combination of genes. Bus et al. (2019) have described multiple approaches to achieve durable resistance, including pyramiding resistance alleles at multiple loci, thereby decreasing the likelihood of occurrence of a single mutation in the pathogen that will overcome the host's resistance (Mundt 1990). Phenotypic identification of seedlings that combine two or more genes for resistance to the same pathogen is labor intensive, costly, and relies on a collection of differential races of the pathogen. However, MAB can be used to precisely and efficiently pyramid resistance alleles at multiple resistance/susceptibility loci, although lack of predictive DNA tests for multiple resistance/susceptibility loci in breeding relevant genotypes remains a limiting factor.

Most sources of resistance to diseases and pests have been identified in wild *Malus* species/genetic backgrounds bearing crabapple fruits that are small in size, astringent, and of low eating quality, thus they are not amenable for immediate use in apple scion improvement. Although introgression of resistance alleles from wild sources is possible and has been done, this is both challenging and time-consuming. Examples of successful introgression of wild resistance sources (typically single major gene resistance) include the apple scab resistance locus *Rvi6* (formerly known as *Vf*; Gessler et al. 2006). In recent years, introgressing and combining such sources of resistance along with fruit quality could be accelerated via rapid cycle breeding techniques (e.g., Le Roux et al. 2012).

5.2.2.2 Apple Scab

Apple scab, caused by the hemibiotrophic fungal pathogen *Venturia inaequalis* (Cooke) G. Winter, is one of the most economically important diseases of apple worldwide (Bowen et al. 2011). The disease is named for the characteristic leaf and fruit lesions caused by conidia that are produced by stromata (Bowen et al. 2011). Severe infections can cause tree defoliation along with undesirable scabby fruit lesions. In fact, even minor fruit lesions can result in loss of marketable fruit. There are several reviews that cover the disease, management, host–pathogen interactions, and breeding for resistance (Bowen et al. 2011; Bus et al. 2011; Gessler et al. 2006; Khajuria et al. 2018).

In earlier breeding programs, such as the PRI program, screening for apple scab involved controlled greenhouse inoculations (Crosby et al. 1992), while in many current breeding programs, phenotyping for selection for scab resistance relies more on natural infection in unsprayed field/orchard blocks. Meanwhile, most genetic studies have focused on understanding the mechanisms of host resistance to apple scab, as well as phenotyping seedlings/accessions under greenhouse/growth chamber conditions by artificially inoculating (e.g., spray or droplet methods) plant material (e.g., seedlings or replicated propagules) with either single isolates or mixtures of isolates (e.g., Bastiaanse et al. 2016; Caffier et al. 2015; Cova et al. 2015; Patocchi et al. 2005). Following inoculation, scab symptoms are typically rated on an ordinal scale (e.g., Chevalier et al. 1991) that evaluates levels of infection on inoculated leaves. As there are several known *V. inaequalis* races and multiple host resistance factors (e.g., *R* genes), observed symptoms vary with the specific host–pathogen interaction. To date, 18 gene-for-gene relationships in the *Malus–V. inaequalis* plant pathosystem have been identified and are summarized in recent reviews (Bus et al. 2011; Khajuria et al. 2018).

Due to the economic importance of apple scab, most apple breeding programs have been engaged in developing resistant cultivars. For example, in the USA, the now retired PRI cooperative apple breeding program was focused on developing disease-resistant (specifically apple scab resistant) cultivars, and on elucidating genetic control of disease resistance (Janick 2006). Several resistant cultivars have been released, however, but only a few have been grown in limited regions and on a small scale, such as 'Juliet' (Europe), 'Florina Querina' (Europe), 'Topaz' (Europe), 'WineCrisp' (U.S.), 'GoldRush' (U.S.), and 'CrimsonCrisp' (U.S.) (Bowen et al. 2011; Kellerhals 2017). Coincidentally, 'Honeycrisp', an economically important cultivar in the USA, developed at the University of Minnesota and selected for its high fruit quality, has also demonstrated resistance to apple scab (Table 5.2). A list of current apple cultivars with resistance to apple scab is cataloged by Kellerhals (2017).

As of now, a group of 20 loci associated with resistance/susceptibility to apple scab have been identified and characterized (Table 5.2). *Rvi6* (formerly known as *Vf*), the first locus associated with scab resistance to be identified and characterized (reviewed in Gessler et al. 2006) has been extensively deployed by breeding programs worldwide (Bus et al. 2011). This gene has been overcome by *V. inaequalis* race 6 in the 1990s in some regions in Europe, and it has been further reported to be overcome in the USA in some scab-resistant genotypes (Papp et al. 2020). Several predictive DNA tests for various *Rvi* loci exist that breeders can utilize in MAB for apple scab resistance (Evans and Peace 2017; Khajuria et al. 2018).

5.2.2.3 Fire Blight

Fire blight, a bacterial disease caused by *Erwinia amylovora* (Burrill) Winslow et al. (*Ea*) infects both young developing shoots and fruits, as well as attacking root tissues (Norelli et al. 2003). In some cases, it can result in severe tissue damage, and ultimately tree death. In recent decades, apple orchards have become more vulnerable to fire blight epidemics due to the large-scale production of susceptible apple cultivars, as well as lack of sustainable control methods effective against all stages of the fire blight disease (Norelli et al. 2003). Thus, breeding cultivars resistant to fire blight is an important goal in many apple scion breeding programs (Kellerhals et al. 2017). Peil et al. (2009) and van der Zwet et al. (2012) have provided thorough reviews of breeding for resistance to fire blight in both apple and pear.

Phenotyping resistance/susceptibility to fire blight is challenging due to intermittent disease occurrence, strong influence of various environmental (e.g., temperature, humidity, and precipitation) and host (e.g., vigor and growth status) factors on susceptibility, quantitative host resistance, *Ea* strain virulence, and host × *Ea* strain interactions (Brown 2012; Lee et al. 2010; Silva et al. 2019; van der Zwet et al. 2012). Additionally, results from different phenotyping methods are often uncorrelated (Peil et al. 2009). As fire blight is listed as a quarantine disease in many countries, most programs rely on artificial inoculation of actively growing shoots under greenhouse conditions. Although greenhouse inoculation often increases the likelihood of infection and identification of highly susceptible individuals, it can overestimate susceptibility and might not be predictive of responses under field conditions (Hampson and Sholberg 2008). Recently, high-throughput methods (e.g., stomatal conductance, imaging using RGB and multispectral cameras, and visible-near spectral reflectance) for assessing fire blight symptoms have been reported (Jarolmasjed et al. 2019). Although promising, more research must be conducted before these can be implemented for routine phenotyping of fire blight resistance/susceptibility.

Although most apple cultivars are susceptible to fire blight, there is a variation for levels of susceptibility with cultivars such as 'Aurora Golden Gala', 'Delicious', 'Empire', 'Enterprise', 'Fiesta', 'Frostbite', 'Liberty, 'Keepsake', and 'Wildung' ('SnowSweet'®) reported to demonstrate very low to moderate susceptibility (Aldwinckle et al. 1999; Kostick et al. 2019; Lezek et al. 1987; Mohan et al. 2002; van der Zwet et al. 2012). Furthermore, several QTLs for

Table 5.2. Scab resistance loci, phenotypes, and resistance sources in apple (*Malus*) (updated from Bus et al. 2011 and Khajuria et al. 2018)

Locus	Chr.	Isolate(s)	Phenotype	Source	Population(s)	Reference(s)
Rvi1 (*Vg*)	12	101, EU-B05	Necrosis	Golden Delicious	Golden Delicious × Idared Golden Delicious × *M. floribunda* 821 Discovery × TN10-8 Prima × Fiesta Fiesta × Discovery Discovery × Prima Durello di Forli × Fiesta	Benaouf and Parisi (2000), Calenge et al. (2004), Durel et al. (2000, 2004)
Rvi2 (*Vh2*)	2	1639[a]	Stellate necrosis	TSR34T15	F_2 of Russian apple R12740-7A: TSR34T15 and TSR33T239	Bus et al. (2005a)
Rvi3 (*Vh3.1*)	4	855 and 856	Stellate necrosis	Geneva	Geneva × Braeburn Elstar × Geneva	Bastiaanse et al. (2016)
Rvi4 (*Vh4/Vx/Vr1*)	2	1797–9	HR	TSR33T239	F_2 of Russian apple R12740-7A: TSR34T15 and TSR33T239	Bus et al. (2005a)
Rvi5 (*Vm*)	17	1810–1	HR	9-AR2T196	Golden Delicious × Murray	Patocchi et al. (2005)
Rvi6 (*Vf*)	1	EU-D42	Chlorosis	Priscilla (*M. floribunda* 821)	–	Vitnatzer et al. (2004)
Rvi7 (*Vfh*)	8	EU-NL05 and 1066	HR	*M. floribunda* 821	–	Parisi et al. (2004)
Rvi8 (*Vh8*)	2	NZ188B.2 and NZ193B.2	Stellate necrosis	B45	Royal Gala × W193B	Bus et al. (2005b)
Rvi9 (*Vdg*)	2	1639	Stellate necrosis	K2	Gala × Dolgo Elstar × Dolgo	Broggini et al. (2011)
Rvi10 (*Va*)	1	EU-NL19 and EU-B04	HR	A723-6	Fortune × PRI 1841–11 NY 489 × PRI 1841–11	Hemmat et al. (2003)
Rvi11 (*Vbj*)	2	–	Stellate necrosis/chlorosis	A722-7	A722-7 × Golden Delicious	Gygax et al. (2004)
Rvi12 (*Vb*)	12	–	Chlorosis	Hansen's baccata #2	Golden Delicious × Hansen's baccata #2	Dayton and Williams (1968) Erdin et al. (2006)
Rvi13 (*Vd*)	10		Stellate necrosis	Durello di Forli	Durello di Forli × Fiesta	Tartarini et al. (2004)

(continued)

Table 5.2. (continued)

Locus	Chr.	Isolate(s)	Phenotype	Source	Population(s)	Reference(s)
		174C, 1066 and EU-NL05[b]				
Rvi14 (*Vdr1*)	6	–	Chlorosis	Dülmener Rosenapfel	Gala × Dülmener Rosenapfel	Soufflet-Freslon et al. (2008)
Rvi15 (*Vr2*)	2	–	HR	GMAL 2473	Golden Delicious × GMAL 2473 GMAL 2473 × Idared	Galli et al. (2010), Patocchi et al. (2004)
Rvi16 (*Vmis*)	3	–	HR	MIS, OP 93.051, G07-098	Splendour × MIS OP 93.051 G07-098 Scired × AK617Scired × AK653	Bus et al. (2010)
Rvi17 (*Va1*)	1	–	Chlorosis	Antonovka APF22	Golden Delicious × Antonovka APF22	Dunemann and Egerer (2010)
Rvi18 (*V25*)	11	–	HR	1980-015-25	1980-015-025 × 1973–001-041	Soriano et al. (2014)
Rvi19	1	–	All classes	Honeycrisp	Honeycrisp × Twin Bee Gala	Clark (2014), Clark et al. (2014a)
Rvi20	15	–	All classes	Honeycrisp	Honeycrisp × Twin Bee Gala	Clark (2014), Clark et al. (2014b)

Chr.: Chromosome
HR: Hypersensitive response
MIS OP: Open-pollinated mildew immune selection
[a]Putative; AvrRvi2 segregated 3:1 instead of 1:1
[b]Further evaluation needed

resistance to fire blight have been identified in *Malus* species and in apple cultivars (Table 5.3; Calenge et al. 2005a; Desnoues et al. 2018; Durel et al. 2009; Emeriewen et al. 2014, 2017; Gardiner et al. 2012; Khan et al. 2006, 2013; Le Roux et al. 2010; Peil et al. 2007, 2019; van de Weg et al. 2018, and reviewed in Emeriewen et al. 2019). In addition, a single fire blight resistance gene, *FB_MR5,* on chromosome 3 of *M. robusta* 5 has been functionally characterized, as well as other candidate resistance genes have been proposed (reviewed in Emeriewen et al. 2019). Most QTLs and all candidate resistance genes have been identified in wild genetic backgrounds, bearing astringent crabapple-like fruit (e.g., *M. robusta* 5, *M. floribunda* 821, *M. fusca,* and 'Evereste'), thereby limiting their breeding utility for apple scion improvement (Emeriewen et al. 2019; Peil et al. 2020). It is important to note that strong sources of resistance (e.g., *FB_MR5*) have been overcome by more aggressive *Ea* strains, and thus these sources alone may not be sustainable for the long term. Although several QTLs have been identified, few predictive DNA tests for breeding relevant germplasm have been developed, thus limiting a breeder's ability to employ MAB for resistance to fire blight (Evans and Peace 2017).

5.2.2.4 Powdery Mildew

Apple powdery mildew is incited by the obligate biotrophic fungus *Podosphaera leucotricha* (Ell. and Ev.) Salm., which can infect actively growing vegetative tissues, blossoms, and fruits,

Table 5.3. QTLs reported for apple (*Malus*) resistance/susceptibility to fire blight

Chr.	Strain(s)	Population(s)	Source(s)	% VE	Reference(s)
2	Ea273	Co-op 16 × Co-op 17	–	11.2–14.7	Khan et al. (2013)
3	AFRS273	Royal Gala × *M. sieversii* Kaz 95 18–07	–	–W	Desnoues et al. (2018)
	CFBP1430	Fiesta × Discovery Prima × Fiesta	Fiesta Prima	4.4–4.9 5.1–7.5	Calenge et al. (2005a)
	Ea222	Idared × *M. robusta* 5	*M. robusta* 5	67.0–83.0	Peil et al. (2007)
		Idared × *M. robusta* 5 Malling 9 × *M. robusta* 5	*M. robusta* 5	80.0	Peil et al. (2008)
	E2002a Ea222_JKI Ea273 ICMP11176	Ottawa3 × *M. robusta* 5 Idared × *M. robusta* 5 Ottawa3 × *M. robusta* 5 Malling 9 × *M. robusta* 5	*M. robusta* 5	–	Gardiner et al. (2012)
	Ea222_JKI Ea236	Idared × *M. robusta* 5	*M. robusta* 5	–	Peil et al. (2019)
5	CFBP1430	Florina × Nova Easygro	Florina Nova Easygro	6.1–10.1 7.2–7.9	Le Roux et al. (2010)
	LP101	Royal Gala × *M. sieversii* Kaz 95 18–07	–	–	Desnoues et al. (2018)
6	Ea273	Co-op 16 × Co-op 17	–	16.8–20.1	Khan et al. (2013)
7	CFBP1430	Prima × Fiesta Fiesta × Discovery	Fiesta	34.3–46.6	Calenge et al. (2005a)
		Gala × Enterprise X-6398 × X-6683	EnterpriseX-6398	–	van de Weg et al. (2018)
	Ea273E2002a	Ottawa3 × *M. robusta* 5	–	–	Gardiner et al. (2012)
	Ea610	Fiesta × Discovery	Fiesta	34.0–38.6	Khan et al. (2006)
8	AFRS273 Ea273	Royal Gala × *M. sieversii* Kaz 95 18–07	–	– 12.5	Desnoues et al. (2018)
	CFBP1430	Gala × Enterprise[a]	Gala; Enterprise	25.0	van de Weg et al. (2018)
9	AFRS273 Ea273	Royal Gala × *M. sieversii* Kaz 95 18–07	–	– 15.2	Desnoues et al. (2018)
	CFBP1430	Florina × Nova Easygro	Nova Easygro	6.6–7.3	Le Roux et al. (2010)
10	CFBP1430	Florina × Nova Easygro	Florina	12.5–17.9	Le Roux et al. (2010)
	Ea222	*M. fusca* MAL0045 × Idared	*M. fusca*	60.0	Emeriewen et al. (2014)
	Ea273 LP101	Royal Gala × *M. sieversii* Kaz 95 18–07	–	15.4 –	Desnoues et al. (2018)
11	AFRS273	Royal Gala × *M. sieversii* Kaz 95 18–07	–	–	Desnoues et al. (2018)

(continued)

Table 5.3. (continued)

Chr.	Strain(s)	Population(s)	Source(s)	% VE	Reference(s)
12	AFRS273	Royal Gala × *M. sieversii* Kaz 95 18–07	–	–	Desnoues et al. (2018)
	CFBP1430	Fiesta × Discovery Golden Delicious × *M. floribunda* 821 MM106 × Evereste	Discovery *M. floribunda* 821 Evereste	5.0 56.5–57.2 39.5–40.4	Calenge et al. (2005a) and Durel et al. (2009)
	Ea222_JKI Ea3049	Idared × *M. arnoldiana*	*M. arnoldiana*	>50%	Emeriewen et al. (2017)
13	CFBP1430	Fiesta × Discovery Gala × Enterprise[a] Gala × Enterprise	Discovery Enterprise Enterprise, Gala[b]	8.0 34.0 –	Calenge et al. (2005a) and van de Weg et al. (2018)
15	CFBP1430	MM106 × Evereste	Evereste	6.2–6.9	Durel et al. (2009)
	Ea273	Co-op 16 × Co-op 17	–	17.4	Khan et al. (2013)
	LP101	Royal Gala × *M. sieversii* Kaz 95 18–07	–	–	Desnoues et al. (2018)

Chr.: Chromosome; %VE: Percent variance explained
[a]Subset of population
[b]Although inconclusive, a putative resistance allele was identified in 'Gala'

thereby resulting in reduced fruit set, fruit quality, yield, and tree vigor (Amiri and Gañán 2019; Butt et al. 1983; Turechek et al. 2004). Management of powdery mildew typically requires multiple fungicide applications during the growing season (Butt et al. 1983). Amiri and Gañán (2019) have provided an overview of the biology, management, cultivar susceptibility, and host–pathogen interactions of powdery mildew in tree fruit systems.

As an obligate biotroph, *P. leucotricha* is difficult to culture for use in artificial inoculation. For phenotyping resistance/susceptibility to powdery mildew, most breeding programs and genetic studies rely on infection via natural inoculum or by gently brushing conidia from infected leaves onto uninfected leaves under either greenhouse, growth chamber, and/or field conditions (e.g., Bus et al. 2010; Calenge and Durel 2006; James et al. 2004; Luo et al. 2019). As most studies rely on natural infection or artificial inoculations with composite conidia rather than single spore isolates, it can be difficult to characterize specific interactions between the host and different *P. leucotricha* races.

Although there is variation among current commercial cultivars for susceptibility to powdery mildew, no commercial cultivar is completely immune (Amiri and Gañán 2019; Kellerhals 2017). Multiple loci associated with resistance/susceptibility to powdery mildew have been reported with most being qualitative/Mendelian (Table 5.4, also summarized by Amiri and Gañán 2019). Most sources of resistance have been characterized in wild genetic backgrounds with poor fruit quality, thus reducing their breeding utility. Unfortunately, a breakdown of resistance from multiple sources has been reported. For example, Caffier and Laurens (2005) have reported that *Pl-2*, a qualitative/Mendelian trait locus derived from *M. zumi* MAL68/5, has been overcome by particular virulent isolates in France. Nevertheless, several predictive DNA tests for powdery mildew resistance/susceptibility loci are available that can be utilized in MAB for powdery mildew resistance (Evans and Peace 2017; Luo et al. 2019).

5.2.2.5 Other Diseases and Pests

In addition to the three primary diseases described above, there are several other pre- and postharvest diseases, as well as pests that impact productivity and sustainability of apple production systems including rosy leaf-curling aphid (*Dysaphis* cf. *devecta* Wlk.), rosy apple aphid (*Dysaphis plantaginea* Passerini), woolly apple aphid (*Eriosoma lanigerum* Hausm.), alternaria blotch (*Alternia alternata*), Valsa canker (*Valsa ceratosperma* [Tode ex Fr.] Maire), and blue mold (*Penicillium expansum* Link.). Jurick (2017) and Norelli (2017) have described some of these fungal and bacterial diseases, respectively, that can infect apples. For several of these other diseases and pests, loci associated with resistance/susceptibility have been identified, and predictive DNA tests have been developed for use in MAB (Table 5.5; Evans and Peace 2017).

5.3 Recent Advances and Future Directions

5.3.1 Genotyping: SNP Arrays and GBS

Three apple single nucleotide polymorphism (SNP) genotyping arrays, namely 8K (Chagné et al. 2012), 20K (Bianco et al. 2014), and 480K (Bianco et al. 2016), have been reported. Peace et al. (2019) have described the implementation of SNP arrays and their associated advantages and disadvantages. As high-quality genotypic data are a fundamental requirement for genetic studies, Vanderzande et al. (2019) have offered a curation workflow (https://www.youtube.com/watch?v=bxC8nPcH2BM&t=7s) for genotypic data of pedigreed germplasm using the 8K SNP array. Thus far, apple genetic and genomic studies have typically relied on using 8K and 20K SNP arrays for multiple applications, including developing SNP-based linkage maps (e.g., Clark et al. 2014b; Di Pierro et al. 2016; Howard et al. 2017), relatedness studies (e.g., Howard et al. 2017), and QTL identification (e.g., Allard et al. 2016; Chagné et al. 2016; Guan et al. 2015; Howard et al. 2018, 2019; van de Weg et al. 2018; Verma et al. 2019).

In addition to SNP arrays, genotyping-by-sequencing (GBS) has been used as an alternative approach to generate markers (Ban and Choi 2018; Gardner et al. 2014; Larsen et al. 2019; Lee et al. 2017; McClure et al. 2016, 2018, 2019; Norelli et al. 2017). Gardner et al. (2014) pioneered the first development of a GBS pipeline in apple, demonstrating its cost-effectiveness in processing vast amounts of DNA samples for use

Table 5.4. QTLs reported for apple (*Malus*) resistance to powdery mildew

Locus	Chromosome	Population	Source	Reference(s)
Pl-1	12	Idared × *M. robusta* 5	*M. robusta* MAL59/9	Dunemann et al. (2007)
Pl-2	11	Royal Gala × A689-24	*M. zumi* MAL68/5	Jänsch et al. (2015)
Pl-d	12	Fiesta × A871-14	D12	James et al. (2004)
Pl-m	11	Idared × *M. robusta* 5	MIS[a]	Bus et al. (2010)
Pl-w	8	Fiesta × E295-4 Discovery × TN10-8	White Angel	Evans and James (2003) and Calenge et al. (2005b)
Pl-bj	2	Pinova × *M. baccata jackii* 419	*M. baccata jackii* 419	Dunemann and Schuster (2009)
–	2	Discovery × TN10-8	–	Calenge and Durel (2006)
–	13		–	
–	12	Idared × U 211	U 211	Stankiewicz-Kosyl et al. (2005)

[a]MIS: Mildew immune selection

Table 5.5. QTLs reported for apple (*Malus*) resistance to other diseases and pests

Pathogen/Pest	Locus	Chr	Population(s)	Source(s)	Reference(s)
Alternaria alternata (Alternaria blotch)	*Alt*	11	Starking Delicious × Jonathan	Jonathan	Moriya et al. (2013)
Penicillium expansum (Blue mold)	qM-*Pe*3.1 qM-*Pe*10.1	3 10	Royal Gala × PI613981	PI613981 Royal Gala	Norelli et al. (2017)
Glomerella cingulata (Glomerella leaf spot)	*Rgls*	15	Fuji × [various parents]	Fuji	Dantas et al. (2009) and Liu et al. (2016)
Dysaphis cf. *devecta* (Rosy leaf-curling aphid)	–	7	Fiesta × Discovery	Fiesta	Stoeckli et al. (2008)
Dysaphis plantaginea (Rosy apple aphid)	*Dp-fl*	8	Florina × [various parents]	Florina	Dapena et al. (2009) and Pagliarani et al. (2016)
	–	17	Fiesta × Discovery	Fiesta	Stoeckli et al. (2008)
Valsa ceratosperma (Valsa canker)	*Vc.xc59.z9* *Vc.03–8.f16*	9 16	Zisai Pearl × Red Fuji	Zisai Pearl	Tan et al. (2017)
Eriosoma lanigerum (Woolly apple aphid)	*Er1* *Er2* *Er3*	8 17 8	Northern Spy × [various parents] *M. robusta* 5 × [various parents] Aotea 1 × [various parents]	Northern Spy *M. robusta* 5 Aotea 1	Bus et al. (2008)
	Er4	7	MIS OP 93.051 G02-54[a]	–	Bus et al. (2010)

Chr.: Chromosome
[a]MIS OP: Open-pollinated mildew immune selection

in linkage mapping and for QTL discovery. In addition to reports on GBS-based linkage maps (Ban and Choi 2018; McClure et al. 2016; Norelli et al. 2017) and QTL identification (McClure et al. 2016; Norelli et al. 2017), GBS has been used in combination with genome-wide association studies (GWAS) for marker–trait association (Larsen et al. 2019; Lee et al. 2017; McClure et al. 2018, 2019).

5.3.2 DNA-informed Breeding

Advances in genetic maps, DNA markers, and availability of whole genome sequences have resulted in a better understanding of the inheritance and physiology of certain traits, as well as the development of practical applications in breeding (e.g., DNA-informed breeding or MAB). As described previously in this chapter, DNA-informed breeding can have several advantages, including potential cost savings, more efficient backcrossing, early generation selection, and more efficient pyramiding of desirable alleles (Collard and Mackill 2008; Kellerhals et al. 2009; Peace 2017; Wannemuehler et al. 2019; Xu and Crouch 2008). In many apple scion breeding programs, the use of DNA datasets and knowledge to inform breeding decisions has become relatively routine. Evans and Peace (2017) describe in detail how DNA-informed breeding has been applied in the WABP. Specifically, DNA information in the WABP has been used for characterization (e.g., identity verification and parentage confirmation, among others) and evaluation (e.g., MAPS and

MASS; Evans and Peace 2017). In the University of Minnesota (UMN) apple breeding program, DNA tests for scab resistance, skin color, acidity, and texture have been used for MAPS and MASS resulting in significant resource savings (Teh et al. 2018). However, promises of DNA-informed breeding in many crops, including apple have not been fully realized (Moose and Mumm 2008; Peace 2017).

Although >1000 QTLs and Mendelian trait loci (MTLs) associated with various traits in apple have been reported in the Genome Database for Rosaceae (https://www.rosaceae.org/; Jung et al. 2019), few DNA tests for breeding-relevant germplasm are available (examples described by Luo et al. 2020). With the increased availability of genomic data, lack of high-quality phenotypic data has become a bottleneck in predictive DNA test development (Evans and Peace 2017). Significant research efforts are being focused on the development and evaluation of high-throughput precise phenotyping methodologies (e.g., Coupel-Ledru et al. 2019; Jarolmasjed et al. 2019). As described by Evans and Peace (2017), Peace (2017), and Ru et al. (2015), other challenges hampering routine application of DNA-informed breeding in apple include lack of access to DNA-based diagnostic services and costs associated with DNA extraction (especially for SNP-based assays).

Looking forward, as QTL discovery continues, the development of multi-marker (e.g., multi-SNP) arrays, enabling simultaneous genotyping of multiple trait loci, will be likely (Evans and Peace 2017). Multi-marker arrays are unlikely to fully replace single-locus tests as not all trait loci are relevant to a given breeding program. However, with declining genotyping costs, breeder(s) might develop a program-specific multi-marker assay that addresses trait priorities tailored to their breeding program(s). The use of single DNA tests will likely continue to increase and complement multi-marker arrays, as new MTLs are identified, this will enable a breeder to choose tests with increased specificity to their program and germplasm.

Given the higher benefit-to-cost ratio compared to seedling testing, DNA-informed parent selection continues to be the most efficient use of DNA information (Evans and Peace 2017). Additionally, DNA information will continue to be used for pyramiding of favorable alleles (e.g., resistance alleles) at multiple loci and is essential for early selection in rapid cycle breeding. As the chasm between phenotyping and genotyping costs widen, genomic selection (see Sect. 5.3.3) would likely be a more cost-effective alternative to target highly quantitative traits even for a perennial tree crop, such as apple.

5.3.3 Genomic Selection

Given that QTL mapping is often limited to detecting only major-effect QTLs that can be used in MAB, genomic (or genome-wide) selection (GS) has been introduced as a genomic-assisted breeding strategy that incorporates all markers distributed across the genome (Meuwissen et al. 2001), thus making this especially relevant for highly quantitative traits. In GS, a training population with both phenotypic and genotypic data is used to construct a prediction model (Meuwissen et al. 2001). The prediction model is subsequently used to predict genomic estimated breeding values (GEBVs) of individuals in a test population with only genotypic data (Meuwissen et al. 2001), obviating the need for phenotyping the test population. The seminal work of Meuwissen et al. (2001) has been discussed in numerous reviews (Crossa et al. 2017; Heffner et al. 2009; Lorenz et al. 2011).

The use of GS in accelerating apple breeding is even more appealing, in view of the crop's long generation time and high phenotyping costs. To date, there are several reports on GS in apple, with fruit quality traits being the primary targets (Kumar et al. 2012, 2015b; McClure et al. 2018; Migicovsky et al. 2016; Muranty et al. 2015; Roth et al. 2020), as summarized in Table 5.6. Relatedness between training and test populations, population sizes, methods for estimating accuracies, trait phenotypic distributions, and linkage disequilibrium impact prediction accuracy; thus, prediction accuracies are quite variable across studies even for the same trait

Table 5.6. Prediction accuracies for traits reported in genomic selection studies for apple

Trait	Prediction accuracy, r	Reference(s)
Fruit firmness	0.83	Kumar et al. (2012)
SSC	0.89	
TA	0.81	
Russet	0.82	
Weighted cortex intensity	0.82	
Astringency	0.67	
Crispness	0.20	Kumar et al. (2015b)
Flavor intensity	0.21	
Fruit firmness	0.25	
Fruit weight	0.35	
Greasiness	0.20	
Juiciness	0.15	
Harvest date	0.45–0.50	McClure et al. (2018)
Fruit weight	0.19–0.20	
Color	0.37–0.41	
Firmness at harvest	0.20–0.21	
Firmness after storage	0.15–0.27	
Change in firmness	0.08–0.40	
TA	0.39–0.42	
SSC	0.18–0.19	
SSC:TA	0.41–0.45	
Various polyphenols	−0.18–0.49	McClure et al. (2019)
Flower size	0.33	Migicovsky et al. (2016)[a]
Fruit flesh oxidation	0.38	
Fruit juiciness	0.40	
Fruit length	0.47	
Fruit overcolor	0.32	
Fruit stem length	0.35	
Fruit weight	0.36	
Fruit width	0.48	
Overcolor intensity	0.41	
Harvest season	0.57	
Leaf margin	0.32	
Attractiveness	0.21	Muranty et al. (2015)
Fruit cropping	0.08	
Fruit size	0.23	
Percent russet	0.20	
Fruit cracking	0.01	
Preharvest dropping	−0.02	
Percent overcolor	0.37	
Overcolor	0.35	
Ground color	0.06	
Color type	−0.14	
Acoustic linear distance	0.10–0.72	Roth et al. (2020)
Number of force peak	0.16–0.78	
PC1[b]	0.19–0.81	
PC2[b]	0.01–0.40	

[a]Traits with $r \geq 0.30$ are reported
[b]Obtained from principal component analysis of acoustic measurements
SSC: Soluble solids content
TA: Titratable acidity
PC: Principal component

(Heffner et al. 2009; Lorenz et al. 2011; Muranty et al. 2015). As a result, standardized and improved GS models are needed before these can be routinely used in breeding programs.

5.3.4 Non-destructive Phenotyping

Given the drawbacks of traditional phenotyping methods (e.g., destructive, time-consuming, and laborious), non-destructive phenotyping tools are garnering considerable attention as alternatives to estimate fruit ripening, quality, and postharvest traits (Nicolai et al. 2007; Porep et al. 2015; Wang et al. 2015). Setting up a model to predict postharvest traits requires correlating non-destructive measurements (typically near-infrared spectroscopy (NIR)) with destructive measurements using a statistical model (e.g., linear regression). The accuracy of a statistical model is computed and expressed in three metrics, including coefficient of determination (R^2), root mean square error of prediction (RMSEP), and the ratio of performance to deviation (RPD). Generally, an RPD value > 1.5 is deemed reliable for use in quantitative prediction (Nicolai et al. 2007; Porep et al. 2015). Concepts, strategies, and applications of NIR spectroscopy have been extensively covered in several reviews (Cortés et al. 2019; Nicolai et al. 2007; Porep et al. 2015; Saeys et al. 2019; Wang et al. 2015).

In many studies, soluble solids content (SSC) has been reported to be reliably predicted with NIR (Beghi et al. 2013; Fan et al. 2016; Kumar et al. 2015a; Luo et al. 2018; McGlone et al. 2002; Travers et al. 2014; Zhang et al. 2019). Dry matter concentration, a trait highly correlated with post-storage SSC and used as a quality index for apples, is also reliably predicted with NIR (Kumar et al. 2015a; McGlone et al. 2003; Palmer et al. 2010; Teh et al. 2020b; Travers et al. 2014). Other postharvest traits, such as firmness and titratable acidity, have not been correlated with NIR prediction models (Beghi et al. 2013; Kumar et al. 2015a; McGlone et al. 2003; Zude et al. 2006).

Newer technologies, such as hyperspectral and vision imaging, are emerging as alternative non-destructive tools to evaluate fruit quality (Li et al. 2019; Lu et al. 2016a; Lu and Lu 2018; Peng and Lu 2008; Sofu et al. 2016). In particular, structured-illumination reflectance imaging along with three-dimensional geometric reconstruction of fruit can detect fresh bruises in apples (Lu et al. 2016b; Lu and Lu 2018). Additionally, the decline of chlorophyll content during fruit ripening is being utilized as a harvest index in non-destructive evaluations (Betemps et al. 2012; Cocetta et al. 2017; Greer 2005; Nyasordzi et al. 2013; Sadar and Zanella 2019; Solovchenko et al. 2005; Zude-Sasse et al. 2002). While chlorophyll content of the skin can be masked by anthocyanin content in red apple cultivars, measuring absorbances at 670 nm and 720 nm wavelengths using a hand-held delta absorbance (DA) meter (Sintéleia, Bologna, Italy) provides an index of absorbance difference ($I_{AD} = A_{670nm} - A_{720nm}$) for fruit ripeness (Ziosi et al. 2008). This approach was first implemented in stone fruit (Ziosi et al. 2008) and has since been implemented in various apple studies (Cocetta et al. 2017; DeLong et al. 2016; Nyasordzi et al. 2013; Sadar and Zanella 2019; Toivonen and Hampson 2014).

Non-destructive phenotyping would be especially useful during the early stages of a breeding program, wherein the number of available fruit on young seedling trees is always limiting. Future work should focus on bridging the chasm between instrumentally derived traits and traits targeted for selection.

5.4 Concluding Remarks

Genetic and genomic advances have resulted in improved understanding of inheritance and physiology of certain traits, as well as for developing practical applications for breeders (e.g., MAB). Apple scion breeding has become more accurate and efficient due to the availability of DNA-based tools, although in limited numbers, for both selection and germplasm characterization. Phenotyping continues to be a bottleneck limiting our ability to efficiently dissect complex quantitative traits and make genetic gains in breeding programs. As most traits targeted for selection are influenced by both environment and management practices, standardized phenotyping is critical. In addition, current phenotypic measures are often destructive, resource-intensive, time-consuming, and indirect estimates of selection traits. For the promise of genomic tools to be fully realized, future work should focus on enabling more standardized, objective, efficient, non-destructive, and direct phenotyping for breeding selection and/or genetic dissection of traits.

References

Aldwinckle HS, Gustafson HL, Forsline PL (1999) Evaluation of the core subset of the USDA apple germplasm collection for resistance to fire blight. Acta Hortic 489:269–272. https://doi.org/10.17660/ActaHortic.1999.489.46

Allard A, Bink MCAM, Martinez S, Kelner J-J, Legave J-M, di Guardo M, Di Pierro EA, Laurens F, van de Weg WE, Costes E (2016) Detecting QTLs and putative candidate genes involved in budbreak and flowering time in an apple multiparental population. J Exp Bot 67:2875–2888. https://doi.org/10.1093/jxb/erw130

Amiri A, Gañán L (2019) Powdery mildew: biology, epidemiology, and management of *Podosphaera* spp. of tree fruit. In: Xu X, Fountain M (eds) Integrated management of diseases and insect pests of tree fruit. Burleigh Dodds Science Publ, pp 21–38. https://doi.org/10.19103/AS.2019.0046.02

Amyotte B, Bowen AJ, Banks T, Rajcan I, Somers DJ (2017) Mapping the sensory perception of apple using descriptive sensory evaluation in a genome wide association study. PLoS ONE 12. https://doi.org/10.1371/journal.pone.0171710

Ban SH, Choi C (2018) Development of an apple F1 segregating population genetic linkage map using genotyping-by-sequencing. Plant Breed Biotechnol 6:434–443. https://doi.org/10.9787/PBB.2018.6.4.434

Bastiaanse H, Bassett HCM, Kirk C, Gardiner SE, Deng C, Groenworld R, Chagné D, Bus VGM (2016) Scab resistance in 'Geneva' apple is conditioned by a resistance gene cluster with complex genetic control. Mol Plant Pathol 17:159–172. https://doi.org/10.1111/mpp.12269

Beghi R, Spinardi A, Bodria L, Mignani I, Guidetti R (2013) Apples nutraceutic properties evaluation through a visible and near-infrared portable system. Food Bioprocess Tech 6:2547–2554. https://doi.org/10.1007/s11947-012-0824-7

Ben Sadok I, Tiecher A, Galvez-Lopez D, Lahaye M, Lasserre-Zuber P, Bruneau M, Hanteville S, Robic R, Cournol R, Laurens F (2015) Apple fruit texture QTLs: year and cold storage effects on sensory and instrumental traits. Tree Genet Genomes 11:119. https://doi.org/10.1007/s11295-015-0947-x

Benaouf G, Parisi L (2000) Genetics of host-pathogen relationships between *Venturia inaequalis* races 6 and 7 and *Malus* species. Phytopathology 90:236–242. https://doi.org/10.1094/PHYTO.2000.90.3.236

Betemps DL, Fachinello JC, Galarça SP, Portela NM, Remorini D, Massai R, Agati G (2012) Non-destructive evaluation of ripening and quality traits in apples using a multiparametric fluorescence sensor. J Sci Food Agr 92:1855–1864. https://doi.org/10.1002/jsfa.5552

Bianco L, Cestaro A, Sargent DJ, Banchi E, Derdak S, Di Guardo M, Salvi S, Jansen J, Viola R, Gut I, Laurens F, Chagné D, Velasco R, van de Weg E, Troggio M (2014) Development and validation of a 20K single nucleotide polymorphism (SNP) whole genome genotyping array for apple (*Malus* × *domestica* Borkh.). PLoS ONE 9:e110377. https://doi.org/10.1371/journal.pone.0110377

Bianco L, Cestaro A, Linsmith G, Muranty H, Denancé C, Théron A, Poncet C, Micheletti D, Kerschbamer E, Di Pierro EA, Larger S, Pindo M, van de Weg E, Davassi A, Laurens F, Velasco R, Durel C-E, Troggio M (2016) Development and validation of the Axiom® Apple 480K SNP genotyping array. Plant J 86:62–74. https://doi.org/10.1111/tpj.13145

Bowen JK, Mesarich CH, Bus VGM, Beresford RM, Plummer KM, Templeton MD (2011) *Venturia inaequalis*: the causal agent of apple scab. Mol Plant Pathol 12:105–122. https://doi.org/10.1111/J.1364-3703.2010.00656.X

Broggini GAL, Bus VGM, Parrivicini G, Kumar S, Groenwold R, Gessler C (2011) Genetic mapping of 14 avirulence genes in an EU-B04 × 1639 progeny of *Venturia inaequalis*. Fungal Genet Biol 48:166–176. https://doi.org/10.1016/j.fgb.2010.09.001

Brown S (2012) Apple. In: Badenes ML, Byrne DH (eds) Fruit breeding, handbook of plant breeding, Springer-Verlag, pp 329–367

Brown SK, Maloney KE (2003) Genetic improvement of apple: breeding, markers, mapping and biotechnology. In: Ferree D, Warrington I (eds) Apples: botany, production, and uses. CAB International, pp 31–59

Brown SK, Terry DE (1997) The Cornell University apple breeding program: past, present, and future. Fruit Var J 51:199–204

Bus VGM, Rikkerink EHA, van de Weg WE, Rusholme RL, Gardiner SE, Bassett HCM, Kodde LP, Parisi L, Laurens FND, Meulenbroek EJ, Plummer KM (2005a) The *Vh2* and *Vh4* scab resistance genes in two differential hosts derived from Russian apple R12740-7A map to the same linkage group of apple. Mol Breed 15:103–116. https://doi.org/10.1007/s11032-004-3609-5

Bus VGM, Rikkerink EHA, van de Weg WE, Rusholme RL, Rikkerink E, Gardiner S (2005b) The *Vh8* locus of a new gene for gene interaction between *Venturia inaequalis* and the wild apple *Malus sieversii* is closely linked to the *Vh2* locus in *Malus pumila* R12740–7A. New Phytol 166:1035–1049. https://doi.org/10.1111/j.1469-8137.2005.01395.x

Bus VGM, Chagné D, Bassett HCM, Bowatte D, Calenge F, Celton JM, Durel C-E, Malone MT, Patocchi A, Ranatunga AC, Rikkerink EHA, Tustin DS, Zhou J, Gardiner SE (2008) Genome mapping of three major resistance genes to woolly aphid (*Eriosoma lanigerum* Hausm.). Tree Genet Genomes 4:223–236. https://doi.org/10.1007/s11295-007-0103-3

Bus VGM, Bassett HCM, Bowatte D, Chagné D, Ranatunga CA, Ulluwishewa D, Wiedow C, Gardiner SE (2010) Genome mapping of an apple scab, a powdery mildew and a woolly apple aphid resistance gene from open-pollinated Mildew Immune Selection. Tree Genet Genomes 6:477–487. https://doi.org/10.1007/s11295-009-0265-2

Bus VGM, Rikkerink EHA, Caffier V, Durel C-E, Plummer KM (2011) Revision of the nomenclature of the differential host-pathogen interactions of *Venturia inaequalis* and *Malus*. Annu Rev Phytopathol 49:391–413. https://doi.org/10.1146/annurev-phyto-072910-095339

Bus VGM, Bowen JK, Broggini GAL, Laurens FND (2019) Breeding fruit cultivars with durable disease resistance. In: Xu X, Fountain M (eds) Integrated management of diseases and insect pests of tree fruit. Burleigh Dodds Science Publishing Limited, pp 233–274. https://doi.org/10.19103/AS.2019.0046.12

Buti M, Poles L, Caset D, Magnago P, Fernandez FF, Colgan RJ, Velasco R, Sargent DJ (2015) Identification and validation of a QTL influencing bitter pit symptoms in apple (*Malus* × *domestica*). Mol Breed 35:29. https://doi.org/10.1007/s11032-015-0258-9

Butt DJ, Martin KM, Swait AAJ (1983) Apple powdery mildew: damage, loss and economic injury level. In: Proceedings of the 10th international congress of plant protection, Brighton, p 118

Caffier V, Laurens F (2005) Breakdown of *Pl2*, a major gene of resistance to apple powdery mildew, in a French experimental orchard. Plant Pathol 54:116–124. https://doi.org/10.1111/j.1365-3059.2005.01147.x

Caffier V, Patocchi A, Expert P, Bellanger MN, Durel CE, Hilber-Bodmer M, Broggini GAL, Groenwold R, Bus VGM (2015) Virulence characterization of *Venturia inaequalis* reference isolates on the differential set of *Malus* hosts. Plant Dis 99:370–375. https://doi.org/10.1094/PDIS-07-14-0708-RE

Calenge F, Durel C-E (2006) Both stable and unstable QTLs for resistance to powdery mildew are detected in apple after four years of field assessments. Mol Breed 17:329–339. https://doi.org/10.1007/s11032-006-9004-7

Calenge F, Faure A, Goerre M, Gebhardt C, van de Weg WE, Parisi L, Durel CE (2004) Quantitative trait loci (QTL) analysis reveals both broad-spectrum and isolate-specific QTL for scab resistance in an apple progeny challenged with eight isolates of *Venturia inaequalis*. Phytopathology 94:370–379. https://doi.org/10.1094/PHYTO.2004.94.4.370

Calenge F, Drouet D, Denancé C, van de Weg WE, Brisset MN, Paulin JP, Durel CE (2005a) Identification of major QTL together with several minor additive or epistatic QTLs for resistance to fire blight in apple in two related progenies. Theor Appl Genet 111:128–135. https://doi.org/10.1007/s00122-005-2002-z

Calenge F, van der Linden CG, van de Weg E, Schouten HJ, van Arkel G, Denance C, Durel CE (2005b) Resistance gene analogues identified through the NBS-profiling method map close to major genes and QTL for disease resistance in apple. Theor Appl Genet 110:660–668. https://doi.org/10.1007/s00122-004-1891-6

Cao K, Chang Y, Sun R, Shen F, Wu T, Wang Y, Zhang X, Han Z (2015) Candidate gene prediction via quantitative trait locus analysis of fruit shape index traits in apple. Euphytica 206:381–391. https://doi.org/10.1007/s10681-015-1488-y

Cappellin L, Costa F, Aprea E, Betta E, Gasperi F, Biasioli F (2015a) Double clustering of PTR-ToF-MS data enables the mapping of QTLs related to apple fruit volatilome. Scient Hortic 197:24–32. https://doi.org/10.1016/j.scienta.2015.10.043

Cappellin L, Farneti B, Di Guardo M, Busatto N, Khomenko I, Romano A, Velasco R, Costa G, Biasioli F, Costa F (2015b) QTL analysis coupled with PTR-ToF-MS and candidate gene-based association mapping validate the role of Md-AAT1 as a major gene in the control of flavor in apple fruit. Plant Mol Biol Rep 33:239–252. https://doi.org/10.1007/s11105-014-0744-y

Chagné D, Crowhurst RN, Troggio M, Davey MW, Gilmore B, Lawley C, Vanderzande S, Hellens RP, Kumar S, Cestaro A, Velasco R, Main D, Rees JD, Iezzoni A, Mockler T, Wilhelm L, van de Weg E, Gardiner SE, Bassil N, Peace C (2012) Genome-wide SNP detection, validation, and development of an 8K SNP array for apple. PLoS ONE 7. https://doi.org/10.1371/journal.pone.0031745

Chagné D, Kirk C, How N, Whitworth C, Fontic C, Reig G, Sawyer G, Rouse S, Poles L, Gardiner SE, Kumar S, Espley R, Volz RK, Troggio M, Iglesias I (2016) A functional genetic marker for apple red skin coloration across different environments. Tree Genet Genomes 12. https://doi.org/10.1007/s11295-016-1025-8

Chevalier M, Lespinasse Y, Renaudin S (1991) A microscopic study of the different classes of symptoms coded by the *Vf* gene in apple for resistance to scab (*Venturia inaequalis*). Plant Pathol 40:249–256. https://doi.org/10.1111/j.1365-3059.1991.tb02374.x

Christeller JT, McGhie TK, Johnston JW, Carr B, Chagné D (2019) Quantitative trait loci influencing pentacyclic triterpene composition in apple fruit peel. Sci Rep 9:18501. https://doi.org/10.1038/s41598-019-55070-5

Clark MD (2014) Characterizing the host response and genetic control in 'Honeycrisp' to apple scab (*Venturia inaequalis*). Retrieved from the University of Minnesota Digital Conservancy. http://hdl.handle.net/11299/172140

Clark MD, Bus VG, Luby JJ, Bradeen JM (2014a) Characterization of the defence response to *Venturia inaequalis* in 'Honeycrisp' apple, its ancestors, and progeny. Eur J Plant Pathol 140:69–81. https://doi.org/10.1007/s10658-014-0444-3

Clark MD, Schmitz CA, Rosyara UR, Luby JJ, Bradeen JM (2014b) A consensus 'Honeycrisp' apple (*Malus* × *domestica*) genetic linkage map from three full-sib progeny populations. Tree Genet Genomes 10:627–639. https://doi.org/10.1007/s11295-014-0709-1

Cocetta G, Beghi R, Mignani I, Spinardi A (2017) Nondestructive apple ripening stage determination using the delta absorbance meter at harvest and after storage. HortTechnol 27:54–64. https://doi.org/10.21273/HORTTECH03495-16

Collard BCY, Mackill DJ (2008) Marker-assisted selection: an approach for precision plant breeding in the twenty-first century. Phil Trans Roy Soc B 363:557–572. https://doi.org/10.1098/rstb.2007.2170

Cortés V, Blasco J, Aleixos N, Cubero S, Talens P (2019) Monitoring strategies for quality control of agricultural products using visible and near-infrared spectroscopy: A review. Trends Food Sci Tech 85:138–148. https://doi.org/10.1016/j.tifs.2019.01.015

Costa F, Stella S, van de Weg E, Guerra W, Cecchinel M, Dallavia J, Koller B, Sansavini S (2005) Role of the genes *Md-ACO1* and *Md-ACS1* in ethylene production and shelf-life of apple (*Malus domestica* Borkh.). Euphytica 141:181–190. https://doi.org/10.1007/s10681-005-6805-4

Costa FA, van de Weg WE, Stella S, Dondini L, Pratesi D, Musacchi S, Sansavini S (2008) Map position and functional allelic diversity of Md-Exp7, a new putative expansin gene associated with fruit softening in apple (*Malus* × *domestica* Borkh.) and pear (*Pyrus communis*). Tree Genet Genomes 4:575–586. https://doi.org/10.1007/s11295-008-0133-5

Costa F, Cappellin L, Longhi S, Guerra W, Magnago P, Porro D, Soukoulis C, Salvi S, Velasco R, Biasioli F, Gasperi F (2011) Assessment of apple (*Malus* × *domestica* Borkh.) fruit texture by a combined acoustic-mechanical profiling strategy. Postharv Biol Tech 61:21–28. https://doi.org/10.1016/j.postharvbio.2011.02.006

Costa F, Cappellin L, Zini E, Patocchi A, Kellerhals M, Komjanc M, Gessler C, Biasioli F (2013) QTL validation and stability for volatile organic compounds (VOCs) in apple. Plant Sci 21:1–7. https://doi.org/10.1016/j.plantsci.2013.05.018

Coupel-Ledru A, Pallas B, Delalande M, Boudon F, Carrié E, Martinez S, Regnard J-L, Costes E (2019) Multi-scale high-throughput phenotyping of apple architectural and functional traits in orchard reveals genotypic variability under contrasted watering regimes. Hortic Res 6:52. https://doi.org/10.1038/s41438-019-0137-3

Cova V, Bandara NL, Liang W, Tartarini S, Patocchi A, Troggio M, Velasco R, Komjanc M (2015) Fine mapping of the *Rvi5* (*Vm*) apple scab resistance locus in the 'Murray' apple genotype. Mol Breed 35:200. https://doi.org/10.1007/s11032-015-0396-0

Crossa J, Pérez-Rodríguez P, Cuevas J, Montesinos-López O, Jarquín D, de los Campos G, Burgueño J, González-Camacho JM, Pérez-Elizalde S, Beyene Y, Dreisigacker S, (2017) Genomic selection in plant breeding: methods, models, and perspectives. Trends Plant Sci 22:961–975. https://doi.org/10.1016/j.tplants.2017.08.011

Crosby JA, Janick J, Pecknold PC, Korban SS, O'Connor PA, Ries SM, Goffreda J, Voordeckers A (1992) Breeding apples for scab resistance: 1945–1990. Acta Hortic 317:43–70. https://doi.org/10.17660/ActaHortic.1992.317.5

Daccord N, Celton J-M, Linsmith G, Becker C, Choisne N, Schijlen E, van de Geest H, Bianco L, Micheletti D, Velasco R, Di Pierro EA, Gouzy J, Rees DJG, Guérif P, Muranty H, Durel C-E, Laurens F, Lespinasse Y, Gaillard S, Aubourg S, Quesneville H, Weigel D, van de Weg E, Troggio M, Bucher E (2017) High-quality *de novo* assembly of the apple genome and methylome dynamics of early fruit development. Nat Genet 49:1099–1108. https://doi.org/10.1038/ng.3886

Dantas ACM, Silva MF, Nodari RO (2009) Avanços genéticos da macieira no controle de doenças (Genetic advances in the control of apple diseases). In: Stadnik, MJ (ed) Manejo integrado de doenças da macieira. Florianópolis: CCAUFSC, pp 127–152

Dapena E, Miñarro M, Blasquez MD (2009) Evaluation of the resistance to the rosy apple aphid using a genetic marker. Acta Hortic 814:787–790. https://doi.org/10.17660/ActaHortic.2009.814.133

Dayton DF, Williams EB (1968) Independent genes in *Malus* for resistance to *Venturia inaequalis*. Proc Am Soc Hortic Sci 92:89–94

DeLong JM, Harrison PA, Harkness L (2016) Determination of optimal harvest boundaries for 'Ambrosia' apple fruit using a delta-absorbance meter. J Hortic Sci Biotech 91:243–249. https://doi.org/10.1080/14620316.2016.1148369

Desnoues E, Norelli JL, Aldwinckle HS, Wisniewski ME, Evans KM, Malnoy M, Khan A (2018) Identification of novel strain-specific and environment-dependent minor QTLs linked to fire blight resistance in apple. Plant Mol Biol Rep 36:247–256. https://doi.org/10.1007/s11105-018-1076-0

Di Guardo M, Bink MCAM, Guerra W, Letschka T, Lozano L, Busatto N, Poles L, Tadiella A, Bianco L, Visser RGF, van de Weg E, Costa F (2017) Deciphering the genetic control of fruit texture in apple by multiple family-based analysis and genome-wide association. J Exp Bot 68:1451–1466. https://doi.org/10.1093/jxb/erx017

Di Pierro EA, Gianfranceschi L, Di Guardo M, Koehorst-van Putten H, Kruisselbrink JW, Longhi S, Troggio M, Bianco L, Muranty H, Pagliarani G, Tartarini S, Letschka T, Luis LL, Garkava-Gustavsson L, Micheletti D, Bink MCAM, Voorrips R, Aziz E, Velasco R, Laurens F, van de Weg WE (2016) A high-density, multi-parental SNP genetic map on apple validates a new mapping approach for outcrossing species. Hortic Res 3:16057. https://doi.org/10.1038/hortres.2016.57

Dixon J, Hewett EW (2000) Factors affecting apple aroma/flavor volatile concentration: a review. NZ J Crop Hort 28:155–173. https://doi.org/10.1080/01140671.2000.9514136

Dunemann F, Egerer J (2010) A major resistance gene from Russian apple 'Antonovka' conferring field immunity against apple scab is closely linked to the *Vf* locus. Tree Genet Genomes 6:627–633. https://doi.org/10.1007/s11295-010-0278-x

Dunemann F, Peil A, Urbanietz A, Garcia-Libreros T (2007) Mapping of the apple powdery mildew resistance gene *Pl1* and its genetic association with an NBS-LRR candidate resistance gene. Plant Breed 126:476–481. https://doi.org/10.1111/j.1439-0523.2007.01415.x

Dunemann F, Schuster M (2009) Genetic characterization and mapping of the major powdery mildew resistance gene *Plbj* from *Malus baccata jackii*. Acta Hortic 814:791–798. https://doi.org/10.17660/ActaHortic.2009. 814.134

Dunemann F, Ulrich D, Boudichevskaia A, Grafe C, Weber WE (2009) QTL mapping of aroma compounds analysed by headspace solid-phase microextraction gas chromatography in the apple progeny 'Discovery' × 'Prima.' Mol Breed 23:501–521. https://doi.org/10.1007/s11032-008-9252-9

Durel CE, van de Weg E, Venisse JS, Parisi L (2000) Localisation of a major gene for scab resistance on the European genetic map of the Prima × Fiesta cross. OILB/WPRS Bull 23:245–248

Durel CE, Calenge F, Parisi L, van de Weg WE, Kodde LP, Liebhard R, Gessler C, Thiermann M, Dunemann F, Gennari F, Tartarini S (2004) An overview of the position and robustness of scab resistance QTLs and major genes by aligning genetic maps of five apple progenies. Acta Hortic 663:135–140. https://doi.org/10.17660/ActaHortic.2004.663.17

Durel C-E, Denancé C, Brisset MN (2009) Two distinct major QTL for resistance to fire blight co-localize on linkage group 12 in apple genotypes 'Evereste' and *Malus floribunda* clone 821. Genome 52:139–147. https://doi.org/10.1139/g08-111

El Hadi MAM, Zhang F-J, Wu F-F, Zhou C-H, Tao J (2013) Advances in fruit aroma volatile research. Molecules 18:8200–8229. https://doi.org/10.3390/molecules18078200

Emeriewen OF, Malnoy M, Richter K, Kilian A, Hanke M-V, Peil A (2014) Evidence of a major QTL for fire blight resistance in apple wild species *Malus fusca*. Acta Hortic 1056:289–294. https://doi.org/10.17660/ActaHortic.2014.1056.49

Emeriewen OF, Peil A, Richter K, Zini E, Hanke MV, Malnoy M (2017) Fire blight resistance of *Malus* × *arnoldiana* is controlled by a quantitative trait locus located at the distal end of linkage group 12. Eur J Plant Pathol 148:1011–1018. https://doi.org/10.1007/s10658-017-1152-6

Emeriewen OF, Wöhner T, Flachowsky H, Peil A (2019) *Malus* hosts–*Erwinia amylovora* interactions: strain pathogenicity and resistance mechanisms. Front Plant Sci 10:551. https://doi.org/10.3389/fpls.2019.00551

Erdin N, Tartarini S, Broggini GA, Gennari F, Sansavini S, Gessler C, Patocchi A (2006) Mapping of the apple scab-resistance gene *Vb*. Genome 49:1238–1245. https://doi.org/10.1139/g06-096

Evans K (2013) Apple breeding in the Pacific Northwest. Acta Hortic 976:75–78. https://doi.org/10.17660/ActaHortic.2013.976.6

Evans KM, James CM (2003) Identification of SCAR markers linked to *Pl-w* mildew resistance in apple. Theor Appl Genet 106:1178–1183. https://doi.org/10.1007/s00122-002-1147-2

Evans K, Peace C (2017) Advances in marker-assisted breeding of apples. In: Evans K (ed) Achieving sustainable cultivation of apples. Burleigh Dodds Science Publ, pp 165–191. https://doi.org/10.19103/AS.2016.0017.09

Evans K, Brutcher L, Konishi B, Barritt B (2010) Correlation of sensory analysis with physical textural data from a computerized penetrometer in the Washington State University apple breeding program. HortTechnol 20:1026–1029. https://doi.org/10.21273/HORTSCI.20.6.1026

Evans K, Barritt BH, Konishi BS, Brutcher LJ, Ross CF (2012) 'WA 38' apple. HortScience 47:1177–1179. https://doi.org/10.21273/HORTSCI.47.8.1177

Fan S, Zhang B, Li J, Huang W, Wang C (2016) Effect of spectrum measurement position variation on the robustness of NIR spectroscopy models for soluble solids content of apple. Biosyst Eng 143:9–19. https://doi.org/10.1016/j.biosystemseng.2015.12.012

FAOSTAT (2017) Food and Agriculture Organization of the United States http://www.fao.org/faostat/en/#data/QC. Accessed 4 Dec 2019

Farneti B, Di Guardo M, Khomenko I, Cappellin L, Biasioli F, Velasco R, Costa F (2017) Genome-wide association study unravels the genetic control of the apple volatilome and its interplay with fruit texture. J Exp Bot 68:1467–1478. https://doi.org/10.1093/jxb/erx018

Flachowsky H, Peil A, Sopanen T, Elo A, Hanke M -V (2007) Overexpression of *BpMADS4* from silver birch (*Betula pendula* Roth.) induces early-flowering in apple (*Malus* × *domestica* Borkh.) Plant Breed 126:137–145. https://doi.org/10.1111/j.1439-0523.2007.01344.x

Flachowsky H, Hanke M-V, Peil A, Strauss SH, Fladung M (2009) A review of transgenic approaches to accelerate breeding of woody plants. Plant Breed 128:217–226. https://doi.org/10.1111/j.1439-0523.2008.01591.x

Flachowsky H, Le Roux P-M, Peil A, Patocchi A, Richter K, Hanke M-V (2011) Application of a high-speed breeding technology to apple (*Malus* × *domestica*) based on transgenic early flowering plants and marker-assisted selection. New Phytol 192:364–377. https://doi.org/10.1111/j.1469-8137.2011.03813.x

Fuleki T, Pelayo E, Palabay RB (1994) Sugar composition of varietal juices produced from fresh and stored apples. J Agric Food Chem 42:1266–1275. https://doi.org/10.1021/jf00042a003

Galli P, Broggini GAL, Kellerhals M, Gessler C, Patocchi A (2010) High-resolution genetic map of the *Rvi15* (*Vr2*) apple scab resistance locus. Mol Breed 26:561–572. https://doi.org/10.1007/s11032-010-9391-7

Gardiner SE, Bus VGN, Rusholme RL, Chagńe D, Rikkerink E (2007) Apples. In: Kole C (ed) Genome mapping and molecular breeding in plants: fruits and nuts. Springer, pp 1–62

Gardiner SE, Norelli JL, de Silva N, Fazio G, Peil A, Malnoy M, Horner M, Bowatte D, Carlisle C, Wiedow C, Wan Y, Bassett CL, Baldo AM, Cleton J-M, Richter K, Aldwinckle HS, Bus VGM (2012) Putative resistance gene markers associated with quantitative trait loci for fire blight resistance in *Malus* 'Robusta 5' accessions. BMC Genet 13:25. https://doi.org/10.1186/1471-2156-13-25

Gardner KM, Brown P, Cooke TF, Cann S, Costa F, Bustamante C, Velasco R, Troggio M, Myles S (2014) Fast and cost-effective genetic mapping in apple using next-generation sequencing. G3-Genes Genom Genet 4:1681–1687. https://doi.org/10.1534/g3.114.011023

Garris A (2013) SnapDragon and RubyFrost are new apple varieties. Cornell Chronicle http://news.cornell.edu/stories/2013/08/snapdragon-and-rubyfrost-are-new-apple-varieties. Accessed 06 Jan 2020

Gessler C, Patocchi A, Sansavini S, Tartarini S, Gianfranceschi L (2006) *Venturia inaequalis* in apple. Crit Rev Plant Sci 25:473–503. https://doi.org/10.1080/07352680601015975

Gianfranceschi L, Soglio V (2004) The European project HiDRAS: innovative multidisiplinary approaches to breeding high quality disease resistant apples. Acta Hortic 663:327–330. https://doi.org/10.17660/ActaHortic.2004.663.55

Greer DH (2005) Non-destructive chlorophyll fluorescence and colour measurements of 'Braeburn' and 'Royal Gala' apple (*Malus domestica*) fruit development throughout the growing season. NZ J Crop Hort 33:413–421. https://doi.org/10.1080/01140671.2005.9514378

Guan Y, Peace C, Rudell D, Verma S, Evans K (2015) QTLs detected for individual sugars and soluble solids content in apple. Mol Breed 35:334. https://doi.org/10.1007/s11032-015-0334-1

Gygax M, Gianfranceschi L, Liebhard R, Kellerhals M, Gessler C, Patocchi A (2004) Molecular markers linked to the apple scab resistance gene *Vbj* derived from *Malus baccata jackii*. Theor Appl Genet 109:1702–1709. https://doi.org/10.1007/s00122-004-1803-9

Hampson CR, Sholberg PL (2008) Estimating combining ability for fire blight resistance in apple progenies. Acta Hortic 793:337–343. https://doi.org/10.17660/ActaHortic.2008.793.50

Hancock JF, Luby JJ, Brown SK, Lobos GA (2008) Apples. In: Hancock JF (ed) Temperate fruit crop breeding: germplasm to genomics. Springer, pp 1–38

Harker FR, Amos RL, Echeverria G, Gunson FA (2006) Influence of texture on taste: insights gained during studies of hardness, juiciness, and sweetness of apple fruit. J Food Sci 71:77–82. https://doi.org/10.1111/j.1365-2621.2006.tb08925.x

Harker FR, Kupferman EM, Marin AB, Gunson FA, Triggs CM (2008) Eating quality standards for apples based on consumer preferences. Postharv Biol Technol 50:70–78. https://doi.org/10.1016/j.postharvbio.2008.03.020

Harshman JM (2015) Quantifying efficiency and accuracy in the Washington State University apple breeding program. Dissertation, Washington State University. http://hdl.handle.net/2376/6242

Heffner EL, Sorrells ME, Jannink JL (2009) Genomic selection for crop improvement. Crop Sci 49:1–12. https://doi.org/10.2135/cropsci2008.08.0512

Hemmat M, Brown SK, Aldwinckle HS, Weeden NF (2003) Identification and mapping of markers for resistance to apple scab from 'Antonovka' and 'Hansen's baccata #2'. Acta Hortic 622:153–161. https://doi.org/10.17660/ActaHortic.2003.622.13

Howard NP, van de Weg E, Bedford DS, Peace CP, Vanderzande S, Clark MD, Teh SL, Cai L, Luby JJ (2017) Elucidation of the 'Honeycrisp' pedigree through haplotype analysis with a multi-family integrated SNP linkage map and a large apple (*Malus* × *domestica*) pedigree-connected SNP data set. Hortic Res 4:17003. https://doi.org/10.1038/hortres.2017.3

Howard NP, van de Weg E, Tillman J, Tong CBS, Silverstein KAT, Luby JJ (2018) Two QTL characterized for soft scald and soggy breakdown in apple (*Malus* × *domestica*) through pedigree-based analysis of a large population of interconnected families. Tree Genet Genomes 14:2. https://doi.org/10.1007/s11295-017-1216-y

Howard NP, Tillman J, Vanderzande S, Luby JJ (2019) Genetics of zonal leaf chlorosis and genetic linkage to a major gene regulating skin anthocyanin production (*MdMYB1*) in the apple (*Malus* × *domestica*) cultivar Honeycrisp. PLoS ONE 14. https://doi.org/10.1371/journal.pone.0210611

Iezzoni A, Weebadde C, Luby J, Yu C, van de Weg E, Fazio G, Main D, Peace CP, Bassil NV, McFerson J (2010) RosBREED: enabling marker-assisted breeding in Rosaceae. Acta Hortic 859:389–394. https://doi.org/10.17660/ActaHortic.2010.859.47

Iezzoni A, Peace C, Main D, Bassil N, Coe M, Finn C, Gasic K, Luby J, Hokanson S, McFerson J, Norelli J, Olmstead Mm Whitaker V, Yue C (2017) RosBREED2: progress and future plans to enable DNA-informed breeding in the *Rosaceae*. Acta Hortic 1172:115–118. https://doi.org/10.17660/ActaHortic.2017.1172.20

Iwanami H (2011) Breeding for fruit quality in apple. In: Jenks MA, Bebeli PJ (eds) Breeding for fruit quality. John Wiley & Sons, pp 175–200

James CM, Clarke JB, Evans KM (2004) Identification of molecular markers linked to the mildew resistance gene *Pl-d* in apple. Theor Appl Genet 110:175–181. https://doi.org/10.1007/s00122-004-1836-0

Janick J (2002) History of the PRI apple breeding program. Acta Hortic 595:55–60. https://doi.org/10.17660/ActaHortic.2002.595.7

Janick J (2006) The PRI apple breeding program. HortScience 41:8–9. https://doi.org/10.21273/HORTSCI.41.1.8

Jänsch M, Broggini GA, Weger J, Bus VG, Gardiner SE, Bassett H, Patocchi A (2015) Identification of SNPs linked to eight apple disease resistance loci. Mol Breed 35:45. https://doi.org/10.1007/s11032-015-0242-4

Jarolmasjed S, Sankaran S, Marzougui A, Kostick S, Si Y, Vargas JJQ, Evans K (2019) High-throughput phenotyping of fire blight disease symptoms using sensing techniques in apple. Front Plant Sci 10:576. https://doi.org/10.3389/fpls.2019.00576

Jung S, Lee T, Cheng C-H, Buble K, Zheng P, Yu J, Humann J, Ficklin SP, Gasic K, Scott K, Frank M, Ru S, Hough H, Evans K, Peace C, Olmstead M, DeVetter LW, McFerson J, Coe M, Wegrzyn JL, Staton ME, Abbott AG, Main D (2019) 15 years of GDR: New data and functionality in the Genome Database for Rosaceae. Nucleic Acids Res 47:D1137–D1145. https://doi.org/10.1093/nar/gky1000

Jurick WM (2017) Pre- and postharvest fungal apple diseases. In: Evans K (ed) Achieving sustainable cultivation of apples. Burleigh Dodds Science Publ, pp 371–382. https://doi.org/10.19103/AS.2016.0017.17

Kellerhals M (2017) Advances in pest- and disease-resistant apple varieties. In: Evans K (ed) Achieving sustainable cultivation of apples. Burleigh Dodds Science Publ, pp 461–481. https://doi.org/10.19103/AS.2016.0017.22

Kellerhals M, Spuhler M, Duffy B, Patocchi A, Frey JE (2009) Selection efficiency in apple breeding. Acta Hortic 177–184. https://doi.org/10.17660/ActaHortic.2009.814.22

Kellerhals M, Schütz S, Patocchi A (2017) Breeding for host resistance to fire blight. J Plant Pathol 99:37–43. https://doi.org/10.4454/jpp.v99i0.3923

Kenis K, Keulemans J, Davey MW (2008) Identification and stability of QTLs for fruit quality traits in apple. Tree Genet Genomes 4:647–661. https://doi.org/10.1007/s11295-008-0140-6

Khajuria YP, Kaul S, Wani AA, Dhar MK (2018) Genetics of resistance in apple against *Venturia inaequalis* (Wint.) Cke. Tree Genet Genomes 14:16. https://doi.org/10.1007/s11295-018-1226-4.

Khan MA, Duffy B, Gessler C, Patocchi A (2006) QTL mapping of fire blight resistance in apple. Mol Breed 17:299–306. https://doi.org/10.1007/s11032-006-9000-y

Khan SA, Chibon PY, de Vos RC, Schipper BA, Walraven E, Beekwilder J, van Dijk T, Finkers R, Visser RG, van de Weg EW, Bovy A (2012) Genetic analysis of metabolites in apple fruits indicates an mQTL hotspot for phenolic compounds on linkage group 16. J Exp Biol 63:2895–2908. https://doi.org/10.1093/jxb/err464

Khan MA, Zhao Y, Korban SS (2013) Identification of genetic loci associated with fire blight resistance in *Malus* through combined use of QTL and association mapping. Physiol Plant 148:344–353. https://doi.org/10.1111/ppl.12068

King GJ, Lynn JR, Dover CJ, Evans KM, Seymour GB (2001) Resolution of quantitative trait loci for mechanical measures accounting for genetic variation in fruit texture of apple (*Malus pumila* Mill.). Theor Appl Genet 102:1227–1235. https://doi.org/10.1007/s001220000530

Kostick SA, Norelli JL, Evans KM (2019) Novel metrics to classify fire blight resistance of 94 apple cultivars. Plant Pathol 68:985–996. https://doi.org/10.1111/ppa.13012

Kumar S, Chagné D, Bink MC, Volz RK, Whitworth C, Carlisle C (2012) Genomic selection for fruit quality traits in apple (*Malus* × *domestica* Borkh.). PLoS ONE 7: e36674. https://doi.org/10.1371/journal.pone.0036674

Kumar S, Garrick DJ, Bink MCAM, Whitworth C, Chagné D, Volz RK (2013) Novel genomic approaches unravel genetic architecture of complex traits in apple. BMC Genomics 14:393. https://doi.org/10.1186/1471-2164-14-393

Kumar S, Rowan D, Hunt M, Chagné D, Whitworth C, Souleyre E (2015c) Genome-wide scans reveal genetic architecture of apple flavor volatiles. Mol Breed 35:118. https://doi.org/10.1007/s11032-015-0312-7

Kumar S, McGlone A, Whitworth C, Volz R (2015a) Postharvest performance of apple phenotypes predicted by near-infrared (NIR) spectral analysis. Postharv Biol Tec 100:16–22. https://doi.org/10.1016/j.postharvbio.2014.09.021

Kumar S, Molloy C, Muñoz P, Daetwyler H, Chagné D, Volz R (2015b) Genome-enabled estimates of additive and nonadditive genetic variances and prediction of apple phenotypes across environments. G3-Genes Genom Genet 5:2711–2718. https://doi.org/10.1534/g3.115.021105

Larsen B, Migicovsky Z, Jeppesen AA, Gardner KM, Toldam-Andersen TB, Myles S, Ørgaard M, Petersen MA, Pedersen C (2019) Genome-wide association studies in apple reveal loci for aroma volatiles, sugar composition, and harvest date. Plant Genome 12:2. https://doi.org/10.3835/plantgenome2018.12.0104

Laurens F (1999) Review of the current apple breeding programmes in the world: objectives for scion cultivar improvement. Acta Hortic 484:163–170. https://doi.org/10.17660/ActaHortic.1998.484.26

Laurens F, Aranzana MJ, Arús P, Bonany J, Corelli L, Patocchi A, Peil A, Quilot B, Salvi S, van de Weg E, Vecchietti A (2010) FruitBreedomics: a new European initiative to bridge the gap between scientific research and breeding *Rosaceae* fruit tree crops. Book of Abstracts, v 2. IHC Lisbon, p 242

Laurens F, Aranzana MJ, Arus P, Bassi D, Bink M, Bonany J, Caprera A, Corelli-Grappadelli L, Costes E, Durel C-E, Mauroux J-B, Muranty H, Nazzicari N, Pascal T, Patocchi A, Peil A, Quilot-Turion B, rossini L, Stella A, Troggio M, Velasco R, van de Weg E, (2018) An integrated approach for increasing breeding efficiency in apple and peach in Europe. Hortic Res 5:11. https://doi.org/10.1038/s41438-018-0016-3

Le Lezek M, Paulin JP, Lecomte P (1987) Shoot and blossom susceptibility to fireblight of apple cultivars. Acta Hortic 217:311–315. https://doi.org/10.17660/ActaHortic.1987.217.54

Le Roux P-MF, Khan MA, Broggini GAL, Duffy B, Gessler C, Patocchi A (2010) Mapping of quantitative trait loci for fire blight resistance in the apple cultivars 'Florina' and 'Nova Easygro.' Genome 53:710–722. https://doi.org/10.1139/g10-047

Le Roux P-MF, Flachowsky H, Hanke M-V, Gessler C, Patocchi A (2012) Use of a transgenic early flowering approach in apple (*Malus* × *domestica* Borkh.) to introgress fire blight resistance from cultivar Evereste. Mol Breed 30:857–874. https://doi.org/10.1007/s11032-011-9669-4

Lee SA, Ngugi HK, Halbrendt NO, O'Keefe G, Lehman B, Travis JW, Sinn JP, McNellis TW (2010) Virulence characteristics accounting for fire blight disease severity in apple trees and seedlings. Phytopathology 100:539–550. https://doi.org/10.1094/PHYTO-100-6-0539

Lee SJ, Ban SH, Kim GH, Kwon SI, Kim JH, Choi C (2017) Identification of potential gene-associated major traits using GBS-GWAS for Korean apple germplasm collections. Plant Breed 136:977–986. https://doi.org/10.1111/pbr.12544

Li J, Luo W, Wang Z, Fan S (2019) Early detection of decay on apples using hyperspectral reflectance imaging combining both principal component analysis and improved watershed segmentation method. Postharv Biol Tec 149:235–246. https://doi.org/10.1016/j.postharvbio.2018.12.007

Liebhard R, Kellerhals M, Pfammatter W, Jertmini M, Gessler C (2003) Mapping quantitative physiological traits in apple (*Malus* × *domestica* Borkh.) Plant Mol Biol 52:511–526. https://doi.org/10.1023/A:1024886500979

Liu YX, Li BH, Wang CH, Liu CX, Kong XH, Zhu J, Dai HY (2016) Genetics and molecular marker identification of a resistance to *Glomerella* leaf spot in apple. Hortic Plant J 2:121–125. https://doi.org/10.1016/j.hpj.2016.06.002

Longhi S, Moretto M, Viola R, Velasco R, Costa F (2012) Comprehensive QTL mapping survey dissects the complex fruit texture physiology in apple (*Malus* x *domestica* Borkh.). J Exp Bot 63:1107–1121. https://doi.org/10.1093/jxb/err326

Longhi S, Hamblin MT, Trainotti L, Peace CP, Velasco R, Costa F (2013) A candidate gene based approach validates *Md-PG1* as the main responsible for a QTL impacting fruit in apple (*Malus* x *domestica* Borkh.). BMC Plant Biol 13:37. https://doi.org/10.1186/1471-2229-13-37

Lorenz AJ, Chao S, Heffner AFG, EL, Hayashi T, Iwata H, Smith KP, Sorrells ME, Jannink JL, (2011) Genomic selection in plant breeding: knowledge and prospects. Adv Agron 110:77–123. https://doi.org/10.1016/B978-0-12-385531-2.00002-5

Lu Y, Lu R (2018) Structured-illumination reflectance imaging coupled with phase analysis techniques for surface profiling of apples. J Food Eng 232:11–20. https://doi.org/10.1016/j.jfoodeng.2018.03.016

Lu Y, Li R, Lu R (2016) Structured-illumination reflectance imaging (SIRI) for enhanced detection of fresh bruises in apples. Postharv Biol Tech 117:89–93. https://doi.org/10.1016/j.postharvbio.2016.02.005

Luby JJ (1991) Breeding cold-hardy fruit crops in Minnesota. HortScience 26:507–512. https://doi.org/10.21273/HORTSCI.26.5.507

Luo X, Ye Z, Xu H, Zhang D, Bai S, Ying Y (2018) Robustness improvement of NIR-based determination of soluble solids in apple fruit by local calibration. Postharv Biol Tech 139:82–90. https://doi.org/10.1016/j.postharvbio.2018.01.019

Luo F, Sandefur P, Evans K, Peace C (2019) A DNA test for routinely predicting mildew resistance in descendants of crabapple 'White Angel.' Mol Breed 39:33. https://doi.org/10.1007/s11032-019-0933-3

Luo F, Evans K, Norelli JL, Zhang Z, Peace C (2020) Prospects for achieving durable disease resistance with elite fruit quality in apple breeding. Tree Genet Genomes 16:21. https://doi.org/10.1007/s11295-020-1414-x

Ma B, Zhao S, Wu B, Wang D, Peng Q, Owiti A, Fang T, Liao L, Ogutu C, Korban SS, Li S (2016) Construction of a high density linkage map and its application in the identification of QTLs for soluble sugar and organic acid components in apple. Tree Genet Genomes 12:1. https://doi.org/10.1007/s11295-015-0959-6

Maliepaard C, Sillanpaa MJ, van Ooijen JW, Jansen RC, Arjas E (2001) Bayesian versus frequentist analysis of multiple quantitative trait loci with an application to an outbred apple cross. Theor Appl Genet 103:1243–1253. https://doi.org/10.1007/s001220100720

McClure KA, Gardner KM, Toivonen PM, Hampson CR, Song J, Forney CF, DeLong J, Rajcan I, Myles S (2016) QTL analysis of soft scald in two apple populations. Hortic Res 3:16043. https://doi.org/10.1038/hortres.2016.43

McClure KA, Gardner KM, Douglas GM, Song J, Forney CF, DeLong J, Fan L, Du L, Toivonen P, Somers DJ, Rajcan I (2018) A genome-wide association study of apple quality and scab resistance. Plant Genome 11. https://doi.org/10.3835/plantgenome2017.08.0075

McClure KA, Gong Y, Song J, Vinqvist-Tymchuk M, Palmer LC, Fan L, Burgher-MacLellan K, Zhang Z, Celton JM, Forney CF, Migicovsky Z (2019) Genome-wide association studies in apple reveal loci of large effect controlling apple polyphenols. Hortic Res 6:107. https://doi.org/10.1038/s41438-019-0190-y

McGlone VA, Jordan RB, Martinsen PJ (2002) Vis/NIR estimation at harvest of pre-and post-storage quality indices for 'Royal Gala' apple. Postharv Biol Tec 25:135–144. https://doi.org/10.1016/S0925-5214(01)00180-6

McGlone VA, Jordan RB, Seelye R, Clark CJ (2003) Dry-matter—a better predictor of the post-storage soluble solids in apples? Postharv Biol Tec 28:431–435. https://doi.org/10.1016/S0925-5214(02)00207-7

Meuwissen THE, Hayes BJ, Goddard ME (2001) Prediction of total genetic value using genome-wide dense marker maps. Genetics 157:1819–1829. https://doi.org/10.1093/genetics/157.4.1819

Migicovsky Z, Gardner KM, Money D, Sawler J, Bloom JS, Moffett P, Chao CT, Schwaninger H, Fazio G, Zhong GY, Myles S (2016) Genome to phenome mapping in apple using historical data. Plant Genome 9:2. https://doi.org/10.3835/plantgenome2015.11.0113

Mohan SK, Fallahi E, Bijman VP (2002) Evaluation of apple varieties for susceptibility to *Erwinia amylovora* by artificial inoculation under field conditions. Acta Hortic 590:373–5. https://doi.org/10.17660/ActaHortic.2002.590.56

Moose SP, Mumm RH (2008) Molecular plant breeding as the foundation for 21st century crop improvement. Plant Physiol 147:969–977. https://doi.org/10.1104/pp.108.118232

Moriya S, Terakami S, Iwanami H, Haji T, Okada K, Yamamoto T, Abe K (2013) Genetic mapping and marker-assisted selection of the gene conferring susceptibility to Alternaria blotch caused by *Alternaria alternata* apple pathotype in apple. Acta Hortic 976:555–560. https://doi.org/10.17660/ActaHortic.2013.976.79

Moriya S, Kunihisa M, Okada K, Iwanami H, Iwata H, Minamikawa M, Katayose Y, Matsumoto T, Mori S, Sasaki H, Matsumoto T (2017) Identification of QTLs for flesh mealiness in apple (*Malus* × *domestica* Borkh.). Hortic J 86:159–170. https://doi.org/10.2503/hortj.MI-156

Mundt CC (1990) Probability of mutation to multiple virulence and durability of resistance gene pyramids: further comments. Phytopathology 81:240–242. https://doi.org/10.1094/Phyto-80-221

Muranty H, Troggio M, Sadok IB, Al Rifaï M, Auwerkerken A, Banchi E, Velasco R, Stevanato P, Van De Weg WE, Di Guardo M, Kumar S (2015) Accuracy and responses of genomic selection on key traits in apple breeding. Hortic Res 2:15060. https://doi.org/10.1038/hortres.2015.60

Nicolai BM, Beullens K, Bobelyn E, Peirs A, Saeys W, Theron KI, Lammertyn J (2007) Nondestructive measurement of fruit and vegetable quality by means of NIR spectroscopy: A review. Postharv Biol Tech 46:99–118. https://doi.org/10.1016/j.postharvbio.2007.06.024

Norelli J (2017) Bacterial diseases affecting apples. In: Evans K (ed) Achieving sustainable cultivation of apples. Burleigh Dodds Science Publ, pp 403–424. https://doi.org/10.19103/AS.20160017.20

Norelli JL, Jones AL, Aldwinckle HS (2003) Fire blight management in the twenty-first century: Using new technologies that enhance host resistance in apple. Plant Dis 87:756–765. https://doi.org/10.1094/PDIS.2003.87.7.756

Norelli JL, Wisniewski M, Fazio G, Burchard E, Gutierrez B, Levin E, Droby S (2017) Genotyping-by-sequencing markers facilitate the identification of quantitative trait loci controlling resistance to *Penicillium expansum* in *Malus sieversii*. PLoS ONE 12. https://doi.org/10.1371/journal.pone.0172949

Nyasordzi J, Friedman H, Schmilovitch Z, Ignat T, Weksler A, Rot I, Lurie S (2013) Utilizing the I_{AD} index to determine internal quality attributes of apples at harvest and after storage. Postharvest Biol Tec 77:80–86. https://doi.org/10.1016/j.postharvbio.2012.11.002

Pagliarani G, Dapena E, Miñarro M, Denancé C, Lespinasse Y, Rat-Morris E, Troggio M, Durel CE, Tartarini S (2016) Fine mapping of the rosy apple aphid resistance locus *Dp-fl* on linkage group 8 of the apple cultivar 'Florina.' Tree Genet Genomes 12:56. https://doi.org/10.1007/s11295-016-1015-x

Palmer JW, Harker FR, Tustin DS, Johnston J (2010) Fruit dry matter concentration: A new quality metric for apples. J Sci Food Agr 90:2586–2594. https://doi.org/10.1002/jsfa.4125

Papp D, Singh J, Gadoury D, Khan MA (2020) New North American isolates of *Venturia inaequalis* can overcome apple scab resistance of *Malus floribunda* 821. Plant Dis 104:649–655. https://doi.org/10.1094/PDIS-10-19-2082-RE

Parisi L, Fouillet V, Schouten HJ, Groenwold R, Laurens F, Didelot F, Evans K, Fischer C, Gennari F, Kemp H, Lateur M, Patocchi A, Thissen J, Tsipouridis C (2004) Variability of the pathogenicity of *Venturia inaequalis* in Europe. Acta Hortic 663:107–114. https://doi.org/10.17660/ActaHortic.2004.663.13

Patocchi A, Bigler B, Koller B, Kellerhals M, Gessler C (2004) *Vr2*: A new apple scab resistance gene. Theor Appl Genet 109:1087–1092. https://doi.org/10.1007/s00122-004-1723-8

Patocchi A, Walser M, Tartarini S, Broggini GAL, Gennari F, Sansavini S, Gessler C (2005) Identification by genome scanning approach (GSA) of a microsatellite tightly associated with the apple scab resistance gene *Vm*. Genome 48:630–636. https://doi.org/10.1139/G05-036

Peace C (2017) DNA-informed breeding of rosaceous crops: Promises, progress and prospects. Hortic Res 4:17006. https://doi.org/10.1038/hortres.2017.6

Peace CP, Bianco L, Troggio M, van de Weg E, Howard NP, Comille A, Durel C-E, Myles S, Migicovsky Z, Schaffer RJ, Costes E, Fazio G, Yamane H, van Nocker S, Gottschalk C, Costa F, Chagné D, Zhang X, Patocchi A, Gardiner SE, Hardner C, Kumar S, Laurens F, Bucher E, Main D, Jung S,

Vanderzande S (2019) Apple whole genome sequences: recent advances and new prospects. Hortic Res 6:59. https://doi.org/10.1038/s41438-019-0141-7

Peil A, Garcia-Libreros T, Richter K, Trognitz C, Trognitz B, Hanke M-V, Flachowsky H (2007) Strong evidence for a fire blight resistance gene of *Malus robusta* located on linkage group 3. Plant Breed 126:470–475. https://doi.org/10.1111/j.1439-0523.2007.01408.x

Peil A, Hanke MV, Flachowsky H, Richter K, Garcia-Libreros T, Celton JM, Gardiner S, Horner M, Bus V (2008) Confirmation of the fire blight QTL of *Malus* × *robusta* 5 on linkage group 3. Acta Hortic 793:297–303. https://doi.org/10.17660/ActaHortic.2008.793.44

Peil A, Bus VMG, Geider K, Richter K, Flachowsky H, Hanke MV (2009) Improvement of fire blight resistance in apple and pear. Int J Plant Breed 3:1–27

Peil A, Hübert C, Wensing A, Horner M, Emeriewen OF, Richter K, Wöhner T, Chagné D, Orellana-Torrejon C, Saeed M, Troggio M, Stefani E, Gardiner SE, Hanke M-V, Flachowsky H, Bus VGM (2019) Mapping of fire blight resistance in *Malus* × *robusta* 5 flowers following artificial inoculation. BMC Plant Biol 19:532. https://doi.org/10.1186/s12870-019-2154-7

Peil A, Emeriewen OF, Khan A, Kostick S, Malnoy M (2020) Status of fire blight resistance breeding in *Malus*. J Plant Pathol. https://doi.org/10.1007/s42161-020-00581-8

Peng Y, Lu R (2008) Analysis of spatially resolved hyperspectral scattering images for assessing apple fruit firmness and soluble solids content. Postharv Biol Tech 48:52–62. https://doi.org/10.1016/j.postharvbio.2007.09.019

Porep JU, Kammerer DR, Carle R (2015) On-line application of near infrared (NIR) spectroscopy in food production. Trends Food Sci Tech 46:211–230. https://doi.org/10.1016/j.tifs.2015.10.002

Potts SM, Khan MA, Han Y, Kushad MM, Korban SS (2014) Identification of quantitative trait loci (QTLs) for fruit quality traits in apple. Plant Mol Biol Rep 32:109–116. https://doi.org/10.1007/s11105-013-0634-8

Roth M, Muranty H, Di Guardo M, Guerra W, Patocchi A, Costa F (2020) Genomic prediction of fruit texture and training population optimization towards the application of genomic selection in apple. Hortic Res 7:148. https://doi.org/10.1038/s41438-020-00370-5

Rowan DD, Hunt MB, Dimouro A, Alspach PA, Weskett R, Volz RK, Gardiner SE, Chagné D (2009) Profiling fruit volatiles in the progeny of a 'Royal Gala' × 'Granny Smith' apple (*Malus* × *domestica*) cross. J Agr Food Chem 57:7953–7961. https://doi.org/10.1021/jf901678v

Ru S, Main D, Evans K, Peace C (2015) Current applications, challenges, and perspectives of marker-assisted seedling selection in Rosaceae tree fruit breeding. Tree Genet Genomes 11:8. https://doi.org/10.1007/s11295-015-0834-5

Rymenants M, van de Weg E, Auwerkerken A, De Wit I, Czech A, Nijland B, Heuven H, De Storme N, Keulemans W (2020) Detection of QTL for apple fruit acidity and sweetness using sensorial evaluation in multiple pedigreed full-sib families. Tree Genet Genomes 16:71. https://doi.org/10.1007/s11295-020-01466-8

Sadar N, Zanella A (2019) A study on the potential of IAD as a surrogate index of quality and storability in cv. 'Gala' apple fruit. Agronomy 9:642. https://doi.org/10.3390/agronomy9100642

Saeys W, Do Trong NN, van Beers R, Nicolaï BM (2019) Multivariate calibration of spectroscopic sensors for postharvest quality evaluation: A review. Postharv Biol Tech 158. https://doi.org/10.1016/j.postharvbio.2019.110981

Schlathölter I, Jänsch M, Flachowsky H, Broggini GAL, Hanke M-V, Patocchi A (2018) Generation of advanced fire blight-resistant apple (*Malus* × *domestica*) selections of the fifth generation within 7 years of applying the early flowering approach. Planta 247:1475–1488. https://doi.org/10.1007/s00425-018-2876-z

Schmitz CA, Clark MD, Luby JJ, Bradeen JM, Guan Y, Evans K, Orcheski B, Brown S, Verma S, Peace C (2013) Fruit texture phenotypes of the RosBREED U. S. apple reference germplasm set. HortScience 48:296–303. https://doi.org/10.21273/HORTSCI.48.3.296

Silva KJP, Singh J, Bednarek R, Fei Z, Khan MA (2019) Differential gene regulatory responses to fire blight infection in two apple cultivars (*Malus* × *domestica*). Hortic Res 6:35. https://doi.org/10.1038/s41438-019-0120-z

Sofu MM, Er O, Kayacan MC, Cetişli B (2016) Design of an automatic apple sorting system using machine vision. Comput Electron Agr 127:395–405. https://doi.org/10.1016/j.compag.2016.06.030

Solovchenko AE, Chivkunova OB, Merzlyak MN, Gudkovsky VA (2005) Relationships between chlorophyll and carotenoid pigments during on-and off-tree ripening of apple fruit as revealed non-destructively with reflectance spectroscopy. Postharv Biol Tech 38:9–17. https://doi.org/10.1016/j.postharvbio.2005.05.004

Song J, Forney CF (2008) Flavour volatile production and regulation in fruit. Can J Plant Sci 88:537–550. https://doi.org/10.4141/CJPS07170

Soriano JM, Madduri M, Schaart JG, van der Burgh A, van Kaauwen MPW, Tomic L, Groenwold R, Velasco R, van de Weg E, Schouten HJ (2014) Fine mapping of the gene Rvi18 (V25) for broad-spectrum resistance to apple scab, and development of a linked SSR marker suitable for marker-assisted breeding. Mol Breed 34:2021–2032. https://doi.org/10.1007/s11032-014-0159-3

Soufflet-Freslon V, Gianfranceschi L, Patocchi A, Durel CE (2008) Inheritance studies of apple scab resistance and identification of Rvi14, a new major gene that acts together with other broad-spectrum QTL. Genome 51:657–667. https://doi.org/10.1139/G08-046

Souleyre EJ, Bowen JK, Matich AJ, Tomes S, Chen X, Hunt MB, Wang MY, Ileperuma NR, Richards K,

Rowan DD, Chagné D (2019) Genetic control of α-farnesene production in apple fruit and its role in fungal pathogenesis. Plant J 100:1148–1162. https://doi.org/10.1111/tpj.14504

Stankiewicz-Kosyl M, Pitera E, Gawronski S (2005) Mapping QTL involved in powdery mildew resistance of the apple clone U 211. Plant Breed 124:63–66. https://doi.org/10.1111/j.1439-0523.2004.01049.x

Stoeckli S, Mody K, Gessler C, Patocchi A, Jermini M, Dorn S (2008) QTL analysis for aphid resistance and growth traits in apple. Tree Genet Genomes 4:833–847. https://doi.org/10.1007/s11295-008-0156-y

Sun R, Chang Y, Yang F, Wang Y, Li H, Zhao Y, Chen D, Wu T, Zhang X, Han Z (2015) A dense SNP genetic map constructed using restriction site-associated DNA sequencing enables detection of QTLs controlling apple fruit quality. BMC Genomics 16:747. https://doi.org/10.1186/s12864-015-1946-x

Tan Y, Lv S, Liu X, Gao T, Li T, Wang Y, Wu T, Zhang X, Han Y, Korban SS, Han Z (2017) Development of high-density interspecific genetic maps for identification of QTLs conferring resistance to *Valsa ceratosperma* in apple. Euphytica 213:10. https://doi.org/10.1007/s10681-016-1790-3

Tartarini S, Gennari F, Pratesi D, Palazetti C, Sansavini S, Parisi L, Fouillet A, Durel CE (2004) Characterization and genetic mapping of a major scab resistance gene from the old Italian apple cultivar 'Durello di Forlì'. Acta Hortic 663:129–133. https://doi.org/10.17660/ActaHortic.2004.663.16

Teh SL, Clark MD, Luby JJ (2018) Marker development and applications in the University of Minnesota apple and grape breeding programs. Acta Hortic 1203:93–98. https://doi.org/10.17660/ActaHortic.2018.1203.14

Teh SL, Coggins JL, Kostick SA, Evans KM (2020b) Location, year, and tree age impact NIR-based postharvest prediction of dry matter concentration for 58 apple accessions. Postharv Biol Tech 166. https://doi.org/10.1016/j.postharvbio.2020.111125

Teh SL, Brutcher L, Schonberg B, Evans K (2020a) Eleven-year correlation of physical fruit texture traits between computerized penetrometers and sensory assessment in an apple breeding program. HortTechnol 30: 719–724. https://doi.org/10.21273/HORTTECH04698-20

Toivonen PM, Hampson CR (2014) Relationship of IAD index to internal quality attributes of apples treated with 1-methylcyclopropene and stored in air or controlled atmospheres. Postharv Biol Tech 91:90–95. https://doi.org/10.1016/j.postharvbio.2013.12.024

Travers S, Bertelsen MG, Kucheryavskiy SV (2014) Predicting apple (cv. Elshof) postharvest dry matter and soluble solids content with near infrared spectroscopy. J Sci Food Agr 94:955–962. https://doi.org/10.1002/jsfa.6343

Turechek WW, Carroll JE, Rosenberger DA (2004) Powdery mildew of apple. New York State IPM Program. https://hdl.handle.net/1813/43120

van de Weg E, Di Guardo M, Jänsch M, Socquet-Juglard D, Costa F, Baumgartner I, Broggini GAL, Kellerhals M, Troggio M, Laurens F, Durel C-E, Patocchi A (2018) Epistatic fire blight resistance QTL alleles in the apple cultivar 'Enterprise' and selection X-6398 discovered and characterized through pedigree-informed analysis. Mol Breed 38:5. https://doi.org/10.1007/s11032-017-0755-0

van der Zwet T, Orolaza-Halbrendt N, Zeller W (2012) Fire blight: history, biology and management. APS Press, Saint Paul, MN, USA

Vanderzande S, Howard NP, Cai L, De Silva LC, Antanaviciute L, Bink MCAM, Kruisselbrink JW, Bassil N, Gasic K, Iezzoni A, van de Weg E, Peace C (2019) High-quality, genome-wide SNP genotypic data for pedigreed germplasm of the diploid outbreeding species apple, peach, and sweet cherry through a common workflow. PLoS ONE 14. https://doi.org/10.1371/journal.pone.0210928

Velasco R, Zharkikh A, Affourtit J, Dhingra A, Cestaro A, Kalyanaraman A, Fontana P, Bhatnagar SK, Troggio M, Pruss D, Salvi S, Pindo M, Baldi P, Catelletti S, Cavaiuolo M, Coppola G, Costa F, Cova V, Dal Ri A, Goremykin V, Komjanc M, Longhi S, Magnago P, Malacarne G, Malnoy M, Micheletti D, Moretto M, Perazzolli M, Si-Ammour A, Vezzulli S, Zini E, Eldredge G, Fitzgerald LM, Gutin N, Lanchbury G, Macalma T, Mitchell JT, Reid J, Wardell B, Kodira C, Chen Z, Desany B, Niazi F, Palmer M, Koepke T, Jiwan D, Schaeffer S, Krishnan V, Wu C, Chu VT, King ST, Vick J, Tao Q, Mraz A, Stormo A, Stormo K, Bogden R, Ederle D, Stella A, Vecchietti A, Kater MM, Masiero S, Lasserre P, Lespinasse Y, Allan AC, Bus V, Chagné D, Crowhurst RN, Gleave AP, Lavezzo E, Fawcett JA, Proost S, Rouzé P, Sterck L, Toppo S, Lazzari B, Hellens RP, Durel C-E, Gutin A, Bumgarner R, Gardiner SE, Skolnick M, Egholm M, Van de Peer Y, Salamini F, Viola R (2010) The genome of the domesticated apple (*Malus* × *domestica* Borkh.). Nat Genet 42:833–841. https://doi.org/10.1038/ng.654

Verma S, Evans K, Guan Y, Luby JJ, Rosyara UR, Howard NP, Bassil N, Bink MCAM, van de Weg WE, Peace CP (2019) Two large-effect QTLs, *Ma* and *Ma3*, determine genetic potential for acidity in apple fruit: breeding insights from a multi-family study. Tree Genet Genomes 15. https://doi.org/10.1007/s11295-019-1324-y

Vinatzer BA, Patocchi A, Tartarini S, Gianfranceschi L, Sansavini S, Gessler C (2004) Isolation of two microsatellite markers from BAC clones of the *Vf* scab resistance region and molecular characterization of scab-resistant accessions in *Malus* germplasm. Plant Breed 123:321–326. https://doi.org/10.1111/j.1439-0523.2004.00973.x

Vogt J, Schiller D, Ulrich D, Schwab W, Dunemann F (2013) Identification of lipoxygenase (LOX) genes putatively involved in fruit flavour formation in apple (*Malus* × *domestica*). Tree Genet Genomes 9:1493–1511. https://doi.org/10.1007/s11295-013-0653-5

Wang H, Peng J, Xie C, Bao Y, He Y (2015) Fruit quality evaluation using spectroscopy technology: A review.

Sensors 15:11889–11927. https://doi.org/10.3390/s150511889

Wannemuehler SD, Luby JJ, Changyan Y, Bedford DS, Gallardo RK, McCracken VA (2019) A cost-benefit analysis of DNA informed apple breeding. HortScience 54:1998–2004. https://doi.org/10.21273/HORTSCI14173-19

Way RD, Aldwinckle HS, Lamb RC, Rejman A, Sansavini S, Shen T, Watkins R, Westwood MN, Yoshida Y (1991) Apples (*Malus*). Acta Hortic 290:3–46. https://doi.org/10.17660/ActaHortic.1991.290.1.

Watkins CB (2017) Advances in postharvest handling and storage of apples. In: Evans K (ed) Achieving sustainable cultivation of apples. Burleigh Dodds Science Publ, pp 337–367. https://doi.org/10.19103/AS.2016.0017.09

Weigl K, Wenzel S, Flachowsky H, Peil A, Hanke M-V (2015) Integration of *BPMADS4* on various linkage groups improves the utilization of the rapid cycle breeding system in apple. Plant Biotechnol J 13:246–258. https://doi.org/10.1111/pbi.12267

Xu Y, Crouch JH (2008) Marker-assisted selection in plant breeding: from publications to practice. Crop Sci 48:391–407. https://doi.org/10.2135/cropsci2007.04.0191

Xu K, Wang A, Brown S (2012) Genetic characterization of the *Ma* locus with pH and titratable acidity in apple. Mol Breed 30:899–912. https://doi.org/10.1007/s11032-011-9674-7

Yue C, Gallardo RK, Luby J, Rihn A, McFerson JR, McCracken V, Bedford D, Brown S, Evans K, Weebadde C, Sebolt A, Iezzoni AF (2013) An investigation of U.S. apple producers' trait prioritization—evidence from audience surveys. HortScience 48:1378–1384. https://doi.org/10.21273/HORTSCI.48.11.1378

Zhang Y, Nock JF, Al Shoffe Y, Watkins CB (2019) Non-destructive prediction of soluble solids and dry matter contents in eight apple cultivars using near-infrared spectroscopy. Postharv Biol Tech 151:111–118. https://doi.org/10.1016/j.postharvbio.2019.01.009

Ziosi V, Noferini M, Fiori G, Tadiello A, Trainotti L, Casadoro G, Costa G (2008) A new index based on vis spectroscopy to characterize the progression of ripening in peach fruit. Postharv Biol Tech 49:319–329. https://doi.org/10.1016/j.postharvbio.2008.01.017

Zude M, Herold B, Roger JM, Bellon-Maurel V, Landahl S (2006) Non-destructive tests on the prediction of apple fruit flesh firmness and soluble solids content on tree and in shelf life. J Food Eng 77:254–260. https://doi.org/10.1016/j.jfoodeng.2005.06.027

Zude-Sasse M, Truppel I, Herold B (2002) An approach to non-destructive apple fruit chlorophyll determination. Postharv Biol Tech 25:123–133. https://doi.org/10.1016/S0925-5214(01)00173-9

Genetics, Breeding, and Genomics of Apple Rootstocks

6

Gennaro Fazio

Abstract

The ability to graft genetically diverse apple cultivars onto diverse root systems (rootstocks) has increased the breeding efficiency of the whole apple system by dividing the combinatorial complexity into two systems—one focused on scion traits (e.g., fruit color, acidity, and crispness, among others) and the other one on rootstock traits (e.g., productivity, disease resistance, and growth habit, among others). While apple rootstocks may have originated to facilitate the propagation of desired fruit, this has since transitioned to the selection of superior root systems that fit certain needs of apple growers. Efforts to identify and improve traits associated with whole apple tree performance by purposeful hybridization are relatively recent compared with the initial selection of apple rootstocks. Several breeding programs have developed breeding protocols depending on the regional needs (cold hardiness in both Canada and Russia, fire blight resistance in the U.S.A., and drought tolerance in China, among others) resulting in the development of diverse new apple rootstocks and complex traits that go well beyond the initial dwarfing growth habit for which several of the founding 'Malling' generation of rootstocks are known for. Advances in apple genetics, phenomics, and genomics are increasing the ability to customize the performance of apple rootstocks to specific cultivation requirements, including scion cultivar, soil type, water availability, disease pressure, and tree training system, thus contributing to significant gains in sustainable yield of the final product; i.e., high-quality and nutritious apples that customers will purchase recurrently.

6.1 Introduction

In the fruit tree industry, a rootstock is the basal part of a tree that extends from the union where the scion is grafted to all underground parts of the tree. As such, a rootstock usually comprises a stem (trunk) and the root system of a tree. In some instances, a bridge stem is grafted between the scion and the rootstock, referred to as an 'interstem', that may be used to either improve graft compatibility between a scion and a rootstock or to further modulate traits of the scion. To discuss the genetics, genomics, and breeding of apple rootstocks, it is important to define the origins of rootstocks and the different genetic classes they belong to.

The art and science of grafting was used in ancient times to vegetatively propagate desirable

G. Fazio (✉)
Plant Genetic Resources Unit, USDA ARS,
Geneva, NY 14456, USA
e-mail: gennaro.fazio@usda.gov

S. S. Korban (ed.), *The Apple Genome*, Compendium of Plant Genomes,
https://doi.org/10.1007/978-3-030-74682-7_6

fruit tree phenotypes. Initially, this practice might have relied on existing trees as recipients of uniform graft wood collected from desirable trees. Over time, seed collected from the fruit of local heterogeneous populations was used to generate seedlings that were later grafted with desirable tree phenotypes. A further development from this practice was to collect seeds from a single mother tree to generate a more uniform seedling population as the recipient of graft wood. As expected, different sources of seedlings would influence genetic properties of rootstocks depending on their relatedness (full-sibs, half-sibs, or feral heterogeneous populations) and population origin (geographical or functional; e.g., cider vs. fresh eating types).

The practice of using apple seedlings as rootstocks remains common today for various reasons as described herein. (1) Seedlings derived from the particular open-pollinated maternal parents (half-sibs), such as 'Antonovka', are valued as the maternal parent that possesses desirable traits (such as cold hardiness, productivity), and used to propagate highly vigorous trees. (2) Seedlings from selected apomictic populations are chosen due to their uniformity, as well as for other desirable traits, such as drought tolerance or productivity. However, it is likely that apomixis is not complete, thus resulting in some heterogeneity of the seedling rootstocks. (3) As seedlings are free of most viruses and viroids, they are used to maintain virus-free clones of desirable apple genotypes for certification and distribution purposes.

While seedlings have been the first form of rootstocks used to vegetatively propagate apples, at some point in apple grafting history, fruit fanciers have noted that some mother trees are very good at yielding seedlings of better rootstock performance than others. Thus, the practice of using rooted suckers (adventitious shoots from root tissues at the base of a tree) as rootstocks has been launched for propagating new trees. This has represented a paradigm shift in the world of rootstocks, as it has served as the genesis of clonal rootstocks. As it is now known, the genetics of a clonal rootstock (root sucker) is the same as that of the 'maternal tree' except for rare, but notable mutations that may occur in vegetative tissues of plants, as discussed later in this chapter. Except for apomictic seedlings, apple seedling rootstocks are highly heterogeneous, and quite often, their performance is unpredictable. This is in contrast to clonal rootstocks, as they maintain their uniform genetics and subsequent uniform field performance.

The origin of the first set of distinct clonal apple rootstocks goes back to the beginning of the last century where researchers in East Malling, England have collected clonal material that has already been circulating and sold in western Europe (mostly in France, England, Belgium, and Germany), and have conducted rigorous experiments describing morphology and performance of these rootstocks (Hatton 1917, 1919, 1920). This first set of clonal apple rootstocks comprised 16 selections (Malling I-XVI) that have exhibited several desirable nurseries and agronomic rootstock traits such as dwarfing growth habit, precocity (induction of early bearing), and ease of propagation. These rootstocks are deemed the genetic founders of clonal apple rootstocks.

6.2 Genetic Origin of the Apple Rootstock Founding Germplasm

Where did Malling rootstocks come from, and what were they related to? A major objective of East Malling scientists was to characterize apple rootstock clones that were being commercially traded in western Europe under the names of 'French Paradise', 'English Paradise', 'Jaune de Metz', and 'Doucin' in the late 1800 s and early 1900 s. This characterization was likely to be conducted in response to some confusion regarding the horticultural properties of rootstocks being sold under the same name (Webster et al. 2000).

While no molecular genetic analyses of all 16 original selections have been conducted, several studies have assessed genetic relationships among some of these selections. Genetic fingerprinting using seven microsatellite (or simple sequence repeats [SSRs]) loci of the apple rootstock germplasm, including Malling (M) 1, 4, 7, 8, 9, and 13, as well as other apple rootstock and scion cultivars, have revealed closer relationships between M.8 and 9 than with M.4, 1, and 13, although all are found to belong to the same group as that of the apple cultivar 'Cox's Orange Pippin', with M.1 deemed as the most closely related to 'Cox's Orange Pippin' (Oraguzie et al. 2005). The placement order of these initial rootstock selections within the apple germplasm has been further confirmed using 83 SSR markers distributed across all 17 apple chromosomes indicating that these foundation rootstocks, in fact, belong to the domesticated apple germplasm (Fig. 6.1).

While acknowledging the role of the Persian region as an important conduit for the domestication of apples through the Silk Road routes, a Persian origin of the 'Paradise' ('Jaune de Metz') stocks is not ascertained in a study comparing Malling apple rootstocks to 159 Iranian landraces and accessions derived from *Malus sieversii* and *M. sylvestris* (Gharghani et al. 2010). It is observed that the founding rootstock germplasm has very limited genetic diversity, as Malling rootstocks have familial similarities. This is not surprising as these stocks have been selected for the following reason. Dwarfed fruit trees including apples have been popular in royal gardens, abbeys, and with fruit fanciers, as evidenced in writings dating back to the sixteenth century, thus rendering it more likely that the

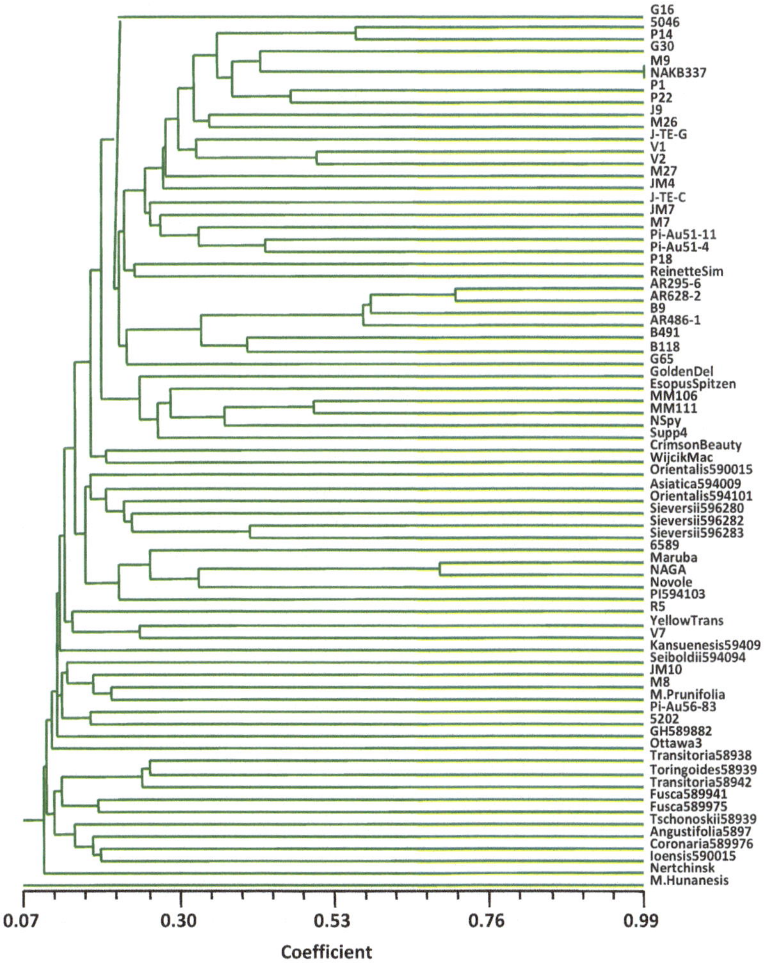

Fig. 6.1 An unweighted pair-group method with arithmetic mean (UPGMA) phenogram of genetic similarity (Jaccard's coefficient) among apple rootstocks, as well as with domesticated and wild apples

most fruitful dwarfed apples along with any resulting seedlings have exchanged hands and maintained, as the unique dwarfing growth habit is retained even with grafting (Tukey 1964). Therefore, dwarfing is the trait that initially characterized much of this founding germplasm. While this germplasm is important for the establishment of dwarfed apple trees during the twentieth century, with M.9 (and clones) being the most successful and most propagated rootstock in the world, these stocks are far from perfect for all apple-growing regions in the world. This is due to their levels of adaptability and response to various abiotic (primarily cold tolerance) and biotic (disease and insect) pressures. Therefore, new rootstocks breeding has become a priority in several countries that required enhanced stocks to address these biotic and abiotic stresses.

6.3 Breeding Apple Rootstocks

Apple-growing regions in the world have highly varied environments depending on climate, soil, water, biological factors, and management practices. These differences have highlighted the need for better suited/adapted rootstocks that maximize production in each environment. Moreover, this has spurred the launch of several breeding programs, beginning with those at each of East Malling and Malling-Merton in England, releasing rootstocks such as M.26 and Malling-Merton (MM) 106, respectively. This is followed by other breeding programs in North America, Europe, Asia, New Zealand, and South Africa that have pursued various goals and achieved several successes (Bonany et al. 2004; Carlson 1981; Casavela 1977; Cummins and Aldwinckle 1974; Fazio et al. 2015b; Fischer et al. 1993; Hutchinson 1977; Jakubowski et al. 2000; Khanizadeh et al. 2005; Marini and Fazio 2018; Mong 1991; Pieniazek 1971; Sandanayaka et al. 2003; Tamai et al. 2003; Trajkovski and Andersson 1980; Tsuchiya 1988; Wan et al. 2011; Wertheim 1998; Xiang et al. 1995).

Breeding of apple rootstocks is a lengthy process, not only because of long generation times due to juvenility ($\sim$2–7 years of vegetative growth from seed germination to flowering) but also due to the long evaluation time required to assess the performance of selections across multiple locations. A modern apple orchard's expected productive life is between 12 and 25 years. A rootstock selection should have at least two full evaluation cycles of 10–12 years at multiple locations in order to be deemed reliable as a choice rootstock candidate. Even after such lengthy trials, new and significant genotype $\times$ environment interactions may be encountered following commercial use, perhaps generating new knowledge about unforeseen traits. Such is the case with Geneva (G) 213 rootstock, bred and selected in Geneva, New York, when a new reduced chill requirement trait has surfaced when this rootstock is tested under low-chill environments in Brazil (Macedo et al. 2018), far away from the original North American breeding ground where chilling requirements are readily met.

A breeding cycle normally comprises the period from the first parental selection and hybridization to the next generation of parental selection and hybridization. For apple rootstocks, this period usually takes between 20 and 30 years as this process requires several cultural management steps commonly involved in fruit tree production. An overall breeding scheme timetable begins with sexual hybridizations of selected parents (designated as year 0). Then, in the following year (designated as year 1), seeds are extracted, stratified, and germinated. Subsequently (years 2–5), young seedlings are subjected to marker-assisted selection (MAS) for target traits, evaluation of propagation properties, screening for disease/pest resistance, and identification of desirable and promising candidates serving as first clones for use in grafting to scion apple cultivars. This is followed (years 5–7) by the production of the first grafted apple trees (of scions onto rootstock clones), and these grafted trees are then evaluated in a regular orchard (years 7–17) for various agronomic traits, including productivity, disease/pest resistance, and other pomological characteristics. From this first orchard test, selections are then transferred to multiple orchards under different environments/geographical locations (years 15–25). During years 20–30, evaluation is completed by national and international rootstock testing organizations to provide critical and independent assessments of the performance of these rootstocks. The above-described scheme of the various rootstock breeding stages is outlined in several publications (Fazio and Robinson 2019; Fazio et al. 2015b). As can be readily noted, the above timeline includes some overlaps among the different stages due to instances of identifying early selections based on field observations and marker-genotype predictions, with the availability of new relevant genomic knowledge, but this relies highly on independent 'in-field' testing by apple growers and rootstock testing projects (Cowgill et al. 2017).

The above outlined time-consuming breeding cycle is demonstrated in the development of current commercial rootstocks from early founder rootstocks such as M.9 (or M.8). For example, the rootstock G.213, the latest release from the Geneva® (New York) breeding program, is only two generations away from M.9 (generation 0), which is a parent of Ottawa 3 (generation 1), and a parent of G.213 (generation 2). Other apple rootstock breeding programs have generally followed similar timetables in developing rootstocks following two generations of breeding. It is likely that the third-generation rootstocks will become available in the next few years.

In recent years, apple rootstock breeding has become more involved, compared to 20–50 years ago, due to our expanded knowledge of component traits critical for the production of high-quality apples. Among these component traits for dwarfing, early bearing induction, and increased sylleptic branching are extrinsic properties that are either imparted or induced on a grafted scion. Among other component traits, root tolerance to replant disease, resistance to woolly apple aphids, root morphology, and propensity for vegetative propagation are intrinsic properties to the rootstock itself as these are not measured on the scion but are directly measured on the rootstock. Furthermore, there are some traits that may overlap, and they may be either conferred or observed on both the rootstock and the scion, such as resistance to fire blight disease and nutrient foraging. A discussion of the genetics and genomics of apple rootstocks will follow these target traits along with their component trait distinctions.

6.4 Genetics of Extrinsic Traits Imparted to Scions or Whole Trees

6.4.1 Dwarfing

While some genetically dwarf apple rootstocks induce dwarfing phenotypes on grafted scions, others do not, thereby allowing grafted scions to grow into standard-sized trees. It is important to point out that dwarfing apple rootstocks are also somewhat dwarfed when observed as non-grafted trees. Such a phenotype is not related to lengths of internodes, but rather to the cumulative numbers of developing internodes per year, thus suggesting an early cessation of apical meristematic growth during a growing season.

A bulked-segregant analysis (BSA) for a progeny of M.9 along with random amplified polymorphic DNA (RAPD), and subsequently SSR markers, were used in the identification of the first locus for dwarfing, *Dw1*, located on chromosome 5 (Pilcher et al. 2008; Rusholme et al. 2004). A quantitative trait locus (QTL) analysis of two segregating populations using first-generation derivatives of M.9 and M.8, Ottawa 3 (O.3) and Budagovsky 9 (Bud9 or B.9) as parents has revealed that the majority of observed variations for the dwarfing phenotype can be explained by interactions of two loci, *Dw1* and a newly identified *Dw2*, located on chromosome 11 (Fazio et al. 2014b). This two-locus model is later confirmed in another QTL analysis identifying these same loci along with additional minor effectors on chromosomes 6, 9, 10, and 12 (Foster et al. 2015). Furthermore, the presence of molecular markers linked to *Dw1* and *Dw2* in a set of dwarfing rootstocks, derived from multiple breeding programs, has also been confirmed, thus asserting the importance of the founders Malling dwarfing germplasm (Foster et al. 2015). Interestingly, markers linked to *Dw1* have also been identified in a syntenic region in pear conferring some level of dwarfing (Knabel et al. 2015). Subsequently, yet another dwarfing locus, *Dw3*, has been identified on chromosome 13, and added to the *Dw1/Dw2* model, using tree size and bark thickness as parameters to investigate dwarfing in a population derived from M.27 (a super-dwarfing phenotype derived from M.9 × M.13) and M.116 (a semi-vigorous phenotype derived from MM.106 × M.27) (Harrison et al. 2016). This backcross population, while seemingly problematic due to inbreeding (shared *S*-alleles likely eliminated half of the potential gametes), has provided some level of support to earlier observations that bark thickness in some dwarfing rootstocks can be used as an early phenotypic selection tool for dwarfing (Beakbane et al. 1973; Elfving et al. 1993).

Levels of vigor control imparted by a rootstock are likely to be influenced by various loci associated with root interactions with soil, pH, biology (a rootstock in replant soil may be dwarfed due to interactions with replant biological factors), and water availability that are likely to be influenced by factors other than *Dw1/Dw2/Dw3*, as observed in several multi-location field trials (Autio et al. 2017; Cowgill et al. 2017; Marini et al. 2013, 2017). Regardless of these interactions, these dwarfing loci are reported to have pleiotropic effects on other horticultural traits, including leaf and fruit nutrition (Fazio et al. 2017). While other sources of dwarfing factors may have been identified in germplasm unrelated to the original Malling selections, there is very limited published information on the genetic inheritance of these other factors to date (Gao et al. 2011; Han et al. 2011; Jin et al. 2012; Rong et al. 2011; Yang et al. 2011; Zhang et al. 2011).

6.4.2 Early Bearing

Early bearing, also referred to as precocity induction, is the ability of a rootstock to overcome long juvenility periods and promote the early development of flowers in scions of young grafted trees. This is in contrast to the standard (wild-type) long juvenility period observed in most apple trees. Rootstock induction of early bearing (Eb) influences partitioning of resources to fruit production instead of vegetative growth; therefore, it is related to dwarfing. This observation is supported by the co-location of *Eb1* and *Eb2* loci for the early bearing trait with QTLs associated with the dwarfing loci *Dw1* and *Dw2* (Fazio et al. 2014b). Furthermore, a third factor *Eb3*, located on chromosome 16, has also been identified that likely modulates the number of fruit per tree in the first bearing season (Fazio et al. 2014b). The relatedness between early bearing and dwarfing is found to be significant, wherein flower clusters counted in the first bearing season have an effect on tree diameter measured two seasons later, thus indicating that rootstock genotypes inducing early flowering in trees display overall significant negative effects on tree growth (Fig. 6.2). Moreover, the diversion of resources from vegetative growth to fruit production, particularly when trees are young, has long-lasting implications on the overall growth potential of such trees.

6.4.3 Nutrient Uptake and Translocation to Scions

The phenotypic effects of nutrient uptake by an apple rootstock on a scion, as well as the potential for new rootstock breeding for varying levels of nutrient uptake and translocation have been known for several decades (Kennedy et al. 1980). The first report on the genetic underpinnings of this set of traits has indicated that there is a complex landscape of major and minor effect QTLs related to scion leaf content for potassium (K), sodium (Na), phosphorous (P), calcium (Ca), copper (Cu), sulfur (S), zinc (Zn), magnesium (Mg), nickel (Ni), and molybdenum (Mo) (Fazio et al. 2013). In a subsequent study wherein mineral nutrient contents in fruits and leaves of field-planted 'Brookfield Gala' trees grafted onto a replicated full-sibling rootstock

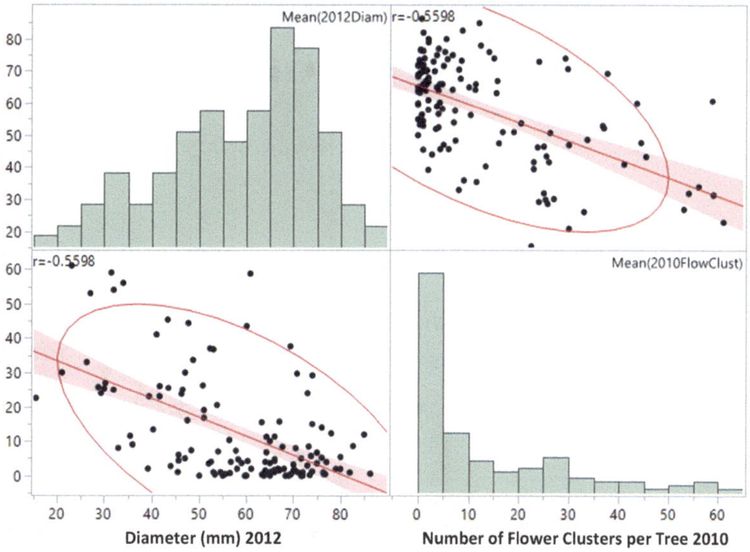

Fig. 6.2 Scatterplot matrices showing mean tree early flowering (Season 2) and mean tree diameter (Season 4) data for a rootstock breeding population segregating for *Dw1* and *Dw2* that was grafted with 'Gala' scion. Early bearing rootstocks are those bearing more than 5 flower clusters per tree

population, measured for two consecutive growing seasons, have found significant genotypic mean correlations between years in pairs of same mineral concentration values and between fruit and leaf values for the same mineral per year data (Fazio, Pers. comm). Furthermore, QTL analyses have found significant effects linked to loci for leaf K values on chromosomes 5 and 11, and leaf P values on chromosomes 5, 11, and 16. Moreover, minor QTLs have been identified in single-year data for Ca, Cu, Mg, Na, S, Zn, and N, wherein some of these QTLs are found to be co-located, thus suggesting the presence of either similar pathways or effector loci for several nutrients (Fig. 6.3). In addition, significant correlations are detected between soluble solids and fruit Ca (0.35 $p = 0.0001$), soluble solids and fruit Cu (0.4226, $p < 0.0001$), red peel color, and fruit Mg (0.29, $p = 0.0017$), and fruit S with firmness (0.26, $p = 0.0051$). While the inheritance of nutrient-related traits is quite complex and dependent on environmental conditions, it is still highly relevant to fruit tree performance and fruit quality, thus it should be deemed as a prime target for marker-assisted breeding (MAB) (Fazio et al. 2018a, b, 2020b).

Variations in the fertility of the soil environment, differences in pH, differential availability of water, and interactions with soil biological factors all contribute to the error term when measuring nutrient levels in scion tissues (Fazio et al. 2015a). Therefore, one of the strategies to overcome these variations is to conduct experiments aimed at measuring rootstock nutrient uptake using uniform either sterile soil, soil-less media, or hydroponics (or yet another form, aeroponics). Recently, aeroponics has been successfully used to assess apple rootstock genotype × pH interactions relevant to nutrient uptake measured in leaves, revealing yet another dimension to the nutrient genetics landscape (Al Farqani 2019). For now, rootstock breeding has focused on targets such as increased Ca or enhanced K/Ca balance that contributes to decreased sensitivity to bitter-pit (Fazio et al. 2018b, 2020b).

6.4.4 Genetics of Water Use Efficiency and Drought Tolerance in Apple

Water use efficiency (WUE) and drought tolerance are related to whole-tree traits, wherein both the rootstock and the scion contribute to the achievement of a specific status. Thus, different scion/rootstock combinations can yield different outcomes for these two traits. Often, WUE is the

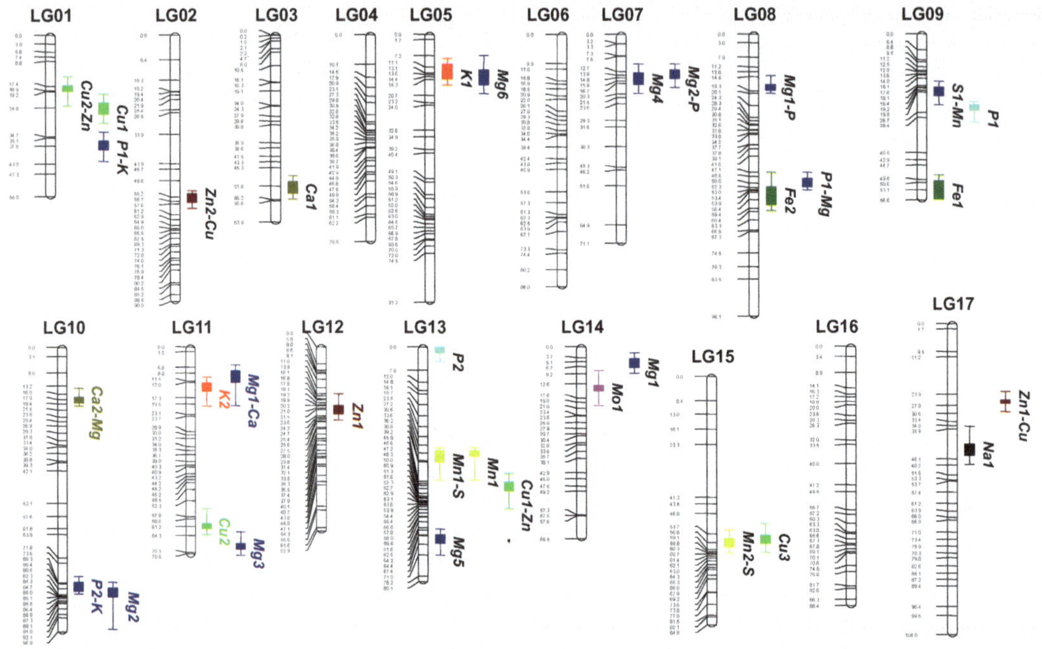

Fig. 6.3 A consensus genetic map of the O3R5 population highlighting locations of QTLs for leaf nutrient concentrations with peak LOD > 3.2. QTLs are named according to the nutrient symbol and are numbered sequentially in decreasing order of significance (e.g., K1, K2). QTLs obtained with covariate QTL analysis are labeled with the covariate written right after the dash symbol (e.g., P2-K)

ratio of photosynthesized carbon to units of transpired water, and it is influenced by water uptake potential in the roots (related to root morphology, vessel composition, water sensing apparatus, and soil exploration index, among others) and transpiration efficiency in the scion (related to leaf morphology, sun exposure, and canopy architecture, among others) (Fernandez et al. 1997; Tworkoski et al. 2016; Zhang et al. 2014). WUE is determined by either calculating units of seasonal dry matter increase/units of water (H_2O) or by measuring carbon dioxide (CO_2), oxygen (O_2), and H_2O flux of tree canopies during short intervals (Glenn 2014).

Wild apple trees are found thriving in very disparate environments with regards to water availability (temporal and quantity). It is likely that the process of natural selection has shifted allele frequencies in contrasting environments (river valleys vs. high plains) as reported in studies, wherein WUE physiological and morphological components are compared in both domesticated and related wild species of apples (Bassett et al. 2011; Maguylo and Bassett 2014). Interestingly, new expressed sequence tags (ESTs) related to water-deficit response have been identified in apple roots and may provide opportune targets for selection (Bassett et al. 2014). Furthermore, carbon isotope discrimination ($\Delta^{13}C$) (synonymous with WUE) measurements are found to be associated with three QTLs, located on linkage group (LG)8, LG15, and LG16, in a 'Honeycrisp' × 'Qinguan' breeding population (Wang et al. 2018). It has also been found that 'Qinguan', a Chinese apple cultivar, exhibits upregulation of a protein associated with WUE (MdAGo4.1—*Malus* argonaute protein) during water stress (Zhou et al. 2016). In another study, two genes, *MhYTP1* and *MhYTP2*, are identified in a wild *Malus* species, *M. hupehensis*, often used as an apomictic rootstock exhibiting differential expression in multiple tissues of plants subjected to either well-watered or water-stressed (six-day period) treatments (Wang et al. 2017). It has been proposed that these changes in gene expression may be associated with DNA methylation patterns observed when comparing drought-sensitive and drought-tolerant apple cultivars in a water-deficit environment (Xu et al. 2018).

Therefore, the key to developing drought-tolerant and WUE apple trees that maintain productivity in a water-challenged environment is highly dependent on the use of high-throughput phenotyping that can detect real-time changes in water stress status in field trees that feature a combined scion/rootstock genetic contribution to this trait complex (Virlet et al. 2015).

6.4.5 Phytohormone Flux

There are multiple ways by which a rootstock can influence levels of phytohormones in a scion (and vice versa), as these are synthesized and produced in different tissues (cytokinins in roots vs. auxins in areal organs), under different environmental conditions (abscisic acid [ABA] under water deficit), or at different times during a growing season. This complexity is compounded by the fact that phytohormones interact with each other to generate diverse phenotypic outcomes influencing various plant functions such as dormancy, flowering, vegetative growth, tree architecture, and photosynthesis.

An analysis of endogenous phytohormones in apple rootstocks has demonstrated heritable differences as more diverse apple rootstocks have become available (Adams et al. 2018; Tworkoski and Fazio 2016; Tworkoski and Miller 2007). It has been reported that endogenous production of phytohormones in a rootstock may be a factor in changes in hormone levels detected in a grafted scion (Lordan et al. 2017; van Hooijdonk et al. 2011). However, these changes may also be due to other rootstock-mediated phenotypes such as water status, crop load, nutrient status, and long-distance gene expression modifications, among others (Samuoliene et al. 2019).

Breeding of rootstocks for hormone flux modifications in scions requires laboratory-intensive (and expensive) methods that cannot yet be applied in a high-throughput phenotyping platform necessary to measure each of the target phytohormones in a breeding population. Therefore, an alternative approach has been proposed, relying on tracking gene expression levels of phytohormone-related genes to indirectly monitor these target phytohormone traits (Chen et al. 2020; Feng et al. 2019; Foster et al. 2018; Hao et al. 2019; Li et al. 2016; Song et al. 2016), and then identify differentially expressed genes as candidates for both MAB and genomic selection (GS). In grafted systems like apple trees, phytohormones acting as communication intermediaries between distant tree tissues (e.g., cytokinin produced in root-tips affecting scion tissue development) can be interpreted differently depending on the specific genetic composition of the scion related to hormone signaling. Thus, while high levels of cytokinin have been detected in 'Honeycrisp' scion grafted onto a G.11 rootstock, a different scion cultivar grafted onto the same rootstock may display different levels due to a scion's inherent cytokinin signaling apparatus (Lordan et al. 2017). Therefore, breeding of rootstocks for hormone flux modification of scions should be associated with specific desirable scion phenotypes that can be modified by rootstocks, and that are not easily amenable to available solutions that involve treatments with exogenous phytohormones.

6.4.6 Sylleptic Branching

The ability of apple rootstocks to modify the number of axillary shoots within the same season's growth has been observed in millions of nursery trees grafted with the same scion cultivar (Fazio and Robinson 2008a, b). Within comparable rootstock vigor classes, some rootstocks tend to have the capacity to develop higher numbers of axillary buds that grow into shoots during nursery tree development. This observed tendency to induce more branches or feathers in nursery trees is desirable as most apple trees are graded by the number of developing branches when sold by nurseries, and they are valued by apple growers due to the long-term implications of planting well-feathered young trees on the economic potential of an orchard (Hooijdonk et al. 2015; Reig et al. 2019).

The genetic inheritance of this axillary branching/feathering trait is quite complex as individual component traits inducing growth of vegetative axillary buds are likely related to water status, hormone signaling, nutrient availability, and a scion's inherent bud dormancy apparatus (Foster et al. 2018; Neri et al. 2001; Vanderzande et al. 2016). The legacy vigor controlling apple rootstocks such as M.9 and B.9 does not promote good sylleptic branching in nursery trees (van Hooijdonk et al. 2011); however, most of the novel genetic variation for sylleptic branching induction has been reported in families that have 'Robusta 5' (*M.* × *robusta* clone 5) in their ancestries such as G.935, G.41, and G.213 (Bartish et al. 1999). Genetic variation for sylleptic branching has been observed in the architecture of own-rooted apple seedlings derived from half-sib populations of *M. sieversii* mother plants collected from various regions west of the Tian-Shan mountain range (China), thus indicating the presence of wide genetic diversity for this trait in the wild apple germplasm, which is highly useful for breeding for sylleptic branching in rootstocks (Fazio et al. 2014a).

Rootstocks that promote sylleptic and proleptic branch formation can have major influences on fruit-bearing branch renewal, as this is an important component of high-density apple orchards wherein a highly rigorous pruning scheme is used (Lordan et al. 2018; Reig et al. 2020; Robinson et al. 2007). Often, this pruning scheme involves the removal of branches that are 2.5 cm in diameter at the base to promote the development of small fruiting branches to increase the productivity and fruit quality of apple trees (Lordan et al. 2018; Reig et al. 2020; Robinson et al. 2007).

6.4.7 Endodormancy and Chilling Requirement

Apple tree buds require a period of winter cold temperatures above freezing and below 7°C followed by a period of warm temperatures in order to break dormancy and resume growth. Endodormancy refers to the period of cold (chilling) temperature requirement, and the chilling requirement is the accumulated number of hours specific to each cultivar needed for release from dormancy (Cook and Jacobs 1999; Rufato et al. 2010). Failure to achieve this requirement causes a lack of uniform flowering and loss of productivity even with the best applications of exogenous bud-break promoting chemical compounds or cultural practices (Hawerroth et al. 2009).

Genetic variation for rootstock induced reduction of the chilling requirement has been characterized in the Malling and Geneva germplasm series, wherein M.7 and G.213 have been shown to consistently promote uniform budbreak of high-chill cultivars Fuji and Gala in a low-chill environment (Couvillon et al. 1984; Macedo et al. 2018; Young and Werner 1984a, b). This effect is quantitatively inherited, and has been evaluated in breeding populations in Geneva, NY (Fazio, Pers. comm.). It is important to note that as climate change transforms some of the current apple-growing regions from high–medium chill to low chill, apple rootstocks will play an important role in maintaining the productivity of medium–high chill cultivars in these regions.

6.4.8 Cold Tolerance

Similar to other temperate tree fruit crops, apple trees can display sensitivity to cold stress, whereby apple rootstocks have been demonstrated to serve as components in either tolerance or sensitivity to various cold events.

There are three forms of cold injury that apple rootstocks have shown sensitivity to, depending on the timing of these cold events. Early winter/late fall cold events tend to injure trees that have not yet achieved full hardiness status and may still be actively growing. Mid-winter events can occur when apple trees are subjected to short durations of warm temperatures that deharden some tissues of trees, and then are subjected to severe cold temperatures within a matter of either hours or days after de-hardening; moreover, trees can be subjected to temperatures below −30°C that can damage most above-ground apple tissues even for cold, hardy rootstocks. Furthermore, apple root tissues can be severely damaged by soil temperatures below −10° C that can occur when there is no snow cover, primarily to the segment of the rootstock that is exposed above-ground (referred to as the rootstock shank), as well as to the segment just below the soil surface that can also be injured, thus resulting in slow tree death that may take a whole season before it can be noticed (Embree and Mcrae1991; Prive et al. 2001). Finally, early de-hardening in the springtime and early bud-break can damage newly developing tender tissues if subjected to drastic cold temperature events.

It has been reported that genetic variation for late fall/early winter cold temperature events is present in the Geneva rootstock breeding program (Moran et al. 2018). Genotypic variation for mid-winter cold hardiness has also been observed in rootstock germplasm, wherein M.7 is reported to be quite sensitive to winter damage at moderately cold temperatures of −20°C. Moreover, winter cold hardiness has been investigated in several rootstocks, wherein M.7 is reported to be very tender. However, M.2, M.4, M.9, MM.106, and P.16 (a selection from the Poland [P]-series) are deemed as tender, whereas M.26, MM.111, MM.104, P.1, and J.9 (a selection from the Jork series) as moderately hardy, and Antonovka seedling, A.2 (a selection from the Alnarp [A] series), Beautiful Arcade, O.3 (a selection from the Ottawa [O] series), O.8, B.9, P.2, P.22, and P.18 as hardy (Quamme 1990; Quamme et al. 1997). Mid-winter survival under severe cold conditions such as those prevailing in Poland, Canada, or some regions of the northern U.S. provides some additional estimates of a rootstock's genetic potential to tolerate extremely cold temperature events that have classified M.9 as more sensitive than either M.26 or B.9 (Czynczyk and Zagaja 1984), while O.3, V.1 (a selection from the Vineland [V] series in Ontario, Canada), V.3, G.16, G.30, and MARK (an open-pollinated seedling of M.9 introduced by Michigan State University) as less sensitive (survival > 90%) than those of B.118, M.9T337 (a certified virus-free clone of M.9 selected at East Malling), B.9, M.9, Nic 29 (another M.9 selection identified in Belgium), Supporter 4 (a selection from a cross between M.9 and M.4), M.26, and MM.111 (survival ~ 50%), or those of M.7 and MM.106 (survival < 25%) (Robinson et al. 2006). In other studies using controlled freezing of non-grafted rootstocks, it is found that G.41, G.11, G.30, B.9, P.2, and M.26 have similar levels of cold hardiness, while G.935 has a higher level of root cold hardiness than that of M.26 (Moran et al. 2011a, b). Recently, Moran et al. (2018) have evaluated fall, winter, and spring hardiness of rootstocks developed from the Vineland and Geneva breeding programs and have noted genotypic variations for all cold temperature events.

Genetic mechanisms for apple rootstock tolerance to cold temperatures have been investigated, and these likely involve dehydrin-related proteins (Du et al. 2015) and MYB genes (Wu et al. 2018). Moreover, ESTs of apple tissues exposed to cold stress have also been identified (Wisniewski et al. 2008). While some of these studies have provided some insights into the diverse mechanisms for apple cold tolerance, there is paucity with regards to the genetic inheritance of such factors in a breeding population of apple rootstocks. For additional information, please see the chapter on cold hardiness in this book.

6.4.9 Graft Union Strength

Ideally, joining two genetically distinct individuals by grafting implies that at the physical interface between the two genotypes there is cell-to-cell and tissue-to-tissue adhesion, interlacing, as well as uniform development (Tworkoski and Fazio 2011). Sometimes, an ideal situation occurs resulting in the development of a continuum of lignified vascular tissue spanning both genotypes as noted in Fig. 6.4. However, some genotype combinations result in a chaotic

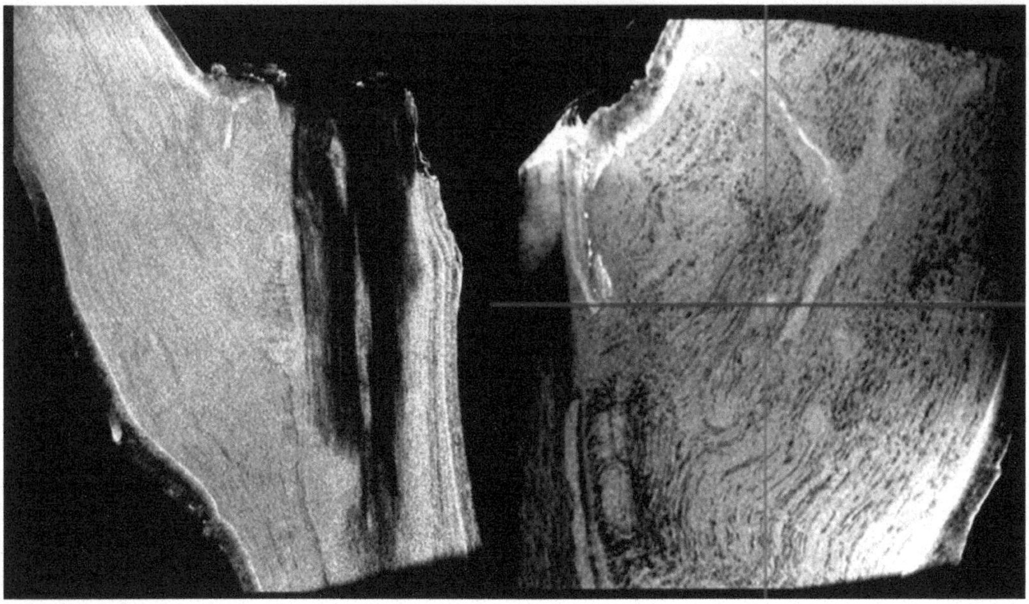

Fig. 6.4 Graft unions of two different scions, cv. 'Fuji' (on the left) and cv. 'Scilate' (on the right) grafted onto a G.41 apple rootstock as visualized by micro X-ray tomography (BRC Imaging Facility, Cornell University). The 'Fuji'/G.41 combination shows an organized development of vascular tissue between the two genotypes signifying a stronger graft union. In contrast, a weaker 'Scilate'/G.41 combination displays a disorganized vascular vessel arrangement that is interspersed with weak pockets of parenchyma cells

arrangement of vascular tissues within the region around the union, sometimes interspersed by low strength parenchyma cells (Fig. 6.4), thus producing a weak graft union (Adams et al. 2017; Basedow and Crassweler 2017). Such weak graft unions can result in breakage and tree death under severe wind conditions, as well as when a trellis support system is not properly installed (Wallis et al. 2017).

It is not unusual to observe that related rootstocks may behave differently when grafted with the same scion cultivar, thus suggesting the presence of a genetic component to this graft union compatibility/incompatibility. For example, a combination of G.214 rootstock and 'Honeycrisp' scion cultivar displays a very strong graft union when compared to that of its half-sib G.41 rootstock with 'Honeycrisp' (Fazio et al. 2020a). The underlying genetic factors for lack of graft union strength or graft compatibility are not understood as the graft union is the conduit for many hormonal and metabolic compounds that may be differentially processed by the adjoining genotypes (Zhou et al. 2020). However, exogenous applications of auxin compounds or compounds involved in auxin flux are reported to "normalize" uniformity and strength of a graft union (Adams et al. 2017). As graft union strength or graft compatibility trait is the result of interactions between two different genotypes, it is possible to breed for and select for optimal graft compatibility combinations in both scions and rootstocks.

6.4.10 Biennial Bearing

Alternate or biennial bearing is the tendency of a scion cultivar to set a heavy fruit crop in one season, but then a low fruit crop for the following season, due to poor/low flowering (return bloom). It has been proposed that this phenomenon hinges on carbohydrate balance and hormone sensing by developing meristems (Racsko and Miller 2011; Veal et al. 2011). Furthermore, QTLs have been identified that explain some of the observed genetic variations for biennial bearing in scions (Durand et al. 2017; Guitton et al. 2012). While genetic determinism in scions is prevalent, apple rootstocks also seem to have strong and significant effects on alternate bearing of scions that are likely to exhibit such a biennial bearing habit (Kviklys and Samuoliene 2020; Lordan et al. 2017, 2019; Reig et al. 2019; Serra et al. 2016).

It is observed that rootstock-modulated crop load in bearing years is likely to be correlated with alternate bearing. While initial reports have indicated that the more dwarfing rootstocks reduce the likelihood of scion cultivars going into biennial bearing (Barritt et al. 1997), successive replicated field trials have demonstrated that even dwarfing rootstocks, such as B.9, could display severe biennial tendencies in scion cultivars (Lordan et al. 2019; Reig et al. 2019). Although genetic variability in apple rootstocks for the biennial bearing trait has been investigated, breeding and selection of rootstocks that would decrease alternate bearing tendencies of scion cultivars are complex. In fact, this complex trait would require high numbers of field replications, thus limiting for QTL discovery; moreover, this trait can only be evaluated when other breeding requirements have been met.

6.5 Genetics of Intrinsic Traits to Apple Rootstocks

6.5.1 Propagation

With the exception of seedling and apomictic rootstocks (Ferree 1998; Inoue et al. 2002; Ma et al. 2012; Wang et al. 2019), the majority of apple rootstocks are propagated vegetatively by either layering, softwood cuttings, or micropropagation (Musacchi and Neri 2019). As to be expected, apple rootstocks that are obligate apomicts cannot be improved by conventional genetics and breeding approaches, as any hybridization efforts would result in the regeneration of the mother plant phenotype (Olien 1987). Interestingly, some apomictic apple rootstocks are only partial apomicts, and these can be identified by a series of phenotypic and molecular methods such as leaf color and morphology or by molecular markers, including SSRs and single nucleotide polymorphisms (SNPs), that can discern hybridized vs. apomictic plants (Bisognin et al. 2009; Ma et al. 2012; Sha et al. 2011). However, early research on apomictic breeding has revealed that frequencies of apomixis in hybrid progeny are much lower than those of their apomictic parents (Cummins et al. 1983; Schmidt et al. 1974). A gene encoding for fertilization-independent endosperm (FIF) and involved in reproductive tissue development in *M. hupehensis*, *MhFIF*, is found to be associated with the display of apomixis in apple rootstocks (Liu et al. 2012a,b). However, it remains to be seen if significant progress can be made in the development of yield-efficient and disease-resistant apple rootstocks using apomictic genotypes.

As for conventional vegetative propagation methods such as cuttings and micro-cuttings, adventitious rooting and induction of axillary shoots are two of the most important traits of interest. It has been found that an adventitious root formation (*ARRO-1*) gene in apple rootstock M.26 is associated with the formation of new roots (Smolka et al. 2009). Using a proteomic approach to compare indole-3-butyric acid (IBA)-induced and non-induced control tissues of the M.9-T337 rootstock has revealed the presence of 3,335 differentially expressed proteins during a 72 h period following induction, thus setting up a framework for pursuing research studies in rooting (Lei et al. 2018).

While most apple cuttings are obtained using softwood (same year growth), the vigorous *M. prunifolia* rootstock 'Marubakaido' has shown a very good propensity for hardwood cutting propagation. This trait is inherited by dwarfing rootstock 'JM7' ('M.9' × 'Marubakaido'). By using 120 seedling progeny of 'JM7', two QTLs for hardwood cutting propensity have been detected on linkage groups 13 and 17 (Moriya et al. 2015). Moreover, it has also been proposed that several auxin-related genes co-located with a QTL on linkage group 17, and that these genes serve as possible candidates for effectors for this hardwood cutting propagation trait (Moriya et al. 2015).

Fig. 6.5 Burr knot formation on an M.9 rootstock around the region of the graft union

It is interesting to point out that both M.9 and M.26 tend to develop aerial root formations, referred to as burr knots (Fig. 6.5). These burr knots would be deemed desirable for vegetative propagation of rootstocks; however, they have negative effects in the orchard as they serve as entry points for insects followed by fungal and bacterial pathogens (Bergh and Leskey 2003). A phenotypic assessment of domesticated and wild apples has revealed wide variations in both suckering (adventitious shoots developed on roots) and burr knot formation, thus suggesting the presence of underlying genetic variations for both traits (Kumar et al. 2018). Furthermore, it has been observed that burr knot formation in scions can be influenced by rootstocks (Marini et al. 2003). Rootstocks released by the Geneva breeding program have demonstrated that it is possible to breed against both traits, suckering and burr knot development, while maintaining acceptable rooting in layering beds and stool cuttings using exogenous plant growth regulator treatments (Adams 2010). Efforts to breed rootstocks for in vitro micro-cutting propagation have been challenging due to variabilities in culture media, as well as in growing culture conditions.

6.5.2 Resistance to Fire Blight

Fire blight, caused by the anaerobic gram-negative bacterium *Erwinia amylovora*, is one of the most devastating diseases that affect apple rootstocks, as it can result in economic losses upwards of $100 million per year (Busdieker-Jesse et al. 2016). The rootstock phase of fire blight is very damaging as entire orchards may have to be removed because susceptible rootstocks, such as M.9, M.27, M.26, and most other commercial rootstocks, can be girdled near the graft union, eventually leading to tree decline and death. Current disease control methods include treatments with copper compounds, antibiotics, and competitive biological agents; however, it is best to develop genetic resistance against fire blight.

Genetic resistance to *E. amylovora* has been identified in several wild apple species, and these have been utilized to breed for a new series of fire-blight-resistant rootstocks and scions (Emeriewen et al. 2020; Forsline et al. 2002; Kostick et al. 2021; Peil et al. 2020; Tegtmeier et al. 2020). It has been reported that there are two forms of resistance present in apple rootstocks. In one form, multigenic resistance has been identified in *M.* × *robusta* clone 5, commonly referred to as 'Robusta 5', and in particular accessions of *M. fusca,* wherein all plant tissues are likely to be either resistant or immune to *E. amylovora* (Aldwinckle et al. 1974; Cummins and Aldwinckle 1974; Gardiner et al. 2012; Peil et al. 2019). In another form, ontogenic resistance has been identified in the apple rootstock B.9, wherein one-year-old shoots demonstrate high levels of susceptibility to fire blight, but then as they grow older (two-year-old and older), these tissues are found to be resistant to *E. amylovora* (Russo et al. 2008a,b). Although inheritance of ontogenic fire blight resistance has not yet been investigated, multiple major and minor genetic factors have been identified in *M.* × *robusta* derivatives characterized by strain-specific reactions (Fazio et al. 2008; Gardiner et al. 2012), resulting in isolation and testing of an NBS-LRR type resistance gene, *FB_MR5,* found on chromosome 3 using transgenic approaches (Broggini et al. 2013, 2014a, b; Fahrentrapp et al. 2013; Kost et al. 2015). Moreover, another major fire blight resistance factor, a QTL, also identified in an apple rootstock derived from 'Robusta 5', is found on chromosome 7, but this does not seem to be affected by strain specificity (Gardiner et al. 2012).

Thus far, only the commercial rootstocks intentionally bred to provide some level of protection against fire blight have been developed by the Geneva breeding program, including G.65, G.11, G.16, G.30, G.202, G.41, G.935, G.214, G.814, G.213, G.969, G.890, G.222, and G.210. These rootstocks have been developed using field (fully-grown grafted trees) and greenhouse inoculation techniques that feature multiple bacterial strains. This approach is more efficient in selecting high-resistant plant material than molecular markers as the latter would require multiple assays to address the multigenic nature of this resistance (Fazio et al. 2015b; Russo et al. 2007).

Furthermore, indirect effects of resistant rootstocks on tolerance of scion cultivars have been evaluated, and it is reported that this is likely to be mediated by complex long-distance mechanisms that transform global gene expression of grafted scions (Jensen et al. 2003, 2010, 2012).

6.5.3 Rootstock Tolerance to Replant Disease and Crown Gall

Apple replant disease (ARD) is observed when young trees of sensitive apple rootstocks are planted in an orchard ground previously planted with trees of either the same or related genus, thus allowing for the buildup of strong biological components, including saprophytic fungi, oomycetes, bacterial, and nematode factors (Mazzola 1998). Symptoms of ARD include tree stunting, poor root growth, and loss of productivity (Auvil et al. 2011). Genetic tolerance to this complex disease has been identified in apple germplasm, including rootstocks (Isutsa and Merwin 2000; White and Tustin 2000). Natural tolerance is quite complex as multiple causative agents identified in the etiology of this disease may have specific gene-for-gene interactions with the host, competitive interactions with host pseudo-symbionts, or interactions with either root morphology or metabolic compounds (Atucha et al. 2013; Emmett et al. 2014; Gu and Mazzola 2003; Leisso et al. 2017; Rumberger et al. 2007; Yao et al. 2006). It is reported that tolerant rootstocks are capable of modifying the composition of negative replant factors and the rhizo-microbial community at large, thereby resulting in either absence or reduction of symptomatology in soils planted with these genotypes (Deakin et al. 2019; St. Laurent et al. 2010).

Transcriptomic analysis of the 'Robusta 5'-derived tolerance in the G.935 apple rootstock has revealed the presence of several major factors involved in defense activation compared to that of the replant sensitive B.9 when challenged by the replant component *Pythium ultimum* (Shin et al. 2016; Zhu and Saltzgiver 2020). Similarly, phytoalexin biosynthesis and defense gene activation have been observed when comparing replant bioactive soils to sterilized soils in four apple genotypes (Reim et al. 2020). Among the upregulated genes, four genes are found to be significantly upregulated in roots and are likely to be related to host plant response to ARD. These genes include *MNL2* (putative mannosidase), *ALF5* (multi-antimicrobial extrusion protein), *UGT73B4* (uridine diphosphate (UDP)-glycosyltransferase 73B4), and *ECHI* (chitin-binding) (Reim et al. 2020). Moreover, another six highly upregulated genes belong to the phytoalexin biosynthesis pathway, as their genotype-specific gene expression pattern is found to be consistent with the phytoalexin content in roots (Reim et al. 2020).

The complex genetic relationships expected with ARD and the difficulty in phenotyping large breeding populations render this disease as a prime target for molecular marker development for breeding new tolerant rootstocks (Fazio and Mazzola 2004; Mazzola and Manici 2012). Interestingly, Reim et al. (2020) have reported that biphenyl synthase (*BIS*) genes are useful as early biomarkers for ARD as expression patterns of these genes have correlated well with phenotypic reactions of investigated *Malus* genotypes. Moreover, inoculation with *Pythium* and *Rhizoctonia* spp. of clean roots derived from sterile micropropagation of 'Robusta 5' derivatives has identified QTLs playing roles in the observed tolerance derived from 'Robusta 5' (Zhu and Saltzgiver 2020; Fazio, Pers. comm.). This latter finding seems to hold up in many different world apple-growing regions (Costa and Stassen 2011; Macedo et al. 2019; Manici et al. 2013; Robinson et al. 2012; Spornberger et al. 2020) indicating the presence of common etiological and tolerance components that can be leveraged for new rootstock development.

Crown and root galls are caused by strains of *Agrobacterium tumefaciens* in some sensitive apple rootstocks. Severe infections can cause significant losses to nursery apple tree production and some yield losses in an orchard setting (Scheer 1978; Drahorad 1978). Inoculation tests performed with three strains of *A. tumefaciens* on M.9, M.26, M.7, MM.111, and MARK apple rootstocks indicate that they are all susceptible, although at different levels of severity and in different tissues (Stover and Walsh 1998). Efforts to develop resistant rootstocks in Japan have resulted in the identification of resistance in *M. sieboldii* (Sanashi 63), which was hybridized with apple rootstock JM7. QTL analysis of a full-sib progeny from such a cross yielded a major QTL from *M. sieboldii* found on linkage group (LG) 02, accounting for 75–85% of the variation (Moriya et al. 2008, 2010).

6.5.4 Root Morphology

Root morphology can have a major influence on apple tree performance, ranging from water and nutrient foraging to tree anchorage. Phenotypic and genetic variations of root morphology traits have been observed in the wild apple, *M. sieversii*, and in apple rootstock breeding populations (Fazio et al. 2014a; Fazio and Volder 2009). Variations in fine-root development, root vigor, volume explored by roots, root gravitropism, and root longevity have all been explored in apple roots using either rhizotrons or aeroponic systems (Al Farqani 2019; An et al. 2017; Atucha et al. 2013; Emmett et al. 2014). The development of useful markers for these difficult-to-phenotype traits is dependent on reliable description of the traits in breeding and mapping populations over several seasons and under different growing conditions. Gene expression analysis can be leveraged to discover differentially expressed genes (DEGs) between diametrically contrasting phenotypes, which in turn can reveal candidate genes associated with observed phenotypic variations (Fig. 6.6).

6.5.5 Interactions with Soil Type, pH, and Salinity

Apple root systems respond to various physical and chemical properties of the soil depending on their genetic compositions. While there seems to be relative paucity of reports on genetic variations of interactions with soil types, preliminary experiments conducted by Fazio et al. (2012) have detected differential responses of 33 apple rootstock genotypes to soil type and pH for growth and nutrient variables measured over two soil types and five pH levels (Fig. 6.7). These findings have been further confirmed by recent multiple experiments conducted both in soil and aeroponics (Al Farqani 2019).

Gene expression analysis of salt-stressed apple roots has revealed the presence of a complex landscape of DEGs associated with osmoprotectant pathways (Li et al. 2013). Moreover, upregulation of genes associated with sucrose, amino acid, alkaloid, flavonoid, and carotenoid metabolism has also been observed in apple roots grown under saline-alkali stress conditions in addition to genes associated with reactive oxygen species (ROS) protectants (Jia et al. 2019; Zhu

Fig. 6.6 Differences in root branching and fine-root development observed in apple rootstocks visualized using aeroponic growing systems

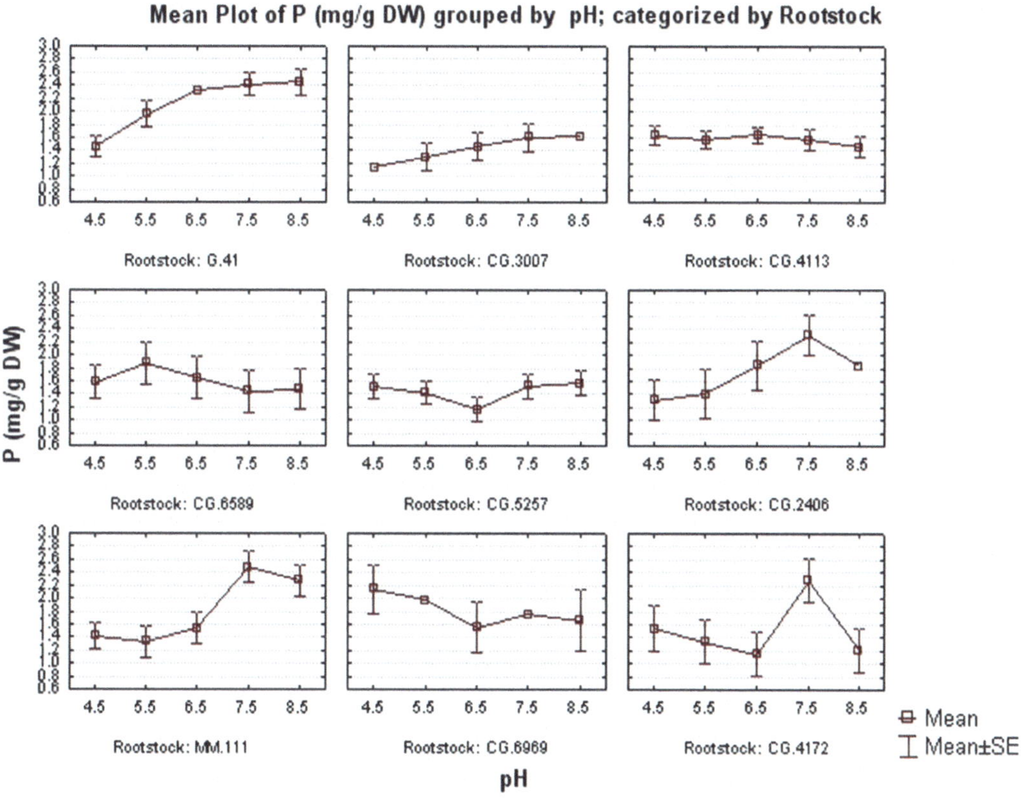

Fig. 6.7 Effects of pH on absorbance and translocation of phosphorous (P) measured in leaves of grafted Gala scions onto nine different rootstocks

et al. 2020). These major effect genes identified in these experiments may be probed for differential expression in apple rootstock breeding populations as means for identifying tolerant selections.

6.5.6 Sensitivity to Viruses and Phytoplasmas

Apple viruses and phytoplasmas cause a loss in productivity by either killing the rootstock outright, such as apple rootstock G.16 (Robinson et al. 2003), or by interfering with normal tissue development (roots, branches, leaves, and fruit), as well as with normal plant physiological processes (Howell et al. 1996; Wood 1996). These pathogens are mostly spread by grafting using infected wood of either scions or rootstocks. While genotypic variations of rootstock sensitivity to viruses have been observed in genetically related rootstocks (Wright et al. 2020), there is no reported information on genetic resistance.

Breeding efforts in both Germany and Italy have been undertaken to generate phytoplasma (*Candidatus* Phytoplasma mali, causal agent of apple proliferation disease or apple witches' broom disease) resistant rootstocks derived from wild apple species *M. sieboldii* and *M. sargentii* (Bisognin et al. 2009; Jarausch et al. 2007, 2008; Seemuller et al. 2007, 2008). Other efforts to evaluate apple rootstocks for sensitivity to viruses (Apple Stem Grooving Virus, Apple Chlorotic Leaf Virus, Apple Stem Pitting Virus, and Tomato Ringspot Virus) are currently in progress with advanced selections available in the Geneva (NY) breeding program.

6.5.7 Rootstock Resistance to Woolly Apple Aphids (WAA)

The Malling-Merton breeding program was initiated primarily to introduce resistance to woolly apple aphid (WAA), *Eriosoma lanigerum* (Hausmann), in rootstocks planted in the southern hemisphere where insect pressure caused major losses. The source of WAA resistance was the apple cultivar 'Northern Spy' (popular for pie-filling) resulting in the selection of apple rootstocks MM.106 and MM.111.

Genetic analysis of the 'Northern Spy' resistance has revealed that it is largely monogenic (*Er1*) and located on chromosome 8. An additional monogenic resistance gene, *Er2*, has been identified on chromosome 17 of 'Robusta 5', and this has been utilized extensively in the Geneva (NY) and the New Zealand breeding programs (Bus et al. 2008). Additional resistance loci, *Er3* and *Er4*, from other sources have been identified in progeny derived from 'Aotea' and the 'Mildew Immune Selection', and these loci are mapped on chromosomes 8 and 7, respectively (Bus et al. 2010; Sandanayaka et al. 2003, 2005). The largely monogenic inheritance of this trait readily lends itself to marker-assisted breeding (Bassett et al. 2015).

However, variations and mutations of WAA ecotypes have resulted in loss of effectiveness of some of the above WAA resistance loci in certain parts of the world; therefore, novel sources of resistance must be identified in available genetic resources (Sharma and Pramanick 2012).

6.6 Conclusions

The portfolio of desirable scion cultivars available for cultivation by apple growers is becoming increasingly diversified with new and high-value apples being released every year. On the other hand, apple rootstocks have until recently served as a virtual monoculture but are now also becoming more diversified with more specialized characteristics, beyond dwarfing and early bearing. In particular, major gains in productivity can be made if growers and nurseries are able to match the unique properties of scion cultivars to unique properties of rootstocks to promote synergies in production efficiency and fruit quality. Optimal pairings between scions and rootstocks have been attained for just a few standard scion cultivars such as 'Fuji', 'Gala', 'Honeycrisp', and 'Golden Delicious' because of empirical knowledge gained by several field tests.

Although in some cases we can extrapolate scion/rootstock pairings based on similarities, additional research must be conducted with new scion cultivars and new rootstocks. Perhaps, optimal pairings in the future can also be guided by a deeper understanding of molecular genetic rootstock/scion interactions using the tools of genomics, proteomics, and phenomics, among others, that are currently under study.

References

Adams RR (2010) Increasing the Rooting in Apple Rootstock Stoolbeds. M.S. Thesis, Cornell University, Ithaca, NY, 138 p

Adams S, Black BL, Fazio G, Roberts NA (2017) The effect of plant growth regulators on apple graft union flexural strength and flexibility. J Am Pomol Soc 71:8–18

Adams S, Lordan J, Fazio G, Bugbee B, Francescatto P, Robinson TL, Black B (2018) Effect of scion and graft type on transpiration, hydraulic resistance and xylem hormone profile of apples grafted on Geneva (R) 41 and M.9-NIC (TM) 29 rootstocks. Scient Hortic 227:213–222

Al Farqani ASA (2019) Evaluation of root's architecture, growth performance and fruiting of Honeycrisp™ apple scion grafted on 8 rootstocks in response to soil and solution's pH using field, aeroponics, and minirhizotron growing systems. Ph.D. Dissertation, Cornell University, Ithaca, NY, 303 p

Aldwinckle HS, Cummins JN, Antoszewski R, Harrison L, Zych CC (1974) Inheritance of fire blight susceptibility in some apple rootstock families. In: Proceedings of the XIX international horticultural congress. IA. Section VII. Fruits (Int Soc Hortic Sci), pp 309–339

An HS, Dong HQ, Wu T, Wang Y, Xu XF, Zhang XZ, Han ZH (2017) Root growth angle: an important trait that influences the deep rooting of apple rootstocks. Scient Hortic 216:256–263

Atucha A, Emmett B, Bauerle T (2013) Growth rate of fine root systems influences rootstock tolerance to replant disease. Plant Soil 376:1–10

Autio W, Robinson T, Black B, Crassweller R, Fallahi E, Parker M, Quezada RP, Wolfe D (2017) Budagovsky, Geneva, Pillnitz, and Malling apple rootstocks affect 'Fuji' performance over the first five years of the 2010 NC-140 'Fuji' apple rootstock trial. J Am Pomol Soc 71:167–182

Auvil TD, Schmidt TR, Hanrahan I, Castillo F, McFerson JR, Fazio G (2011) Evaluation of dwarfing rootstocks in Washington apple replant sites. Acta Hortic 903:265–271

Barritt BH, Konishi BS, Dilley MA, Kappel F (1997) Tree size, yield and biennial bearing relationships with 40 apple rootstocks and three scion cultivars. Acta Hortic 451:105–112

Bartish IV, Weeden NF, Tobutt KR, Alston FH (1999) The use of interspecific crosses in *Malus* to map the genes of characters important for apple rootstock breeding. Acta Hortic 484:319–323

Basedow M, Crassweler R (2017) Potential anatomical methods for the determination of weak wood in apple. J Am Pomol Soc 71:19–28

Bassett CL, Baldo AM, Moore JT, Jenkins RM, Soffe DS, Wisniewski ME, Norelli JL, Farrell RE (2014) Genes responding to water deficit in apple (*Malus × domestica* Borkh.) roots. BMC Plant Biol 14:182. https://doi.org/10.1186/1471-2229-14-182

Bassett CL, Glenn DM, Forsline PL, Wisniewski ME, Farrell RE (2011) Characterizing water use efficiency and water deficit responses in apple (*Malus × domestica* Borkh. and *Malus sieversii* Ledeb.) M. Roem. HortScience 46:1079–1084

Bassett H, Malone M, Ward S, Foster T, Chagné D, Bus V (2015) Marker-assisted selection in an apple rootstock breeding family. Acta Hortic 1100:25–28

Beakbane AB, Thompson EC, Schmidt H, Brown AG, Watkins R, Alston F (1973) Selection problems in rootstock breeding. In: Proceedings of Eucarpia fruit section symposium V. Top fruit breeding. Canterbury, Sept. 11–14 1973, pp 115–124

Bergh JC, Leskey TC (2003) Biology, ecology, and management of dogwood borer in eastern apple orchards. Can Entomol 135:615–635

Bisognin C, Seemuller E, Citterio S, Velasco R, Grando MS, Jarausch W (2009) Use of SSR markers to assess sexual vs apomictic origin and ploidy level of breeding progeny derived from crosses of apple proliferation-resistant *Malus sieboldii* and its hybrids with *Malus × domestica* cultivars. Plant Breed 128:507–513

Bonany J, Carbo J, Casals M, Iglesias I, Montserrat R (2004) Performance of some clonal apple rootstocks in Girona and Lleida (Catalonia, NE-Spain) with special reference to fruit quality. Acta Hortic 658:333–339

Broggini GAL, Durel CE, Vergne E, Chevreau E, Fahrentrapp J, Vanblaere T, Peil A, Flachowsky H, Hanke MV, Krens FA, Schouten HJ, Gessler C (2013) Cisgenic approach for improved disease resistance in apple. Acta Hortic 974:117–121

Broggini GAL, Kost T, Fahrentrapp J, Patocchi A, Wohner T, Flachowsky H, Peil A, Hanke MV, Gessler C (2014a) FB-MR5 is an apple gene providing resistance to fire blight. Acta Hortic 1056:273–276

Broggini GAL, Wohner T, Fahrentrapp J, Kost TD, Flachowsky H, Peil A, Hanke MV, Richter K, Patocchi A, Gessler C (2014b) Engineering fire blight resistance into the apple cultivar 'Gala' using the FB_MR5 CC-NBS-LRR resistance gene of *Malus × robusta* 5. Plant Biotech J 12" 728–733

Bus VGM, Bassett HCM, Bowatte D, Chagné D, Ranatunga CA, Ulluwishewa D, Wiedow C,

Gardiner SE (2010) Genome mapping of an apple scab, a powdery mildew and a woolly apple aphid resistance gene from open-pollinated Mildew Immune Selection. Tree Genet Genomes 6:477–487

Bus VGM, Chagne D, Bassett HCM, Bowatte D, Calenge F, Celton JM, Durel CE, Malone MT, Patocchi A, Ranatunga AC, Rikkerink EHA, Tustin DS, Zhou J, Gardiner SE (2008) Genome mapping of three major resistance genes to woolly apple aphid (*Eriosoma lanigerum* Hausm.). Tree Genet Genomes 4:223–236

Busdieker-Jesse NL, Nogueira L, Onal H, Bullock DS (2016) The economic impact of new technology adoption on the US apple industry. J Agric Resour Econ 41:549–569

Carlson RF (1981) The Mark apple rootstock. Fruit Var J 35:78–79

Casavela S (1977) Research on the selection of new vegetative rootstocks for apples. Lucr Stiint Inst Cerc Pomicult 6:159–167

Chen Y, An X, Zhao D, Li E, Ma R, Li Z, Cheng C (2020) Transcription profiles reveal sugar and hormone signaling pathways mediating tree branch architecture in apple (*Malus domestica* Borkh.) grafted on different rootstocks. PLoS ONE 15:e0236530–e0236530

Cook NC, Jacobs G (1999) Suboptimal winter chilling impedes development of acrotony in apple shoots. HortScience 34:1213–1216

Costa C, Stassen PJC (2011) Overview of apple rootstocks in South Africa. Acta Hortic 903:385–390

Couvillon GA, Finardi N, Magnani M, Freire C (1984) Rootstock influences the chilling requirement of 'Rome Beauty' apple in Brazil. HortScience 19:255–256

Cowgill WP Jr, Autio WR, Hoover EE, Marini RP, Domoto PA (2017) NC-140 multi-state research project: improving economic and environmental sustainability in tree-fruit production through changes in rootstock use. J Am Pomol Soc 71:34–46

Cummins JN, Aldwinckle HS (1974) Breeding apple rootstocks. HortScience 9:367–372

Cummins JN, Aldwinckle HS, Janick J (1983) Breeding apple rootstocks. Plant Breed Rev 1:294–394

Czynczyk A, Zagaja SW (1984) Evaluation of growth and cropping of apple trees grafted on dwarf rootstocks and interstems. Comp Fruit Tree 17:37–49

Deakin G, Fernandez-Fernandez F, Bennett J, Passey T, Harrison N, Tilston EL, Xu X (2019) The effect of rotating apple rootstock genotypes on apple replant disease and rhizosphere microbiome. Phytobiomes J 3:273–285

Du F, Xu J, Li D, Wang X, Du F, Xu JN, Li D, Wang XY (2015) The identification of novel and differentially expressed apple-tree genes under low-temperature stress using high-throughput Illumina sequencing. Mol Biol Rep 42:569–580

Drahorad W (1978) Crown gall. Obstbau Weinbau 15:107–109

Durand J-B, Allard A, Guitton B, van de Weg E, Bink MCAM, Costes E (2017) Predicting flowering behavior and exploring its genetic determinism in an apple multi-family population based on statistical indices and simplified phenotyping. Front Plant Sci 8:858. https://doi.org/10.3389/fpls.2017.0085

Elfving DC, Schecter I, Hutchinson A (1993) The history of the Vineland (V.) apple rootstocks. Fruit Var J 47:52–58

Embree CG, Mcrae KB (1991) An Exploratory study of reciprocal apple rootstock and scion hardiness with 2 methods of assessment. HortScience 26:1523–1525

Emeriewen OF, Richter K, Berner T, Keilwagen J, Schnable PS, Malnoy M, Peil A (2020) Construction of a dense genetic map of the *Malus fusca* fire blight resistant accession MAL0045 using tunable genotyping-by-sequencing SNPs and microsatellites. Sci Rep 10:10

Emmett B, Nelson E, Kessler A, Bauerle T (2014) Fine-root system development and susceptibility to pathogen colonization. Planta 239:325–340

Fahrentrapp J, Broggini GAL, Gessler C, Kellerhals M, Peil A, Malnoy M, Richter K (2013) Fine mapping of the fire blight resistance locus in *Malus* × *robusta* 5 on linkage group 3. Acta Hortic 976:499–500

Fazio G, Adams S, Lordan J, Roberts N, Robinson TL, Black B (2020a) Rootstock, scion, and graft type influence graft union flexural strength of apple trees. J Am Pomol Soc 74:24–44

Fazio G, Cheng L, Grusak MA, Robinson TL (2015a) Apple rootstocks influence mineral nutrient concentration of leaves and fruit. NY Fruit Quart 25:11–15

Fazio G, Cheng L, Lordan J, Francescatto P, Grusak MA, Robinson TL (2018a) Breeding apple rootstocks for modulation of mineral nutrients in scions. Acta Horitc 1217:29–37

Fazio G, Forsline P, Chao CT, Richards C, Volk G (2014a) Tree and root architecture of *Malus sieversii* seedlings for rootstock breeding. Acta Horitc 1058:585–594

Fazio G, Grusak M, Robinson TL (2017) Apple rootstocks' dwarfing loci relationships with mineral nutrient concentration in scion leaves and fruit. Acta Hortic 1177:93–102

Fazio G, Kviklys D, Grusak MA, Robinson T (2013) Phenotypic diversity and QTL mapping of absorption and translocation of nutrients by apple rootstocks. Asp Appl Biol 119:37–50

Fazio G, Kviklys D, Grusak MA, Robinson TL (2012) Soil pH, soil type and replantdisease affect growth and nutrient absorption in apple rootstocks. NY Fruit Quart 20:22–28

Fazio G, Lordan J, Francescatto P, Cheng L, Wallis A, Grusak MA, Robinson TL (2018b) 'Honeycrisp' apple fruit nutrient concentration affected by apple rootstocks. Acta Hortic 1228:223–227

Fazio G, Lordan J, Grusak MA, Francescatto P, Robinson TL (2020b) I. Mineral nutrient profiles and relationships of 'Honeycrisp' grown on a genetically diverse set of rootstocks under Western New York climatic conditions. Scient Hortic 266: 108477. https://doi.org/10.1016/j.scienta.2019.05.004

Fazio G, Mazzola M (2004) Target traits for the development of marker assisted selection of apple rootstocks - prospects and benefits. Acta Hortic 663 (823):827

Fazio G, Robinson T (2008a) Modification of nursery tree architecture by apple rootstocks. HortScience 43:1271

Fazio G, Robinson TL (2008b) Modification of nursery tree architecture with apple rootstocks: A breeding perspective. NY Fruit Quart 16:13–16

Fazio G, Robinson T (2019) Advances in the development and utilization of fruit tree rootstocks: a case study for apple, In: Lang GA (ed) Achieving sustainable cultivation of temperate zone tree fruits and berries, volume 1: physiology, genetics and cultivation. Burleigh Dodds Sci Publ Ltd, Cambridge, UK, pp 31–72

Fazio G, Robinson TL, Aldwinckle HS (2015b) The Geneva apple rootstock breeding program. Plant Breed Rev 39:379–424

Fazio G, Volder A (2009) Genetic control of root architecture, physiology and disease resistance traits. HortScience 44:986

Fazio G, Wan Y, Kviklys D, Romero L, Adams R, Strickland D, Robinson T (2014b) *Dw2*, a new dwarfing locus in apple rootstocks and its relationship to induction of early bearing in apple scions. J Am Soc Hortic Sci 139:87–98

Fazio G, Wan Y, Russo NL, Aldwinckle HS (2008) Investigation on the inheritance of strain specific resistance to *Erwinia amylovora* in an apple rootstock segregating population. Acta Hortic 793:331–335

Feng Y, Sun QR, Zhang GF, Wu T, Zhang XZ, Xu XF, Han ZH, Wang Y (2019) Genome-wide identification and characterization of ABC transporters in nine Rosaceae species identifying *MdABCG28* as a possible cytokinin transporter linked to dwarfing. Int J Mol Sci 20:5783. https://doi.org/10.3390/ijms20225783

Fernandez RT, Perry RL, Flore JA (1997) Drought response of young apple trees on three rootstocks. II. Gas exchange, chlorophyll fluorescence, water relations, and leaf. J Am Soc Hortic Sci 122:841–848

Ferree DC (1998) Performance of eight apomictic selections as apple rootstocks. HortScience 33:641–643

Fischer M, Schmidt H, Kellerhals M (1993) The Pillnitz apple rootstock breeding programme results: Progress in temperate fruit breeding. In: Proceedings of Eucarpia fruit breed sect meet. Kluwer Acad Publ, Dordrecht, Netherlands, pp 141–145

Forsline PL, Aldwinckle HS, Hale C, Mitchell R (2002) Natural occurrence of fire blight in USDA apple germplasm collection after 10 years of observation. Acta Hortic 590:351–357

Foster TM, Celton JM, Chagné D, Tustin DS, Gardiner SE (2015) Two quantitative trait loci, *Dw1* and *Dw2*, are primarily responsible for rootstock-induced dwarfing in apple. Hortic Res 2:15001. https://doi.org/10.1038/hortres.2015.1

Foster TM, Ledger SE, Janssen BJ, Luo Z, Drummond RSM, Tomes S, Karunairetnam S, Waite CN, Funnell KA, van Hooijdonk BM, Saei A, Seleznyova AN, Snowden KC (2018) Expression of *MdCCD7* in the scion determines the extent of sylleptic branching and the primary shoot growth rate of apple trees. J Exp Bot 69:2379–2390

Gao H, Dang Z, Li D, Mei L, Wan Y, Wang L, Zhao Z, Fazio G (2011) The history of apple breeding in People's Republic of China. Acta Hortic 903:199–205

Gardiner SE, Norelli JL, de Silva N, Fazio G, Peil A, Malnoy M, Horner M, Bowatte D, Carlisle C, Wiedow C, Wan YZ, Bassett CL, Baldo AM, Celton JM, Richter K, Aldwinckle HS, Bus VGM (2012) Putative resistance gene markers associated with quantitative trait loci for fire blight resistance in *Malus* 'Robusta 5' accessions. BMC Genet 13:25. https://doi.org/10.1186/1471-2156-13-25

Gharghani A, Zamani Z, Talaie A, Fattahi R, Hajnajari H, Oraguzie NC, Wiedow C, Gardiner SE (2010) The role of Iran (Persia) in apple (*Malus* × *domestica* Borkh.) domestication, evolution and migration via the Silk Trade Route. Acta Hortic 229–236

Glenn DM (2014) An analysis of ash and isotopic carbon discrimination (Δ13C) methods to evaluate water use efficency in apple. Scient Hortic 171:32–36. https://doi.org/10.1016/j.scienta.2014.03.031

Gu YH, Mazzola M (2003) Modification of fluorescent pseudomonad community and control of apple replant disease induced in a wheat cultivar-specific manner. Appl Soil Ecol 24:57–72

Guitton B, Kelner JJ, Velasco R, Gardiner SE, Chagné D, Costes E (2012) Genetic control of biennial bearing in apple. J Exp Bot 63:131–149

Han ZH, Xu XF, Shen T, Wang Y, Li TZ, Kong J, Zhang XZ (2011) *Malus xiaojinensis* - a promising apple rootstock with resistance to Fe chlorosis. Acta Hortic 903:163–167

Hao S, Lu Y, Liu J, Bu Y, Chen Q, Ma N, Zhou Z, Yao Y (2019) GIBBERELLIN INSENSITIVE DWARF1 plays an important role in the growth regulation of dwarf apple rootstocks. HortScience 54:416–422

Harrison N, Harrison RJ, Barber-Perez N, Cascant-Lopez E, Cobo-Medina M, Lipska M, Conde-Ruiz R, Brain P, Gregory PJ, Fernandez-Fernandez F (2016) A new three-locus model for rootstock-induced dwarfing in apple revealed by genetic mapping of root bark percentage. J Exp Bot 67:1871–1881

Hatton RG (1917) Paradise apple stocks. J Roy Hortic Soc 42:361–399

Hatton RG (1919) Paradise apple stocks their fruit and blossom described. J Roy Hortic Soc 44:89–94

Hatton RG (1920) Suggestion for the right selection of apple stocks. J Roy Hortic Soc 45:257–268

Hawerroth FJ, Petri JL, Herter FG, Leite GB, Leonetti JF, Marafon AC, Simoes F (2009) Phenology, budbreak and apple fruit production by hydrogen cyanamide and mineral oil application. Bragantia 68:961–971

Hooijdonk BMV, Tustin DS, Dayatilake D, Oliver M, van Hooijdonk BM (2015) Nursery tree design modifies

annual dry matter production of newly grafted 'Royal Gala' apple trees. Scient Hortic 197:404–410
Howell WE, Mink GI, Hurtt SS, Fuster JA, Postman JD (1996) Select *Malus* clones for rapid detection of apple stem grooving virus. Plant Dis 80:1200–1202
Hutchinson A (1977) Rootstock research with new clones and future possibilities. Comp Fruit Tree 10:8–12
Inoue E, Sakuma F, Kasumi M, Hara H, Tsukihashi T (2002) Production of apomict-like seedling through intergeneric cross between Japanese pear and apple by means of pollen irradiation. Acta Hortic 587:211–215
Isutsa DK, Merwin IA (2000) *Malus* germplasm varies in resistance or tolerance to apple replant disease in a mixture of New York orchard soils. HortScience 35:262–268
Jakubowski T, Zagaja SW, Geibel M, Fischer M, Fischer C (2000) 45 years of apple rootstocks breeding in Poland. Acta Hortic 538:723–727
Jarausch W, Bisognin C, Peccerella T, Schneider B, Seemuller E (2008) Development of resistant rootstocks for the control of apple proliferation disease. Acta Hortic 381–385
Jarausch W, Bisognin C, Schneider B, Grando S, Velasco R, Seemuller E (2007) Breeding of apple rootstocks resistant to ' Candidatus Phytoplasma mali'. Bull Insect 60:299–300
Jensen PJ, Halbrendt N, Fazio G, Makalowska I, Altman N, Praul C, Maximova SN, Ngugi HK, Crassweller RM, Travis JW, McNellis TW (2012) Rootstock-regulated gene expression patterns associated with fire blight resistance in apple. BMC Genom 13:9. https://doi.org/10.1186/1471-2164-13-9
Jensen PJ, Makalowska I, Altman N, Fazio G, Praul C, Maximova SN, Crassweller RM, Travis JW, McNellis TW (2010) Rootstock-regulated gene expression patterns in apple tree scions. Tree Genet Genomes 6:57–72
Jensen PJ, Rytter J, Detwiler EA, Travis JW, McNellis TW (2003) Rootstock effects on gene expression patterns in apple tree scions. Plant Mol Biol 53:493–511
Jia XM, Wang H, Svetla S, Zhu YF, Hu Y, Cheng L, Zhao T, Wang YX (2019) Comparative physiological responses and adaptive strategies of apple *Malus halliana* to salt, alkali and saline-alkali stress. Scient Hortic 245:154–162
Jin WM, Zhang Q, Liu SZ, Wei QP, Cheng ZM, Xue XH, Yang TZ (2012) Genetic Diversity of 41 apple rootstocks based on simple sequence repeat markers. J Am Soc Hortic Sci 137:51–56
Kennedy AJ, Rowe RW, Samuelson TJ (1980) The effects of apple rootstock genotypes on mineral content of scion leaves. Euphytica 29:477–482
Khanizadeh S, Groleau Y, Levasseur A, Granger R, Rousselle GL, Davidson C (2005) Development and evaluation of St Jean-Morden apple rootstock series. HortScience 40:521–522
Knabel M, Friend AP, Palmer JW, Diack R, Wiedow C, Alspach P, Deng C, Gardiner SE, Tustin DS, Schaffer R, Foster T, Chagné D (2015) Genetic control of pear rootstock-induced dwarfing and precocity is linked to a chromosomal region syntenic to the apple *Dw1* loci. BMC Plant Biol 15:230. https://doi.org/10.1186/s12870-015-0620-4
Kost TD, Gessler C, Jansch M, Flachowsky H, Patocchi A, Broggini GA (2015) Development of the first cisgenic apple with increased resistance to fire blight. PLoS ONE 10: https://doi.org/10.1371/journal.pone.0143980
Kostick SA, Norelli JL, Teh SL, Evans KM (2021) Quantitative variation and heritability estimates of fire blight resistance in a pedigree-connected apple germplasm set. J Plant Pathol 8:28. https://doi.org/10.1007/s42161-020-00543-0
Kumar C, Singh SK, Pramanick KK, Verma MK, Srivastav M, Singh R, Bharadwaj C, Naga KC (2018) Morphological and biochemical diversity among the *Malus* species including indigenous Himalayan wild apples. Scient Hortic 233:204–219
Kviklys D, Samuoliene G (2020) Relationships among the rootstock, crop load, and sugar hormone signaling of apple tree, and their effects on biennial bearing. Front Plant Sci 11:1213. https://doi.org/10.3389/fpls.2020.01213
Lei C, Fan S, Li K, Meng Y, Mao J, Han M, Zhao C, Bao L, Zhang D (2018) iTRAQ-based proteomic analysis reveals potential regulation networks of IBA-induced adventitious root formation in apple. Int J Mol Sci 19(3):667. https://doi.org/10.3390/ijms19030667
Leisso R, Rudell D, Mazzola M (2017) Metabolic composition of apple rootstock rhizodeposits differs in a genotype-specific manner and affects growth of subsequent plantings. Soil Biol Biochem 113:201–214
Li Q, Liu J, Tan DX, Allan AC, Jiang Y, Xu X, Han Z, Kong J, Li QT, Liu J, Jiang YZ, Xu XF, Han ZH, Kong J (2013) A genome-wide expression profile of salt-responsive genes in the apple rootstock *Malus zumi*. Int J Mol Sci 14:21053–21070
Li G, Ma J, Tan M, Mao J, An N, Sha G, Zhang D, Zhao C, Han M, Li GF, Ma JJ, Tan M, Mao JP, An N, Sha GL, Zhang D, Zhao CP, Han MY (2016) Transcriptome analysis reveals the effects of sugar metabolism and auxin and cytokinin signaling pathways on root growth and development of grafted apple. BMC Genom 17:150. https://doi.org/10.1186/s12864-016-2484-x
Liu D, Dong Q, Fang M, Chen K, Hao Y, Liu DD, Dong QL, Fang MJ, Chen KQ, Hao YJ (2012a) Ectopic expression of an apple apomixis-related gene *MhFIE* induces co-suppression and results in abnormal vegetative and reproductive development in tomato. J Plant Physiol 169:1866–1873
Liu D, Dong Q, Sun C, Wang Q, You C, Yao Y, Hao Y, Liu DD, Dong QL, Sun C, Wang QL, You CX,

Yao YX, Hao YJ (2012b) Functional characterization of an apple apomixis-related *MhFIE* gene in reproduction development. Plant Sci 185(186):105–111

Lordan J, Fazio G, Francescatto P, Robinson T (2017) Effects of apple (*Malus* × *domestica*) rootstocks on scion performance and hormone concentration. Scient Hortic 225:96–105

Lordan J, Fazio G, Francescatto P, Robinson TL (2019) II. Horticultural performance of 'Honeycrisp' grown on a genetically diverse set of rootstocks under Western New York climatic conditions. Scient Hortic 257: 108686. https://doi.org/10.1016/j.scienta.2019.108686

Lordan J, Francescatto P, Dominguez LI, Robinson TL (2018) Long-term effects of tree density and tree shape on apple orchard performance, a 20 year study - Part 1, agronomic analysis. Scient Hortic 238:303–317

Ma R, Huang Y, Sha G, Shi Y, Li J, Gong X, Li M, Ma RQ, Huang Y, Sha GL, Shi YJ, Li J, Gong XH, Li M (2012) Identifying hybrids of apomictic apple rootstocks with SSR and flow cytometry. J Fruit Sci 29:461–465

Macedo TA, Sander GF, Michelon MF, Carminatti JF, Rufato AR, Rufato L, Robinson TL (2018) Chilling requirement and budburst uniformity of cultivar 'Maxi Gala' grafted on different rootstocks. Acta Hortic 1228:241–245

Macedo TA, Silva PS, Sander GF, Welter JF, Rufato L, Rossi A (2019) Productivity and quality of'Fuji Suprema' apple fruit in different rootstocks and growing conditions. Scient Hortic 256:108651

Maguylo K, Bassett C (2014) Phenotyping *M. sieversii* collections from Kazakhstan for leaf traits and tree architecture. Acta Horitc 1058:335–341

Manici LM, Kelderer M, Franke-Whittle IH, Ruhmer T, Baab G, Nicoletti F, Caputo F, Topp A, Insam H, Naef A (2013) Relationship between root-endophytic microbial communities and replant disease in specialized apple growing areas in Europe. Appl Soil Ecol 72:207–214

Marini RP, Autio WR, Black B, Cline J, Cowgill WR Jr, Crassweller RM, Domoto PA, Hampson C, Moran R, Quezada RA, Robinson T, Ward DL, Wolfe D (2013) Return bloom on 'Golden Delicious' apple trees as affected by previous season's crop density on three rootstocks at 11 locations. J Am Pomol Soc 67:72–79

Marini RP, Crassweller RM, Autio W, Moran R, Robinson TL, Cline J, Wolfe D, Parra-Quezada R (2017) The interactive effects of early season temperatures, crop density and rootstock on average fruit weight of 'Golden Delicious' apple. Acta Hortic 1177:189–194

Marini RP, Fazio G (2018) Apple rootstocks: history, physiology, management, and breeding. Hortic Rev 45:197–312

Marini RP, Parker ML, Barden JA, Unrath CR (2003) The effect of eight dwarf rootstocks on burrknot development on 'Gala' apple trees at two locations. J Am Pomol Soc 57:93–96

Mazzola M (1998) Elucidation of the microbial complex having a causal role in the development of apple replant disease in Washington. Phytopathology 88:930–938

Mazzola M, Manici LM (2012) Apple replant disease: role of microbial ecology in cause and control. Annu RevPhytopathol 50:45–65

Mong QY (1991) A hardy, dwarfing apple rootstock germplasm 'Zhaai 76'. China Fruits, p 42

Moran RE, Sun YP, Geng F, Zhang DL, Fazio G (2011a) Cold temperature tolerance of trunk and root tissues in one- or two-year-old apple rootstocks. HortScience 46:1460–1464

Moran RE, Zhang D, Sun Y (2011b) Cold temperature tolerance of G.16 and G.935 apple roots. Acta Hortic 903:289–293

Moran RE, Peterson BJ, Fazio G, Cline J (2018) Genotypic variation in apple rootstock low temperature tolerance during Spring and Fall. J Am Soc Hortic Sci 143:319–332

Moriya S, Iwanami H, Takahashi S, Kotoda N, Suzaki K, Abe K (2008) Evaluation and inheritance of crown gall resistance in apple rootstocks. J Jpn Soc Hortic Sci 77:236–241

Moriya S, Iwanami H, Haji T, Okada K, Yamada M, Yamamoto T, Abe K (2015) Identification and genetic characterization of a quantitative trait locus for adventitious rooting from apple hardwood cuttings. Tree Genet Genomes 11:59. https://doi.org/10.1007/s11295-015-0883-9

Moriya S, Iwanami H, Takahashi S, Kotoda N, Suzaki K, Yamamoto T, Abe K (2010) Genetic mapping of the crown gall resistance gene of the wild apple *Malus sieboldii*. Tree Genet Genomes 6:195–203

Musacchi S, Neri D (2019) Optimizing production of quality nursery plants for fruit tree cultivation. In: Lang GA (ed) Achieving sustainable cultivation of temperate zone tree fruits and berries. Burleigh Dodds Sci Publisher, Cambridge, UK, pp 183–242. https://doi.org/10.19103/as.2018.0040.06

Neri D, Papachatzis G, Sabbatini P, Zucconi F (2001) Lateral shoot formation in apple plants. Rivista Frutticolt Ortofloricol 63:31–36

Olien WC (1987) Apomictic crabapples and their potential for research and fruit production. HortScience 22:541–546

Oraguzie NC, Yamamoto T, Soejima J, Suzuki T, Silva HND (2005) DNA fingerprinting of apple (*Malus* spp.) rootstocks using simple sequence repeats. Plant Breed 124:197–202

Peil A, Emeriewen OF, Khan A, Kostick S, Malnoy M (2020) Status of fire blight resistance breeding in *Malus*. J Plant Pathol. https://doi.org/10.1007/s42161-020-00581-8

Peil A, Huebert C, Wensing A, Horner M, Emeriewen OF, Richter K, Woehner T, Chagné D, Orellana-Torrejon C, Saeed M, Troggio M, Stefani E, Gardiner SE, Hanke M-V, Flachowsky H, Bus VGM (2019) Mapping of fire blight resistance in *Malus* x *robusta* 5 flowers following artificial inoculation. BMC Plant Biol 19(1):532. https://doi.org/10.1186/s12870-019-2154-7

Pieniazek SA (1971) Russian breeding of clonal apple rootstocks. Am Nurseryman 134: 32/34/36

Pilcher RLR, Celton JM, Gardiner SE, Tustin DS (2008) Genetic markers linked to the dwarfing trait of apple rootstock 'Malling 9'. J Am Soc Hortic Sci 133:100–106

Prive JP, Zhang MIN, Embree CG, Hebb D, Palmer JW, Wunsche JN (2001) The influence of freeze-thaw cycling in cold hardiness studies on apple rootstocks. Acta Hortic 557:123–130

Quamme HA (1990) Cold hardiness of apple rootstocks. Comp Fruit Tree 23:11–16

Quamme HA, Brownlee RT, Barritt BH, Kappel F (1997) Cold hardiness evaluation of apple rootstocks. Acta Hortic 451:187–193

Racsko J, Miller DD (2011) Effect of different growth inducing rootstocks on alternate bearing of 'Royal Gala' and 'Vista Bella' apples. Acta Hortic 903:411–418

Reig G, Lordan J, Sazo MM, Hoying SA, Fargione MJ, Reginato GH, Donahue DJ, Francescatto P, Fazio G, Robinson TL (2019) Effect of tree type and rootstock on the long-term performance of 'Gala', 'Fuji' and 'Honeycrisp' apple trees trained to Tall Spindle under New York State climatic conditions. Scient Hortic 246:506–517

Reig G, Lordan J, Hoying S, Fargione M, Donahue DJ, Francescatto P, Acimovic D, Fazio G, Robinson T (2020) Long-term performance of 'Delicious' apple trees grafted on Geneva® rootstocks and trained to four high-density systems under New York State climatic conditions. HortScience 55:1538–1550. https://doi.org/doi.org/10.21273/HORTSCI14904-21220

Reim S, Rohr A-D, Winkelmann T, Weiss S, Liu B, Beerhues L, Schmitz M, Hanke M-V, Flachowsky H (2020) Genes involved in stress response and especially in phytoalexin biosynthesis are upregulated in four *Malus* genotypes in response to apple replant disease. Front Plant Sci 10:1724. https://doi.org/10.3389/fpls.2019.01724

Robinson T, Aldwinckle H, Fazio G, Holleran T, Janick J (2003) The Geneva series of apple rootstocks from Cornell: performance, disease resistance, and commercialization. Acta Hortic 622:513–520

Robinson TL, Fazio G, Aldwinckle HS, Hoying SA, Russo N (2006) Field performance of Geneva apple rootstocks in the Eastern USA. Sodininkyste-ir-Darzininkyste 25:181–191

Robinson TL, DeMarree AM, Hoying SA (2007) An economic comparison of five high density apple planting systems. Acta Hortic 732:481–489

Robinson T, Autio W, Clements J, Cowgill W, Embree C, Gonzalez V, Hoying S, Kushad M, Parker M, Parra R, Schupp J (2012) Rootstock tolerance to apple replant disease for improved sustainability of apple production. Acta Hortic 940:521–528

Rong Z, Yi K, Yang F, Liu Z, Li X, Wan D, Wan Y (2011) A dwarfing apple rootstock: 'Liaozhen 2'. Acta Hortic 903:169–176

Rufato L, Kretzschmar AA, Brighenti AF, Macedo TA, Mendes M, Silva LC (2010) Bud break in different cultivars of apple trees in two regions of Santa Catarina State, Brazil. Acta Hortic 884:643–646

Rumberger A, Merwin IA, Thies JE (2007) Microbial community development in the rhizosphere of apple trees at a replant disease site. Soil Biol Biochem 39:1645–1654

Rusholme RL, Gardiner SE, Bassett HCM, Tustin DS, Ward SM, Didier A (2004) Identifying genetic markers for an apple rootstock dwarfing gene. Acta Hortic 663:405–409

Russo NL, Robinson TL, Fazio G, Aldwinckle HS (2007) Field evaluation of 64 apple rootstocks for orchard performance and fire blight resistance. HortScience 42:1517–1525

Russo NL, Aldwinckle HS, Robinson TL, Fazio G (2008a) Budagovsky 9 rootstock: uncovering a novel resistance to fire blight. Acta Hortic 793:321–324

Russo NL, Robinson TL, Fazio G, Aldwinckle HS (2008b) Fire blight resistance of Budagovsky 9 apple rootstock. Plant Dis 92:385–391

Samuoliene G, Viskeliene A, Sirtautas R, Kviklys D (2019) Relationships among the rootstock and morphogenetic behaviour of apple tree in aspect for fruit quality. Acta Hortic 1242:115–122

Sandanayaka WRM, Bus VGM, Connolly P, Newcomb R (2003) Characteristics associated with Woolly Apple Aphid *Eriosoma lanigerum*, resistance of three apple rootstocks. Entomol Experim Applic 109:63–72

Sandanayaka WRM, Bus VGM, Connolly P (2005) Mechanisms of woolly aphid [*Eriosoma lanigerum* (Hausm.)] resistance in apple. J Appl Entomol 129:534–541

van der Scheer HAT (1978) Research on crown gall of fruit trees. Gewasbescherming 9:97–101

Schmidt H, Antoszewski R, Harrison L, Zych CC (1974) The inheritance of apomixis in polyploid *Malus* rootstock material. In: Proceedings of XIX Int Hortic Congress. Section VII. Fruits, pp 340–456

Seemuller E, Moll E, Schneider B (2007) *Malus sieboldii*-based rootstocks mediate apple proliferation resistance to grafted trees. Bull Insectol 60:301–302

Seemuller E, Moll E, Schneider B (2008) Apple proliferation resistance of *Malus sieboldii*-based rootstocks in comparison to rootstocks derived from other *Malus* species. Euro J Plant Pathol 121:109–119

Serra S, Leisso R, Giordani L, Kalcsits L, Musacchi S (2016) Crop load influences fruit quality, nutritional balance, and return bloom in 'Honeycrisp' apple. HortScience 51:236–244

Sha GL, Gong XH, Huang Y, Shao YC, Yin T, Hao YJ, Shu HR (2011) 'Qingzhen 1' and 'Qingzhen 2', two apple rootstocks with high apomitic fruit setting ability. Acta Hortic 903:159–162

Sharma YP, Pramanick KK (2012) Utilization of plant genetic resources for the improvement of temperate fruit crops. Indian J Genet Plant Breed 72:130–135

Shin S, Zheng P, Fazio G, Mazzola M, Main D, Zhu Y (2016) Transcriptome changes specifically associated with apple (*Malus domestica*) root defense response during Pythium ultimum infection. Physiol Mol Plant Pathol 94:16–26

Smolka A, Welander M, Olsson P, Holefors A, Zhu L-H (2009) Involvement of the *ARRO-1* gene in adventitious root formation in apple. Plant Sci 177:710–715

Song C, Zhang D, Zhang J, Zheng L, Zhao C, Ma J, An N, Han M, Song CH, Zhang D, Zhang J, Zheng LW, Zhao CP, Ma JJ, An N, Han MY (2016) Expression analysis of key auxin synthesis, transport, and metabolism genes in different young dwarfing apple trees. Acta Physiol Plant 38:43. https://doi.org/10.1007/s11738-016-2065-2

Spornberger A, Schueller E, Noll D (2020) The influence of Geneva rootstocks on the vegetative and generative characteristics of the apple cultivar 'Topaz' in an organically managed replanted orchard. Int J Fruit Sci 20(S3):S1436–S1444. https://doi.org/10.1080/15538362.2020.1799475

St. Laurent A, Merwin IA, Fazio G, Thies JE, Brown MG (2010) Rootstock genotype succession influences apple replant disease and root-zone microbial community composition in an orchard soil. Plant Soil 337:259–272

Stover E, Walsh C (1998) Crown gall in apple rootstocks: inoculation above and below soil and relationship to root mass proliferation. HortScience 33:92–95

Tamai H, Komatsu H, Koike H, Ono T, Shigehara I (2003) Performance of 'Rakuraku Fuji' apple trees on JM.7 and M.9 Nagano rootstocks, and M.9 Nagano/Marubakaido interstem combination in Japan. J Am Pomol Soc 57:157–160

Tegtmeier R, Pompili V, Singh J, Micheletti D, Silva KJP, Malnoy M, Khan A (2020) Candidate gene mapping identifies genomic variations in the fire blight susceptibility genes *HIPM* and *DIPM* across the *Malus* germplasm. Sci Rep 10:1637. https://doi.org/10.1038/s41598-020-73284-w

Trajkovski V Andersson G (1980) Rootstock breeding. In: Report for 1978–1979. Division of fruit breeding, Swedish University of Agricultural Science, Balsgard, pp 36–48

Tsuchiya S (1988) Breeding of dwarfing rootstocks for fruit trees. Res J Food Agric 11:3–11

Tukey HB (1964) Dwarfed fruit trees for Orchard, garden and home. Macmillan, NY

Tworkoski T, Fazio G (2011) Physiological and morphological effects of size-controlling rootstocks on 'Fuji' apple scions. Acta Hortic 903:865–872

Tworkoski T, Fazio G (2016a) Hormone and growth interactions of scions and size-controlling rootstocks of young apple trees. Plant Grow Regul 78:105–119

Tworkoski T, Fazio G, Glenn DM (2016) Apple rootstock resistance to drought. Scient Hortic 204:70–78

Tworkoski T, Miller S (2007) Endogenous hormone concentrations and bud-break response to exogenous benzyl adenine in shoots of apple trees with two growth habits grown on three rootstocks. J Hortic Sci Biotech 82:960–966

van Hooijdonk B, Woolley D, Warrington I, Tustin S (2011) Rootstocks modify scion architecture, endogenous hormones, and root growth of newly grafted 'Royal Gala' apple trees. J Am Soc Hortic Sci 136:93–102

Vanderzande S, Hias N, Edge-Garza D, Costes E, Davey MW, Keulemans J (2016) Sylleptic branching in winter-headed apple (*Malus* × *domestica*) trees: Accession-dependent responses and their relationships with other tree architectural characteristics. Tree Genet Genomes 12:87. https://doi.org/10.1007/s11295-016-1046-3

Veal D, Damerow L, Blanke MM (2011) Selective mechanical thinning to regulate fruit set, improve quality and overcome alternate bearing in fruit crops. Acta Hortic 903:775–781

Virlet N, Costes E, Martinez S, Kelner JJ, Regnard JL (2015) Multispectral airborne imagery in the field reveals genetic determinisms of morphological and transpiration traits of an apple tree hybrid population in response to water deficit. J Exp Bot 66:5453–5465

Wallis A, Harshman JM, Butler B, Price D, Fazio G, Walsh C (2017) Performance of Geneva (R) apple rootstock selections with 'Brookfield Gala' and 'Cripps Pink' in a tall spindle system. J Am Pomol Soc 71:137–148

Wan Y, Li D, Zhao Z, Mei L, Han M, Schwaninger H, Fazio G (2011) The distribution of wild apple germplasm in Northwest China and its potential application for apple rootstock breeding. Acta Hortic 903:123–141

Wang N, Guo TL, Sun X, Jia X, Wang P, Shao Y, Liang BW, Gong XQ, Ma FW (2017) Functions of two *Malus hupehensis* (Pamp.) Rehd. *YTPs* (*MhYTP1* and *MhYTP2*) in biotic- and abiotic-stress responses. Plant Sci 261:18–27

Wang H, Zhao S, Mao K, Dong Q, Liang B, Li C, Wei Z, Li M, Ma F (2018) Mapping QTLs for water-use efficiency reveals the potential candidate genes involved in regulating the trait in apple under drought stress. BMC Plant Biol 18:136. https://doi.org/10.1186/s12870-018-1308-3

Wang Y, Li W, Xu X, Qiu C, Wu T, Wei Q, Ma F, Han Z (2019) Progress of apple rootstock breeding and its use. Hortic Plant J 5:183–191

Webster T, Tobutt K, Evans K (2000) Breeding and evaluation of new rootstocks for apple, pear and sweet cherry. Comp Fruit Tree 33:100–104

Wertheim SJ (1998) Rootstock guide: apple, pear, cherry, European plum. Proefstation voor de Fruitteelt (Fruit Research Station), Wilhelminadorp, Netherlands

White M, Tustin S (2000) New apple rootstock alternatives for the Southern Hemisphere. Comp Fruit Tree 33:112–115

Wisniewski M, Bassett C, Norelli J, Macarisin D, Gasic K, Korban S (2008) Expressed-sequence tag (EST) analysis of the response of apple (*Malus* × *domestica*) to low temperature and water deficit. Physiol Plant 133:298–317. https://doi.org/10.1111/j.1399-3054.2008.01063.x

Wood GA (1996) Past and present status of virus and phytoplasma diseases in apple rootstocks in New Zealand. NZ J Crop Hortic Sci 24:133–141

Wright AA, Cross AR, Harper SJ (2020) A bushel of viruses: identification of seventeen novel putative viruses by RNA-Seq in six apple trees. PLoS ONE 15:

Wu RG, Wang Y, Wu T, Xu XF, Han ZH (2018) Functional characterisation of *MdMYB44* as a negative regulator in the response to cold and salt stress in apple calli. J Hortic Sci Biotech 93:347–355

Xiang B, Cheng M, Li X, Jin Q, Liu Y, Xie X, Xiang BX, Cheng MH, Li XL, Jin Q, Liu YQ, Xie XL (1995) A study of drought resistance in the rootstock germplasms of apple. J Southwest Agric Univ 17:381–385

Xu JD, Zhou SS, Gong XQ, Song ., van Nocker S, Ma FW, Guan QM (2018) Single-base methylome analysis reveals dynamic epigenomic differences associated with water deficit in apple. Plant Biotech J 16:672–687

Yang T, Wang Q, Gao J, Li K, Han F, Tian J (2011) Cold-and-drought resistant dwarfing apple rootstocks from *Malus baccata*. Acta Hortic 903:177–182

Yao SR, Merwin IA, Brown MG (2006) Root dynamics of apple rootstocks in a replanted orchard. HortScience 41:1149–1155

Young E, Werner DJ (1984a) Effects of rootstock and scion chilling during rest on resumption of growth in apple and peach. J Am Soc Hortic Sci 109:548–551

Young E, Werner DJ (1984b) Rootstock and scion chilling effects on apple and peach bud break. Comp Fruit Tree 17:32–36

Zhang LS, Li XW, Zhang LX, Li BZ, Han MY, Liu FT, Zheng P, Alva AK (2014) Role of abscisic acid (ABA) in modulating the responses of two apple rootstocks to drought stress. Pak J Bot 46:117–126

Zhang MJ, Ding LH, Wang Q, Li YB, Yan XK, Xing GJ (2011) Development of cold resistant apple rootstocks in China. Acta Hortic 903:183–186

Zhou SS, Ma SX, Li MJ, Li CY, Gong XQ, Guan QM, Tan YX, Shao Y, Li C, Ma FW (2016) Comprehensive genomic analysis and expression profiling of Argonaute gene family and examination of their regulatory roles in water-use efficiency and abiotic stress responses in apple. Acta Physiol Plant 38:231. https://doi.org/10.1007/s11738-016-2245-0

Zhou YM, Tian X, Yao JJ, Zhang ZF, Wang Y, Zhang XZ, Li W, Wu T, Han ZH, Xu XF, Qiu CP (2020) Morphological and photosynthetic responses differ among eight apple scion-rootstock combinations. Scient Hortic 261: https://doi.org/10.1016/j.scienta.2019.108981

Zhu Y, Jia X, Wu Y, Hu Y, Cheng L, Zhao T, Huang Z, Wang Y (2020) Quantitative proteomic analysis of *Malus halliana* exposed to salt-alkali mixed stress reveals alterations in energy metabolism and stress regulation. Plant Grow Reg 90:205–222

Zhu YM, Saltzgiver M (2020) A systematic analysis of apple root resistance traits to *Pythium ultimum* infection and the underpinned molecular regulations of defense activation. Hortic Res 7:62. https://doi.org/10.1038/s41438-020-0286-4

Genetic and Physical Mapping of the Apple Genome

7

Yuepeng Han and Schuyler S. Korban

Abstract

In apple genome mapping, two distinctive types of maps have been developed—a genetic map, also known as a linkage map, and a physical map. Linkage maps have been established for many important horticultural traits in apple. This involves identifying molecular markers linked to major genes and quantitative trait loci (QTL). Such genetic maps have been developed for various desirable economic traits for the apple, including those for disease resistance, pest resistance, growth habit, flowering, and budbreak, as well as for various fruit quality traits. In addition, genome-wide physical maps have also been developed for the apple. Such maps not only serve as platforms for large-scale genome sequencing efforts, but are also essential for studying the genetic basis of complex traits, as well as for pursuing genomics research studies. Strategies for establishing genetic and physical maps of the apple genome along with the details of these findings will be presented.

7.1 Introduction

Genome mapping involves the identification of locations of genes and/or positioning of genetic markers on each chromosome of a genome, as well as the determination of relative distances between genes and/or of genetic markers along these chromosomes. In genome mapping, genetic markers can take on various forms, and genes may be deemed as one of such particular forms of genetic markers. Overall, there are two types of genome mapping efforts that can be undertaken. These include genetic mapping and physical mapping. Genetic mapping involves the calculation of the relative distance between any two genes or genetic markers on a chromosome, while physical mapping involves the determination of the absolute position of genes or genetic markers on each chromosome. Therefore, there are two distinctive types of "maps": a genetic map (also known as a linkage map) and a physical map, which can be generated in the field of genome mapping. Although both maps display positions of a collection of genetic markers and gene loci, distances of a genetic map are estimated by calculating frequencies of recombination between

Y. Han (✉)
Key Laboratory of Plant Germplasm Enhancement and Specialty Agriculture, Wuhan Botanical Garden of the Chinese Academy of Sciences, Wuhan 430074, China
e-mail: yphan@wbgcas.cn

Y. Han
Sino-African Joint Research Center, Chinese Academy of Sciences, Wuhan 430074, China

S. S. Korban
Department of Natural Resources and Environmental Sciences, University of Illinois at Urbana-Champaign, Urbana, IL 61801, USA

S. S. Korban (ed.), *The Apple Genome*, Compendium of Plant Genomes,
https://doi.org/10.1007/978-3-030-74682-7_7

either genes or genetic markers during crossover of homologous chromosomes, whereas those of a physical map are actual physical distances, ideally measured in nucleotide bases.

The unit of relative distance in genetic maps is the centiMorgan (cM), named in honor of the early geneticist Thomas Hunt Morgan. It is common knowledge that recombination events do not occur evenly along a chromosome. Thus, some sections of a chromosome would exhibit either recombination hotspots or coldspots, yielding either higher or lower recombination frequencies, respectively, than otherwise would be expected (Nachman 2002). Hence, a recombination frequency is an approximate estimation of how physically far apart genes or markers are located along a chromosome. Overall, two adjacent genes or markers with a low recombination frequency are likely to be closer together, while those with a higher recombination frequency are likely to be farther apart. Thus, a physical map is a more accurate representation of a genome.

Overall, a genetic (linkage) map serves as a platform to identify molecular markers linked to major genes/quantitative trait loci (QTL) contributing to desirable economic traits, whereas a physical map is very useful both for the development of DNA markers for a genomic region of interest and effective positional cloning of genes. Therefore, establishing reliable and dense linkage and physical maps are critical for pursuing genetic studies.

Apple (*Malus* × *domestica* Borkh.) is one of the most important temperate fruit tree crops grown in the world. It belongs to the sub-family Maloideae in the Rosaceae family. The domesticated apple has an autopolyploidy origin, but its genome is diploidized, with a basic chromosome number of x = 17 (Velasco et al. 2010). Overall, the apple genome exhibits high levels of heterozygosity due to genetic barriers, including self-incompatibility and lack of inbreeding. In addition, the apple displays juvenile periods ranging between 6 and 10 years or even longer, making it difficult to pursue genetic improvement using conventional breeding methods. Thus, early identification of young seedlings carrying economically desirable traits using marker-assisted selection (MAS) is highly valuable in apple breeding programs. Furthermore, knowledge of genome-wide DNA-sequence variations among segregating populations is a fundamental step toward dissecting those genetic controls of complex traits in plant species of large and/or complex genomes. During the past two decades, various efforts have been undertaken to characterize the genetic basis of desirable horticultural traits in apples. These efforts have significantly contributed to the development of genomic tools and resources for the apple, including robust molecular markers, high-density linkage maps, an integrated physical and genetic map, and drafts of complete genome sequences. In this chapter, we report on efforts and progress made in undertaking genetic and physical mapping of the apple genome.

7.2 A Double Pseudo-Testcross Strategy and Genetic Markers

The first steps involved in constructing a genetic map include the development of a mapping population and robust genetic markers. The primary types of mapping populations are F_2 and backcross (BC) populations, as well as recombinant inbred lines (RILs). An F_2 population is developed by selfing F_1 hybrids derived by crossing two parents, while BC populations are developed by crossing selected F_1 hybrids with one of the two parents used in the original cross. RILs are developed by single-seed selections from individual plants of an F_2 population.

As apple is self-incompatible and has a long juvenile period, this highly prevents the development of the above-listed primary types of mapping populations. However, as apples are highly heterozygous, an F_1 generation displays wide segregation for various traits. As many loci consist of two alleles, a segregating F_1 generation of apple would in fact exhibit either a typical F_2 segregation pattern with a 1:2:1 ratio (referred to as intercross loci) or a backcross pattern with a 1:1 ratio (referred to as testcross loci). Using testcross markers segregating in one parent but not in the other, Grattapaglia and Sederoff (1994) have proposed undertaking a double pseudo-testcross strategy for pursuing linkage mapping using an F_1 apple progeny derived from a cross between two outbred parents. The pseudo-testcross mapping strategy is simple to implement, and in principle can be exploited using any type of molecular marker (Fig. 7.1). Using this pseudo-testcross strategy, two parent-specific linkage maps are then constructed. Subsequently, these two maps are integrated based on genetic markers present in both linkage maps. This pseudo-testcross strategy has greatly facilitated conducting genetic mapping studies of outcrossed species, and it has been widely used in constructing linkage maps of various fruit tree crops, including apple.

7.2.1 Development of Molecular Markers

Over the past three decades, multiple technologies have been used to develop molecular markers for apples in order to genotype germplasm and/or segregate populations. Earlier, isozymes, restriction fragment length polymorphisms (RFLPs), and random amplified polymorphic DNAs (RAPDs) have been initially used in linkage mapping studies (Hemmat et al. 1994; Conner et al. 1997). Although isozymes are robust and informative, they are influenced by environmental conditions and exhibit variations during plant development. RAPDs are quick and easy to assay; moreover, those RAPD loci found to be heterozygous in the two parents can also be used for map alignment. However, RAPDs present a drawback due to their low degree of transferability across multiple mapping populations. Although RFLPs are highly reliable and reproducible, their assays are time-consuming, labor-intensive, and expensive.

Subsequently, amplified fragment length polymorphisms (AFLPs) and simple sequence repeats (SSRs), also known as microsatellites, have been developed to construct linkage maps in apples (Maliepaard et al. 1998). As AFLPs are dominant markers and their assays are time-consuming and expensive, SSRs, frequently abundant in genomes, have been deemed preferable for use in constructing linkage maps. SSR markers are either non-expressed/genomic SSRs (G-SSRs) or expressed sequence tag (EST) SSRs (EST-SSRs). Furthermore, they are co-dominant, informative, reliable, and highly reproducible. Microsatellites in coding regions may influence gene expression and/or functionality, rendering them as valuable resources for genetic studies. A major advantage of developing EST-SSRs is the likelihood of identifying associations between functional genes and phenotypes. In addition, EST-SSRs are highly transferable across species. However, the

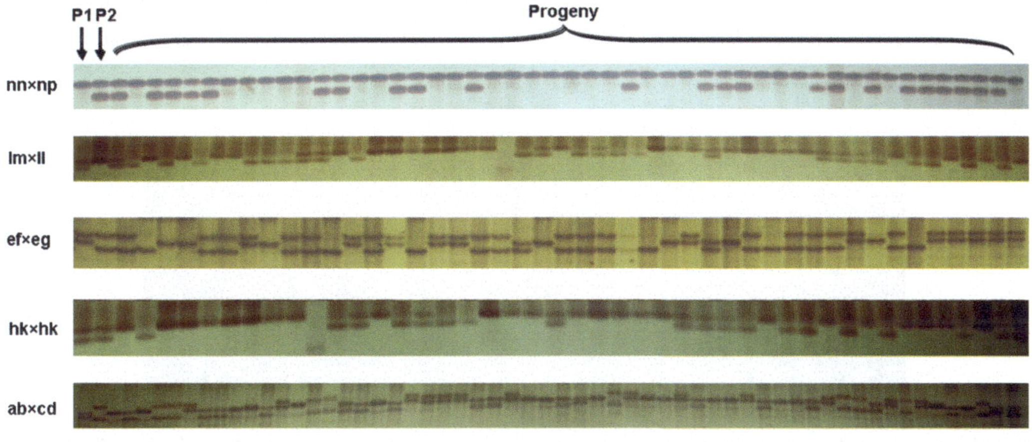

Fig. 7.1 Different types of molecular markers used for a pseudo-testcross mapping strategy. P1 and P2 correspond to the two parents of a cross, while the other letters correspond to different molecular marker combinations

mutation rate of transcribed sequences is lower than that detected in non-transcribed sequences. Thus, the frequency of EST-SSRs is expected to be lower than that of G-SSRs.

A critical step in developing SSR markers is the recovery of sequence information. In apple, SSR markers have been first developed from a genomic library of the cultivar 'Royal Gala' enriched for 'GA' repeats (Guilford et al. 1997). A total of 28 SSRs have been developed by sequencing some of the generated DNA fragments. Later, more than 400 SSRs have been developed using this SSR-enriched genomic library approach (Guilford et al. 1997; Gianfranceschi et al. 1998; Hokanson et al. 1998; Liebhard et al. 2002; Hemmat et al. 2003; Vinatzer et al. 2004; Silfverberg-Dilworth et al. 2006). Liebhard et al. (2002) and Silfverberg-Dilworth et al. (2006) have developed ~300 SSRs that have been mostly used to build a reference map for a 'Fiesta' × 'Discovery' progeny. Subsequently, SSRs associated with expressed sequences have also been extensively exploited, and a total of 1,160 non-redundant EST sequences containing SSR motifs have been identified (Han et al. 2011). Of 1,160 EST-SSRs, 323 are found to be polymorphic between two parents of a segregating population, and these have been successfully mapped in apple (Fig. 7.2). While investigating genome-wide SSRs in the apple draft genome, Zhang et al. (2012) have identified a total of 28,538 SSRs, with an overall density of 40.8 SSRs per Mb (Table 7.1). Di-nucleotide repeats are the most frequent SSRs in the apple genome, accounting for 71.9% of all SSRs. Interestingly, it has been reported that AT/TA repeats are the most frequent in genomic regions, accounting for 38.3% of all G-SSRs; however, AG/GA dimers prevail in transcribed sequences, accounting for 59.4% of all EST-SSRs. It is reported that ~79.0% of SSRs in the apple genome are polymorphic among cultivars and wild relatives (Zhang et al. 2012).

With the availability of large amounts of DNA sequence information for apples, single nucleotide polymorphisms (SNPs) have attracted much more attention in the last decade. An SNP corresponds to a variation in a single nucleotide that occurs at a specific position within a genome. Thus, SNPs are major contributors to genetic variations and account for approximately 80% of all known polymorphisms in organisms. For example, approximately 200 SNPs have been developed from bacterial artificial chromosome (BAC) end sequences (BESs), and these SNPs have been genetically mapped using a bin mapping strategy (Han et al. 2009). Furthermore, a total of 37,807 SNPs have been identified by mining an EST database of 14 apple genotypes, with an average of one SNP per 187 bp (Khan et al. 2012a). Recently, high-throughput SNP discovery through whole-genome re-sequencing has been widely reported in apple. For example, the International RosBREED SNP Consortium (IRSC) has selected 27 apple cultivars from worldwide breeding programs for re-sequencing at low-coverage using an Illumina Genome

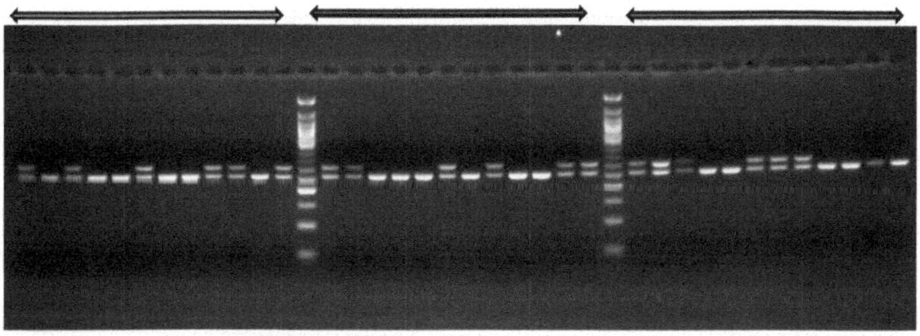

Fig. 7.2 Validation of three EST-SSRs using 12 apple seedlings derived from a cross between 'Coop 17' and 'Coop 16' using metaphor gel electrophoresis. Each of the colored double-arrow bars corresponds to a different EST-SSR marker

Table 7.1 SSRs identified in the apple genome

Repeat	Region	Repeat length (nucleotides)				Total	Frequency (%)
		<30	30–40	40–50	>50		
Dimer	CDS	76	0	0	0	76	0.27
	UTR	2098	598	158	153	3007	10.54
	Genomic	10860	3817	1755	1010	17442	61.12
Trimer	CDS	260	5	0	0	265	0.93
	UTR	425	22	3	2	452	1.58
	Genomic	2411	211	69	95	2786	9.76
Tetramer	CDS	10	0	0	0	10	0.04
	UTR	158	11	1	1	171	0.60
	Genomic	1472	107	22	11	1612	5.65
Pentamer	CDS	16	0	0	0	16	0.06
	UTR	223	7	0	0	230	0.81
	Genomic	1658	19	1	2	1680	5.89
Hexamer	CDS	136	8	1	1	146	0.51
	UTR	248	10	2	0	260	0.91
	Genomic	355	20	9	1	385	1.35

CDS, coding sequence; UTR, untranslated region

Analyzer II (Chagné et al. 2012a, b). Alignments of these sequences to the whole-genome sequence of 'Golden Delicious' have revealed a total of 2,113,120 SNPs, with an average frequency of one SNP per 288 bp of the apple genome. Later, re-sequencing and variant calling have been conducted for 16 accessions, including 13 apple (*M.* × *domestica*) cultivars, one accession of *M.* × *micromalus,* and two di-haploid (DH) accessions derived from 'Golden Delicious', revealing the presence of 16,614,171 variants such as SNPs and indels (Bianco et al. 2014). Of these variants, 750,763 SNPs are of high quality with SNP_Scores above 0.7, thus serving as valuable resources for developing an SNP array. In addition, whole-genome re-sequencing, at high depth of coverage, has been carried out for 63 different cultivars covering most of the genetic diversity of the cultivated apple (Bianco et al. 2016). Following the alignments of these sequences against the Apple Genome Reference v.3 (https://www.rosaceae.org/), a total of 15,499,525 variants have been identified with multi-sample SNP calling.

7.2.2 High-Throughput Genotyping

Screening DNA variants in a segregating population is a key step in undertaking genetic mapping efforts. In line with the ever-expanding genome sequence information, cost-effective genotyping technologies have been developed that can simultaneously assay different genetic markers.

Initially, a fluorescence microsatellite genotyping method was developed to facilitate high-throughput genotyping of SSRs. Multiple microsatellite products of different sizes and/or labeled with different fluorescence dyes were mixed and then subjected to capillary electrophoresis. This fluorescence-labeled microsatellite genotyping method allowed for unambiguous allele identification (Fig. 7.3), and it was successfully used in genotyping of a segregating population of apples (Han et al. 2011).

Subsequently, as SNP arrays offer highly multiplexed assays at a relatively low cost, high-density SNP arrays have been developed to evaluate whether or not a genome-wild

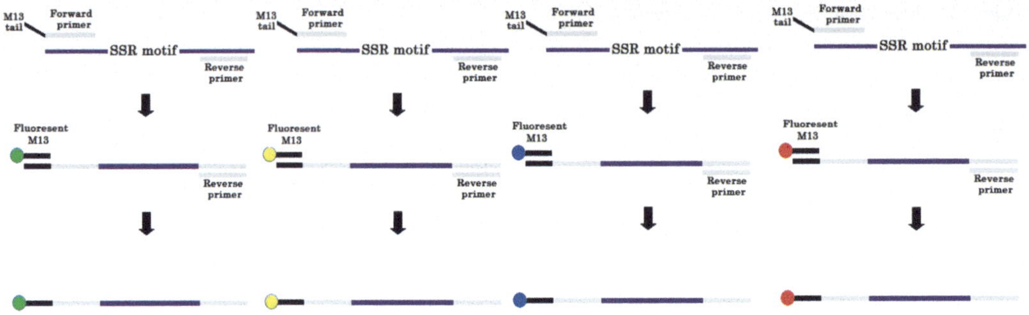

Fig. 7.3 Fluorescent microsatellite genotyping using a three-primer strategy

assessment of allelic variation is feasible in apples. The first apple SNP array, composed of 1,536 SNPs and labeled as the Illumina GoldenGate™ assay, has been developed in the United States by Khan et al. (2012a). During the same year, an 8 K Infinium II array has been also developed by the Plant & Food Research Institute in New Zealand (Chagné et al. 2012a, b). These two apple SNP arrays represent the first generation of high-throughput SNP genotyping platforms. However, the amount and distribution of SNPs present in these first-generation arrays have not been sufficient to perform pedigree-based analysis at high levels of precision. Thus, two additional apple SNP arrays, the 20 K whole-genome genotyping (WGG) array (Bianco et al. 2014) and the Axiom® Apple480K (Bianco et al. 2016), have been subsequently developed.

Furthermore, restriction-site associated DNA sequencing (RAD-Seq) has been developed to scan SNPs at the genome-wide level in apple, and this has proven to be a fast and cost-effective approach for pursuing genetic mapping in apple (Gardner et al. 2014). For example, a double-digest RAD-Seq (ddRAD-Seq) was used to genotype an apple segregating population of 1,733 F_1 seedlings derived from 'Jonathan' × 'Golden Delicious'. This resulted in ~15,000 and 38,000 SNPs for *Eco*RI and *Hin*dIII restriction sites, respectively, and this was used to develop a dense SNP-based genetic linkage map (Sun et al. 2015). Moreover, an efficient genotyping method of a specific-locus amplified fragment sequencing (SLAF-seq), similar to ddRAD-Seq, was developed to screen 30 diverse apple accessions, consisting of 20 cultivars and 10 wild relatives (Fig. 7.4; Ma et al. 2017). This screening resulted in rapid observation of linkage disequilibrium (LD) decays in both cultivated and wild apples, with r^2-values below 0.2 at both 440 bp and 280 bp, respectively (Ma et al. 2017). With the advent of next-generation sequencing (NGS), whole-genome re-sequencing has been used for genotyping limited numbers of samples in apple (Duan et al. 2017). However, this strategy has been deemed cost-prohibitive for large populations consisting of thousands of individuals.

7.3 Genetic Mapping of the Apple Genome

7.3.1 Linkage Map Construction

Earlier, the lack of availability of molecular markers has been one of the main reasons that hindered the development of genetic maps for apple. Earlier, allozymes have been the only form of genetic markers available for apples. However, allozymes are not suitable for genome mapping as they are limited in number and low in polymorphisms. With the availability of molecular markers such as RAPDs, Hemmat et al. (1994) have developed the first genetic map for apples using an F_1 population derived from a cross between 'Rome Beauty' and 'White Angel'. The linkage map for 'Rome Beauty' consists of 156 markers spread across 21 linkage groups, while the 'White Angel' map contains 253 markers

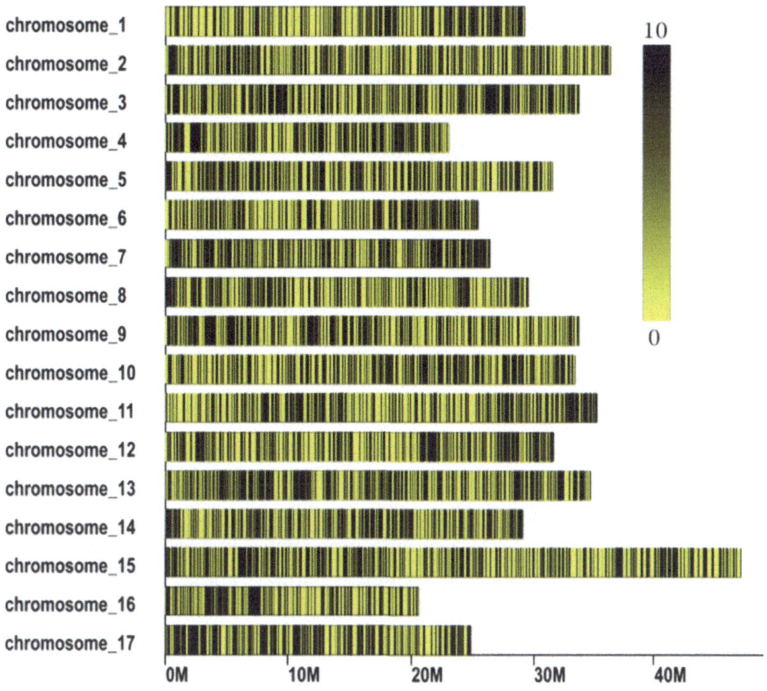

Fig. 7.4 Distribution of 39,635 SNPs along an apple reference genome

arranged across 24 linkage groups. Later, linkage maps have been constructed for three parents of two segregating populations, 'Wijcik McIntosh' NY75441-67 and 'Wijcik McIntosh' × NY75441-58 (Conner et al. 1997). The linkage maps for 'Wijcik McIntosh', NY75441-67, and NY75441-58 consist of 238, 110, and 183 markers mapped onto 19, 16, and 18 linkage groups, respectively. These earliest linkage maps have been primarily based on RAPD markers, along with several isozymes and morphological markers. As RAPDs have a very low degree of transferability across species and mapping populations, the usefulness of the earliest linkage maps of apple has been rather limited. Moreover, the number of linkage groups identified does not correspond to the number of chromosomes. For the above reasons, an integrated linkage map based on a number of co-dominant transferable markers, mostly RFLPs along with a small number of microsatellites, distributed over all 17 linkage groups have been developed using a 'Prima' × 'Fiesta' population consisting of 152 seedlings, and is deemed as the first apple reference genetic map (Maliepaard et al. 1998).

As mentioned above, SSRs are informative, reliable, and easily transferable across species. Therefore, co-dominant SSRs are superior markers for genome mapping. An SSR-based map can serve as an ideal reference map as SSRs are easily transferable to different progenies. A large-scale application of SSRs in linkage map construction has been initially conducted by both Liebhard et al. (2003) and Silfverberg-Dilworth et al. (2006). A set of approximately 300 SSRs have been used to build a reference map for the 'Fiesta' × 'Discovery' progeny. This reference linkage map also contains large numbers of dominant AFLPs and RAPDs that have filled in gaps or marker-rare regions (Liebhard et al. 2003). Such robust polymerase chain reaction (PCR)-based saturated reference maps are essential for genome-wide identification of quantitative trait loci (QTLs) controlling complex traits.

To integrate physical and genetic maps, a linkage map primarily based on SSRs was developed by Han et al. (2011). A total of 442 SSRs (329 EST-SSRs and 113 gSSRs) along with 12 sequence tag sites (STS) were used to generate a saturated reference map for the 'Co-op

17' × 'Co-op 16' progeny. Later, Zhang et al. (2012) investigated genome-wide SSRs based on the draft whole-genome sequence of apple, and a total of 299 SSRs were selected to construct an SSR-based reference linkage map for the 'Jonathan' × 'Golden Delicious' progeny.

Although the number of mapped SSRs is an important determinant of the value of a reference map, it is more important that such mapped SSRs have good coverage of the entire genome along with a relatively even distribution along the chromosomes. However, as SSRs are not evenly distributed across the genome, the marker density of SSR-based linkage maps is usually low and insufficient for fine-mapping of QTLs of interest. Besides SSRs, SNPs are also excellent markers for genome mapping as SNPs not only represent the most common type of DNA sequence variation in a genome, but they are also amenable for high-throughput genotyping using a highly robust array-based platform. In apple, SNPs are notably abundant as the apple is an outcross species. To develop highly saturated linkage maps of apple, SNPs have been extensively exploited by whole-genome re-sequencing to develop SNP arrays (Chagné et al. 2012a, b; Bianco et al. 2014, 2016). A couple of high-density SNP-based linkage maps have been constructed using SNP arrays in apple (Antanaviciute et al. 2012; Khan et al. 2012b; Falginella et al. 2015; Ma et al. 2016). An example of SNP-based linkage map is presented in Fig. 7.5 (Ma et al. 2016). An ultra-high-density linkage map of apple comprising 15,417 SNP markers spanning 1267 cM, with an average distance of 0.37 cM and a maximum distance of 3.29 cM between adjacent markers, has been developed by Di Pierro et al. (2016). These high-density linkage maps not only facilitate the identification of candidate genes of interest present within assembled sequence data of the apple genome, but they are also useful for identifying genomic structural variations. With the development of advanced sequencing technologies, genotyping of mapping populations can be conducted by direct genomic DNA sequencing at a low cost. Genome-wide screening of genomic DNA variations will accelerate studies of genome mapping.

7.3.2 Mapping of Genes for Disease and Pest Resistance

Most gene mapping studies have focused on the identification of markers closely linked to genes responsible for resistance to diseases and pests. Many resistance traits against diseases or pests in apples are simply inherited, and they are mostly controlled by single genes with recognizable effects. Until now, significant progress has been made in identifying genes conferring resistance to various diseases and pests in apples, such as apple scab, powdery mildew, and aphids (Han and Korban 2010).

7.3.2.1 Apple Scab Resistance

Apple scab, incited by *Venturia inaequalis* Cke. (Wint.), is a serious fungal disease in all apple-growing regions of the world. Early efforts have focused on identifying genes controlling apple scab from different *Malus* germplasm to develop cultivars with resistance to this disease. Although different genes for scab resistance have been identified, the *Vf* gene form of the wild crab apple species *M. floribunda* clone 821 has been the focus of major research efforts (Gessler et al. 2006; Gessler and Pertot 2012). In recent years, the nomenclature of the *Vf* gene has been changed to *Rvi6*. Therefore, we will include this new nomenclature for all sub-headings; however, as the work in all previously published articles cited in the sections below have used the *Vf* nomenclature, we will retain the use of this older nomenclature as cited in these articles to avoid confusion.

The *Vf* (*Rvi6*) Gene

The *Vf* gene has been widely introgressed into the cultivated apple, and dozens of scab-resistant cultivars have been developed from different breeding programs around the world (Crosby

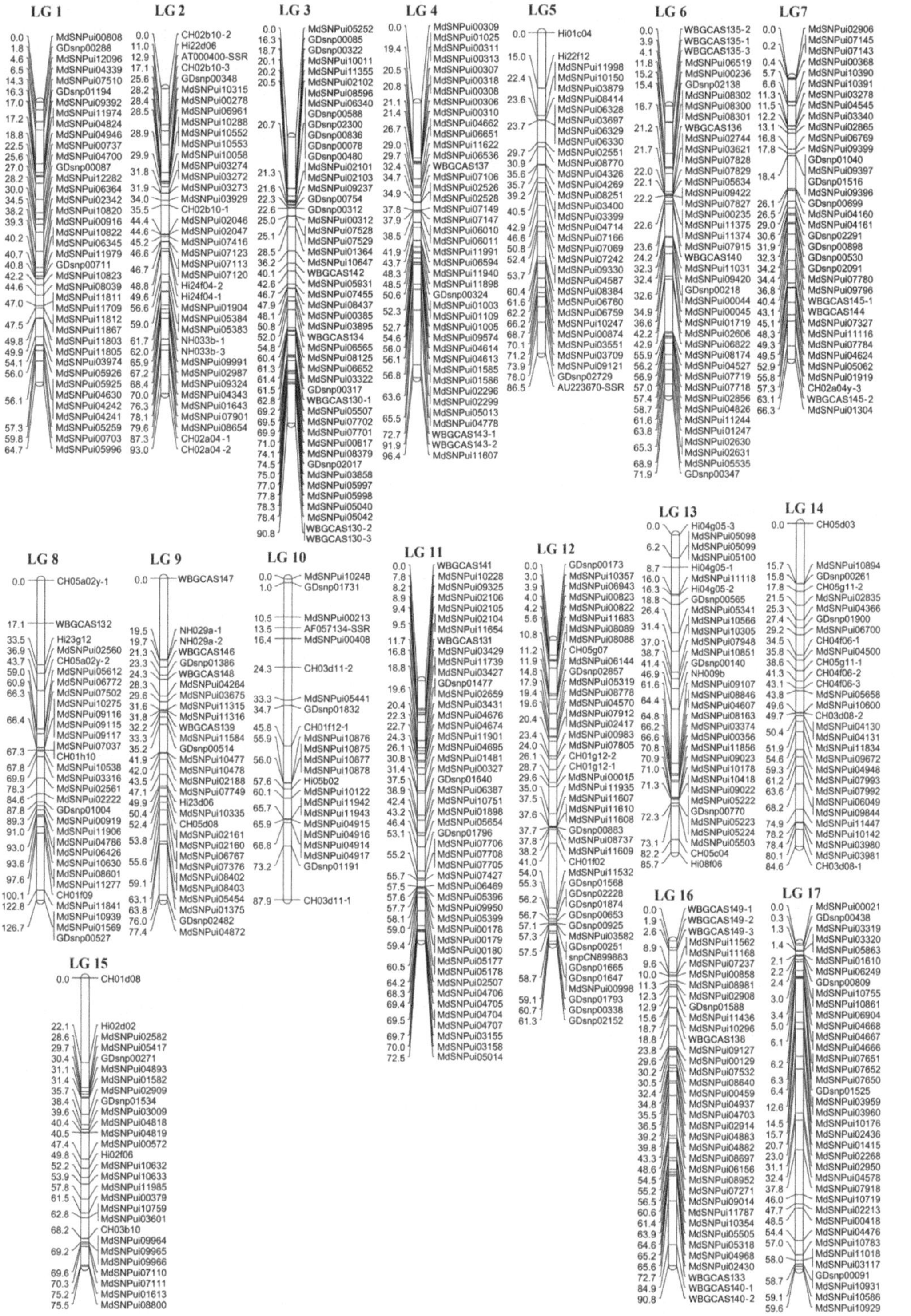

Fig. 7.5 Genetic linkage maps of the 'JG' × 'WSH' apple mapping population developed using an apple SNP array. The genetic distance is shown in centimorgan (cM) along the left-hand side. LG, linkage group; JG, cv. 'Jiguan' (female parent); WSH, cv. 'Wangshanhong' (male parent)

et al. 1992). Early on, the *Vf* gene has been found to confer field immunity to races 1 to 5 of the fungal pathogen *V. inaequalis* (Lespinasse 1989), but in later years, it has been found that this gene can be overcome by races 6 and 7 in some apple/ *Malus* genotypes (Roberts and Crute 1994; Parisi and Lespinasse 1996; Papp et al. 2019). The *Vf* locus has been mapped onto LG1 of the apple genome (Maliepaard et al. 1998). Interestingly, a team of Swiss/Italian researchers in Europe and a group at the University of Illinois at Urbana-Champaign in the United States have separately isolated and cloned the *Vf* gene using chromosome walking and chromosome landing strategies, respectively. These two cloning strategies have been previously described by Han and Korban (2010); therefore, a synopsis of these strategies will be provided herein.

Cloning of the *Vf* (*Rvi6*) Gene Using a Chromosome Walking Strategy

In this strategy, RAPD markers linked to the *Vf* gene were identified and converted into either sequence characterized amplified regions (SCARs) or cleaved amplified polymorphic sequences (CAPS) (Yang and Krüger 1994; Koller et al. 1994; Tartarini 1996; Gianfranceschi et al. 1996; Gardiner et al. 1996; Yang et al. 1997; Hemmat et al. 1998; Gianfranceschi et al. (1996); Yang and Korban (1996); Tartarini et al. (1999)). Identifying reliable and reproducible makers facilitated the launch of efforts for pursuing map-based cloning of the *Vf* gene using a chromosome walking strategy, as described in the following sections.

In order to identify markers closely flanking the *Vf* gene, segregating populations for this gene were used to identify a CAPS marker M18 located on the right-hand side of the *Vf* gene at an interval of 0.5 cM, while two other markers AL07 and AM19, at a genetic distance of 0.3 cM, were mapped along the left-hand side of the *Vf* gene at an interval of 0.9 cM (Tartarini et al. 2019; Patocchi et al. 1999b). This allowed for the construction of a BAC contig spanning the *Vf* gene region, wherein Patocchi et al. (1999a) estimated the physical distance between the markers M18 and AL07 to be ~550 kb, while the physical distance between M18 and AM19 was found to be less than 870 kb, thus suggesting that AM19 was closer to the *Vf* gene than that of AL07 (Patocchi et al. 1999a). Chromosome walking was then initiated from the two flanking markers AM19 and M18 toward the *Vf* gene using a genomic BAC library of apple cv. 'Florina', with an average insert size of 100–150 kb (Patocchi et al. 1999a; Vinatzer et al. 1998). Following nine rounds of chromosome walking steps, a total of 13 BAC clones spanning the *Vf* gene region between the two markers M18 and AM19 were identified (Patocchi et al. 1999b). A fine-mapping effort narrowed the *Vf* gene to a shorter region of ~350 kb, consisting of a minimum of five clones corresponding to the minimum tiling path (Patocchi et al. 1999b). Inserts of these five BAC clones were then used as probes to screen a cDNA library from scab-inoculated leaves of cv. 'Florina'. Hybridizing cDNA sequences were compared to sequences available in Genbank resulting in the identification of three cDNA clones that were similar to the *Cladosporium fulvum* (*Cf*) resistance gene cluster from tomato (Vinatzer et al. 2001). A gene cluster of several *Cf* homologs was found in the *Vf* region, and genomic DNA sequences of apple scab resistance *HcrVf* genes, *HcrVf* 1 to 4, were found and physically mapped to the *Vf* region (Vinatzer et al. 2001). Of these, *HcrVf* 1, *HcrVf2*, and *HcrVf4* were expressed and co-segregated with the *Vf* resistance reaction, while *HcrVf3* was not expressed and designated as a pseudogene. Furthermore, the deduced proteins of *HcrVf1*, *HcrVf2*, and *HcrVf4* belonged to the leucine-rich repeat (LRR) class of proteins (Vinatzer et al. 2001). These *HcrVf* genes were deemed similar to *Cf* resistance genes, as they are likely to encode extracytoplasmic glycoproteins anchored to a cell membrane, with the majority of the extracytoplasmic domain consisting of LRR motifs (Vinatzer et al. 2001).

Cloning of the *Vf* (*Rvi6*) Gene Using a Chromosome Landing Strategy

In the second strategy for positional cloning of this *Vf* gene, Xu and Korban (2002) used chromosome landing to isolate this gene. They capitalized on a narrow-down bulk segregant strategy to screen scab-resistant individuals carrying the *Vf* gene, each of two bulks, and each consisting of two individuals, carrying a short introgressed region either left or right of the *Vf* gene. These bulks were used to identify AFLP markers tightly linked to the *Vf* gene (Xu and Korban 2000). During the process of this "narrow-down" bulking strategy, scab-resistant apple selections were pooled to generate a resistant bulk (RB), while scab-susceptible apple cultivars were used to prepare a susceptible bulk (SB). These RB and SB, as well as the original source of the *Vf* gene, *M. floribunda* 821, were used to screen AFLP markers linked to the *Vf* gene. Mapping results demonstrated that eight AFLP markers were linked to the *Vf* gene, but a single marker was found to be tightly linked to the *Vf* gene (Xu and Korban 2000). Subsequently, these eight AFLP markers were used to estimate the extent of the introgressed region covering the *Vf* gene. Within the RB bulk, a group of selections was found to have a short introgressed genomic region located left (designated as L-RB) of the *Vf* gene, while another group of selections was found to have a relatively short introgressed genomic region located to the right (designated as R-RB) of the *Vf* gene. Therefore, four DNA samples from *M. floribunda* 821, L-RB, R-RB, and SB were then used to screen for those AFLP markers closest to the *Vf* gene. A total of 31 putative AFLP markers were identified as these were present in *M. floribunda 821*, L-RB, and R-RB, but not in SB. These AFLP markers, likely to be linked to the *Vf* gene, were mapped using all scab-resistant selections individually. Subsequently, this narrow-down strategy led to estimating the extent of the introgressed region present in each scab-resistant selection in these sub-bulks, and a total of 15 AFLP markers that co-segregated with the *Vf* gene were identified.

This was subsequently followed by a fine-mapping of the *Vf* gene using scab-resistant F1 seedling populations from each of a cross 'Co-op 17' × 'Co-op 16' and a cross of 'Jonafree' × 'Ill. Del. No.1'. These seedling populations identified seven AFLP markers that were very close to the *Vf* gene, wherein no recombinant was identified in the two resistant seedling populations, while an eighth AFLP marker yielded two recombinants and was placed on the right side of the *Vf* gene at an interval of 0.4 cM; however, the remaining seven AFLP markers yielded a single recombinant and were anchored to the left side of the *Vf* gene at an interval of 0.2 cM. Therefore, of all 15 *Vf*-linked AFLP markers, 11 were converted into SCAR markers (Xu et al. 2001a). These SCARs were then used to screen two apple BAC libraries, one from *M. floribunda* 821 while the second from apple cv. 'Goldrush' (scab-resistant, carrying *Vf*), to identify BAC clones containing the *Vf* gene (Xu et al. 2001b; Xu and Korban 2001). A 200-kb BAC contig spanning the *Vf* locus was constructed (Xu et al. 2002). BAC clones containing the *Vf* gene were subcloned, and these were hybridized with total cDNAs prepared from salicylic acid-treated leaves. A cluster of four resistance paralogs, designated as *Vfa1*, *Vfa2*, *Vfa3*, and *Vfa4*, was identified. Following analysis of this cluster, *Vfa1*, *Vfa2*, and *Vfa4* were found to encode proteins characterized with extracellular LRRs and transmembrane (TM) domains and were differentially expressed in leaf tissues, with *Vfa4* highly expressed in mature leaves. However, *Vfa3* was truncated and found to be non-functional. Differential expression profiles among the four paralogs were observed during leaf development. When pairwise comparisons of these *Vf* paralogs were conducted, it was revealed that these must have evolved from four ancient *Vf* members following two sequential duplication events of a single *Vf* progenitor initially present in the *Malus* genome (Xu and Korban 2004).

Other Genes Conferring Resistance to Apple Scab

In addition to the *Vf* (*Rvi6*) gene described above, a total of 15 other major apple scab resistance genes have been identified (Bus et al. 2011; Han and Korban 2010). Of these, a group of six genes, including *Vh2* (also known as *Vr*), *VT57*, *Vh4* (also known as *Vr1* or *Vx*), *Vr2* (also known as *Rvi15*, identified from apple accession GMAL 2473 at PGRU in Geneva, New York), *Vbj* (from

M. baccata jackii), and *Vh8* (from *M. sieversii* W193B) have been mapped onto LG2 (Schouten et al. 2014; Gyax et al. 2004); moreover, *Vh2*, *VT57*, *Vbj*, and *Vh8* are reported to be closely co-located to each other (Bus et al. 2005b; Patocchi et al. 2004; Galli et al. 2010). It is important to point out that *Vh2*, *VT57*, and *Vh4* are identified in the Russian apple seedling R12740-4A (*Malus* Mill. sp.), and these contribute to the high level of scab resistance observed in this particular selection (Dayton and Williams 1968; Bus et al. 2005a; Boudichevskaia et al. 2004, 2006). Furthermore, another two genes, *Vb* (from Hansen's *baccata* #2) and *Vg* (from 'Golden Declicious'), are mapped onto LG12 (Hemmat et al. 2003; Erdin et al. 2006; Bénaouf et al. 1997; Bénaouf and Parisi 2000). Furthermore, genes *Va* (from 'Antonovka' PI 172633), *Vd* (from an Italian apple cultivar 'Durello di Forli'), and *Vm* (from *M.* × *micromalus* clone no. 804) are mapped onto LGs 1, 10, and 17, respectively (Dayton and Williams 1970; Cheng et al. 1998; Patocchi et al. 2005; Tartarini et al. 2004; Calenge et al. 2005a; Soriano et al. 2009), whereas *Vmis* (from a Mildew Immune Seedling [MIS], an open-pollinated seedling of apple cv. 'Starking Delicious') has not yet been mapped (Dayton 1977; Gardiner et al. 2001). Overall, it appears that LG2 must serve as a hotspot for apple scab resistance genes. Additional details of these genes and their linkages to molecular markers are provided by Han and Korban (2010).

Bastiaanse et al. (2016) have also identified five additional loci conferring scab resistance in apple cv. 'Geneva', located at the bottom of LG4, and designated as *Rvh3.1* to *Rvh3.5*. It is reported that *Rvh3.1*, *Rvh3.2,* and *Rvh3.3* confer dominant scab resistance, while *rvh3.4* and *rvh3.5* display recessive scab resistance. Five nucleotide-binding site–leucine-rich repeat (NBS–LRR) candidate genes are found to be tightly linked to the dominant *R* genes *Rvh3.1*, *Rvh3.2*, and *Rvh3.3* (Bastiaanse et al. 2016).

7.3.2.2 Powdery Mildew Resistance

The fungal disease powdery mildew is incited by *Podosphaera leucotricha,* and it is deemed as the second most important economic disease of apple. Mapping of genes for resistance to powdery mildew has been difficult, and thus far, five major genes for resistance have been identified. These include *Pl-1* from *M.* × *robusta* (Markussen et al. 1995; Dunemann et al. 2004, 2007), *Pl-2* from *M.* × *zumi* (Knight and Alston 1968; Seglias and Gessler 1997), *Pl-w* from the crab apple 'White Angel' (Gallot et al. 1985), *Pl-d* from the accession 'D12' (Visser and Verhaegh 1976), and *Pl-m* from MIS (Dayton 1977).

The *Pl-1* gene has been mapped onto LG12 near the Hi07f01 SSR marker and close to an NBS–LRR candidate resistance gene (Dunemann et al. 2007). Recently, comparative transcriptome analysis has revealed the presence of 24 genes that are differentially expressed between trees resistant to and susceptible to powdery mildew (Jensen et al. 2014). These genes are physically clustered within the vicinity of the previously reported major QTL for powdery mildew resistance on LG12. Therefore, this region on LG12 represents a particular region of interest for mildew resistance genes (Han and Korban 2010). The *Pl-2* gene from *M.* × *zumi* is located on LG6, later referred to as LG11 following Maliepaard et al. (1998), and an EST close (0.9 cM) to the *Pl-2* gene has been identified (Dunemann et al. 1999; Gardiner et al. 2003; Kellerhals et al. 2000). The *Pl-w* gene, from the ornamental crab apple cv. 'White Angel', has been mapped to LG8 (Korban anad Dayton 1983; Gallot et al. 1985; Manganaris 1989; Battle and Alston 1996; Hemmat et al. 1994; Maliepaard et al. 1998; Evans and James 2003). The *Pl-d,* a gene identified from a seedling (D12) of the 'D series' of wild apples grown in South Tyrol (Visser and Verhaegh 1976), has been mapped to LG12 at a similar position to that of the *Pl-1* gene (James et al. 2004). These findings suggest that there is a clustering of genes for resistance to both scab and mildew on LG12, thus highlighting the importance of this linkage group (Han and Korban 2010). As for the *Pl-m* gene from MIS, it is proposed that this gene is mapped onto LG11 (Gardiner et al. 2003; Han and Korban 2010).

Interestingly, seven QTLs for powdery mildew resistance have been identified in a clone of an open-pollinated apple cv. 'Primula', designated as 'U211', in Poland (Stankiewicz et al. 2005). At least two of these QTLs are reported to be different from those of *Pl-1*, *Pl-2*, *Pl-w*, and *Pl-d*, thus are likely to serve as two additional genes for mildew resistance (Calenge and Durel 2006).

Recently, knocking out specific members of the *Mildew Locus O* (*MLO*) gene family, responsible for powdery mildew susceptibility (*S*-genes), has been used to identify and map additional candidate genes for mildew resistance (Pessina et al. 2014). The *MLO* genes can be divided into seven clades, and *S*-genes are usually clustered with clades IV and V. The apple genome contains four *MLO* genes that belong to clade V. Of these four genes, *MdMLO11* and *MdMLO19* are up-regulated during powdery mildew infection, while *MdMLO5* and *MdMLO7* are not transcriptionally responsive to powdery mildew infection (Pessina et al. 2014). Silencing of *MdMLO19* reduces powdery mildew disease severity by 75%, while knock-down of *MdMLO11*, either alone or in combination with *MdMLO19*, does not cause any reduction or additional reduction of susceptibility compared with *MdMLO19* alone (Pessina et al. 2016). Hence, it has been deemed that *MdMLO19* plays a critical role in apple powdery mildew susceptibility.

7.3.2.3 Fire Blight Resistance

Fire blight, incited by the bacterium *Erwinia amylovora*, is a very serious economic disease of apple, as well as for other members of the Rosaceae family. This disease, first discovered on European pear, has become widespread to most apple and European pear-growing regions around the world (Bonn and van der Zwet 2000; Jock et al. 2002; Peil et al. 2009). Following assessment of various wild and cultivated apples, several numbers of species such as *M.* × *robusta* 5 and *M. fusca* clone H-12, along with various apple cultivars such as 'Delicious', 'Winesap', 'Prima', 'Enterprise', 'Liberty', and 'Priscilla', have been found to be either immune or highly resistant to fire blight (Aldwinkle et al. 1976; Korban et al. 1988; Janick et al. 1996). The mechanism(s) of pathogen–host interactions has been reviewed by Khan et al. (2012c).

Overall, genetic resistance to fire blight is reported to be quantitatively controlled, and several major resistance genes have been identified in apples (Korban et al. 1988; Lespinasse and Aldwinckle 2000). Using two progenies, 'Fiesta' × 'Discovery' and 'Prima' × 'Fiesta', Calenge et al. (2005b) have identified a major QTL for fire blight resistance located on LG7 of 'Fiesta', and this QTL is flanked by the two SSR markers (Silfverberg-Dilworth et al. 2006; N'Diaye et al. 2008). Furthermore, four minor QTLs have been also identified on LGs 3, 12, and 13. Khan et al. (2006) have also detected a major QTL on LG7 of 'Fiesta' in the same region as that reported by Calenge et al. (2005b). A subsequent study has revealed that this major QTL on LG7 was co-located with two candidate genes associated with fire blight resistance—an HSP90 gene-family member and a WRKY transcription factor (Baldo et al. 2010).

A second major QTL for fire blight resistance was identified on LG3 of *M.* × *robusta* clone 5 (Peil et al. 2007). A whole-genome scanning approach using SSR markers allowed for mapping of a fire blight resistance gene, and a QTL on LG3 from *M.* × *robusta* 5 was located in the same genomic region of the QTL from 'Fiesta' (Calenge et al. 2005b). As the frequency of phenotypic variation explained by this QTL was very high (80%), a major gene was likely to be involved and that different alleles conferring different levels of resistance must be present in this region of LG3 (Peil et al. 2007). The QTL on LG3 was also confirmed in different mapping populations having *M.* × *robusta* 5 in their pedigrees and following inoculations with different *E. amylovora* strains (Gardiner et al. 2012; Peil et al. 2007 and 2008). Later, Fahrentrapp et al. (2013) identified the Fb_MR5 candidate resistance gene within the QTL region on LG3 of *M.* × *robusta* 5 using a map-based cloning approach. The FB_MR5 gene belonged to the CC-NBS-LRR resistance gene family, and it was linked to two SSR markers, Ch03e03 and Fem18. This gene was confirmed to be the resistance determinant in *M.* × *robusta* 5, as FB_MR5-carrying transgenic plants of the fire blight susceptible cultivar 'Gala' demonstrated resistance to fire blight (Broggini et al. 2014).

Durel et al. (2009) conducted QTL analysis using two F_1 progenies derived from two crosses, a cross between the fire-blight-susceptible rootstock MM106 and a fire-blight-resistant ornamental cv. 'Evereste' and another cross between the moderately susceptible cv. 'Golden Delicious' and the wild species *M. floribunda* 821 of unknown level of resistance to fire blight. Following greenhouse inoculation of seedlings of these two progenies with *E. amylovora*, a major QTL, designated as Fb_E, was identified on the distal region of LG12 of both cv. 'Evereste' and *M. floribunda* 821, explaining 40–70% of the observed phenotypic variation. Subsequently, high-resolution mapping was performed for the major QTL Fb_E on LG12 of cv. 'Evereste' using a chromosome landing strategy, and the genomic region covering the resistance locus was delimited to 78 kb (Parravicini et al. 2011). Two candidate genes, with homologies to the *Pto/Prf* complex in tomato, were deemed the most likely candidates of fire blight resistance genes.

Using an F_1 progeny derived from a cross between two resistant apple cultivars 'Florina' and 'Nova Easygro', Le Roux et al. (2010) identified two QTLs for fire blight resistance in 'Florina'. One QTL, located on LG10, accounted for approximately 15% of the phenotypic variation, and a second QTL, located on LG5, explained approximately 10% of the phenotypic variation. In addition, Khan et al. (2013) detected three minor QTLs for fire blight resistance using an F_1 pedigree of 'Coop 16' × 'Coop 17', and these were located on LG2, LG6, and LG15.

More recently, a total of 13 novel marker-trait associations linked to fire blight resistance were identified in a mapping population of *M. sieversii* with *M.* × *domestica* (Desnoues et al. 2018). These QTLs were distributed over eight linkage groups, but several QTLs were identified on the same linkage groups as those of previously identified QTLs; however, these were located farther away. Interestingly, QTLs identified on LG10 at 15.9 cM, with bacterial strain Ea273 in 2011, were likely the same as the QTL identified with strain CFBP1430 for lesion length during the first two weeks following inoculation in apple cv. 'Florina' (Le Roux et al. 2010), while QTLs identified on LG15 for bacterial strain LP101 could be the same QTLs identified previously with strain CFBP1430 in cv. 'Florina' and in the mapping population of 'Coop 16' × 'Coop 17' (Khan et al. 2013).

7.3.2.4 Aphid Resistance

As various groups of aphids attack apples, resulting in significant damage, efforts have been undertaken to identify sources for genetic resistance. Fortunately, monogenic resistance has been identified in cultivated as well as wild apples for rosy leaf-curling aphid (*Dysaphis devecta*), woolly apple aphid (*Eriosoma lanigerum*), and rosy apple aphid (*Dysaphis plantaginea*). Efforts to map genes for resistance to *D. devecta* and *E. lanigerum* have been successful, while limited efforts have been undertaken to map genes for resistance to *D. plantaginea*.

Genes for Resistance to the Rosy Leaf-Curling Aphid

A total of three genes, *Sd-1* (identified in apple cv. 'Cox's Orange Pippin'), *Sd-2* (identified in apple cv. 'Northern Spy'), and *Sd-3* (identified in both *M.* × *robusta* and *M.* × *zumi),* conferring resistance to different biotypes of the rosy leaf-curling aphid (*D. devecta*) have been identified (Alston and Briggs 1968, 1977). The *Sd-1* gene confers resistance to biotypes 1 and 2 of *D. devecta*, while *Sd-2* and *Sd-3* confer resistance to biotypes 1 and 3, respectively.

Genetic mapping of the *Sd-1* gene was first conducted using the scab-resistant cultivar 'Fiesta', developed from a cross of 'Cox's Orange Pippin' × 'Idared' (Roche et al. 1997a), and the *Sd-1* gene was mapped to a single locus on LG7 (Roche et al. 1997b). Subsequently, a detailed fine-mapping of the *Sd1* locus was conducted using a chromosome walking strategy as described above for the positional cloning of the *Vf* gene (Cevik and King 2002a). However, as the BAC library was developed from cv. 'Florina' (Vinatzer et al. 1998), later found to be susceptible to rosy apple aphid, this BAC library did not contain the *Sd-1* gene. Nevertheless, the *Sd-1* locus was mapped within a small region, ~180 kb, and covered by two overlapping BAC clones (Cevik and King 2002b). The ends of these two BAC clones were isolated and sequenced, and it was found that one BAC-end sequence was highly similar to NBS–LRR resistance gene sequences. As this BAC library did not contain the *Sd-1* resistance gene, a BAC library from an aphid-resistant genotype must be developed in order to identify and clone the *Sd-1* gene. Furthermore, based on a fluorescent in situ hybridization (FISH) analysis, the *Sd-1* region was likely to be located very close to the telomere, and this suggested that further efforts would be necessary to map and clone the *Sd-1* gene (Cevik and King 2002b).

The *Sd-2* resistance locus was first reported as being distinct from the *Sd-1* locus (Alston and Briggs 1977). However, Cevik and King (2002a) mapped the *Sd-2* gene using a cross of 'Double Red Northern Spy' (*Sd-2sd-2*) × 'Totem' (*sd-2sd-2*) and found that both *Sd-1* and *Sd-2* loci were in fact located within the same genomic region. Thus, it was deemed that these two genes were likely to be allelic.

7.3.2.5 Genes for Resistance to the Woolly Apple Aphid

Genetic sources of woolly apple aphid (WAA) resistance have been identified in *Malus*, and at least three different major genes have been reported. One WAA resistance gene was identified in apple cv. 'Northern Spy' (Crane et al. 1936), designated as the *Er* gene (Knight et al. 1962), and then renamed as *Er1* (Alston et al. 2000). A second WAA resistance gene, *Er2*, was identified in *M.* × *robusta* clone 5, often referred to as 'Robusta 5' (Alston et al. 2000), while a third WAA resistance gene, *Er3*, was identified in *M. sieboldii* selection 'Aotea 1' (Bus et al. 2008).

Following gene mapping efforts, *Er1* and *Er3* genes were found to be located at the top of LG8 (Bus et al. 2008); however, after a new WAA biotype was identified that was capable of overcoming only *Er3*, it was suggested that these were either two closely linked genes or different alleles with different functions (Sandanayaka et al. 2003; Bus et al. 2008). As for the *Er2* gene, this was mapped onto LG17 of 'Robusta 5' (Bus et al. (2008). Furthermore, three NBS-LRR genes were mapped to the *Er2* region (Calenge et al. 2005a). Subsequently, five additional genes within the vicinity of the *Er2* region were found to be differentially expressed between trees resistant and susceptible to WAA (Jensen et al. 2014).

Genes for Resistance to the Rosy Apple Aphid

Although a single dominant gene, Sm_h, for resistance to rosy apple aphid, *D. plantaginea,* was identified in an open-pollinated selection of *M.* × *robusta* (MAL59/9), no molecular studies have been conducted. However, limited genetic studies have been conducted on the scab-resistant cultivar 'Florina' that was also found to be either resistant or tolerant to *D. plantaginea* (Lespinasse et al. 1985). At this time, it has been reported that at least two independent genes have been proposed to be involved in resistance to rosy apple aphid detected in the scab-resistant apple cultivar 'Florina'. Interestingly, this genetic resistance was independent of the *Vf* scab resistance gene in 'Florina' (Minarro and Dapena 2004).

In a different study, transcriptional changes were observed following *D. plantaginea* infestation, thereby identifying 21 DNA fragments that were differentially expressed in 'Florina' compared to either apple cultivar 'Topaz' (susceptible to rosy apple aphid), mechanical wounding, or non-infested leaves (Qubbaj et al. 2005).

7.3.3 Tree Architecture

7.3.3.1 Columnar Growth Habit

The columnar growth habit is one of the most important traits of fruit trees as it facilitates the establishment of high-density planted orchards. In apple, the columnar growth habit is primarily controlled by a single dominant gene, *Co* (columnar). The *Co* locus has been first mapped onto LG10 of apple cv. 'Wijcik McIntosh' (Conner et al. 1997). Later, genetic mapping of the columnar trait has been performed using other columnar genotypes, including 'Totem', 'Telamon', and 'Tuscan', and all results further confirmed that the *Co* locus is located on LG10 (Kim et al. 2003; Tian et al. 2005; Fernandez-Fernandez et al. 2008; Kenis and Keulemans 2005, 2007). Tian et al. (2005) have reported on fine-mapping of the *Co* region by adding nine new markers using a 'Spur Fuji' × 'Telamon' progeny. These markers have been identified by a bulked segregant analysis, and the *Co* gene is mapped between the two SSRs CH3d11 and COL.

Subsequently, significant progress in fine-mapping of the *Co* gene has been achieved with the aid of the draft genome sequence of 'Golden Delicious'. Moriya et al. (2012) conducted fine-mapping of *Co* using 1000 F_1 seedlings derived from 31 populations. The *Co* locus was delimited to a 196 kb region of the apple genome, between the two SSR markers Mdo.chr10.11 and Mdo.chr10.15. Bai et al. (2012a) developed five SSR markers to screen 528 F_1 seedlings derived from four populations along with 290 different germplasm or selections and delimited the *Co* locus to a 193 kb region. This 193 kb region was found to contain 26 predicted genes in the 'Golden Delicious' genome, and three putative *Lateral Organ Boundaries Domain* (*LBD*) genes were deemed as strong candidates for the *Co* gene.

Baldi et al. (2013) conducted fine-mapping of *Co* using 1551 F_1 seedlings derived from different crosses. The *Co* locus was delimited to a region of 0.56 cM flanked by the two SSR markers Co04R11 and Co04R13, and corresponding to 393 kb of the 'Golden Delicious' genome sequence. A total of 36 genes were predicted in this region, with seven genes deemed as potential candidates for their involvement in controlling shoot development. This 393 kb region overlapped with the 196 kb region previously identified by Moriya et al. (2012), thus further confirming the reliability of the *Co* fine-mapping efforts. DNA sequence comparisons of BAC clones covering the *Co* region between the mutant 'Wijcik' and the wild-type 'McIntosh' revealed an insertion of a novel non-coding DNA element of 1956 bp specific to the Pyreae taxonomical tribe (Wolters et al. 2013). This insertion was present in 'Wijcik', but absent in 'McIntosh'. A total of six genes were present in the 50 kb region surrounding the insertion, and with only one gene encoding a putative 2OG-Fe(II) oxygenase, designated as *MdCo31*, which was differentially expressed between 'McIntosh' and 'Wijcik' shoot buds. Ectopic expression of *MdCo31* in *Arabidopsis* yielded compact plants with shortened floral internodes, similar to the phenotype observed in 'Wijcik'. This indicated that *MdCo31* was a strong candidate gene for the columnar growth habit.

7.3.3.2 Dwarfing Growth Habit

A gene for dwarfing growth habit, *Dw1,* has been identified in the apple rootstock 'Malling 9', and this has been mapped onto the top of LG5 (Rushlomé-Pilcher et al. 2008). The 'M9' dwarfing effect is related to an increase in the proportion of floral buds and the reduction of the number and length of branches in the first year of growth after grafting (Seleznyova et al. 2008; Foster et al. 2014). This gene is a major determinant of dwarfing, as most dwarfing and semi-dwarfing rootstocks carry this dwarfing allele.

Fazio et al. (2014) have conducted QTL mapping of scion vigor suppression using a genetic map of a segregating population from a cross between 'Ottawa 3' and 'Robusta 5'. Their finding has not only confirmed the presence of previously identified *Dw1* locus, but has also revealed the presence of a new dwarfing locus on LG11, designated as *Dw2*. Later, Foster et al. (2015) have also conducted QTL mapping of the dwarfing trait using a rootstock population derived from the cross between 'M9' and 'Robusta 5' (non-dwarfing). Their results have confirmed that *Dw1* exhibits a significant influence on dwarfing of the scion, while *Dw2* has a smaller effect on dwarfing. In addition, four minor-effect QTLs have also been detected on LGs 6, 9, 10, and 12. Therefore, *Dw1* and *Dw2* are primarily responsible for rootstock-induced dwarfing, and rootstocks carrying both *Dw1* and *Dw2* show the highest degree of dwarfing.

Recently, Harrison et al. (2016) have revealed the presence of a third QTL for dwarfing on LG13 of apple, designated as *Dw3*, following genetic mapping of root bark percentage (percentage of the whole root area constituted by the root cortex). It is reported that there is a strong link between root bark percentage and rootstock-induced dwarfing of the scion. This, in turn, leads to a proposed three-locus model that is capable of explaining levels of dwarfing from the dwarf 'M.27' to the semi-invigorating rootstock 'M.116'. The QTL on LG13 (*Rb3*) is likely to be analogous to the dwarfing QTL *Dw3*.

7.3.4 Budbreak, Flowering Time, and Fruit Ripening

Allard et al. (2016) conducted QTL mapping of budbreak (release from dormancy) and flowering time using an integrated genetic map containing 6849 SNPs. A total of four major QTLs on LG7, LG10, LG12, and LG9 were detected, accounting for 5.6–21.3% of the trait variance. Two homologs of the flowering genes *AGL24* and the florigen *FLOWERING LOCUS T* (*FT*) were found within the vicinity of QTLs on LG9 and LG12, respectively. Besides these major QTLs, two minor QTLs were also identified on LG15 and LG8. These minor QTLs co-localized with three of the four *Dormancy Associated MADs-box* (*DAM*) genes in the apple genome, and representing a cluster of genes unique to Rosaceae. This finding suggested that mechanisms underlying chilling perception might be common among perennial and annual plants (Allard et al. 2016). Subsequently, a genome-wide association (GWAS) study utilized an apple core collection and confirmed the position of the major QTL for budbreak on LG9, and narrowed the confidence interval to ~ 360 kb (Urrestazaru et al. 2017). This QTL highlights the importance of two major candidate genes encoding transcription factors containing NAC or MADS-box (putative flowering repressor *FLOWERING LOCUS C* [putative *FLC*]) domains (Urrestarazu et al. 2017).

Recently, Miotto et al. (2019) used an Infinium II SNP platform for genotyping, containing 8,788 SNPs, to construct linkage maps of a cross between a genotype for low chilling requirement (CR), 'M13/91', with a genotype with a moderate CR, cv. 'Fred Hough'. They detected a major QTL for budbreak at the beginning of LG9, and this QTL was found to be stable throughout seven growing seasons at two different field sites, accounting for 24.5–45.6% and 31–48% of the phenotypic variation in these two sites. It is proposed that several candidate genes, such as *MdoICE1* (mediating cold and abscisic acid [ABA] responsiveness) and *MdoMPT* (a mitochondrial phosphate transporter (MPT) gene involved in accelerating bud dormancy release), among others are likely to be present in this QTL region (Moito et al. 2019).

As for flowering time, Urrestarazu et al. (2017) pursued GWAS mapping to identify QTLs for flowering time, as well as for fruit-ripening period. They used 1,168 different apple genotypes grown over six locations, phenotyped for phenological traits, and genotyped at a high-density of SNPs using the Axiom®Apple 480 K SNP array. It was found that GWAS retained two SNPs as co-factors on chromosome 9 for the flowering period and six SNPs for the ripening period (four on chromosome 3, one on chromosome 10, and one on chromosome 16), which together accounted for 8.9 and 17.2% of the observed phenotypic variance, respectively. For both traits, SNPs in weak linkage disequilibrium were also detected nearby, thus suggesting the presence of allelic heterogeneity. Within this region, genes coding for transcription factors containing either NAC or MADS domains were identified as major candidates, within small confidence intervals, and computed for associated genomic regions (Urrestarazu et al. 2017).

7.3.5 Fruit Quality Traits

Fruit quality traits encompass numerous parameters and various characters, such as fruit color, fruit size, flesh firmness, acidity, and shelf life, among others. All these traits are rather complex and controlled predominantly by QTLs with some major genes playing some important roles for some traits.

7.3.5.1 Red Coloration of Fruit

Red coloration plays an important role in determining quality in pome fruits, particularly in apples. Cheng et al. (1996) have identified a RAPD marker, BC226, that is closely linked (< 2 cM) to the *Rf* gene on LG9 controlling fruit skin color in apples. This RAPD marker has been converted into a SCAR marker, BC226SCAR, and used to investigate segregation of red fruit color in a seedling population of 'Rome Beauty' × 'White Angel'. It has been revealed that the genetic control of apple fruit skin color is rather simple, involving two alleles of the *Rf* locus deemed responsible for red and yellow coloration.

Subsequently, two apple MYB transcription factors, *MdMYB1* and *MdMYB10*, were reported to be responsible for apple skin or cortex coloration (Takos et al. 2006; Espley et al. 2006). In a later study, Chagné et al. (2007) mapped *MdMYB10* onto LG9 and reported that, in fact, *MdMYB10*, *MdMYB1*, and *Rf* were all alleles of the same gene, wherein *MdMYB10* was responsible for red coloration of fruit flesh and foliage, while *MdMYB1* controlled fruit skin color. Although several *MYB10* genes involved in anthocyanin accumulation have been characterized in various species of Rosaceae (Takos et al. 2006; Ban et al. 2007; Espley et al. 2007), a long terminal repeat (LTR) retrotransposon insertion upstream of *MdMYB1* was found to be associated with the red-skinned phenotype (Zhang et al. 2019).

Chagné et al. (2013) have reported on a novel red fruit-flesh phenotype of apple cv. 'Sangrado', having green foliage, while the fruit develops red flesh in the cortex, during late maturity. Using an F_1 population derived from a cross between the white-fleshed cv. 'Sciros' and a red-fleshed selection from 'Sangrado', a major QTL explaining 31.0% of the observed phenotypic variation has been identified on LG17. Using 450 F_1 seedlings from three populations, this QTL is delimited to an 800 kb region using a fine-mapping strategy. Moreover, this region is reported to contain 74 predicted genes. Within this region, a transcription factor, designated as *MdMYB110a* which is a paralog of *MdMYB10*, has been confirmed to be responsible for red coloration in the apple cortex. It is suggested that *MdMYB10* and *MdMYB110a* must have evolved from a whole-genome duplication event that must have occurred within the Maloideae family, which would account for observed differences in expression patterns and responses to fruit maturity.

7.3.5.2 Non-anthocyanin Phenolic Compounds in Fruit

Apple serves as one of the major sources of phenolic compounds in the human diet. Using a segregating F_1 population derived from a cross between 'Prima' and 'Fiesta', Khan et al. (2012d) have pursued a QTL mapping effort for phenolic compounds in apple fruit. A total of 418 and 254 metabolites have been identified in fruit peel (skin) and flesh, respectively. Furthermore, QTL mapping has revealed the presence of 488 and 181 QTLs in peel and flesh tissues, respectively. QTL hotspots regulating metabolites belonging to the phenylpropanoid pathway are primarily located on LG1, LG8, LG13, and LG16. The most important QTL hotspot with the largest number of metabolites is detected on LG16 with QTLs for 33 peel-related and 17 flesh-related phenolic compounds. A structural gene encoding leucoanthocyanidin reductase (LAR1) in the QTL hotspot on LG16 has been reported to exert a large effect on the metabolism of phenolic compounds in apple fruits (also see Chap. 18 in this volume).

Chagné et al. (2012a, b) mapped 79 QTLs associated with 17 polyphenolic compounds in apple fruits using an F_1 progeny derived from a cross between 'Royal Gala' and 'Braeburn'. In silico alignment of parental genetic maps with the whole draft, apple genome sequence revealed that a single gene, *LAR1*, co-located with a QTL cluster of flavanol compounds, including catechin, epicatechin, and procyanidin dimers, as well as five unknown procyanidin oligomers were identified close to the top of LG16. Yet, another gene encoding hydroxy cinnamate/quinate transferase (HCT/HQT), co-located with a QTL for chlorogenic acid content, was mapped close to the bottom of LG 17. Therefore, *LAR1* and *HCT/HQT* were likely to be involved in influencing levels of phenolic compounds in apple fruit.

There are a number of reports on the involvement of transcription factors (TFs) in the regulation of phenolic compounds in apples. Verdu et al. (2014) have identified ~70 QTLs controlling levels of phenolic compounds and the average degree of polymerization of flavanols in a cider apple progeny derived from a cross between two hybrids, 'X5210' and 'X8402'. This finding further confirms the importance of the *HCT/HQT* gene on LG17 that is involved in the biosynthesis of phenolic compounds in apple fruit. Moreover, this has also revealed that four transcription factors, including *MdTTG1*, *bHLH33*, *MYB110a*, and *MYB110b*, are located in QTL regions controlling the content of several phenolic groups, such as hydroxycinnamic acids and flavonols. Hence, the biosynthesis of phenolic compounds in apple is regulated not only by proanthocyanidin (PA) pathway structural genes but also by TFs such as *MYB* and *WD40* genes.

In a recent study, Busatto et al. (2019) investigated the levels of phenolic compounds between wild and domesticated accessions, particularly in flesh tissues, and found that differences observed could be attributed to higher growth rates of domesticated apples. This was further confirmed with observed higher contents of procyanidin B2 + B4 and phloridzin in russet-skinned apples, known to have higher levels of these compounds. Moreover, expression analysis of 16 polyphenolic genes demonstrated higher levels of expression for nine of these genes at harvest time compared to that during fruit development. McClure et al. (2019) used GWAS consisting of 136 apple cultivars, and identified candidate genes for the production of quercitrin, epicatechin, catechin, chlorogenic acid, 4-*O*-caffeoylquinic acid, and procyanidins B1, B2, and C1. These candidate genes were confirmed to be present within the QTL hotspot on LG16, and as reported previously (Khan et al. 2012d; Chagné et al. 2012a, b) although there were slight differences in the location of the most significant SNP across phenotypes used in this GWAS, all were within the boundaries of this QTL hotspot.

7.3.5.3 Concentrations of Vitamin C in Apple Fruit

Vitamin C or L-ascorbic acid (AsA) is closely related to maintaining fruit quality during post-harvest storage, as well as susceptibility to internal browning in apple fruits. Using an F_1 progeny derived from 'Telamon' × 'Braeburn', Mellidou et al. (2012) have conducted QTL mapping for Vitamin C content in apple fruits. Stable QTLs controlling AsA levels have been detected on LGs 10, 11, 16, and 17. Furthermore, metabolic genes encoding GDP-L-galactose phosphorylase (GGP), dehydroascorbate reductase (DHAR), and nucleobase-ascorbate transporter are located within confidence intervals of these QTL map locations. Of particular interest are three paralogs of *MdGGP,* as they are co-located within QTL regions for AsA levels in fruit.

Allelic variants of *MdGGP1* and *MdGGP3,* derived from apple cv. 'Braeburn', were associated with high levels of AsA in fruits collected from the mapping population, as well as from other apple germplasm (Mellidou et al. 2012). Furthermore, significant differences in expression of *MdGGP1* alleles were observed in fruit from high- and low-AsA genotypes, thus suggesting that *MdGGP1* played a critical role in the regulation of AsA biosynthesis in apple fruit. Moreover, *MdDHAR3-3* is co-located with a stable QTL for browning in apple cv. 'Telamon'. These findings suggested that the redox status of the AsA pool was related to susceptibility to flesh browning in apples.

7.3.5.4 Fruit Acidity and Sweetness

Acidity and sugar content are the two main factors determining fruit eating quality. The major organic acid in apple fruit is malic acid. A major QTL or the *Ma* gene, responsible for fruit acidity, has been initially mapped onto LG16, accounting for ~30% of the observed variance (Maliepaard et al. 1998; Liebhard et al. 2003; Kenis et al. 2008). The *Ma* locus is flanked by two SSRs, Hi0h08 and Hi22f, present in a BAC contig, and located at a distance of 428 kb apart from each other (Fig. 7.6). Subsequently, Xu et al. (2012) have developed several SSR markers within the *Ma* region using the whole draft genome sequence of 'Golden Delicious', and these SSR markers have been used to screen two half-sib apple populations. This has revealed that the *Ma* locus is delimited to a physical region no longer than 150 kb on chromosome 16 in the 'Golden Delicious' genome. Fine-mapping has narrowed down the *Ma* locus to a region of 65–82 kb in length containing 12–19 predicted genes, depending on the haplotypes (Bai et al. 2012b). Of these, two aluminum-activated malate transporter-like genes, designated *Ma1* and *Ma2*, are identified as strong candidates of the *Ma* gene. Analysis of expression profiling has further indicated that *Ma1* is the major determinant controlling fruit acidity, as *Ma1* expression is significantly correlated with fruit titratable acidity. The low acidity allele of *Ma1* is mainly characterized by a mutation at base 1455 in the open reading frame (also see Chap. 14 in this volume).

Recently, Verma et al. (2019) determined numbers and locations of QTLs for acidity variation using a large apple breeding population consisting of 16 full-sib families, representing nine major breeding parents, and genotyped using an 8 K SNP array. They determined titratable acidity at harvest, and following 10- and 20-week of cold storage treatments, for three successive seasons. Using pedigree-based QTL mapping software, FlexQTL™, they detected two QTLs, on LG16 (*Ma* locus) and LG8 (designated as *Ma3*) that together accounted for 66 ± 5% of the phenotypic variation. They found that an additive allele dosage model for the two QTLs was capable of explaining most acidity variation, with an average of +1.8 mg/L at harvest per high-acidity allele. They reported that due to the presence of high-acidity alleles, the faster is the depletion with storage, with all combinations likely to converge to a common baseline. Furthermore, it was reported that each QTL had a rare second high-acidity allele with either a stronger or a weaker effect.

In addition to the *Ma* gene QTLs, six other QTLs for apple fruit acidity have also been detected on LGs 2, 8, 10, 13, 15, and 17 (Liebhard et al. 2003; Kenis et al. 2008). A QTL on LG8 accounts for up to 46% of the observed variance in an F_1 population derived from a cross between 'Fiesta' and 'Discovery' (Liebhard et al. 2003). This QTL is also found to be linked to SSR markers CH05a02y on LG8 of apple cv. Jonathan (Zhang et al. 2012) and Hi01f09 on LG8 of cv. Jiguan (Ma et al. 2016). Therefore, it

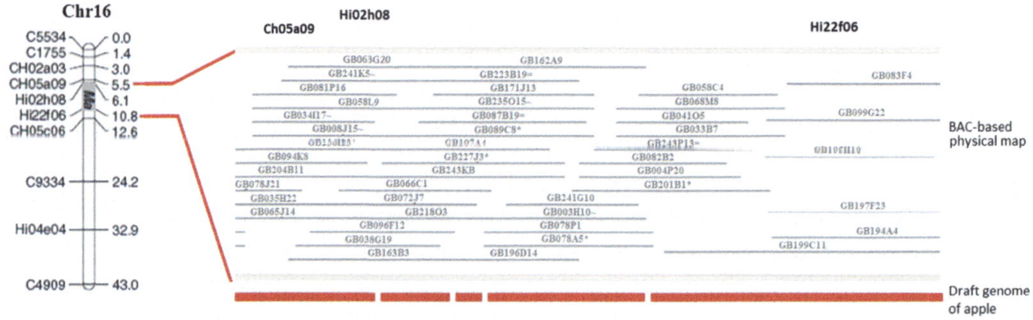

Fig. 7.6 A physical map covering the *Ma* locus in apple

is likely that fruit acidity is also controlled by a QTL on LG8 in apple.

There are several studies on QTL mapping for sugar content in apple fruits. For example, Liebhard et al. (2003) reported on QTL mapping for sugar content in apple fruits using a cross between 'Fiesta' and 'Discovery'. Two minor QTLs for fruit sugar content were detected on LG6 and LG8 of both 'Fiesta' and 'Discovery', and these accounted for 10–17% of the observed variance. Later, Kenis et al. (2008) investigated the inheritance of fruit sugar content using a cross between apple cvs. Telamon and Braeburn. A major QTL was identified on LG10 in both parents over two successive years and accounting for up to 30% of the phenotypic variation. In addition, a minor QTL was also detected on LG2 in both parents over two successive years, and accounting for ~8% of the observed phenotypic variation. Subsequently, Guan et al. (2015) used fruits from 15 apple breeding populations, analyzed at harvest, as well as after 10 and 20 weeks of cold storage followed by 1 week at room temperature in two successive years, and subjected these fruit samples to a total of 1416 polymorphic SNPs filtered from the RosBreed Apple SNP Infinium® array for QTL analysis using Flex QTL™ software. They were able to identify QTLs associated with sweetness. Interestingly, these populations were derived from eight elite cultivars, including 'Arlet', 'Aurora Golden Gala', 'Cripps Pink', 'Delicious', 'Enterprise', 'Granny Smith', 'Honeycrisp', and 'Splendour', along with two advanced Washington apple selections, 'WA 5' and 'WA 7'. They identified QTLs for individual sugars, including those of fructose, glucose, sucrose, and sorbitol on LGs 1, 2, 3, 4, 5, 9, 11, 12, 13, 15, and 16, as well as QTLs for soluble solid content (SSC) on LGs 2, 3, 12, 13, and 15 (Guan et al. 2015). Interestingly, a QTL region on LG1 was consistently identified for both fructose and sucrose, from harvest through storage in both years, accounting for 34–67% and 13–41% of total phenotypic variation, respectively. It was reported that these stable QTLs explaining phenotypic variation on LG1 for fructose content point to a promising genomic region for developing DNA-based markers that can be useful in marker-assisted breeding for sweetness selection in apple breeding programs (Guan et al. 2015).

The above findings have confirmed that genetic control of fruit sweetness traits is polygenic or quantitatively controlled (Kenis et al. 2008). Therefore, future positional cloning efforts of some of these major QTLs responsible for fruit sweetness could be pursued, although this would be rather challenging (also see Chap. 14 in this volume).

Recently, Zhen et al. (2018) used gene-tagged markers to identify 25 *SWEET* sugar transporter genes, involved in sugar accumulation, in the apple genome. Of these genes, 24 were located on five homologous pairs of chromosomes (3–11, 5–10, 4–12, 6–14, 13–16), and one on chromosome 17. Nine of these genes were found to be highly expressed throughout fruit development; therefore, molecular markers were developed for these nine *MdSWEET* genes, and used for genotyping of 188 apple cultivars. An association of polymorphic *MdSWEET* genes with soluble sugar content in mature fruit was analyzed. It was found that three genes, *MdSWEET2e*, *MdSWEET9b*, and *MdSWEET15a*, were significantly associated with fruit sugar content, wherein *MdSWEET15a* and *MdSWEET9b* accounting for a relatively large proportion of observed phenotypic variations in sugar content. Furthermore, both *MdSWEET9b* and *MdSWEET15a* were determined to be located on chromosomal regions harboring QTLs for sugar content. It was concluded that *MdSWEET9b* and *MdSWEET15a* were likely candidates regulating fruit sugar accumulation in apples (Zhen et al. 2018).

7.3.5.5 Volatile Compounds in Ripe Fruits

Aromas are very important components of organoleptic qualities of fruits. It is well known that fruits accumulate a variety of volatile compounds, such as esters and phenylpropenes. Esters are synthesized as a result of fatty acid degradation or from amino acid precursors, while phenylpropenes are produced via a side branch of the general phenylpropanoid pathway. To date, there are few reports on the inheritance of volatile compounds in apple fruits. Using proton transfer reaction mass spectrometry (PTR-MS), Zini et al. (2005) have analyzed the headspace composition of volatile organic compounds (VOCs) emitted by fruits of an F_1 progeny of 'Fiesta' × 'Discovery'. QTL analysis for all PTR-MS peaks has identified 10 QTLs associated with VOCs with LOD scores > 2.5. This finding has demonstrated that the contents of volatile compounds are quantitatively controlled.

Souleyre et al. (2014) have identified 46 QTLs for the production of esters and alcohols on 15 linkage groups of apples. Of these QTLs, one major QTL on LG2 co-locates with the alcohol acyl-transferase 1 (AAT1) encoding gene that catalyzes the final step of the biosynthetic pathway of volatile esters. Later, AAT1 has been determined to be linked to phenylpropene production in apples (Yauk et al. 2017). In addition, another QTL on LG1 is found to co-locate with six candidate genes encoding phenylpropene *O*-methyltransferases, *MdoOMT1 to 6* (Yauk et al. 2015). Of these genes, only *MdoOMT1* is strongly associated with estragole production in ripening fruit. Following a functional analysis study, it is demonstrated that *MdoOMT1* is required for the production of phenylpropenes, including estragole that may contribute to ripe apple fruit aroma.

Recently, Larsen et al. (2019) conducted a GWAS, using 145 apple cultivars, and was able to associate genotype and phenotype information by combining genotyping-by-sequencing (GBS) generated SNP markers with fruit flavor volatile data. They were able to detect associations for acetate esters, particularly of butyl acetate and hexyl acetate on chromosome 2 within a 2- to 3-Mb region, wherein several alcohol acyl-transferases, including AAT1, were clustered.

7.3.6 Fruit Texture

Fruit firmness plays an important role in determining storage life. Loss of firmness is mainly due to enzyme activities that depolymerize cell wall pectin. Fruit texture parameters have been recently dissected into two groups of sub-traits, mechanical and acoustic, to distinguish between firmness (based on mechanical sub-traits) and crispness (based on acoustic sub-traits) of apples. These parameters have been used in QTL mapping studies.

Using a cross between 'Prima' and 'Fiesta', King et al. (2000) first reported on QTL mapping of fruit firmness in apple. Phenotypic data were collected at six different sites over two successive years. A total of three major QTLs were identified on LG1, LG10, and LG16, respectively. Those QTLs on LG10 and LG16 were later confirmed by Kenis et al. (2008). Moreover, Kenis et al. (2008) investigated the inheritance of fruit quality traits using a population derived from a cross between 'Telamon' and 'Braeburn'. The strongest QTL was identified on LG10 of both parents over two successive seasons, accounting for up to 31% of the observed phenotypic variation. Later, the region of the QTL on LG10 was further narrowed down by Costa et al. (2010) using a full-sib population of 'Fuji' × 'Mondial Gala', wherein fruit firmness data were collected at different maturity stages, from harvest to post-harvest storage. QTL mapping revealed that at least two major QTLs controlling fruit firmness and softening were located on LG10. One QTL, designated as *Md-PG1*, encoded polygalacturonase and depolymerized cell wall pectin, while another QTL, designated as *Md-ACO1*, was involved in ethylene biosynthesis. *Md-PG1* has been reported to be related to changes in fruit firmness occurring during ripening under ambient conditions, whereas *Md-ACO1* was likely to be associated with changes in fruit firmness during and after cold storage. Therefore, *MdPG1* and *MdACO1*, located at the center and bottom of chromosome 10, respectively, have been deemed as candidate genes controlling apple firmness and storability.

A novel apple gene, *Md-Exp7*, sharing high sequence similarities with plant-specific ripening expansin genes, was cloned by Costa et al. (2008). A functional marker (Md-$Exp7_{SSR}$), based on an SSR motif (CT)n located within the untranslated region of *Md-Exp7*, was first developed. This marker was then used to map the homolog *Md-Exp7* gene onto LG1 of the apple genome within a region carrying a major apple QTL for fruit firmness that has been previously identified (Costa et al. 2008). An association analysis was conducted using a collection of 31 apple cultivars revealing that fruit softening was correlated with the CT repeat number, wherein the longer the Md-Exp7 SSR marker fragment, the higher the level of fruit softening (Costa et al. 2008).

Longhi et al. (2012) have conducted a thorough QTL mapping effort of apple fruit texture. QTLs have been identified on LGs 5, 8, 10, 11, 12, and 16, accounting for 15.6–49% of the total phenotypic variance. This highly significant QTL cluster is mapped onto chromosome 10, and it is co-located with *Md-PG1,* coding for polygalacturonase (PG) which depolymerizes cell wall pectin (Costa et al. 2010). In a subsequent study, *Md-PG1*, has been confirmed to be mapped onto chromosome 10 and co-localized within a statistical interval of a major hotspot QTL associated with several fruit texture sub-phenotypes (Longhi et al. 2013). Moreover, other candidate genes related to *Md-NOR* and *Md-RIN* TFs, including *Md-Pel* (pectate lyase), and *Md-ACS1,* involved in ethylene biosynthesis, have also been mapped within statistical intervals. All these findings suggest that fruit texture is a complex trait that must be controlled by several genes in apples. Amyotte et al. (2017) have identified a significant sensory texture locus on chromosome 13 (Chr13: 6,049,060) accounting for 23% of the variation for *crisp* and 25% of the variation for *juicy* texture; thus, further supporting associations between sensory juiciness and chromosome 13 detected in diverse germplasm.

Subsequently, Di Guardo et al. (2017) used two QTL mapping approaches, a pedigree-based analysis (PBA) using six full-sib pedigreed families, and a GWAS using 233 apple accessions, to further elucidate the genetic control of fruit texture. All plant materials were genotyped using a 20K SNP array and phenotyped using a sophisticated high-resolution texture analyzer. The QTL findings indicated that chromosome 10 controls the mechanical properties, while chromosomes 2 and 14 were associated with the acoustic response.

7.3.6.1 Apple Skin Russeting

Apple fruit skin russeting is attributed to the formation of a plastic periderm in response to microcracking on stiff cuticles. It is proposed that damaged cuticles of the epidermis are shed, while the peridermal layer undergoes suberization to seal the fruit (Lashbrooke et al. 2015). Fruit of russeted apple cultivars has epidermal cells of irregular sized along with irregular arrangements and variable cuticle thickness (Knoche and Lang 2017). Russeted apple cultivars are rich in procyanidin B2 + B4 and phloridzin in the pulp at harvest (Gutierrez et al. 2018; Busatto et al. 2019). The russeting trait is genetically controlled by two or more genes, with one deemed as a major gene (Alston et al. 1975).

Falginella et al. (2015) have reported on a major QTL for russeting development on LG12 of apple. This major QTL is located in an approximately 400 kb region, containing 58 predicted genes, based on the reference genome sequence of 'Golden Delicious'. Of 58 genes, a putative ATP-BINDING CASSETTE G (ABCG) family transporter, designated as *Ru_RGT*, has been identified as a candidate for russet development. The *Ru_RGT* gene encodes a G subfamily ABC half-transporter protein that is involved in cuticle development. The major gene *Ru_RGT* likely controls russeting development via regulation of cuticle organization.

Lashbrooke et al. (2015) conducted QTL mapping for apple russeting using a full-sib mapping population and identified two QTLs detected on LG2 and LG15. Genomic analysis of these two QTL regions identified a set of candidate genes putatively involved in cuticle biosynthesis. Gene expression profiling revealed that an SHN1/WIN1 transcription factor, designated *MdSHN3* and located within the vicinity of the QTL on LG15, was expressed at low levels in apple accessions with improper cuticle formation. Orthologs of SHN1/WIN1 have been well known to regulate cuticle formation in different plants such as *Arabidopsis*, tomato, and barley.

Hence, the *MdSHN3* gene was proposed as a critical regulator of cuticle biosynthesis in apple fruits. Recently, Legay et al. (2016) proposed that the transcription factor MdMYB 93, associated with some key suberin biosynthetic genes, was found to be involved in the regulation of suberin deposition in russeted apple skins. This was proposed to begin with the synthesis of monomeric precursors, their transport, polymerization, and all the ways to deposition of suberin in the primary cell wall (Legay et al. 2017).

7.3.7 Fruit Size

Unlike many fruits derived from carpel tissues, both apple and pear fruits are derived from the hypanthium, consisting of fused sepals, petals, and anther-derived tissues. Therefore, it is interesting to explore the mechanisms underlying fruit size development. To date, there are only a few studies that have specifically addressed fruit size development in apples. Kenis et al. (2008) have investigated the inheritance of fruit size using a population of 'Telamon' × 'Braeburn'. Two QTLs controlling fruit diameter and fruit length have been detected on LG10 and LG15, respectively, and accounting for 23–33% of the observed phenotypic variation. Later, Devoghalaere et al. (2012) have reported on QTL mapping of apple fruit size over multiple seasons using F_1 progenies derived from two crosses, 'Royal Gala' × 'Braeburn' and 'Starkrimson' × 'Granny Smith'. Two QTLs controlling fruit weight, *fw1* and *fw2*, have been identified on LG8 and LG15 for all four parents. In a subsequent study, these QTLs *fw1* and *fw2* were found to harbor 11 and 7 genes in selective regions of *M.* × *domestica*, respectively (Duan et al. 2017). However, QTL regions of LG5 and LG10 have auxin response factors *MdARF6* and *MdARF106*, respectively. *MdARF106* is found to be expressed during stages of fruit cell division and cell expansion, which is consistent with their potential roles in controlling fruit size (Devoghalaere et al. 2012).

Yao et al. (2015) conducted genetic mapping for apple fruit size using a 'Royal Gala' (RG) × 'A689-24' segregating population, and four QTLs were detected on LG2, LG10 (two QTLs), and LG11. The QTL on LG11 accounted for 13.9% of fruit weight variation, and a microRNA gene (miRNA172) was identified as the causative gene. A transposon insertional allele of miRNA172 caused a reduction in levels of expression, thus resulting in an increase in fruit size. In constrast, over-expression of miRNA172 in transgenic apple lines significantly reduced fruit size.

7.4 Physical Mapping of the Apple Genome

Genome-wide physical mapping serves as a platform that facilitates large-scale genome sequencing efforts, as well as an essential tool for determining the genetic basis of complex traits, as well as for pursuing genomics research efforts.

In order to construct a physical map, there are various approaches that have been developed, such as BAC restriction-based fingerprinting, fluorescent in situ hybridization (FISH), and sequence-tagged site (STS) mapping. BAC restriction-based fingerprinting locates relative positions of recognition sequences for restriction endonucleases along a DNA molecule, while FISH utilizes information obtained by hybridizing a probe, carrying a marker, to intact chromosomes. On the other hand, STS mapping relies on mapping positions of short sequences via PCR and/or hybridization analysis of genome fragments. Recently, whole-genome sequencing has become available to generate sequence-based physical maps wherein distances are measured based on numbers of base pairs.

For apple, BAC restriction-based fingerprinting and whole-genome sequencing mapping have been used to develop genome-wide physical maps (Han et al. 2007; Velasco et al. 2010).

7.4.1 Large-Insert Genomic Library Construction

The construction of large-insert genomic libraries is a necessary step in pursuing physical mapping using the BAC restriction-based fingerprinting method. Early on, a yeast artificial chromosome (YAC) cloning system has been developed to clone large DNA fragments (up to 1000 kb). However, the YAC cloning system is found to be unstable in yeast; moreover, it is difficult to purify. Most importantly, it is found that chimeric YACs account for as high as 40% of a whole library. Therefore, these disadvantages have limited the utility of YAC libraries. Subsequently, alternative cloning systems for large DNA inserts have been developed, such as the BAC, transformation-competent artificial chromosome (TAC), and the bacteriophage P1-derived artificial chromosome (PAC) (Han and Korban 2010). Of these cloning systems, BACs have been widely used to develop genomic libraries for plants due to their capabilities of carrying large DNA inserts, up to 300 kb, high cloning efficiency, as well as their stability for maintaining foreign DNA (Han and Korban 2010).

Thus far, all large-insert genomic libraries developed for apple have been constructed using the BAC system. For example, Vinatzer et al. (1998) constructed a BAC library for 'Florina' using the BAC cloning vector pECBAC1, consisting of 36,864 recombinant clones with an average insert size of ~120 kb, and representing ~5 × apple haploid-genome equivalents. Later, Xu et al. (2001b) constructed a BAC library for *M. floribunda* 821 in the BAC cloning vector pBeloBAC11, consisting of 31,584 BAC clones with an average insert size of 125 kb, and representing approximately ~5 × *M. floribunda* 821 haploid-genome equivalents. In addition, another BAC library was constructed for 'Goldrush', consisting of 35,328 clones with an average insert size of ~110 kb, using a similar strategy (Xu et al. 2002). All the above-described apple BAC libraries have been first used to isolate and clone the *Vf (Rvi6)* gene for resistance to apple scab, as described above (Vinatzer et al. 2001; Xu and Korban 2002; Han and Korban 2010).

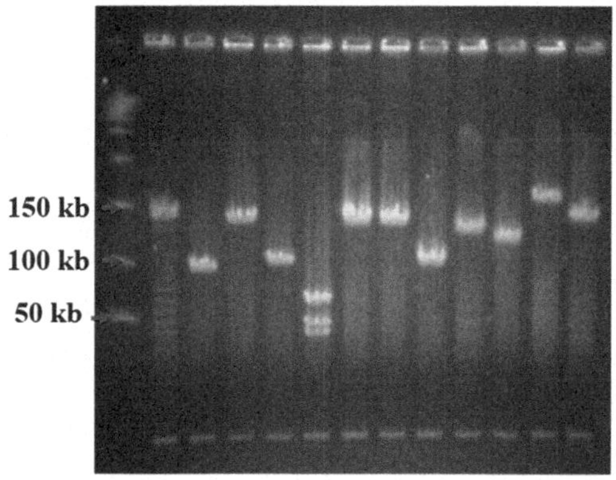

Fig. 7.7 BACs randomly selected from a 'GoldRush' BAC library constructed with *Hin*dIII

Subsequently, in an effort to develop a genome-wide physical map for the apple using BAC fingerprinting, a second BAC library for 'Goldrush' was also constructed, wherein high molecular weight (HMW) apple DNA was restricted with *Hin*dIII instead of *Bam*HI (Han et al. 2007). This additional 'Goldrush' library consisted of 46,791 clones, representing ~7 × haploid genome equivalents and with an average insert size of 125 kb (Fig. 7.7).

Later, a BAC library for the apple rootstock Geneva 41 (G.41), a cross of 'Malling 27' × 'Robusta 5', was constructed consisting of 41,472 clones with an average insert size of ~120 kb cloned into a pECBAC1 vector (G. Fazio, Pers. Comm.). Moreover, a BAC library from 'Pinkie' has also been constructed at the New Zealand Institute of Plant and Food Research, consisting of 56,400 clones, corresponding to ~7 × genome coverage and with an average insert size of 110 kb, cloned in a pCLD04541 vector (E. Rikkerink, Pers. Comm.). In addition, a cosmid library from 'Pinkie' was also constructed by the New Zealand Institute of Plant and Food Research in SuperCos1, consisting of 170,000 clones (with an average insert size of 35 kb), and representing ~7x genome coverage (E. Rikkerink, Pers. Comm.).

7.4.2 A BAC-Based Physical Map of the Apple Genome

7.4.2.1 BAC Fingerprinting

A total of 82,503 BAC clones from two BAC libraries, developed from *M.* × *domestica* cv. 'GoldRush' using *Bam*HI and *Hind*III, respectively, were fingerprinted using an agarose gel-based restriction fingerprinting protocol (Han et al. 2007). DNA fragments were stained using SYBR Green I, visualized by fluorescence using the Typhoon 8600 Imager, and the captured gel images were analyzed using Image version 3.10b (http://www.sanger.ac.uk/Software/Image/). Bands of 702 to 29,950 bp in size were collected for contig assembly. All cleaned band data were standardized using DNA markers II and V, converted into migration rates, ranging from 891 to 3463, yielding a total of 74,281 clones that were successfully fingerprinted, and entered into the FPC database for contig assembly. These clones represented ~10.5× haploid genome equivalents, among which 44.4% were equivalent to 4.4× haploid genomes (clones of an average of 23.6 bands), derived from the *Bam*HI library, and 55.6% were equivalent to 6.1× haploid genomes (clones of an average of 26.0 bands), derived from the *Hind*III library (Han et al. 2007).

7.4.2.2 Fingerprint Analysis and Contig Assembly

The assembly of contigs was conducted using the FPC version 7.2 program (see also http://www.agcol.arizona.edu/software/fpc) (Fig. 7.8). Following several rounds of cleaning of overlapping clones, a tolerance of 7 and a cutoff value of 3×10^{-9} were used for automatic contig assembly. A DQer function was run for several rounds at a final cutoff of 3×10^{-16} (by setting at 3×10^{-13}). As a result, a total of 68,058 BAC clones (92%) were assembled into 3,943 contigs, wherein the physical length of automated contigs was estimated to be 943.8 Mb, based on 242,001 unique bands, and each band was equivalent to 3.9 kb.

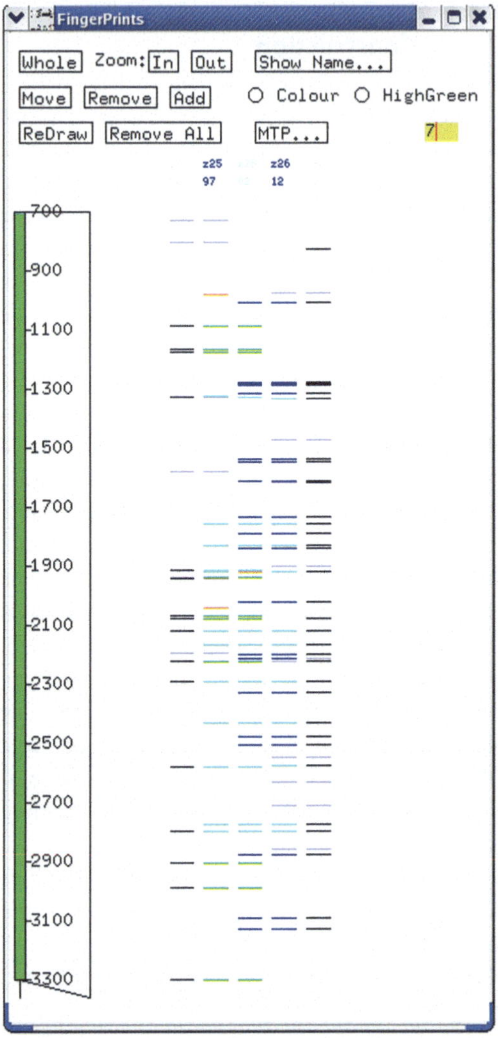

Fig. 7.8 A check of the reliability of an apple contig assembly

Following the automated map assembly, a manual review of the assembly was undertaken to assess and enhance relative orders of clones within contigs, as well as to identify joints between contigs, as well as to disassemble larger chimeric contigs. Furthermore, individual singleton clones were then added to contigs as necessary to increase the coverage of sparse regions within contigs. This detailed and iterative process yielded a total number of 2702 contigs,

Table 7.2 Status of the apple physical map before and after manual editing

Parameters	Automated contig assembly	After manual editing
Number of clones in FPC database	74,281	74,281
Number of singletons	6,223	5,953
Number of contigs	3,943	2,702
Contigs containing		
>200 clones	2	5
101–200 clones	21	67
51–100 clones	212	266
26–50 clones	522	537
10–25	1,407	945
3–9	1,534	754
2	245	128
Unique bands of the contigs	242,001	237,763
Physical length of contigs in megabase pairs	943.8	927.3

consisting of 237,763 unique bands and spanning 927.3 Mb in physical length (Table 7.2). It was found that the longest contig consisted of 287 clones, covering 702 unique bands and spanning 2.7 Mb in physical length (Han et al. 2007).

7.4.2.3 Validation of Contig Reliability

Contig assembly was validated for reliability using variable stringencies (increasing the stringency of contig assembly), by assembling contigs using fingerprints from individual libraries (checking fingerprint patterns of every contig), by checking consensus maps of contigs, as well as by using DNA markers (three contigs were checked using either DNA markers or PCR probes). Overall, the assembly of the contigs was verified and validated (Han et al. 2007; Han and Korban 2010).

7.5 Integration of Physical and Genetic Maps in Apple

The physical map was anchored to the genetic map using two approaches. In the first approach, 52 BAC-end sequence-based SNPs were developed, and these were used to anchor 51 contigs onto the apple genetic map (Han et al. 2009). However, in the second approach, a PCR-based approach was used to screen the BAC library with 649 genetic markers, consisting of 637 SSRs and 15 STSs. This approach allowed for single-locus markers to identify ~7.0 positive BAC clones, ranging from 1 to 16, whereas multilocus markers detected 5 to 25 positive BAC clones, on average of 9.5 clones. Thus, 308 single-locus markers were identified in single contigs, and 291 single-locus markers were identified in two or three contigs, and therefore, all 539 single-locus markers were anchored onto single contigs. Moreover, a total of 23 multilocus markers were identified in either two or three contigs, and these were anchored onto linkage groups based on allele sizes of either these multilocus markers and/or their neighboring molecular markers located on the same contig. Furthermore, a total of 33 sequence-tagged connectors (STCs) were used to screen the apple BAC library, and these were anchored onto different contigs (Han et al. 2009).

Overall, a total of 664 contigs have been merged to yield a genome-wide physical map for the apple genome, and this map consisted of

Table 7.3 Number of SSR markers and cumulative physical lengths of BAC contigs genetically mapped onto different linkage groups in apple

Chromosome	Linkage group		Physical map		
	No. of markers	Length (cM)	No. of contigs	No. of markers	Length (Mb)
1	22	83.7	24	32	26.96
2	31	80.0	39	50	30.01
3	25	63.9	23	38	22.21
4	21	73.0	21	37	22.59
5	43	83.6	37	64	30.36
6	20	81.9	17	29	15.51
7	17	58.5	19	20	17.00
8	26	75.9	23	34	18.52
9	36	72.0	24	43	20.35
10	34	82.0	42	55	33.78
11	27	87.2	42	46	34.23
12	25	70.2	31	40	23.64
13	21	70.7	28	35	23.82
14	22	60.9	19	39	18.17
15	51	116.0	40	62	37.73
16	21	80.7	24	38	22.26
17	25	74.7	20	37	23.62
Total	467	1314.9	473	699	420.76

1,806 contigs, estimated to span 828.03 Mb in physical length. Lengths of most of the contigs (47%) are 500–1000 kb, while 30% of these contigs are longer than 1 Mb with an average size of anchored contigs of 897.4 kb in length, ranging from 97 kb to 4.0 Mb. Fortunately, 470 contigs, spanning 420.76 Mb in cumulative physical length, have been anchored onto the 17 linkage groups. Thus, based on the estimated apple genome size of 700 Mb (Velasco et al. 2010), the physical length of anchored contigs covers 60% of the apple genome (Table 7.3). Moreover, the average physical length of this integrated map on each linkage group of the apple is approximately 24.8 Mb, and ranging between 17.0 Mb and 37.73 Mb (Han et al. 2009).

7.6 A Sequence-Based Physical Map of the Apple Genome

7.6.1 The First Draft Genome Sequence for Apple

The genome of apple cv. 'Golden Delicious' was sequenced using a whole-genome shotgun approach by Velasco et al. (2010). This sequencing strategy incorporated a mixture of Sanger sequencing and Roche 454 sequencing. The former yielded a large sequence read scaffold-genome, while the latter will be filled in gaps, and built coverage depth. The size of the apple genome was estimated to be 742.3 Mb. Sequence coverage was 16.9-fold total, with 26%

and 74% provided by a Sanger dye primer sequencing and 454 sequencing, respectively. The sequence reads were assembled using an iterative assembly approach, yielding 122,146 contigs. Subsequently, 103,076 out of 122,146 contigs were assembled into 1,629 metacontigs, with a total length of 598.3 Mb accounting for 71.2% of the apple genome. The total contig length was 603.9 Mb covering approximately 81.3% of the apple genome (Velasco et al. 2010).

The metacontigs were then anchored onto a high-quality genetic map with 1,643 markers, resulting in a sequence-based physical map of the apple genome. The total size of repetitive elements, including 31,678 transposable elements, was 500.7 Mb, accounting for 67% of the apple genome. Nearly all the unassembled parts of the genome, corresponding to 138.4 Mb in size, consisted of repetitive sequences. It was reported that ~71.2% of genetically anchored sequences belonged to the gene-rich part of the genome, covering as many as 90.2% of genes assigned to chromosomes. A total of 57,386 genes (11,444 genes specific for apple) were predicted in the apple genome, including some genes found only in one of the homoeologous chromosome pairs (Velasco et al. 2010; also see Chap. 8 in this volume).

Subsequently, Li et al. (2016) have constructed yet another draft genome of apple cv. 'Golden Delicious'. This genome sequence has been developed using a combination of 76 Gb (~102 × genome coverage) Illumina HiSeq data and 21.7 Gb (~29× genome coverage) PacBio data. The draft genome is approximately 632.4 Mb in size, corresponding to ~90% of the estimated genome, consisting of predicted 53,922 protein-coding genes and 2,765 non-coding RNA genes (also see Chap. 8 in this volume).

7.6.2 High-Quality *de Novo* Assembly of the Apple Genome

The genome of a doubled-haploid 'Golden Delicious' line, designated as 'GDDH13', has been sequenced and assembled using a mixture of short-read sequencing, long-read sequencing, and optical mapping (Daccord et al. 2017). Sequence coverage was 235-fold in total, consisting of ~120-fold coverage of Illumina paired-end reads (72 Gb), 80-fold coverage of Illumina Nextera mate-pair reads (58 Gb) at three different insert sizes (2, 5, and 10 kb), as well as ~35-fold coverage of PacBio sequencing data. Although initially 2,150 contigs were produced, with an N50 of 620 kb, these were later modified and N50 was increased from 620 kb to 699 kb, for a total contig length of 625.2 Mb. Subsequently, a ~600-fold-coverage BioNano optical map was used to generate a consensus map consisting of an assembly of 649.7 Mb. Following sequence annotation, it was determined that this draft genome consisted of 42,140 protein-coding genes, accounting for 23.3% of the genome assembly, and 1,965 non-protein-coding genes (Daccord et al. 2017; also see Chap. 8 in this volume).

7.7 Comparative Mapping Between Apple and Other Rosaceae Species

Comparative mapping reveals the degree of conservation of genome structure between related genera. This enhances our understanding of those genome evolutionary forces that shape current chromosomal configurations of various species in Rosaceae. Furthermore, comparative mapping studies will facilitate in identifying those genes controlling common traits in different genera, and will also allow for information gained in one species to aid in our pursuit of similar studies in related species.

The first comparative study among Rosaceous genomes was reported by Dirlewanger et al. (2004). It was revealed that high levels of orthologous marker order conservation were observed between *Prunus* (peach) linkage groups (1, 3, and 4) and *Malus* (apple) linkage groups (5, 8, 9, 10, 13, 16, and 17), with one linkage group of *Prunus* usually corresponding to a pair of homeologous linkage groups in apple. Approximately half of linkage group 1 of *Prunus* aligned with the two homologous linkage groups 13 and 16 in apple. These results suggested that the two genera, *Prunus* and *Malus*, share large conserved syntenic blocks, and the whole-genome duplication in apple must have occurred the following divergence of the two genera.

Whole-genome sequence comparisons between the peach and apple genomes have identified a total of 349 conserved regions, with the largest region being 6.1 Mb in size (Jung et al. 2012). Overall, each chromosome of *Prunus* matches up with two or more in *Malus*. For example, the peach chr.3 and chr.5 match up with two pairs of homologous chromosomes, chr.9/chr.17 and chr.14/chr.6, respectively. The peach chr.1 aligns with two pairs of homologous chromosomes, chr.13/chr.16 and chr.8/chr.15, in apple. The peach chr.4 matches up with two pairs of homologous chromosomes in apple, chr.3/chr.11 and chr.5/chr.10. The peach chr.6 aligns with three pairs of homologous chromosomes in apple, chr.2/chr.15, chr.3/chr.11, and chr.4/chr.12. The peach chr.8 aligns with two pairs of homologous chromosomes in apple, chr.5/chr.10 and chr.3/chr.11. These results further confirm that whole-genome duplication must have occurred in the apple genome.

Alignment of the *Malus* genetic map with the *Pyrus* (pear) genetic map using 102 SSRs in common, including 64 pear SSR markers and 38 apple SSR markers, has revealed that 90 SSRs show complete co-linearity between the two genomes (Celton et al. 2009). This demonstrates the presence of a high degree of co-linearity between the genomes of *Malus* and *Pyrus*. Later, comparative mapping using molecular markers has also revealed a number of small chromosomal translocations between *Malus* and *Pyrus*, suggested to play an important role in the evolution of these two genera (Illa et al. 2011). Whole-genome sequence comparisons have further revealed that the apple and pear genomes are similar in genic regions, but differ in non-genic regions (Wu et al. 2013). These observed DNA sequence differences between apple and pear genomes are mainly attributed to the presence of repetitive sequences, predominantly those of transposable elements.

7.8 Conclusions

Major advances have been made in developing genetic and physical maps for the apple. These advances have been critical in identifying and locating various important loci controlling highly desirable traits involved in growth habit, biotic resistance, and fruit quality traits, among others. Moreover, these maps along with the integration of genetic and physical maps have now been supplemented with the release of complete genome sequences for the apple genome due to the availability of next-generation sequencing technologies. These advances will continue to have a significant impact on the identification and analysis of not only various genes of interest but also in identifying and delineating genetic and biological networks and pathways that will contribute to rapid advances in the genetic enhancement efforts of the apple.

References

Aldwinkle HS, Way RD, Livermore KG, Preczewski JL, Beer SV (1976) Fire blight in the Geneva apple collection. Fruit Var J 30:42–55

Allard A, Bink MC, Martinez S, Kelner JJ, Legave JM, di Guardo M, Di Pierro EA, Laurens F, van de Weg EW, Costes E (2016) Detecting QTLs and putative candidate genes involved in budbreak and flowering time in an apple multiparental population. J Exp Bot 67:2875–288

Alston FH, Briggs JB (1968) Resistance to *Sappaphis devecta* (Wlk.) in apple. Euphytica 17:468–472

Alston FH, Briggs JB (1977) Resistance genes in apple and biotypes of *Dysaphis devecta*. Ann Appl Biol 87:75–81

Alston FH, Watkins R (1975) Apple breeding at East Malling. In: Brown AG, Watkins R, Alston F (eds) Proceedings Eucarpia symposium on tree fruit breeding: 11–14 September 1973; Canterbury, pp 14–29

Alston FH, Phillips KL, Evans KM (2000) A Malus gene list. Acta Hortic 538:561–570

Amyotte B, Bowen AJ, Banks T, Rajcan I, Somers DJ (2017) Mapping the sensory perception of apple using descriptive sensory evaluation in a genome wide association study. PLoS ONE 12:

Antanaviciute L, Fernández-Fernández F, Jansen J, Banchi E, Evans KM, Viola R, Velasco R, Dunwell JM, Troggio M, Sargent DJ (2012) Development of a dense SNP-based linkage map of an apple rootstock progeny using the *Malus* Infinium whole genome genotyping array. BMC Genom 13:203. https://doi.org/10.1186/1471-2164-13-203

Bai T, Zhu Y, Fernández-Fernández F, Keulemans J, Brown S, Xu K (2012a) Fine genetic mapping of the *Co* locus controlling columnar growth habit in apple. Mol Genet Genomics 287:437–450

Bai Y, Dougherty L, Li M, Fazio G, Cheng L, Xu K (2012b) A natural mutation-led truncation in one of the two aluminum-activated malate transporter-like genes at the *Ma* locus is associated with low fruit acidity in apple. Mol Genet Genomics 287:663–678

Baldi P, Wolters PJ, Komjanc M, Viola R, Velasco R, Salvi S (2013) Genetic and physical characterisation of the locus controlling columnar habit in apple (*Malus* × *domestica* Borkh.). Mol Breed 31:429–440

Baldo A, Norelli JL, Farrell RE Jr, Bassett CL, Aldwinckle HS, Malnoy M (2010) Identification of genes differentially expressed during interaction of resistant and susceptible apple cultivars (*Malus* × *domestica*) with *Erwinia amylovora*. BMC Plant Biol 10:1. https://doi.org/10.1186/1471-2229-10-1

Ban Y, Honda C, Hatsuyama Y, Igarashi M, Bessho H, Moriguchi T (2007) Isolation and functional analysis of a MYB transcription factor gene that is a key regulator for the development of red coloration in apple skin. Plant Cell Physiol 48:58–970

Bastiaanse H, Bassett HC, Kirk C, Gardiner SE, Deng C, Groenworld R, Chagné D, Bus VG (2016) Scab resistance in 'Geneva' apple is conditioned by a resistance gene cluster with complex genetic control. Mol Plant Pathol 17:159–172

Bénaouf G, Parisi L (2000) Genetics of host-pathogen relationships between *Venturia inaequalis* races 6 and 7 and *Malus* species. Phytopathology 90:236–242

Bénaouf G, Parisi L, Laurens F (1997) Inheritance of *Malus floribunda* clone 821 resistance to *Venturia inaequalis*. IOBC/WPRS Bull 20:1–7

Bianco L, Cestaro A, Sargent DJ, Banchi E, Derdak S, Di Guardo M, Salvi S, Jansen J, Viola R, Gut I, Laurens F, Chagné D, Velasco R, van de Weg E, Troggio M (2014) Development and validation of a 20 K single nucleotide polymorphism (SNP) whole genome genotyping array for apple (*Malus* × *domestica* Borkh.). PLoS ONE 9: e110377

Bianco L, Cestaro A, Linsmith G, Muranty H, Denancé C, Théron A, Poncet C, Micheletti D, Kerschbamer E, Di Pierro EA, Larger S, Pindo M, Van de Weg E, Davassi A, Laurens F, Velasco R, Durel CE, Troggio M (2016) Development and validation of the Axiom(®) Apple480K SNP genotyping array. Plant J 86:62–74

Bonn WG, van der Zwet T (2000) Distribution and economic importance of fire blight. In: Vanneste JL (ed) Fire Blight: the disease and its causative agent, Erwinia amylovora. CABI Publish, Wallingford, UK, pp 37–55

Boudichevskaia A, Fischer C, Flachowsky H, Hanke V, Dunemann F (2004) Development of molecular markers for *Vr1*, a scab resistance factor from R12740-7A apple. Acta Hortic 663:171–176

Boudichevskaia A, Flachowsky H, Peil A, Fischer C, Dunemann F (2006) Development of a multiallelic SCAR marker for the scab resistance gene *Vr1/Vh4/Vx* from R12740-7A apple and its utility for molecular breeding. Tree Genet Genomes 2:186–195

Broggini GA, Wöhner T, Fahrentrapp J, Kost TD, Flachowsky H, Peil A, Hanke MV, Richter K, Patocchi A, Gessler C (2014) Engineering fire blight resistance into the apple cultivar 'Gala' using the FB_MR5 CC-NBS-LRR resistance gene of *Malus* × *robusta* 5. Plant Biotechnol J 12:728–733

Bus VGM, Rikkerink EHA, van de Weg WE, Rusholme RL, Gardiner SE, Bassett HCM, Kodde LP, Parisi L, Laurens FND, Meulenbroek EJ, Plummer KM (2005a) The *Vh2* and *Vh4* scab resistance genes in two differential hosts derived from Russian apple R12740-7A map to the same linkage group of apple. Mol Breed 15:103–116

Bus VGM, Laurens FND, van de Weg WE, Rusholme RL, Rikkerink EHA, Gardiner SE, Bassett HCM, Kodde LP, Plummer KM (2005b) The *Vh8* locus of a new gene-for gene interaction between *Venturia inaequalis* and the wild apple *Malus sieversii* is closely linked to the *Vh2* locus in *Malus pumila* R12740-7A. New Phytol 166:1035–1049

Bus VGM, Chagné D, Bassett HCM, Bowatte D, Calenge F, Celton J-M, Durel CE, Malone MT, Patocchi A, Ranatunga AC, Rikkerink EHA, Tustin DS, Zhou J, Gardiner SE (2008) Genome mapping of three major resistance genes to woolly apple aphid (*Eriosoma lanigerum* Hausm). Tree Genet Genomes 4: 233–236

Bus VGM, Rikkerink EHA, Caffier V, Durel CE, Plummer KM (2011) Revision of the nomenclature of the differential host–pathogen interactions of *Venturia inaequalis* and *Malus*. Annu Rev Phytopathol 49:391–413

Busatto N, Matsumoto D, Tadiello A, Vrhovsek U, Costa F (2019) Multifaceted analyses disclose the role of fruit size and skin-russeting in the accumulation pattern of phenolic compounds in apple. PLoS ONE 14(7): https://doi.org/10.1371/journal.pone.0219354

Calenge F, Durel CE (2006) Both stable and unstable QTLs for resistance to powdery mildew are detected in apple afler four years of field assessments. Mol Breed 17:329–339

Calenge F, Van der Linden CG, Van de Weg E, Schouten HJ, Van Arkel G, Denancé C, Durel CE (2005a) Resistance gene analogues identified through the NBS-profiling method map close to major genes and QTL for disease resistance in apple. Theor Appl Genet 110:660–668

Calenge F, Drouet D, Denancé C, Van de Weg WE, Brisset MN, Paulin JP, Durel CE (2005b) Identification of a major QTL together with several minor additive or epistatic QTLs for resistance to fire blight in apple in two related progenies. Theor Appl Genet 111:128–135

Celton JM, Chagné D, Tustin SD, Terakami S, Nishitani C, Yamamoto T, Gardiner SE (2009) Update on comparative genome mapping between *Malus* and *Pyrus*. BMC Res Notes 2:182. https://doi.org/10.1186/1756-0500-2-182

Cevik V, King J (2002a) High-resolution genetic analysis of the *Sd*-1 aphid resistance locus in *Malus* spp. Theor Appl Genet 105:346–354

Cevik V, King GJ (2002b) Resolving the aphid resistance locus *Sd*-1 on a BAC contig within a sub-telomeric region of *Malus* linkage group 7. Genome 45:939–945

Chagné D, Carlisle CM, Blond C, Volz RK, Whitworth CJ, Oraguzie NC, Crowhurst RN, Allan AC, Espley RV, Hellens RP, Gardiner SE (2007) Mapping a candidate gene (*MdMYB10*) for red flesh and foliage colour in apple. BMC Genom 8:212. https://doi.org/10.1186/1471-2164-8-212

Chagné D, Krieger C, Rassam M, Sullivan M, Fraser J, André C, Pindo M, Troggio M, Gardiner SE, Henry RA, Allan AC, McGhie TK, Laing WA (2012a) QTL and candidate gene mapping for polyphenolic composition in apple fruit. BMC Plant Biol 12:12

Chagné D, Crowhurst RN, Troggio M, Davey MW, Gilmore B, Lawley C, Vanderzande S, Hellens RP, Kumar S, Cestaro A, Velasco R, Main D, Rees JD, Iezzoni A, Mockler T, Wilhelm L, Van de Weg E, Gardiner SE, Bassil N, Peace C (2012b) Genome-wide SNP detection, validation, and development of an 8 K SNP array for apple. PLoS ONE 7:

Chagné D, Lin-Wang K, Espley RV, Volz RK, How NM, Rouse S, Brendolise C, Carlisle CM, Kumar S, De Silva N, Micheletti D, McGhie T, Crowhurst RN, Storey RD, Velasco R, Hellens RP, Gardiner SE, Allan AC (2013) An ancient duplication of apple MYB transcription factors is responsible for novel red fruit-flesh phenotypes. Plant Physiol 161:225–239

Cheng FS, Weeden NF, Brown SK (1996) Identification of co-dominant RAPD markers tightly linked to fruit skin color in apple. Theor Appl Genet 93:222–227

Cheng FS, Weeden NF, Brown SK, Aldwinckle HS, Gardiner SE, Bus VG (1998) Development of a DNA marker for *Vm*, a gene conferring resistance to apple scab. Genome 41:208–214

Conner PJ, Brown SK, Weeden NF (1997) Randomly amplified polymorphicDNA-based genetic linkage maps of three apple cultivars. J Am Soc Hort Sci 122:350–359

Costa F, Van de Weg WE, Stella S, Dondini L, Pratesi D, Musacchi S, Sansavini S (2008) Map position and functional allelic diversity of *Md-Exp7*, a new putative expansin gene associated with fruit softening in apple (*Malus* × *domestica* Borkh.) and pear (*Pyrus communis*). Tree Genet Genomes 4:575–586

Costa F, Peace CP, Stella S, Serra S, Musacchi S, Bazzani M, Sansavini S, Van de Weg WE (2010) QTL dynamics for fruit firmness and softening around an ethylene-dependent polygalacturonase gene in apple (*Malus* x *domestica* Borkh.). J Exp Bot 61:3029–3039

Crane MB, Greenslade RM, Massee AM, Tydemans HM (1936) Studies on the resistance and immunity of apples to the wooly apple aphid, *Eriosoma lanigerum* (Hausm.). J Pomol Hort Sci 14:137–163

Crosby JA, Janick J, Pecknold PC, Korban SS, O'Connor PA, Ries SM, Goffreda J, Voordeckers A (1992) Breeding apples for scab resistance: 1945–1990. Fruit Var J 46:145–166

Daccord N, Celton JM, Linsmith G, Becker C, Choisne N, Schijlen E, van de Geest H, Bianco L, Micheletti D, Velasco R, Di Pierro EA, Gouzy J, Rees DJG, Guérif P, Muranty H, Durel CE, Laurens F, Lespinasse Y, Gaillard S, Aubourg S, Quesneville H, Weigel D, van de Weg E, Troggio M, Bucher E (2017) High-quality de novo assembly of the apple genome and methylome dynamics of early fruit development. Nat Genet 49:1099–1106

Dayton DF (1977) Genetic immunity to apple mildew incited by *Podosphaera leucotricha*. HortScience 12:225–226

Dayton DF, Williams EB (1968) Independent genes in *Malus* for resistance to *Venturia inaequalis*. Proc Amer Soc Hort Sci 92:89–94

Dayton DF, Williams EB (1970) Additional allelic genes in *Malus* for scab resistance of two reaction types. J Am Soc Hort Sci 95:735–736

Desnoues E, Norelli JL, Aldwinckle HS, Wisniewski ME, Evans KM, Malnoy M, Khan A (2018) Identification of novel strain-specific and environment-dependent minor QTLs linked to fire blight resistance in apples. Plant Mol Biol Rep 36:247–256

Devoghalaere F, Doucen T, Guitton B, Keeling J, Payne W, Ling TJ, Ross JJ, Hallett IC, Gunaseelan K, Dayatilake GA, Diak R, Breen KC, Tustin DS, Costes E, Chagné D, Schaffer RJ, David KM (2012) A genomics approach to understanding the role of auxin in apple (*Malus* × *domestica*) fruit size control. BMC Plant Biol 12:7. https://doi.org/10.1186/1471-2229-12-7

Di Guardo M, Bink MCAM, Guerra W, Letschka T, Lozano L, Busatto N, Poles L, Tadiello A, Bianco L, Visser RGF, van de Weg E, Costa F (2017) Deciphering the genetic control of fruit texture in apple by multiple family-based analysis and genome-wide association. J Expt Bot 68:1451–1466

Di Pierro EA, Gianfranceschi L, Di Guardo M, Koehorst-van Putten HJ, Kruisselbrink JW, Longhi S, Troggio M, Bianco L, Muranty H, Pagliarani G, Tartarini S, Letschka T, Lozano Luis L, Garkava-Gustavsson L, Micheletti D, Bink MC, Voorrips RE, Aziz E, Velasco R, Laurens F, van de Weg WE (2016) A high-density, multi-parental SNP genetic map on apple validates a new mapping approach for outcrossing species. Hortic Res 3:16057

Duan N, Bai Y, Sun H, Wang N, Ma Y, Li M, Wang X, Jiao C, Legall N, Mao L, Wan S, Wang K, He T, Feng S, Zhang Z, Mao Z, Shen X, Chen X, Jiang Y, Wu S, Yin C, Ge S, Yang L, Jiang S, Xu H, Liu J, Wang D, Qu C, Wang Y, Zuo W, Xiang L, Liu C, Zhang D, Gao Y, Xu Y, Xu K, Chao T, Fazio G, Shu H, Zhong GY, Cheng L, Fei Z, Chen X (2017) Genome re-sequencing reveals the history of apple and supports a two-stage model for fruit enlargement. Nat Commun 8:249. https://doi.org/10.1038/s41467-017-00336-7

Dunemann F, Bräcker G, Markussen T, Roche P (1999) Identification of molecular markers for the major mildew resistance gene *Pl2* in apple. Acta Hort 484:411–416

Dunemann F, Urbanietz A, Gardiner S, Bassett H, Legg W, Rusholme R, Bus V, Ranatunga C (2004) Marker assisted selection for Pl-1 powdery mildew resistance in apple - old markers for a new resistance gene? Acta Hortic 663:757–762

Dunemann F, Peil A, Urbanietz A, Garcia-Libreros T (2007) Mapping of the apple powdery mildew resistance gene *Pl1* and its genetic association with an NBS-LRR candidate resistance gene. Plant Breed 126:476–481

Durel CE, Denancé C, Brisset MN (2009) Two distinct major QTL for resistance to fire blight co-localize on linkage group 12 in apple genotypes 'Evereste' and *Malus floribunda* clone 821. Genome 52:139–147

Erdin N, Tartarini S, Broggini GAL, Gennari F, Sansavini S, Gessler C, Patocchi A (2006) Mapping of the apple scab resistance gene *Vb*. Genome 49:1238–1245

Espley RV, Hellens RP, Puterill J, Kutty-Amma S, Allan AC (2006) Red colouration in apple fruit is due to the activity of a MYB transcription factor, MdMYB10. Plant J 49:414–427

Espley RV, Hellens RP, Putterill J, Stevenson DE, Kutty-Amma S, Allan AC (2007) Red colouration in apple fruit is due to the activity of the MYB transcription factor, MdMYB10. Plant J 49:414–427

Evans KM, James CM (2003) Identification of SCAR markers linked to *Pl-w* mildew resistance in apple. Theor Appl Genet 106:1178–1183

Fahrentrapp J, Broggini GAL, Kellerhals M, Peil A, Richter K, Zini E, Gessler C (2013) A candidate gene for fire blight resistance in *Malus* × *robusta* 5 is coding for a CC–NBS–LRR. Tree Genet Genome 9:237–251

Falginella L, Cipriani G, Monte C, Gregori R, Testolin R, Velasco R, Troggio M, Tartarini S (2015) A major QTL controlling apple skin russeting maps on the linkage group 12 of 'Renetta Grigia di Torriana'. BMC Plant Biol 15:150

Fazio G, Wan Y, Kviklys D, Romero L, Adams R, Strickland D, Robinson T (2014) *Dw2*, a new dwarfing locus in apple rootstocks and its relationship to induction of early bearing in apple scions. J Am Soc Hortic Sci 139:87–98

Fernandez-Fernandez F, Evans KM, Clarke JB, Govan CL, James CM, Maric S, Tobutt KR (2008) Development of an STS map of an interspecific progeny of *Malus*. Tree Genet Genom 4:469–479

Foster TM, Watson AE, van Hooijdonk BM, Schaffer RJ (2014) Key flowering genes including FT-like genes are upregulated in the vasculature of apple dwarfing rootstocks. Tree Genet Genomes 10:189–202

Foster TM, Celton J-M, Chagné D, Tustin DS, Gardiner SE (2015) Two quantitative trait loci, *Dw1* and *Dw2*, are primarily responsible for rootstock-induced dwarfing in apple. Hortic Res 2:15001

Galli P, Patocchi A, Broggini GA, Gessler C (2010) The *Rvi15* (*Vr2*) apple scab resistance locus contains three TIR-NBS-LRR genes. Mol Plant-Microbe Interact 23:608–617

Gallot JC, Lamb RC, Aldwinkle HS (1985) Resistance to powdery mildew from some small fruited *Malus* cultivars. HortScience 20:1085–1087

Gardiner SE, Bassett HCM, Noiton DAM, Bus VG, Hofstee ME, White AG, Ball RD, Forster RL, Rikkerink EHA (1996) A detailed linkage map around an apple scab resistance gene demonstrates that class 3A and 3B progeny both carry the *Vf* gene. Theor Appl Genet 93:485–493

Gardiner S, Bassett H, Murdoch J, Meech S, Cook M, Bus V, Ranatunga C, Rikkerink E (2001) Major pest and disease resistance loci in apple available to breeders. In: Plant & animal Genome IX conferences, San Diego, CA, USA

Gardiner S, Murdoch J, Meech S, Rusholme R, Bassett H, Cook M, Bus V, Rikkerink E, Gleave A, Crowhurst R, Ross G, Warrington I (2003) Candidate resistance genes from an EST database prove a rich source of markers for major genes conferring resistance to important apple pests and diseases. Acta Hortic 622:141–151

Gardiner SE, Norelli JL, de Silva N, Fazio G, Peil A, Malnoy M, Horner M, Bowatte D, Carlisle C, Wiedow C, Wan Y, Bassett CL, Baldo AM, Celton JM, Richter K, Aldwinckle HS, Bus VG (2012) Putative resistance gene markers associated with quantitative trait loci for fire blight resistance in *Malus* 'Robusta 5' accessions. BMC Genet 13:25. https://doi.org/10.1186/1471-2156-13-25

Gardner RG, Cummins JN, Aldwinckle HS (1980) Inheritance of fire blight resistance in *Malus* in relation to rootstock breeding. J Am Soc Hort Sci 105:912–916

Gardner KM, Brown P, Cooke TF, Cann S, Costa F, Bustamante C, Velasco R, Troggio M, Myles S (2014) Fast and cost-effective genetic mapping in apple using next-generation sequencing. G3 (Bethesda) 4:1681–1687

Gessler C, Pertot I (2012) *Vf* scab resistance of *Malus*. Trees 26:95–108. https://doi.org/10.1007/s00468-011-0618-y

Gessler C, Patocchi A, Sansavini S, Tartarini S, Gianfranceschi L (2006) *Venturia inaequalis* resistance in apple. Crit Rev Plant Sci 25:473–503

Gianfranceschi L, Koller B, Seglias N, Kellerhals M, Gessler C (1996) Molecular selection in apple for resistance to scab caused by *Venturia inaequalis*. Theor Appl Genet 93:199–204

Gianfranceschi L, Seglias N, Tarchini R, Komjanc M, Gessler C (1998) Simple sequence repeats for the genetic analysis of apple. Theor Appl Genet 96:1069–1076

Grattapaglia D, Sederoff RR (1994) Genetic linkage maps of *Eucalyptus grandis* and *Eucalyptus urophylla* using a pseudo-testcross: mapping strategy and RAPD markers. Genetics 137:1121–1137

Guan Y, Peace C, Rudell D, Verma S, Evans KM (2015) QTLs detected for individual sugars and soluble solids content in apple. Mol Breed 35:135. https://doi.org/10.1007/s11032-015-0334-1

Guilford P, Prakash S, Zhu JM, Rikkerink E, Gardiner S, Bassett H, Forster R (1997) Microsatellites in *Malus* × *domestica* (apple): abundance, polymorphism and cultivar identification. Theor Appl Genet 94:249–254

Gutierrez BL, Zhong G-Y, Brown SK (2018) Increased phloridzin content associated with russeting in apple (*Malus domestica* (Suckow) Borkh.) fruit. Genet Resour Crop Evol 65:2135–3149

Han Y, Korban SS (2010) Strategies for map-based cloning in apple. Crit Rev Plant Sci 29:265–284

Han Y, Gasic K, Marron B, Beever JE, Korban SS (2007) A BAC-based physical map of the apple genome. Genomics 89:630–637

Han Y, Chagné D, Gasic K, Rikkerink EH, Beever JE, Gardiner SE, Korban SS (2009) BAC-end sequence-based SNPs and Bin mapping for rapid integration of physical and genetic maps in apple. Genomics 93:282–288

Han Y, Zheng D, Vimolmangkang S, Khan MA, Beever JE, Korban SS (2011) Integration of physical and genetic maps in apple confirms whole genome and segmental duplications in the apple genome. J Exp Bot 62:5117–5130

Harrison N, Harrison RJ, Barber-Perez N, Cascant-Lopez E, Cobo-Medina M, Lipska M, Conde-Ruíz R, Brain P, Gregory PJ, Fernández-Fernández F (2016) A new three-locus model for rootstock-induced dwarfing in apple revealed by genetic mapping of root bark percentage. J Exp Bot 67:1871–1881

Hemmat M, Weeden NF, Manganaris AG, Lawson DM (1994) Molecular marker linkage map for apple. J Heredity 85:4–11

Hemmat M, Weeden NF, Aldwinckle HS, Brown SK (1998) Molecular markers for the scab resistance (*Vf*) region in apple. J Am Soc Hort Sci 123:992–996

Hemmat M, Brown SK, Aldwinckle HS, Weeden NF (2003) Identification and mapping of markers for resistance to apple scab from 'Antonovka' and 'Hansen's baccata #2'. Acta Hort 622:153–161

Hokanson SC, Szewc-McFadden AK, Lamboy WF, McFerson JR (1998) Microsatellite (SSR) markers reveal genetic identities, genetic diversity and relationships in a *Malus* × *domestica* Borkh. core subset collection. Theor Appl Genet 97:671–683

Illa E, Sargent DJ, Lopez Girona E, Bushakra J, Cestaro A, Crowhurst R, Pindo M, Cabrera A, van der Knaap E, Iezzoni A, Gardiner S, Velasco R, Arús P, Chagné D, Troggio M (2011) Comparative analysis of rosaceous genomes and the reconstruction of a putative ancestral genome for the family. BMC Evol Biol 11:9

James CM, Clarke JB, Evans KM (2004) Identification of molecular markers linked to the mildew resistance gene *Pl-d* in apple. Theor Appl Genet 110:175–181

Janick J, Cummins JN, Brown SK, Hemmat M (1996) Apples. In: Janick J, Moore JN (eds.) Fruit breeding: tree and tropical fruits,. vol I. Wiley, New York, pp 177

Jensen PJ, Fazio G, Altman N, Praul C, McNellis TW (2014) Mapping in an apple (*Malus* × *domestica*) F1 segregating population based on physical clustering of differentially expressed genes. BMC Genom 15:261

Jock S, Donat V, Lopez MM, Bazzi C, Geider K (2002) Following spread of fire blight in Western and southern Europe by molecular differentiation of *Erwinia amylovora* strains with PFGE analysis. Environ Microbiol 4:106–114

Jung S, Cestaro A, Troggio M, Main D, Zheng P, Cho I, Folta KM, Sosinski B, Abbott A, Celton JM, Arús P, Shulaev V, Verde I, Morgante M, Rokhsar D, Velasco R, Sargent DJ (2012) Whole genome comparisons of *Fragaria*, *Prunus* and *Malus* reveal different modes of evolution between Rosaceous subfamilies. BMC Genom 13:129

Kellerhals M, Gianfranceschi L, Seglias N, Gessler C (2000) Marker-assisted selection in apple breeding. Acta Hortic 521:255–265

Kenis K, Keulemans J (2005) Genetic linkage maps of two apple cultivars (*Malus* × *domestica* Borkh.) based on AFLP and microsatellite markers. Mol Breed 15:205–219

Kenis K, Keulemans J (2007) Study of tree architecture of apple (*Malus* × *domestica* Borkh.) by QTL analysis of growth traits. Mol Breed 19:193–208

Kenis K, Keulemans J, Davey MW (2008) Identification and stability of QTLs for fruit quality traits in apple. Tree Genet Genomes 4:647–661

Khan MA, Duffy B, Gessler C, Patocchi A (2006) QTL mapping of fire blight resistance in apple. Mol Breed 17:299–306

Khan MA, Han Y, Zhao YF, Korban SS (2012a) A high-throughput apple SNP genotyping platform using the GoldenGate™ assay. Gene 494:196–201

Khan MA, Han Y, Zhao YF, Troggio M, Korban SS (2012b) A multipopulation consensus genetic map reveals inconsistent marker order among maps likely attributed to structural variations in the apple genome. PLoS ONE 7:

Khan MA, Zhao Y, Korban SS (2012c) Molecular mechanisms of fire blight resistance in Rosaceae. Plant Mol Biol Rep 30:247–260

Khan SA, Chibon PY, de Vos RC, Schipper BA, Walraven E, Beekwilder J, van Dijk T, Finkers R, Visser RG, van de Weg EW, Bovy A, Cestaro A, Velasco R, Jacobsen E, Schouten HJ (2012d) Genetic analysis of metabolites in apple fruits indicates an mQTL hotspot for phenolic compounds on linkage group 16. J Exp Bot 63:2895–2908

Khan MA, Zhao YF, Korban SS (2013) Identification of genetic loci associated with fire blight resistance in

Malus through combined use of QTL and association mapping. Physiol Plant 148:344–353

Kim MY, Song KJ, Hwang JH, Shin YU, Lee HJ (2003) Development of RAPD and SCAR markers linked to the *Co* gene conferring columnar growth habit in apple (*Malus pumila* Mill.). J Hort Sci Bio 78:512–517

King GJ, Maliepaard C, Lynn JR, Alston FH, Durel CE, Evans KM,·Griffon B, Laurens F, Manganaris AG,· Schrevens E, Tartarini S, Verhaegh J (2000) Quantitative genetic analysis and comparison of physical and sensory descriptors relating to fruit flesh firmness in apple (Malus pumila Mill.). Theor Appl Genet 100:1074–1084

Knight RL, Alston FH (1968) Sources of field immunity to mildew *(Podosphaera leucotricha)* in apple. Can J Genet Cytol 10:294–298

Knight RL, Briggs JB, Massee AM, Tydeman HM (1962) The inheritance of resistance to woolly aphid, *Eriosoma lanigerum* (Hsmnn.), in the apple. J Hort Sci 37:207–218

Knoche M, Lang A (2017) Ongoing growth challenges fruit skin integrity. Crit Rev Plant Sci 36:190–215

Koller B, Gianfranceschi L, Seglias N, McDermott J, Gessler C (1994) DNA markers linked to *Malus floribunda* 821 scab resistance. Plant Mol Biol 26:597–602

Korban SS, Dayton DF (1983) Evaluation of *Malus* germplasm for resistance to powdery mildew. HortScience 18:219–220

Korban SS, Ries SM, Klopmeyer MJ, Morissey JF, Hattermann DR (1988) Genotypic responses of scab-resistant apple cultivar/selections to two strains of *Erwinia amylovora* and the inheritance of resistance to fire blight. Ann Appl Biol 113:101–105

Larsen B, Migicovsky Z, Jeppesen AA, Gardner KM, Toldam-Andersen TB, Myles S, Ørgaard M, Mikael Petersen MA, Pedersen C (2019) Genome-wide association studies in apple reveal loci for aroma volatiles, sugar composition, and harvest date. Plant Genome 12: https://doi.org/10.3835/plantgenome2018.12.0104

Lashbrooke J, Aharoni A, Costa F (2015) Genome investigation suggests *MdSHN3*, an APETALA2-domain transcription factor gene, to be a positive regulator of apple fruit cuticle formation and an inhibitor of russet development. J Exp Bot 66:6579–6589

Le Roux PM, Khan MA, Broggini GA, Duffy B, Gessler C, Patocchi A (2010) Mapping of quantitative trait loci for fire blight resistance in the apple cultivars 'Florina' and 'Nova Easygro'. Genome 53:710–722

Legay S, Guerriero G, André C, Guignard C, Cocco E, Charton S, Boutry M, Rowland O, Hausman J-F (2016) *MdMyb93* is a regulator of suberin deposition in russeted apple fruit skins. N Phytol 212:977–991. https://doi.org/10.1111/nph.14170

Legay S, Cocco E, André CM, Guignard C, Hausman J-F, Guerriero G (2017) Differential lipid composition and gene expression in the semi-russeted 'Cox Orange Pippin' apple variety. Front Plant Sci 8:1656. https://doi.org/10.3389/fpls.2017.01656

Lespinasse Y (1989) Breeding pome fruits with stable resistance to diseases: genes, resistance mechanisms, present work and prospects. IOBC/WPRS Bull 2:100–115

Lespinasse Y, Aldwinckle H (2000) Breeding for resistance to fire blight. In: Vanneste JL (ed) Fire blight: the disease and its causative agent, *Erwinia amylovora*. CAB International, Wallingford, UK, pp 253–273

Lespinasse Y, Olivier JM, Lespinasse JM, Le Lezec M (1985) 'Florina- Quérina'®: la résistance du pommier à la tavelure. Arboricult Fruit 378:43–47

Li X, Kui L, Zhang J, Xie Y, Wang L, Yan Y, Wang N, Xu J, Li C, Wang W, van Nocker S, Dong Y, Ma F, Guan Q (2016) Improved hybrid de novo genome assembly of domesticated apple (*Malus* × *domestica*). Gigascience 5:35

Liebhard R, Gianfranceschi L, Koller B, Ryder CD, Tarchini R, Van DeWeg E, Gessler C (2002) Development and characterization of 140 new microsatellites in apple (*Malus* × *domestica* Borkh.). Mol Breed 10:217–241

Liebhard R, Koller B, Gianfranceschi L, Gessler C (2003) Creating a saturatedreference map for the apple (*Malus* × *domestica* Borkh.) genome. Theor Appl Genet 106:1497–1508

Longhi S, Moretto M, Viola R, Velasco R, Costa F (2012) Comprehensive QTL mapping survey dissects the complex fruit texture physiology in apple (*Malus* × *domestica* Borkh.). J Exp Bot 63:1107–1121

Longhi S, Hamblin MT, Trainotti L, Peace CP, Velasco R, Costa F (2013) A candidate gene based approach validates Md-PG1 as the main responsible for a QTL impacting fruit texture in apple (*Malus* × *domestica* Borkh.). BMC Plant Biol 13:37. https://doi.org/10.1186/1471-2229-13-37

Ma B, Zhao S, Wu B, Wang D, Peng Q, Owiti A, Fang T, Liao L, Ogutu C, Korban SS, Li S, Han YP (2016) Construction of a high density linkage map and its application in the identification of QTLs for soluble sugar and organic acid components in apple. Tree Genet Genomes 12:1. https://doi.org/10.1007/s11295-015-0959-6

Ma B, Liao L, Peng Q, Fang T, Zhou H, Korban SS, Han Y (2017) Reduced representation genome sequencing reveals patterns of genetic diversity and selection in apple. J Integr Plant Biol 59:190–204

Maliepaard C, Alston FH, van Arkel G, Brown LM, Chevreau E, Dunemann F, Evans KM, Gardiner S, Guilford P, van Heusden AW, Janse J, Laurens F, Lynn JR, Manganaris AG, den Nijs APM, Periam N, Rikkerink E, Roche P, Ryder C, Sansavini S, Schmidt H, Tartarini S, Verhaegh JJ, Vrielinkvan Ginkel M, King GJ (1998) Aligning male and female linkage maps of apple (*Malus pumila* Mill.) using multiallelic markers. Theor Appl Genet 97:60–73

Manganaris AG (1989) Isoenzymes as genetic markers in apple breeding. PhD thesis, University of London, London, UK

Markussen T, Kruger J, Schmidt H, Dunemann F (1995) Identification of PCR-based markers linked to the powdery-mildew-resistance gene *Pl1* from *Malus robusta* in cultivated apple. Plant Breed 114:530–534

McClure KA, Gong Y, Song J, Vinqvist-Tymchuk M, Palmer LC, Fan L, Burgher-MacLellan K, Zhang ZQ, Celton J-M, Forney CF, Migicovsky Z, Myles S (2019) Genome-wide association studies in apple reveal loci of large effect controlling apple polyphenols. Hortic Res 6:107. https://doi.org/10.1038/s41438-019-0190-y

Mellidou I, Chagné D, Laing WA, Keulemans J, Davey MW (2012) Allelic variation in paralogs of GDP-L-galactose phosphorylase is a major determinant of vitamin C concentrations in apple fruit. Plant Physiol 160:1613–1629

Minarro M, Dapena E (2004) Inheritance of tolerance to the rosy apple aphid of the cv. 'Florina'. Acta Hortic 663: 261–264

Miotto YE, Tessele C, Czermainski ABC, Porto DD, Falavigna VdS, Sartor T, Cattani AM, Delatorre CA, de Alencar SA, da Silva-Junior OB, Togawa RC, do Carmo Costa MM, Pappas Jr. GJ, Grynberg P, de Oliveira PRD, Kvitschal MV, Denardi F, Buffon V, Revers LF (2019) Spring is coming: Genetic analyses of the bud break date locus reveal candidate genes from the cold perception pathway to dormancy release in apple (*Malus* × *domestica* Borkh.). Front Plant Sci 10:33 https://doi.org/10.3389/fpls.2019.00033

Moriya S, Okada K, Haji T, Yamamoto T, Abe K (2012) Fine mapping of *Co*, a gene controlling columnar growth habit located on apple (*Malus* × *domestica* Borkh.) linkage group 10. Plant Breed 131:641–647

N'Diaye A, Van de Weg WE, Kodde LP, Koller B, Dunemann F, Thiermann M, Tartarini S, Gennari F, Durel CE (2008) Construction of an integrated consensus map of the apple genome based on four mapping populations. Tree Genet Genomes 4:727–743

Nachman MW (2002) Variation in recombination rate across the genome: evidence and implications. Curr Opin Genet Dev 12:657–663

Papp D, Singh J, Gadoury DM, Khan MA (2019) New North American isolates of *Venturia inaequalis* can overcome apple scab resistance of *Malus floribunda* 821. Plant Dis 104(3):649–55

Parisi L, Lespinasse Y (1996) Pathogenicity of *Venturia inaequalis* race 6 strains on apple clones (*Malus* spp.). Plant Dis 80:1179–1183

Parravicini G, Gessler C, Denancé C, Lasserre-Zuber P, Vergne E, Brisset MN, Patocchi A, Durel CE, Broggini GA (2011) Identification of serine/threonine kinase and nucleotide-binding site-leucine-rich repeat (NBS-LRR) genes in the fire blight resistance quantitative trait locus of apple cultivar 'Evereste'. Mol Plant Pathol 12:493–505

Patocchi A, Gianfranceschi L, Gessler C (1999a) Towards the map-based cloning of *Vf*: Fine and physical mapping of the *Vf* region. Theor Appl Genet 99:1012–1017

Patocchi A, Vinatzer BA, Gianfranceschi L, Tartarini S, Zhang H-B, Sansavini S, Gessler C (1999b) Construction of a 550 kb BAC contig spanning the genomic region containing the apple resistance gene *Vf*. Mol Gen Genet 262:884–891

Patocchi A, Bigler B, Koller B, Kellerhals M, Gessler C (2004) *Vr2*: A new apple scab resistance gene. Theor Appl Genet 109:1087–1092

Patocchi A, Walser M, Tartarini S, Broggini GAL, Gennari F, Sansavini S, Gessler C (2005) Identification by genome scanning approach (GSA) of a microsatellite tightly associated with the apple scab resistance gene *Vm*. Genome 48:630–663

Peil A, Garcia-Libreros T, Richter K, Trognitz FC, Trognitz B, Hanke MV, Flachowsky H (2007) Strong evidence for a fire blight resistance gene of *Malus robusta* located on linkage group 3. Plant Breed 126:470–475

Peil A, Bus VGM, Geider K, Richter K, Flachowsky H, Hanke MV (2009) Improvement of fire blight resistance in apple and pear. Int J Plant Breed 3:1–27

Pessina S, Pavan S, Catalano D, Gallotta A, Visser RG, Bai Y, Malnoy M, Schouten HJ (2014) Characterization of the *MLO* gene family in Rosaceae and gene expression analysis in *Malus domestica*. BMC Genom 15:618

Pessina S, Angeli D, Martens S, Visser RG, Bai Y, Salamini F, Velasco R, Schouten HJ, Malnoy M (2016) The knock-down of the expression of *MdMLO19* reduces susceptibility to powdery mildew (*Podosphaera leucotricha*) in apple (*Malus domestica*). Plant Biotechnol J 14:2033–2044

Qubbaj T, Reineke A, Zebitz CPW (2005) Molecular interactions between rosy apple aphids, *Dysaphis plantaginea*, and resistant and susceptible cultivars of its primary host *Malus domestica*. Entomol Exper Applicat 115:145–152

Roberts AL, Crute IR (1994) Apple scab resistance from *Malus floribunda* 821 (*Vf*) is rendered ineffective by isolates of *Venturia inaequalis* from *Malus floribunda*. Norw J Agr Sci 17:403–406

Roche P, Alston FH, Maliepaard C, Evans KM, Vrielink R, Dunemann F, Markussen T, Tartarini S, Brown LM, Ryder C, King GJ (1997a) RFLP and RAPD markers linked to the rosy leaf curling aphid resistance gene (*Sd-1*) in apple. Theor Appl Genet 94:528–533

Roche PG, Van Arkel AW, Heusden Van (1997b) A specific PCR assay based on an RFLP marker closely linked to the *Sd1* gene for resistance to biotypes 1 and 2 of the rosy leaf curling aphid in apple. Plant Breed 116:567–572

Rushlomé-Pilcher RL, Celton J-M, Gardiner SE (2008) Genetic markers linked to the dwarfing trait of apple rootstock 'Malling 9'. J Am Soc Hort Sci 133:100–106

Sandanayaka WRM, Bus VGM, Connolly P, Newcomb R (2003) Characteristics associated with woolly apple aphid *Eriosoma lanigerum*, resistance of three apple rootstocks. Entomol Exp Appl 109:63–72

Schouten HJ, Brinkhuis J, van der Burgh A, Schaart JG, Groenwold R, Broggini GAL, Gessler C (2014) Cloning and functional characterization of the *Rvi15* (*Vr2*) gene for apple scab resistance. Tree Genet Genomes 10:251–260

Seglias NP, Gessler C (1997) Genetics of apple powdery mildew resistance from *Malus zumi* (Pl2). IOBC/WPRS Bull 20:195–208

Seleznyova AN, Tustin DS, Thorp TG (2008) Apple dwarfing rootstocks and interstocks affect the type of growth units produced during the annual growth cycle: precocious transition to flowering affects the composition and vigour of annual shoots. Ann Bot 101:679–87

Silfverberg-Dilworth E, Matasci CL, Van de Weg WE, Van Kaauwen MPW, Walser M, Kodde LP, Soglio V, Gianfranceschi L, Durel CE, Costa F, Yamamoto T, Koller B, Gessler C, Patocchi A (2006) Microsatellite markers spanning the apple (*Malus × domestica* Borkh.) genome. Tree Genet Genomes 2:202–224

Soriano JM, Joshi SG, Van Kaauwen W, Noordijk Y, Groenwold R, Henken B, van de Weg WE, Schouten HJ (2009) Identification and mapping of the novel apple scab resistance gene *Vd3*. Tree Genet Genomes 5:475–482

Souleyre EJ, Chagné D, Chen X, Tomes S, Turner RM, Wang MY, Maddumage R, Hunt MB, Winz RA, Wiedow C, Hamiaux C, Gardiner SE, Rowan DD, Atkinson RG (2014) The *AAT1* locus is critical for the biosynthesis of esters contributing to 'ripe apple' flavour in 'Royal Gala' and 'Granny Smith' apples. Plant J 78:903–915

Sun R, Chang Y, Yang F, Wang Y, Li H, Zhao Y, Chen D, Wu T, Zhang X, Han Z (2015) A dense SNP genetic map constructed using restriction site-associated DNA sequencing enables detection of QTLs controlling apple fruit quality. BMC Genomics 16:747

Takos AM, Jaffé FW, Jacob SR, Bogs J, Robinson SP, Walker AR (2006) Light-induced expression of a MYB gene regulates anthocyanin biosynthesis in red apples. Plant Physiol 142:1216–1232

Tartarini S (1996) RAPD markers linked to the *Vf* gene for scab resistance in apple. Theor Appl Genet 92:803–810

Tartarini S, Gianfranceschi L, Sansavini S, Gessler C (1999) Development of reliable PCR markers for the selection of the *Vf* gene conferring scab resistance in apple. Plant Breed 118:183–186

Tartarini S, Gennari F, Pratesi D, Palazzetti C, Sansavini S, Parisi L, Fouillet A, Fouillet V, Durel CE (2004) Characterisation and genetic mapping of a major scab resistance gene from the old Italian apple cultivar 'Durello di Forli'. Acta Hortic 663:129–133

Tian YK, Wang CH, Zhang JS, James C, Dai HY (2005) Mapping *Co*, a gene controlling the columnar phenotype of apple, with molecular markers. Euphytica 145:181–188

Urrestarazu J, Muranty H, Denancé C, Leforestier D, Ravon E, Guyader A, Guisnel R, Feugey L, Aubourg S, Celton J-M, Daccord N, Dondini L, Gregori R, Lateur M, Houben P, Ordidge M, Paprstein F, Sedlak J, Nybom H, Garkava-Gustavsson L, Troggio M, Bianco L, Velasco R, Poncet C, Théron A, Moriya S, Bink MCAM, Laurens F, Tartarini S, Durel C-E (2017) Genome-wide association mapping of flowering and ripening periods in apple. Front Plant Sci 8:1923 https://doi.org/10.3389/fpls.2017.01923

Velasco R, Zharkikh A, Affourtit J, Dhingra A, Cestaro A, Kalyanaraman A, Fontana P, Bhatnagar SK, Troggio M, Pruss D, Salvi S, Pindo M, Baldi P, Castelletti S, Cavaiuolo M, Coppola G, Costa F, et al (2010) The genome of the domesticated apple (Malus × domestica Borkh.). Nat Genet 42:833–839

Verdu CF, Guyot S, Childebrand N, Bahut M, Celton JM, Gaillard S, Lasserre-Zuber P, Troggio M, Guilet D, Laurens F (2014) QTL analysis and candidate gene mapping for the polyphenol content in cider apple. PLoS ONE 9:

Verma S, Evans K, Guan Y, Luby JJ, Rosyara UR, Howard NP, Bassil N, Bink MCAM, van De Weg WE, Peace CP (2019) Two large-effect QTLs, *Ma* and *Ma3*, determine genetic potential for acidity in apple fruit: breeding insights from a multi-family study. Tree Genet Genomes 15:18. https://doi.org/10.1007/s11295-019-1324-y

Vinatzer BA, Zhang H-B, Sansavini S (1998) Construction and characterization of a bacterial artificial chromosome library of apple. Theor Appl Genet 97:1183–1190

Vinatzer BA, Patocchi A, Gianfranceschi L, Tartarini S, Zhang H-B, Gessler C, Sansavini S (2001) Apple (*Malus sp.*) contains receptor-like genes homologous to the *Cf* resistance gene family of tomato with a cluster of such genes co-segregating with *Vf* apple scab resistance. Mol Plant-Microbe Interact 14:508–515

Vinatzer BA, Patocchi A, Tartarini S, Gianfranceschi L, Sansavini S, Gessler C (2004) Isolation of two microsatellite markers from BAC clones of the *Vf* scab resistance region and molecular characterization of scab-resistant accessions in *Malus* germplasm. Plant Breed 123:321–326

Visser T, Verhaegh JJ (1976) Review of tree fruit breeding carried out at the Institute for Horticultural Plant Breeding at Wageningen from 1951 to 1976. In: Proceedings of Eucarpia meeting of tree fruit breeding, Wageningen, The Netherlands, pp 113–132

Wolters PJ, Schouten HJ, Velasco R, Si-Ammour A, Baldi P (2013) Evidence for regulation of columnar habit in apple by a putative 2OG-Fe(II) oxygenase. New Phytol 200:993–999

Wu J, Wang Z, Shi Z, Zhang S, Ming R, Zhu S, Khan MA, Tao S, Korban SS, Wang H et al (2013) The genome of the pear (*Pyrus bretschneideri* Rehd.). Genome Res 23:396–408
Xu M, Korban SS (2000) Saturation mapping of the apple scab resistance gene *Vf* using AFLP markers. Theor Appl Genet 101:844–851
Xu M, Korban SS (2001) AFLP-derived SCARs facilitate construction of a 1.1 Mb sequence-ready map of a region that spans the *Vf* locus in the apple genome. Plant Mol Biol 50:803–818
Xu M, Korban SS (2002) A cluster of four receptor-like genes resides in the *Vf* locus that confers resistance to apple scab disease. Genetics 162:1995–2006
Xu M, Korban SS (2004) Somatic variation plays a key role in the evolution of the *Vf* gene family residing in the *Vf* locus that confers resistance to apple scab disease. Mol Phylogenet Evol 32:57–65
Xu M, Huaracha E, Korban SS (2001a) Development of sequence characterized amplified regions (SCARs) from amplified fragment length polymorphism (AFLP) markers tightly linked to the *Vf* gene in apple. Genome 44:63–70
Xu M, Song J, Cheng Z, Jiang J, Korban SS (2001b) A bacterial artificial chromosome (BAC) library of *Malus floribunda* 821 and contig construction for positional cloning of the apple scab resistance gene *Vf*. Genome 44:1104–1113
Xu M, Song J, Jiang J, Korban SS (2002) Constructing a bacterial artificial chromosome library of the apple cultivar GoldRush. Acta Hortic 595:103–112
Xu K, Wang A, Brown S (2012) Genetic characterization of the *Ma* locus with pH and titratable acidity in apple. Mol Breed 30:899–912
Yang H, Korban SS (1996) Screening apples for OPD20/600 using sequence-specific primers. Theor Appl Genet 92:263–266
Yang HY, Krüger J (1994) Identification of an RAPD markers linked to the *Vf* gene for scab resistance in apples. Plant Breed 112:323–329
Yang HY, Korban SS, Krüger J, Schmidt H (1997) The use of a modified bulk segregant analysis to identify a molecular marker linked to a scab resistance gene in apple. Euphytica 94:175–182
Yao JL, Xu J, Cornille A, Tomes S, Karunairetnam S, Luo Z, Bassett H, Whitworth C, Rees-George J, Ranatunga C, Snirc A, Crowhurst R, de Silva N, Warren B, Deng C, Kumar S, Chagné D, Bus VG, Volz RK, Rikkerink EH, Gardiner SE, Giraud T, MacDiarmid R, Gleave AP (2015) A microRNA allele that emerged prior to apple domestication may underlie fruit size evolution. Plant J 84:417–427
Yauk YK, Chagné D, Tomes S, Matich AJ, Wang MY, Chen X, Maddumage R, Hunt MB, Rowan DD, Atkinson RG (2015) The O-methyltransferase gene MdoOMT1 is required for biosynthesis of methylated phenylpropenes in ripe apple fruit. Plant J 82:937–950
Yauk YK, Souleyre EJF, Matich AJ, Chen X, Wang MY, Plunkett B, Dare AP, Espley RV, Tomes S, Chagné D, Atkinson RG (2017) Alcohol acyl transferase 1 links two distinct volatile pathways that produce esters and phenylpropenes in apple fruit. Plant J 91:292–305
Zhang Q, Ma B, Li H, Chang Y, Han Y, Li J, Wei G, Zhao S, Khan MA, Zhou Y, Gu C, Zhang X, Han Z, Korban SS, Li S, Han Y (2012) Identification, characterization, and utilization of genome-wide simple sequence repeats to identify a QTL for acidity in apple. BMC Genom 13:537. https://doi.org/10.1186/1471-2164-13-537
Zhang L, Hu J, Han X, Li J, Gao Y, Richards CM, Zhang C, Tian Y, Liu G, Gul H, Wang D, Tian Y, Yang C, Meng M, Yuan G, Kang G, Wu Y, Wang K, Zhang H, Wang D, Con P (2019) A high-quality apple genome assembly reveals the association of a retrotransposon and red fruit colour. Nat Commun 10:1494. https://doi.org/10.1038/s41467-019-09518-x
Zhen Q, Fang T, Peng Q, Liao L, Zhao L, Owiti A, Han Y (2018) Developing gene-tagged molecular markers for evaluation of genetic association of apple *SWEET* genes with fruit sugar accumulation. Hortic Res 5:14. https://doi.org/10.1038/s41438-018-0024-3
Zini E, Biasioli F, Gasperi F, Mott D, Aprea E, Märk TD, Patocchi A, Gessler C, Komjanc M (2005) QTL mapping of volatile compounds in ripe apples detected by proton transfer reaction-mass spectrometry. Euphytica 145:269–279

8 The Apple Genome and Epigenome

Jean-Marc Celton, Luca Bianco, Gareth Linsmith, Sandrine Balzergue, and Michela Troggio

Abstract

Within a span of the last two decades, sequencing technologies have rapidly evolved, allowing for the production of genome sequences from multiple angiosperm clades, including major crops and botanical models. Apple (*Malus* × *domestica*) is among the first crop genomes to be fully sequenced, thus contributing to expanded knowledge of genome structure, biological functions, trait physiology and inheritance, and leading to practical applications for crop improvement. Access to full genome sequences has also allowed for the development of new fields of investigation, such as epigenetics. With the advent of accurate and cheaper new sequencing technologies, together with high-throughput phenotyping methods, the next decade will probably see the development of projects based on whole epi/genome sequencing or resequencing, providing researchers with a firm foundation to investigate the dynamics involved in the development of traits of interest.

J.-M. Celton (✉) · S. Balzergue
Univ Angers, Institut Agro, INRAE, IRHS, SFR QUASAV, F-49000 Angers, France
e-mail: jean-marc.celton@inrae.fr

L. Bianco · G. Linsmith · M. Troggio
Research and Innovation Center, Fondazione Edmund Mach, 38010 San Michele all'Adige, Italy

G. Linsmith
Earlham Institute, Norwich Research Park, Norwich, UK

G. Linsmith
Center for Plant Systems Biology, VIB, Technologiepark 71, 9052 Gent, Belgium

G. Linsmith
Department of Plant Biotechnology and Bioinformatics, Ghent University, Technologiepark 71, 9052 Ghent, Belgium

8.1 Introduction

Structural genomics is key to understanding a plant genome via the determination of the DNA sequence order. Knowledge of sequences of a plant genome enables the exploration of genome structures and allows for inferring molecular phylogeny. It also allows for the study of genes controlling an organism's growth, development, and adaptation to its environment.

Prior to the advent of Next Generation Sequencing (NGS) technologies in 2005 (Morozova and Marra 2008), DNA sequencing was based on Sanger's sequencing method. This latter technique was used to sequence the *Arabidopsis thaliana* genome (The Arabidopsis Genome Initiative 2000), as well as those of *Oryza sativa* (rice) in 2002, *Carica papaya* (papaya) in 2008, and *Zea mays* (maize) in 2009. Back then, it was presumed that sequencing of a single species was representative of all other

S. S. Korban (ed.), *The Apple Genome*, Compendium of Plant Genomes,
https://doi.org/10.1007/978-3-030-74682-7_8

related plant species and that information gathered on a single genotype could be useful and applied to all related organisms. However, as pointed out by Tagu et al. (2014), model organisms were often not archetypal, and they did not replicate the biology of their close relatives. Nevertheless, these initial sequencing and assembly efforts served as critical milestones in the study of many species for which no genomic resources were available until then.

In addition to costs involved in sequencing projects, there were major factors that hindered obtaining high-quality genome assembly from many non-model species, including genome size, the extent of genome duplication, heterozygosity, repetitive sequence composition, and ploidy level. Moreover, the early sequencing technology did not provide sufficient support for the assembly algorithms to overcome some of the above-mentioned challenges. Most of these factors were certainly very limiting to sequencing the apple (*Malus* × *domestica* Borkh.) genome at that time. However, major advances have been made on several fronts, including sequencing technologies, bioinformatics tools, and cost reductions, among others, that all have allowed for the pursuit of large sequencing projects for the apple genome, and these will be presented herein in this chapter.

8.2 De Novo Sequencing of the Apple Genome

The advent of NGS technologies has enabled researchers to bypass most of the biological and associated technological constraints listed above. It has allowed for the successful generation of the first de novo high-quality draft genome assembly of the apple (Velasco et al. 2010). This first Whole-Genome Sequence (WGS) for apple has resulted from the collaborative efforts of scientists from 14 institutions. Due to its popularity in commercial orchards across the world and its prominence in the recent ancestry of many modern apple cultivars, the heterozygous 'Golden Delicious' apple genotype has been chosen to represent this crop from a genomic perspective.

The assembly of this first draft apple genome sequence resulted from the use of sequence reads generated from first-generation Sanger sequencing technology, as well as from the recent 454 sequencing by synthesis technology. Anchoring of 1,629 meta-contigs was performed using a high-quality genetic map, carrying 1,643 markers, in order to reconstruct all 17 linkage groups, or chromosomes, of the apple. Altogether, a 603.9 Mb assembled sequence data covered about 81.3% of the estimated genome size, with 500.7 Mb (67%) representing repetitive elements. Altogether, genome annotation identified 63,541 genes (Table 8.1).

The first genome sequence of the apple has provided highly valuable information on gene sequence and order, thereby providing the scientific community with a valuable tool to "initiate a new era" (Peace et al. 2019). The key basic discoveries that have been made as a result include the following: (1) All 17 chromosomes of the apple are derived from a relatively recent Whole-Genome Duplication (WGD) event of a Rosaceae ancestor having nine chromosomes; (2) the gene pool of the cultivated apple, *M.* × *domestica*, is primarily derived from the wild species *M. sieversii*; (3) the large size of the pome fruit of apple is likely to be the result of an expansion of a MADS-box fruit development gene family; and (4) gene families involved in sorbitol metabolism have been expanded in the apple genome compared to those of non-Rosaceae genomes (Peace et al. 2019).

The genome sequence of 'Golden Delicious' was further improved by Li et al. (2016). In this effort, the apple genome sequence was complemented with a 76 Gb sequence data generated from an Illumina HiSeq (high-throughput sequencing) and 21.7 Gb sequence data generated from PacBio (long reads) sequencing systems. Using these collective sequence data, the new genome assembly increased the N50 of contigs (sequence length of the shortest *contig* at 50% of the total genome length) by seven-fold to 112 Kb, yielding a final genome size of

Table 8.1 Comparisons of genome sequences and annotations of five different apple genotypes

Overall genome parameters/technologies/genome sequencing	Genome parameters analyzed	'Golden Delicious'			'Hanfu'	*Malus baccata*
		Heterozygous		Doubled-haploid GDDH13	HFTH1	Y-B094
		Velasco et al. (2010)	Li et al. (2016)	Daccord et al. (2017)	Zhang et al. (2019)	Chen et al. (2019)
	Estimated genome size (Mb)	742.3	701	651	708.5	779
	Seq assembly size (Mb)	603.9	632.4	643.2	658.8	665
	N50 (Kb)	16.17	112	5,558	6,990	452
	Nb prot-coding genes	63,141	53,922	42,140	44,677	46,114
	TE (%)	42.4	NA	57.3	59.8	58.5
Technology	Sanger (Gb)	3.26				
	454 seq (Gb)	9.28				
	Illumina seq (Gb)		76	120	43.3	173.6
	Hi-C seq (Gb)				95.5	
	PacBio (Gb)		21.7	24	77	
	BioNano			600 × coverage	224 × coverage	
	Ref. genetic map	Yes	No	Yes	No	Yes
Genome completeness	CEGMA %	88.7	93.1			90.6
	BUSCO*%	86.7	51.5	94.9	95.9	93.2

BUSCO* = complete gene

632.4 Mb, and covering 90% of the estimated genome size (701 Mb). Furthermore, new annotation analyses predicted the apple genome carried 53,922 protein-coding genes and 2,765 non-coding genes (Table 8.1). At the time, further progress in the apple genome assembly was deemed difficult due to the highly heterozygous status of the 'Golden Delicious' genotype.

Often during de novo assembly, heterozygosity and repetitive regions of a genome complicate the assembly graph, resulting in fragmented assemblies that would join together contigs belonging either to different alleles or to recently duplicated genes, as it is the case for apple (Kajitani et al. 2014). This is mostly attributed to a high single nucleotide variant (SNV) density and structural variations present within a genome (Kajitani et al. 2014).

In order to overcome such constraints, it was necessary to sequence and assemble a simplified version of the apple genome. As a suitable biological material, a doubled-haploid derivative line of 'Golden Delicious' known as "GDDH13" (Golden Delicious Doubled Haploid #13), has been available since the 1980s at the Institut National de Recherche pour l'Agriculture, l'Alimentation et l'Environnement (INRAE) in Angers (France). The genotype of GDDH13 was sequenced to an unprecedented depth using Illumina short reads and PacBio long reads and then assembled de novo in combination with optical maps (BioNano) and a high-density consensus genetic map (Daccord et al. 2017). The total length of this new assembly was 643.2 Mb (genome size estimated at 651 Mb), with 42,140 annotated protein-coding genes and 1,965 non-protein-coding genes (Table 8.1). As already reported by Velasco et al. (2010), transposable elements (TEs) represented the majority of the sequence (57.3%), while protein-coding genes represented 23.3% of the assembly. Compared to previous versions of the 'Golden Delicious' genome, the N50 of scaffolds was increased to 5.5 Mb, which, in association with the integrated linkage map, yielded highly contiguous pseudomolecules. Moreover, the new gene prediction analysis reduced the estimated number of genes from 63,541 to 41,140, closer to the number of genes reported for the pear genome, as well as from recent filtering out of overlapping genes (Wu et al. 2013). Finally, the 'GDDG13' genome sequence analysis permitted a better assessment of whole-genome duplication of the apple (Fig. 8.1), and indicated that only very few regions had no synteny to other parts of the genome (i.e., mid-region of Chromosome 4).

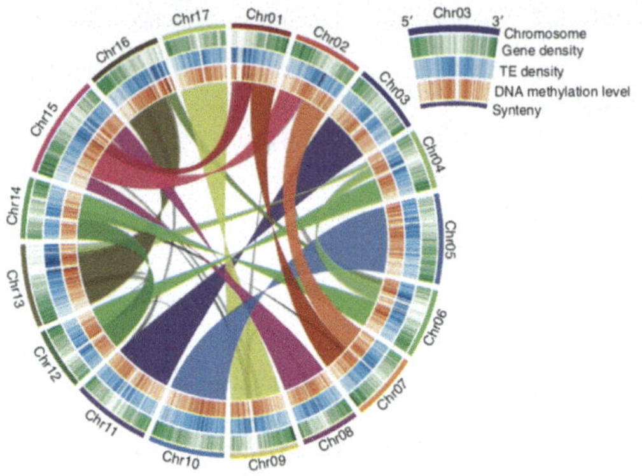

Fig. 8.1 Synteny and distribution of genomic and epigenomic features of the "GDDH13" apple genome (Daccord et al. 2017). Coloured lines link collinearity blocks that represent syntenic regions. Rings correspond to chromosomes (Chr), while heat maps represent gene density (green), transposable element (TE) density (blue), and DNA methylation levels (orange)

More recently, an apple genome sequence of an anther-derived trihaploid homozygous 'Hanfu' line, "HFTH1", has been released (Zhang et al. 2019). Similar to 'GDDH13', this "HFTH1" line has an advantage in simplifying genome assembly, and it is derived from an important Chinese cv. 'Hanfu', a hybrid of 'Dongguang' x 'Fuji'. In this effort, a combination of single-molecular real-time (SMRT) sequencing (PacBio), chromosome conformation capture (Hi-C) sequencing, and optical mapping was used to generate a 658.9 Mb assembly, with a contig N50 of 6.99 Mb. The "HFTH1" genome size assembly was close to that of "GDDH13", but represented 93% of the estimated genome size (708.54 Mb). Genome annotation yielded presence of 44,677 protein-coding genes (Table 8.1), of which 4.29% were located within gap regions of the "GDDH13" genome. Once again, TE elements accounted for the majority (59.8%) of the apple genome sequence.

Genome sequence comparisons between "GDDH13" and "HFTH1" lines yielded interesting hints about the genetic diversity in apple. Zhang and colleagues (2019) identified a significant SNP density, as well as 18,047 deletions and 12,101 insertions of lengths greater than 100 bp. Interestingly, SNP density was not evenly distributed among coding genes, with a significant enrichment identified in disease-resistance domains, indicating incidence of differential selection pressures in these two genotypes. Additional SNP analyses revealed significant average SNP density among chromosomes, as well as between homologous chromosome pairs, thus indicating potential introgression and fixation during domestication and breeding following the WGD event (Zhang et al. 2019).

Recently, the genome sequence of a wild apple species, *M. baccata*, was released (Chen et al. 2019). The total length of this assembly was 665 Mb (genome size estimated at 779 Mb), with 46,114 annotated protein-coding genes (Table 8.1). Sequencing of this species is particularly interesting. It is one of the four wild apple species that has contributed to the early hybridizations of the cultivated apple (*M.* × *domestica*), and represents a valuable breeding resource because of its cold tolerance and resistance to various diseases.

8.3 Apple Genome Assembly Quality Assessment

Comparisons of assembled sequences to Illumina data, both by alignment of reads back to the assembly and by analysis of oligonucleotides (k-mer) spectra, provide insights into the quality of the assembled sequences. Alignments of Illumina data of a genotype to existing apple assemblies reveal that all assemblies are broadly consistent with read data. With the exception of the 'Golden Delicious' assembly of Li et al. (2016), over 90% of reads align to all contig level assemblies, with the highest proportion of reads (97.9%) mapping to the "GDDH13" assembly (Daccord et al. 2017) (Table 8.2). Assemblies with the lowest Illumina alignment rates are the DBG2OLC assembly of Li et al. (2016) and the scaffold level assembly of Velasco et al. (2010) (Table 8.2).

Since the publication of the first 'Golden Delicious' assembly (Velasco et al. 2010), technological and algorithmic developments along with judicious use of homozygous genotypes have greatly decreased the collapse of repeats and separation of haplotypes present in apple assemblies. However, as demonstrated by the quality metrics (Table 8.2), improvements in contiguity have sometimes come at the cost of increased rates of misassembly and lower quality consensus sequences. When considering all of the metrics, the highest quality apple assemblies produced to date are those produced from homozygous genotypes, the "HFTH1" (Zhang et al. 2019) and "GDDH13" (Daccord et al. 2017) assemblies.

The "GDDH13" assembly of Daccord et al. (2017) has a higher proportion of Illumina reads mapped back to the assembly than any other apple assembly. It also has a lower proportion of paired Illumina reads mapping in the incorrect pairing orientation than any other apple assembly. It does, however, display a lower proportion of the assembly covered by Illumina read alignments than other assemblies assessed here. The "HFTH1" assembly of Zhang et al. (2019) has a higher proportion of the sequence covered by read alignments than any other apple assembly with a high proportion of reads mapping to the assembly and mapping in the correct pairing orientation. Kmer spectrum copy-number analysis indicates that the "GDDH13" and "HFTH1" assemblies are also the first apple assemblies, wherein neither display evidence of significant missing content or of significant erroneously duplicated content (Fig. 8.2).

8.4 Whole-Genome Duplication: An Event Shared by the Malinae Subtribe

Several studies have reported on the incidence of a common WGD event within the Malinae subtribe (Velasco et al. 2010; Chagné et al. 2014; Daccord et al. 2017; Linsmith et al. 2019; Chen et al. 2019). In fact, distribution of synonymous substitutions per synonymous site (KS) calculated for whole paranomes of *M.* × *domestica*, *Pyrus communis* (European pear), and *Pyrus* × *bretschneideri* (Asian pear or Chinese white pear) support incidence of a common WGD. Signature WGD peaks in KS plots for these three species occur at almost identical KS values of $\sim$ 0.16 (Fig. 8.3; Linsmith et al. 2019). Comparisons of these WGD KS peaks with KS peaks of ortholog distributions between pears and apple and between pears/apple and rose (*Rosa chinensis*) suggest that the WGD must have occurred quite a long time following the divergence of Amygdaloideae and Rosoideae,

Table 8.2 Comparisons of apple genomes assembly quality metrics

Cultivar/ Line/Species	Ref. assembly	Nb sequences	Min (bp)	Max (kb)	Collapsed repeats	Total collapsed sequence	% Read mapped	% Properly paired	Error rate	% Mapping insertion	% Mapping deletion	% Indel homopolymer	% > 1x
'Golden Delicious'	Velasco et al. (2010)	121,751	500	154.4	25,612	335,351,996	93.55	87.49	0.0196	4.51	4.31	64.90	97.72
'Golden Delicious'	Li et al. (2016)	9675	677	958.4	1684	82,578,891	88.75	81.08	0.0253	29.57	18.29	58.39	90.51
GDDH13	Daccord et al. (2017) (Chr + LG0)	819	1119	50,230	NA	NA	97.86	95.90	0.0027	1.59	0.67	58.81	88.25
GDDH13	Daccord et al. (2017) (Scaffolds)	1062	928	19,920	137	6,186,461	97.86	95.90	0.0027	1.59	0.67	60.18	88.25
Malus baccata	Chen et al. (2019)	15,414	500	7025.7	2563	19,260,718	96.61	91.35	0.0074	1.71	2.03	59.11	91.75
'Hanfu'	Zhang et al. (2019)	207	10,114	56,450	NA	NA	95.41	90.85	0.0074	0.74	0.48	64.35	99.51

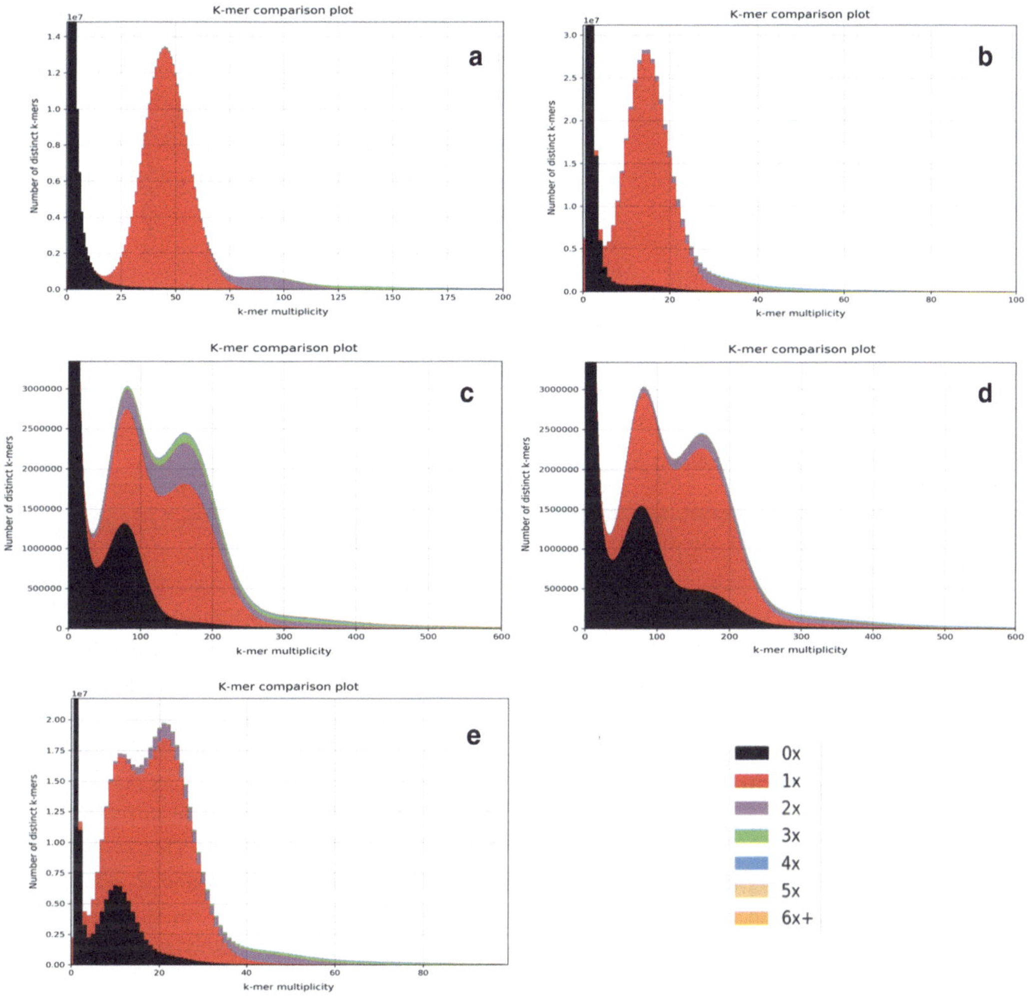

Fig. 8.2 K-mer spectrum comparison plots for the following genotypes: **a** Triple-haploid "HFTH1" (Zhang et al. 2019); **b** doubled-haploid "GDDH13" (Daccord et al. 2017); **c** heterozygous 'Golden Delicious' (Velasco et al. 2010); **d** heterozygous 'Golden Delicious' (Li et al. 2016); and **e** heterozygous *Malus baccata* (Chen et al. 2019)

but well prior to the divergence of pear and apple unless substantial substitution rate acceleration/deceleration has occurred in these lineages.

8.5 Orthology Analysis

Predicted protein sequences from apple genotype "GDDH13" (Daccord et al. 2017) were compared with those from eight other species, including *P. communis* (BartlettDHv2.0) (Linsmith et al. 2019), *P.* × *bretschneideri* (Wu et al. 2013), *Fragaria vesca* (Edger et al. 2018), *Prunus persica* (Verde et al. 2013), *Rosa chinensis* (Raymond et al. 2018), *Rubus occidentalis* (VanBuren et al. 2016), *Vitis vinifera* (The French-Italian Public Consortium for Grapevine Genome Characterization 2007), and *Arabidopsis thaliana* (The Arabidopsis Genome Initiative Proteins 2000). All sequences were clustered into 20,677 orthologous groups (≥ 2 members), 8,877 of which (43%) were common to all nine genomes (Fig. 8.4). A set of 414 gene clusters

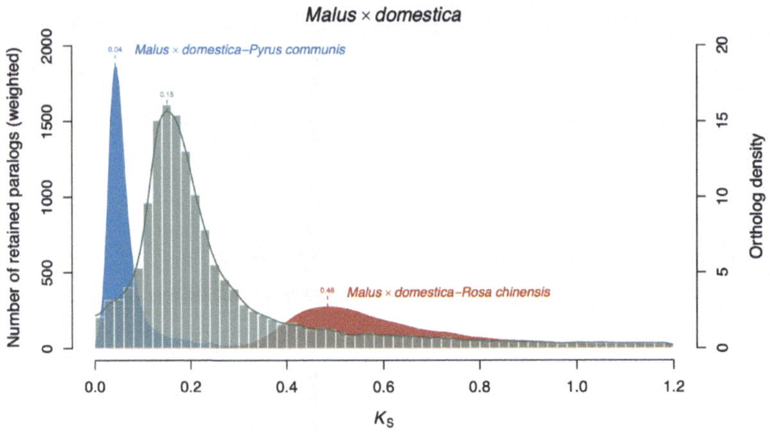

Fig. 8.3 Paralog K_s distributions of *Malus* × *domestica* genotype "GDDH13" (grey histogram and line, left-hand y-axis; a peak represents a WGD event) and one-to-one ortholog K_s distributions between sets of two species, *M.* × *domestica–Pyrus communis* (blue curve) and *M.* × *domestica—Rosa chinensis* (red curve) (kernel-density estimates, right-hand y-axis; a peak represents a species divergence event)

were identified as specific to the three pome fruits analyzed (i.e., to apple and to the two pear species). Among these, five gene clusters were found to be specific to apple, while 30 gene clusters were specific to the Asian (22) and the European (8) pears. All remaining gene clusters could be identified in one of the other Malinae genomes. Interestingly, 85 gene clusters were identified as being shared by the two pear species but not by apple (Linsmith et al. 2019 supplementary data).

8.6 Whole-Genome Resequencing Projects for Apple

Following publication of the first apple genome sequence, resequencing of various apple genotypes and accessions were launched in order to identify genetic variants for molecular marker development. As opposed to de novo sequencing, resequencing studies rely on the use of a reference genome for both alignment and analysis of newly generated sequences. Using NGS technologies, Chagné and colleagues (2012) resequenced a set of 27 cultivars, ancestors, and founders, selected to represent the pedigrees of worldwide apple breeding programmes. The sequenced genome coverage was low, varying from 1.3× to 12.8×. Alignment of sequences onto the then reference apple genome led to the identification of single nucleotide polymorphisms (SNPs) that were then used, in turn, to develop an 8K SNP array for high-throughput genotyping. Altogether, over 2,000,000 SNPs were identified, with an average of one SNP and every 288 bp of the genome. This original SNP array was subsequently expanded wherein 14 additional genotypes were resequenced to an average depth of 38.7×, allowing for identification of high-quality SNPs and developing a 20K SNP array (Bianco et al. 2014). This set of SNPs was further complemented in a follow-up study by Bianco et al. (2016). Using sequencing data generated from 52 additional cultivars covering the majority of the genetic diversity of the cultivated apple, over 12,700,000 high-confidence SNP variants were identified, and following further filtering, and these were used to generate a robust 487K SNP array (Bianco et al. 2016).

Boocock et al. (2015) extended resequencing of additional apple genotypes to identify copy-number variations (CNVRs) in the apple genome. To achieve this, they used part of the data generated by Chagné et al. (2012) and complemented it with data generated from resequencing of 11 cultivars and hybrids. In this study, a total of 876 CNVRs were identified, indicating the

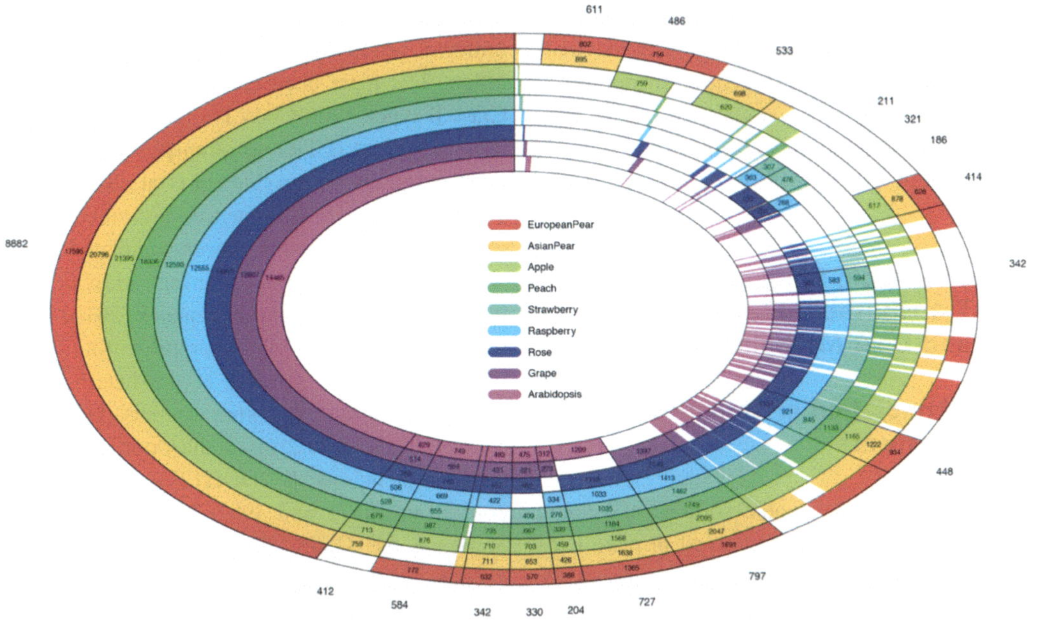

Fig. 8.4 Plot of protein clusters shared by nine species, including *Malus* × *domestica* (GDDH13), *Pyrus communis* (BartlettDHv2.0), *P.* × *bretschneideri*, *Fragaria vesca*, *Prunus persica*, *Rosa chinensis*, *Rubus occidentalis*, *Vitis vinifera*, and *Arabidopsis thaliana* (Linsmith et al. 2019)

potential significance of such variations in apple diversity. Furthermore, this study confirmed the highly heterozygous nature of apple, with the identification of SNP variants every 208 bp of the genome.

Later, two additional apple genotypes, 'Nagafu No. 2' and 'Qinguan', were resequenced at a genome coverage of 31× and 28×, respectively (Xing et al. 2016). In addition to the identification of over 2,700,000 SNPs, 82,600 structural variants and 1,572,000 INDELs, locations of 190 genes likely involved in flowering were determined. In another study, Lee et al. (2016) resequenced four somatic variants of 'Fuji', including 'Danhong', 'Hangawi', 'Benishogun' and 'Yataka Fuji', along with the original 'Fuji' (at an average coverage of 16.8×), and proposed that two INDEL markers would allow for the differentiation of 'Fuji' from its somatic variant 'Benishogun'. Simultaneously, the 'Fuji' genome was also sequenced in another study to develop SNP markers for each allele at 1 cM intervals and identify haplotypes of 'Fuji' and determine their inheritance using 115 related apple accessions (Kunihisa et al. 2016). Over 2,800,000 variants, SNPs, and INDELS were identified between 'Fuji' and the heterozygous 'Golden Delicious' genome; moreover, QTLs were identified by ANOVA based on detected 'Fuji' haplotypes (Kunihisa et al. 2016).

More recently, Duan et al. (2017) reported on genome variations following resequencing, at an average depth of 12.2× genome coverage, of 117 diverse accessions from 24 *Malus* species, and identified over 7,200,000 SNPs and 431,000 INDELs. As a result, they proposed a comprehensive model for apple domestication along the Silk Road, and presented new evidence supporting a model for fruit size evolution comprised of two major events (Duan et al. 2017). Overall, genome resequencing allowed for investigations of the history of apple speciation and domestication, and could be used in the future as a tool for assisted breeding.

As demonstrated in the above studies, genome resequencing has yielded massive amounts of data and information on genome diversity and structure. This has allowed researchers to develop powerful SNP arrays that are being used today in many studies, including pedigree analysis (Muranty et al. 2020), genetic mapping (Di Pierro et al. 2016), and genome-wide association studies (GWAS) (Urrestarazu et al. 2017; McClure et al. 2018, 2019; Larsen et al. 2019).

8.7 The Apple Epigenome: Transposable Elements and DNA Methylation

8.7.1 Annotation of Transposable Elements

Mutations serve as valuable sources of genetic diversity, and are heavily involved in plant evolution. Transposable elements (TEs) account for significant portions of all prokaryote and eukaryote genomes, representing most of the interspersed repeats (Piégu et al. 2015). As such, it is proposed that TEs could contribute significantly to genetic variations among *Malus* species and within genotypes. In recent years, evidence supporting the importance of TEs in genome evolution, genome structure, regulation of gene expression, and epigenetics has been mounting (Encode Project Consortium 2012; Fedoroff 2012; Chénais et al. 2012). In plants, transcription and mobility of TEs are limited by DNA methylation (Mirouze et al. 2009). Interestingly, TE-related DNA methylation can also influence the expression of nearby genes resulting in phenotypic variability even in the absence of genetic changes (Slotkin and Martienssen 2007). Therefore, given the potential importance of TEs in regulating gene expression along with their prevalence in the apple genome, unravelling their sequence and distribution is necessary.

Piégu et al. (2015) have proposed that TEs are defined as discrete segments of DNA capable of moving within a host genome from one chromosome to another. Furthermore, the status of TE copies within a genome varies depending on the age and activity of the element, leading to the identification of many fossil TE sequences alongside a few active copies. TEs can be classified in multiple ways in order to generate uniform groups based on particular features. Often, TEs have been classified as Class I and Class II types, with Class I containing both long terminal repeat (LTR)-retrotransposons and non-LTR retrotransposons (Sahebi et al. 2018).

In apple genome sequences presented in the above section, TEs have been annotated using various pipeline and differing classification parameters. Velasco and colleagues (2010) identified a total of 314.5 Mb TEs, representing 42.4% of the sequenced genome in the heterozygous 'Golden Delicious' genome; whereas, TEs represent 365.7 Mb (57.2%) of the doubled-haploid 'Golden Delicious' ("GDDH13") genome (Daccord et al. 2017), and 399.9 Mb (59.8%) of the homozygous 'Hanfu' line ("HFTH1") genome sequence (Zhang et al. 2019). Therefore, this observed global increase in TEs could be mostly attributed to the improved quality of apple genome sequence assemblies due to advances in read sequence length and optical mapping technologies. However, it is rather challenging to conduct comparisons of TE annotations, as different TE annotation pipelines have been used. While Velasco et al. (2010) have identified repetitive elements by conducting searches (BLASTN and BLASTX) against RepBase14.01, NCBI databases, and the Uniprot database, Daccord et al. (2017) have annotated these elements using consensus sequences provided by the TE*denovo* detection pipeline (Flutre et al. 2011). On the other hand, Zhang et al. (2019) have used a combination of both de novo and homology-based approaches to annotate repetitive sequences. In order to compare TE composition in "GDDH13" and "HFTH1", Zhang et al. (2019) have re-annotated the "GDDH13" TEs. Nevertheless, as a result of these different annotation strategies, comparisons among various apple genomes remain challenging, and this is analyzed in Tables 8.3, 8.4, 8.5, and 8.6.

Using currently available data, it can be assumed that TE contents of the total nucleotide space in the apple genome accounts for ~ 60% of the assembly.

Features of the genome and its TEs	'Golden Delicious' (Velasco et al. 2010)	
	Mb	%
Genome size	742.3	100.0
LTR/Copia	40.6	5.5
LTR/Gypsy	187.1	25.2
LTR/TRIM/Cassandra	3.2	0.4
LINE/RTE	48.1	6.5
TIR/hAT	2.1	0.3
TIR/CACTA	0.1	0.0
TIR/MITE/Spring	4.4	0.6
Unclassified TEs	28.9	3.9
Total TEs	314.5	42.4

Table 8.3 TE annotation of the 'Golden Delicious' genome (Velasco et al. 2010). Class I TEs are highlighted in light grey, while Class II TEs are highlighted in dark grey

It is noteworthy to point out that during TE annotation of "GDDH13", a particular TE sequence unique to the apple genome, accounting for 3.6% of the whole-genome assembly, has been identified. This TE, termed *HODOR* (high-copy Golden Delicious repeat) is classified as a 9,716 bp LTR-RT. Interestingly, high densities of *HODOR* coincide with regions of chromosomes with reduced levels of recombination, thus suggesting that centromeric regions in the apple genome may be located within regions characterized by an over-representation of this

Features of the genome and its TEs	"GDDH13" (Daccord et al. 2017)	
	Mb	%
Genome size	651.0	100.0
LTR Retrotransposon (RLX)	235.4	36.2
LINE (RIX)	28.0	4.3
DIRS (RYX)	10.4	1.6
SINE (RSX)	1.7	0.3
Unclassified or non-autonomous retrotransposons (RXX)	3.1	0.5
TIR Transposon (DTX)	65.5	10.1
MITE (DXX)	16.7	2.6
Helitron (DHX)	4.8	0.7
Maverick (DMX)	0.2	0.0
non-annotated	116.0	17.9
Potential Host Gene	6.3	1.0
No repeat category	6.1	0.9
Non protein coding genes	4.0	0.6
Total TEs	365.8	56.2

Table 8.4 TE annotation of the "GDDH13" genome (Daccord et al. 2017). Class I TE are in light grey, while Class II TE are in dark grey

Features of the genome and its TEs	'"HFTH1" Zhang et al. (2019)		"GDDH13" Daccord et al. (2017)	
	Mb	%	Mb	%
Genome size	658.9	100.0	651.0	100.0
LTR Copia	104.5	15.9	90.1	13.8
LTR Gypsy	167.4	25.4	153.5	23.6
LTR Caulimovirus	3.6	0.5	3.5	0.5
LTR Cassandra	2.5	0.4	2.4	0.4
LTR Bel-Pao	0.6	0.1	0.5	0.1
LTR Others	25.6	3.9	24.8	3.8
SINE	0.8	0.1	0.8	0.1
LINE	30.3	4.6	27.7	4.3
DNA	64.8	9.8	65.5	10.1
Simple repeat	12.4	1.9	11.1	1.7
RC	3.5	0.5	3.3	0.5
Others	13.4	2.0	13.8	2.1
Total TEs	400.0	60.7	368.7	56.6

Table 8.5 TE annotations of both "HFTH1" and "GDDH13" genomes performed by Zhang et al. (2019). Class I TEs are in light grey, while Class II TEs are in dark grey

Features of the genome and its TEs	***Malus baccata*** (Chen et al. 2019)	
	Mb	%
Genome size	665.0	100
LTR/Gypsy	194.2	29.2
LTR/Copia	106.4	16
LTR/other	39.9	6
LINE/RTE	53.2	8
LINE/L1	8.6	1.3
LINE/other	1.3	0.2
SINEs	10.0	1.5
CMC	6.0	0.9
DNA	21.9	3.3
hAT	25.9	3.9
PIF	14.0	2.1
Other	10.0	1.5
Total TEs	491.4	73.9

Table 8.6 TE annotation of the *Malus baccata* genome (Chen et al. 2019). Class I TEs are in light grey, while Class II TEs are in dark grey

particular TE. Furthermore, this LTR element has also been identified in the BartlettDHv2.0 genome (Linsmith et al. 2019). A BLAST analysis has revealed the presence of 232 full-length HODOR copies in the pear genome, accounting for approximately one-third of the number of full-length copies identified in the apple genome. To date, the HODOR element has only been identified in apple and pear genomes. However, it is important to point out that these two genomes have been reassembled using the latest long-read technology to produce chromosome scale assemblies. As HODOR is a transposable element, it may not have been completely assembled in previous Rosaceae genomes that have relied on short-read data. However, sequenced genomes of *F. vesca* (Edger et al. 2018), *R. chinensis* (Raymond et al. 2018), and *R. occidentalis* (Van Buren et al. 2016) have been assembled using long-read data, but a BLAST analysis has not detected any trace of the HODOR element in these assemblies. Therefore, future in-depth studies of repeat element contents of Rosaceae genomes should be conducted to identify the time point in Rosaceae evolution at which this element has first emerged, and to assess whether or not this element is involved in or relates to observed phenotypic differences within the Rosaceae family.

8.7.2 Role(s) of TEs in Apple

It is known that TEs play important roles in a genome structure, and they can be used as markers for a genome's evolution. As demonstrated by Daccord et al. (2017), TE annotation analysis has indicated that a major TE burst is estimated to have occurred around 21 million years ago (Mya). This burst must have affected all TE types. It is suggested that the ancestor of modern apples must have undergone environmental changes, presumably an uplifting event that has likely occurred at the Tian Shan mountains, a form of stress leading to the activation of TEs. It is hypothesized that this TE burst may have contributed to the diversification and possibly to the speciation of the predecessor of both apple and pear.

In a recent study by Zhang et al. (2019), the presence of a TE has been associated with a major agronomical trait in apple. In fact, they have identified an LTR retrotransposon insertion upstream of the *MdMYB1* gene (Fig. 8.5), a core transcriptional activator of anthocyanin biosynthesis. This TE insertion is considered as an enhancer, and it controls red colour development by lowering the threshold value of the light response.

As mentioned earlier, TEs serve as major sources of genetic diversity in apple, and can influence genome structure and expression of nearby genes. Therefore, particular attention should be paid to undertaking high-quality annotation of TE sequences in all apple genome sequences, possibly using similar TE annotation pipelines in order to allow comparisons among genomes. Future studies using apple as a model plant, having undergone a rather recent WGD event, would aid in a better

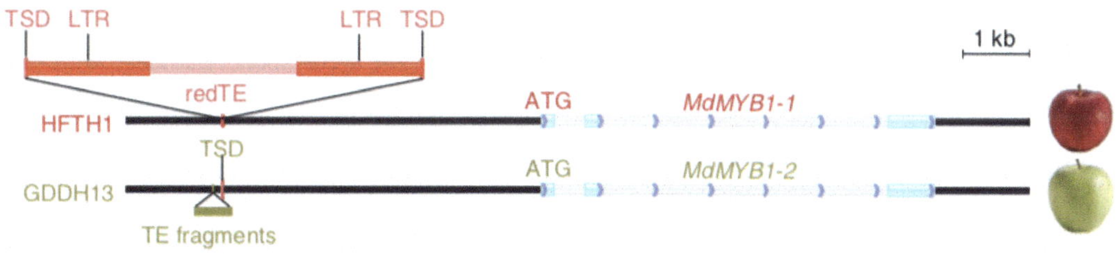

Fig. 8.5 The molecular structure of *MdMYB1-1* and *MdMYB1-2* alleles along with flanking sequences. The insertion sites upstream of *MdMYB1-1* and *MdMYB1-2* are indicated by a red line (for "HFTH1") and golden-yellow line (for "GDDH13"), respectively (from Zhang et al. 2019)

understanding of the role(s) of TEs by comparing the evolution of homologous gene sequences within syntenic blocks, and confirm the importance of TEs in the evolutionary pathway of a plant genome.

8.8 The Apple Methylome

The advent of whole-genome sequencing (WGS) technologies has also provided opportunities for investigating an organism's methylome. It is known that cytosine methylation in plants exists in all sequence contexts, but primarily occurs at CG or CHG sequences (where H is any base, except for a G), and less often at CHH sites. Changes in methylation patterns of genomic DNA have been reported in various developmental processes, and have been shown to be influenced by environmental stresses in plants (Seymour et al. 2014). Various studies have recently demonstrated that local DNA methylation variants, represented by differential cytosine methylation at particular loci, can have major effects on transcription of nearby genes, and can be inherited over generations (Becker et al. 2011; Zhang et al. 2006). Thus, access to near-complete WGS provides the necessary foundation for pursuing epigenetic studies.

Epigenetic studies in apple were initiated following sequencing of the "GDDH13" genome. In fact, the main reason for generating this high-quality genome was to enable an understanding of the epigenetic mechanisms involved in fruit size regulation. This analysis was based on the methylome sequencing of two "GDDH" clones derived from independent chromosomal duplications originating from the same haploid line. The first clone ("GDDH13") was found to produce large-sized fruits, while the second clone ("GDDH18") produced small-sized fruits. Globally, DNA methylation levels in "GDDH" leaves were 49%, 39%, and 12% in CG, CHG, and CHH contexts, respectively. As presented in Fig. 8.1, DNA methylation was not evenly spread across all apple chromosomes, with peaks of methylation coinciding with recombination cold spots and the *HODOR* sequence. As expected, from previous studies on model plants (Law and Jacobsen 2010), reduced DNA methylation levels were identified in gene sequences; whereas, TEs were hypermethylated. Further analysis using previously generated transcriptome datasets indicated that highly-methylated genes showed the lowest levels of expression, while genes with either low to very low methylation levels had higher transcription levels (Daccord et al. 2019). Therefore, methylome comparisons of the two genetically identical clones, "GDDH13" and "GDDH18", allowed for identification of differentially methylated regions (DMRs) located nearby genes likely involved in fruit size regulation (Daccord et al. 2017).

The study of epigenetic mechanisms influencing phenotypic changes in apple has been the subject of various studies. Several research groups have investigated the effects of the methylome on expression of genes involved in fruit skin colouration. Ma et al. (2018) took advantage of the "GDDH13" apple genome to investigate the effects of promoter methylation on *MdMYB1* expression, and found that differential methylation profiles influenced gene transcription levels, and in turn, fruit skin colouration. Similarly, several research groups investigated the effects of DNA methylation in various apple sports, including 'Red Delicious', 'Yanfu 3', 'Yanfu 8', 'Nagafu 2', 'Ralls', 'Shannonghong', and 'Fuji' on fruit colour development (Li et al. 2019; Jiang et al. 2019; Cho et al. 2020). They identified linkages between promoter methylation of several genes involved in the anthocyanin biosynthesis pathway and fruit skin colouration. Moreover, El-Sharkawy et al. (2015) compared methylomes of the yellow-skin apple sport 'Blondee' with that of the original 'Gala'. It was reported that there was a differential methylation level in the promotor of *MdMYB10*, involved in the anthocyanin pigment synthesis pathway, and that identified DMRs influenced transcript expression of MdMYB10, thereby leading to differential phenotypes.

Thus, knowledge of the epigenetic landscape of apple cultivars could provide new tools to study and differentiate among somatic variants. Furthermore, this could lead to the future development of epigenetic markers for marker-assisted selection useful in apple breeding programmes.

In addition to fruit skin colour, other characters of agronomic importance have also been investigated. Xu et al. (2018) investigated the effects of water deficit on the methylome of the drought-tolerant apple cv. 'Qinguan' and the drought-sensitive cv. 'Honeycrisp'. They identified significant DMRs following stress treatment, and these were associated with multiple genes, including transcription factors, and TEs. This indicated that the apple methylome was dynamic, and capable of responding to abiotic stress in order to adapt the transcriptional response of the tree. Furthermore, gene priming following biotic stress was also recently investigated using the "GDDH13" genotype (Gully et al. 2019). Based on their findings, it was hypothesized that stress could lead to long-lasting transcriptional memory (via gene priming), and that DNA methylation was required for appropriate biotic stress response (Gully et al. 2019).

Methylome changes during flower bud formation and winter dormancy release were also investigated by Xing et al. (2019) (using cv. 'Nagafu No.2') and Kumar et al. (2016) (using cv. 'Royal Delicious'), respectively. In both studies, it was suggested that epigenetic modifications played critical roles on these major agronomical traits in apple. Furthermore, DNA methylation levels of particular genes in apple rootstocks were also investigated (Feng et al. 2017). It was reported that differential methylation levels in the *IPT5b* promoter region were detected between the dwarfing rootstock 'M.9' and the vigorous rootstock '*M.* × *robusta*'. It was proposed that this differential methylation level was responsible for lower levels of expression of the *IPT5b* gene (involved in the cytokinin biosynthetic pathway) in 'M.9', thus leading to the dwarfing effect of this rootstock (Feng et al. 2017). Finally, Perrin et al. (2020) investigated the transmission of DNA methylation patterns using apple as a model perennial plant. This study demonstrated that significant methylation changes resulted from both sexual (via seeds) and asexual (via grafting) reproduction and indicated that newly grafted plants might inherit part of the epigenomic history of their donor tree.

As mentioned above, epigenetic studies will continue to benefit from the availability of high-quality apple WGS in the future. Such studies will likely lead to the identification of genes or epimutations involved in important agronomic traits that have not yet been investigated, such as fruit shape, disease resistance, and tree architecture. As mentioned by Peace et al. (2019), controlled epigenetic changes could also be used to induce targeted DNA methylation changes or to induce the mobilization of TEs, thus leading to the development of new phenotypes, and serving as a powerful set of tools for gene discovery.

8.9 Data and Tools

The reference portal for genomic data for apple is the Genome Database for Rosaceae (GDR), as described by Jung et al. (2019). This resource currently (September 2020) hosts eight different *Malus* reference genomes, each of which can be browsed through with the JBrowse tool (Buels et al. 2016). Such a search reports not only the sequence of a genome, but it also allows for overlaying of other available information, such as genes, mRNAs, TEs, and markers, among others, to a genome sequence. From the GDR, it is also possible to BLAST sequences against each of the sequenced genomes. Furthermore, comprehensive search/browse features are also provided for all information available for each genome (including markers, genetic maps, quantitative trait loci, publicly available breeding data, and publications, among others). Other available tools include a Synteny Viewer based on MCScanX (Wang et al. 2012) to conduct comparative analyses among different genomes, as well as a Breeder Toolbox, labelled as a

Breeding Information Management System (BIMS).

In addition, GDR provides the GDRCyc, a database of metabolic and signalling pathways for several Rosaceae species, including apple, built with PathwayTools (Caspi et al. 2013), beginning with gene models for each of the hosted genomes. Such pathways are used by downstream analysis tools, also available through GDR, such as the Pathway Inspector (Bianco et al. 2017) that performs pathway enrichment analyses, beginning with RNA-Seq data.

8.9.1 Conclusions

Advances in the development and release of full genome sequences for the apple have been made, and these have contributed to the availability of highly reliable references genomes. These genome sequences have allowed for the pursuit of resequencing efforts as well as the development of a new field of investigation, including epigenetics. Currently, research efforts have been undertaken to investigate epigenetic mechanisms influencing various phenotypic traits and changes in apple that will further expand the scope of our fundamental knowledge of the apple and aid in pursuing effective strategies for genetic enhancement efforts of the apple.

References

Becker C, Hagmann J, Müller J, Koenig D, Stegle O, Borgwardt K, Weigel D (2011) Spontaneous epigenetic variation in the *Arabidopsis thaliana* methylome. Nature 480:245–251

Bianco L, Cestaro A, Linsmith G, Muranty H, Denance C, Théron A, Poncet C, Micheleti D, Kerschbamer E, Di Piero EA, Larger S, Pindo M, Van de Weg E, Davassi A, Laurens F, Velasco R, Durel CE, Troggio M (2016) Development and validation of the Axiom®Apple480K SNP genotyping array. Plant J 86:62–74

Bianco L, Cestaro A, Sargent DJ, Banchi E, Derdak S, Di Guardo M, Salvi S, Jansen J, Viola R, Gut I, Laurens F, Chagné D, Velasco R, Van de Weg E, Troggio M (2014) Development and validation of a 20K single nucleotide polymorphism (SNP) whole genome genotyping array for apple (*Malus* x *domestica* Borkh). PloS ONE 9:e110377

Bianco L, Riccadonna S, Lavezzo E, Falda M, Formentin E, Cavalieri D, Toppo S, Fontana P (2017) Pathway Inspector: a pathway based web application for RNAseq analysis of model and non-model organisms. Bioinformatics 33:453–455

Boocock J, Chagné D, Merriman TR, Black MA (2015) The distribution and impact of common copy-number variation in the genome of the domesticated apple *Malus* x *domestica* Borkh. BMC Genomics 16:848

Buels R, Yao E, Diesh CM, Hayes RD, Munoz-Torrez M, Helt G, Goodstein DM, Elsik CG, Lewis SE, Stein L, Holmes IH (2016) JBrowse: a dynamic web platform for genome visualization and analysis. Genome Biol 17:66. https://doi.org/10.1186/s13059-016-0924-1

Caspi R, Dreher K, Karp PD (2013) The challenge of constructing classifying and representing metabolic pathways. FEMS Microbiol Lett 345:85–93

Chagné D, Crowhurst RN, Troggio M, Davey MW, Gilmore B, Lawley C, Vanderzande S, Hellens RP, Kumar S, Cestaro A, Velasco R, Main D, Rees JD, Iezzoni A, Mockler T, Wilhelm L, Van de Wew E, Gardiner SE, Bassil N, Peace C (2012) Genome-wide SNP detection validation and development of an 8K SNP array for apple. PloS ONE 7:e31745

Chen X, Li S, Zhang D, Han M, Jin X, Zhao C, Wang S, Xing L, Ma J, Ji J, An N (2019) Sequencing of a wild apple (*Malus baccata*) genome unravels the differences between cultivated and wild apple species regarding disease resistance and cold tolerance G3: GENES GENOMES GENET 9:2051–2060

Chénais B, Caruso A, Hiard S, Casse N (2012) The impact of transposable elements on eukaryotic genomes: from genome size increase to genetic adaptation to stressful environments. Gene 509:7–15

Cho HJ, Kim GH, Choi C (2020) Differential gene expression and epigenetic analyses between striped and T blushed skinned sports of 'Fuji' apple. Scient Hortic 261. https://doiorg/doiorg/101016/jscienta2019108944

Daccord N, Celton J-M, Linsmith G, Becker C, Choisne N, Schijlen E, van de Geest H, Bianco L, Micheletti D, Velasco R, Di Pierro EA, Gouzy J, Rees DJG, Guérif P, Muranty H, Durel CE, Laurens F, Lespinasse Y, Gaillard S, Aubourg S, Quesneville H, Weigel D, Van de Weg E, Troggio M, Bucher E (2017) High-quality de novo assembly of the apple genome and methylome dynamics of early fruit development. Nat Genet 49:1099. https://doiorg/doi:101038/ng3886

Di Pierro EA, Gianfranceschi L, Di Guardo M, Koehorst-van Putten HJJ, Kruisselbrink JW, Longhi S, Troggio M, Bianco L, Muranty H, Pagliarani G, Tartarini S,

Letschka T, Luis LL, Garkava-Gustavsson L, Micheletti D, Bink MCAM, Voorrips R, Aziz E, Velasco R, Laurens F, Van de Weg WE (2016) A high-density, multi-parental SNP genetic map on apple validates a new mapping approach for outcrossing species. Hort Res 3:16057. https://doi.org/10.1038/hortres.2016.57

Duan N, Bai Y, Sun H, Wang N, Ma Y, Li M, Wang X, Jiao C, Legall N, Mao L, Wan S, Wang K, He T, Feng S, Zhang Z, Mao Z, Shen X, Chen X, Jiang Y, Wu S, Yin C, Ge S, Yang L, Jiang S, Xu H, Liu J, Wang D, Qu C, Wang Y, Zuo W, Xiang L, Liu C, Zhang D, Gao Y, Xu Y, Xu K, Chao T, Fazio G, Shu H, Zhong G-Y, Cheng L, Fei Z, Chen X (2017) Genome re-sequencing reveals the history of apple and supports a two-stage model for fruit enlargement. Nat Commun 8:249. https://doi.org/10.1038/s41467-017-00336-7

Edger PP, VanBuren R, Colle M, Poorten TJ, Wai CM, Niederhuth CE, Alger EI, Ou S, Acharya CB, Wang J, Callow P, McKain MR, Shi J, Collier C, Xiong Z, Mower JP, Slovin JP, Hytonen T, Jiang N, Childs KL, Knapp SJ (2018) Single-molecule sequencing and optical mapping yields an improved genome of woodland strawberry (*Fragaria vesca*) with chromosome-scale contiguity. GigaSci 7(2):1–7. https://doi.org/10.1093/gigascience/gix124

El-Sharkawy I, Liang D, Xu K (2015) Transcriptome analysis of an apple (*Malus* × *domestica*) yellow fruit somatic mutation identifies a gene network module highly associated with anthocyanin and epigenetic regulation. J Exp Bot 66:7359–7376. https://doi.org/10.1093/jxb/erv433

ENCODE Project Consortium (2012) An integrated encyclopedia of DNA elements in the human genome. Nature 489:57–74

Fedoroff NV (2012) Transposable elements epigenetics and genome evolution. Science 338:758–767

Feng Y, Zhang X, Wu T, Xu X, Han Z, Wang Y (2017) Methylation effect on IPT5b gene expression determines cytokinin biosynthesis in apple rootstock Biochem Biophys Res Commun 482:604–609

Flutre T, Duprat E, Feuillet C, Quesneville H (2011) Considering transposable element diversification in de novo annotation approaches. PloS ONE 6:e16526

Gully K, Celton J-M, Degrave A, Pelletier S, Brisset M-N, Bucher E (2019) Biotic stress-induced priming and de-priming of transcriptional memory in arabidopsis and apple. Epigenomes 3(1):3. https://doi.org/10.3390/epigenomes3010003

Jiang S-H, Sun Q-G, Chen M, Wang N, Xu H-F, Fang H-C, Wang Y-C, Zhang Z-Y, Chen X-S (2019) Methylome and transcriptome analyses of apple fruit somatic mutations reveal the difference of red phenotype. BMC Genomics 20:117. https://doi.org/10.1186/s12864-019-5499-2

Jung S, Lee T, Cheng C-H, Buble K, Zheng P, Yu J, Humann J, Ficklin SP, Gasic K, Scott K, Frank M, Ru S, Hough H, Evans K, Peace C, Olmstead M, DeVetter LW, McFerson J, Coe M, Wegrzyn JL, Staton ME, Abbott AG, Main D (2019) 15 years of GDR: new data and functionality in the Genome Database for Rosaceae. Nucleic Acids Res 47:D1137–D1145

Kajitani R, Toshimoto K, Noguchi H, Toyoda A, Ogura Y, Okuno M, Yabana M, Harada M, Nagayasu E, Maruyama H, Kohara Y, Fujiyama A, Hayashi T, Itoh T (2014) Efficient *de novo* assembly of highly heterozygous genomes from whole-genome shotgun short reads. Genome Res 24:1384–1395

Kumar G, Rattan UK, Singh AK (2016) Chilling-mediated DNA methylation changes during dormancy and its release reveal the importance of epigenetic regulation during winter dormancy in apple (*Malus* × *domestica* Borkh). PloS ONE 11:e0149934

Kunihisa M, Moriya S, Abe K, Okada K, Haji T, Hayashi T, Kawahara Y, Itoh R, Itoh T, Katayose Y, Kanamori H, Matsumoto T, Mori S, Sasaki H, Matsumoto T, Nishitani C, Terakami S, Yamamoto T (2016) Genomic dissection of a 'Fuji' apple cultivar: re-sequencing SNP marker development definition of haplotypes and QTL detection. Breed Sci 66(4):499–515. https://doiorg/doi:101270/jsbbs16018

Larsen B, Migicovsky Z, Jeppesen AA, Gardner KM, Toldam-Andersen TB, Myles S, Orgaard M, Agerlin Petersen M, Pedersen C (2019) Genome-wide association studies in apple reveal loci for aroma volatiles sugar composition and harvest date. Plant Genome 12 (2). https://doi.org/10.3835/plantgenome2018.12.0104

Law JA, Jacobsen SE (2010) Establishing maintaining and modifying DNA methylation patterns in plants and animals. Nat Rev Genet 11:204–220

Lee HS, Kim GH, Kwon SI, Kim JH, Kwon YS, Choi C (2016) Analysis of 'Fuji' apple somatic variants from next-generation sequencing. Genet Mol Res 15(3). https://doi.org/10.4238/gmr.15038185

Li W-F, Ning G-X, Mao J, Guo Z-H, Zhou Q, Chen B-H (2019) Whole-genome DNA methylation patterns and complex associations with gene expression associated with anthocyanin biosynthesis in apple fruit skin. Planta 250:1833–1847

Li X, Kui L, Jing Zhang J, Xie Y, Wang L, Yan Y, Wang N, Xu J, Li C, Wang W, van Nocker S, Dong Y, Ma F, Guan Q (2016) Improved hybrid de novo genome assembly of domesticated apple (*Malus* × *domestica*). GigaSci 5:35. https://doi.org/10.1186/s13742-016-0139-0

Linsmith G, Rombauts S, Montanari S, Deng CH, Celton JM, Guérif P, Liu C, Rolf L, Zurn JD, Cestaro A, Bassil NV, Bakker LV, Schijlen E, Gardiner SE, Lespinasse Y, Durel CE, Velasco R, Neale DB, Chagné D, Van de Peer Y, Troggio M, Bianco L (2019) Pseudo-chromosome length genome assembly of a double haploid 'Bartlett' pear (*Pyrus communis* L.). BioRxiv 8 (12):giz138. https://doi.org/10.1093/gigascience/giz138

Ma C, Jing C, Chang B, Yan J, Liang B, Liu L, Yang Y, Zhao Z (2018) The effect of promoter methylation on MdMYB1 expression determines the level of anthocyanin accumulation in skins of two non-red apple cultivars. BMC Plant Biol 18:108. https://doi.org/10.1186/s12870-018-1320-7

McClure KA, Gardner KM, Douglas GM, Song J, Forney CF, DeLong J, Fan L, Du L, Toivonen PMA, Somers DJ, Rajcan I, Myles S (2018) A genome-wide association study of apple quality and scab resistance. Plant Genome 11(1). https://doi.org/10.3835/plantgenome2017.08.0075

McClure KA, Gong Y, Song J, Vinqvist-Tymchuk M, Campbell Palmer L, Fan L, Burgher-MacLellan K, Zhang Z, Celton J-M, Forney CF, Migicovsky Z, Myles S (2019) Genome-wide association studies in apple reveal loci of large effect controlling apple polyphenols. Hortic Res 6:107. https://doi.org/10.1038/s41438-019-0190-y

Mirouze M, Reinders J, Bucher E, Nishimura T, Schneeberger K, Ossowski S, Cao J, Weigel D, Paszkowski J, Mathieu O (2009) Selective epigenetic control of retrotransposition in *Arabidopsis*. Nature 461:427–430

Morozova O, Marra MA (2008) Applications of next-generation sequencing technologies in functional genomics. Genomics 92:255–264

Muranty H, Denancé C, Feugey L, Crépin JL, Barbier Y, Tartarini S, Ordidge M, Troggio M, Lateur M, Nybom H, Paprstein F, Laurens F, Durel CE (2020) Using whole-genome SNP data to reconstruct a large multi-generation pedigree in apple germplasm. BMC Plant Biol 20:2. https://doi.org/10.1186/s12870-019-2171-6

Peace CP, Bianco L, Troggio M, Van de Weg E, Howard NP, Cornille A, Durel CE, Myles S, Migicovsky Z, Schaffer R, Costes E, Fazio G, Yamane H, van Nocker S, Gottschalk C, Costa F, Chagné D, Zhang X, Patocchi A, Gardiner S, Hardner C, Kumar S, Laurens F, Bucher E, Main D, Jung S, Vanderzande S (2019) Apple whole genome sequences: recent advances and new prospects. Hortic Res 6:59. https://doi.org/10.1038/s41438-019-0141-7

Perrin A, Daccord N, Roquis D, Celton JM, Vergne E, Bucher E (2020) Divergent DNA methylation signatures of juvenile seedlings, grafts and adult apple trees. Epigenomes 4(1):4. https://doi.org/10.3390/epigenomes4010004

Piégu B, Bire S, Arensburger P, Bigot Y (2015) A survey of transposable element classification systems – A call for a fundamental update to meet the challenge of their diversity and complexity. Mol Phylogenet Evol 86:90–109

Raymond O, Gouzy J, Just J, Badouin H, Verdenaud M, Lemainque A, Vergne P, Moja S, Choisne N, Pont C, Carrère S, Caissard JC, Couloux A, Cottret L, Aury J, Szécsi J, Latrasse D, Madoui M-A, François L, Fu X, Yang S-H, Dubois A, Piola F, Larrieu A, Perez M, Labadie K, Perrier L, Govetto B, Labrousse Y, Villand P, Bardoux C, Boltz V, Lopez-Roques C, Heitzler P, Vernoux T, Vandenbussche M, Quesneville H, Boualem A, Bendahmane A, Liu C, Le Bris M, Salse J, Baudino S, Benhamed M, Wincker P, Bendahmane M (2018) The Rosa genome provides new insights into the domestication of modern roses. Nat Genet 50:772–777

Sahebi M, Hanafi MM, van Wijnen AJ, Rice D, Rafii MY, Azizi P, Osman M, Taheri S, Abu Bakar MF, Mat Isa MN, Noor YM (2018) Contribution of transposable elements in the plant's genome. Gene 665:155–166

Seymour DK, Koenig D, Hagmann J, Becker C, Weigel D (2014) Evolution of DNA Methylation Patterns in the Brassicaceae is Driven by Differences in Genome Organization. PLOS Genet 10ºe1004785. https://doiorg/doi:101371/journalpgen1004785

Slotkin RK, Martienssen R (2007) Transposable elements and the epigenetic regulation of the genome. Nature 8:272–285

The Arabidopsis Genome Initiative (2000) Analysis of the genome sequence of the flowering plant *Arabidopsis thaliana*. Nature 408:796–815

The French-Italian Public Consortium for Grapevine Genome Characterization - Jaillon O, Aury J-M, Noel B, Policriti A, Clepet C, Casagrande A, Choisne N, Aubourg S, Vitulo N, Jubin C, Vezzi A, Legeai F, Hugueney P, Dasilva C, Horner D, Mica E, Jublot D, Poulain J, Bruye`re C, Billault A, Segurens B, Gouyvenoux M, Ugarte E, Cattonaro F, Anthouard V, Vico V, Del Fabbro C, Alaux M, Di Gaspero G, Dumas V, Felice N, Paillard S, Juman I, Moroldo M, Scalabrin S, Canaguier A, Le Clainche I, Malacrida G, Durand E, Pesole G, Laucou V, Chatelet P, Merdinoglu D, Delledonne M, Pezzotti M, Lecharny A, Scarpelli C, Artiguenave F, Pe` ME, Giorgio Valle G, Morgante M, Caboche M, Adam-Blondon A-F, Weissenbach J, Quétier F, Wincker P (2007) The grapevine genome sequence suggests ancestral hexaploidization in major angiosperm phyla. Nature 449:463–467

Urrestarazu J, Muranty H, Denancé C, Leforestier D, Ravon E, Guyader A, Guisnel R, Feugey L, Aubourg S, Celton J-M, Daccord N, Dondini L, Gregori R, Lateur M, Houben P, Ordidge M, Paprstein F, Sedlak J, Nybom H, Garkava-Gustavsson L, Troggio M, Bianco L, Velasco R, Poncet C, Théron A, Moriya S, Bink MCAM, Laurens F, Tartarini S, Durel C-E (2017) Genome-wide association mapping of flowering and ripening periods in apple. Front Plant Sci 8:1923. https://doiorg/103389/fpls201701923

Vanburen R, Bryant D, Bushakra JM, Vining KJ, Edger PP, Rowley ER, Priest HD, Michae TP, Lyons E, Filichkin SA, Dossett M, Finn CE, Bassil NV, Mockler TC, (2016) The genome of black raspberry (*Rubus occidentalis*). Plant J 87:535–547

Velasco R, Zharkikh A, Affourtit J, Dhingra A, Cestaro A, Kalyanaraman A, Fontana P, Bhatnagar SK, Troggio M, Pruss D, Dhingra, A, Cestaro A, Kalyanaraman A, Fontana P, Bhatnagar SK, Troggio M, Pruss D, Salvi S, Pindo M, Baldi P, Castelletti S, Cavaiuolo M, Coppola G, Costa F, Cova V, Ri AD, Goremykin V, Komjanc M, Longhi S, Magnago P, Malacarne G, Malnoy M, Micheletti D, Moretto M,

Perazzolli M, Si-Ammour A, Vezzulli S, Zini E, Eldredge G, Fitzgerald LM, Gutin N, Lanchbury G, Macalma T, Mitchell JT, Reid J, Wardell B, Kodira C, Chen Z, Desany B, Niazi F, Palmer M, Koepke T, Jiwan D, Schaeffer S, Krishnan V, Wu C, Chu VT, King ST, Vick J, Tao Q, Mraz A, Stormo A, Stormo K, Bogden R, Ederle D, Stella A, Vecchietti A, Kater MM, Masiero S, Lasserre P, Lespinasse Y, Allan AC, Bus V, Chagné D, Crowhurst RN, Gleave AP, Lavezzo E, Fawcett JA, Proost S, Rouzé P, Sterck L, Toppo S, Lazzari B, Hellens RP, Durel C-E, Gutin A, Bumgarner R, Gardiner SE, Skolnick M, Egholm M, Van de Peer Y, Salamini F, Viola R (2010) The genome of the domesticated apple (*Malus* × *domestica* Borkh.). Nat Genet 42:833–839. https://doiorg/101038/ng654

Wang Y, Tang H, DeBarry JD, Tan X, Li J, Wang X, Lee TH, Jin H, Marler B, Guo H, Kissinger JC, Paterson AH (2012) MCScanX: a toolkit for detection and evolutionary analysis of gene synteny and collinearity. Nucleic Acids Res 40:e49

Wu J, Wang Z, Shi Z, Zhang S, Ming R, Zhu S, Khan MA, Tao S, Korban SS, Wang H, Chen NJ, Nishio T, Xu X, Cong L, Qi K, Huang X, Wang Y, Zhao X, Wu J, Deng C, Gou C, Zhou W, Yin H, Qin G, Sha Y, Tao Y, Chen H, Yang Y, Song Y, Zhan D, Wang J, Li L, Dai M, Gu C, Wang Y, Shi D, Wang X, Zhang H, Zeng L, Zheng D, Wang C, Chen M, Wang G, Xie L, Sovero V, Sha S, Huang W, Zhang S, Zhang M, Sun J, Xu L, Li Y, Liu X, Li Q, Shen J, Wang J, Paull RE, Bennetzen JL, Wang J, Zhang S (2013) The genome of the pear (*Pyrus bretschneideri* Rehd.). Genome Res 23:396–408

Xing L, Li Y, Qi S, Zhang C, Ma W, Zuo X, Liang J, Gao C, Jia P, Shah K, Zhang D, An N, Zhao C, Han M, Zhao J (2019) Comparative RNA-sequencing and DNA methylation analyses of apple (*Malus domestica* Borkh.) buds with diverse flowering capabilities reveal novel insights into the regulatory mechanisms of flower bud formation. Plant Cell Physiol. https://doiorg/doi:101093/pcp/pcz080

Xing L, Zhang D, Song X, Weng K, Shen Y, Li Y, Zhao C, Ma J, An N, Han M (2016) Genome-wide sequence variation identification and floral-associated trait comparisons based on the re-sequencing of the 'Nagafu No 2' and 'Qinguan' varieties of apple (*Malus domestica* Borkh). Front Plant Sci 7:908. https://doi.org/10.3389/fpls.2016.00908

Xu J, Zhou S, Gong X, Song Y, van Nocker S, Ma F, Guan Q (2018) Single-base methylome analysis reveals dynamic epigenomic differences associated with water deficit in apple. Plant Biotech J 16:672–687

Zhang L, Hu J, Han X, Li J, Gao Y, Richards CM, Zhang C, Tian Y, Liu G, Gul H, Wang D, Tian Y, Yang C, Meng M, Yuan G, Kang G, Wu Y, Wang K, Zhang H, Wang D, Cong P (2019) A high-quality apple genome assembly reveals the association of a retrotransposon and red fruit colour. Nat Commun 10:1494. https://doi.org/10.1038/s41467-019-09518-x

Zhang X, Yazaki J, Sundaresan A, Cokus S, Chan SW-L, Chen H, Henderson IR, Shinn P, Pellegrini M, Jacobsen SE, Ecker JR (2006) Genome-wide high-resolution mapping and functional analysis of DNA methylation in *Arabidopsis*. Cell 126:1189–1201

Regulatory Sequences in Apple

9

Xiao-Fei Wang and Yu-Jin Hao

Abstract

Apple (*Malus* × *domestica* Borkh.) is one of the most important cultivated fruit trees in the world. However, studies on apple gene regulation have long lagged behind model plants due to incomplete information on the genome sequence. This has limited identification and functional characterization of regulatory sequences, that is, numbers, types, and functional roles of genes involved in the regulation of various biological processes in the apple. In general, regulatory sequences can either promote or repress the expression of downstream genes through different regulatory mechanisms, such as protein–DNA binding or protein–protein interactions. Gene expression is highly regulated at the transcriptional level. As a primary type of regulatory sequences, transcription factors (TFs) play critical roles in this process by either positively or negatively regulating the expression of multiple genes, including both downstream structural genes and other regulatory sequences. TFs regulate gene expression by interacting with specific *cis*-elements in target genes and by forming protein complexes with other TFs to either promote or suppress target gene expression in regulatory modules. Over the past decade, apple whole-genome sequencing efforts have enabled the collection of genomic information on apple TF families and their specific roles in regulation of fruit quality traits, abiotic and biotic stress resistance, and various other important biological traits. In this chapter, we will provide an introduction of apple TFs, and review recent studies on their roles in the regulation of apple growth and development.

9.1 Introduction

The genus *Malus* belongs to the Rosaceae family. It includes the cultivated orchard apple, *Malus* × *domestica* Borkh., an important horticultural crop whose global production has increased in the last decade to more than 80 megatons on a cultivated area of nearly 3,000,000 ha. The apple has been cultivated for thousands of years in China.

The first draft of the whole-genome sequence (WGS) of the cultivated apple, of 'Golden Delicious', was published and labelled v1.0 (Velasco et al. 2010). The apple was the third fruit crop to be sequenced, following grape and papaya (Bolger et al. 2014). The heterozygosity of the 'Golden Delicious' genome was relatively high, and approximately 81% of the genome was

X.-F. Wang (✉) · Y.-J. Hao
National Key Laboratory of Crop Biology, Shangdong Agricultural University, Shangdong, China
e-mail: xfwang2004@163.com

Y.-J. Hao
e-mail: haoyujin@sdau.edu.cn

S. S. Korban (ed.), *The Apple Genome*, Compendium of Plant Genomes,
https://doi.org/10.1007/978-3-030-74682-7_9

sequenced, with a total contig length of 604 Mb compared to the estimated genome size of 742 Mb and 63,141 annotated genes (Velasco et al. 2010). Although incomplete, this draft sequence was beneficial to the apple research community, as it demonstrated that the 17 apple chromosomes were derived from a genome-wide duplication of nine original chromosomes in a Rosaceae ancestor (Velasco et al. 2010). Subsequently, a high-quality apple WGS from a double-haploid 'Golden Delicious', line "GDDH13", was released, and designated as GDDH13 v1.1. This genome was 643 Mb in size, with 42,140 annotated genes and a scaffold N50 of 5.558 Mb (Daccord et al. 2017). This new WGS provided additional genomic resources for apple research (Peace et al. 2019). Recently, apple genome re-sequencing of 117 diverse accessions was conducted and used to demonstrate that the cultivated apple probably originated from *M. sieversii* in Kazakhstan via deep introgression with *M. sylvestris,* and that *M. sieversii* from Xinjiang in China might not have directly contributed to apple domestication. This study supported a two-stage process for the evolution of the enlarged apple fruit size and generated molecular markers useful in apple breeding programmes (Duan et al. 2017). More recently, a high-quality genome of an apple anther-derived homozygous line, "HFTH1", was assembled with a contig N50 of 6.99 Mb. This study identified numerous genomic variations due mainly to activities of transposable elements and demonstrated that a retrotransposon insertion in the *MdMYB1* promoter was associated with the red-skinned apple phenotype (Zhang et al. 2019).

Overall, these whole-genome sequencing and re-sequencing efforts have contributed to our knowledge of apple genetics and functional genomics, and have provided useful guidance for future molecular breeding efforts. In particular, apple WGSs have spurred major advances in apple functional genetics. Numerous studies have identified candidate genes related to important horticultural traits such as flowering, fruit ripening, fruit quality characteristics, and stress resistance (Gu et al. 2019; Peace et al. 2019).

In general, genomic sequences include both non-coding and coding sequences. Non-coding sequences are comprised of multiple types of regulatory elements, such as promoters, enhancers, silencers, and insulators, and these regulate spatial and temporal transcriptions of targeted protein-coding sequences (Orenstein and Shamir 2016). While coding sequences are those DNAs that are transcribed into RNAs, and in turn, these RNAs are translated into proteins. Among these coding sequences, TFs function to serve as major regulatory sequences. TFs contain DNA-binding domains, and function as either activators or repressors by binding to *cis*-acting elements in target gene promoters, thereby increasing or decreasing transcription of numerous genes controlling multiple biological processes, including growth, development and stress tolerance (Orenstein and Shamir 2016). Often, TFs have additional functional domains allowing them to interact with various other regulatory sequences (Boeva 2016).

Only a handful of TF genes have been functionally characterized in apple and in other woody plants until recently. In this chapter, we will focus on advances in identifying apple TFs and exploring their specific functions in regulating apple physiological processes. We will present an overview of apple TF families and then proceed to cover recent advances in functional characterizations of these TF families.

9.2 Transcription Factor Families in Apple

Based on the availability of multiple plant WGSs, it is now relatively easy to identify genes encoding plant TFs. These TFs account for a considerable portion of a genome and can be grouped into different gene families based on their conserved protein–protein and DNA-binding domains, as well as various other completed genome sequences. *Arabidopsis* is the most extensively characterized plant from a genomic perspective. The *Arabidopsis* genome is approximately 130 Mb in size, and contains approximately 30,000 protein-coding genes, over 2,000 of which are transcription factors, belonging to more than 50 families. Similar to genomes of other model organisms, such as *Caenorhabditis* and *Drosophila*, the *Arabidopsis* genome exhibits broad duplications (Riechmann and Ratcliffe 2000). Furthermore, major TF families of *Arabidopsis* include the following: MYB, basic helix-loop-helix (bHLH), APETALA2/ETHYLENE RESPONSE FACTOR (AP2/ERF), NAM/ATAF1,2/CUC2 (NAC), Teosinte branched1/cycloidea/proliferating cell factor 1 (TCP), basic leucine zipper (bZIP), and MADS families (Table 9.1). Many of these families are plant-specific, such as AP, NAC, WRKY, and DNA-binding one zinc finger (Dof) (Riechmann and Ratcliffe 2000). As these multiple TF families regulate a variety of target genes, these likely allow plants to respond to various biotic and abiotic stresses. For example, MYB TFs play key roles in plant secondary metabolites, AP2 TFs play specific roles in the regulation of floral development, both TCP and NAC TFs influence hormone-mediated growth and development of leaves and roots, WRKY proteins regulate multiple stress responses, and SQUAMOSA promoter binding protein-like (SPL) TFs participate in regulating the transition from juvenile to adult growth (Samad et al. 2017).

Previously, general characteristics of apple TFs have been inferred from homology-based cloning of homologous sequences from other model plants. In particular, the *Arabidopsis* genome has been used as a model to identify TFs in apple and in other fruit tree species. Following the release of the apple whole-genome sequence, it has become relatively straightforward to identify apple TF-encoding genes (Jin et al. 2017). A listing of major TF families of those of apple and *Arabidopsis*, based on information from the Plant TFDB (http://planttfdb.cbi.pku.edu.cn/), with *Arabidopsis* used as a reference, is presented in Table 9.1 (Davuluri et al. 2003; Velasco et al. 2010; Jin et al. 2017). Interestingly, some of these TF families have been

Table 9.1 A listing of major transcription factor (TF) families, along with numbers of genes in each of these families, identified in both *Arabidopsis* and apple through PlantTFDB

TF family	PlantTFDB in *Arabidopsis*	PlantTFDB in apple	Genomic identification in apple	Reference(s)
SBP	30	42	27	Li et al. (2013)
WRKY	90	139	127/119	Meng et al. (2016), Lui et al. (2017)
ARF	37	30	31	Luo et al. (2014)
LBD	50	80	58	Wang et al. (2013)
Dof	47	60	60	Yang et al. (2018), Zhang et al. (2018b)
AP2/ERF	30	45	259	Girardi et al. (2013)
bHLH	225	250	188/175	Mao et al. (2017), Yang et al. (2017)
bZIP	127	114	114	Li et al. (2016b)
CO-like	22	17	64	Liu et al. (2018)
GATA	41	35	35	Chen et al. (2017b)
GRF	9	12	16	Zheng et al. (2018a)
MADS	146	142	146	Tian et al. (2015)
MYB	168	238	229	Cao et al. (2013)
NAC	138	253	180	Su et al. (2013)
TCP	33	58	52	Xu et al. (2014)

identified only in plant species, including those of AP2, AS2, Dof, EIL, GARP, LFY, NAC, Nin-like, SBP, TCP, WRKY, and YABBY gene families. These families may play specific roles in plant biological processes and in response to external environmental changes.

Multiple transcription factor families have been identified in apple, including those of MYB (Cao et al. 2013), AP2/ERF (Girardi et al. 2013), SBP (Li et al. 2013), NAC (Su et al. 2013), LATERAL ORGAN BOUNDARIES domain (LBD) (Wang et al. 2013), Auxin responsive factor (ARF) (Luo et al. 2014), TCP (Xu et al. 2014), MADS (Tian et al. 2015), bZIP (Li et al. 2016b), WRKY (Meng et al. 2016; Lui et al. 2017), GATA (Chen et al. 2017b), bHLH (Mao et al. 2017; Yang et al. 2017), Dof (Yang et al. 2018; Zhang et al. 2018b), BBX (Liu et al. 2018), and Growth-regulating factor (GRF) (Zheng et al. 2018a) gene families (Table 9.1). It is important to point out that most of these analyses have been conducted prior to 2017 using the 2010 v1.0 apple reference genome (Velasco et al. 2010). This has provided an overview of apple TFs, including their chromosomal locations, subfamily classifications, intron–exon organisations, sequence features, limited spatio-temporal expression patterns, and transcriptional responses in response to different environmental and hormonal stimuli.

In the next section, a review along with a description of detailed information about apple TFs that have been functionally characterized will be presented.

9.2.1 MYB TF Family

MYB TFs have been genome-wide isolated and characterized in numerous plant species, including *Arabidopsis* and apples (Dubos et al. 2010; Cao et al. 2013). MYB TFs correspond to a large gene family in plants and known to perform various functions in plant biological processes in apple (Dubos et al. 2010; Allan and Espley 2018). MYB TFs can be divided into three subfamilies based on the MYB domains, including R1R2R3, R2R3, and MYB-related (Dubos et al. 2010). Most MYBs belong to the R2R3 subfamily, having a DNA-binding domain (MYB domain) in the N-terminal and a regulatory domain (for either activation or repression) that is mainly present in the C terminal region (Dubos et al. 2010). In apple, a total of 229 MYB TFs have been identified, including 222 belonging to the R2R3 subfamily, five to the R1R2R3 subfamily, and two to the 4R-like subfamily (Cao et al. 2013). Furthermore, MYB TFs are classified into 45 phylogenetic subgroups based on conserved sequence motifs in their non-DNA-binding domain (DBD) regions (Cao et al. 2013). Many MYB TFs have been functionally characterized in apple, and their roles have been well investigated in multiple biological processes, including secondary metabolism (flavonoids, flavour, and organic acids), hormone signalling, and stress defence (Allan and Espley 2018; Gu et al. 2019), as presented in Table 9.2 and Fig. 9.1.

Studies on MYB TFs have mainly focused on their roles in regulating the phenylalanine metabolic pathway (specifically anthocyanin and proanthocyanidin accumulation) and fruit coloration. Anthocyanins are important secondary metabolites that determine fruit coloration (Allan and Espley 2018; Honda and Moriya 2018). Interestingly, anthocyanin biosynthesis is regulated by a variety of environmental and hormonal factors (An et al. 2020c). In apple, MdMYB1/MYB10/MYBA are found to be homologs of *Arabidopsis* MYB75/PAP1, and are involved as major regulators of anthocyanin biosynthesis and fruit coloration (Borevitz et al. 2000; Espley et al. 2009; Allan and Espley 2018). In fact, MdMYB1/MYB10/MYBA are allelic, and it has been reported that there is an insertion in the *MdMYB10* promoter to promote its auto-activation, leading to over-accumulation of anthocyanin in both leaves and fruits (Espley et al. 2009). Incidentally, the expression of *MdMYB10* also affects the release of volatiles, mainly through the phenylpropanoid metabolic pathway (Yauk et al. 2017). Whereas, MdMYB110a, a paralog of MdMYB1/MdMYB10, is a second apple TF associated with the distinctive phenotype of red-coloured fruit flesh cortex and green leaves (Chagné et al. 2013).

Table 9.2 A listing of MYB transcription factors in apple, their counterparts in *Arabidopsis*, along with their functions

Malus × *domestica*	Corresponding counterparts in *Arabidopsis*	Function(s)	Reference(s)
MdMYB1/MYB10/MYBA	AtMYB75/AtPAP1, AtMYB90/AtPAP2	Anthocyanin biosynthesis Organic acid accumulation Osmotic stress tolerance	Espley et al. (2009) Hu et al. (2016a) Gao et al. (2011)
MdMYB110a	AtMYB113, AtMYB114	Anthocyanin biosynthesis	Chagné et al. (2013)
MdMYB9, MdMYB11	AtTT2	Anthocyanin and PA biosynthesis	An et al. (2015)
MdMYB24L	AtMYB24	Anthocyanin biosynthesis	Wang et al. (2019a)
MdMYBPA1	MYBPA1	Anthocyanin accumulation	Wang et al. (2018b)
MdMYB23	AtMYB10	PA biosynthesis	An et al. (2018c)
MdMYBDL1	AtMYB58	Anthocyanin biosynthesis	Xu et al. (2017), Liu et al. (2019a)
MdMYB16	AtMYB3	Anthocyanin biosynthesis	Xu et al. (2017, Liu et al. (2019a)
MdMYB308	AtMYB7	Anthocyanin biosynthesis	Xu et al. (2017), Liu et al. (2019a)
MdMYB12	AtMYB32	PA biosynthesis	Wang et al. (2017)
MdMYB22	AtMYB4	PA biosynthesis	Wang et al. (2017)
MdMYB3	MdMYB3	Anthocyanin biosynthesis	Vimolmangkang et al. (2013)
MdMYBL2	AtMYBL2	Anthocyanin biosynthesis	Wang et al. (2019b)
MdMYB73	AtMYB44	Organic acid accumulation	Hu et al. (2017)
MdMYB44	AtMYB44	Organic acid accumulation	Jia et al. (2018b)
MdSIMYB1	AtMYB112	Abiotic stresses tolerance	Wang et al. (2014)
MdoMYB121	AtMYB5	Abiotic stresses tolerance	Cao et al. (2013)
MdMYB4	AtMYB4	Abiotic stresses tolerance	Wu et al. (2017b)
MdMYB15L	AtMYB15	Cold stress tolerance	Xu et al. (2018a, b)
MdMYB46	AtMYB46	Salt and osmotic stress tolerance	Chen et al. (2019)
MdMYB88, MdMYB124	AtFLP; AtMYB88	Cold stress tolerance; Root development	Xie et al. (2018) Wang et al. (2019c)
MdMYB308L	AtMYB3	Cold stress tolerance	An et al. (2020d)
MdMYB39L	AtMYB9	Sucrose uptake; Stamen development	Meng et al. (2018a), Li et al. (2020)
MdMYB58	AtMYB15	Iron homeostasis	Wang et al. (2018a)
MdMYB2	AtMYB2	Phosphate utilisation	Yang et al. (2020)
MdMYB93	AtMYB93	Suberin synthesis	Legay et al. (2016)

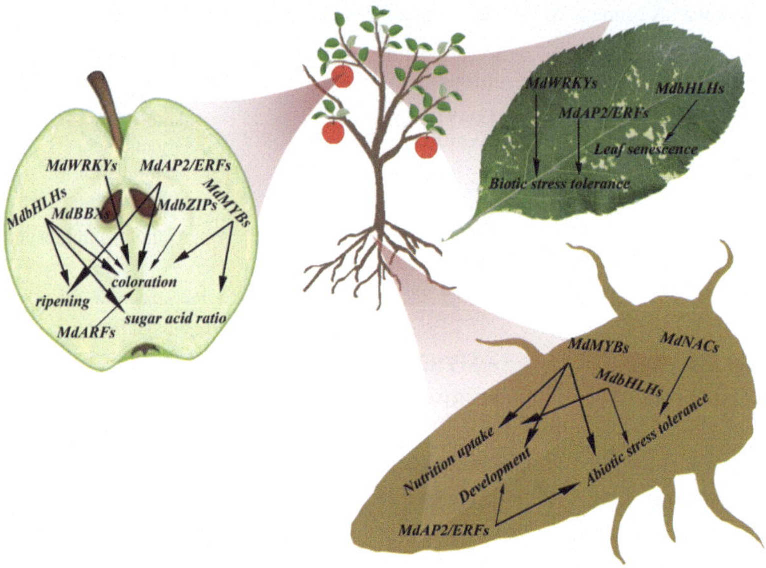

Fig. 9.1 A model depicting the identified TFs in regulating apple growth and development

It has been reported that in most horticultural plants, including apple, light and UVB (ultraviolet-B, 280–315 nm) play key roles in regulating the phenylpropanoid metabolism of leaves, flowers, and fruits (Li et al. 2012). In particular, it has been observed that transcript levels of the apple *MdMYB1* gene is induced by light, and this gene transcriptionally activates other structural genes that, in turn, regulate anthocyanin biosynthesis (Li et al. 2012). Moreover, as a critical repressor in the light signalling pathway, a gene encoding for the constitutive photomorphogenic1 (COP1) is found to regulate light-mediated anthocyanin accumulation by ubiquitination and degradation of MdMYB1 (Li et al. 2012). In another study, expression of the apple TF *MdHY5,* coding for a bZIP protein, is reported to be induced by light, and it promotes expression of *MdMYB1* (An et al. 2017b). On the other hand, an apple MdSIZ1, a SIZI sumo E3 ligase, directly interacts with and sumoylates MdMYB1 to modulate anthocyanin accumulation at low temperatures (Zhou et al. 2017a). Whereas, MdEIL1, an ethylene-responsive regulator, binds directly to the promoter of *MdMYB1*, thereby regulating ethylene-modulated anthocyanin accumulation. Furthermore, MdMYB1 interacts with the *MdERF3* promoter, a positive regulator of ethylene biosynthesis, promoting ethylene biosynthesis and anthocyanin accumulation in a positive feedback pathway (An et al. 2018d).

In recent studies, it has been reported that transcript levels of *MdbZIP44* (An et al. 2018f), *MdWRKY40* (An et al. 2019c), *MdERF38* (An et al. 2020e), and *MdTCP46* (An et al. 2020a) can be induced by abscisic acid (ABA), wounding, drought, and high-light intensity, respectively. All of these TFs interact with *MdMYB1* and promote its transcriptional activity, thereby inducing environmentally and hormonally modulated anthocyanin accumulation and fruit coloration (An et al. 2020c). Interestingly, it has been observed that nitrate plays a negative role in fruit coloration as an apple *MdBT2* gene, a member of the "bric-à-brac, tramtrack, broad-complex" (BTB) domain gene family and homologous to an *Arabidopsis* nitrate-responsive gene, *AtBT2*, is induced by exogenous nitrate (Wang et al. 2018c). In fact, MdBT2 interacts with both MdMYB1 and MdMYB9, thus negatively regulating their protein stabilities, and inhibiting both anthocyanin and proanthocyanidin (PA) accumulation and fruit coloration (An et al. 2018a; Wang et al. 2018c). Furthermore, an apple MdMIEL1, encoding a RING E3 ligase that responds to multiple abiotic stresses, is found to interact with MdMYB1 to modulate its ubiquitination and protein stability, thereby negatively regulating MdMYB1-mediated anthocyanin biosynthesis (An et al. 2017a). Transparent testa2 (TT2) and PA1 types of MYB TFs are involved in PAs biosynthesis, and ectopic expression of MdMYB9 and MdMYB11, homologs of *Arabidopsis* TT2 proteins, promote both anthocyanin and PA accumulation. Furthermore, MdMYB9 and MdMYB11 interact with a bHLH TF, MdbHLH3, to regulate the transcription of structural genes participating in anthocyanin and PA biosynthesis (An et al. 2015). Moreover, *MdMYB24L* responds to jasmonic acid (JA), and interacts with JA signalling factors (MdJAZs and MdMYC2) to promote anthocyanin accumulation (Wang et al. 2019a).

On the other hand, *MdMYBPA1*, a PA1-type MYB TF, is found to respond to low temperature and promotes the conversion of PA to anthocyanin (Wang et al. 2018b). In fact, cold stress induces expression of the apple *MdMYB23*, and that overexpression of *MdMYB23* promotes increased cold tolerance by activating transcription of C-repeat binding transcription factors (CBFs) *MdCBF1* and *MdCBF2*. Furthermore, MdMYB23 also modulates PA accumulation by promoting the expression of *MdANR*, the key enzyme of PA biosynthesis (An et al. 2018c). Whereas, overexpression of *MdMYBDL1*, a light-inducible MYB-like gene, enhances anthocyanin accumulation by inhibiting expression of both *MdMYB16* and *MdMYB308*, two negative regulators of anthocyanin biosynthesis (Xu et al. 2017; Liu et al. 2019a). In another group of MYB TFs, *MdMYB12* and *MdMYB22* are reported to regulate the phenylalanine metabolic pathway in apple (Wang et al. 2017). Whereas, overexpression of *MdMYB3* is found to be high in red-skinned rather than in either yellow- or green-skinned apples, as it promotes accumulation of both anthocyanins and flavonols, and regulates flower development (Vimolmangkang et al. 2013). Moreover, in red-fleshed apples, MdMYBL2 acts as a positive regulator of cytokinin-induced anthocyanin biosynthesis (Wang et al. 2019b).

Another functional property of MYB TFs is their participation in organic acid accumulation in apple. For example, MdMYB1 can not only control anthocyanin biosynthesis, but it also can influence organic acid accumulation by modulating the vacuolar acidity via vacuolar H^+-ATPases (VHAs) and vacuolar H^+-pyrophosphatases (VHPs) (Hu et al. 2016a). In another TF, MdMYB73 interacts with MdCIbHLH1 influencing organic acid accumulation by promoting expression of apple *aluminium-activated malate transporter 9* (*MdALMT9*), *vacuolar ATPase subunit A* (*MdVHA-A*), and *vacuolar pyrophosphatase 1* (*MdVHP1*) (Hu et al. 2017). While, *MdMYB44*, identified by bulk segregant analysis, has an A/G SNP in its promoter region that contributes to accumulation of different levels of acidity in apple fruit (Jia et al. 2018b).

It has been well documented that MYB TFs influence tolerance to biotic and abiotic stresses in various plants, including apple (Dubos et al. 2010). For example, transcript levels of the apple *MdSIMYB1*, an R2R3-MYB TF, are influenced by various abiotic stresses, and overexpression of *MdSIMYB1* promotes apple tolerance to various abiotic stresses, such as salinity, drought, and cold stresses (Wang et al. 2014). While overexpression of *MdMYB10* in *Arabidopsis* promotes osmotic stress tolerance (Gao et al. 2011). In another study, transgenic apple and tomato plants over-/expressing *MdoMYB121*, induced by multiple stresses, demonstrate enhanced tolerance to salinity, drought, and cold stresses (Cao et al. 2013). Furthermore, MdMYB4 is found to be a positive regulator of cold and salt tolerance (Wu et al. 2017b), while MdMYB15L is a negative regulator of the cold stress response (Xu et al. 2018a, b). In a recent study, transgenic apple overexpressing *MdMYB46* demonstrates enhanced tolerance to both salt and osmotic stress (Chen et al. 2019). In another study, transgenic plants expressing *MdMYB88* and *MdMYB124*, two homologs induced by cold, demonstrate enhanced cold tolerance via elevated expression of cold-related genes (Xie et al. 2018). Yet another TF, MdMYB308L promotes cold tolerance through upregulation of transcription of an *MdCBF2* gene (An et al. 2020d).

Another reported role of MYB TFs is their involvement in regulating plant organ development (Dubos et al. 2010). As reduced levels of accumulation of the sugar alcohol sorbitol contribute to abnormal stamen development, it has been observed that a stamen-specific TF, MdMYB39L, plays an essential role in sorbitol-mediated growth and development of stamens and pollen tubes (Meng et al. 2018a). Recently, MdMYB39L is shown to bind to the promoter of a sugar transporter, *MdSTP13a*, to promote sucrose uptake and pollen tube growth (Li et al. 2020). Furthermore, an apple FOUR LIPS (MdFLP), also known as MdMYB124, is also found to regulate adventitious root development by modulating the auxin response (Wang et al. 2019c).

In other roles, some MYB TFs are found to regulate nutrient uptake and utilisation in apple. For example, *MdMYB58* is induced by iron deficiency and transcriptionally represses *MdMATE43* expression, thus negatively regulating Fe homeostasis (Wang et al. 2018a). While, ectopic expression of *MdMYB2,* found to be responsive to phosphate deficiency, regulates both phosphate utilization and gibberellin (GA)-regulated growth (Yang et al. 2020).

It is quite likely that apple MYB TF may also have additional functions. For example, *MdMYB93* is highly expressed in russeted suberised skin of apple fruits, and it has been found that transient expression of *MdMYB93* promotes suberin synthesis (Legay et al. 2016).

9.2.2 The bHLH TF Family

Similar to MYBs, bHLH proteins are also among the largest TF families. These proteins consist of a conserved bHLH domain of about 60 amino acids that includes two functional motifs, a basic (b) motif and an 'HLH' motif (Toledo-Ortiz et al. 2003). The 'b' motif is located at the N-terminus, consists of 13–17 amino acids, and recognizes an E-box DNA sequence (CANNTG) (Kavas et al. 2016), while the 'HLH' motif functions in dimerisation with other interacting proteins (Massari and Murre 2000).

In plants, bHLH TFs act as key regulators of metabolic, physiological, and developmental processes, such as light signalling, hormone signal transduction, abiotic and biotic stress responses, and nutrient utilization. Yang et al. (2017) have identified 175 bHLH proteins in apple using multiple sequence alignments, and have classified them into 23 subfamilies. Mao et al. (2017) have also characterized bHLH TFs in apple, and have identified 188 bHLH TFs, and these have been subdivided into 18 groups. Their regulatory roles in secondary metabolism, abiotic stress responses, as well as nutrient uptake and utilisation have been functionally identified (Table 9.3 and Fig. 9.1).

Among these apple bHLH TFs, MdbHLH3 and its homolog MdbHLH33 have been well documented to be involved in regulating anthocyanin accumulation and in cold stress tolerance. It has been observed that as low temperature promotes biosynthesis of anthocyanin, MdbHLH3 is found to be cold-induced, and it interacts with MdMYB1 (allelic to MdMYB10) to regulate low-temperature enhanced anthocyanin accumulation and fruit coloration in apple (Xie et al. 2012). In addition to its interaction with MdMYB1, MdbHLH3 interacts with MdMYB9/11 to activate the expression of *MdMYB1*, *MdMYB9*, and *MdMYB11*, thereby promoting anthocyanin and PA accumulation (An et al. 2015). It has been reported that the apple hexokinase, MdHXK1, is a glucose sensor, and it directly phosphorylates and stabilizes MdbHLH3 to regulate glucose-modulated anthocyanin accumulation (Hu et al. 2016b). Furthermore, MdbHLH3 can also promote ethylene release by promoting transcription of genes involved in ethylene biosynthesis (Hu et al. 2019). Interestingly, MdbHLH33 positively regulates anthocyanin accumulation and cold stress tolerance by promoting transcription of related regulatory and structural genes (Xie et al. 2012; Xu et al. 2017, 2018a).

Yet another bHLH TF, MdMYC2, is an important regulator of the phytohormone jasmonate (JA)-induced anthocyanin accumulation (An et al. 2016). While expression of another apple bHLH TF, *MdCIbHLH1*, a homolog of the

Table 9.3 A listing of bHLH transcription factors in apple, their counterparts in *Arabidopsis*, along with their functions

Malus × domestica	Corresponding counterparts in *Arabidopsis*	Function(s)	Reference(s)
MdbHLH3; MdbHLH33	AtTT8	Flavonol biosynthesis; Ethylene biosynthesis; Cold stress tolerance	Xie et al. (2012), An et al. (2015), Hu et al. (2019), Xu et al. (2017, 2018a)
MdMYC2	AtMYC2	Anthocyanin biosynthesis; Ethylene biosynthesis; Al tolerance	An et al. (2016), Li et al. (2017b), An et al. (2017c)
MdCIbHLH1	AtICE1	Cold stress tolerance; Malic acid biosynthesis	Feng et al. (2012), Zhang et al. (2020)
MdbHLH104	AtbHLH104	Fe uptake	Zhao et al. (2016)
MxFIT	AtFIT	Fe uptake	Yin et al. (2014)
MdSAT1	AtbHLH19	Fe uptake	Wang et al. (2018a)
MdbHLH93	AtbHLH93	Leaf senescence	An et al. (2019a, b, c)
MdPIF1	AtPIF1	Seed germination; hypocotyl elongation	Zhou et al. (2017b)

Arabidopsis ICE1, is apparently induced by cold treatment. In fact, overexpression of *MdCIbHLH1* contributes to cold stress tolerance by inducing transcription of *MdCBF2* (Feng et al. 2012). It has also been observed that MdCIbHLH1 interacts with MdMYB73 to regulate malic acid biosynthesis and vacuolar acidity (Hu et al. 2017; Zhang et al. 2020).

As mentioned above, some bHLH TFs are functional in apple fruit ripening, which is promoted by both ethylene and JA. It has been reported that MdMYC2 encodes a bHLH TF whose transcript is induced by JA, wherein MdMYC2 binds directly to promoters of apple genes encoding 1-aminocyclopropane-1-carboxylic acid (ACC) synthase (ACS1), *MdACS1*, ACC oxidase, *MdACO1*, and the secondary TF ETHYLENE RESPONSE FACTOR 3 (ERF3), *MdERF3*, thereby promoting ethylene biosynthesis. Furthermore, MdMYC2 also interacts with MdERF2, a suppressor of ethylene biosynthesis, thereby enhancing the expression of *MdACS1*, and ultimately modulating both ethylene biosynthesis and fruit ripening through crosstalk between and ethylene and JA signalling (Li et al. 2017b).

In addition to the above-described functional roles, bHLH TFs play essential roles in both nutrient uptake and utilization. As iron (Fe) deficiency affects the growth and development of fruit trees, some bHLH TFs belonging to the Ib subgroup have been shown to regulate iron utilization in apple (Yin et al. 2014; Zhao et al. 2016; Wang et al. 2018a). For example, MdbHLH104 is a key regulator of the Fe deficiency response, and transgenic apple plants expressing *MdbHLH104* have demonstrated increased H^+-ATPase activity in response to Fe deficiency (Zhao et al. 2016). MdbHLH104 interacts with the promoter of the apple gene coding for the *Arabidopsis* PM H^+-ATPase (AHA), *MdAHA8,* to positively regulate its expression, and thereby promoting Fe uptake. In addition, MdbHLH104 directly regulates transcription of other Fe-responsive apple genes, such as *MdbHLH38*, *MdbHLH39*, and *POPEYE* (*MdPYE*), an apple gene coding for a bHLH protein induced by iron deficiency. It is observed that MdbHLH104 interacts with subgroup IVc bHLH proteins to co-regulate *MdAHA8* transcription, thereby promoting both PM H^+-ATPase activity and Fe utilization (Zhao et al. 2016). In another study, an Fe deficiency-induced transcription factor (FIT) bHLH gene, *MxFIT*, has been isolated from the species *M.* x *iaojinensis*, a promising apple rootstock with resistance to iron chlorosis (Yin et al. 2014). It is found that overexpression of *MxFIT* promotes tolerance to iron deficiency stress (Yin et al. 2014). Furthermore, an apple member of the IVa subfamily of bHLH TFs, MdSAT1 (an apple symbiotic ammonium transporter1) interacts with MdMYB58, and competitively weakens its transcriptional activity, thereby promoting Fe utilisation (Wang et al. 2018a).

It is important to point out that bHLH TFs also participate in other biological processes allowing for plants to acclimate to multiple environments. For example, it has been found that expression of *MdbHLH93* is induced during leaf senescence (An et al. 2019b). Furthermore, overexpression of *MdbHLH93* promotes expression of a *SENESCENCE-ASSOCIATED GENE (SAG), MdSAG18,* and other apple senescence-related genes, and that this regulates senescence through an ABA-dependent pathway (An et al. 2019b). In another study, it is reported that ectopic expression of an apple *MdPIF1*, coding for PHYTOCHROME INTERACTING FACTOR1 (PIF1)-a bHLH protein similar to AtPIF1, inhibits seed germination and promotes hypocotyl elongation (Zhou et al. 2017b). In another example, it is reported that a MdMYC2 gene interacts with *MdERF3* promoter, thereby negatively regulating aluminium (Al) tolerance via regulation of ethylene biosynthesis mediated by MdERF3 (An et al. 2017c).

9.2.3 The WRKY TF Family

WRKY TFs belong to a large family. To date, 74 and 109 WRKY TFs have been isolated from *Arabidopsis* and rice, respectively (Phukan et al. 2016). WRKY TFs contain a conserved WRKY domain consisting of a highly conserved DNA-binding WRKYGQK motif and a zinc-finger motif (C2H2 or C2HC) (Eulgem et al. 2000). In general, this domain binds to a W-box *cis*-acting element (C/T)TGAC(C/T) (Brand et al. 2013; Chen et al. 2017a). Overall, WRKY TFs can be classified into four subfamilies based on numbers of WRKY domains and types of zinc finger motifs. Type I proteins possess two WRKY domains; type II proteins contain one WRKY domain and a C2H2 zinc finger motif; while type III proteins contain one WRKY domain and a C2HC zinc finger motif; and type IV proteins contain one WRKY domain with an incomplete zinc-finger motif (Eulgem et al. 2000).

It is reported that WRKY TFs are involved in various plant biological processes, including tolerance to biotic and abiotic stresses, secondary metabolites, and phytohormone responses (Chen et al. 2017a). In one recent study, 127 MdWRKYs have been identified in apple. These are classified into four subfamilies (Meng et al. 2016), while in another study, 119 WRKY candidates have been identified in the apple genome. These have been classified into three types, Types I to III, based on their conserved WRKY and zinc-finger domains (Lui et al. 2017).

Most studies on apple WRKY TFs have primarily focused on their roles in responses to abiotic and biotic stresses (Table 9.4 and Fig. 9.1). For example, it has been found that MdWRKY79 induces expression of the *NUCLEOTIDE BINDING/LEUCINE-RICH REPEAT (NLR) 16, MdNLR16*, thereby modulating sorbitol-mediated resistance to the fungal pathogen *Alternaria alternata* that incites Alternaria blotch disease of apple (Meng et al. 2018b). Whereas, *MdWRKY15* enhances apple resistance to the fungal pathogen *Botryosphaeria dothidea,* inciting bot or white rot, by directly regulating salicylic acid (SA) biosynthesis via *Isochorismate Synthase* (*ICS*), *MdICS1* (Zhao et al. 2020).

Likewise, *MdWRKY46* activates the expression of an apple gene encoding for an auxin-responsive GH3 family protein, also known as *AVR PPHB SUSCEPTIBLE 3* (*PBS3*), *MdPBS3.1*, another key regulator of SA biosynthesis (Zhao et al. 2019a). Expression of *MdPBS3.1* results in SA signalling that enhances plant resistance to *B. dothidea* (Zhao et al. 2019a). Yet another WRKY TF member, *MdWRKY31*, also responds to SA, and its ectopic expression increases resistance to biotic stress via increased transcription of SA-related genes (Zhao et al. 2019b). Furthermore, expression of both *MdWRKYN1* and *MdWRKY26* is induced in apple leaves inoculated with *A. alternaria* f. sp. *mali* (ALT1), and overexpression of these two genes enhances disease resistance by promoting transcription of pathogenesis-related (PR) genes mediated by these WRKY proteins (Zhang et al. 2017). In another study, transcription of *MdWRKY1* is also reported to be induced by ALT1, and ectopic expression of *MdWRKY1* also promotes plant resistance to the fungal pathogen *Phytophthora parasitica* var. *nicotianae*, which incites black shank (root and crown rot) disease (Liu et al. 2019b). Furthermore, *MdWRKY11* is reported to pathogen-induced, and hence it is also involved in disease resistance (Fan et al. 2011).

Recent studies have demonstrated that apple WRKY TFs also play essential roles in anthocyanin accumulation and phytohormone response. As wounding induces anthocyanin accumulation, it has been observed that overexpression of *MdWRKY40* promotes wounding-mediated anthocyanin biosynthesis (An et al. 2019c). Interestingly, it has been found that in red-fleshed apples, expression of *MdWRKY11* is associated with anthocyanin synthesis, and that MYB TFs and *MdHY5* (a master regulator that integrates signals from multiple pathways) are involved in this process (Liu et al. 2019b). Whereas, it has been observed that *MdWRKY31* is induced by the phytohormone ABA, and that MdWRKY31 binds directly to the promoter of the TF gene related to *AB13/viviparous1* (*RAV1*), *MdRAV1*, to regulate plant sensitivity to ABA (Zhao et al. 2019b). Furthermore, expression of *MdWRKY9* is found to be relatively high in dwarfing apple rootstocks (highly valued for establishing high-density plantings and for promoting precocious fruiting), and additional studies have demonstrated that MdWRKY9 modulates dwarfing through direct repression of brassinosteroids (BRs) via inhibition of expression of an apple *DWARFING* (*DWF*) gene *MdDWF4* (Zheng et al. 2018b).

9.2.4 The AP2/ERF Gene Family

The AP2/ERF TF family is characterized by the presence of an AP2/ERF domain, a region of approximately 60–70 amino acids that are involved in DNA binding. AP2/ERF TFs can be classified into three subgroups, including AP2 proteins with two repeated AP2/ERF domains, ERF proteins with single AP2/ERF domains, and RAV proteins with B3 domains (Nakano et al. 2006). AP2/ERF proteins regulate a variety of processes, including plant growth and development, as well as responses to multiple environmental stimuli (Riechmann et al. 2000). In apple, a total of 259 AP2/ERF proteins, each containing at least one ERF domain, have been identified (Girardi et al. 2013). It is important to point out that the ERF family can also be subdivided into two subfamilies, an ERF subfamily and a CBF/DREB subfamily (Sakuma et al. 2002). The apple genome has been found to contain 68 *MdDREB* genes that can be further classified into six subgroups (Zhao et al. 2012). Their roles in fruit growth and development, as well as in abiotic and biotic stress responses, have been identified (Table 9.5 and Fig. 9.1).

9.2.4.1 ERF TFs

ERF TFs participate in both fruit ripening and anthocyanin accumulation in apple fruit. It has been reported that both *MdERF1* and *MdERF2* are expressed during fruit ripening (Wang et al. 2007), wherein transcription of *MdERF2* is inhibited by ethylene during ripening, and MdERF2 interacts with MdERF3 to repress *MdACS1* transcription, thereby negatively influencing fruit ripening (Li et al. 2016a,

Table 9.4 A listing of WRKY transcription factors in apple, their counterparts in *Arabidopsis*, along with their functions

Malus × *domestica*	Corresponding counterparts in *Arabidopsis*	Function(s)	Reference(s)
MdWRKY79	AtWRKY72	Disease resistance	Meng et al. (2018b)
MdWRKY15	AtWRKY7	Disease resistance	Zhao et al. (2020)
MdWRKY46	AtWRKY53	SA biosynthesis	Zhao et al. (2019a)
MdWRKY31	AtWRKY6	Disease resistance	Zhao et al. (2019b)
MdWRKYN1	AtDSC1	Disease resistance	Zhang et al. (2017)
MdWRKY1	AtWRKY15	Disease resistance	Liu et al. (2019b)
MdWRKY11	AtWRKY11	Disease resistance	Fan et al. (2011)
MdWRKY40	AtWRKY40	Anthocyanin biosynthesis	An et al. (2019c)
MdWRKY9	AtWRKY11	BRs biosynthesis	Zheng et al. (2018b)

Table 9.5 A listing of AP2/ERF transcription factors in apple, their counterparts in *Arabidopsis*, along with their functions

Malus × *domestica* / other *Malus sp.*	Corresponding counterparts in *Arabidopsis*	Function(s)	Reference(s)
MdERF2	AtERF105	Fruit ripening	Li et al. (2016a, 2017b)
MdERF3	AtERF1A	Fruit ripening	Li et al. (2016a, 2017b)
MdERF17	AtERF17	Fruit peel degreening	Han et al. (2018)
MdERF38	AtERF38	Anthocyanin biosynthesis	An et al. (2020e)
MdERF1B	AtERF1	Anthocyanin and PA biosynthesis	Zhang et al. (2018a)
MdERF11	AtERF008	Disease resistance	Wang et al. (2020)
MdERF4	AtERF4	Salt stress tolerance	An et al. (2018g)
MsDREB6.2; MpDREB2A	AtDREB2A	Drought tolerance	Liao et al. (2017)
MdDREB76	AtERF060	Salt and drought stress tolerance	Li et al. (2019), Sharma et al. (2019)
MbDREB1	AtDREB1	Abiotic stress tolerance	Yang et al. (2011)
MdSHN3	AtSHN3	Cuticle formation, russet development	Lashbrooke et al. (2015)
MdANT1; MdANT2	AtANT	Cell proliferation	Dash and Malladi (2012)

2017b). Moreover, MdMYC2 promotes ethylene biosynthesis and fruit coloration by binding directly to the promoter of *MdERF3*, thereby promoting *MdACS1* transcription (Li et al. 2017b). It has been observed that serine repeat insertions in the coding region of the *MdERF17* gene enhance its transcriptional activity, promoting the expression of chlorophyll-degradation-related genes and contributing to fruit peel degreening (Han et al. 2018). Furthermore, the apple ERF protein MdERF38 plays a positive role in drought-modulated anthocyanin accumulation mediated by MdMYB1 (An et al. 2020e). Moreover, MdERF1B interacts with specific MYB proteins, including MdMYB1, MdMYB9, and MdMYB11 to activate ethylene-modulated anthocyanin and PA biosynthesis (Zhang et al. 2018a).

It is known that ERF TFs also play important roles in stress tolerance. For example, *ERF11* expression is induced following infection by *B. dothidea*, and overexpression of *MdERF11* confers resistance to *B. dothidea* by modulating the SA biosynthesis pathway (Wang et al. 2020). An et al. (2018g) have isolated a salt-responsive ERF gene, *MdERF4,* and have demonstrated that ectopic expression of *MdERF4* reduces salt stress tolerance, which is in contrast to transgenic plants expressing *MdERF3* that are found to be more tolerant to salt stress. It is reported that *MdERF4* binds directly to the promoter of *MdERF3* and suppresses its expression, thereby regulating ethylene-modulated salt stress tolerance (An et al. 2018g).

9.2.4.2 Dehydration Response Element Binding (DREB) TFs

DREB TFs operate in response to abiotic stress. It has been reported that transgenic apple plants carrying *MsDREB6.2,* a gene isolated from *M. sieversii,* demonstrate cytokinin-deficient phenotypes due to expression of *MdCKX4a*, a cytokinin (CK) catabolism gene that degrades endogenous CK, and thereby promoting drought tolerance (Liao et al. 2017). While another apple *DREB* gene, *MdDREB76*, promotes salt and drought stress tolerance by inducing transcription of stress-related genes (Sharma et al. 2019).

A gene for dwarf growth habit isolated from *M. baccata*, *MbDREB1,* was found to promote abiotic stress tolerance via ABA-dependent and ABA-independent pathways (Yang et al. 2011). In another study, it was reported that the methylation level of the promoter of an apple *DREB2A* gene, *MpDREB2A*, from *M. prunifolia* was shown to influence its transcription when ectopically expressed in transgenic *Arabidopsis*, and enhancing drought resistance (Li et al. 2019).

9.2.4.3 AP2 TF Subfamily

Within the AP2 subfamily, an apple MdSHN3 is found to contain an APETALA2-domain, and that overexpression of an apple *MdSHN3,* a WAX Inducer1/SHINE1 TF, promotes cuticle formation and represses russet development on apple fruit (Lashbrooke et al. 2015). Moreover, apple *AINTEGUMENTA* genes, *MdANT1* and *MdANT2*, are found to modulate cell production during fruit development by balancing the transcription of cell proliferation-related genes (Dash and Malladi 2012).

9.2.5 The MADS-Box TF Family

Members of the MADS-box gene family play critical roles in a variety of physiological and developmental processes in plants. All MADS proteins contain a highly conserved N-terminal MADS-box domain with approximately 59 amino acids that function in DNA binding (Riechmann and Ratciliffe 2000). Based on phylogenetic analysis, the plant MADS-box family can be divided into two major subfamilies, type I (serum response factor [SRF]) and type II (myocyte enhancer factor 2 [MEF2]) (Alvarez-Buylla et al. 2000). Type I genes can be classified into different MADS (M) protein including Ma, Mb, and Mr subgroups, while type II genes can be classified into MIKCC and MIKC* subgroups, wherein these consist of a MADS (M) domain, an intervening (I) domain, a keratin-like (K) domain, and a C-terminal (C) region, thus referred to as MIKC-type proteins. The MIKCC can be further subdivided into 12 subfamilies (Henschel et al. 2002; Becker and Theissen 2003).

Within the apple genome, a total of 146 MADS-box genes have been identified, and these are clustered into six subfamilies (Tian et al. 2015). Yet, in another study, a total of 142 apple MADS-box proteins have been identified, including six Dormancy-Associated MADS-box (DAM) proteins and four Flowering Locus C-like (FLC-like) proteins (Kumar et al. 2016). Among these, the functions of three *MdDAMs* and two *SHORT VEGETATIVE PHASE-like* (*MdSVPs*) MADS-box genes have been investigated in apple, and it is found that ectopic expression of these genes contributes to delayed budbreak and altered architecture changes (Wu et al. 2017a). In another study, it is reported that *MdMADS8* and *MdMADS9* bind to promoters of *MdACS1* and *MdACO1*, promoting their transcription and ultimately regulating fleshy fruit development (Ireland et al. 2013). While another gene, *FRUITFULL* (*FUL*), is associated with the regulation of fruit development and dehiscence, and it is reported that an orthologous *FUL-like* candidate apple gene, *MdMADS2.1*, influences fruit flesh firmness (Cevik et al. 2010). Furthermore, ectopic expression of *MdMADS2* in tobacco results in earlier flowering and shorter bolts, suggesting that *MdMADS2* may regulate flower development in apple (Sung et al. 1999). Moreover, overexpression of an apple *MdMADS5*, a putative homolog of *Arabidopsis APETALA1* (*AP1*), leads to early flowering (Kotoda et al. 2002). Whereas in apetalous mutants of apple, expression of an apple *PISTILLATA* (*MdPI*) gene, is abolished by a retrotransposon insertion in the intron position, and this mutation confers parthenocarpic fruit development (Yao et al. 2001).

9.2.6 The bZIP TF Family

The bZIP TF family consists of important regulators of various plant biological processes. The bZIP TFs are named based on their possession of a conserved bZIP domain with 60–80 amino acids, consisting of a basic region and a leucine zipper region (Hu and Sauer 1992).

In apple, a total of 114 bZIP TFs have been identified and classified into ten subfamilies based on their sequences. These TFs have different expression profiles, wherein some MdbZIPs are highly expressed during fruit ripening, while others are differentially expressed in vegetative organs (Li et al. 2016b).

Among these TFs, the ELONGATED HYPOCOYTL5 (HY5) bZIP TF plays multiple roles in plant photomorphogenesis and in stress tolerance. In apple, MdHY5 is involved in regulating anthocyanin accumulation, and it is reported that various environmental and hormonal factors promote its expression (An et al. 2017b). It has been observed that MdHY5 positively regulates the expression of *MdMYB10* by binding to the G-box *cis*-element. Furthermore, expression of *MdNRTs* and *MdNIRs*, involved in nitrate assimilation and uptake, are also induced by MdHY5, thereby highlighting the role of MdHY5 in nutrient utilization (An et al. 2017b). Interestingly, an apple B-box MdBBX22, belonging to a family of zinc-finger TFs, interacts with MdHY5 and promotes its transcriptional activity to regulate anthocyanin accumulation and fruit coloration (An et al. 2019a). Another study reported that MdHY5 binds to the promoter of the light-inducible apple gene *MdMYBDL1* to activate its expression, negatively regulating transcription of both *MdMYB16* and *MdMYB308*, two negative regulators of anthocyanin biosynthesis, thereby promoting anthocyanin accumulation (Liu et al. 2019a). Furthermore, MdHY5 also acts as a positive regulator of cold tolerance (An et al. 2017d).

Among other bZIP TFs, an ABA-responsive transcription factor, AREB2, is found to promote soluble sugar accumulation by upregulating transcription of related genes, including amylase and sugar transporter genes (Ma et al. 2017). Similarly, the bZIP TF MdbZIP44 regulates ABA-modulated anthocyanin accumulation by enhancing binding of MdMYB1 to promoters of downstream target genes (An et al. 2018f).

9.2.7 The BBX TF Family

Plant BBX proteins belong to a zinc-finger TF family, and they contain a B-box domain (either one or two conserved B-box motifs) at their N-termini along with a CCT (CONSTANS, CO-like, and TOC1) domain at their C-termini (Gangappa and Botto 2014). These BBX proteins are known to be involved in regulating photomorphogenesis, shade avoidance, and stress response (Gangappa and Botto 2014).

In apple, a total of 64 *MdBBXs* have been identified, all of which are classified into five subfamilies based on their phylogenetic relationships (Liu et al. 2018). MdBBX22 responds to UV-B, and as mentioned above, it interacts with MdHY5 to promote its transcriptional activity and positively to regulate photomorphogenesis (An et al. 2019a). On the other hand, the BBX protein MdBBX37 regulates photomorphogenesis through synergistic interactions with MdMYBs and MdHY5 (An et al. 2020b). It is reported that UV-B radiation enhances anthocyanin accumulation and fruit coloration, and an apple B-box protein-encoding gene, *MdCOL11,* positively regulates UV-B enhanced anthocyanin biosynthesis (Bai et al. 2014). Interestingly, expression of *MdBBX20* is also significantly induced by UV-B, and MdBBX20 can regulate anthocyanin accumulation in response to both ultraviolet radiation and low temperature (Fang et al. 2019b). Furthermore, another BBX protein, *MdCOL4*, links UV-B and high-temperature signalling to regulate apple fruit coloration (Fang et al. 2019a). Moreover, ectopic expression of *MdBBX10* in *Arabidopsis* is found to enhance tolerance to both drought and salt stresses (Liu et al. 2019c).

9.2.8 The NAC TF Family

The plant-specific NAC (NAM, ATAF, and CUC) TF family is recognized by a conserved NAC domain that contains approximately 160 amino acids. The NAC TF family can be classified into five subgroups (A to E). NAC TFs play diverse roles in regulating plant biological processes, particularly in responses to both biotic and abiotic stresses (Olsen et al. 2005).

A total of 180 NAC genes have been identified in apple, and these have been grouped into six subfamilies (I-VI) based on their phylogenetic relationships (Su et al. 2013). Functional characterization of apple NAC TFs has mainly revealed their participation in plant abiotic stress tolerance. For example, *MdNAC1* is a NAC transcription factor, and its overexpression leads to a dwarfing phenotype, likely attributed to reduced levels of endogenous ABA and brassinosteroid (BR) (Jia et al. 2018a). Moreover, overexpression of *MdNAC1* confers drought stress tolerance by promoting photosynthesis and reactive oxygen species (ROS) scavenging (Jia et al. 2019).

Interestingly, *MdNAC029* responds to low-temperature, and overexpression of *MdNAC029* reduces cold tolerance by directly repressing the expression of *MdCBFs* (An et al. 2018b). Whereas it is found that salt treatment induces *MdNAC047* expression, and ectopic expression of *MdNAC047* facilitates the release of ethylene by regulating *MdERF3* transcription, thereby promoting ethylene-modulated salt stress tolerance (An et al. 2018e).

On the other hand, it has been reported that levels of expression of *MdNAC52* increase during apple colour development and that overexpression of *MdNAC52* in apple calli promotes anthocyanin accumulation (Sun et al. 2019). Interestingly, *MdNAC52* also regulates PA accumulation by controlling transcription of *MdMYB9*, *MdMYB11* , and *LAR* genes (Sun et al. 2019).

9.2.9 The ARF TF Family

Auxin response factor (ARF) TFs regulate auxin response by binding to auxin-responsive elements (AuxREs, TGTCTC) in promoters of auxin-regulated genes (Guilfoyle et al. 1998). ARF TFs contain three conserved domains, including a DNA-binding domain (DBD), a middle region (MR), and a protein–protein interaction domain (conserved domains III/IV, CTD) (Tiwari et al. 2003).

A total of 31 ARF proteins have been identified in apple, and these have been classified into three subgroups (groups I, II, and III) (Luo et al. 2014). Studies on apple ARF TFs have focused primarily on their roles in fruit coloration and ripening. For example, *MdARF19* influences anthocyanin accumulation by regulating expression of *MdLOB52*, a negative regulator of anthocyanin accumulation (Wang et al. 2018d). Whereas, MdARF13 interacts with MdMYB10, and negatively regulates the anthocyanin metabolic pathway (Wang et al. 2018a, b, c, d). On the other hand, expression of *MdARF5* is induced by auxin treatment, and MdARF5 binds to the *MdERF2* promoter to induce ethylene biosynthesis, thereby participating in auxin-modulated ethylene biosynthesis and fruit ripening (Yue et al. 2020).

9.2.9.1 Other TF Families

In addition to the above-described list of transcription factor families, other TFs have also been investigated through genome-wide analyses and functional characterization in apple. A total of 58 LATERAL ORGAN BOUNDARIES DOMAIN (LBD) genes (Wang et al. 2013), 60 putative DNA-binding one zinc-finger (Dof) genes (Zhang et al. 2018b), 52 TEOSINTE BRANCHED/CYCLOIDEA/PROLIFERATING CELL FACTOR (TCP) genes (Xu et al. 2014), 27 SQUAMOSA promoter binding protein (SBP) genes (Li et al. 2013), 16 GROWTH-REGULATING FACTOR (GRF) genes (Zheng et al. 2018a), and 35 GATA genes, evolutionarily conserved TFs that interact with the WGATAR (W = T or A; R = G or A) sequence motif and are ubiquitous in eukaryotes (Chen et al. 2017b), have been identified in apple through genome-wide surveys.

As for functional characterization of some of these TFs, it has been reported that an apple MdLBD13 plays a key role in nitrate-mediated anthocyanin biosynthesis and nitrate utilization (Li et al. 2017a) while MdLOB52 has been observed to influence MdARF19-modulated anthocyanin accumulation (Wang et al. 2018d). On the other hand, *LEAFY* transgenic apple plants exhibit columnar phenotypes with shorter internodes (Flachowsky et al. 2010); whereas, an MDH1, a homolog of BELL1 (BEL1), plays a specific role in the regulation of apple ovule development (Dong et al. 2000).

9.3 Conclusions

Overall, TFs play important roles in plant growth and development by activating and/or repressing the transcription of genes involved in a variety of biological processes. The ongoing accumulation of completed plant genome sequences and the continued development of bioinformatics tools have allowed for extensive identification, functional characterization, and evolutionary investigation of various plant TF families (Jin et al. 2017).

The completion of multiple draft apple genome sequences (Velasco et al. 2010; Daccord et al. 2017; Zhang et al. 2019) has allowed for the identification and characterization of transcription factors at the whole-genome level. Members of multiple TF families, including those of MYB (Cao et al. 2013), bZIP (Li et al. 2016b), WRKY (Meng et al. 2016; Lui et al. 2017), bHLH (Mao et al. 2017; Yang et al. 2017), and BBX (Liu et al. 2018) families have been identified in the apple genome, and the functions of some TFs have been clearly demonstrated. For example, more than 30 apple MYB TFs have been functionally characterized and shown to participate in secondary metabolism, abiotic and biotic stress tolerance, nutrient utilisation, and organ development, among other biological processes (Fig. 9.1).

Transgenic apple plants with altered expression of *MdMYB10* (Espley et al. 2009), *MdoMYB121* (Cao et al. 2013), *MdSIMYB1* (Wang et al. 2014), *MdMYB88/124* (Liu et al. 2018), *MdMYB46* (Chen et al. 2019), *MdbHLH3* (Xie et al. 2012), *MdCIbHLH1* (Feng et al. 2012), *MdbHLH104* (Zhao et al. 2016), *MdMADS8/9* (Ireland et al. 2013), *MdDAMb and MdSVPa* (Wu et al. 2017), *MdAREB2* (Ma et al. 2017), *MsDREB6.2* (Liao et al. 2017), *MdWRKY9* (Zheng et al. 2018b), and *MdNAC1* (Jia et al. 2019) have been developed. These transgenic lines have allowed for pursuing detailed characterization of TFs and determine their roles in the regulation of apple growth and development. Recently, the CRISPR-Cas9 gene-editing technology can be used to pursue targeted edits in important genes, as well as to generate novel alleles that confer important changes in apple fruit traits (Osakabe et al. 2018). Such new technologies will provide scientists with convenient means for the rapid introgression of TF-derived improvements in consumer traits based on known functions of apple TFs.

9.4 Dedication

The Senior author, Prof. Xiao-Fei Wang, and this book volume editor, Prof. Schuyler S. Korban, would like to dedicate this chapter on "Regulatory Sequences in Apple" in memory of Prof. Yu-Jin Hao who passed away on March 15, 2021, as he will leave behind an indelible void in the plant molecular biology community, agricultural science community, and the scientific community at large, both in his native, China, and around the world, and he will be sorely missed. Prof. Hao has had a highly distinguished career. He was the Cheung Kong Scholar Distinguished Professor at Shangdong Agricultural University (SDAU) and winner of several awards, including the Chinese National Outstanding Youth Science Foundation, the National Millions of Talents Project, the Chinese Ministry of Education "Changjiang Scholars and Innovative Research" innovation team leader, and the "Taishan scholar in Shandong Province" professor. He also served as vice-president of SDAU, director of the National Engineering and Technology Research Center of Apple, and as an expert reviewer for the National Natural Science Foundation of China, among other critical roles. He also served as an Associate Editor of the journals of "Horticulture Research," "Plant Cell, Tissue & Organ Culture," and "Scientia Horticultureae". He published over 237 scientific articles in top-tier journals, focusing on functional and regulatory roles of various transcription factors and related transcription regulators of genes associated with fruit quality traits, such as anthocyanin and organic acids, among others, as well as those genes involved in biotic and abiotic stress tolerance in apple. A comprehensive list, although not complete, of his publications is available on https://www.researchgate.net/profile/Yu-Jin-Hao.

References

Allan AC, Espley RV (2018) MYBs drive novel consumer traits in fruits and vegetables. Trends Plant Sci 23(8):693–705

Alvarez-Buylla ER, Liljegren SJ, Pelaz S, Gold SE, Burgeff C, Ditta GS, Vergara-Silva F, Yanofsky MF (2000) MADS-box gene evolution beyond flowers: expression in pollen, endosperm, guard cells, roots and trichomes. Plant J 24(4):457–466

An X-H, Tian Y, Chen K-Q, Liu X-J, Liu D-D, Xie X-B, Cheng C-G, Cong P-H, Hao Y-J (2015) MdMYB9 and MdMYB11 are involved in the regulation of the JA-induced biosynthesis of anthocyanin and proanthocyanidin in apples. Plant Cell Physiol 56(4):650–662

An J-P, Li H-H, Song L-Q, Su L, Liu X, You C-X, Wang X-F, Hao Y-J (2016) The molecular cloning and functional characterization of MdMYC2, a bHLH transcription factor in apple. Plant Physiol Biochem 108:24–31

An J-P, Liu X, Li H-H, You C-X, Wang X-F, Hao Y-J (2017a) Apple RING E3 ligase MdMIEL1 inhibits anthocyanin accumulation by ubiquitinating and degrading MdMYB1 protein. Plant Cell Physiol 58 (11):1953–1962

An J-P, Qu F-J, Yao J-F, Wang X-N, You C-X, Wang X-F, Hao Y-J (2017b) The bZIP transcription factor MdHY5 regulates anthocyanin accumulation and nitrate assimilation in apple. Hortic Res 4(1):1–9

An J-P, Wang X-N, Yao J-F, Ren Y-R, You C-X, Wang X-F, Hao Y-J (2017c) Apple MdMYC2 reduces aluminum stress tolerance by directly regulating *MdERF3* gene. Plant Soil 418(1–2):255–266

An J-P, Yao J-F, Wang X-N, You C-X, Wang X-F, Hao Y-J (2017d) MdHY5 positively regulates cold tolerance via CBF-dependent and CBF-independent pathways in apple. J Plant Physiol 218:275–281

An J-P, An X-H, Yao J-F, Wang X-N, You C-X, Wang X-F, Hao Y-J (2018a) BTB protein MdBT2 inhibits anthocyanin and proanthocyanidin biosynthesis by triggering MdMYB9 degradation in apple. Tree Physiol 38(10):1578–1587

An J-P, Li R, Qu F-J, You C-X, Wang X-F, Hao Y-J (2018b) An apple NAC transcription factor negatively regulates cold tolerance via CBF-dependent pathway. J Plant Physiol 221:74–80

An J-P, Li R, Qu F-J, You C-X, Wang X-F, Hao Y-J (2018c) R2R3-MYB transcription factor MdMYB23 is involved in the cold tolerance and proanthocyanidin accumulation in apple. Plant J 96(3):562–577

An J-P, Wang X-F, Li Y-Y, Song L-Q, Zhao L-L, You C-X, Hao Y-J (2018d) EIN3-LIKE1, MYB1, and ETHYLENE RESPONSE FACTOR3 act in a regulatory loop that synergistically modulates ethylene biosynthesis and anthocyanin accumulation. Plant Physiol 178(2):808–823

An J-P, Yao J-F, Xu R-R, You C-X, Wang X-F, Hao Y-J (2018e) An apple NAC transcription factor enhances salt stress tolerance by modulating the ethylene response. Physiol Plant 164(3):279–289

An J-P, Yao J-F, Xu R-R, You C-X, Wang X-F, Hao Y-J (2018f) Apple bZIP transcription factor MdbZIP44 regulates abscisic acid-promoted anthocyanin accumulation. Plant Cell Environ 41(11):2678–2692

An J-P, Zhang X-W, Xu R-R, You C-X, Wang X-F, Hao Y-J (2018g) Apple MdERF4 negatively regulates salt tolerance by inhibiting MdERF3 transcription. Plant Sci 276:181–188

An J-P, Wang X-F, Zhang X-W, Bi S-Q, You C-X, Hao Y-J (2019a) MdBBX22 regulates UV-B-induced anthocyanin biosynthesis through regulating the function of MdHY5 and is targeted by MdBT2 for 26S proteasome-mediated degradation. Plant Biotechnol J 17(12):2231–2233

An J-P, Zhang X-W, Bi S-Q, You C-X, Wang X-F, Hao Y-J (2019b) MdbHLH93, an apple activator regulating leaf senescence, is regulated by ABA and MdBT2 in antagonistic ways. New Phytol 222(2):735–751

An J-P, Zhang X-W, You C-X, Bi S-Q, Wang X-F, Hao Y-J (2019c) MdWRKY40 promotes wounding-induced anthocyanin biosynthesis in association with MdMYB1 and undergoes MdBT2-mediated degradation. New Phytol 224(1):380–395

An J-P, Liu Y-H, Zhang X-W, Bi S-Q, Wang X-F, You C-X, Hao Y-J (2020a) Dynamic regulation of different light intensity-modulated anthocyanin biosynthesis by BT2-TCP46-MYB1 in apple. J Exp Bot 71(10):3094–3109

An J-P, Wang X-F, Espley RV, Lin-Wang K, Bi S-Q, You C-X, Hao Y-J (2020b) An apple B-Box protein MdBBX37 modulates anthocyanin biosynthesis and hypocotyl elongation synergistically with MdMYBs and MdHY5. Plant Cell Physiol 61(1):130–143

An J-P, Wang X-F, Hao Y-J (2020c) BTB/TAZ protein MdBT2 integrates multiple hormonal and environmental signals to regulate anthocyanin biosynthesis in apple. J Integr Plant Biol 62(11):1643–1646

An J-P, Wang X-F, Zhang X-W, Xu H-F, Bi S-Q, You C-X, Hao Y-J (2020d) An apple MYB transcription factor regulates cold tolerance and anthocyanin accumulation and undergoes MIEL1-mediated degradation. Plant Biotechnol J 18(2):337–353

An J-P, Zhang X-W, Bi S-Q, You C-X, Wang X-F, Hao Y-J (2020e) The ERF transcription factor MdERF38 promotes drought stress-induced anthocyanin biosynthesis in apple. Plant J 101(3):573–589

Bai S, Saito T, Honda C, Hatsuyama Y, Ito A, Moriguchi T (2014) An apple B-box protein, MdCOL11, is involved in UV-B-and temperature-induced anthocyanin biosynthesis. Planta 240 (5):1051–1062

Becker A, Theissen G (2003) The major clades of MADS-box genes and their role in the development and evolution of flowering plants. Mol Phylogenet Evol 29 (3):464–489

Boeva V (2016) Analysis of genomic sequence motifs for deciphering transcription factor binding and transcriptional regulation in eukaryotic cells. Front Genet 7:24

Bolger ME, Weisshaar B, Scholz U, Stein N, Usadel B, Mayer KF (2014) Plant genome sequencing-applications for crop improvement. Curr Opin Biotechnol 26:31–37

Borevitz JO, Xia Y, Blount J, Dixon RA, Lamb C (2000) Activation tagging identifies a conserved MYB regulator of phenylpropanoid biosynthesis. Plant Cell 12 (12):2383–2393

Brand LH, Fischer NM, Harter K, Kohlbacher O, Wanke D (2013) Elucidating the evolutionary conserved DNA-binding specificities of WRKY transcription factors by molecular dynamics and in vitro binding assays. Nucleic Acids Res 41(21):9764–9778

Cao Z-H, Zhang S-Z, Wang R-K, Zhang R-F, Hao Y-J (2013) Genome wide analysis of the apple MYB transcription factor family allows the identification of MdoMYB121 gene conferring abiotic stress tolerance in plants. PLoS ONE 8(7):e69955

Cevik V, Ryder CD, Popovich A, Manning K, King GJ, Seymour GB (2010) A FRUITFULL-like gene is associated with genetic variation for fruit flesh firmness in apple (*Malus domestica* Borkh.). Tree Genet Genomes 6(2):271–279

Chagné D, Lin-Wang K, Espley RV, Volz RK, How NM, Rouse S, Brendolise C, Carlisle CM, Kumar S, De Silva N (2013) An ancient duplication of apple MYB transcription factors is responsible for novel red fruit-flesh phenotypes. Plant Physiol 161(1):225–239

Chen F, Hu Y, Vannozzi A, Wu K, Cai H, Qin Y, Mullis A, Lin Z, Zhang L (2017a) The WRKY transcription factor family in model plants and crops. Crit Rev Plant Sci 36(5–6):311–335

Chen H, Shao H, Li K, Zhang D, Fan S, Li Y, Han M (2017b) Genome-wide identification, evolution, and expression analysis of GATA transcription factors in apple (*Malus* × *domestica* Borkh.). Gene 627:460–472

Chen K, Song M, Guo Y, Liu L, Xue H, Dai H, Zhang Z (2019) MdMYB46 could enhance salt and osmotic stress tolerance in apple by directly activating stress-responsive signals. Plant Biotechnol J 17(12):2341–2355

Daccord N, Celton J-M, Linsmith G, Becker C, Choisne N, Schijlen E, Van de Geest H, Bianco L, Micheletti D, Velasco R (2017) High-quality de novo assembly of the apple genome and methylome dynamics of early fruit development. Nat Genet 49 (7):1099–1106

Dash M, Malladi A (2012) The *AINTEGUMENTA* genes, *MdANT1* and *MdANT2*, are associated with the regulation of cell production during fruit growth in apple (*Malus* × *domestica* Borkh.). BMC Plant Biol 12(1):98. https://doi.org/10.1186/1471-2229-12-98

Davuluri RV, Sun H, Palaniswamy SK, Matthews N, Molina C, Kurtz M, Grotewold E (2003) AGRIS:

Arabidopsis gene regulatory information server, an information resource of *Arabidopsis* cis-regulatory elements and transcription factors. BMC Bioinformatics 4(1):25. https://doi.org/10.1186/1471-2105-4-25

Dong Y-H, Yao J-L, Atkinson RG, Putterill JJ, Morris BA, Gardner RC (2000) *MDH1*: an apple homeobox gene belonging to the BEL1 family. Plant Mol Biol 42(4):623–633

Duan N, Bai Y, Sun H, Wang N, Ma Y, Li M, Wang X, Jiao C, Legall N, Mao L (2017) Genome re-sequencing reveals the history of apple and supports a two-stage model for fruit enlargement. Nat Commun 8(1):249. https://doi.org/10.1038/s41467-017-00336-7

Dubos C, Stracke R, Grotewold E, Weisshaar B, Martin C, Lepiniec L (2010) MYB transcription factors in *Arabidopsis*. Trends Plant Sci 15(10):573–581

Espley RV, Brendolise C, Chagné D, Kutty-Amma S, Green S, Volz R, Putterill J, Schouten HJ, Gardiner SE, Hellens RP (2009) Multiple repeats of a promoter segment causes transcription factor autoregulation in red apples. Plant Cell 21(1):168–183

Eulgem T, Rushton PJ, Robatzek S, Somssich IE (2000) The WRKY superfamily of plant transcription factors. Trends Plant Sci 5(5):199–206

Fang H, Dong Y, Yue X, Chen X, He N, Hu J, Jiang S, Xu H, Wang Y, Su M (2019a) MdCOL4 interaction mediates crosstalk between UV-B and high temperature to control fruit coloration in apple. Plant Cell Physiol 60(5):1055–1066

Fang H, Dong Y, Yue X, Hu J, Jiang S, Xu H, Wang Y, Su M, Zhang J, Zhang Z (2019b) The B-box zinc finger protein MdBBX20 integrates anthocyanin accumulation in response to ultraviolet radiation and low temperature. Plant Cell Environ 42(7):2090–2104

Fan H, Wang F, Gao H, Wang L, Xu J, Zhao Z (2011) Pathogen-induced MdWRKY1 in 'Qinguan' apple enhances disease resistance. J Plant Biol 54(3):150–158

Feng X-M, Zhao Q, Zhao L-L, Qiao Y, Xie X-B, Li H-F, Yao Y-X, You C-X, Hao Y-J (2012) The cold-induced basic helix-loop-helix transcription factor gene *MdCIbHLH1* encodes an ICE-like protein in apple. BMC Plant Biol 12(1):22. https://doi.org/10.1186/1471-2229-12-22

Flachowsky H, Hättasch C, Höfer M, Peil A, Hanke M-V (2010) Overexpression of *LEAFY* in apple leads to a columnar phenotype with shorter internodes. Planta 231(2):251–263

Gangappa SN, Botto JF (2014) The BBX family of plant transcription factors. Trends Plant Sci 19(7):460–470

Gao J-J, Zhang Z, Peng R-H, Xiong A-S, Xu J, Zhu B, Yao Q-H (2011) Forced expression of *Mdmyb10*, a myb transcription factor gene from apple, enhances tolerance to osmotic stress in transgenic *Arabidopsis*. Mol Biol Rep 38(1):205–211

Girardi CL, Rombaldi CV, Dal Cero J, Nobile PM, Laurens F, Bouzayen M, Quecini V (2013) Genome-wide analysis of the AP2/ERF superfamily in apple and transcriptional evidence of ERF involvement in scab pathogenesis. Sci Hortic 151:112–121

Guilfoyle TJ, Ulmasov T, Hagen G (1998) The ARF family of transcription factors and their role in plant hormone-responsive transcription. Cell Mol Life Sci 54(7):619–627

Gu K-D, Wang C-K, Hu D-G, Hao Y-J (2019) How do anthocyanins paint our horticultural products? Sci Hortic 249:257–262

Han Z, Hu Y, Lv Y, Rose JK, Sun Y, Shen F, Wang Y, Zhang X, Xu X, Wu T (2018) Natural variation underlies differences in ETHYLENE RESPONSE FACTOR17 activity in fruit peel degreening. Plant Physiol 176(3):2292–2304

Henschel K, Kofuji R, Hasebe M, Saedler H, Münster T, Theissen G (2002) Two ancient classes of MIKC-type MADS-box genes are present in the moss *Physcomitrella patens*. Mol Biol Evol 19(6):801–814

Honda C, Moriya S (2018) Anthocyanin biosynthesis in apple fruit. Hortic J 87(3):305–314

Hu JC, Sauer R (1992) The basic-region leucine-zipper family of DNA binding proteins. Nucleic acids and molecular biology. Springer, New York, NY, pp 82–101

Hu D-G, Sun C-H, Ma Q-J, You C-X, Cheng L, Hao Y-J (2016a) MdMYB1 regulates anthocyanin and malate accumulation by directly facilitating their transport into vacuoles in apples. Plant Physiol 170(3):1315–1330

Hu D-G, Sun C-H, Zhang Q-Y, An J-P, You C-X, Hao Y-J (2016b) Glucose sensor MdHXK1 phosphorylates and stabilizes MdbHLH3 to promote anthocyanin biosynthesis in apple. PLoS Genet 12(8):e1006273

Hu D-G, Li Y-Y, Zhang Q-Y, Li M, Sun C-H, Yu J-Q, Hao Y-J (2017) The R2R3-MYB transcription factor MdMYB73 is involved in malate accumulation and vacuolar acidification in apple. Plant J 91(3):443–454

Hu D-G, Yu J-Q, Han P-L, Xie X-B, Sun C-H, Zhang Q-Y, Wang J-H, Hao Y-J (2019) The regulatory module MdPUB29-MdbHLH3 connects ethylene biosynthesis with fruit quality in apple. New Phytol 221(4):1966–1982

Ireland HS, Yao J-L, Tomes S, Sutherland PW, Nieuwenhuizen N, Gunaseelan K, Winz RA, David KM, Schaffer RJ (2013) Apple *SEPALLATA1/2*-like genes control fruit flesh development and ripening. Plant J 73(6):1044–1056

Jia D, Gong X, Li M, Li C, Sun T, Ma F (2018a) Overexpression of a novel apple NAC transcription factor gene, *MdNAC1*, confers the dwarf phenotype in transgenic apple (*Malus domestica*). Genes 9(5):229

Jia D, Shen F, Wang Y, Wu T, Xu X, Zhang X, Han Z (2018b) Apple fruit acidity is genetically diversified by natural variations in three hierarchical epistatic genes: *MdSAUR37*, *MdPP2CH* and *MdALMTII*. Plant J 95(3):427–443

Jia D, Jiang Q, van Nocker S, Gong X, Ma F (2019) An apple (*Malus domestica*) NAC transcription factor enhances drought tolerance in transgenic apple plants. Plant Physiol Biochem 139:504–512

Jin J, Tian F, Yang D-C, Meng Y-Q, Kong L, Luo J, Gao G (2017) PlantTFDB 4.0: toward a central hub for transcription factors and regulatory interactions in plants. Nucleic Acids Res 45(D1):D1040-D1045. https://doi.org/10.1093/nar/gkw982

Kavas M, Baloğlu MC, Atabay ES, Ziplar UT, Daşgan HY, Ünver T (2016) Genome-wide characterization and expression analysis of common bean bHLH transcription factors in response to excess salt concentration. Mol Genet Genomics 291(1):129–143

Kotoda N, Wada M, Kusaba S, Kano-Murakami Y, Masuda T, Soejima J (2002) Overexpression of *MdMADS5*, an APETALA1-like gene of apple, causes early flowering in transgenic *Arabidopsis*. Plant Sci 162(5):679–687

Kumar G, Arya P, Gupta K, Randhawa V, Acharya V, Singh AK (2016) Comparative phylogenetic analysis and transcriptional profiling of MADS-box gene family identified DAM and FLC-like genes in apple (*Malus* × *domestica*). Sci Rep 6(1):1–13

Lashbrooke J, Aharoni A, Costa F (2015) Genome investigation suggests *MdSHN3*, an APETALA2-domain transcription factor gene, to be a positive regulator of apple fruit cuticle formation and an inhibitor of russet development. J Exp Bot 66 (21):6579–6589

Legay S, Guerriero G, André C, Guignard C, Cocco E, Charton S, Boutry M, Rowland O, Hausman JF (2016) MdMyb93 is a regulator of suberin deposition in russeted apple fruit skins. New Phytol 212(4):977–991

Liao X, Guo X, Wang Q, Wang Y, Zhao D, Yao L, Wang S, Liu G, Li T (2017) Overexpression of *MsDREB6.2* results in cytokinin-deficient developmental phenotypes and enhances drought tolerance in transgenic apple plants. Plant J 89(3):510–526

Liu X, Li R, Dai Y, Chen X, Wang X (2018) Genome-wide identification and expression analysis of the B-box gene family in the apple (*Malus domestica* Borkh.) genome. Mol Genet Genomics 293(2):303–315

Liu W, Wang Y, Sun J, Jiang H, Xu H, Wang N, Jiang S, Fang H, Zhang Z, Wang Y-L (2019a) *MdMYBDL1* employed by *MdHY5* increases anthocyanin accumulation via repression of *MdMYB16/308* in apple. Plant Sci 283:32–40. https://doi.org/10.1016/j.plantsci.2019.01.016

Liu W, Wang Y, Yu L, Jiang H, Guo Z, Xu H, Jiang S, Fang H, Zhang J, Su M (2019b) MdWRKY11 participates in anthocyanin accumulation in red-fleshed apples by affecting MYB transcription factors and the photoresponse factor MdHY5. J Agric Food Chem 67(32):8783–8793

Liu X, Li R, Dai Y, Yuan L, Sun Q, Zhang S, Wang X (2019c) A B-box zinc finger protein, MdBBX10, enhanced salt and drought stresses tolerance in *Arabidopsis*. Plant Mol Biol 99(4–5):437–447

Li Y-Y, Mao K, Zhao C, Zhao X-Y, Zhang H-L, Shu H-R, Hao Y-J (2012) MdCOP1 ubiquitin E3 ligases interact with MdMYB1 to regulate light-induced anthocyanin biosynthesis and red fruit coloration in apple. Plant Physiol 160(2):1011–1022

Li J, Hou H, Li X, Xiang J, Yin X, Gao H, Zheng Y, Bassett CL, Wang X (2013) Genome-wide identification and analysis of the SBP-box family genes in apple (*Malus* × *domestica* Borkh.). Plant Physiol Biochem 70:100–114

Li T, Jiang Z, Zhang L, Tan D, Wei Y, Yuan H, Li T, Wang A (2016a) Apple (*Malus domestica*) MdERF2 negatively affects ethylene biosynthesis during fruit ripening by suppressing MdACS1 transcription. Plant J 88(5):735–748

Li Y-Y, Meng D, Li M, Cheng L (2016b) Genome-wide identification and expression analysis of the bZIP gene family in apple (*Malus domestica*). Tree Genet Genomes 12(4):82. https://doi.org/10.1007/s11295-016-1043-6

Li H-H, Liu X, An J-P, Hao Y-J, Wang X-F, You C-X (2017a) Cloning and elucidation of the functional role of apple MdLBD13 in anthocyanin biosynthesis and nitrate assimilation. Plant Cell Tiss Org Cult (PCTOC) 130(1):47–59

Li T, Xu Y, Zhang L, Ji Y, Tan D, Yuan H, Wang A (2017b) The jasmonate-activated transcription factor MdMYC2 regulates ETHYLENE RESPONSE FACTOR and ethylene biosynthetic genes to promote ethylene biosynthesis during apple fruit ripening. Plant Cell 29(6):1316–1334

Li X, Xie Y, Lu L, Yan M, Fang N, Xu J, Wang L, Yan Y, Zhao T, van Nocker S (2019) Contribution of methylation regulation of MpDREB2A promoter to drought resistance of *Mauls prunifolia*. Plant Soil 441:15–32. https://doi.org/10.1007/s11104-019-04149-z

Li C, Meng D, Pineros MA, Mao Y, Dandekar AM, Cheng L (2020) A sugar transporter takes up both hexose and sucrose for sorbitol-modulated in vitro pollen tube growth in apple. Plant Cell 32(2):449–469

Lui S, Luo C, Zhu L, Sha R, Qu S, Cai B, Wang S (2017) Identification and expression analysis of WRKY transcription factor genes in response to fungal pathogen and hormone treatments in apple (*Malus domestica*). J Plant Biol 60(2):215–230

Luo X-C, Sun M-H, Xu R-R, Shu H-R, Wang J-W, Zhang S-Z (2014) Genomewide identification and expression analysis of the ARF gene family in apple. J Genet 93 (3):785–797

Mao K, Dong Q, Li C, Liu C, Ma F (2017) Genome wide identification and characterization of apple bHLH transcription factors and expression analysis in response to drought and salt stress. Front Plant Sci 8:480. https://doi.org/10.3389/fpls.2017.00480

Massari ME, Murre C (2000) Helix-loop-helix proteins: regulators of transcription in eucaryotic organisms. Mol Cell Biol 20(2):429–440

Ma Q-J, Sun M-H, Lu J, Liu Y-J, Hu D-G, Hao Y-J (2017) Transcription factor AREB2 is involved in soluble sugar accumulation by activating sugar transporter and amylase genes. Plant Physiol 174(4):2348–2362

Meng D, Li Y, Bai Y, Li M, Cheng L (2016) Genome-wide identification and characterization of WRKY transcriptional factor family in apple and analysis of their responses to waterlogging and drought stress. Plant Physiol Biochem 103:71–83

Meng D, He M, Bai Y, Xu H, Dandekar AM, Fei Z, Cheng L (2018a) Decreased sorbitol synthesis leads to abnormal stamen development and reduced pollen tube growth via an MYB transcription factor, MdMYB39L, in apple (*Malus domestica*). New Phytol 217(2):641–656

Meng D, Li C, Park H-J, González J, Wang J, Dandekar AM, Turgeon BG, Cheng L (2018b) Sorbitol modulates resistance to *Alternaria alternata* by regulating the expression of an NLR resistance gene in apple. Plant Cell 30(7):1562–1581

Nakano T, Suzuki K, Fujimura T, Shinshi H (2006) Genome-wide analysis of the ERF gene family in *Arabidopsis* and rice. Plant Physiol 140(2):411–432

Olsen AN, Ernst HA, Leggio LL, Skriver K (2005) NAC transcription factors: structurally distinct, functionally diverse. Trends Plant Sci 10(2):79–87

Orenstein Y, Shamir R (2016) Modeling protein-DNA binding via high-throughput in vitro technologies. Brief Funct Genomics 16(3):171–180

Osakabe Y, Liang Z, Ren C, Nishitani C, Osakabe K, Wada M, Komori S, Malnoy M, Velasco R, Poli M (2018) CRISPR-Cas9-mediated genome editing in apple and grapevine. Nat Protoc 13(12):2844–2863

Peace CP, Bianco L, Troggio M, Van de Weg E, Howard NP, Cornille A, Durel C-E, Myles S, Migicovsky Z, Schaffer RJ (2019) Apple whole genome sequences: recent advances and new prospects. Hortic Res 6(1):1–24

Phukan UJ, Jeena GS, Shukla RK (2016) WRKY transcription factors: molecular regulation and stress responses in plants. Front Plant Sci 7:760. https://doi.org/10.3389/fpls.2016.00760

Riechmann JL, Ratcliffe OJ (2000) A genomic perspective on plant transcription factors. Curr Opin Plant Biol 3(5):423–434

Sakuma Y, Liu Q, Dubouzet JG, Abe H, Shinozaki K, Yamaguchi-Shinozaki K (2002) DNA-binding specificity of the ERF/AP2 domain of *Arabidopsis* DREBs, transcription factors involved in dehydration-and cold-inducible gene expression. Biochem Biophys Res Commun 290(3):998–1009

Samad AF, Sajad M, Nazaruddin N, Fauzi IA, Murad A, Zainal Z, Ismail I (2017) MicroRNA and transcription factor: key players in plant regulatory network. Front Plant Sci 8:565. https://doi.org/10.3389/fpls.2017.00565

Sharma V, Goel P, Kumar S, Singh AK (2019) An apple transcription factor, MdDREB76, confers salt and drought tolerance in transgenic tobacco by activating the expression of stress-responsive genes. Plant Cell Rep 38(2):221–241

Sung S-K, Yu G-H, An G (1999) Characterization of MdMADS2, a member of the SQUAMOSA subfamily of genes, in apple. Plant Physiol 120(4):969–978

Sun Q, Jiang S, Zhang T, Xu H, Fang H, Zhang J, Su M, Wang Y, Zhang Z, Wang N (2019) Apple NAC transcription factor MdNAC52 regulates biosynthesis of anthocyanin and proanthocyanidin through MdMYB9 and MdMYB11. Plant Sci 289:110286. https://doi.org/10.1016/j.plantsci.2019.110286

Su H, Zhang S, Yuan X, Chen C, Wang X-F, Hao Y-J (2013) Genome-wide analysis and identification of stress-responsive genes of the NAM-ATAF1,2-CUC2 transcription factor family in apple. Plant Physiol Biochem 71:11–21

Tian Y, Dong Q, Ji Z, Chi F, Cong P, Zhou Z (2015) Genome-wide identification and analysis of the MADS-box gene family in apple. Gene 555(2):277–290

Tiwari SB, Hagen G, Guilfoyle T (2003) The roles of auxin response factor domains in auxin-responsive transcription. Plant Cell 15(2):533–543

Toledo-Ortiz G, Huq E, Quail PH (2003) The *Arabidopsis* basic/helix-loop-helix transcription factor family. Plant Cell 15(8):1749–1770

Velasco R, Zharkikh A, Affourtit J, Dhingra A, Cestaro A, Kalyanaraman A, Fontana P, Bhatnagar SK, Troggio M, Pruss D, Dhingra, A, Cestaro A, Kalyanaraman A, Fontana P, Bhatnagar SK, Troggio M, Pruss D, Salvi S, Pindo M, Baldi P, Castelletti S, Cavaiuolo M, Coppola G, Costa F, Cova V, Ri AD, Goremykin V, Komjanc M, Longhi S, Magnago P, Malacarne G, Malnoy M, Micheletti D, Moretto M, Perazzolli M, Si-Ammour A, Vezzulli S, Zini E, Eldredge G, Fitzgerald LM, Gutin N, Lanchbury G, Macalma T, Mitchell JT, Reid J, Wardell B, Kodira C, Chen Z, Desany B, Niazi F, Palmer M, Koepke T, Jiwan D, Schaeffer S, Krishnan V, Wu C, Chu VT, King ST, Vick J, Tao Q, Mraz A, Stormo A, Stormo K, Bogden R, Ederle D, Stella A, Vecchietti A, Kater MM, Masiero S, Lasserre P, Lespinasse Y, Allan AC, Bus V, Chagné D, Crowhurst RN, Gleave AP, Lavezzo E, Fawcett JA, Proost S, Rouzé P, Sterck L, Toppo S, Lazzari B, Hellens RP, Durel C-E, Gutin A, Bumgarner R, Gardiner SE, Skolnick M, Egholm M, Van de Peer Y, Salamini F, Viola R (2010) The genome of the domesticated apple (*Malus* × *domestica* Borkh.). Nat Genet 42:833–839. https://doiorg/10.1038/ng.654

Vimolmangkang S, Han Y, Wei G, Korban SS (2013) An apple MYB transcription factor, MdMYB3, is involved in regulation of anthocyanin biosynthesis and flower development. BMC Plant Biol 13(1):176. https://doi.org/10.1186/1471-2229-13-176

Wang X, Zhang S, Su L, Liu X, Hao Y (2013) A genome-wide analysis of the LBD (LATERAL ORGAN BOUNDARIES domain) gene family in *Malus domestica* with a functional characterization of *MdLBD11*. PLoS ONE 8(2):e57044. https://doi.org/10.1371/journal.pone.0057044

Wang R-K, Cao Z-H, Hao Y-J (2014) Overexpression of a R2R3 MYB gene MdSIMYB1 increases tolerance to multiple stresses in transgenic tobacco and apples. Physiol Plant 150(1):76–87

Wang A, Tan D, Takahashi A, Zhong Li T, Harada T (2007) MdERFs, two ethylene-response factors involved in apple fruit ripening. J Exp Bot 58 (13):3743–3748

Wang N, Xu H, Jiang S, Zhang Z, Lu N, Qiu H, Qu C, Wang Y, Wu S, Chen X (2017) MYB12 and MYB22 play essential roles in proanthocyanidin and flavonol synthesis in red-fleshed apple (*Malus sieversii f. niedzwetzkyana*). Plant J 90(2):276–292

Wang F-P, Wang X-F, Zhang J, Ma F, Hao Y-J (2018a) MdMYB58 modulates Fe homeostasis by directly binding to the MdMATE43 promoter in plants. Plant Cell Physiol 59(12):2476–2489

Wang N, Qu C, Jiang S, Chen Z, Xu H, Fang H, Su M, Zhang J, Wang Y, Liu W, Zhang Z, Lu N, Chen X (2018b) The proanthocyanidin-specific transcription factor MdMYBPA1 initiates anthocyanin synthesis under low-temperature conditions in red-fleshed apples. Plant J 96(1):39–55

Wang X-F, An J-P, Liu X, Su L, You C-X, Hao Y-J (2018c) The nitrate-responsive protein MdBT2 regulates anthocyanin biosynthesis by interacting with the MdMYB1 transcription factor. Plant Physiol 178 (2):890–906

Wang Y, Wang N, Xu H, Jiang S, Fang H, Zhang T, Su M, Xu L, Zhang Z, Chen X (2018d) Nitrogen affects anthocyanin biosynthesis by regulating MdLOB52 downstream of MdARF19 in callus cultures of red-fleshed apple (*Malus sieversii f. niedzwetzkyana*). J Plant Growth Regul 37(3):719–729

Wang Y, Liu W, Jiang H, Mao Z, Wang N, Jiang S, Xu H, Yang G, Zhang Z, Chen X (2019a) The R2R3-MYB transcription factor MdMYB24-like is involved in methyl jasmonate-induced anthocyanin biosynthesis in apple. Plant Physiol Biochem 139:273–282

Wang Z, Li J, Mao Y, Zhang M, Wang R, Hu Y, Mao Z, Shen X (2019b) Transcriptional regulation of *MdPIN3* and *MdPIN10* by MdFLP during apple self-rooted stock adventitious root gravitropism. BMC Plant Biol 19(1):229. https://doi.org/10.1186/s12870-019-1847-2

Wang Y, Sun J, Wang N, Xu H, Qu C, Jiang S, Fang H, Su M, Zhang Z, Chen X (2019c) MdMYBL2 helps regulate cytokinin-induced anthocyanin biosynthesis in red-fleshed apple (*Malus sieversii f. niedzwetzkyana*) callus. Funct Plant Biol 46(2):187–196

Wang J-H, Gu K-D, Han P-L, Yu J-Q, Wang C-K, Zhang Q-Y, You C-X, Hu D-G, Hao Y-J (2020) Apple ethylene response factor MdERF11 confers resistance to fungal pathogen *Botryosphaeria dothidea*. Plant Sci 291:110351

Wu R, Tomes S, Karunairetnam S, Tustin SD, Hellens RP, Allan AC, Macknight RC, Varkonyi-Gasic E (2017a) *SVP*-like MADS box genes control dormancy and budbreak in apple. Front Plant Sci 8:477. https://doi.org/10.3389/fpls.2017.00477

Wu R, Wang Y, Wu T, Xu X, Han Z (2017b) MdMYB4, an R2R3-Type MYB transcription factor, plays a crucial role in cold and salt stress in apple calli. J Am Soc Hortic Sci 142(3):209–216

Xie X-B, Li S, Zhang R-F, Zhao J, Chen Y-C, Zhao Q, Yao Y-X, You C-X, Zhang X-S, Hao Y-J (2012) The bHLH transcription factor MdbHLH3 promotes anthocyanin accumulation and fruit colouration in response to low temperature in apples. Plant Cell Environ 35 (11):1884–1897

Xie Y, Chen P, Yan Y, Bao C, Li X, Wang L, Shen X, Li H, Liu X, Niu C (2018) An atypical R2R3 MYB transcription factor increases cold hardiness by CBF-dependent and CBF-independent pathways in apple. New Phytol 218(1):201–218

Xu R, Sun P, Jia F, Lu L, Li Y, Zhang S, Huang J (2014) Genomewide analysis of TCP transcription factor gene family in *Malus domestica*. J Genet 93(3):733–746

Xu H, Wang N, Liu J, Qu C, Wang Y, Jiang S, Lu N, Wang D, Zhang Z, Chen X (2017) The molecular mechanism underlying anthocyanin metabolism in apple using the *MdMYB16* and *MdbHLH33* genes. Plant Mol Biol 94(1–2):149–165

Xu H, Wang N, Wang Y, Jiang S, Fang H, Zhang J, Su M, Zuo W, Xu L, Zhang Z (2018a) Overexpression of the transcription factor MdbHLH33 increases cold tolerance of transgenic apple callus. Plant Cell Tiss Org Cult (PCTOC) 134(1):131–140

Xu H, Yang G, Zhang J, Wang Y, Zhang T, Wang N, Jiang S, Zhang Z, Chen X (2018b) Overexpression of a repressor MdMYB15L negatively regulates anthocyanin and cold tolerance in red-fleshed callus. Biochem Biophys Res Commun 500(2):405–410

Yang W, Liu X-D, Chi X-J, Wu C-A, Li Y-Z, Song L-L, Liu X-M, Wang Y-F, Wang F-W, Zhang C (2011) Dwarf apple MbDREB1 enhances plant tolerance to low temperature, drought, and salt stress via both ABA-dependent and ABA-independent pathways. Planta 233(2):219–229

Yang J, Gao M, Huang L, Wang Y, van Nocker S, Wan R, Guo C, Wang X, Gao H (2017) Identification and expression analysis of the apple (*Malus* × *domestica*) basic helix-loop-helix transcription factor family. Sci Rep 7(1):1–14

Yang Q, Chen Q, Zhu Y, Li T (2018) Identification of *MdDof* genes in apple and analysis of their response to biotic or abiotic stress. Funct Plant Biol 45(5):528–541

Yang Y-Y, Ren Y-R, Zheng P-F, Qu F-J, Song L-Q, You C-X, Wang X-F, Hao Y-J (2020) Functional identification of apple *MdMYB2* gene in phosphate-starvation response. J Plant Physiol 244:153089

Yao J-L, Dong Y-H, Morris BA (2001) Parthenocarpic apple fruit production conferred by transposon insertion mutations in a MADS-box transcription factor. Proc Natl Acad Sci 98(3):1306–1311

Yauk YK, Souleyre EJ, Matich AJ, Chen X, Wang MY, Plunkett B, Dare AP, Espley RV, Tomes S, Chagné D (2017) Alcohol acyl transferase 1 links two distinct volatile pathways that produce esters and phenylpropenes in apple fruit. Plant J 91(2):292–305

Yin L, Wang Y, Yuan M, Zhang X, Xu X, Han Z (2014) Characterization of MxFIT, an iron deficiency induced transcriptional factor in *Malus xiaojinensis*. Plant Physiol Biochem 75:89–95

Yue P, Lu Q, Liu Z, Lv T, Li X, Bu H, Liu W, Xu Y, Yuan H, Wang A (2020) Auxin-activated MdARF5 induces the expression of ethylene biosynthetic genes to initiate apple fruit ripening. New Phytol 226 (6):1781

Zhang Q, Li Y, Zhang Y, Wu C, Wang S, Hao L, Wang S, Li T (2017) Md-miR156ab and Md-miR395 target WRKY transcription factors to influence apple resistance to leaf spot disease. Front Plant Sci 8:526. https://doi.org/10.3389/fpls.2017.00526

Zhang J, Xu H, Wang N, Jiang S, Fang H, Zhang Z, Yang G, Wang Y, Su M, Xu L (2018a) The ethylene response factor MdERF1B regulates anthocyanin and proanthocyanidin biosynthesis in apple. Plant Mol Biol 98(3):205–218

Zhang Z, Yuan L, Liu X, Chen X, Wang X (2018b) Evolution analysis of Dof transcription factor family and their expression in response to multiple abiotic stresses in *Malus domestica*. Gene 639:137–148

Zhang L, Hu J, Han X, Li J, Gao Y, Richards CM, Zhang C, Tian Y, Liu G, Gul H (2019) A high-quality apple genome assembly reveals the association of a retrotransposon and red fruit colour. Nat Commun 10 (1):1494. https://doi.org/10.1038/s41467-019-09518-x

Zhang Q-Y, Gu K-D, Cheng L, Wang J-H, Yu J-Q, Wang X-F, You C-X, Hu D-G, Hao Y-J (2020) BTB-TAZ domain protein MdBT2 modulates malate accumulation and vacuolar acidification in response to nitrate. Plant Physiol 183(2):750–764

Zhao T, Liang D, Wang P, Liu J, Ma F (2012) Genome-wide analysis and expression profiling of the DREB transcription factor gene family in *Malus* under abiotic stress. Mol Genet Genomics 287(5):423–436

Zhao Q, Ren Y-R, Wang Q-J, Yao Y-X, You C-X, Hao Y-J (2016) Overexpression of *MdbHLH104* gene enhances the tolerance to iron deficiency in apple. Plant Biotechnol J 14(7):1633–1645

Zhao X-Y, Qi C-H, Jiang H, Zhong M-S, You C-X, Li Y-Y, Hao Y-J (2019a) MdHIR4 transcription and translation levels associated with disease in apple are regulated by MdWRKY31. Plant Mol Biol 101(1–2):149–162

Zhao X-Y, Qi C-H, Jiang H, Zhong M-S, Zhao Q, You C-X, Li Y-Y, Hao Y-J (2019b) MdWRKY46-enhanced apple resistance to *Botryosphaeria dothidea* by activating the expression of MdPBS3.1 in the salicylic acid signaling pathway. Mol Plant Microbe Interact 32 (10):1391–1401

Zhao X-Y, Qi C-H, Jiang H, Zhong M-S, You C-X, Li Y-Y, Hao Y-J (2020) MdWRKY15 improves resistance of apple to *Botryosphaeria dothidea* via the salicylic acid-mediated pathway by directly binding the MdICS1 promoter. J Integr Plant Biol 62(4):527–543

Zheng L, Ma J, Song C, Zhang L, Gao C, Zhang D, An N, Mao J, Han M (2018a) Genome-wide identification and expression analysis of *GRF* genes regulating apple tree architecture. Tree Genet Genomes 14(4):54. https://doi.org/10.1007/s11295-018-1267-8

Zheng X, Zhao Y, Shan D, Shi K, Wang L, Li Q, Wang N, Zhou J, Yao J, Xue Y (2018b) *MdWRKY9* overexpression confers intensive dwarfing in the M26 rootstock of apple by directly inhibiting brassinosteroid synthetase *MdDWF4* expression. New Phytol 217(3):1086–1098

Zhou L-J, Li Y-Y, Zhang R-F, Zhang C-L, Xie X-B, Zhao C, Hao Y-J (2017a) The small ubiquitin-like modifier E3 ligase MdSIZ1 promotes anthocyanin accumulation by sumoylating MdMYB1 under low-temperature conditions in apple. Plant Cell Environ 40 (10):2068–2080

Zhou L-J, Mao K, Qiao Y, Jiang H, Li Y-Y, Hao Y-J (2017b) Functional identification of MdPIF1 as a phytochrome interacting factor in apple. Plant Physiol Biochem 119:178–188

Genome Editing in Apple

10

Chikako Nishitani, Keishi Osakabe, and
Yuriko Osakabe

Abstract

Genome editing with artificially engineered nucleases is an advanced molecular technology for pursuing precise and effective genetic engineering. In this technology, engineered nucleases induce DNA double-strand breaks at targeted sites in a genome, stimulating the DNA repair system in cells, thus enabling site-directed mutagenesis. Genome editing using CRISPR (clustered regularly interspaced short palindromic repeats)/CRISPR-associated protein9 (Cas9), originating from a defence system of prokaryotes, is a powerful technology that is now being widely utilized in molecular research studies, as well as in breeding programmes of various plant species, including fruit trees, to impart either novel or enhanced traits to established commercial cultivars or to new cultivars/genotypes. Recently, several reports have demonstrated successful apple genome editing and the introduction of important traits, such as those for early flowering and reduced fire blight susceptibility, to popular commercial cultivars, such as 'Gala' and 'Golden Delicious'. It is important to point out that these reports reveal that such genome-edited/mutant apple plants or cell lines do not carry foreign genes. Nevertheless, during the process of precise genome editing, the coexistence of various types of mutations referred to as "mosaic mutations" and off-target effects are major concerns. Therefore, to minimize such effects, selection of target sequences and estimation of off-target effects for CRISPR/Cas9 has been developed for many organisms, and these have also been employed for apple by using in silico analysis based on genome information. On the other hand, apple genome heterozygosity has led to difficulties in genome editing, as the complex genome of apple precludes the use of some of these basic techniques for genome editing. Therefore, further studies focused on genome information and culture techniques tailored for apple are needed. It will be highly critical for each apple cultivar in developing precise and efficient genome editing for apple. This chapter will provide an overview of current studies of genome editing in apple and will discern and explore how this strategy will provide insights into molecular breeding technologies for genetic improvement of the apple.

C. Nishitani
NARO Institute of Fruit Tree and Tea Science, 2-1 Fujimoto, Tsukuba 305-8605, Ibaraki, Japan

K. Osakabe
Faculty of Bioscience and Bioindustry, Fujii Memorial Institute of Medical Sciences, Tokushima University, 3-18-15, Kuramoto-cho, Tokushima 770-8503, Japan

Y. Osakabe (✉)
School of Life Science and Technology, Tokyo Institute of Technology, 4259 Nagatsuda-cho, Midori-ku, Yokohama 226-8502, Japan
e-mail: osakabe.y.ab@m.titech.ac.jp

S. S. Korban (ed.), *The Apple Genome*, Compendium of Plant Genomes,
https://doi.org/10.1007/978-3-030-74682-7_10

10.1 Introduction

Genome editing technologies using artificially engineered nucleases are effective for site-directed mutagenesis in various organisms (Osakabe and Osakabe 2015). These technologies have introduced opportunities for precise and rapid genetic improvement of various organisms, including plants (Limera et al. 2017). CRISPR (clustered regularly interspaced short palindromic repeats)/CRISPR-associated protein9 (Cas9), a defence system of prokaryotes, has been utilized as an efficient and easy-to-use genome-editing tool that is now widely applied to various crop plants (Kamburova et al. 2017).

In fruit trees, the first successful application of genome editing targeting an endogenous gene by CRISPR/Cas9 was reported in apple (Nishitani et al. 2016). Since then, additional studies of apple genome editing by CRISPR/Cas9 have been reported (Malnoy et al. 2016; Charrier et al. 2019; Pompili et al. 2020). Furthermore, genome editing studies in other fruit trees, such as grape (Ren et al. 2016; Nakajima et al. 2017), citrus (Peng et al. 2017), banana (Kaur et al. 2018), kiwi (Wang et al. 2018), European pear (Charrier et al. 2019), and blueberry (Ohmori et al. 2020) have also been reported. All these reports confirm the applicability and usefulness of genome editing for fruit trees. In this chapter, we will first cover the fundamental basis of genome editing, its advantages, and associated technical issues of importance. Subsequently, we will review progress and technological advances in pursuing apple genome editing and then discuss future opportunities and challenges of genome editing for apple and other fruit tree crops.

10.2 Genome Editing and Its Advantages

10.2.1 Genome Editing Tools and Mechanisms

In general, genome editing involves the following outlined molecular and biological processes. First, an artificial nuclease is designed and optimized for target DNA recognition and digestion. Following the delivery of this nuclease into cells, digestion of the double-stranded DNA occurs at the target site, that is, DNA double-strand breaks (DSBs). Subsequently, this is followed by the repair of these digested DNAs via endogenous DNA repair pathways. There are two major DNA repair pathways in eukaryotes, non-homologous end joining (NHEJ) and homology-directed repair (HDR) (Osakabe and Osakabe 2015). NHEJ is an endogenous error-prone repair system that induces insertions and deletions (indels) at DSB sites, thereby generating altered sequences. Quite often, these mutations disrupt gene function. On the other hand, digested DNA can be precisely repaired using homologous DNA templates in the HDR pathway. Both NHEJ and HDR pathways have been demonstrated in various plant species, and studies have suggested that overall, the frequency of HDR repair is less than that of the NHEJ pathway (Bortesi and Fischer 2015; Steinert et al. 2016). This is because activity of the HDR pathway is likely to be limited to specific cell cycle stages, such as that of the S/G2 stage and to other more complex processes (Rozov et al. 2019). Moreover in homologous recombination (HR), the homologous chromosome can be used as a template (Rong and Golic 2003; Hayut et al. 2017). Recently, advanced HR technologies using genome editing have been reported (Čermák et al. 2015; Miki et al. 2018).

To date, three types of artificial nucleases have been developed as genome editing tools. These include a zinc finger nuclease (ZFN), transcription activator-like effector nuclease (TALEN), and CRISPR/Cas9. All these artificial nucleases consist of a DNA-recognition molecule along with a DNA cleavage molecule, and based on DNA-recognition molecules, they are classified into two types, protein-based and RNA-based genome editing tools (Osakabe and Osakabe 2015). ZFN and TALEN have protein-based DNA-recognition domains. The DNA-recognition domain of either ZFN or TALEN is fused to a restriction enzyme, *Fok*I nuclease, to cleave DNA at the target site. As *Fok*I functions as a dimer, the formation of a heterodimer of either ZFN or TALEN is necessary to induce DSB. On the other hand, CRISPR/Cas9 utilizes an RNA-based DNA-recognition system to form a ribonucleoprotein (RNP) complex of a Cas9 nuclease and a guide RNA (gRNA). CRISPR/Cas9 specifically binds the target DNA through RNA–DNA pairing via a 20-nt gRNA. Adjacent to gRNA target sequences, 5′-NGG-3′ sequences, referred to as protospacer adjacent motifs (PAMs), are necessary for recognition by SpCas9 (*Streptococcus pyogenes* Cas9). The gRNA can be engineered more easily than other protein-based tools, and due to its high efficiency and ease of design, CRISPR/Cas9 has been used widely in various organisms, including fruit trees.

10.2.2 Efficiency of Mutation, Mosaicism, and Off-Target Effects in Genome Editing

In general, the mutation efficiency depends on target sequences; therefore, the design of a gRNA target sequence is key for CRISPR/Cas9-mediated genome editing (Liu et al. 2016). Any insufficient or non-specific activity of genome editing tools can sometimes yield secondary consequences, such as mosaicism and off-target effects.

Mosaicism, that is, the incidence of various mutation patterns within an individual plant or line, is sometimes observed due to different mutation profiles from individual cells. In brief, its mechanism involves a non-simultaneous or stochastic DSB followed by repair in cells under different conditions (Mehravar et al. 2019). In order to obtain genome-edited plants with stable traits, mosaicism should be avoided. In traditional mutagenesis, several rounds of in vitro regeneration and propagation have been effectively utilized to achieve chimeral separation to get rid of mosaicism in tree plants (Abu-Qaoud et al. 1990; Spiegel-Roy 1990). Therefore, this strategy can also be utilized to minimize the incidence of mosaic mutations in genome editing.

It has been observed that off-target effects are influenced by gRNA specificity, as gRNA sequences used to target DNA can highly impact the efficiency of genome editing, and that mutations may sometimes occur in genes of similar sequences (Fu et al. 2014; Liu et al. 2016). To minimize off-target effects, *in silico* software programs for gRNA design using genomic information with reference sequences such as CasOT (Xiao et al. 2014), CRISPOR (Haeussler et al. 2016), and CRISPR-P (Liu et al. 2017) have been developed, and are widely used in various plant species. It has been reported that no clear off-target effects are detected in Arabidopsis, tomato, and apple when using in silico designed gRNAs (Osakabe et al. 2016; Pan et al. 2016; Pompili et al. 2020).

An alternative approach to reduce off-target effects would involve the use of truncated gRNAs. Truncated gRNAs can induce mutations in both mammalian and plant cells with low off-target effects (Fu et al. 2014; Osakabe et al. 2016). Successful genome editing with an 18-bp gRNA has been demonstrated in both apple and tomato (Nishitani et al. 2016; Ueta et al. 2017). As target specificity is a critical factor in the utilization of genome editing, fully exploiting the above-described technologies to minimize off-target effects is highly necessary.

10.2.3 Advantages of Accuracy in Genome Editing

Although genome editing can introduce mutations via molecular mechanisms similar to those of traditional methods for mutagenesis following the DSB process, it can induce mutagenesis specifically at a targeted genome position. Compared with other traditional methods, this observed precision and efficiency of mutagenesis using genome editing are deemed important features of genetic engineering for many crops. Furthermore, with genome editing, there is a low risk of generating unexpected traits due to mutations in other loci.

To enhance the specificity of genome editing, genome sequence information is essential for pursuing an experimental design. As genome sequence information has become available for many plant species, including that for apple (Velasco et al. 2010; Kunihisa et al. 2016; Daccord et al. 2017; Chen et al. 2019), target sequences for genome editing can be widely utilized. Further bioinformatic analysis to complete high-quality and comprehensive sequences will continue to contribute to successful genome editing in various crops. As fruit trees are widely heterogeneous with high ploidy levels, precise and efficient mutagenesis using genome editing is also applicable to specific gene modifications, despite the presence of sequences similar to those target genes that are frequently detected within a genome. In apple, it has been reported that a genome-wide duplication (GWD) event must have occurred in relatively recent times (over 50 million years ago) that has resulted in polyploidization of the genome size, from 9 ancestral chromosomes to 17 chromosomes (Velasco et al. 2010).

The complexity of the apple genome is one of the major issues in genome engineering of this crop. For example, there might be two loci from an originally single-copy gene in a diploid apple genome (Mimida et al. 2009); therefore, considering the presence of heterozygous alleles in the apple, this suggests that there are up to four homologous genes present in this genome. Hence, due to sequence similarities among these genes, it is rather difficult to conduct functional analysis for each of the homologous genes to discern their likely role(s). However, genome editing can be employed to determine how each of the genes within a gene family is involved in a trait of interest or a plant developmental/biological process. Recently, the CRISPR/Cas9-mediated disruption of a single gene in a gene family of apple has been reported (Pompili et al. 2020). However, as apple reference genome sequences are currently available for *Malus* × *domestica* cv. 'Golden Delicious' (Velasco et al. 2010; Daccord et al. 2017), *M.* × *domestica* cv. 'Hanfu' (Zhang et al. 2019), and *M. baccata* (Chen et al. 2019), further studies of genome sequences will enhance precision and efficiency of genome editing in apple.

10.2.4 Genome Editing Without Foreign DNA Insertion

Due to the nature of molecular mechanisms involved in genome editing, insertion of foreign DNA is not necessary, thereby rendering this technology as highly desirable. Although stable transformation via *Agrobacterium* is one of the most widely popular methods used for the delivery of artificial nucleases into plant cells, transient expression of artificial nucleases in plant cells is sufficient to induce mutations (Clasen et al. 2016). Moreover, elimination of engineered nuclease genes that are stably transformed into plant cells following genome editing has been reported (Pompili et al. 2020). Such transgene-free mutagenesis will enhance future applications when considering social acceptance and regulations of genome-edited cultivars based on the Cartagena Protocol for LMO (living modified organisms) (Gao 2019). Furthermore, recovery of null segregants offers yet an effective strategy for eliminating foreign genes. However, it is important to point out that the next generation of a genome-edited genotype is no longer deemed as the original cultivar for heterozygous plants, such as that of the apple.

Recent studies on the development of transgene-free genome editing have revealed the applicability of genome editing for genetic engineering of various crops (Gao 2019). Direct delivery of RNPs into plant cells has been developed in apple and grape (Malnoy et al. 2016); Arabidopsis, tobacco, lettuce, and rice (Woo et al. 2015); petunia (Subburaj et al. 2016); potato (Andersson et al. 2018); soybean and tobacco (Kim et al. 2017); maize (Svitashev et al. 2016); and wheat (Liang et al. 2017). Mutated lettuce, maize, and wheat plants have been regenerated using direct delivery of RNPs due to ease of redifferentiation from either protoplasts or callus (Woo et al. 2015; Svitashev et al. 2016; Liang et al. 2017). As long as mutated cells can grow into plantlets following RNP-mediated mutagenesis, this technology is deemed as one of the most robust methods to generate foreign DNA-free genome-edited plants.

10.3 Progress in Apple Genome Editing Technology

10.3.1 CRISPR/Cas9 Vector-Mediated Genome Editing

Currently, there are a few reports on genome editing (Table 10.1). Among these, CRISPR/Cas9-mediated genome editing targeting apple endogenous genes has been pursued (Malnoy et al. 2016; Nishitani et al. 2016; Charrier et al. 2019; Pompili et al. 2020). Within these studies, *Agrobacterium*-mediated transformation have been used to develop genome-edited plantlets (Nishitani et al. 2016; Charrier et al. 2019; Pompili et al. 2020). Whereas, Malnoy et al. (2016) have attempted RNP genome editing using a polyethylene glycol (PEG) gene transfer method, they have not been able to recover regenerated plantlets. In an earlier study, ZFN-mediated genome editing, targeting a reporter β-glucuronidase (GUS) gene, has been reported (Peer et al. 2015).

In our laboratory, an endogenous apple *phytoene desaturase* (*PDS*) gene, a key gene in carotenoid biosynthesis, has been targeted for gene editing (Nishitani et al. 2016). Disruption of the *PDS* gene results in an albino phenotype in plants. This allows for readily visual observation of such gene disruption; thus, the *PDS* gene is commonly used as a target gene for early attempts at plant genome editing. In this study, a CRISPR/Cas9 vector, Cas9 and a single gRNA have been used to target the *PDS* gene and to stably transfer it into the apple rootstock 'JM2' (Table 10.1) (Nishitani et al. 2016). This vector harbours a plant-codon-optimized *Cas9* fused to a selection marker gene coding for the green fluorescent protein, *GFP,* via a 2A peptide under the control of a 2× *35SCaMV* promoter, along with a gRNA expression cassette driven by an Arabidopsis *AtU6-1* promoter. Kanamycin is used as a selection marker for T-DNA transformation. Transformation efficiency of 2.5–4.5% is achieved using both kanamycin and PCR detection of the transgenes. Approximately 8 months following transformation, albino phenotypes have been observed in 13.6% of regenerated T0 transformants when using a gRNA having the highest activity (Fig. 10.1). Following DNA sequencing, biallelic mutations of the apple *PDS* gene have been detected, with low mosaic levels detected. It is important to point out that visible albino shoots developing from basal regions of green regenerated shoots have been observed around 3 months following the transfer of these shoot explants onto a propagation medium (Osakabe et al. 2018). Furthermore, the majority of mutations of small deletions detected in these mutated plantlets are presumably repaired via the NHEJ pathway. Interestingly, involvement of the HR pathway is also proposed in this study (Nishitani et al. 2016). Analysis of alleles distinguished by their sequence polymorphisms in biallelic mutations has demonstrated that a single minor mutant sequence in the two types of biallelic sequences in one of the alleles is identical to a major type of mutation in biallelic sequences of the second allele. This 'mirror' phenomenon is frequently observed in regenerated mutants, suggesting the presence of a likely copying mechanism to repair the target locus using the HR pathway. Together, these results have suggested that activation of the HR pathway is elicited without an exogenous template (Nishitani et al. 2016).

In another study, *PDS* gene disruption has been undertaken using stable transformation of apple cv. 'Gala' (Charrier et al. 2019). The CRISPR/Cas9 vector, wherein a *SpCas9* is driven by the *PcUbi4-2* promoter from parsley (*Petroselinum crispum*) and two gRNAs is driven by two apple promoters *MdU3* and *MdU6,* is used for multiplex genome editing, while kanamycin is used for the selection of mutated regenerants. Although simultaneous mutations of both targets are frequently detected, in this study, a long deletion, 272 bp, between these two targets is not observed. Moreover, repeated transfer of selected albino sectors to fresh proliferation

Table 10.1 Studies on apple genome editing

Nuclease	Genotype(s)	Target(s)	Software for target search	Reference genome	Expected phenotype	Cas9/ZFN promoter	gRNA promoter	Mutation efficiency** (efficiency of 100% somatic mutations)***	Off-target	Detection method	Reference
SpCas9	JM2	*MdPDS*	CasOT	*Malus* × *domestica* v1.0 genome	Albino	2X35S	AtU6-1	31.8% (42.8%)	n.d.****	Average of more than 40 amplicons were analyzed per line	Nishitani et al. (2016)
SpCas9	Gala	*MdPDS*	CRISPOR	Apple genome assembly GDDH13 v1.1	Albino	PcUbi4-2	MdU3, U6	92.60%	n.d.	A total of 3–8 amplicons sequenced per line	Charrier et al. (2019)
		MdTFL1.1			Early flowering			90%	One off-target site identical sequence to an on-target was mutated. No off-target effects in other two sites with three mismatches each	Based on the early flowering phenotype	
	Conference (pear)*							9%	n.d.		
SpCas9	Gala	*MdDIPM4*	CRISPOR	Apple genome assembly GDDH13 v1.1	Resistance to fire blight	AtUbi10	AtU6	77.8% (70.4%)	No off-target effects in five potential sites (four mismatches each)	Average of 3,000 read counts analyzed per line by MiSeq	Pompili et al. (2020)
	Golden Delicious							73.3% (70%)			
SpCas9	Golden Delicious	*MdDIPM-1*	CRISPR RGEN Tools		Resistance to fire blight	RNP infection	RNP infection	6.7%	n.d.	At least 17,000 read counts analyzed per experiment by MiSeq	Malnoy et al. (2016)
		MdDIPM-2						3.3%			
		MdDIPM-4						6.9%			
	Chardonnay (grape)*	*MLO-7*			Resistance to powdery mildew			0.10%			
ZFN	Galaxy (a sport of 'Gala')	*GUS*				Hsp18.2		10–40%	n.d.	GUS staining	Peer et al. 2015

*; Genome editing in other plant species were also estimated in the same study
**; Ratio of transformants with mutations or ratio of mutated sequences from infected protoplasts (Malnoy et al. 2016)
***; Ratio of transformants (100% somatic mutations including biallelic mutations) in edited transformants
****; Not cetermined

Fig. 10.1 Transformation of CRISPR/Cas9 vector into 'JM2' apple rootstock. **a** Leaflet explants incubated on a kanamycin-containing redifferentiation medium after 1-week of co-culture with *Agrobacterium tumefaciens* harbouring a CRISPR/Cas9 vector. **b** Redifferentiating explants following 2 to 3 months of incubation. **c** Transformant genome-edited shoots (albino phenotypes) detected along basal regions of regenerated shoots; bar = 1 cm

medium has reduced mosaic mutations in mutant shoots; 36 lines with mosaic mutations are obtained from a total of 41 transgenic lines (87.8%).

These observed differences in mutation efficiencies between the two studies may be attributed to gRNA activity, as well as to other factors, such as apple cultivars, culture methods, and vector systems (Nishitani et al. 2016; Charrier et al. 2019) (Table 10.1). Moreover, the observed reduction of incidence of mosaics by repeated transfer of shoots to fresh proliferation media is demonstrated by both Nishitani et al. (2016) and Charrier et al. (2019). Therefore, further optimization of the protocol and associated factors is required to develop highly efficient genome editing and recovery of low mosaic mutations in apple.

10.3.2 Genome Editing for Useful Traits

10.3.2.1 Early Flowering

Charrier et al. (2019) have reported on genome editing of the apple *Terminal Flower* (*TFL1.1*) gene using the same vector as that for *PDS*. The *TFL* gene is a floral repressor, and knockout plants exhibit accelerated flowering (Shannon and Meeks-Wanger 1991; Kotoda et al. 2006). Previously, it has been reported that two *TFL* genes, *TFL1.1* and *TFL1.2*, have been expressed in vegetative tissues of apple (Mimida et al. 2009). Recently, a CRISPR/Cas9 vector harbouring two gRNAs specifically targeting *TFL1.1* has been used to transform apple cv. 'Gala' (Charrier et al. 2019). Most of the CRISPR/Cas9 transgenic lines for *TFL1.1* flowered within a period of 12 months following the transfer of transformed lines onto a proliferation medium. Interestingly, flower buds are observed on terminal buds of shoots, and opened flowers have demonstrated the presence of all floral organs, often in irregular numbers. If these flowers are fertile, then null segregants could be rapidly generated.

It is important to point out that the production of early flowering phenotypes via genetic engineering is more difficult than that of producing albino phenotypes, as proliferation efficiency is lower during culture due to cessation of vegetative growth. In fact, within the same study, ratios of recovery of apple CRISPR/Cas9 transformants for *TFL1.1* range between 0.75 and 5.49% compared to 14.25% for transformants carrying the *PDS* gene (Charrier et al. 2019). Interestingly, based on early flowering phenotypes, the *TFL1.1* mutation ratio within transformants is 90%. When five transformants with early flowering have been subjected to sequence analysis, biallelic mutations are confirmed. Furthermore, when a vector harbouring the same gRNAs is also used for the *TFL1.1* gene in pear, which is closely related to apple, wherein a single nucleotide mismatch in the gRNA to target sequences in pear is located at 18 bases upstream of the PAM sequence, a mutation is induced in the pear genome using this gRNA. Moreover, a transformation efficiency of 24% is obtained, and 9% of transformants have demonstrated early flowering. Similarly, when five transformants with early flowering are sequenced, biallelic mutations in all these transformants are detected. It is important to point out that that the observed low mutation efficiency in pear, compared to that in apple, is attributed to the presence of a mismatch at the target site (Charrier et al. 2019).

10.3.2.2 Disease Resistance

Disease resistance is one of the most critical targets in apple breeding programmes. For example, fire blight and apple scab are major diseases of both apple and pear, incited by a bacterium and a fungus, respectively (see other chapters). To overcome these serious diseases, genome editing targeting relevant susceptibility genes would be one of the most powerful strategies to deal with plant and microbial interactions and thereby enhance disease resistance.

The bacterium *Erwinia amylovora* is the causal pathogen of fire blight disease (Milčevičová et al. 2010). In order to reduce susceptibility to fire blight in apple, a gene encoding for the disease-specific protein of *E. amylovora,* DspE-interacting protein 4, the susceptibility gene *DIPM4* is knocked out with CRISPR/Cas9 using a *DIPM4*-specific gRNA (Pompili et al. 2020). In particular, an FLP (Flippase)/*FRT* (*FLP Recombination Target*) site-specific recombination system in the CRISPR/Cas9 allows for removal of the T-DNA insertion following stable transformation (Pompili et al. 2020). The vector has been designed with the *Flippase (FLP)* gene under the control of an *Hsp17.5-E* promoter from soybean and two *Flippase Recombination Target* (*FRT*) sites adjacent to each left and right borders. This vector system is used for *Agrobacterium*-mediated transformation of both 'Gala' and 'Golden Delicious' and yielding transformation efficiencies of 1.55% and 2.55%. A total of 57 transformants (27 from 'Gala' and 30 from 'Golden Delicious') have been analyzed, and mutation efficiencies of 77.8% and 73.3%, respectively, have been obtained. Among edited transformants, it is observed that 7.4% ('Gala') and 20% ('Golden Delicious') yield single-type mutations. Furthermore, all tested homozygous lines have demonstrated significant reduction in susceptibility to *E. amylovora*, on average by 50%, compared with controls for both cultivars. This finding has demonstrated that the *DIPM4* gene is associated with fire blight susceptibility.

In an earlier study, it has been reported that using RNA interference (RNAi) of four *DIPM* genes, resistance to fire blight increased in apple cv. 'Galaxy'; however, the functionality of each of the individual *DIPM* genes has not been clearly discerned as two or more *DIPM* genes are simultaneously silenced by RNAi (Borejsza-Wysocka et al. 2004).

Therefore, using an FLP/FRT-mediated T-DNA elimination system has allowed for confirmation of the functionality of one of these *DIPM* genes (Pompili et al. 2020). Moreover, mutant plants exhibiting stable removal of the selection marker gene for kanamycin (*NPTII*) have been obtained from four lines out of six lines tested. Together, the above findings have confirmed the usefulness of genome editing for evaluating individual loci for *DIPM* genes (Pompili et al. 2020).

10.3.3 Foreign-DNA-Free Genome Editing in Apple

It has long been reported that the efficiency of stable transformation of apple fluctuates depending on the cultivar/genotype tested (James et al. 1989; Norelli et al. 1994; Puite and Schaart 1996; Seong and Song 2008; Wada et al. 2009). Therefore, it is difficult to exploit the genome editing technology for apple cv. 'Fuji' due to its low transformation efficiency (Seong and Song 2008). Thus, it is expected that either transient expression or direct delivery of a genome editing tool would be more applicable to a wider group of cultivars than that of stable transformation.

Transient expression of CRISPR/Cas9 has been utilized to target an endogenous *PDS* gene in apple cv. 'Gala' (Charrier et al. 2019). Among 747 regenerated shoot buds induced from 229 leaf explants, three shoots are albino, and of these, two shoots have demonstrated the absence of integration of *Cas9* and selection marker (*nptII*) genes. Therefore, transient expression is a simple and promising approach for assessing utility of the CRISPR/Cas9 system; however, mutation rates are very low (0.27%). Furthermore, removal of selection marker genes and CRISPR/Cas9 cassettes using either Cre/loxP or FLP/FRT has also been reported for apple (see above Sect. 10.3.2) (Pompili et al. 2020).

On the other hand, direct delivery of RNPs into apple and grape protoplasts has been conducted (Malnoy et al. 2016; Osakabe et al. 2018). This approach is based on the transient introduction of RNP into protoplasts. By using PEG-mediated transformation, RNPs targeting *DIPM-1*, *2*, and *4* genes have been separately delivered into 'Golden Delicious' protoplasts, prepared from leaf tissues (Malnoy et al. 2016). It is observed that one day following the introduction of RNPs, mutations are detected in all targets. Based on the results of targeted deep sequencing for all protoplasts, it is found that mutation efficiencies range between 3.3 and 6.9%. Moreover, deletion mutations are preferentially observed for genes *DIPM-1* and *4*; whereas, insertion mutations are more frequent than deletion mutations for *DIPM-2*, thus suggesting that both mutations can be induced depending on target sequences or genes (Malnoy et al. 2016). Furthermore, the mutation efficiency for apple is influenced by the ratio (w/w) of Cas9 protein to gRNA, suggesting that a high amount of Cas9 protein is important for this system; whereas, the Cas9/gRNA ratio has no effect on the mutation efficiency for grape (Malnoy et al. 2016). Overall, the regeneration systems for mutated apple and grape protoplasts require further improvements and should be optimised.

10.4 Future Goals

The genome-editing technology is becoming a useful technology for pursuing genetic engineering efforts for apple. Recent studies have provided promising results in using various genome editing technologies for apple, such as for purposes of trait modifications, identification of functionality of distinct family genes, as well as of foreign gene-free technologies using transient expression of RNP transfection. It is also worth noting that genome editing has been applied to apple rootstocks (Nishitani et al. 2016), as rootstocks are highly critical in promoting fruit quality and cultivation of apples. These collective efforts will provide further opportunities for the genetic engineering of apple for economically useful traits.

Thus far, recent studies in apple genome editing have demonstrated the incidence of small mutations resulting in gene disruptions. Indeed, useful traits for apple would be improved by gene disruption, as there are several studies of using either antisense or RNAi technologies for modifying apple traits (Murata et al. 2001; Broothaerts et al. 2004). However, multiple mutagenesis efforts should be undertaken to collectively improve these various critical economic traits. Recently, gene targeting (GT) has been developed in mammalian species as a precise genome editing technology for a knock-in system (Čermák et al. 2015); however, HDR-mediated GT is highly inefficient in plants. As engineered nucleases efficiently induce targeted DSBs in a genome, GT frequencies are dramatically enhanced (Puchta and Fauser 2013). Recently, several studies have demonstrated efficient GT techniques using CRISPR/Cas9 and a combination of virus-based replicons in plants (Gil-Humanes et al. 2017; Wang et al. 2017; Dahan-Meir et al. 2018). All these efforts are likely to be useful in future plant breeding programmes. In apple breeding, it is likely that several targets could be modified using GT technology. For example, using traditional cross-hybridisation, it has taken a minimum of six generations (at least 25 to 30 years) to transfer the scab resistance gene, *Vf* (*Rvi6*), from the wild species *M. floribunda* 821 into elite lines of the cultivated apple, *M.* × *domestica* (Crosby et al. 1994). However, transfer of such a gene can now be introduced into *M.* × *domestica* more readily, and in a relatively short period of time via GT technology.

It is also important to point out that both efficient tissue culture regeneration and genetic transformation methods are required for efficient genome editing in plants. So far, the number of apple cultivars that have been used in genome editing are limited to 'JM2', 'Golden Delicious', and 'Gala', for which both efficient regeneration systems and genetic transformation systems are already established. Although regeneration protocols have been developed for many apple cultivars/genotypes, these protocols are widely variable for each cultivar/genotype. Moreover, these regeneration protocols should be optimized to increase regeneration efficiency from somatic tissues, particularly following the selection of transformants, so that they can be amenable for use in genome editing studies and applications.

Interestingly, genome-wide prediction technology, which can select parents that would produce superior progenies, has recently been explored in fruit tree crops (Iwata et al. 2016). While genome editing is generally appropriate for use in modifying traits controlled by a small number of known genes, genome prediction is suitable for traits influenced by unknown genes. Therefore, it is worthwhile to consider the use of both genome editing and prediction to accelerate genetic engineering and breeding of apple. Together, cross-hybridization of genome-edited parents to introduce novel superior traits would be one of the promising strategies in fruit tree breeding programmes. As we continue to better understand the whole apple genome, these efforts will also enable the development of future technologies in apple breeding to generate novel and valuable economic traits.

References

Abu-Qaoud H, Skirvin RM, Chevreau E (1990) *In vitro* separation of chimeral pears into their component genotypes. Euphytica 48(2):189–196

Andersson M, Turesson H, Olsson N, Fält AS, Ohlsson P, Gonzalez MN, Samuelsson M, Hofvander P (2018) Genome editing in potato via CRISPR-Cas9 ribonucleoprotein delivery. Physiol Plant 164(4):378–384

Borejsza-Wysocka EE, Malnoy M, Meng X, Bonasera JM, Nissinen RM, Kim JF, Beer SV, Aldwinckle HS (2004) Silencing of apple proteins that interact with DspE, a pathogenicity effector from *Erwinia amylovora*, as a strategy to increase resistance to fire blight Acta Hortic 663:469–474

Bortesi L, Fischer R (2015) The CRISPR/Cas9 system for plant genome editing and beyond. Biotech Adv 33 (1):41–52

Broothaerts W, Keulemans J, Van Nerum I (2004) Self-fertile apple resulting from *S*-RNase gene silencing. Plant Cell Rep 22(7):497–501

Čermák T, Baltes NJ, Čegan R, Zhang Y, Voytas DF (2015) High-frequency, precise modification of the tomato genome. Genome Biol 16(1):232. https://doi.org/10.1186/s13059-015-0796-9

Charrier A, Vergne E, Dousset N, Richer A, Petiteau A, Chevreau E (2019) Efficient targeted mutagenesis in apple and first time edition of pear using the CRISPR-Cas9 system. Front Plant Sci 10.40. https://doi.org/10.3389/fpls.2019.00040

Chen X, Li S, Zhang D, Han M, Jin X, Zhao C, Wang S, Xing L, Ma J, Ji J, An N (2019) Sequencing of a wild apple (*Malus baccata*) genome unravels the differences between cultivated and wild apple species regarding disease resistance and cold tolerance. G3: Genes Genomes Genet 9(7):2051–2060

Clasen BM, Stoddard TJ, Luo S, Demorest ZL, Li J, Cedrone F, Tibebu R, Davison S, Ray EE, Daulhac A, Coffman A, Yabandith A, Retterath A, Haun W, Baltes NJ, Mathis L, Voytas DF, Zhang F (2016) Improving cold storage and processing traits in potato through targeted gene knockout. Plant Biotech J 14 (1):169–176

Crosby JA, Janick J, Pecknold PC, Goffreda JC, Korban SS (1994) 'GoldRush' apple. HortScience 29 (7):827–828

Daccord N, Celton JM, Linsmith G, Becker C, Choisne N, Schijlen E, van de Geest H, Bianco L, Micheletti D, Velasco R, Pierro EAD, Gouzy J, Rees DJG, Guérif P, Muranty H, Durel CE, Laurens F, Lespinasse Y, Gaillard S, Aubourg S, Quesneville H, Weigel D, van de Weg E, Troggio M, Bucher E (2017) High-quality de novo assembly of the apple genome and methylome dynamics of early fruit development. Nat Genet 49(7):1099–1106. https://doi.org/10.1038/ng.3886

Dahan-Meir T, Filler-Hayut S, Melamed-Bessudo C, Bocobza S, Czosnek H, Aharoni A, Levy AA (2018) Efficient *in planta* gene targeting in tomato using geminiviral replicons and the CRISPR/Cas9 system. Plant J 95(1):5–16

Fu Y, Sander JD, Reyon D, Cascio VM, Joung JK (2014) Improving CRISPR-Cas nuclease specificity using truncated guide RNAs. Nat Biotech 32(3):279–284

Gao C (2019) Precision plant breeding using genome editing technologies. Transgenic Res 28:53–55

Gil-Humanes J, Wang Y, Liang Z, Shan Q, Ozuna CV, Sánchez-León S, Baltes NJ, Starker C, Barro F, Gao C, Voytas DF (2017) High-efficiency gene targeting in hexaploid wheat using DNA replicons and CRISPR/Cas9. Plant J 89(6):1251–1262

Haeussler M, Schönig K, Eckert H, Eschstruth A, Mianné J, Renaud JB, Schneider-Maunoury S, Shkumatava A, Teboul L, Kent J, Joly JS, Concordet JP (2016) Evaluation of off-target and on-target scoring algorithms and integration into the guide RNA selection tool CRISPOR. Genome Biol 17(1):148. https://doi.org/10.1186/s13059-016-1012-2

Hayut SF, Bessudo CM, Levy AA (2017) Targeted recombination between homologous chromosomes for precise breeding in tomato. Nat Commun 8(1):15605. https://doi.org/10.1038/ncomms15605

Iwata H, Minamikawa MF, Kajiya-Kanegae H, Ishimori M, Hayashi T (2016) Genomics-assisted breeding in fruit trees. Breed Sci 66(1):100–115

James DJ, Passey AJ, Barbara DJ, Bevan M (1989) Genetic transformation of apple (*Malus pumila* Mill.) using a disarmed Ti-binary vector. Plant Cell Rep 7 (8):658–661

Kamburova VS, Nikitina EV, Shermatov SE, Buriev ZT, Kumpatla SP, Emani C, Abdurakhmonov IY (2017) Genome editing in plants: an overview of tools aapplications. Int J Agron 2017: Article ID 7315351. https://doi.org/10.1155/2017/7315351

Kaur N, Alok A, Shivani KN, Pandey P, Awasthi P, Tiwari S (2018) CRISPR/Cas9-mediated efficient editing in *phytoene desaturase* (*PDS*) demonstrates precise manipulation in banana cv. Rasthali Genome Funct Integr Genomics 18(1):89–99

Kim H, Kim ST, Ryu J, Kang BC, Kim JS, Kim SG (2017) CRISPR/Cpf1-mediated DNA-free plant genome editing. Nat Commun 8(1):14406. https://doi.org/10.1038/ncomms14406

Kotoda N, Iwanami H, Takahashi S, Abe K (2006) Antisense expression of *MdTFL1*, a *TFL1*-like gene, reduces the juvenile phase in apple. J Am Soc Hortic Sci 131(1):74–81

Kunihisa M, Moriya S, Abe K, Okada K, Haji T, Hayashi T, Kawahara Y, Itoh R, Itoh T, Katayose Y, Kanamori H, Matsumoto T, Mori S, Sasaki H, Matsumoto T, Nishitani C, Terakami S, Yamamoto T (2016) Genomic dissection of a 'Fuji' apple cultivar: re-sequencing, SNP marker development, definition of haplotypes, and QTL detection. Breed Sci 66(4):499–515

Liang Z, Chen K, Li T, Zhang Y, Wang Y, Zhao Q, Liu J, Zhang H, Liu C, Ran Y, Gao C (2017) Efficient DNA-free genome editing of bread wheat using CRISPR/Cas9 ribonucleoprotein complexes. Nat Commun 8(1):14261. https://doi.org/10.1038/ncomms14261

Limera C, Sabbadini S, Sweet JB, Mezzetti B (2017) New biotechnological tools for the genetic improvement of major woody fruit species. Front Plant Sci 8:1418. https://doi.org/10.3389/fpls.2017.01418

Liu H, Ding Y, Zhou Y, Jin W, Xie K, Chen LL (2017) CRISPR-P 2.0: an improved CRISPR-Cas9 tool for genome editing in plants. Mol Plant 10(3):530–532

Liu X, Homma A, Sayadi J, Yang S, Ohashi J, Takumi T (2016) Sequence features associated with the cleavage efficiency of CRISPR/Cas9 system. Sci Rep 6:19675. https://doi.org/10.1038/srep19675

Malnoy M, Viola R, Jung MH, Koo OJ, Kim S, Kim JS, Velasco R, Nagamangala Kanchiswamy C (2016) DNA-free genetically edited grapevine and apple protoplast using CRISPR/Cas9 ribonucleoproteins. Front Plant Sci 7:1904. https://doi.org/10.3389/fpls.2016.01904

Mehravar M, Shirazi A, Nazari M, Banan M (2019) Mosaicism in CRISPR/Cas9-mediated genome editing. Dev Biol 445(2):156–162

Miki D, Zhang W, Zeng W, Feng Z, Zhu JK (2018) CRISPR/Cas9-mediated gene targeting in *Arabidopsis* using sequential transformation. Nat Commun 9 (1):1967. https://doi.org/10.1038/s41467-018-04416-0

Milčevičová R, Gosch C, Halbwirth H, Stich K, Hanke MV, Peil A, Flachowsky H, Rozhon W, Jonak C, Oufir M, Hausman JF, Matušíková I, Fluch S, Wilhelm E (2010) *Erwinia amylovora*-induced defense mechanisms of two apple species that differ in susceptibility to fire blight. Plant Sci 179 (1–2):60–67

Mimida N, Kotoda N, Ueda T, Igarashi M, Hatsuyama Y, Iwanami H, Moriya S, Abe K (2009) Four *TFL1/CEN*-like genes on distinct linkage groups show different expression patterns to regulate vegetative and reproductive development in apple (*Malus× domestica* Borkh.). Plant Cell Physiol 50(2): 394–412

Murata M, Nishimura M, Murai N, Haruta M, Homma S, Itoh Y (2001) A transgenic apple callus showing reduced polyphenol oxidase activity and lower browning potential. Biosci Biotech Biochem 65(2):383–388

Nakajima I, Ban Y, Azuma A, Onoue N, Moriguchi T, Yamamoto T, Toki S, Endo M (2017) CRISPR/Cas9-mediated targeted mutagenesis in grape. PLoS ONE 12(5):e0177966. https://doi.org/10.1371/journal.pone.0177966

Nishitani C, Hirai N, Komori S, Wada M, Okada K, Osakabe K, Yamamoto T, Osakabe Y (2016) Efficient genome editing in apple using a CRISPR/Cas9 system. Sci Rep 6:31481. https://doi.org/10.1038/srep31481

Norelli JL, Aldwinckle HS, Destéfano-Beltrán L, Jaynes JM (1994) Transgenic 'Mailing 26'apple expressing the attacin E gene has increased resistance to *Erwinia amylovora*. Euphytica 77(1–2):123–128

Ohmori M, Yamane H, Osakabe K, Osakabe Y, Tao R (2020) Targeted mutagenesis of CENTRORADIALIS using CRISPR/Cas9 system through the improvement of genetic transformation efficiency of tetraploid highbush blueberry. J Hortic Sci Biotech. https://doi.org/10.1080/14620316.2020.1822760

Osakabe Y, Osakabe K (2015) Genome editing with engineered nucleases in plants. Plant Cell Physiol 56 (3):389–400

Osakabe Y, Watanabe T, Sugano SS, Ueta R, Ishihara R, Shinozaki K, Osakabe K (2016) Optimization of CRISPR/Cas9 genome editing to modify abiotic stress responses in plants. Sci Rep 6:26685. https://doi.org/10.1038/srep26685

Osakabe Y, Liang Z, Ren C, Nishitani C, Osakabe K, Wada M, Komori S, Malnoy M, Velasco R, Poli M, Jung MH, Koo OJ, Viola R, Kanchiswamy CN (2018) CRISPR–Cas9-mediated genome editing in apple and grapevine. Nat Protoc 13(12):2844–2863

Pan C, Ye L, Qin L, Liu X, He Y, Wang J, Chen L, Lu G (2016) CRISPR/Cas9-mediated efficient and heritable targeted mutagenesis in tomato plants in the first and later generations. Sci Rep 6:24765. https://doi.org/10.1038/srep24765

Peer R, Rivlin G, Golobovitch S, Lapidot M, Gal-On A, Vainstein A, Tzfira T, Flaishman MA (2015) Targeted mutagenesis using zinc-finger nucleases in perennial fruit trees. Planta 241(4):941–951

Peng A, Chen S, Lei T, Xu L, He Y, Wu L, Yao L, Zou X (2017) Engineering canker-resistant plants through CRISPR/Cas9-targeted editing of the susceptibility gene *CsLOB1* promoter in citrus. Plant Biotech J 15 (12):1509–1519

Pompili V, Dalla Costa L, Piazza S, Pindo M, Malnoy M (2020) Reduced fire blight susceptibility in apple cultivars using a high-efficiency CRISPR/Cas9-FLP/FRT-based gene editing system. Plant Biotech J 18(3):845–858. https://doi.org/10.1111/pbi.13253

Puchta H, Fauser F (2013) Gene targeting in plants: 25 years later. Int J Dev Biol 57:629–637

Puite KJ, Schaart JG (1996) Genetic modification of the commercial apple cultivars Gala, Golden Delicious and Elstar via an *Agrobacterium tumefaciens*-mediated transformation method. Plant Sci 119:125–133

Ren C, Liu X, Zhang Z, Wang Y, Duan W, Li S, Liang Z (2016) CRISPR/Cas9-mediated efficient targeted mutagenesis in Chardonnay (*Vitis vinifera* L.). Sci Rep 6:32289. https://doi.org/10.1038/srep32289

Rong YS, Golic KG (2003) The homologous chromosome is an effective template for the repair of mitotic DNA double-strand breaks in Drosophila. Genetics 165(4):1831–1842

Rozov SM, Permyakova NV, Deineko EV (2019) The problem of the low rates of CRISPR/Cas9-mediated knock-ins in plants: Approaches and solutions. Int J Mol Sci 20(13):3371. https://doi.org/10.3390/ijms20133371

Seong ES, Song KJ (2008) Factors affecting the early gene transfer step in the development of transgenic 'Fuji' apple plants. Plant Grow Reg 54(2):89–95

Shannon S, Meeks-Wagner DR (1991) A mutation in the Arabidopsis *TFL1* gene affects inflorescence meristem development. Plant Cell 3(9):877–892

Spiegel-Roy P (1990) Economic and agricultural impact of mutation breeding in fruit trees. Mut Breed Rev 5:1–26

Steinert J, Schiml S, Puchta H (2016) Homology-based double-strand break-induced genome engineering in plants. Plant Cell Rep 35(7):1429–1438

Subburaj S, Chung SJ, Lee C, Ryu SM, Kim DH, Kim JS, Bae S, Lee GJ (2016) Site-directed mutagenesis in *Petunia × hybrida* protoplast system using direct delivery of purified recombinant Cas9 ribonucleoproteins. Plant Cell Rep 35(7):1535–1544

Svitashev S, Schwartz C, Lenderts B, Young JK, Cigan AM (2016) Genome editing in maize directed by CRISPR–Cas9 ribonucleoprotein complexes. Nat Commun 7(1):13274. https://doi.org/10.1038/ncomms13274

Ueta R, Abe C, Watanabe T, Sugano SS, Ishihara R, Ezura H, Osakabe Y, Osakabe K (2017) Rapid breeding of parthenocarpic tomato plants using CRISPR/Cas9. Sci Rep 7:507. https://doi.org/10.1038/s41598-017-00501-4

Velasco R, Zharkikh A, Affourtit J, Dhingra A, Cestaro A, Kalyanaraman A, Fontana P, Bhatnagar SK, Troggio M, Pruss D, Salvi S, Pindo M, Baldi P, Castelletti S, Cavaiuolo M, Coppola G, Costa F, Cova V, Ri AD, Goremykin V, Komjanc M, Longhi S, Magnago P, Malacarne G, Malnoy M, Micheletti D, Moretto M, Perazzolli M, Si Ammour A, Vezzulli S, Zini E, Eldredge G, Fitzgerald LM, Gutin N, Lanchbury G, Macalma T, Mitchell JT, Reid J, Wardell B, Kodira C, Chen Z, Desany B, Niazi F, Palmer M, Koepke T, Jiwan D, Schaeffer S, Krishnan V, Wu C, Chu VT, King ST, Vick J, Tao Q, Mraz A, Stormo A, Stormo K, Bogden R, Ederle D, Stella A, Vecchietti A, Kater MM, Masiero S, Lasserre P, Lespinasse Y, Allan AC, Bus V, Chagné D, Crowhurst RN,

Gleave AP, Lavezzo E, Fawcett JA, Proost S, Rouzé P, Sterck L, Toppo S, Lazzari B, Hellens RP, Durel C-E, Gutin A, Bumgarner R, Gardiner SE, Skolnick M, Egholm M, Van de Peer Y, Salamini F, Viola R (2010) The genome of the domesticated apple (*Malus* × *domestica* Borkh.). Nat Genet 42:833–839

Wada M, Ureshino A, Tanaka N, Komori S, Takahashi S, Kudo K, Bessho H (2009) Anatomical analysis by two approaches ensure the promoter activities of apple *AFL* genes. J Jpn Soc Hortic Sc 78(1):32–39

Wang M, Lu Y, Botella JR, Mao Y, Hua K, Zhu JK (2017) Gene targeting by homology-directed repair in rice using a geminivirus-based CRISP)R/Cas9 system. Mol Plant 10(7):1007–1010

Wang Z, Wang S, Li D, Zhang Q, Li L, Zhong C, Liu Y, Huang H (2018) Optimized paired-sgRNA/Cas9 cloning and expression cassette triggers high-efficiency multiplex genome editing in kiwifruit. Plant Biotech J 16(8):1424–1433

Woo JW, Kim J, Kwon SI, Corvalán C, Cho SW, Kim H, Kim SG, Kim ST, Choe S, Kim JS (2015) DNA-free genome editing in plants with preassembled CRISPR-Cas9 ribonucleoproteins. Nat Biotech 33(11):1162–1164

Xiao A, Cheng Z, Kong L, Zhu Z, Lin S, Gao G, Zhang B (2014) CasOT: a genome-wide Cas9/gRNA off-target searching tool. Bioinformatics 30(8):1180–1182

Zhang L, Hu J, Han X, Li J, Gao Y, Richards CM, Zhang C, Tian Y, Liu G, Gul H, Wang D, Tian Y, Yang C, Meng M, Yuan G, Kang G, Wu Y, Wang K, Zhang H, Wang D, Cong P (2019) A high-quality apple genome assembly reveals the association of a retrotransposon and red fruit colour. Nat Commun 10:1494. https://doi.org/10.1038/s41467-019-09518-x

11 Flowering and Juvenility in Apple

Nobuhiro Kotoda

Abstract

Flowering and juvenility are important traits for the cultivation and breeding of apples (*Malus* spp.). As apples are reported to have been cultivated long before ancient Greek civilization, many findings related to morphological and physiological traits, and well-developed cultivation techniques have been accumulated for many cultivated, ornamental, and rootstock apples. During the first half of the 1990s, elucidation of the molecular mechanism(s) of flower development in model plants has received much attention leading to major advances. This is followed by determinations of genetic sequences and functional analyses of genes controlling flowering and juvenility in these plants using genetic transformation technologies during the latter half of the 1990s. Subsequently, genes involved in floral organ development and genes involved in either flower induction or suppression of the apple, such as *MdFT1* and *MdTFL1*, respectively, have been identified. Over the past decade, either stable or transient gene expression systems have been developed to control the flowering time in apple. These advances have had significant impacts on functional analysis of various genes of interest controlling fruit quality traits and pursuing apple breeding efforts with significantly reduced generation cycles. Future efforts should pursue functional analyses of genes, such as *TFL1*/*FT*-like genes; moreover, it is necessary to investigate and delineate relationships between plant hormones/sugars and flowering in apples.

11.1 Introduction

The cultivated apple, *Malus* × *domestica* Borkh., presents unique challenges and constraints in pursuing rapid genetic improvement efforts. This is attributed to its long juvenile period (5–7 years for seedlings), large tree size, self-incompatibility, and inbreeding depression (Brown and Maloney 2003). Of these, shortening the juvenile period directly impacts the efficiency of breeding new cultivars. Furthermore, flowering habits of apple trees also affect the productivity and fruit quality of established apple trees (Dennis 2003). As clonal propagation, via grafting/budding of scion cultivars onto dwarfing rootstocks, is a common practice to maintain genetic/phenotypic identity of desirable cultivars, control tree size, and promote earlier flowering and fruiting, this still requires a period of at least

N. Kotoda (✉)
Faculty of Agriculture, Saga University, Saga 840-8502, Japan
e-mail: koto@cc.saga-u.ac.jp

S. S. Korban (ed.), *The Apple Genome*, Compendium of Plant Genomes,
https://doi.org/10.1007/978-3-030-74682-7_11

36–48 months. Yet, modern apple producers are eager to begin earning return on investment much sooner by shortening the non-fruiting period as much as possible, promote earlier fruiting in newly established orchards, as well as having the flexibility of changing their orchards from old to new promising cultivars within a relatively shorter time frame without suffering significant economic loss while awaiting for these new orchards to come into bearing. Therefore, understanding the mechanism of flowering and juvenility of apples has long been one of the most important objectives of interest.

This chapter provides an overview and description of flower development, juvenility, and those genes controlling flowering in apple. Despite recent progress in genetic engineering/tissue culture techniques, evaluating phenotypes of seedlings derived from sexual hybridizations in a breeding programme require long periods of time. Likewise, conducting genetic analyses of economic traits of interest, particularly those for fruiting and fruit quality traits, require more time due to the length of the juvenile period. However, the development of genetic transformation protocols (James et al. 1989; Wada et al. 2020), molecular markers, such as simple sequence repeats (SSRs) (Gianfranceschi et al. 1998; Hokanson et al. 1998) and single nucleotide polymorphisms (SNPs) (Khan et al. 2012), accumulation of expressed sequence tags (ESTs) (Newcomb et al. 2005; Gasic et al. 2009), and release of whole-genome draft sequences of the domesticated apple (Velasco et al. 2010; Daccord et al. 2017) have facilitated efforts for investigating the molecular mechanism of flowering in apples.

Therefore, morphological and physiological aspects of flowering and juvenility are described herein based on earlier studies (reported before the late 1990s). This will be followed by molecular studies of genes controlling flower development and flower induction (primarily of studies after the late 1990s). Finally, reports on transgenic approaches pursued, over the past decade, to control flowering in apple will be presented.

11.2 Life Cycle of the Apple

The overall life cycle of the apple is schematically presented in Fig. 11.1. In woody plants, such as the apple, acquiring the ability to flower marks the end of the juvenile phase. When actual flowering does not coincide with the end of the juvenile phase, such an intervening period is referred to as an adult vegetative phase or transition phase (Zimmerman 1972; Hanke et al. 2007). In the adult phase, the reproductive organs of an apple tree will produce fruit-bearing shoots (FBS) and enter into a seasonal-flower/fruit formation cycle (reproductive growth cycle). On the other hand, vegetative organs will produce vegetative shoots, such as succulent shoots (SS) (or water sprouts), and enter into a vegetative growth cycle. After flowers bloom on FBS in the spring, a newly budded shoot (bourse shoot) emerges from an axillary meristem along the bourse (Fig. 11.2). Approximately 95% of vegetative meristems on bourse shoots will become committed to floral development, consequently converting into inflorescence meristems and then developing into floral meristems. The remaining 5% of vegetative meristems will continue to produce vegetative shoots (Foster et al. 2003).

The morphology of SS, developing directly from either a branch or trunk, is similar to that of juvenile shoots with vigorous growth. Moreover, SS does not bear floral buds for several years. As the apple fruit ripens during the fall, the apple tree enters into a dormant period beginning in late fall.

It is also important to point out that if seeds from ripened fruit are allowed to germinate, they will enter an ontogenetic growth cycle.

11.3 Flower Development in the Apple

11.3.1 Floral Organ Development

In the apple, floral primordia for the following year is initiated 3–6 weeks after the current bloom season or following the cessation of shoot growth

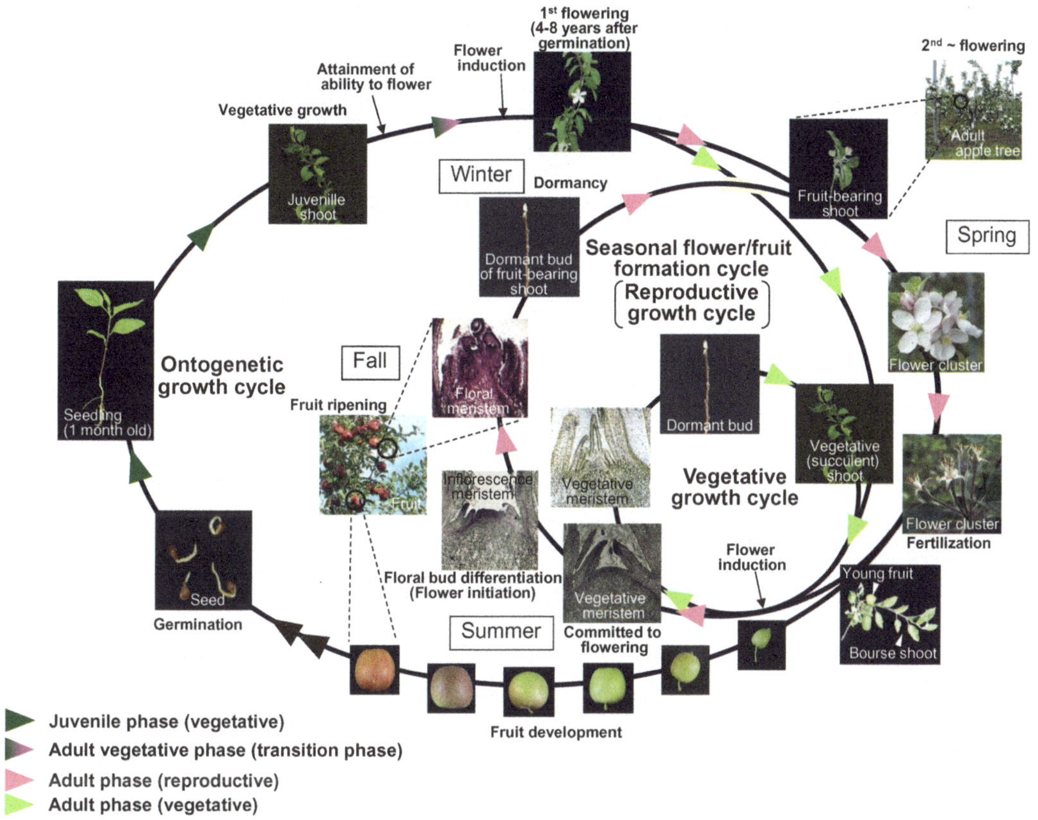

Fig. 11.1 Schematic representation of the life cycle of the apple (Mimida et al. 2009)

(Buban and Faust 1982). In 'Jonathan' apple trees, growth of most current shoots from a cluster base (a bourse) ceases in late June in Morioka, Japan. From this observation, it appears that the transition from the vegetative to the reproductive phase occurs in late June [45–50 days after full bloom (DAFB)] (Kotoda et al. 2000).

In detailed microscopic observations, it is revealed that the apical meristem is a narrow pointed dome on July 3 (52 DAFB), slightly domed on July 17 (66 DAFB), and is clearly expanded by July 31 (80 DAFB). Thus, it appears that floral bud formation is initiated in the middle of July (65–70 DAF) in Morioka (Japan). Furthermore, the apical meristem is considerably domed-shaped in mid-September. In early October, sepal primordia are formed, and this is subsequently followed by the initiation of stamen primordia. By mid-December, several floral primordia are visible (Fig. 11.2). These floral organs will rapidly expand following the end of the dormancy period and are fully differentiated in late April, prior to bloom.

In general, apple inflorescence consists of a terminal flower and four lateral flowers. The terminal flower (king flower) usually grows and opens earlier than those lateral flowers. Scanning electrographs and images of sectioned vegetative and inflorescence meristems of 'Royal Gala' in Havelock North, New Zealand are shown in Fig. 11.3. It is noteworthy to point out that the transition from stage 0 (apex $\leq$ 130 μm in diameter) to stage 1 (apex > 130 μm in diameter) is critical for vegetative meristems to be committed to floral development (Foster et al. 2003). Based on these findings, the first internal cues for flower induction during the seasonal flower/fruit formation cycle may be expressed simultaneously or slightly earlier than the period of cessation of shoot growth in apple.

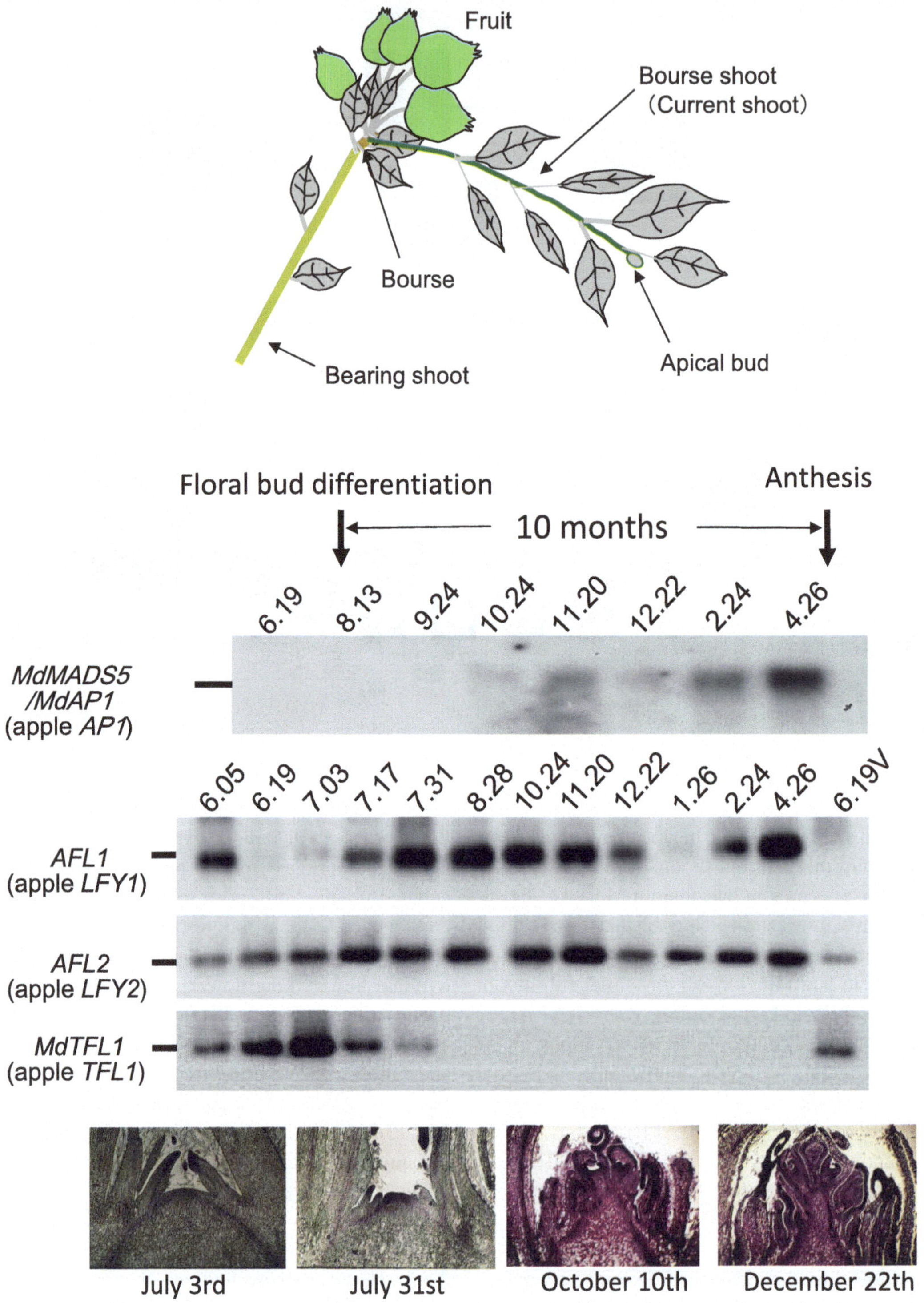

Fig. 11.2 Morphologies of shoots with fruit; floral bud differentiation; and expression of flowering genes in the apple. V, vegetative shoots. The numbers above the panels represent months and days (Kotoda et al. 2000; Kotoda and Wada 2005)

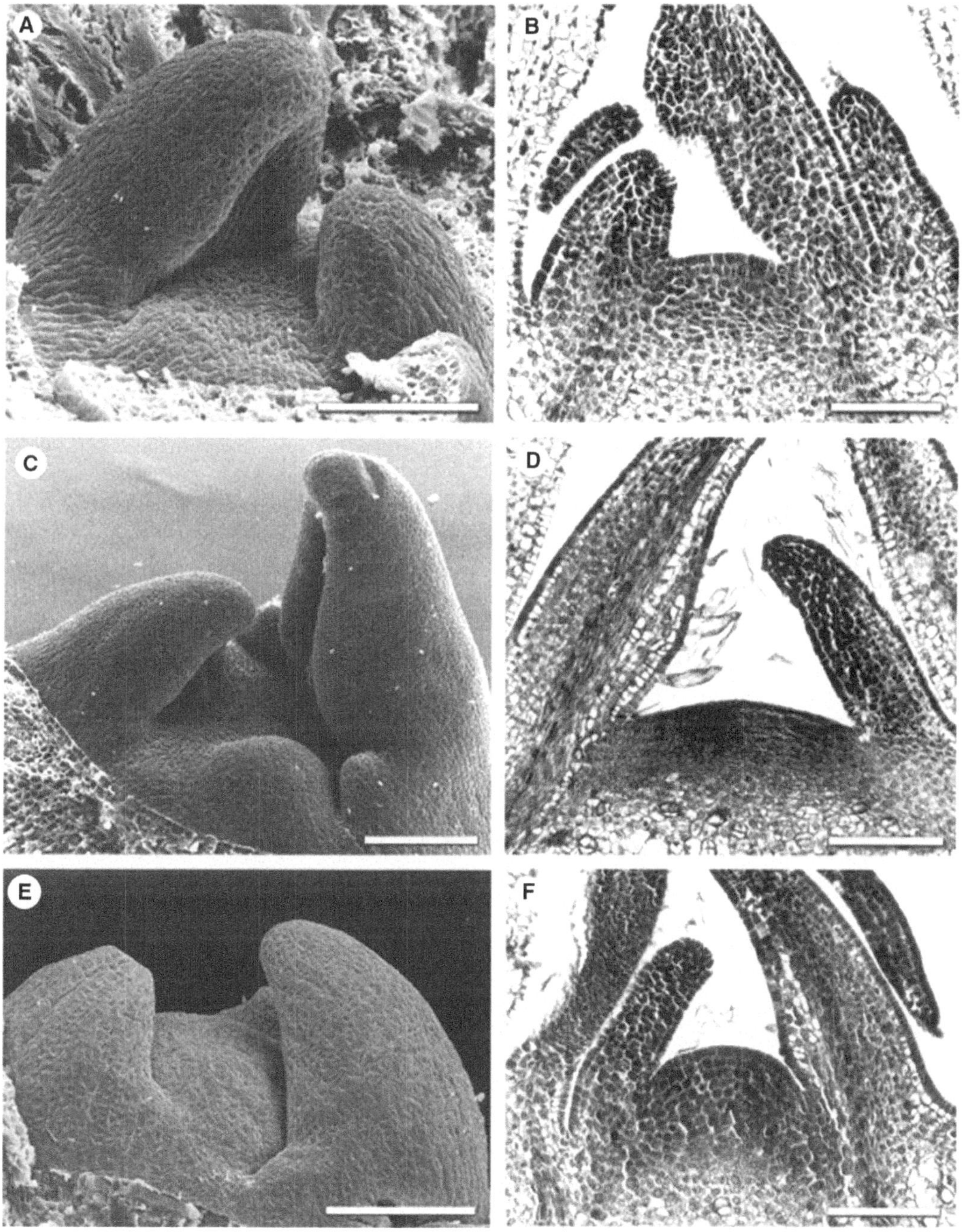

Fig. 11.3 Scanning electron micrographs (SEM) and images of sectioned vegetative and inflorescence meristems of 'Royal Gala' apple. **a** stage 0; **c** stage 1; and **e** stage 2 of SEM; **b** stage 0; **d** stage 1; and **f** stage 2 of median longitudinal sections. Bars = 100 μm (Foster et al. 2003)

11.3.2 Mutants of Flower Development

A typical flower of a dicotyledonous plant is organized into four whorls, from the outermost to the innermost, consisting of sepals (collectively referred to as the calyx), petals (collectively referred to as the corolla), stamens (androecium), and carpels (gynoecium). Some of the most important and compelling studies in plant science are those that have investigated and explained the molecular genetic mechanism of determination of floral organ formation. In the early 1990s, a hypothesis, the ABC model of flower development, explaining the mechanism(s) for determining the formation of the four floral organs is proposed. This ABC model is based on analysis of homeotic mutants of three types of flowers in both Arabidopsis and snapdragon, namely class A (carpel–stamen–stamen–carpel, from the outer to inner whorls), class B (sepal–sepal–carpel–carpel), and class C (sepal–petal–petal–sepal) mutants (Bowman et al. 1991; Coen and Meyerowitz 1991).

In the ABC model, it is proposed that genes involved in homeotic mutations are subdivided into three groups, A, B, and C, and that floral organs are determined by the manner of functionality (presence or absence of expression) of these three genes in each of the four whorls. When only the A gene is functional, then sepals are formed; when both A and B genes are functional, then petals are formed; when both B and C genes are functional, then stamens are formed; when only C genes are functional, then carpels are formed; and when none of these genes are functional, then leaves are formed. Moreover, the A function and the C function are mutually antagonistic. This is the basic concept of the well-known classical ABC model.

In the Arabidopsis model plant, those genes responsible for A, B, and C functions have been identified. All of these genes are found to belong to the *MADS*-box gene family, except for *APETALA2* (*AP2*), and these function as transcription factors (TFs). Genes involved in the formation of floral organs have been isolated and identified in various plants, and the quartet model, consisting of *SEPALLATA* proteins functional in combination with A, B, and C proteins, has since been postulated (Melzer and Theissen 2009). It is important to point out that the ABC model has been found to be applicable to many other plants, and it is still used to explain flower development and evolution (Bowman 2012).

Interestingly, seedless apples, such as 'Spencer Seedless,' 'Rae Ime,' and 'Wellington Bloomless,' also feature a different flower form (Fig. 11.4). This flower form has long been known in ornamental apples. As flowers lacking petals and stamens exhibit a phenotype similar to that of the B class mutant, these are proposed to result from a loss-of-function mutant of the B gene in apple. Such mutants have a retrotransposon-like sequence inserted into the intron of *MdPI,* an apple *PISTILLATA* (*PI*) homologous gene, thereby not allowing for *MdPI* expression in flower buds (Yao et al. 2001). On the other hand, cultivars with double flowers, a highly desirable trait in ornamental plants, may be observed due to a mutation in the homeotic C gene involved in the formation of floral organs (Fig. 11.5).

11.4 Juvenility in the Apple

11.4.1 Developmental Phases in Plants

In general, plant growth consists of three to four developmental phases that are qualitatively different. Although definitions of terms relevant to the growth phase have long been discussed, in recent years, the general growth phase of plants —including perennial plants—has been divided into three developmental phases: the juvenile vegetative phase, adult vegetative phase, and adult reproductive phase (Poethig 1990, 2003). In other words, the developmental phase is first divided into two phases, vegetative and reproductive; moreover, the vegetative phase is further subdivided into a 'state without the ability to respond to external flower induction' (juvenile phase) and a 'state with the ability to respond to external flower induction' (adult phase).

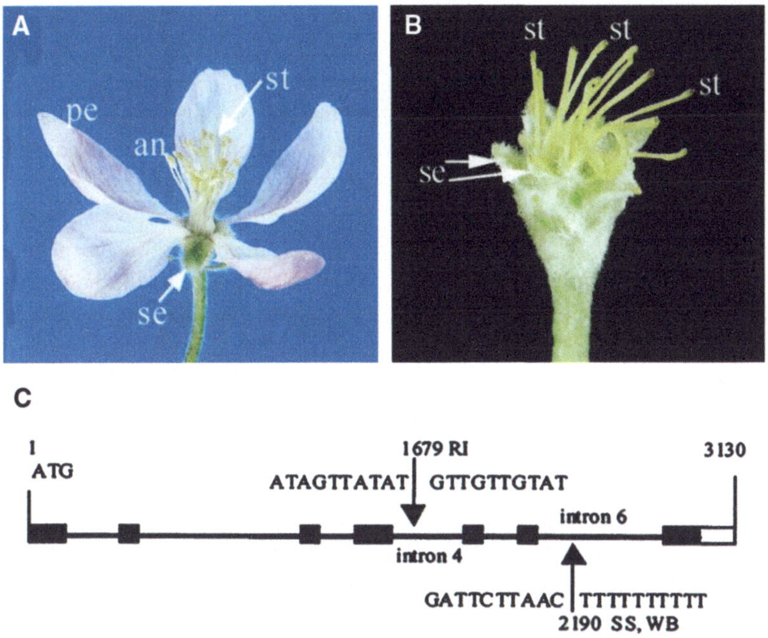

Fig. 11.4 Normal-type and 'Rae Ime' apple flowers. **a** A normal apple flower with sepals (se), petals (pe), anthers (an), and styles (st). **b** A 'Rae Ime' flower with no petals or stamens, but with higher numbers of sepals and styles. **c** Arrows indicate insertion sites in *MdPI* for 'Rae Ime' (RI), 'Spencer Seedless' (SS), and 'Wellington Bloomless' (WB). Numbers indicate nucleotide positions beginning with the ATG codon in the wild-type *MdPI* of 'Granny Smith.' Dark boxes are coding regions, while the white box corresponds to a 39 noncoding region. Lines between boxes represent introns (Yao et al. 2001) (Copyright (2001) National Academy of Sciences, USA)

Fig. 11.5 'Prince Georges' crabapple. Bar = 6 cm

Furthermore, the phase change from the juvenile vegetative to the adult vegetative is progressive, and it is rather difficult to discriminate this transition period, from juvenile to adult, in many plant species. Nevertheless, the change from adult vegetative to reproductive entails a radical change involving floral bud formation; thus it is relatively easy to identify this phase change. In the fields of pomology, arboriculture, and forestry, the developmental phases of woody plants are often classified simply into juvenile and adult phases, while the intermediate growth phase in-between these two phases is referred to as the transition phase (Zimmerman 1972; Hackett 1985).

11.4.2 Juvenility and Rejuvenility

It is relatively easy to estimate the duration of the juvenile phase in plants wherein flower induction conditions have already been delineated, but in plants that have not, it is necessary to confirm their periodic and stable flowering, as it is an indisputable indicator of the end of the juvenile phase (Hackett 1985). The time frame following seed germination during which seedlings are not capable of flowering/fruiting is generally referred to as a juvenile period. The length of the juvenile period varies from 12 months or less, in herbaceous plants, to 50 to 60 years, in woody plants. The juvenile periods (periods required for flowering) for major horticultural and arboricultural crops, including fruit trees and model plants, are listed in Table 11.1.

Woody crops, such as apples, have long juvenile periods, and such juvenility is an obstacle to efficient breeding (Zimmerman 1972; Hackett 1985). Morphological, physiological, and biochemical characteristics of a seedling in the juvenile phase are described as *juvenility*. For example, spines are well-known morphological features of juvenility. As a seedling undergoes a phase change from the juvenile to the adult phase, this is accompanied by likely changes in morphological features, as well as in both physiological and biochemical traits, such as leaf shape, spines, adventitious root formation, biotic stress resistance, anthocyanin content, and photosynthetic efficiency. For example, it has long been known that adventitious root formation in young trees or root induction in vegetative cuttings is readily induced and highly observed.

It is commonly known that after a seedling or tree undergoes a change from the juvenile to the adult phase, it is rather rare to go back to the juvenile phase under normal growing conditions. This is attributed to the fact that each developmental phase is likely maintained by factors that are not directly related to sequence changes of genes (epigenetics) and the role of DNA methylation in these processes (Poethig 1990). However, depending on the plant species, juvenility may be recovered under some conditions, even following phase inversion. This phenomenon is known as *rejuvenation*. For example, it is known that juvenile characteristics, such as the enhanced capability for adventitious root formation and for delayed flowering have been observed during the following: (1) regeneration of plants from adventitious buds and embryos; (2) grafting of scions onto rootstocks with strong juvenility; (3) applications of plant hormones such as gibberellin; (4) severe pruning; and (5) shoot proliferation by cuttings (Hackett 1985).

It is important to point out that rejuvenation by grafting and severe pruning results from vigorous growth, which is in contrast to ontogenetic ageing. Although it is rare to induce complete rejuvenation, it is possible to induce partial rejuvenation for various plants by some of the methods described above.

Table 11.1 Durations of juvenile periods for major horticultural and arboricultural crops

Plant species	Juvenile period
Thale cress/Arabidopsis (*Arabidopsis thaliana*)	1 month
Tomato (*Solanum Lycopersicum*)	2–3 months
Tobacco (*Nicotiana tabacum*)	3 months
Chestnut (*Castanea crenata*)	3–4 years
Stone fruits (*Prunus* spp.)	4–5 years
Apple (*Malus* spp.)	5–7 years
Persimmon (*Diospyros kaki*)	5–7 years
Pear (*Pyrus* spp.)	6–10 years
Citrus (*Citrus* spp.)	7–10 years
Avocado (*Persea americana*)	15 years
Poplar (*Populus* spp.)	20 years

11.4.3 In Vitro Flowering

Although it is rather rare to observe flower bud formation in shoots grown in tissue culture, in vitro flowering has indeed been detected. In vitro flowering is likely to occur when the tissue used for in vitro culture is derived from an adult phase organ, and it is less likely to occur from a juvenile phase organ (Scorza 1982). In pome fruits, it has been reported that sometimes in vitro flowering is observed in regenerated shoot cultures of European pears (*Pyrus communis*), a closely-related crop to apple; however, in vitro flowering of regenerated shoots of apples has not been reported under normal culture conditions (Pers. commun.).

11.4.4 Cultural Methods for Promoting Flowering

As perennial crops such as fruit trees have long juvenile periods, various cultural methods and techniques have been devised and used for shortening these juvenile periods to promote earlier flowering. For example, by optimizing cultural conditions such as growth temperature, light interception, day length, nutrients, and suppression of dormancy (by defoliation), it has been reported that 60–80% of apple seedlings can flower within a period of only 20 months following seed germination in a greenhouse (Aldwinckle 1975). This is in accordance with the proposal that a minimum plant size (minimum numbers of cell divisions) is necessary for a seedling to acquire the capability of developing flower buds. Furthermore, it is well known that grafting of apple scions onto rootstocks is effective in shortening the juvenile period, and promotes earlier flowering (Visser 1964).

In fruit breeding, cultural methods for shortening the juvenile period by grafting scions (budwood/scionwood) onto dwarfing rootstocks or top-grafting (grafting onto upper sections of mature trees) are commonly used. However, the physiological and molecular basis of early flower induction by grafting have not yet been delineated in woody plants such as the apple, although a mobile flowering promoter, FLOWERING LOCUS T (FT) protein, a systemic inducer of flowering may be involved in this mechanism, as will be presented in the next section.

11.5 Genes Controlling Flowering in Apple

Based on knowledge of flowering in *Arabidopsis* (Koornneef et al. 1998; Levy and Dean 1998), studies on the development of an early flowering/fruiting apple have been undertaken. Of course, as it is critical to shorten the breeding time for apple, as well as for other fruit trees, there has been a strong interest in identifying genes related to flower/fruit development in the apple.

Sung and An (1997) first reported on the isolation of an apple *MADS*-box gene, *MdMADS1*, that was expressed in reproductive organs of the 'Fuji' apple. Thereafter, several apple *MADS*-box genes likely related to flowering were isolated and characterized (Sung et al. 1999, 2000; Yao et al. 1999, 2001; Kotoda et al. 2000; van der Linden et al. 2002; Kitahara et al. 2004). The apple gene *MdMADS5* (*MdAP1*), proposed to be involved in promoting flower bud differentiation and calyx formation, was isolated and characterized (Yao et al. 1999; Kotoda et al. 2000). Using genetic transformation, *MdMADS5* (*MdAP1*) was found to be functionally homologous to a class A gene, *APETALA1* (*AP1*), a member of the Arabidopsis *MADS*-box gene family, in a heterologous system, transgenic Arabidopsis (Kotoda et al. 2002).

In addition to *MADS*-box genes, apple *LEAFY* (*LFY*) homologous genes, *AFL1* and *AFL2*, have been isolated and are found to be functionally similar to *LFY* in transgenic Arabidopsis (Wada et al. 2002). Moreover, *MdTFL1*, an apple gene homologous to *TERMINAL FLOWER1* (*TFL1*) that acts to maintain vegetative growth and suppresses flowering, is characterized using both transgenic Arabidopsis and apple (Kotoda et al. 2003; Kotoda and Wada 2005). Subsequently, Esumi et al. (2005) have identified two *TFL1* genes of apple, while Kotoda et al. (2006) have described an early flowering phenotype of transgenic apple using antisense *MdTFL1*. Thereafter, *MdTFL1a*, one of the two apple *TFL1*s, is fully characterized by Mimida et al. (2009).

The *FLOWERING LOCUS T* (*FT*), one of the floral promoters, was characterized in Arabidopsis using molecular genetics tools (Kardailsky et al. 1999; Kobayashi et al. 1999). The characteristics of *FT* attracted a lot of attention as it was proposed to be 'Florigen,' a mobile flower-promoting compound. Later, an apple *MdFT* (GenBank No. AB161112, referred to as *MdFT1*) was identified, together with *MdTFL1* (Kotoda and Wada 2005). Subsequently, two apple *FT*s were characterized in detail using both transgenic Arabidopsis and apple (Kotoda et al. 2010; Tränkner et al. 2010).

11.5.1 Genes Involved in Floral Organ Development

Thus far, many apple *MADS*-box genes (more than 100) have been isolated, and some of these genes have been classified into several classes/groups based on sequence analyses. These include class A, or *AP1/FRUITFULL* (*FUL*)-group (*MdMADS2*, *MdMADS5*, and *MdMADS12*); class B, or *PI/APETALA3* (*AP3*)-group (*MdPI*, *MdMADS13*, and *MdTM6*); class C, or *AGAMOUS* (*AG*)-group (*MdMADS10*, *MdMADS14*, and *MdMADS15*); and *SEPALLATA* (*SEP*)-group (*MdMADS1/MdMADS8*, *MdMADS3/MdMADS7*, *MdMADS4/MdMADS24*, *MdMADS6*, and *MdMADS9*) (Yao et al. 1999; van der Linden et al. 2002; Ireland et al. 2013).

Within class A or the *AP1/FRUITFULL* (*FUL*)-group genes in apple, expression analysis of *MdMADS2*, *MdMADS5* (*MdAP1*), and *MdMADS12* has been conducted. It is observed that *MdMADS2* is highly expressed in both sepals and petals but weakly expressed in both stamens and carpels; however, it is not expressed in leaves (Sung et al. 1999). On the other hand, *MdMADS5* (*MdAP1*) is preferentially expressed in sepals, beginning from the time of sepal formation following floral bud differentiation (Fig. 11.2; Kotoda et al. 2000). Whereas, *MdMADS12* is highly expressed in receptacles and vegetative tissues, such as leaves. Moreover, it is interesting to note that *MdMADS12* accumulates higher amounts of transcripts in mature leaves and shoots than in juvenile tissues (van der Linden et al. 2002). Therefore, as all three genes are expressed in floral organs, their involvements in floral organ development require further investigations.

Within class B or *PI/APETALA3* (*AP3*)-group genes, *MdPI*, *MdTM6*, and *MdMADS13* have been identified and characterized. As mentioned above, Sect. 11.3.2, the function of *MdPI*, responsible for the formation of petals and stamens in the apple, has been elucidated using genetic analysis of apple mutants of a class B gene (Yao et al. 2001). Furthermore, Yao et al. (2018) have developed transgenic 'Bolero' apple lines with either upregulated or downregulated *MdPI* and have confirmed that suppression of *MdPI* confers petal-to-sepal and stamen-to-carpel conversions. Additionally, overexpression of *MdPI* has resulted in the development of flattened apple fruit due to reduced cell growth. In Arabidopsis, PI is found to associate with AP3, thereby generating a PI/AP3 heterodimer that is functional (Riechmann et al. 1998; Honma and Goto 2001). Moreover, using a yeast two-hybrid experiment, it is revealed that MdMADS13 also interacts with MdPI, thus suggesting that these two proteins function cooperatively in apples (Wada et al. 2018).

Within class C or *AGAMOUS* (*AG*)-group genes, *MdMADS10*, *MdMADS14*, *MdMADS15*, and *MdMADS22* have been identified and characterized. Of these four apple genes, *MdMADS10*, similar to the Arabidopsis *SEEDSTICK* (*STK*) (Pinyopich et al. 2003), is found to be expressed within the core of apple fruit. While *MdMADS14*, similar to the Arabidopsis *SHATTERPROOF* (*SHP*) (Ma et al. 1991; Liljegren et al. 2000), is reported to be expressed mainly in carpels; whereas, *MdMADS15*, similar to the Arabidopsis *AG*, is observed to be expressed mainly in sepals, stamens, carpels, and receptacles (van der Linden et al. 2002).

Recently, Klocko et al. (2016) have developed a transgenic 'Galaxy' apple wherein both *MdMADS15* and *MdMADS22* are downregulated by RNA interference (RNAi) and have observed that suppression of *AG*-like genes increased numbers of whorls of petals but lacked anthers, thus enhancing the attractiveness of flowers possessing many petals. Therefore, the expression of *AG*-like genes would be either downregulated or lacking in those ornamental crabapples with double flowers (Fig. 11.5).

11.5.2 Genes Involved in Flower Induction

Over the past three decades, many genes involved in flower induction have been isolated, and their functions have been delineated in model plants, such as Arabidopsis, snapdragon, and rice. In the early 1990s, it has been reported that *FLORICAULA* (*FLO*)/*LFY* and *SQUAMOSA* (*SQUA*)/*AP1* play important roles in the transition from inflorescence to floral meristems (Coen et al. 1990; Irish and Sussex 1990; Huijser et al. 1992; Mandel et al. 1992; Weigel et al. 1992). *LFY* is defined as a floral meristem identity gene, similar to *FLO* of snapdragon; whereas, *AP1* functions as both a floral organ and a meristem identity gene.

Following a report that the juvenile phase of an aspen hybrid of two poplar species was reduced by constitutive expression of an Arabidopsis *LFY* gene (Weigel and Nilsson 1995), an apple *AP1* (*MdAP1*) and *LFY* (*AFL1* and *AFL2*) genes were isolated and characterized (Kotoda et al. 2000; Wada et al. 2002). It was observed that *MdMADS5* (*MdAP1*) was expressed at the developmental stage when sepal primordia would undergo differentiation (Fig. 11.2). Moreover, it was observed that *AFL2* was constitutively expressed throughout the year, but *AFL1* was seasonally expressed, with low levels of expression detected from mid-June to early July, just before floral bud differentiation (Fig. 11.2). Following expression analysis of these three genes, transgenic apples with constitutive expression of either *MdAP1* or *AFL1* were generated to develop an early flowering apple; however, these transgenic apples did not show early flowering (Wada and Kotoda 2003; Kotoda et al. 2006). Thus, it was proposed that apple *LFY* and *AP1* genes might not be sufficient for inducing apple flowering, and that overexpression of either *LFY* or *AP1* was not involved in breaking the juvenile phase in apple.

Subsequently, two genes related to flowering, *TFL1* and *FT*, have been isolated, wherein *TFL1* is a floral repressor (Bradley et al. 1997; Ohshima et al. 1997) and *FT* is a floral promoter (Kardailsky et al. 1999; Kobayashi et al. 1999). These two genes are similar in sequence, and they belong to the *TFL1*/*FT* family that consists of six members, including *ARABIDOPSIS THALIANA CENTRORADIALIS* (*ATC*), *BROTHER OF FT AND TFL1* (*BFT*), *MOTHER OF FT AND TFL1* (*MFT*), and *TWIN SISTER OF FT* (*TSF*) in Arabidopsis, but *TFL1* and *FT* function quite differently from each other (Kobayashi et al. 1999).

Based on *TFL1*, an apple *TFL1* (*MdTFL1*) was isolated, and its expression was investigated. It was found that expression of *MdTFL1* in apical buds of FBS was upregulated just before floral bud differentiation, but was downregulated as these buds approached stage 3 (inflorescence meristem) of development; that is, ~80 DAFB (Fig. 11.2), and was not detectable at stage 4 (floral meristem) of development. These findings clearly demonstrated that *MdTFL1* was developmentally regulated around the time of flower induction. To further delineate the function of *MdTFL1*, transgenic 'Orin' apples expressing antisense RNA of *MdTFL1* (35S:*MdTFL1AS*) were generated. A single transgenic apple plant flowered eight months following transfer to the greenhouse, while the control (non-transgenic) 'Orin' apples did not flower over a period of seven years (Fig. 11.6a–c). These findings further confirmed that *MdTFL1* was functional just like *TFL1*, and that *MdTFL1* was involved in maintaining the juvenile/vegetative phase in apples.

Following the discovery of the *FT* gene, many homologous genes were isolated from a variety of plant species. Endo et al. (2005) reported that

Fig. 11.6 Transgenic apple lines either suppressing *MdTFL1* or overexpressing *MdFT1*. **a** Twelve-month-old transgenic 'Orin' apple with 35S:*MdTFL1* antisense (35S: *MdTFL1AS*) following the break of the first dormancy period. **b** A flower of a transgenic line carrying an 35S:*MdTFL1AS* construct. **c** Fruit of a transgenic line carrying an 35S:*MdTFL1AS* construct. **d** Flower of a transgenic apple 'JM2' (line MdFT1#1-1), overexpressing an *MdFT1*, growing in the greenhouse; a white bar corresponds to 3 cm. **e** A close-up view of this 'JM2' as shown in **d** (Kotoda et al. 2006, 2010)

overexpression of citrus *FT* (*CiFT*, later referred to as *CiFT1*) resulted in highly precocious flowering in transgenic trifoliate oranges. This was the first report on the precocious flowering of transgenic woody plants by using an *FT* gene. Later, a total of three citrus *FT*s were identified, and *CiFT3* was then deemed to be involved in flower induction in citrus (Nishikawa et al. 2007). In the meantime, two apple *FT*s (*MdFT1* and *MdFT2*) were isolated and characterized in both transgenic Arabidopsis and apples (Kotoda and Wada 2005; Kotoda et al. 2010; Tränkner et al. 2010). Interestingly, these two apple *FT*s induced in vitro flowering, the first such observation in tissue culture of a standard apple cultivar. A transgenic 'JM2' apple line, overexpressing an *MdFT1*, with an early flowering phenotype (flowering 8–12 months after *Agrobacterium* infection, or 2–6 months following in vitro shoot regeneration) was subjected to further analysis (Figs. 11.6d and e). It was found that *FT* was expressed in leaves, and FT proteins translated from *FT* mRNA were then transported to apical meristems to induce flowering by forming an FT-FD complex (Abe et al. 2005; Corbesier et al. 2007; Giakountis and Coupland

2008; Zeevaart 2008). Therefore, this mobile flower-promoting substance was proposed to conform to the concept of a 'florigen', a long sought after plant flowering signal (Chailahyan 1968).

In another study, overexpression of *BpMADS4*, a birch (*Betula pendula*) *FUL*-like *MADS*-box gene, in transgenic apple has induced early flowering in apple cv. 'Pinova', with frequent in vitro flowering (Flachowsky et al. 2007). Interestingly, this phenotype resembles that of the *MdFT1*-overexpressing apple in that it produces solitary flower buds along with rounded leaf blades in tissue culture (Kotoda et al. 2010). As *MdMADS12* is a putative ortholog of *FUL*, *MdMADS12* is likely to play an important role in apple flower induction downstream of *MdFT1*.

11.5.3 Genetic Mapping of Apple *TFL1* and *FT*

The genus *Malus*, belonging to the Rosaceae subfamily Amygdaleae, has a higher chromosome number, $x = 17$, than other members of the family Rosaceae, such as peach (*Prunus persica*; $x = 8$), due to its polyploid origin (Sax 1933; Chevreau et al. 1985), resulting from a recent (> 50 Mya) genome-wide duplication (Velasco et al. 2010).

The genes *TFL1* and *FT* in apple have duplicates, *MdTFL1* and *MdTFL1a* for *TFL1* and *MdFT1* and *MdFT2* for *FT*, that are localized at similar positions on likely homoeologous linkage groups, LGs12 and 14 and LGs12 and 4, respectively (Fig. 11.7). As *MdTFL1* and *MdTFL1a* are similarly expressed in vegetative tissues in both adult and juvenile tissues, they could also function redundantly as floral repressors (Mimida et al. 2009). On the other hand, expression patterns of *MdFT1* and *MdFT2* are quite different as *MdFT1* is mainly expressed in apical buds of FBS, while *MdFT2* is mainly in both flower buds and young fruit (Kotoda et al. 2010). This suggests that these genes, *MdFT1* and *MdFT2*, must have undergone differentiation in function during the process of evolution following genome-wide duplication (Kotoda et al. 2010).

11.6 Transgenic Approaches for Controlling Flowering in Apple

It is well known that the apple is a woody perennial crop requiring a period of 5 to 7 years from seed germination to first flowering (juvenile period), and this is considerably longer than that of annual crops such as rice and tomatoes. Therefore, breeding fruit trees, such as apples, often takes at least 15–25 years, from sexual hybridization to selecting a new cultivar. As an example, 'Fuji' apple has been first evaluated as a seedling tree for its fruit characteristics 12 years following sexual hybridization (around 1951) and then has been subsequently released after 23 years from crossing (in 1962) as a named and registered cultivar (Sadamori et al. 1963). Therefore, shortening the juvenile period of fruit trees, such as that for apple, is directly related to the efficiency of breeding any new cultivars.

Earlier studies in Arabidopsis have demonstrated that artificial control of flowering time is possible by either enhancing or suppressing the expression of particular genes involved in flowering. Thus, two approaches, described below, can be considered to develop early flowering plants. In one approach, enhancing expression of genes involved in promoting early flowering depends on *LFY* and *AP1* genes, as they are involved in switching plants to move from vegetative to reproductive growth, as has been demonstrated in *Arabidopsis* (Weigel and Nilsson 1995; Levy and Dean 1998; Liljegren et al. 1999). In a second approach, suppressing expression of genes involved in flowering inhibition focuses on *TFL1* as it suppresses expression of *LFY* and *AP1*. Thus loss of *TFL1* function results in early flowering, and similar to the use of a flowering promoter (Bradley et al. 1997; Ohshima et al. 1997; Levy and Dean 1998; Liljegren et al. 1999).

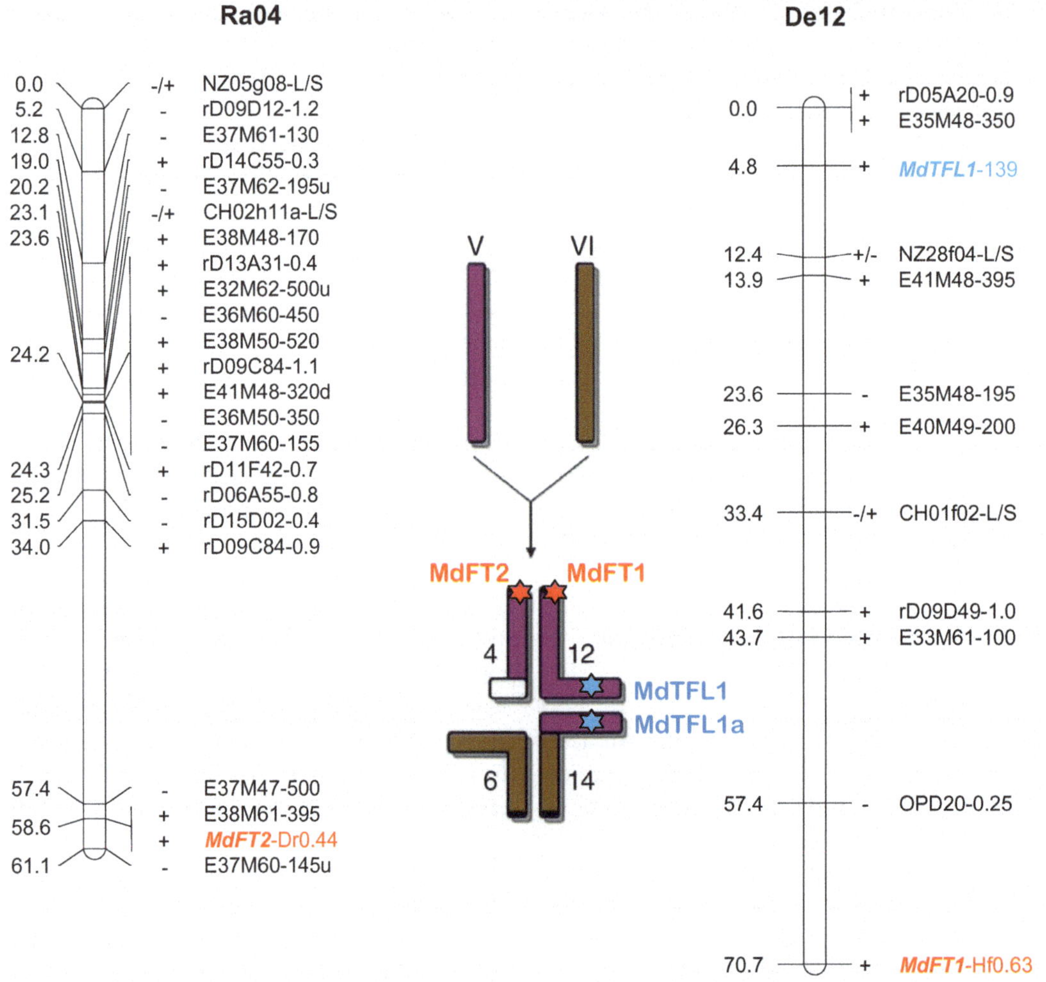

Fig. 11.7 Locations of *TFL1*/*FT*-like genes on linkage maps of apple cultivars 'Ralls Janet' (Ra) and 'Delicious' (De). Loci are listed on the right-hand side of the linkage maps, while genetic distances (cM) are shown on the left-hand side. Chromosomes 4, 6, 12, and 14 have originated from duplications of ancient chromosomes V and VI, followed by translocation and a deletion event. Shared colours in boxes indicate homology between extant chromosomes (Mimida et al. 2009; Kotoda et al. 2010; Velasco et al. 2010)

Over the last decade, it has become abundantly clear that *TFL1*- and *FT*-like genes are key for controlling flowering in woody plants, including apples, as described above (Sect. 11.5). A description of the utilization of *TFL1*- and *FT*-like genes to control flowering time in apples is herein presented.

11.6.1 Overexpression of Genes Promoting Flowering

As mentioned above, overexpression of *BpMADS4* (Flachowsky et al. 2007), *MdFT1* (Kotoda et al. 2010), and *MdFT2* (Tränkner et al. 2010, 2011) have been demonstrated to promote very early flowering, including in vitro flowering in apples. These transgenic approaches have paved the way for the development of an early flowering system for fruit trees. In particular, *BpMADS4* has been used to develop a "high-speed" or "fast-tracking" system for breeding apples. For example, transgenic apples have been developed wherein both scab- and fire-blight resistance genes are pyramided within a period of about two years following hybridization with a wild apple carrying fire-blight resistance, and then with a scab-resistant apple cultivar (Flachowsky et al. 2011; Le Roux et al. 2012). In another fast-track breeding approach, heat-induced expression of poplar (*Populus trichocarpa*) *FT* genes, driven by the heat-inducible *Gmhsp 17.5-E* (HSP)-promoter from soybean (*Glycine max*) is used (Wenzel et al. 2013). In this system, heat treatment of transgenic lines of apple cv. 'Pinova' at 42 °C for 1 h per day for about one month has induced flowering, and these flowers are found to be fertile, thus yielding viable offspring (Wenzel et al. 2013).

Likewise, another revolutionary system has been developed using a virus vector for reducing the generation time in apple seedlings. Yamagishi et al. (2011) have reported on the induction of early flowering in apple seedlings using the Apple Latent Spherical Virus (ALSV) vector expressing an Arabidopsis *FT* gene. It is observed that apple seedlings flowered within 1.5 to 2 months after inoculation of seed cotyledons with ALSV expressing *FT*, and these flowering seedlings have produced a new generation following hybridizations with a cross-compatible cultivar. Thus, this system reduces the generation time of apples to less than one year using a transient expression system of a virus vector without incurring any genetic modification for the next generation.

11.6.2 Development of an Early Flowering System Using Both *TFL1* and *FT* Genes

In another approach, an early flowering system has been developed using both *TFL1* and *FT* genes. In this system, expression of an Arabidopsis *FT* and silencing of *MdTFL1* accelerate flowering time and the life cycle of apple seedlings using an ALSV vector, ALSV-AtFT/MdTFL1 (Yamagishi et al. 2014). It is reported that more than 90% of infected seedlings flower within 1.5 to 3 months. It is also important to point out that in most cases, ALSV is not transmitted via seeds to successive progenies. Again, this transient expression system using a virus vector is different from that of an *Agrobacterium*-mediated transformation, as a transient expression system does not alter the genome of the host plant, while the latter modifies it by integrating T-DNA into the host genome.

11.6.3 Grafting of a Scion onto a Rootstock Expressing *FT* Genes to Induce Early Flowering

Grafting is often used for propagating apple scion cultivars onto rootstocks in order to maintain identity and obtain clonal plants. In addition, top-grafting is often used to promote earlier flowering (and fruit set) of seedlings in traditional apple breeding programmes. In this context, a highly desirable rootstock for promoting the early flowering of scion cultivars has also long been sought after. Thus, the discovery of FT, likely phloem-mobile florigen, sheds light on the development of such a desirable flower-promoting rootstock.

In grafting experiments of Arabidopsis, an *ft-7* rootstock expressing an FT-Green Fluorescent Protein (GFP) fusion encoding gene under the control of a *SUC2* promoter is reported to promote flowering of an *ft-7* mutant scion grafted onto this rootstock, and an FT-GFP protein is detected in the vascular tissue of the *ft-7* shoot (Corbesier et al. 2007). Furthermore, *ft-TWIN SISTER OF FT* (*tsf*) double-mutant scions grafted onto 35S::*FT* plants (*ft tsf*/35S::*FT* plants) have been observed to flower considerably earlier than *ft tsf*/*ft tsf* controls (Jin et al. 2015). In contrast, an apple rootstock expressing either *PtFT1* or *PtFT2* (*P. trichocarpa FT*) under the control of a heat-shock inducible promoter (*HSP*) has not contributed to the early flowering of a grafted wild-type scion apple cultivar following a long-term heat treatment (Wenzel et al. 2013). However, transgenic apple expressing either *PtFT1* or *PtFT2* is observed to flower within six days following a heat treatment (Wenzel et al. 2013).

Similarly, transgenic hybrid aspen (*Populus tremula* × *P. tremuloides* or *P. tremula* × *P. alba*) rootstocks expressing an Arabidopsis *FT* under the control of a heat-shock inducible promoter have not induced flowering of non-transgenic scions, although 66% of non-grafted *FT*-expressing plants have flowered following an inductive heat treatment (Zhang et al. 2010). In trees such as apples and aspens, FT is likely to be either non-transmissible or effective enough to promote flowering by grafting, unlike what is observed in annual plants and in shrubs (Putterill and Varkonyi-Gasic 2016).

11.7 Concluding Remarks and Future Perspectives

The history of cultivated apples is quite long, dating all the way back before the Grecian era (Janick et al. 1996; Luby 2003). Thus, many apple cultivars/strains have been identified and selected. Furthermore, descriptions of various physiological traits for apple have long been accumulated, thereby offering clues that aid in elucidating mechanisms of such physiological and genetic traits, such as those for flowering and juvenility.

Based on ever-accumulating knowledge and recent studies, it has been revealed that *TFL1/FT*-like genes play critical roles in controlling flowering, including that in fruit trees. However, additional physiological traits/factors are yet to be elucidated, both physiologically and genetically, in apples. These include those factors that maintain juvenility, internal and external conditions of flower induction in the seasonal flowering cycle, promotion of flowering by grafting, biennial bearing, and dormancy, although the last two traits have not been discussed herein (see Chaps. 5 and 12 in this volume).

The various traits/factors mentioned above seem to be partially associated with the actions of *TFL1/FT* genes, as well as with their regulators. Therefore, interacting genes with *TFL1/FT* should be of interest in unveiling those multiple functions of *FT*-like genes (Mimida et al. 2011; Putterill and Varkonyi-Gasic 2016). When growing apple seedlings as single stems under appropriate environmental and cultural management conditions, flowering can be induced along the top sections of stems in either the second or third year, thus suggesting that moving inhibitory compounds produced in the roots and moving up along through the stem may prevent flowering during the juvenile phase (Dennis et al. 2003). This reported hypothesis is known as the effect of the distance from the roots on juvenility. Interestingly, likely floral repressors of apples, *MdTFL1*, *MdTFL1a*, and *MdCENa*, are highly expressed in roots (Mimida et al. 2009). Thus, the functions of floral repressors should be thoroughly investigated to further elucidate the mechanism of juvenility in various woody plants.

Although not mentioned here, the involvement of low-molecular signalling compounds, such as plant hormones and sugars, in flower induction or suppression should be further investigated (Xing et al. 2015; Li et al. 2018). Moreover, developing a system that enables control of flowering time, either directly or indirectly, using either *TFL1/FT* or other flowering genes related thereto, would also be of great importance for the fruit industry, including that of the apple industry.

References

Abe M, Kobayashi Y, Yamamoto S, Daimon Y, Yamaguchi A, Ikeda Y, Ichinoki H, Notaguchi M, Goto K, Araki T (2005) FD, a bZIP protein mediating signals from the floral pathway integrator FT at the shoot apex. Science 309:1052–1056

Aldwinckle HS (1975) Flowering of apple seedlings 16–20 months after germination. HortScience 10(2):124–126

Bowman JL (2012) The ABC model of flower development: then and now. Development 139:4095–4098
Bowman JL, Smyth DR, Meyerowitz EM (1991) Genetic interactions among floral homeotic genes of *Arabidopsis*. Development 112:1–20
Bradley D, Ratcliffe O, Vincent C, Carpenter R, Coen E (1997) Inflorescence commitment and architecture in *Arabidopsis*. Science 275:80–83
Brown SK, Maloney KE (2003) Genetic improvement of apple: breeding, markers, mapping and biotechnology. In: Ferree DC, Warrington IJ (eds) Apples: botany, production, and uses. CAB International Publisher, Oxon, pp 31–59
Buban T, Faust M (1982) Flower bud induction in apple trees: Internal control and differentiation. Hortic Rev 4:174–203
Coen ES, Meyerowitz EM (1991) The war of the whorls: genetic interactions controlling flower development. Nature 353:31–37
Coen ES, Romero JM, Doyle S, Elliott R, Murphy G, Carpenter R (1990) *floricaula*: a homeotic gene required for flower development in *Antirrhinum majus*. Cell 63(6):1311–1322
Chailahyan MK (1968) Internal factors of plant flowering. Annu Rev Plant Physiol 19:1–37
Chevreau E, Lespinasse Y, Gallet M (1985) Inheritance of pollen enzymes and polyploidy origin of apple (*Malus* × *domestica* Borkh.). Theor Appl Genet 71:268–277
Corbesier L, Vincent C, Jang S, Fornara F, Fan Q, Searle I, Giakountis A, Farrona S, Gissot L, Turnbull C, Coupland G (2007) FT protein movement contributes to long-distance signaling in floral induction of *Arabidopsis*. Science 316:1030–1033
Daccord N, Celton JM, Linsmith G, Becker C, Choisne N, Schijlen E, van de Geest H, Bianco L, Micheletti D, Velasco R, Di Pierro EA, Gouzy J, Rees DJG, Guérif P, Muranty H, Durel CE, Laurens F, Lespinasse Y, Gaillard S, Aubourg S, Quesneville H, Weigel D, van de Weg E, Troggio M, Bucher E (2017) High-quality de novo assembly of the apple genome and methylome dynamics of early fruit development. Nat Genet 49:1099–1106
Dennis FG Jr (2003) Flowering, pollination and fruit set and development. In: Ferree DC, Warrington IJ (eds) Apples: botany, production, and uses. CAB International Publisher, Oxon, pp 153–166
Endo T, Shimada T, Fujii H, Kobayashi Y, Araki T, Omura M (2005) Ectopic expression of an *FT* homolog from *Citrus* confers an early flowering phenotype on trifoliate orange (*Poncirus trifoliate* L. Raf.). Transgenic Res 14:703–712
Esumi T, Tao R, Yonemori K (2005) Isolation of *LEAFY* and *TERMINAL FLOWER1* homologues from six fruit tree species in the subfamily Maloideae of the family Rosaceae. Sex Plant Reprod 17:277–287
Flachowsky H, Le Roux PM, Peil A, Patocchi A, Richter K, Hanke MV (2011) Application of a high-speed breeding technology to apple (*Malus* × *domestica*) based on transgenic early flowering plants and marker-assisted selection. New Phytol 192:364–377
Flachowsky H, Peil A, Sopanen T, Elo A, Hanke V (2007) Overexpression of *BpMADS4* from silver birch (*Betula pendula*) in apple (*Malus* × *domestica*) induces early flowering. Plant Breed 126:137–145
Foster T, Johnston R, Seleznyova A (2003) A morphological and quantitative characterization of early floral development in apple (*Malus* × *domestica* Borkh.). Ann Bot 92:199–206
Gasic K, Gonzales DO, Thimmapuram J, Liu L, Malnoy M, Gong G, Han Y, Vodkin LO, Liu L, Aldwinckle H, Carroll N, Orvis L, Goldsbrough P, Clifton S, Fulton L, Dante M, Theising B, Martin J, Pape D, Wisniewski ME, Fazio G, Feltus FA, Korban SS (2009) Comparative analysis and functional annotation of an expressed sequence tag collection of apple. Plant Genome 2:23–38
Giakountis A, Coupland G (2008) Phloem transport of flowering signals. Curr Opin Plant Biol 11:687–694
Gianfranceschi L, Seglias N, Tarchini R, Komjanc M, Gessler C (1998) Simple sequence repeats for the genetic analysis of apple. Theor Appl Genet 96 (8):1069–1076
Hackett WP (1985) Juvenility, maturation, and rejuvenility in woody plants. Hortic Rev 7:109–155
Hanke M-V, Flachowsky H, Peil A, Hättasch C (2007) No flower no fruit—Genetic potentials to trigger flowering in fruit trees. Genes Genomes Genomics 1:1–20
Hokanson S, Szewc-McFadden A, Lamboy W, McFerson JR (1998) Microsatellite (SSR) markers reveal genetic identities, genetic diversity and relationships in a *Malus* × *domestica* Borkh. core subset collection. Theor Appl Genet 97(5–6):671–683
Honma T, Goto K (2001) Complexes of MADS-box proteins are sufficient to convert leaves into floral organs. Nature 409:525–529
Huijser P, Klein J, Lonnig W, Meijer H, Saedler H, Sommer H (1992) Bracteomania, an inflorescence anomaly, is caused by the loss of function of the MADS-box gene *squamosa* in *Antirrhinum majus*. EMBO J 11:1239–1249
Ireland HS, Yao J-L, Tomes S, Sutherland PW, Nieuwenhuizen N, Gunasee-lan K, Winz RA, David KM, Schaffer RJ (2013) Apple *SEPALLATA1/2*-like genes control fruit flesh development and ripening. Plant J 73:1044–1056
Irish VF, Sussex IM (1990) Function of the *apetala-1* gene during *Arabidopsis* floral development. Plant Cell 2:741–753
James DJ, Passey AJ, Barbara DJ, Bevan M (1989) Genetic transformation of apple (*Malus pumila* Mill.) using a disarmed Ti-binary vector. Plant Cell Rep 7:658–661
Janick J, Cummins JN, Brown SK, Hemmat M (1996) Apples. In: Janick J, Moor JN (eds) Fruit Breeding. John Wiley & Sons Inc., New York, pp 1–77

Jin S, Jung HS, Chung KS, Lee JH, Ahn JH (2015) FLOWERING LOCUS T has higher protein mobility than TWIN SISTER OF FT. J Exp Bot 66(20):6109–6117

Kardailsky I, Shukla VK, Ahn JH, Dagenais N, Christensen SK, Nguyen JT, Chory J, Harrison MJ, Weigel D (1999) Activation tagging of the floral inducer *FT*. Science 286:1962–1965

Khan MA, Han Y, Zhao YF, Korban SS (2012) A high-throughput apple SNP genotyping platform using the GoldenGate™ assay. Gene 494:196–201

Kitahara K, Ohtsubo T, Soejima J, Matsumoto S (2004) Cloning and characterization of apple class B MADS-box genes including a novel *AP3* homologue *MdTM6*. J Jpn Soc Hortic Sci 73:208–215

Klocko AL, Borejsza-Wysocka E, Brunner AM, Shevchenko O, Aldwinckle H, Strauss SH (2016) Transgenic suppression of *AGAMOUS* genes in apple reduces fertile and increases floral attractiveness. PLoS ONE 11:e0159421. https://doi.org/10.1371/journal.pone.0159421

Kobayashi Y, Kaya H, Goto K, Iwabuchi M. Araki T (1999) A pair of related genes with antagonistic roles in mediating flowering signals. Science 286:1960–1962

Koornneef M, Alonso-Blanco C, Peeters AJM, Soppe W (1998) Genetic control of flowering time in Arabidopsis. Annu Rev Plant Physiol Plant Mol Biol 49:345–370

Kotoda N, Wada M (2005) *MdTFL1*, a *TFL1*-like gene of apple, retards the transition from the vegetative to reproductive phase in transgenic *Arabidopsis*. Plant Sci 168:95–104

Kotoda N, Wada M, Komori S, Kidou S, Abe K, Masuda T, Soejima T (2000) Expression pattern of homologues of floral meristem identity genes *LFY* and *AP1* during flower development in apple. J Am Soc Hortic Sci 125:398–403

Kotoda N, Wada M, Kusaba S, Kano-Murakami Y, Masuda T, Soejima J (2002) Overexpression of *MdMADS5*, an *APETALA1*-like gene of apple, causes early flowering in transgenic *Arabidopsis*. Plant Sci 162:679–687

Kotoda N, Wada M, Masuda T, Soejima J (2003) The break-through in the reduction of juvenile phase in apple using transgenic approaches. Acta Hortic 625:337–343

Kotoda N, Iwanami H, Takahashi S, Abe K (2006) Antisense expression of *MdTFL1*, a *TFL1*-like gene, reduces the juvenile phase in apple. J Am Soc Hortic Sci 131:74–81

Kotoda N, Hayashi H, Suzuki M, Igarashi M, Hatsuyama Y, Kidou S, Igasaki T, Nishiguchi M, Yano K, Shimizu T, Takahashi S, Iwanami H, Moriya S, Abe K (2010) Molecular characterization of *FLOWERING LOCUS T*-like genes of apple (*Malus* × *domestica* Borkh.) Plant Cell Physiol 51 (4):561–575

Le Roux PM, Flachowsky H, Hanke MV, Gessler C, Patocchi A (2012) Use of a transgenic early flowering approach in apple (*Malus* × *domestica* Borkh.) to introgress fire blight resistance from cultivar Evereste. Mol Breed 30:857–874

Levy YY, Dean C (1998) The transition to flowering. Plant Cell 10:1973–1989

Li Y, Zhang D, Zhang X, Xing L, Fan S, Ma J, Zhao C, Du L, Han H (2018) A transcriptome analysis of two apple (*Malus* × *domestica*) cultivars with different flowering abilities reveals a gene network module associated with floral transitions. Sci Hortic 239 (2018):269–281

Liljegren SJ, Gustafson-Brown C, Pinyopich A, Ditta GS, Yanofsky MF (1999) Interactions among *APETALA1*, *LEAFY*, and *TERMINAL FLOWER1* specify meristem fate. Plant Cell 11:1007–1018

Liljegren SJ, Ditta GS, Eshed Y, Savidge B, Bowman JL, Yanofsky MF (2000) *SHATTERPROOF MADS*-box genes control seed dispersal in *Arabidopsis*. Nature 404:766–770

Luby JJ (2003) Taxonomic classification and brief history. In: Ferree DC, Warrington IJ (eds) Apples: botany, production, and uses. CAB International Publisher, Oxon, pp 1–29

Ma H, Yanofsky MF, Meyerowitz EM (1991) *AGL1-AGL6*, an *Arabidopsis* gene family with similarity to floral homeotic and transcription factor genes. Genes Dev 5:484–495

Mandel MA, Gustafson-Brown C, Savidge B, Yanofsky MF (1992) Molecular characterization of the *Arabidopsis* floral homeotic gene *APETALA1*. Nature 360:273–277

Melzer R, Theissen G (2009) Reconstitution of 'floral quartets' *in vitro* involving class B and class E floral homeotic proteins. Nucl Acids Res 37:2723–2736

Mimida N, Kotoda N, Ueda T, Igarashi M, Hatsuyama Y, Iwanami H, Moriya S, Abe K (2009) Four *TFL1/CEN*-like genes on distinct linkage groups show different expression patterns to regulate vegetative and reproductive development in apple (*Malus* × *domestica* Borkh.). Plant Cell Physiol 50:394–412

Mimida N, Kidou S, Iwanami H, Moriya S, Abe K, Voogd C, Varkonyi-Gasic E, Kotoda N (2011) Apple FLOWERING LOCUS T proteins interact with transcription factors implicated in cell growth and organ development. Tree Physiol 31:555–566

Newcomb RD, Crowhurst RN, Gleave AP, Rikkerink EHA, Allan AC, Beuning LL, Bowen JH, Gera E, Jamieson KR, Janssen BJ, Laing WA, McArtney S, Nain B, Ross GS, Snowden KC, Souleyre EJF, Walton EF, Yauk Y-K (2005) Analyses of expressed sequence tags from apple. Plant Physiol 141:147 166

Nishikawa F, Endo T, Shimada T, Fujii H, Shimizu T, Omura M (2007) Increased *CiFT* abundance in the stem correlates with floral induction by low temperature in Satsuma mandarin (*Citrus unshiu* Marc.). J Exp Bot 54:2439–2448

Ohshima S, Murata M, Sakamoto W, Ogura Y, Motoyoshi F (1997) Cloning and molecular analysis

of the *Arabidopsis* gene *Terminal Flower 1*. Mol Gen Genet 254:186–194
Pinyopich A, Ditta GS, Savidge B, Liljegren SJ, Baumann E, Wisman E, Yanofsky MF (2003) Assessing the redundancy of MADS-box genes during carpel and ovule development. Nature 424:85–88
Poethig RS (1990) Phase change and the regulation of shoot morphogenesis in plants. Science 250 (4983):923–930
Poethig RS (2003) Phase change and the regulation of developmental timing in plants. Science 301 (5631):334–336
Putterill J, Varkonyi-Gasic E (2016) FT and florigen long-distance flowering control in plants. Curr Opin Plant Biol 33:77–82
Riechmann JL, Krizek BA, Meyerowitz EM (1998) Dimerization specificity of Arabidopsis MADS domain homeotic proteins APETALA1, APETALA3, PISTILLATA, and AGAMOUS. Proc Natl Acad Sci USA 93:4793–4798
Sadamori S, Yoshida Y, Murakami H, Ishizuka S (1963) New apple variety 'Fuji'. Bull Hortic Res Station. Jpn Ser (C) 1:1–6
Sax K (1933) The origin of the Pomoideae. Proc Am Soc Hortic Sci 30:147–150
Scorza R (1982) *In vitro* flowering. Hortic Rev 4:106–127
Sung S-K, An G (1997) Molecular cloning and characterization of a MADS-box cDNA clone of the Fuji apple. Plant Cell Physiol 38:484–489
Sung S-K, Yu GH, An G (1999) Characterization of *MdMADS2*, a member of the *SQUAMOSA* subfamily of genes, in apple. Plant Physiol 120:969–978
Sung S-K, Yu G-H, Nam J, Jeong D-H, An G (2000) Developmentally regulated expression of two MADS-box genes, *MdMADS3* and *MdMADS4*, in the morphogenesis of flower buds and fruits in apple. Planta 210:519–528
Tränkner C, Lehmann S, Hoenicka H, Hanke M-V, Fladung M, Lenhardt D, Dunemann F, Gau A, Schlangen K, Malnoy M, Flachowsky H (2010) Over-expression of an *FT*-homologous gene of apple induces early flowering in annual and perennial plants. Planta 232:1309–1324
Tränkner C, Lehmann S, Hoenicka H, Hanke M-V, Fladung M, Lenhardt D, Dunemann F, Gau A, Schlangen K, Malnoy M, Flachowsky H (2011) Note added in proof to: Over-expression of an *FT*-homologous gene of apple induces early flowering in annual and perennial plants. Planta 233:217–218
van der Linden CG, Vosman B, Smulders MJM (2002) Cloning and characterization of four apple MADS box genes isolated from vegetative tissues. J Exp Bot 53:1025–1036
Velasco R, Zharkikh A, Affourtit J, Dhingra A, Cestaro A, Kalyanaraman A, Fontana P, Bhatnagar SK, Troggio M, Pruss D, Salvi S, Pindo1 M, Baldi P, Castelletti S, Cavaiuolo M, Coppola G, Costa F, Cova V, Dal Ri A, Goremykin V, Komjanc M, Longhi S, Magnago P, Malacarne1 G, Malnoy M, Micheletti D, Moretto M, Perazzolli M, Si-Ammour A, Vezzulli S, Zini E, Eldredge G, Fitzgerald LM, Gutin N, Lanchbury J, Macalma T, Mitchell JT, Reid J, Wardell B, Kodira C, Chen Z, Desany B, Niazi F, Palmer M, Koepke T, Jiwan D, Schaeffer S, Krishnan V, Wu C, Chu VT, King ST, Vick J, Tao Q, Mraz A, Stormo A, Stormo K, Bogden R, Ederle D, Stella A, Vecchietti A, Kater MM, Masiero S, Lasserre P, Lespinasse Y, Allan AC, Bus V, Chagné D, Crowhurst RN, Gleave AP, Lavezzo E, Fawcett JA, Proost S, Rouzé P, Sterck L, Toppo S, Lazzari B, Hellens RP, Durel C-E, Gutin A, Bumgarner RE, Gardiner SE, Skolnick M, Egholm M, Van de Peer Y, Salamini F, Viola R (2010) The genome of the domesticated apple (*Malus* × *domestica* Borkh.). Nat Genet 42(10):833–841
Visser T (1964) Juvenile phase and growth of apple and pear seedlings. Euphytica 13:119–129
Wada M, Kotoda N (2003) Flowering genes of apple. Flower News Lett 35:18–27
Wada M, Cao Q, Kotoda N, Soejima J, Masuda T (2002) Apple has two orthologues of *FLORICAULA*/ *LEAFY* involved in flowering. Plant Mol Biol 49:567–577
Wada M, Oshino H, Tanaka N, Mimida N, Moriya-Tanaka Y, Honda C, Hanada T, Iwanami H, Komori S (2018) Expression and functional analysis of apple *MdMADS13* on flower and fruit formation. Plant Biotechnol. https://doi.org/10.5511/plantbiotechnology.18.0510a
Wada M, Nishitani C, Komori S (2020) Stable and efficient transformation of apple. Plant Biotechnol 37 (2):163–170. https://doi.org/10.5511/plantbiotechnology.20.0602a
Weigel D, Alvarez J, Smyth DR, Yanofsky MF, Meyerowitz EM (1992) *LEAFY* controls floral meristem identity in *Arabidopsis*. Cell 69:843–859
Weigel D, Nilsson O (1995) A developmental switch sufficient for flower initiation in diverse plants. Nature 377:495–500
Wenzel S, Flachowsky H, Hanke M-V (2013) The Fast-track breeding approach can be improved by heat-induced expression of the *FLOWERING LOCUS T* genes from poplar (*Populus trichocarpa*) in apple (*Malus* × *domestica* Borkh.). Plant Cell Tiss Organ Cult 115:127–137
Xing L-B, Zhang D, Li Y-M, Shen Y-W, Zhao C-P, Ma J-J, An N, Han M-Y (2015) Transcription profiles reveal sugar and hormone signaling pathways mediating flower induction in apple (*Malus domestica* Borkh.). Plant Cell Physiol 56(10):2052–2068
Yamagishi N, Sasaki S, Yamagata K, Komori S, Nagase M, Wada M, Yamamoto T, Yoshikawa N (2011) Promotion of flowering and reduction of a generation time in apple seedlings by ectopical expression of the *Arabidopsis thaliana FT* gene using the Apple latent spherical virus vector. Plant Mol Biol 75:193–204
Yamagishi N, Kishigami R, Yoshikawa N (2014) Reduced generation time of apple seedlings to within a year by means of a plant virus vector: a new plant-breeding technique with no transmission of genetic

modification to the next generation. Plant Biotech J 12:60–68

Yao J-L, Dong Y-H, Kvarnheden A, Morris B (1999) Seven MADS-box genes in apple are expressed in different parts of the fruit. J Am Soc Hortic Sci 124:8–13

Yao J-L, Dong Y-H, Morris BAM (2001) Parthenocarpic apple fruit production conferred by transposon insertion mutations in a MADS-box transcription factor. Proc Natl Acad Sci USA 98:1306–1311

Yao J-L, Xu J, Tomes S, Cui W, Luo Z, Deng C, Ireland HS, Schaffer RJ, Gleave AP (2018) Ectopic expression of the *PISTILLATA* homologous *MdPI* inhibits fruit tissue growth and changes fruit shape in apple. Plant Direct 2018:1–11. https://doi.org/10.1002/pld3.51

Zeevaart JA (2008) Leaf-produced floral signals. Curr Opin Plant Biol 11:541–547

Zhang H, Harry DE, Ma C, Yuceer C, Hsu C-Y, Vikram V, Shevchenko O, Etherington E, Strauss SH (2010) Precocious flowering in trees: the *FLOWERING LOCUS T* gene as a research and breeding tool in *Populus*. J Exp Bot 61(10):2549–2560

Zimmerman RH (1972) Juvenility and flowering in woody plants: a review. HortScience 7:447–455

12 Genetics and Genomics of Cold Hardiness and Dormancy

Jianyang Liu, Timothy S. Artlip,
Sherif M. Sherif, and Michael E. Wisniewski

Abstract

Successful cropping of apple depends on the ability of trees in the orchard to undergo both a period of dormancy and to acclimate to low winter temperatures. These physiological processes are inter-connected and involve cross-talk via shared signal transduction pathway components, phytohormones, and the environment. However, these processes also possess unique components, including activities of downstream regulons that do not appear to interact with each other. Apple dormancy differs from those of many other woody perennials in that it is solely temperature-dependent, rather than relying on both photoperiod and temperature cues. Apple cultivars can also vary considerably in dormancy duration. Dormancy duration is regulated by chilling and heating requirements, that is, numbers of hours between 0 °C and 7 °C, and numbers of hours above 7 °C following chilling, respectively, and both requirements are genetically controlled. Several quantitative trait loci (QTLs) have been identified on Linkage Group (LG) 9 using mapping populations derived from crosses between low- and high-chilling genotypes. Furthermore, regulation and genetics of cold hardiness have also been widely investigated. It is observed that levels of cold tolerance vary among seasons and among apple genotypes. Interestingly, there are several similarities in responses of both herbaceous and woody perennial plants to low temperatures. Notably, C-repeat binding transcription factors (CBFs) appear to serve as central regulators of cold signal transduction and initiation of cold acclimation. However, some regulators in the CBF pathway in apple have not been identified in herbaceous plants, thus suggesting that the CBF-regulatory pathway in apple, and perhaps in other woody perennials, may differ from that in herbaceous plants, such as Arabidopsis. Most low-temperature plant research studies have focused on plant responses to acute low-temperature stress, but not on the abilities of plants, such as apple, to maintain freezing tolerance under chronic and extended periods of low-temperature stress. Therefore, acquiring

Jianyang Liu, Timothy S. Artlip—These authors contributed equally to this chapter.

J. Liu · S. M. Sherif (✉)
School of Plant and Environmental Sciences, Alson H. Smith Jr. Agricultural Research and Extension Center, Virginia Tech, Winchester, VA 22602, USA
e-mail: ssherif@vt.edu

T. S. Artlip (✉)
USDA-ARS-Appalachian Fruit Research Station, Kearneysville, WV 25430, USA
e-mail: tim.artlip@usda.gov

M. E. Wisniewski
Department of Biological Sciences, Virginia Tech, Blacksburg, VA 24061, USA

S. S. Korban (ed.), *The Apple Genome*, Compendium of Plant Genomes,
https://doi.org/10.1007/978-3-030-74682-7_12

a better understanding of apple genetics, molecular biology, and plant responses to complex and dynamic environmental changes is critical for apple genetic improvement efforts and management of apple orchards in response to environmental stress, particularly under current climate change conditions.

12.1 Introduction

As bud dormancy and cold hardiness in perennial plant crops are critical for apple production, related research studies have been ongoing for many decades. Recently increased concerns over climate change and its potential impact on several aspects of plant phenology, particularly those associated with overwintering mechanisms of temperate fruit crops, have highlighted the necessity for increased research.

The apple has originated in temperate regions of Central Asia, where the seasonal temperature fluctuates well above and below 0 °C (Luby et al. 2001; Velasco et al. 2010; Daccord et al. 2017). *Malus* species, both wild and domesticated, have developed adaptive strategies to cope with low-winter temperatures by entering dormancy and undergoing a process of cold acclimation. As active growth of meristematic tissues is inhibited during dormancy and during the process of cold acclimation, living cells undergo biochemical and structural changes that enable them to either avoid or tolerate dehydration stress associated with freezing temperatures. These processes involve several phenological events, including growth cessation, dormancy establishment, cold acclimation, chilling and heat accumulation, dormancy release, and growth resumption in the next growing season. In woody perennials, including temperate fruit trees, endodormancy and survival of plant organs in response to low temperatures (chilling; 0 °C to ca. 7 °C) and/or freezing are linked traits. Cold acclimation is a complex phenomenon involving structural and biochemical alterations, both of which are regulated by changes in gene expression in response to environmental cues. Even when cold hardy, winter injury can still occur in all organs of an apple tree, including roots, despite having undergone cold acclimation. This is due to the extent and duration of freezing tolerance, as well as to the apple tree's ability to maintain maximum freezing tolerance during periods of fluctuating temperatures. These capabilities are under genetic control, and these can vary among apple genotypes/species. Interestingly, studies related to the causes and effects of winter injury on apple trees have been undertaken for over 100 years, as evidenced in a description of the problem by Gardner et al. (1922).

Freezing damage to cells can be attributed to either extracellular or intracellular ice formation. Extracellular ice formation imposes dehydration stress on a cell, as differences in water potential between the cell and the extracellular ice cause water to move out of the cell. In fact, the level of cold hardiness of a genotype is often defined by its ability to withstand a specific level of dehydration stress. In response to the presence of extracellular ice, plant cells undergo the process of cytorrhysis, wherein the whole plant cell compresses and collapses, resulting in permanent damage of the cell wall (Molisch 1897; Wiegand 1906). This process differs from osmotic-induced dehydration, such as salt stress, whereby the plasma membrane pulls away from the cell wall but retains its original shape. It is important to point out that intracellular ice formation is deemed an immediate lethal event wherein all plant cell membranes, organelles, and other structures are disrupted by ice crystals. Upon thawing, these leak out of the cell.

In woody perennials, dormancy is usually referred to as the absence of detectable growth. Based on sources of signals inducing dormancy, it can be categorised into three classes: paradormancy, whereby signals from apical meristems suppress lateral bud growth; endodormancy, whereby growth inhibition is governed by biochemical and molecular signals released from buds; and ecodormancy, whereby bud growth is inhibited by surrounding environmental conditions, such as low temperatures (Lang 1987). In recent years, the definition of bud dormancy has been revised as the "*inability to resume growth from meristems under favorable conditions*" (Rohde and Bhalerao 2007). This definition highlights the status of the meristem, distinguishing it from ecodormancy, and has gained growing recognition (Rohde and Bhalerao 2007; Cooke et al. 2012; Considine and Considine 2016). In this chapter, we will focus on endodormancy, the key mechanism for apple trees to overwinter. Therefore, the term dormancy will be used interchangeably with endodormancy, unless otherwise specified.

Dormancy is a highly regulated and complex process that is subject to influences by various internal and external factors. Regrowth and flowering of apple buds in the spring are dependent on successful bud dormancy and cold acclimation during the winter; thus, highly impacting annual fruit production. Unlike several other woody species, apple trees receive cues from low temperatures in the fall season, instead of short photoperiods, to initiate dormancy (Heide and Prestrud 2005). During dormancy, apple trees must undergo a period of cold exposure, referred to as the chilling requirement (CR), typically expressed as chilling hours/units (CUs), to exit dormancy. Following completion of CR, a period of warm temperature, referred to as the heat requirement (HR), is required to induce uniform budbreak or flowering. Both CR and HR are genetically determined traits, and these are highly variable across apple cultivars (Labuschagné et al. 2002). If CRs or HRs are unmet, this may lead to uneven budbreak or poor floral competency in apple (Hawerroth et al. 2013). CR in apple is reported to have a higher influence than HR on flowering date (Celton et al. 2011). However, HR can counterbalance either insufficient or excess chilling exposure (Ruiz et al. 2007); moreover, exposure to heat wave(s) during endodormancy can partially counterbalance or offset chilling effects (Anzanello et al. 2014), thus suggesting interdependence and overlap between CR and HR. Dormancy parameters are critical in breeding programmes and cultivar selection, as they largely determine the geographical distribution and production of apple worldwide. Recent research has indicated that such parameters are also influenced and shaped by global climate change (Nanninga et al. 2017; Rinne et al. 2018; Wenden et al. 2020).

A major consequence of global climate change is inadequate or disturbed dormancy. A direct demonstration of the effect(s) of such disturbance is an increased risk of frost damage. Therefore, following budbreak, deciduous trees lose cold hardiness, and meristem tissues within buds may suffer cold injuries as a result of a rapid drop in temperature. Global climate change increases frequencies of both spring frosts (Liu et al. 2018) and warm spells in late winter or in early spring (Ma et al. 2019), both of which contribute to increased odds of frost damage. Studies have reported that, over the last three decades, apple bloom has advanced by 6–9 days in Europe (Vitasse et al. 2011; Hoffmann and Rath 2013; Vitasse et al. 2018), thereby exposing fragile floral tissues to damaging spring freeze. Furthermore, the loss of winter chilling conditions required for release from dormancy for many temperate perennials, including apple, is projected to continue for decades to come (Augspurger 2013; Unterberger et al. 2018; Ma et al. 2019). By comparing dormancy dynamics in apple trees grown under contrasting climates, Malagi et al. (2015) have demonstrated that climate dictates entry into and intensity, of dormancy, as a mild winter leads to weakened and shortened endodormancy. As dormancy-related issues are becoming of the general concern in fruit production, various efforts are underway to combat mild winters and late frosts (Liu and Sherif 2019). A key question, which is of considerable interest in apple biology, is how the dormancy-growth cycle is controlled and what key components are involved and orchestrated in dormancy regulation.

12.2 Dormancy

12.2.1 Hormonal Control of Bud Dormancy

Plant hormones are extensively involved in the regulation of dormancy. While functions of some hormones such as abscisic acid (ABA) and gibberellins/gibberellic acid (GA) have been convincingly associated with particular aspects of dormancy, roles of some other hormones, such as ethylene and cytokinin (CK), appear to defy clear generalisation, as their functions are highly context-dependent, and are variable at different stages of plant development or across plant species.

Often, the roles of plant hormones are investigated based on the dynamics of metabolism and signalling. In addition to direct measurements of hormone levels, useful insights can be gained via various molecular tools, such as mutagenesis, analysis of expression of biosynthetic and catabolic genes, and omics. In particular, advances in knowledge of hormone signalling pathways such as perception, signal transduction, and signal interplay for several major hormones has greatly expanded our understanding of hormonal regulation of dormancy.

The phytohormone ABA has been widely recognized as a key regulator of growth cessation and dormancy initiation (Finkelstein 2013; Zheng et al. 2015). In several deciduous woody perennials, ABA content in buds increases at the inception of endodormancy in the fall (Wang et al. 2015; Tuan et al. 2017; Li et al. 2018), and decreases prior to dormancy release with the accumulation of either chilling exposure (Leida et al. 2012) or dormancy-breaking agents; e.g., hydrogen cyanamide (Zheng et al. 2015). During dormancy, bud ABA contents are mainly regulated by in situ synthesis and catabolism, and regulatory steps involved in these two processes are controlled by gene families of nine-*cis*-epoxycarotenoid dioxygennases (*NCEDs*) and cytochrome P450s (*CYP707s*), respectively. In pear, apple's closest relative, ABA *PpNCED-2* and *PpNCED-3* are found to be responsible for increased levels of ABA with chilling accumulation, while *CYP707-3* controls the decline of ABA content towards dormancy release (Li et al. 2018). Exogenous ABA, when applied on apple and pear trees in the fall, induces physiological changes characteristic of endodormancy, confirming ABA's role in promoting dormancy (Guak and Fuchigami 2001; Li et al. 2018). In addition to oxidation, ABA can also be deactivated by conjugating with glucosyl ester (GE) to form an ABA-GE complex, catalysed by glucosyltransferase. ABA-GE is mobile and serves as the storage form of ABA. Some recent reports have indicated that ABA conjugation may participate in the regulation of ABA content during dormancy (Chmielewski et al. 2018; Zhang et al. 2018). However, it still remains unclear to what extent this mechanism supplements the oxidative catabolism of ABA, and how it is spatially and temporally regulated during the bud dormancy cycle.

In addition to ABA metabolism, ABA responses are also regulated at the signalling level. The ABA signalling pathway consists of the following three key components: ABA receptors (RCAR/PYR/PYL), Protein Phosphate 2Cs (PP2Cs), and SNF1-Related Protein Kinase 2s (SnRK2s). The binding of ABA to its receptors releases SnRK2s from suppression of PP2C, thus activating downstream ABA response genes that are mediated by *cis*-regulatory elements (ABREs) (Soon et al. 2012). Genetic studies have suggested that environmental signals and ABA levels modulate transcript levels of these central components. In a transcriptomic study with the Japanese pear *Pyrus pyrifolia* Nakai 'Kosui', Li et al. (2018) have demonstrated that expression of ABA receptor *PYL* and *SnRK2s*, a positive regulator of ABA responses, are upregulated at the point of a tree's entry into dormancy. In contrast, expression of *PP2Cs* has remained low during deep endodormancy, but increases as ABA content decreases towards dormancy release. In the same study, exogenous ABA promotes the expression of *SnRK2s*, but suppresses that of *PP2Cs*. Similar results have also been observed in the Japanese pear

P. pyrifolia 'Suli', wherein *PP2C* genes are upregulated and *SnRK2s* are downregulated after buds exit from dormancy (Bai et al. 2013).

On the other hand, GAs serve as critical regulators of the plant growth/dormancy cycle, playing prominent roles in countering the effects of ABA. Of all naturally occurring forms of GAs, it is GA1, GA3, GA4, and GA7 that are known to possess significant bioactivities, with GA3 being the predominant form in vegetative tissues of apple (Yang et al. 2013). Reduction of GA levels is required for the proper establishment of endodormancy, while the elevation of GA levels is detected towards or soon after the release of endodormancy (Molmann et al. 2005; Cooke et al. 2012). During dormancy, levels of GA and ABA are inversely correlated, with the ABA/GA ratio varying in parallel with the depth of dormancy (Duan et al. 2004). GA is primarily regulated at the metabolic level, with critical steps controlled by biosynthetic genes *GA20ox* and *GA3ox* along with catabolic genes, such as *GA2ox* (Choubane et al. 2012; Wen et al. 2016; Zheng et al. 2018). Transgenic plant studies have revealed that expression of GA metabolic genes is subject to the influence of ABA levels, thus suggesting that there is a very close interdependence of GA and ABA in modulating the progression of dormancy (Seo et al. 2006; Oh et al. 2007; Wen et al. 2016; Yue et al. 2018).

Several mechanisms have been proposed for GA-mediated dormancy release. Firstly, GAs may control the dormancy cycle by restoring intercellular communications via induction of expression of Group 17 1, 3-β-glucanases that hydrolyse callose, a major polysaccharide that contributes to symplastic closure (Rinne et al. 2011). Secondly, GA promotes the production of Reactive Oxygen Species (ROS), known to play central roles in triggering budbreak (Sudawan et al. 2016; Liu et al. 2017; Beauvieux et al. 2018). Finally, GA may induce dormancy release by activating metabolic pathways, particularly those associated with sugar metabolism (Zhuang et al. 2015). Furthermore, soluble sugars not only function as important sources of energy to sustain regrowth but also serve as potential signalling elements that indirectly activate the expression of growth-related genes (Ruan et al. 2010).

In the GA signalling pathway, the GA-GID1-DELLA module is highly conserved in angiosperms (Sun 2011). In this model, GID1 (Gibberellin Insensitive Dwarf1) is a GA receptor, which can, upon forming a GA-GID1 complex, induce degradation of DELLA proteins, as it is a negative regulator of GA responses (Itoh et al. 2003; Ariizumi et al. 2013; Hauvermale et al. 2014). Expression of *GID1* correlates with GA levels during dormancy (Zhu et al. 2015; Yue et al. 2018); whereas, *DELLA* genes appear to respond to dormancy-inducing cues, such as short day (SD) photoperiod (Zawaski and Busov 2014; Yue et al. 2018). Recently, Yue et al. (2018) have proposed that both *GID1* and *DELLA* are subject to the effects of ABA. In apple, expression of *RGL*, a *DELLA* gene, can be induced by overexpression of C-repeat binding transcription factor(s) (*CBFs*) genes, suggesting the involvement of the GA signalling pathway in cold perception (Wisniewski et al. 2015).

Ethylene is also required for both dormancy initiation and dormancy release, as demonstrated by elevated ethylene levels at both stages. Ethylene deficiency or impaired ethylene reception have been reported to result in either abnormal or failed dormancy establishment (Suttle 2004; Ruonala et al. 2006; Sumitomo et al. 2008). The involvement of ethylene in dormancy induction may be ascribed to its ability to confer growth inhibition, which may involve GA deactivation and accumulation of DELLAs (Dubois et al. 2013, 2015). Moreover, ethylene can stabilize DELLAs via CTR1, an ethylene receptor target, without recruiting GA (Achard et al. 2007). The requirement of ethylene during budbreak, as demonstrated by temporary activation of ethylene synthesis in response to dormancy breaking stimuli, enhanced budbreak following exogenous ethylene application, as well as delayed dormancy release via blocking of ethylene signalling, may involve ethylene-mediated ABA degradation (Ophir et al. 2009;

Zheng et al. 2018), repression of CBF genes (Shi et al. 2012), or production of ROS (Ionescu et al. 2017; Beauvieux et al. 2018). However, the effects of ethylene in apple budbreak are rather different from those observed in other plant species, as ethylene production in apple floral buds declines during bud expansion in the spring (Blanpied 1972). Moreover, though fall applications of ethylene-based chemicals, such as ethephon, have been shown to effectively delay the blooming date in *Prunus* species, such effects appear to be insignificant in apple (Gianfagna et al. 1989; Durner et al. 1990; Durner and Gianfagna 1991; Durner 1995).

Both auxins, such as indole-3-acetic acid (IAA) and cytokinins (CKs) are growth-promoting phytohormones that have been primarily implicated in dormancy release. Transcriptomic analysis data from Japanese apricot and poplar have suggested that the majority of auxin-associated genes are downregulated during endodormancy (Zhong et al. 2013; Howe et al. 2015), and IAA levels can be induced by either low temperature or SD photoperiod (Zhang et al. 2012a). In dormant apple buds, overexpression of Dormancy Associated MADS (DAM) appears to lower IAA levels (Yamane et al. 2019), possibly contributing to the maintenance of deeper dormancy. On the other hand, IAA production increases during dormancy release (Nagar and Sood 2006; Qiu et al. 2013; Zhang et al. 2018), preparing for regrowth by restoring symplastic paths (Aloni and Peterson 1997), enhancing GA production (Wolbang and Ross 2001; Wolbang et al. 2004; Frigerio et al. 2006), and switching on the cell cycle machinery (Noriega and Pérez 2017). Indeed, expression analysis has indicated that auxin production and transport may act as essential regulatory nodes in apple dormancy (Porto et al. 2015). This is supported by a recent finding, in which a *CBF*-overexpressing transgenic apple line (T166), with prolonged endodormancy, is characterized by upregulation of auxin catabolism and storage genes, while genes related to auxin transport are downregulated in transgenic plants compared to wild-type M26 rootstock plants (Artlip et al. 2019).

It has been reported that the production of CKs is also subject to environmental signals. That chilling exposure is required for shoot-derived CK increases in apple buds and bark in the spring, which subsequently triggers budburst (Cook et al. 2001). Indeed, genes for CK inactivation or storage are upregulated; whereas, genes for the cell division cycle are downregulated in the aforementioned transgenic apple line T166 displaying delayed exit from dormancy (Artlip et al. 2019). As increased levels of CKs can activate cell cycle genes, thereby increasing rates of cell division and cellular respiration, as both are major events preceding the reactivation of dormant buds (Noriega and Pérez 2017).

12.2.2 Intercellular Regulation of Dormancy

In perennials, the dormancy/growth cycle is highly dependent on the cell-to-cell transport in the symplastic continuum, which in turn relies on the connectivity of specialized channels between adjacent cells known as plasmodesmata (PD) (Cooke et al. 2012). Passage of mobile molecules across the PD is primarily controlled by the deposition and degradation of callose, a 1, 3-β-glucan polymer, and catalyzed by callose synthase (CALS1) and glucanase, respectively (Wu et al. 2018). PD closure plays a central role in plant defence mechanisms, especially against invading viruses, and it has been demonstrated to serve as a critical step in the establishment of dormancy (Rinne and van der Schoot 2003). In a recent study, an ABA-mediated PD constriction in hybrid aspen is demonstrated to prevent dormancy release by limiting the passage of growth factors, such as the Flowering Locus T (*FT*) into dormant buds (Tylewicz et al. 2018). In this study, an ABA-insensitive mutant (*abi1-1*) has been reported to fail in producing PD callose, exhibiting compromised dormancy establishment under dormancy conducive conditions, while PD closure and dormancy are restored via either downregulation of a chromodomain remodelling factor, *PICKLE* (*PKL*), or ectopic expression of *Plasmodesmata Located Protein 1* (*PDLP1*) (Tylewicz et al. 2018). It is noteworthy to point out that PKL is a chromatin remodeller involved in facilitating epigenetic marks (such as histone H3 lysine 27) and repressing expression of tissue-specific genes associated with developmental transitions (Zhang et al. 2012b); whereas, *PDLPs* are important regulators of PD permeability and symplastic transport (Lim et al. 2016). Furthermore, ABA can induce the expression of *callose synthase 1* (*CALS1*) by suppressing the expression of *PKL* (Singh et al. 2019). In addition to regulation by ABA, the connectivity of PD channels can be restored by GA4, which can induce expression of Group 17 1, 3-β-glucanases to hydrolyze callose (Rinne et al. 2011). In a recent RNAi study with aspen, Singh et al. (2019) have proposed a model linking GA to ABA-mediated PD closure. In this model, it is proposed that upon exposure to SD photoperiod, elevated ABA levels suppress expression of *PKL*, releasing suppression of a short vegetative phase (SVP)-Like (SVL), a transcription factor orthologous to an Arabidopsis floral repressor (*SVP*), which activates expression of *CALS1* and the GA catabolic gene *GA2 oxidase*; thus, inducing callose accumulation. An analysis of Differentially Expressed Genes (DEGs) in transgenic apple line T166 with enhanced dormancy has revealed that Group 17 1, 3-β-glucanases are downregulated relative to non-transformed controls, and this may contribute to the observed delay in dormancy release in line T166 (Artlip et al. 2019). Using microscopy and synchrotron techniques, Lee et al. (2017) have demonstrated that PD closure and reduction of pectin pore size limit water movement into dormant buds of spruce. The facts that free-water can only be detected in dormant apple buds that have satisfied the CR (Liu et al. 1993), and that high temperatures favour a quick transition between endodormancy and ecodormancy in apple buds via shortening the rehydration period (Malagi et al. 2015) only reinforce the hypothesis that PD permeability is tightly-linked to the regulation of bud dormancy in apple. This is similar to scenarios observed in other deciduous woody species.

12.2.3 Cell Cycle Progression During Dormancy

When viewing dormancy as a dynamic state of meristems, the cell cycle is of particular importance in understanding the dormancy/growth cycle. In the cell cycle, the G1 phase primes the cell for the S phase, and the G1/S transition phase determines cell cycle progression by integrating environmental signals (Inze and De Veylder 2006). Progression of the cell cycle is governed by cyclins (especially types A, B, and D), in partnership with Cyclin-Dependent Kinases (CDK), that can be repressed by an Inhibitor of a CDK/Kip Related Protein (ICK) (Yamaguchi et al. 2000; Verkest et al. 2005; Lipavska et al. 2011; Torres Acosta et al. 2011). Dormancy is predominantly characterized by an arrested phase at either G1 or G0, and an extended stage of G1 (Devitt and Stafstrom 1995; Gutierrez et al. 2002). Entry into G1 is subject to the regulation of both internal and environmental signals. In a recent study, Vergara et al. (2017) have demonstrated the central role of ABA in modulating the cell cycle during dormancy, in which endogenous ABA suppresses the expression of both cyclins and *CDK*, while upregulating the expression of *ICKs* in dormant grapevine buds. In the same study, an exogenous ABA is reported to suppress expression of cell cycle genes (CCG); however, this observed effect is reversed by the application of hydrogen cyanamide (HC), which reduces the ABA content (Vergara et al. 2017). On the other hand, the cell cycle also participates in the perception of environmental signals by virtue of AP2 transcription factors, AINTEGUMENTALIKE1-4 (AIL1-AIL4). In hybrid aspen, downregulation of AIL1-AIL4 is required for photoperiod-dependent growth cessation and downregulation of D-type cyclins (CYCs) (Karlberg et al. 2011; Azeez et al. 2014).

Therefore, it has been proposed that D-type CYCs (e.g., CYCD3;1) play critical roles in regulating the G1/S transition (Menges et al. 2006). Recently, Artlip et al. (2019) have demonstrated that the apple *AIL1* gene is down-regulated in a transgenic apple line, T166, over-expressing the *CBF* gene. This finding sheds additional light on the mechanism of cold-dependent dormancy induction, which may involve an *AIL1*-mediated cell cycle regulation.

12.2.4 Molecular Controls of Apple Dormancy

It has been long observed that the mechanism(s) regulating dormancy is under strong genetic control and that our knowledge of dormancy has been considerably improved with studies of *Dormancy Associated MADS* genes (*DAMs*) (Bielenberg et al. 2004; Li et al. 2009). It is reported that *DAMs* have been first identified in peach, and that expression of some *DAMs*, including *PpDAM5* and *PpDAM56* are highly correlated with progression of endodormancy (Li et al. 2009; Jimenez et al. 2010). *DAM* genes are highly similar to the *SVP/AGL24* gene family in Arabidopsis, wherein members control vernalization-promoted flowering (Jimenez et al. 2009; Saito et al. 2015). In apple, several *DAM* orthologs have been identified and functionally characterized (Mimida et al. 2015; Porto et al. 2016; Wu et al. 2017); however, there is some inconsistency in the nomenclature of these genes. Among them, two genes, *MdDAMa* and *MdDAMc*, have been proposed as having roles in triggering growth cessation, while three other genes, *MdDAMb*, *MdSVPa*, and *MdSVPb*, are found to participate in the maintenance of bud dormancy (Falavigna et al. 2014; Wu et al. 2017). The functions of some of the *DAM* genes in apple have been assessed in transgenic plants, wherein overexpression of a *Prunus mume* gene, *PmDAM6*, in apple is found to suppress growth and delays bud outgrowth, likely through modulation of contents of the phytohormones ABA and cytokinin in terminal buds (Yamane et al. 2019).

In addition to *DAM* genes, a *MADS* box gene, *FLOWERING LOCUS C* (*FLC*), known to repress flowering in Arabidopsis, was also proposed as a candidate regulator of dormancy. A comparative study of apple cultivars with contrasting CRs revealed differential expression of *FLC*-like genes (Porto et al. 2015). Moreover, expression patterns of *FLC*-like genes in apple were highly correlated with both progression of both dormancy and chilling accumulation (Kumar et al. 2016a; Nishiyama et al. 2019). This finding was highly reminiscent of those observed for *DAM* genes. It should be noted that *FLC* genes have been located at the top of Linkage Group 9 (LG9), a region harbouring major QTLs responsible for both budbreak and flowering in apple (Segura et al. 2007; Allard et al. 2016; Miotto et al. 2019).

In the photoperiod-dependent induction of dormancy, the *CONSTANS* (*CO*)/*FT* regulon has been demonstrated to be the main mediator in the SD signalling pathway, which culminates in arrested cell cycle via regulation of *AIL*, an AP2 transcription factor (Karlberg et al. 2011; Srinivasan et al. 2012). In contrast, various distinct signalling pathways in temperature-induced dormancy are less understood. Recently, the *CBF*-dependent *DAM* gene pathway has been proposed as the main mechanism mediating autumn chilling and dormancy induction. In this pathway, CBF may directly bind to promoter regions (C-repeat/DRE) of *DAM* genes and enhance their transcriptional activities, as confirmed by yeast one-hybrid and transient expression experiments in Japanese pear (Saito et al. 2015; Niu et al. 2016) and Japanese apricot (Zhao et al. 2018). The function(s) of *CBF* in dormancy has been revealed via ectopic expression of a peach *CBF* gene in apple (line T166) as it induces precocious initiation of dormancy, increases freezing tolerance, and delays budbreak (Wisniewski et al. 2011; Artlip et al. 2014). In addition to enhancing expression of *DAM*-like genes, ectopic expression of *CBF* alters *EARLY BUD-BREAK*-like (*EBB*-like) genes in buds (Wisniewski et al. 2015), and that *MdEBB1* is yet another major regulator of dormancy release, first identified in poplar (Yordanov et al. 2014; Busov et al. 2016).

Dormancy has been proposed as a phenomenon driven by epigenetic mechanisms (Liu et al. 2015). Epigenetic modifications could either activate or repress gene expression patterns, either globally or locally, but without altering DNA sequences. In apple, global DNA and histone methylation changes have been reported to be closely associated with chilling, thus highlighting the importance of epigenetic controls mediating environmental cues in dormancy (Kumar et al. 2016b). In particular, *DAM* genes appear to be under strong epigenetic regulation during the dormancy cycle (Kumar et al. 2016b). In peach, a repressed histone marker, H3K27me3, is associated with high-chill cultivars, and that with chilling accumulation, the promoter of a *DAM6* gene is characterized by enriched H3K27me3 and reduced H3K4me3 and H3ac, thus pointing to activated expression (Leida et al. 2012). Similarly, a decrease of H3K4me3 is found to result in reduced expression of a *DAM* homolog (*PpMADS13-1*) in pear (*Pyrus pyrifolia*), and thus contributing to dormancy release (Saito et al. 2015). In addition to histone modification, recent studies have indicated that *DAM* genes can also be inactivated by DNA methylation and small interfering RNAs (siRNAs), both of which target promotors of DAM genes in response to chilling accumulation (Niu et al. 2016; Rothkegel et al. 2017). Altogether, these findings point to the central role of epigenetic modifications in regulating *DAM* genes, thereby confirming the importance of *DAM* genes in regulation of dormancy.

12.2.5 An Overall View of Dormancy

In conclusion, dormancy in perennials is a complex phenomenon involving the orchestration of a variety of processes at the physiological, molecular, and biochemical levels (Fig. 12.1). These processes include but are not limited to biosynthesis, signalling, and transport of phytohormones, modulation of cell cycle machinery, intracellular transport of signalling components, dehydration/hydration of dormant buds, transcriptomic and epigenetic regulation of dormancy-related genes, among others (Fig. 12.1). As such, attempts in deciphering the molecular and biochemical regulation of bud dormancy in fruit trees, including apple, require that concurrent multiple factors are taken into consideration.

Recent research efforts have generally concluded that the apple enters endodormancy by taking cues from low temperatures, and in this process, the transcription factor *CBF* appears to be a prominent regulator in the conduit of low-temperature perception. This in itself involves *DAM* genes and several downstream mechanisms that are shared among photoperiod-dependent pathways. Further molecular data are required to fully characterise key components in regulatory pathways of apple dormancy, particularly those related to cold hardiness.

12.3 Cold Hardiness

12.3.1 Avoidance, Tolerance, and Resistance to Cold Temperatures

Plants rely on three main strategies to deal with low temperatures: avoidance, resistance, and tolerance, all of which are employed by various temperate woody perennials (Levitt 1980). An obvious example of an avoidance mechanism is a life-cycle adaptation, such as shedding leaves prior to freezing temperatures, thus avoiding high vapour pressure deficits/desiccation, and photoinhibition, resulting from low temperatures coupled with high-light levels, as well as ice formation in mesophyll tissues. Some woody perennials, including apple, can also avoid deleterious effects of ice formation in tissues by undercooling (supercooling); i.e., maintaining water in a liquid state at sub-zero temperatures. Cells, tissues, and organs that retain cellular water in a liquid phase at low subfreezing temperatures (–20 to –40 °C), and remain free from internal heterogeneous ice nuclei, as well as isolated from nucleating effects of extracellular ice, are reported to undergo deep undercooling (Wisniewski et al. 2003, 2018). Deep

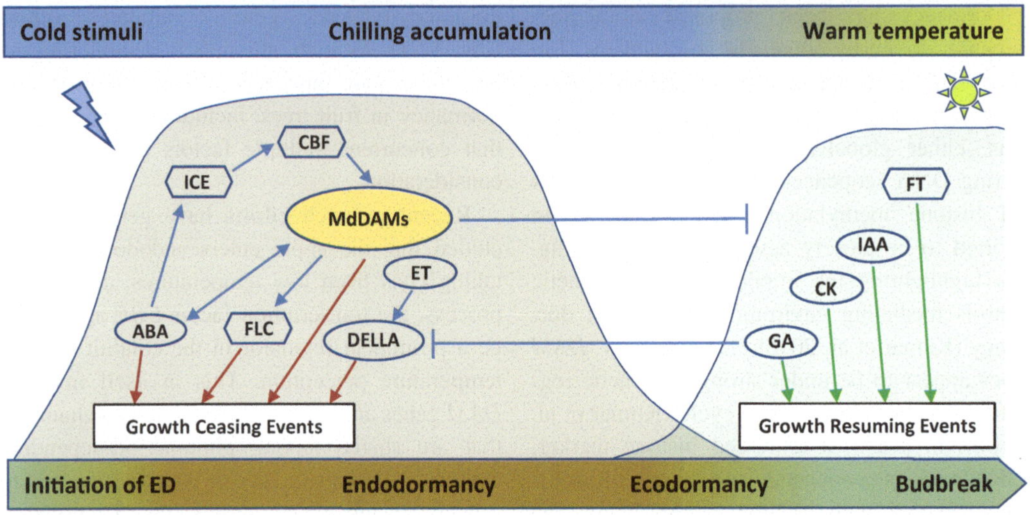

Fig. 12.1 A schematic model of endodormancy progression in apple. Arrows indicate activating actions, with red arrows denoting actions for growth cessation and green arrows denoting actions for growth resumption, while T ended bars denote inhibiting actions. Hexagons represent transcription factors; ovals represent hormones or proteins. Growth Ceasing Events indicate major events related to growth cessation, which include callose deposition, cell cycle arrested at G1/G0 stage, and repressive marks of histones, among others. Growth Resuming Events include callose degradation, cell cycle progressing to G1/S transition, and active marks of histones, among others. Components under the grey dome are upregulated during the initiation of endodormancy, and decline gradually with the chilling accumulation. In contrast, components in the green dome increase as buds transition from endodormancy to ecodormancy, reactivating growth and culminating in budbreak. ABA = abscisic acid, CBF = C-repeat binding transcription factors, CK = cytokinin, DELLA = a negative regulator in GA signalling pathway, containing a DELLA motif, ET = ethylene, GA = gibberellins, IAA = Indole-3-acetic acid (auxin), ICE = Inducer of CBF Expression, FLC = FLOWERING LOCUS C, FT = FLOWERING LOCUS T. This schematic model is adapted from the information presented by Liu and Sherif (2019) and Miotto et al. (2019)

undercooling has been reported to occur in xylem parenchyma of apple twigs (Hong and Sucoff 1982; Sakai and Larcher 1987; Kuroda and Sagisaka 2005; Pramsohler et al. 2012). In contrast to xylem tissues, apple buds cannot deep undercool in the winter (Salazar-Gutierrez 2016), but may slightly undercool following budbreak (Pramsohler and Neuner 2013). Some data suggest that the cytoplasm in cells of vegetative apple buds undergo a glass transition and thus avoid desiccation and intracellular ice formation (Vertucci and Stushnoff 1992; Pramsohler and Neuner 2013). While structural in nature, the ability to deep undercool is brought about through developmental programmes that are induced by changes in gene expression. In this regard, it is similar to various other survival strategies outlined by Levitt (1980).

Tolerance or resistance to low temperatures typically encompasses active changes in gene expression leading to alterations at cellular and sub-cellular levels (Preston and Sandve 2013; Wisniewski et al. 2018). Notably, these include changes in membrane lipid composition, with a bias towards increased unsaturation, thereby allowing membranes to maintain their integrity at sub-freezing temperatures (Uemura and Steponkus 1999). Seasonal changes in various cellular components have been reported in apple (Wang and Faust 1990). Changes in specific oligosaccharides and polyols have also been correlated with cold hardiness in woody plants, including apple (Stushnoff et al. 1999). These compounds contribute to the stabilization of membranes, acting as ROS scavengers or as compatible solutes that provide protection against osmotic stress (Tarkowski and Van den Ende 2015).

Plants must undergo a period of cold acclimation (CA) for their highest levels of tolerance to low temperatures. Cold acclimation in apple is first induced by short days (SDs) followed by exposure to chilling/low temperatures in the fall (Howell and Weiser 1970a; Ketchie 1984). Continued low temperatures are required to maintain maximum levels of cold hardiness, as well as dormancy. Epigenetic modifications, such as methylation, are strongly associated with maintaining dormancy in apple (Kumar et al. 2016b). Significant aspects of the molecular regulatory machinery underlying CA have been reported, particularly those related to plant response to low temperatures. However, details of low-temperature perception are still in debate. The current model proposes that a cold shock causes changes in membrane fluidity, and a rearrangement of the cytoskeleton triggering an influx of Ca^{2+}, which in turn induces transcriptional activation and post-transcriptional modifications of relevant genes and proteins (Fig. 12.2; Zhu 2016). Several regulatory pathways have been documented, with the central pathway involving induction of CBF transcription factor (s).

12.3.2 CBF Transcription Factors—The Central Cold Response Pathway

CBFs are members of the AP2/EREB family of transcription factors, and they bind to a conserved CCGAC promoter motif, termed the C-repeat (CRT) when plants are subjected to low temperature(s) or termed the Dehydration Response Element (DRE) in plants subjected to water deficit (Baker et al. 1994; Jaglo-Ottosen et al. 1998; Liu et al. 1998; Medina et al. 1999). Due to simultaneous studies conducted in different laboratories, dual nomenclatures have been proposed, C-repeat Binding Factor (CBF) and DRE Binding1 (DREB1). As the emphasis in this section is focused on cold hardiness, the CBF term will be used.

It is reported that the CRT is present in promoters of *COld Regulated* (*COR*) genes and/or osmotic stress-related genes (Fig. 12.2). Many of these genes encode functional proteins presumed to provide cryoprotection, such as dehydrins (Graether and Boddington 2014; Falavigna et al. 2015), or active osmotic adjustment via either carbohydrates or quaternary amine accumulation, such as those via various enzyme pathways (see Kozlowski and Pallardy 2002; Welling and Palva 2006). Several amino acid motifs distinguish CBF/DREB1 proteins from other members of the AP2/EREB family, as well as from a related DREB2 group, which is solely responsive to either water deficit or heat shock in dicots, and also binds to the conserved CRT motif (Mizoi et al. 2012).

There are wide variations in the numbers of *CBF* genes present in different species (Wisniewski et al. 2014; Shi et al. 2018). Three of the six *CBF* genes in Arabidopsis are cold-responsive (*AtCBF1,2,3*), while all five apple *CBF*s (*MdCBFs1-5*) are cold-responsive at varying levels (Wisniewski et al. 2011; Feng et al. 2012; Wisniewski et al. 2015). There are relatively few differences between sequences of cold-inducible Arabidopsis *CBF*s. In contrast, notable amino acid sequence differences are present among different apple CBFs (Wisniewski et al. 2011), leading to distinct phylogenetic groupings (Zhao et al. 2012; Wisniewski et al. 2014). Interestingly, there is experimental evidence demonstrating functional differences among *AtCBF*s despite their sequence similarities (Novillo et al. 2004, 2007; Jia et al. 2016; Zhao et al. 2016; Shi et al. 2017; Li et al. 2017). It has been reported that differences in low-temperature adaptation in natural Arabidopsis populations can also be traced to functional *AtCBF* dissimilarities (Gehan et al. 2015). Furthermore, differences in low-temperature induction kinetics (Wisniewski et al. 2011, 2015) and responses to various regulatory factors among *MdCBF*s suggest that there are likely functional differences in apple (Fig. 12.2).

It is known that regulation of *CBF* genes is complex, with most knowledge gained from Arabidopsis studies (Fig. 12.2; Mizoi et al. 2012;

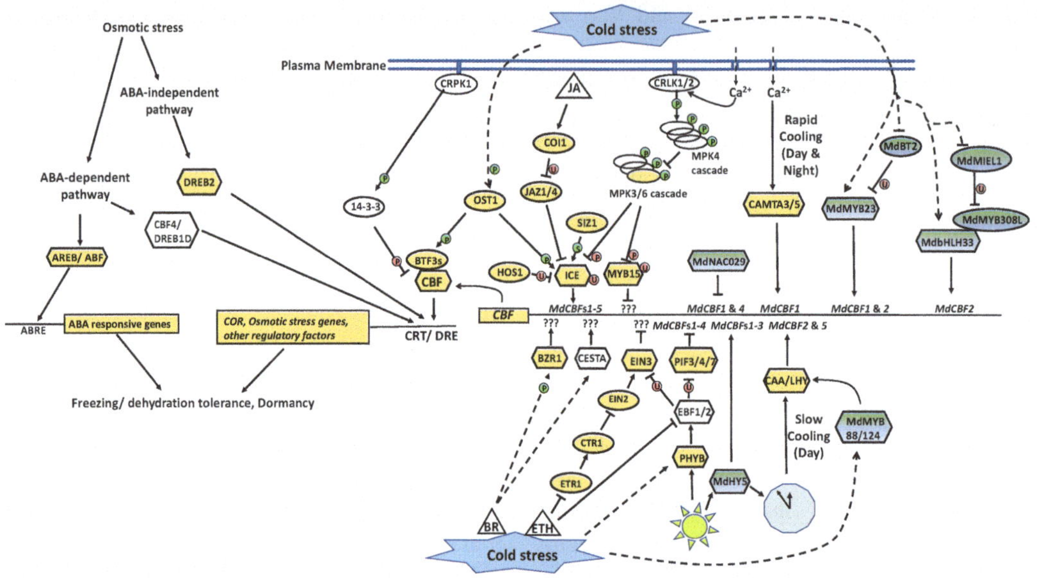

Fig. 12.2 A model of low-temperature signal transduction and transcriptional responses in apple. Lines with arrows denote positive/activation steps. Lines with a "T" denote inhibitory steps. Dashed lines represent unconfirmed pathways. Green circles with a "P" denote phosphorylation, while red circles with a "P" denote dephosphorylation. Green circles with as "S" denote activation/stabilisation of SUMOylation, and red circles with "U" denote deactivation of ubiquitination. Hexagons indicate transcription factors; ovals indicate interacting proteins. Gold shaded hexagons and ovals denote apple homologs of Arabidopsis genes, while green to blue shaded symbols represent apple genes that have not been associated with low-temperature responses in Arabidopsis. The *MdCBF* genes containing promoter-specific transcription factor motifs are indicated, while "???" indicates those relevant promoter motifs that have not yet been identified in apple *CBF* genes. ABA = Abscisic Acid, BR = Brassindole, JA = Jasmonic Acid, ETH = Ethylene. This schematic diagram is based on findings reported by Feng et al. (2012), Mizoi et al. (2012), Du et al. (2015), Wisniewski et al. (2015), Eremina et al. (2016), Zhang et al. (2016), Kidokoro et al. (2017), Wisniewski et al. (2018), An et al. (2017, 2018a, 2018b, 2019), Shi et al. (2018), Xie et al. (2018), and Artlip et al. (2019)

Eremina et al. 2016; Kidokoro et al. 2017; Shi et al. 2018). An Inducer of CBF Expression (ICE), an MYC-like transcription factor, has been demonstrated as a primary regulator of *CBF*s. Moreover, studies in Arabidopsis have confirmed that ICE is constitutively expressed and requires post-translational modification for activity (Chinnusamy et al. 2003). Interestingly, Arabidopsis has two *ICE* genes that largely overlap in function (Kim et al. 2015). In contrast, apple has four *ICE*-like genes (Wisniewski et al. 2014), and among these, it is found that *MdClbHLH1* is cold-inducible and that it induces *MdCBFs1-5* (Feng et al. 2012). The regulation of ICE in Arabidopsis is also complex, with ubiquitination, SUMOylation, and phosphorylation all playing significant regulatory roles (Fig. 12.2). Many of the components of *ICE* regulation have apparent apple homologs that are expressed or active during periods of low temperatures (Du et al. 2015; Zhang et al. 2016; Artlip et al. 2019; Li et al. 2019).

Homologs of other positive and negative *AtCBF*-regulatory factors/pathways are also found in apple, but these have not been characterized to determine whether or not they have either direct or indirect effects on *MdCBF* expression (Fig. 12.2). In addition, promoter elements for BZR1, CESTA, and EIN3 binding have not been found in *MdCBF* genes (Wisniewski et al. 2015). In contrast, evidence is provided for the presence of either promoter elements or regulatory inputs of *MdCBF*s by various elements, including MYB15, PIF3/4/7,

CAMTA3/5, MdHY5, MdMYB88/124 (with CAA/LHY), MdNAC029, MdMYB23, and MdbHLH33 (Du et al. 2015; Wisniewski et al. 2015; Xie et al. 2018; An et al. 2017, 2018a, 2018b, 2019).

In Arabidopsis and in other plant species, an abrupt temperature shift is typically used in *CBF*-regulatory studies. However, few studies have utilized either a gradual drop in temperature or long-term exposure to low temperature when conducting CBF studies. Kidokoro et al. (2017) have suggested that two partially overlapping low-temperature signal-transduction pathways regulating *CBF*s, CAMTA and CCA/LHY, are capable of differentiating between rapid- and slow-cooling of induction of *CBF* gene expression in Arabidopsis. This finding may be highly relevant when considering the perennial nature of apple, particularly during day/night temperature fluctuations in temperate regions, as these fluctuations generally occur gradually, and perennial plants are often exposed to long-term (chronic) low-temperature conditions. Kidokoro et al. (2017) have reported on the involvement of circadian rhythms in the regulation of CBF genes, and in this regard, promoters of *MdCBFs* have numerous light/circadian rhythm elements (Wisniewski et al. 2015), thus suggesting sensitivities to these inputs. In fact, Wisniewski et al. (2015) have demonstrated that under long-term (3 weeks) SD (8 h light) and constant 4 °C conditions, expression levels of *MdCBF1*, *MdCBF2*, and *MdCBF4*, but not of *MdCBF3* and *MdCBF5*, are higher than those of time zero controls. It has been reported that ABA-dependent and -independent pathways may also play important functional roles under chronic low-temperature conditions given the osmotic/dehydration stresses that occur in frozen tissues (Yang et al. 2010; Eremina et al. 2016). Furthermore, many functional genes or regulon(s) for cold/freezing and osmotic stress/dehydration are known to overlap (Fig. 12.2). In addition, bark tissues of temperate trees, including those of apple, are known to undergo active osmotic adjustment during winter months, and this may contribute to the observed freezing resistance of buds as water is withdrawn from buds due to differences in osmotic potential (Pramsohler and Neuner 2013).

Most *MdCBF*s should also be subject to either auto- or cross-regulation as all, but *MdCBF4*, have either a single or more CRTs in their promoter regions (Wisniewski et al. 2015). Interestingly, at least two *MdCBF* genes, both possessing a CRT motif, are not upregulated in a transgenic apple line overexpressing a peach *CBF* gene (*PpCBF1*), even though they are inducible by a shift to low temperature (Wisniewski et al. 2011). Additionally, none of these genes are even differentially upregulated in field-grown transgenic trees relative to non-transgenic trees (Artlip et al. 2019). These findings provide further evidence that multiple regulatory signals are involved in *MdCBF* expression.

It is reported that strict regulation of *CBF*s is warranted as overexpression of *CBF*s elicits various pleiotropic effects. Interestingly, overexpression of *CBF*s results in enhanced freezing tolerance in both herbaceous plants (Liu et al. 1998; Kasuga et al. 1999; Gilmour et al. 2000) and woody perennials, including poplar and apple (Benedict et al. 2006; Wisniewski et al. 2011; Artlip et al. 2014, 2016). This latter finding is reported to be due to the upregulation of genes encoding stress-tolerance-related proteins within the *CBF* regulon (e.g., Wisniewski et al. 2011; Artlip et al. 2019). In fact, Wisniewski et al. (2011) have observed increased freezing tolerance, of –3 °C, in non-acclimated transgenic apple and for –5 °C for cold-acclimated (CA)-transgenic apple trees, growing in a controlled-environment chamber, compared to those of non-transgenic (control) trees. Artlip et al. (2014) have reported an improvement of nearly –2 °C in the month of June in field-grown trees. However, overexpression of *CBF*s frequently leads to undesirable effects, such as growth abnormalities in both herbaceous (e.g., Liu et al. 1998; Kasuga et al. 1999; Gilmour et al. 2000) and woody perennials, including apple (Wisniewski et al. 2011; Artlip et al. 2014; Wisniewski et al. 2015; Artlip et al. 2016). Moreover, overexpression of *PpCBF1* in apple has also resulted in novel sensitivities to SD photoperiods, leading to early bud set and dormancy, as well as to delayed budbreak in the spring (Wisniewski et al. 2011; Artlip et al. 2014; Wisniewski et al. 2015). These SD sensitivities have been further investigated as the onset of dormancy in apple (and pear) is strictly temperature-dependent, and this is in contrast to that of other woody perennials wherein SD regulates dormancy (Heide and Prestrud 2005).

Both Wisniewski et al. (2015) and Artlip et al. (2019) have demonstrated that various genes associated with growth, dormancy, and hormonal activities are altered, in complex manners, in transgenic apple trees overexpressing *PpCBF1*. As noted above in Sect. 12.2.4, *DAM* (*Dormancy Associated MADS-box*) and *EBB* (*Early Bud Break*) exhibit expression patterns in transgenic apple buds that are consistent with delayed budbreak (Wisniewski et al. 2015). Apparently, upregulation of auxin, gibberellic acid, and cytokinin storage/inactivation genes along with downregulation of genes involved in dormancy release likely contribute to delayed-release from dormancy, as observed in *PpCBF1*-overexpressing apple trees (Artlip et al. 2019), and as noted above in Sect. 12.2.1. Furthermore, as mentioned above in Sect. 12.2.4, epigenetic modifications may also play a role in the regulation of dormancy. In this regard, Artlip et al. (2019) have also noted upregulation of genes associated with epigenetic regulation, thus supporting earlier findings wherein methylation changes have been tracked during both dormancy entry and dormancy release (Kumar et al. 2016b). The above collective studies further underscore the need for tight regulation of *CBF* expression, particularly for both deacclimation and dormancy release.

12.3.3 Deacclimation and Reacclimation

Deacclimation (DA), reduction in levels of cold hardiness at any stage of the seasonal cycle of cold hardiness, is a complex process, and not merely a reversal of CA. While CA requires periods of weeks to months, DA requires shorter periods, of days to weeks, and may involve the conversion of carbohydrates and proteins or stored energy reserves, available during cold hardiness, into forms required for growth and increased metabolism (Kalberer et al. 2006). Moreover, reacclimation (RA), the subsequent increase in plant cold hardiness following some level of DA, is not a rapid process. Nevertheless, there is a wide genetic variability for RA capability. For instance, RA for apple bark that is deacclimated by a single day of 15 °C would require a period of three cold days for reversal (Howell and Weiser 1970b).

It is noteworthy to point out that chilling soil temperatures also influence bud phenology and xylem hydraulics, but it is not known as to whether or not this also impacts the cold hardiness of aerial tissues. It is reported that chilled soil temperatures apparently reduce xylem conductivity, thereby delaying tissue rehydration (Greer et al. 2005; Beikircher et al. 2016). Furthermore, decreases in root metabolism and xylem conductivity inhibit both conversion and transport of carbohydrate reserves to sink tissues of aerial portions of trees (Greer et al. 2005).

It has been reported that the duration of endodormancy in temperate fruit trees is genetically controlled and environmentally regulated by the need to accumulate a genetically determined number of chilling units, the number of hours between 0 °C and 7 °C (Naor et al. 2003). Once completed, endodormancy is followed by a period of ecodormancy, also genetically controlled, requiring an accumulation of heat units, duration of warm temperatures for buds to swell, open, and begin to grow (Wisniewski et al. 2018). In a study by Anzanello et al. (2014) on the effects of erratic temperature events on dormancy and CA, they have reported that daily 15 °C temperature fluctuations do not impact endodormancy; however, prolonged elevated temperatures (25 °C) would initiate DA and promote bud development, provided the chilling requirement has been met. Moreover, endodormant tissues are significantly less impacted by warmer temperatures than ecodormant tissues (Anzanello et al. 2014). As ecodormancy progresses, there is increased sensitivity to warmer temperatures culminating in complete DA and a commitment for growth (Howell and Weiser 1970a, b; Kalberer et al. 2006; Arora and Taulavori 2016). Thus, DA and a shift to growth competency would increase hydration of tissues, thereby increasing risks of detrimental ice formation and propagation if freezing temperatures were to return (Kalberer et al. 2006).

It has been well established that widespread changes in gene expression must occur during CA, dormancy, and DA in buds and cambial tissues of poplar (*Populus spp*), often used as a model (Cooke et al. 2012; Petterle et al. 2013; Ding and Nilsson 2016). Transcriptional profiling of apple buds has revealed changes comparable to those observed in poplar, with shifts moving away from genes associated with dormancy and cold hardiness towards growth-related gene expression (Falavigna et al. 2014; Porto et al. 2015; Wisniewski et al. 2015; Falavigna et al. 2015; Kumar et al. 2016b; Miotto et al. 2019). In poplar, the vascular cambium, responsible for stem growth, also undergoes a period of dormancy and exhibits changes in gene expression similar to those observed in bud tissues (Schrader et al. 2004; Bhalerao and Fischer 2016). Furthermore, analogous changes are observed in the expression of apple bark (cambium, phloem, epidermis, and/or phellem) genes, and these are characterised by marked shifts towards cell division and cambial growth-related genes (Wisniewski et al. 2015; Artlip et al. 2019) upon incidence of DA (Howell and Weiser 1970b).

12.3.4 Genetic Diversity in Cold Hardiness

The genetic background of a cultivar plays a critical role in defining capacities for CA, DA, and RA. Cultivars are known to differ in their levels of freezing tolerance or deep undercooling during periods of both CA and DA (e.g., Coleman 1985; Moran et al. 2018). Concomitantly, chilling requirement, heat requirement, and duration of dormancy are also highly genotype-dependent (Haugge and Cummins 1991; Labuschagné et al. 2002; Celton et al. 2011; Anzanello et al. 2014). This has been demonstrated in expression profiles of dormancy-related genes (Falavigna et al. 2014). Moreover, mapping studies using segregating populations have provided information on genetic regulation and inheritance of CA-related traits. For example, Celton et al. (2011) have identified cold hardiness-related QTLs on LGs 01, 03, 09, 10, 12, 15, and 17, with LG 08 and 09 harbouring major QTLs. More specifically, clusters of cell cycle genes have been identified within confidence intervals for a major QTL on LG09. Using a different mapping population, Miotto et al. (2019) have also identified a major QTL on LG09, but have concluded that genes other than those involved in the cell cycle are more closely associated with this QTL. One of the genes in the QTL on LG09, *MdICE1*, is reported to directly influence the regulation of cold hardiness in different apple genotypes. This gene is involved in the control of *MdCBF* gene expression, and it is indirectly involved in the regulation of dormancy through transcriptional activation of dormancy-promoting *MdDAM* genes via *MdCBF* genes (see also Wisniewski et al. 2015; Artlip et al. 2019).

12.3.5 Cold Hardiness—Summary and Future Prospects

In summary, plant cold hardiness (tolerance or avoidance) and seasonal changes in the regulation of cold hardiness are complex, requiring multiple environmental cues that directly influence genetic and epigenetic determinants that are genotype- and species-specific. Both structural and biochemical changes occur in organs and tissues of fruit trees, and these are regulated by changes in gene expression and epigenetic modifications; moreover, these determine the ability of an apple tree to withstand freezing temperatures and the presence of ice in its tissues. Regulation of *CBF* genes and the CBF regulon(s) are vital in this regard. The different phases of cold hardiness (acclimation, deacclimation, and reacclimation) along with stages of dormancy (endodormancy and ecodormancy) are inextricably related processes, with extensive crosstalk between them. Both are subject to hormonal and environmental cues that impact gene expression in a genotype-specific manner. The ability to grow aeconomically-viable apple trees in a changing climate and provide a consistent supply of fruit to consumers will require a better understanding of dormancy and cold hardiness, including the role(s) of microRNAs, non-coding RNAs, and alternative splicing (e.g., Calixto et al. 2019). Rapid cycle breeding of apple (Flachowsky et al. 2011; Luo et al. 2020), along with advanced gene-editing techniques, such as CRISPR/Cas9 in apple (e.g., Osakabe et al. 2018) will increasingly play important roles in cultivar improvement and the development of advanced breeding stocks better adapted to erratic weather conditions and overall winter temperatures. These genetically improved breeding stocks will impact the ability of a specific cultivar to accumulate chilling units in a specific geographical location. These approaches can be complemented by enhanced knowledge of gene expression in the primary apple progenitor, *M. sieversii. Malus sieversii* is known for thriving in harsh climates in Central Asia, disease resistance traits, and potentially valuable fruit quality traits (Luby et al. 2001). Wisniewski et al. (2020) have recently described an apple FOX library in Arabidopsis that overexpresses cDNA inserts taken from winter acclimated *M. sieversii* bark. This forward-genetics approach allows high throughput screening for a variety of environmental stresses such as freezing, osmotic, and salt stresses along with disease tolerance. The apple FOX library has over 4000 lines that have been partially characterised for sequence identity, with numerous inserts having no homologs known in the *M.* × *domestica*, *A. thaliana,* or *Populus trichocarpa* genomes. With tools such as these in hand, the future looks bright for apple breeders to keep pace with the challenges ahead.

References

Achard P, Baghour M, Chapple A, Hedden P, Van Der Straeten D, Genschik P, Moritz T, Harberd NP (2007) The plant stress hormone ethylene controls floral transition via DELLA-dependent regulation of floral meristem-identity genes. Proc Natl Acad Sci USA 104:6484–6489

Allard A, Bink MC, Martinez S, Kelner JJ, Legave JM, Di Guardo M, Di Pierro EA, Laurens F, Van De Weg EW, Costes E (2016) Detecting QTLs and putative candidate genes involved in budbreak and flowering time in an apple multiparental population. J Exp Bot 67:2875–2888

Aloni R, Peterson CA (1997) Auxin promotes dormancy callose removal from the phloem of *Magnolia kobus* and callose accumulation and earlywood vessel differentiation in *Quercus robur*. J Plant Res 110:37–44

An J-P, Yao J-F, Wang X-N, You C-X, Wang X-F, Hao Y-J (2017) MdHY5 positively regulates cold tolerance via CBF-dependent and CBF-independent pathways in apple. J Plant Physiol 218:275–281

An J-P, Li R, Qu F-J, You C-X, Wang X-F, Hao Y-J (2018a) An apple NAC transcription factor negatively regulates cold tolerance via CBF-dependent pathway. J Plant Physiol 221:74–80

An J-P, Li R, Qu F-J, You C-X, Wang X-F, Hao Y-J (2018b) R2R3-MYB transcription factor MdMYB23 is involved in the cold tolerance and proanthocyanidin accumulation in apple Plant J 96: 562–577

An J-P, Wang X-F, Zhang X-W, Xu H-F, Bi S-Q, You C-X, Hao Y-J (2019) An apple MYB transcription factor regulates cold tolerance and anthocyanin accumulation and undergoes MIEL1-mediated degradation. Plant Biotech J. https://doi.org/10.1111/pbi.13201

Anzanello R, Fialho FB, dos Santos HP, Bergamaschi H, Marodin GAB (2014) Bud dormancy in apple trees after thermal fluctuations. Pesquisa Agropecuária Brasileira 49:457–464

Ariizumi T, Hauvermale AL, Nelson SK, Hanada A, Yamaguchi S, Steber CM (2013) Lifting della repression of Arabidopsis seed germination by nonproteolytic gibberellin signaling. Plant Physiol 162:2125–2139

Arora R, Taulavuori K (2016) Increased risk of freeze damage in woody perennials vis-a-vis climate change: Importance of deacclimation and dormancy response. Front Environ Sci 4:44

Artlip TS, Wisniewski ME, Norelli JL (2014) Field evaluation of apple overexpressing a peach *CBF* gene confirms its effect on cold hardiness, dormancy and growth. Environ Exp Bot 106:79–86

Artlip TS, Wisniewski ME, Arora R, Norelli JL (2016) An apple rootstock overexpressing a peach *CBF* gene alters growth and flowering in the scion but does not impact cold hardiness or dormancy. Hortic Res 3:16006

Artlip T, McDermaid A, Ma Q, Wisniewski M (2019) Differential gene expression in non-transgenic and transgenic "M.26" apple overexpressing a peach *CBF* gene during the transition from eco-dormancy to bud break. Hortic Res 6:86

Augspurger CK (2013) Reconstructing patterns of temperature, phenology, and frost damage over 124 years: spring damage risk is increasing. Ecology 94:41–50

Azeez A, Miskolczi P, Tylewicz S, Bhalerao RP (2014) A tree ortholog of APETALA1 mediates photoperiodic control of seasonal growth. Curr Biol 24:717–724

Bai S, Saito T, Sakamoto D, Ito A, Fujii H, Moriguchi T (2013) Transcriptome analysis of Japanese pear (*Pyrus pyrifolia* Nakai) flower buds transitioning through endodormancy. Plant Cell Physiol 54:1132–1151

Baker SS, Wilhelm KS, Thomashow MF (1994) The 5'-region of *Arabidopsis thaliana* cor15a has *cis*-acting elements that confer cold-, drought- and ABA-regulated gene expression. Plant Mol Biol 24:701–713

Beauvieux R, Wenden B, Dirlewanger E (2018) Bud dormancy in perennial fruit tree species: A pivotal role for oxidative cues. Front Plant Sci 9:657

Beikircher B, Mittmann C, Mayr S (2016) Prolonged soil frost affects hydraulics and phenology of apple trees. Front Plant Sci 7:867

Bhalerao RP, Fischer U (2016) Environmental and hormonal control of cambial stem cell dynamics. J Exp Bot 68:79–87

Bielenberg DG, Wang Y, Fan S, Reighard GL, Scorza R, Abbott AG (2004) A deletion affecting several gene candidates is present in the *Evergrowing* peach mutant. J Hered 95:436–444

Benedict C, Skinner JS, Meng R, Chang Y, Bhalerao R, Huner NPA, Finn CE, Chen THH, Hurry V (2006) The CBF1-dependent low temperature signalling pathway, regulon and increase in freeze tolerance are conserved in *Populus* spp. Plant Cell Environ 29:1259–1272

Blanpied GD (1972) A study of ethylene in apple, red raspberry, and cherry. Plant Physiol 49:627–630

Busov V, Carneros E, Yakovlev I (2016) *EARLY BUD-BREAK1* (*EBB1*) defines a conserved mechanism for control of bud-break in woody perennials. Plant Signal Behav 11:

Calixto CPG, Tzioutziou NA, James AB, Hornyik C, Guo W, Zhang R, Nimmo HG. Brown JWS (2019) Cold-dependent expression and alternative splicing of Arabidopsis long non-coding RNAs. Front Plant Sci 10:235

Celton J-M, Martinez S, Jammes M-J, Bechti A, Salvi S, Legave J-M, Costes E (2011) Deciphering the genetic determinism of bud phenology in apple progenies: a new insight into chilling and heat requirement effects on flowering dates and positional candidate genes. New Phytol 192:378–392

Chinnusamy V, Ohta M, Kanrar S, Lee B-h, Hong X, Agarwal M, Zhu J-K (2003) ICE1: a regulator of cold-induced transcriptome and freezing tolerance in *Arabidopsis*. Genes Dev 17:1043–1054

Chmielewski FM, Baldermann S, Götz KP, Homann T, Gödeke K, Schumacher F, Huschek G, Rawe LH (2018) Abscisic acid related metabolites in sweet cherry buds (*Prunus avium* L.). J Hortic 5:1–9

Choubane D, Rabot A, Mortreau E, Legourrierec J, Peron T, Foucher F, Ahcene Y, Pelleschi-Travier S, Leduc N, Hamama L, Sakr S (2012) Photocontrol of bud burst involves gibberellin biosynthesis in *Rosa sp.* J Plant Physiol 169:1271–1280

Coleman WK (1985) Variations in cold resistance among apple cultivars during deacclimation. J Exp Bot 36:1159–1171

Considine MJ, Considine JA (2016) On the language and physiology of dormancy and quiescence in plants. J Exp Bot 67:3189–3203

Cook NC, Bellstedt DU, Jacobs G (2001) Endogenous cytokinin distribution patterns at budburst in Granny Smith and Braeburn apple shoots in relation to bud growth. Sci Hortic 87:53–63

Cooke JEK, Eriksson ME, Junttila O (2012) The dynamic nature of bud dormancy in tree: environmental control and molecular mechanisms. Plant Cell Environ 35:1701–1728

Daccord N, Celton JM, Linsmith G, Becker C, Choisne N, Schijlen E, van de Geest H, Bianco L, Micheletti D, Velasco R, Di Pierro EA, Gouzy J, Rees DJG, Guérif P, Muranty H, Durel CE, Laurens F, Lespinasse Y, Gaillard S, Aubourg S, Quesneville H, Weigel D, van de Weg E, Troggio M, Bucher E (2017) High-quality de novo assembly of the apple genome and methylome dynamics of early fruit development. Nat Genet 49:1099–1106

Devitt ML, Stafstrom JP (1995) Cell cycle regulation during growth-dormancy cycles in pea axillary buds. Plant Mol Biol 29:255–265

Ding J, Nilsson O (2016) Molecular regulation of phenology in trees – because the seasons they are a-changin'. Curr Opin Plant Biol 29:73–79

Du F, X-Y J-N, Li D, Wang X-Y (2015) The identification of novel and differentially expressed apple-tree genes under low-temperature stress using high-throughput Illumina sequencing. Mol Biol Rep 42:569–580

Duan C, Li X, Gao D, Liu H, Li M (2004) Studies on regulations of endogenous ABA and GA3 in sweet cherry flower buds on dormancy. Acta Hortic Sin 31:149–154

Dubois M, Skirycz A, Claeys H, Maleux K, Dhondt S, De Bodt S, Vanden Bossche R, De Milde L, Yoshizumi T, Matsui M, Inze D (2013) Ethylene response factor6 acts as a central regulator of leaf growth under water-limiting conditions in Arabidopsis. Plant Physiol 162:319–332

Dubois M, Van Den Broeck L, Claeys H, Van Vlierberghe K, Matsui M, Inze D (2015) The ETHYLENE RESPONSE FACTORs ERF6 and ERF11 antagonistically regulate mannitol-induced growth inhibition in arabidopsis. Plant Physiol 169:166–179

Durner EF, Gianfagna TJ, Rooney FX, Teiger GS, Seiler MJ, Cantarella MJ (1990) Harvest date and size distribution of peach fruit are altered with fall applied Ethephon. HortScience 25:911–913

Durner EF (1995) Dormant pruning and fall ethephon application influence peach pistil hardiness. J Am Soc Hortic Sci 120:823–829

Durner EF, Gianfagna TJ (1991) Ethephon prolongs dormancy and enhances supercooling in peach flower buds. J Am Soc Hortic Sci 116:500–506

Eremina M, Rozhon W, Poppenberger B (2016) Hormonal control of cold stress responses in plants. Cell Mol Life Sci 73:797–810

Falavigna VS, Porto DD, Buffon V, Margis-Pinheiro M, Pasquali G, Revers LF (2014) Differential transcriptional profiles of dormancy-related genes in apple buds. Plant Mol Biol Rep 32:796–813

Falavigna VS, Miotto YE, Porto DD, Anzanello R, dos Santos HP, Fialho FB, Margis-Pinheiro M, Pasquali G, Revers LF (2015) Functional diversification of the dehydrin family in apple and its contribution to cold acclimation during dormancy. Physiol Plant 155:315–319

Feng X-M, Zhao Q, Zhao L-L, Qiao Y, Xie X-B, Li H-F, Yao Y-X, You C-X, Hao Y-J (2012) The cold-induced basic helix-loop-helix transcription factor gene *MdClbHLH1* encodes an ICE-like protein in apple. BMC Plant Biol 12:22

Finkelstein R (2013) Abscisic acid synthesis and response. Arabidopsis Book 11:

Flachowsky H, Le Roux P-M, Peil A, Patocchi A, Richter K, Hanke M-V (2011) Application of a high-speed breeding technology to apple (*Malus* × *domestica*) based on transgenic early flowering plants and marker-assisted selection. New Phytol 192:364–377

Frigerio M, Alabadi D, Perez-Gomez J, Garcia-Carcel L, Phillips AL, Hedden P, Blazquez MA (2006) Transcriptional regulation of gibberellin metabolism genes by auxin signaling in Arabidopsis. Plant Physiol 142:553–563

Gardner VR, Bradford FC, Hooker HD Jr (1922) Fundamentals of fruit production. McGraw-Hill Book Company Inc, New York, NY, pp 250–387

Guak S, Fuchigami LH (2001) Effects of applied ABA on growth cessation, bud dormancy, cold acclimation, leaf senescence and N mobilization in apple nursery plants. J Hortic Sci Biotechnol 76:459–464

Gehan MA, Park S, Gilmour SJ, An C, Lee C-M, Thomashow MF (2015) Natural variation in the C-repeat binding factor cold response pathway correlates with local adaptation of Arabidopsis ecotypes. Plant J 84:682–693

Gianfagna T, Durner E, Teiger GS (1989) Reducing low temperature injury to peach flower buds with ethephon. Acta Hortic 239:203–206

Gilmour SJ, Sebolt AM, Salazar MP, Everard JD, Thomashow MF (2000) Overexpression of the Arabidopsis CBF3 transcriptional activator mimics multiple biochemical changes associated with cold acclimation. Plant Physiol 124:1854–1865

Graether SP, Boddington KF (2014) Disorder and function: a review of the dehydrin protein family. Front Plant Sci 5:576

Greer DH, Wunsche JN, Norling CL, Wiggins HN (2005) Root-zone temperatures affect phenology of bud break, flower cluster development, shoot extension growth and gas exchange of 'Braeburn' (*Malus domestica*) apple trees. Tree Physiol 26:105–111

Gutierrez C, Ramirez-Parra E, Castellano MM, Del Pozo JC (2002) G1 to S transition: more than a cell cycle engine switch. Curr Opin Plant 5:480–486

Haugge R, Cummins JN (1991) Phenotypic variation of length of bud dormancy in apple cultivars and related *Malus* species. J Am Soc Hortic Sci 116:100–106

Hauvermale AL, Ariizumi T, Steber CM (2014) The roles of the GA receptors GID1a, GID1b, and GID1c in sly1-independent GA signaling. Plant Signal Behav 9:

Hawerroth FJ, Herter FG, Petri JL, Marafon AC, Leonetti JF (2013) Evaluation of winter temperatures on apple budbreak using grafted twigs. Rev Bras de Frutic 35:713–721

Heide OM, Prestrud AK (2005) Low temperature, but not photoperiod, controls growth cessation and dormancy induction and release in apple and pear. Tree Physiol 25:109–114

Hoffmann H, Rath T (2013) Future bloom and blossom frost risk for *Malus domestica* considering climate model and impact model uncertainties. PLoS ONE 8:

Hong S-G, Sucoff E (1982) Rapid increase in deep supercooling of xylem parenchyma. Plant Physiol 69:697–700

Howe GT, Horvath DP, Dharmawardhana P, Priest HD, Mockler TC, Strauss SH (2015) Extensive transcriptome changes during natural onset and release of vegetative bud dormancy in *Populus*. Front Plant Sci 6:989

Howell GS, Weiser CJ (1970a) The environmental control of cold acclimation in apple. Plant Physiol 45:390–394

Howell GS, Weiser CJ (1970b) Fluctuations in the cold resistance of apple twigs during spring dehardening. J Am Soc Hortic Sci 95:190–192

Inze D, De Veylder L (2006) Cell cycle regulation in plant development. Annu Rev Genet 40:77–105

Ionescu IA, Lopez-Ortega G, Burow M, Bayo-Canha A, Junge A, Gericke O, Moller BL, Sanchez-Perez R (2017) Transcriptome and metabolite changes during hydrogen cyanamide-induced floral bud break in sweet cherry. Front Plant Sci 8:1233

Itoh H, Matsuoka M, Steber CM (2003) A role for the ubiquitin-26S-proteasome pathway in gibberellin signaling. Trends Plant Sci 8:492–497

Jaglo-Ottosen KR, Gilmour SJ, Zarka DG, Schabenberger O, Thomashow MF (1998) *Arabidopsis CBF1* overexpression induces *COR* genes and enhances freezing tolerance. Science 280:104–106

Jia Y, Ding Y, Shi Y, Zhang X, Gong Z, Yang S (2016) The *cbfs* triple mutants reveal the essential functions of *CBFs* in cold acclimation and allow the definition of CBF regulons in Arabidopsis. New Phytol 212:345–353

Jimenez S, Lawton-Rauh AL, Reighard GL, Abbott AG, Bielenberg DG (2009) Phylogenetic analysis and molecular evolution of the dormancy associated MADS-box genes from peach. BMC Plant Biol 9:81

Jimenez S, Reighard GL, Bielenberg DG (2010) Gene expression of *DAM5* and *DAM6* is suppressed by chilling temperatures and inversely correlated with bud break rate. Plant Mol Biol 73:157–167

Kalberer SR, Wisniewski M, Arora R (2006) Deacclimation and reacclimation of cold-hardy plants: Current understanding and emerging concepts. Plant Sci 171:3–16

Karlberg A, Bako L, Bhalerao RP (2011) Short day-mediated cessation of growth requires the downregulation of AINTEGUMENTALIKE1 transcription factor in hybrid aspen. PLoS Genet 7:

Kasuga M, Liu Q, Miura S, Yamaguchi-Shinozaki K, Shinozaki K (1999) Improving plant drought, salt, and freezing tolerance by gene transfer of a single stress-inducible transcription factor. Nat Biotechnol 17:287–291

Ketchie DO (1984) Cold resistance of apple trees through the year and its relationship to the physiological stages. Acta Hortic 168:131–137

Kidokoro S, Yoneda K, Takasaki H, Takahashi F, Shinozaki K, Yamaguchi-Shinozaki K (2017) Different cold-signaling pathways function in the responses to rapid and gradual decreases in temperature. Plant Cell 29:760–774

Kim YS, Lee M, Lee J-H, Lee H-J, Park C-M (2015) The unified ICE-CBF pathway provides a transcriptional feedback control of freezing tolerance during cold acclimation in *Arabidopsis*. Plant Mol Biol 89:187–201

Kozlowski TT, Pallardy SG (2002) Acclimation and adaptive responses of woody plants to environmental stresses. Bot Rev 68:270–334

Kumar G, Arya P, Gupta K, Randhawa V, Acharya V, Singh AK (2016a). Comparative phylogenetic analysis and transcriptional profiling of MADS-box gene family identified *DAM* and *FLC*-like genes in apple (*Malus* x *domestica*) Sci Rep 6:20

Kumar G, Rattan UK, Singh AK (2016b) Chilling-mediated DNA methylation changes during dormancy and its release reveal the importance of epigenetic regulation during winter dormancy in apple (*Malus* × *domestica* Borkh.). PLoS ONE 11(2): e0149934

Kuroda H, Sagisaka S (2005) Seasonal ultrastructural changes in xylem parenchyma cells of apple twigs in relation to cold hardiness. J Jpn Soc Hortic Sci 74:181–188

Labuschagné IF, Louw JH, Schmidt K, Sadie A (2002) Genetic variation in chilling requirement in apple progeny. J Am Soc Hortic Sci 127:663–672

Lang GA (1987) Dormancy: a new universal terminology. HortScience 22:817–820

Lee Y, Karunakaran C, Lahlali R, Liu X, Tanino KK, Olsen JE (2017) Photoperiodic regulation of growth-dormancy cycling through induction of multiple bud-shoot barriers preventing water transport into the winter buds of Norway spruce. Front Plant Sci 8:2109

Leida C, Conesa A, Llácer G, Badenes ML, Ríos G (2012) Histone modifications and expression of DAM6 gene in peach are modulated during bud dormancy release in a cultivar-dependent manner. New Phytol 193:67–80

Levitt J (1980) Responses of plants to environmental stress. Chilling, freezing and high temperature, vol 1, 2nd edn. Academic Press, NY, 497pp

Li Z, Reighard GL, Abbott AG, Bielenberg DG (2009) Dormancy-associated MADS genes from the *EVG* locus of peach [*Prunus persica* (L.) Batsch] have distinct seasonal and photoperiodic expression patterns. J Exp Bot 60:3521–3530

Li A, Zhou M, Wei D, Chen H, You C, Lin J (2017) Transcriptome profiling reveals the negative regulation of multiple plant hormone signaling pathways elicited by overexpression of C-repeat binding factors. Front Plant Sci 8:1647

Li J, Xu Y, Niu Q, He L, Teng Y, Bai S (2018) Abscisic acid (ABA) promotes the induction and maintenance of pear (*Pyrus pyrifolia* White Pear Group) flower bud endodormancy. Int J Mol Sci 19:310

Li X, Li M, Zhou B, Yang Y, Wei Q, Zhang J (2019) Transcriptome analysis provides insights into the stress response crosstalk in apple (*Malus* × *domestica*) subjected to drought, cold and high salinity. Sci Rep 9:9071

Lim GH, Kachroo A, Kachroo P (2016) Role of plasmodesmata and plasmodesmata localizing proteins in systemic immunity. Plant Signal Behav 11:

Lipavska H, Maskova P, Vojvodova P (2011) Regulatory dephosphorylation of CDK at G(2)/M in plants: yeast mitotic phosphatase cdc25 induces cytokinin-like effects in transgenic tobacco morphogenesis. Ann Bot 107:1071–1086

Liu J, Sherif SM (2019) Combating spring frost with ethylene. Front Plant Sci 10:1408

Liu Z, Zhu H, Abbott A (2015) Dormancy behaviors and underlying regulatory mechanisms: from perspective of pathways to epigenetic regulation. In: Anderson JV (ed) Advances in plant dormancy. Springer International Publishing, New York, pp 75–105

Liu D, Faust M, Millard M, Line M, Stutte G (1993) States of water in summer-dormant apple buds determined by proton magnetic resonance imaging. J Am Soc Hortic Sci 118:632–637

Liu Q, Kasuga M, Sakuma Y, Abe H, Miura S, Yamaguchi-Shinozaki K, Shinozaki K (1998) Two transcription factors, DREB1 and DREB2, with an EREBP/AP2 DNA binding domain separate two cellular signal transduction pathways in drought- and low temperature responsive gene expression, respectively, in Arabidopsis. Plant Cell 10:1391–1406

Liu B, Zhao S, Tan F, Zhao H, Wang D, Si H, Chen Q (2017) Changes in ROS production and antioxidant capacity during tuber sprouting in potato. Food Chem 237:205–213

Liu Q, Piao S, Janssens IA, Fu Y, Peng S, Lian X, Ciais P, Myneni RB, Penuelas J, Wang T (2018) Extension of the growing season increases vegetation exposure to frost. Nat Commun 9:426

Luby J, Forsline P, Aldwinckle H, Bus V, Geibel M (2001) Silk road apples – collection, evaluation, and utilization of *Malus sieversii* from Central Asia. HortScience 36:225–231

Luo F, Norelli JL, Howard NP, Wisniewski M, Flachowsky H, Hanke M-V, Peace C (2020) Introgressing blue mold resistance into elite apple germplasm by rapid cycle breeding and foreground and background DNA-informed selection. Tree Genet Genomes 16:28

Ma Q, Huang JG, Hanninen H, Berninger F (2019) Divergent trends in the risk of spring frost damage to trees in Europe with recent warming. Glob Chang Biol 25:351–360

Malagi G, Sachet MR, Citadin I, Herter FG, Bonhomme M, Regnard J-L, Legave JM (2015) The comparison of dormancy dynamics in apple trees grown under temperate and mild winter climates imposes a renewal of classical approaches. Trees 29:1365–1380

Medina J, Bargues M, Terol J, Perez-Alonso M, Salinas J (1999) The *Arabidopsis CBF* gene family is composed of three genes encoding AP2 domain-containing proteins whose expression is regulated by low temperature but not by abscisic acid or dehydration. Plant Physiol 119:463–470

Menges M, Samland AK, Planchais S, Murray JA (2006) The D-type cyclin CYCD3;1 is limiting for the G1-to-S-phase transition in Arabidopsis. Plant Cell 18:893–906

Mimida N, Saito T, Moriguchi T, Suzuki A, Komori S, Wada M (2015) Expression of *DORMANCY-ASSOCIATED MADS-BOX (DAM)*-like genes in apple. Biol Plant 59:237–244

Miotto YE, Tessele C, Czermainski ABC, Porto DD, Falavigna VS, Sartor T, Cattani AM, Delatorre CA, de Alencar SA, da Silva-Junior OB, Togawa RC, Costa MMC, Pappas GJ Jr, Grynberg P, de Oliveira PRD, Kvitschal MV, Denardi F, Buffon V, Revers LF (2019) Spring is coming: Genetic analyses of the bud break date locus reveal candidate genes from the cold perception pathway to dormancy release in apple (*Malus* × *domestica* Borkh.). Front Plant Sci 10:33

Mizoi J, Shinozaki K, Yamaguchi-Shinozaki K (2012) AP2/ ERF transcription factors in plant abiotic stress. Biochim Biophys Acta 1819:86–96

Molisch H (1897) Untersuchungen über das erfrieren der Pflanzen (reprinted in English in 1982). Cryo Lett 3:332–390

Molmann JA, Asante DKA, Jensen JB, Krane MN, Ernstsen A, Junttla O, Olsen JE (2005) Low night temperature and inhibition of gibberellin biosynthesis override phytochrome action and induce bud set and cold acclimation, but not dormancy in *PHYA* overexpressors and wild-type of hybrid aspen. Plant Cell Environ 28:1579–1588

Moran RE, Peterson BJ, Fazio G, Cline J (2018) Genotypic variation in apple rootstock low temperature tolerance during spring and fall. J Am Soc Hortic Sci 143:319–332

Nagar PK, Sood S (2006) Changes in endogenous auxins during winter dormancy in tea (*Camellia sinensis* L.) O. Kuntze. Acta Physiol Plant 28:165–169

Nanninga C, Buyarski CR, Pretorius AM, Montgomery RA (2017) Increased exposure to chilling advances the time to budburst in North American tree species. Tree Physiol 37:1727–1738

Naor A, Flaishman M, Stern R, Moshe A, Erez A (2003) Temperature effects on dormancy completion of vegetative buds in apple. J Am Soc Hortic Sci 128:636–641

Nishiyama S, Matsushita MC, Yamane H, Honda C, Okada K, Tamada Y, Moriya S, Tao R (2019) Functional and expressional analyses of apple *FLC*-like in relation to dormancy progress and flower bud development. Tree Physiol tpz111. https://doi.org/10.1093/treephys/tpz0111

Niu Q, Li J, Cai D, Qian M, Jia H, Bai S, Hussain S, Liu G, Teng Y, Zheng X (2016) Dormancy-associated MADS-box genes and microRNAs jointly control dormancy transition in pear (*Pyrus pyrifolia* white pear group) flower bud. J Exp Bot 67:239–257

Noriega X, Pérez FJ (2017) ABA biosynthesis genes are down-regulated while auxin and cytokinin biosynthesis genes are up-regulated during the release of grapevine buds from endodormancy. J Plant Growth Regul 36:814–823

Novillo F, Alonso JM, Ecker JR, Salinas J (2004) CBF2/ DREB1C is a negative regulator of *CBF1/ DREB1B* and *CBF3/ DREB1A* expression and plays a central role in stress tolerance in *Arabidopsis*. Proc Natl Acad Sci USA 101:3985–3990

Novillo F, Medina J, Salinas J (2007) *Arabidopsis* CBF1 and CBF3 have a different function than CBF2 in cold acclimation and define different gene classes in the CBF regulon. Proc Natl Acad Sci USA 104:21002–21007

Oh E, Yamaguchi S, Hu J, Yusuke J, Jung B, Paik I, Lee H-S, Sun T-P, Kamiya Y, Choi G (2007) PIL5, a phytochrome-interacting bHLH protein, regulates gibberellin responsiveness by binding directly to the GAI and RGA promoters in Arabidopsis seeds. Plant Cell 19:1192–1208

Ophir R, Pang X, Halaly T, Venkateswari J, Lavee S, Galbraith D, Or E (2009) Gene-expression profiling of grape bud response to two alternative dormancy-release stimuli expose possible links between impaired mitochondrial activity, hypoxia, ethylene-ABA interplay and cell enlargement. Plant Mol Biol 71:403–423

Osakabe Y, Liang Z, Ren C, Nishitani C, Osakabe K, Wada M, Komori S, Malnoy M, Velasco R, Poli M, Jung M-H, Koo O-J, Viola R, Kanchiswany CN (2018) CRISPR-Cas9-mediated genome editing in apple and grapevine. Nat Prot 13:2844–2863

Petterle A, Karlberg A, Bhalerao RP (2013) Daylength mediated control of seasonal growth in perennial trees. Curr Opin Biotech 16:301–306

Porto DD, Bruneau M, Perini P, Anzanello R, Renou J-P, dos Santos HP, Fialho FB, Revers LF (2015) Transcription profiling of the chilling requirement for bud break in apples: a putative role for *FLC-like* genes. J Exp Bot 66:2659–2672

Porto DD, Da Silveira FV, Arenhart RA, Perini P, Buffon V, Anzanello R, Dos Santos HP, Fialho FB, De Oliveira PRD, Revers LF (2016) Structural genomics and transcriptional characterization of the Dormancy-Associated MADS-box genes during bud dormancy progression in apple. Tree Genet Genomes 12:46

Pramsohler M, Neuner G (2013) Dehydration and osmotic adjustment in apple stem tissue during winter as it relates to the frost resistance of buds. Tree Physiol 33:807–816

Pramsohler M, Hacker J, Neuner G (2012) Freezing pattern and frost killing temperature of apple (*Malus domestica*) wood under controlled conditions and in nature. Tree Physiol 32:819–828

Preston JC, Sandve SR (2013) Adaptation to seasonality and the winter freeze. Front Plant Sci 4:167

Qiu Z, Wan L, Chen T, Wan Y, He X, Lu S, Wang Y, Lin J (2013) The regulation of cambial activity in Chinese fir (*Cunninghamia lanceolata*) involves extensive transcriptome remodeling. New Phytol 199:708–719

Rinne PL, van der Schoot C (2003) Plasmodesmata at the crossroads between development, dormancy, and defense. Can J Bot 81(12):1182–1197

Rinne PL, Welling A, Vahala J, Ripel L, Ruonala R, Kangasjarvi J, van der Schoot C (2011) Chilling of dormant buds hyperinduces *FLOWERING LOCUS T* and recruits GA-inducible 1,3-β-glucanases to reopen signal conduits and release dormancy in *Populus*. Plant Cell 23:130–146

Rinne PLH, Paul LK, van der Schoot C (2018) Decoupling photo- and thermoperiod by projected climate change perturbs bud development, dormancy establishment and vernalization in the model tree *Populus*. BMC Plant Biol 18:220

Rohde A, Bhalerao RP (2007) Plant dormancy in the perennial context. Trends Plant Sci 12:217–223

Rothkegel K, Sanchez E, Montes C, Greve M, Tapia S, Bravo S, Prieto H, Almeida AM (2017) DNA methylation and small interference RNAs participate in the regulation of MADS-box genes involved in dormancy in sweet cherry (*Prunus avium* L.). Tree Physiol 37:1739–1751

Ruan YL, Jin Y, Yang YJ, Li GJ, Boyer JS (2010) Sugar input, metabolism, and signaling mediated by invertase: roles in development, yield potential, and response to drought and heat. Mol Plant 3:942–955

Ruiz D, Campoy JA, Egea J (2007) Chilling and heat requirements of apricot cultivars for flowering. Environ Exp Bot 61:254–263

Ruonala R, Rinne PL, Baghour M, Moritz T, Tuominen H, Kangasjarvi J (2006) Transitions in the functioning of the shoot apical meristem in birch (*Betula pendula*) involve ethylene. Plant J 46:628–640

Saito T, Bai S, Imai T, Ito A, Nakajima I, Moriguchi T (2015) Histone modification and signalling cascade of the dormancy-associated MADS-box gene, *PpMADS*13-1, in Japanese pear (*Pyrus pyrifolia*) during endodormancy. Plant Cell Environ 38:1157–1166

Sakai A, Larcher W (1987) Frost survival of plants: responses and adaptation to freezing stress. Ecological studies, vol 62. Springer, Berlin, 321p

Salazar-Gutierrez MR, Chaves B, Hoogenboom G (2016) Freezing tolerance of apple flower buds. Sci Hortic 198:344–351

Schrader J, Moyle R, Bhalerao R, Hertzberg M, Lundeberg J, Nilsson P, Bhalerao RP (2004) Cambial meristem dormancy in trees involves extensive remodeling of the transcriptome. Plant J 40:173–187

Segura V, Denance C, Durel CE, Costes E (2007) Wide range QTL analysis for complex architectural traits in a 1-year-old apple progeny. Genome 50:159–171

Seo M, Hanada A, Kuwahara A, Endo A, Okamoto M, Yamauchi Y, North H, Marion-Poll A, Sun TP, Koshiba T (2006) Regulation of hormone metabolism in Arabidopsis seeds: phytochrome regulation of abscisic acid metabolism and abscisic acid regulation of gibberellin metabolism. Plant J 48:354–366

Shi Y, Tian S, Hou L, Huang X, Zhang X, Guo H, Yang S (2012) Ethylene signaling negatively regulates freezing tolerance by repressing expression of CBF and type-A ARR genes in Arabidopsis. Plant Cell 24:2578–2595

Shi Y, Huang J, Sun T, Wang X, Zhu C, Ai Y, Gu H (2017) The precise regulation of different *COR* genes by individual CBF transcription factors in *Arabidopsis thaliana*. J Int Plant Biol 59:118–133

Shi Y, Ding Y, Yang S (2018) Molecular regulation of CBF signaling in cold acclimation. Trends Plant Sci 23:623–637

Singh RK, Miskolczi P, Maurya JP, Bhalerao RP (2019) A tree ortholog of *SHORT VEGETATIVE PHASE* floral repressor mediates photoperiodic control of bud dormancy. Curr Biol 29:128–133

Soon FF, Ng LM, Zhou XE, West GM, Kovach A, Tan MH, Suino-Powell KM, He Y, Xu Y,

Chalmers MJ, Brunzelle JS, Zhang H, Yang H, Jiang H, Li J, Yong EL, Cutler S, Zhu JK, Griffin PR, Melcher K, Xu HE (2012) Molecular mimicry regulates ABA signaling by SnRK2 kinases and PP2C phosphatases. Science 335:85–88

Srinivasan C, Dardick C, Callahan A, Scorza R (2012) Plum (*Prunus domestica*) trees transformed with poplar *FT1* result in altered architecture, dormancy requirement, and continuous flowering. PLoS ONE 7:

Stushnoff C, Remmele RL Jr, Essensee V, McNeil M (1999) Low temperature induced biochemical mechanisms: implications for cold acclimation and de-acclimation. In: Jackson MB, Black CR (eds) NATO ASI series, vol 116. Interacting stresses on plants in a changing climate. Springer, Berlin, pp 647–657

Sudawan B, Chang CS, Chao HF, Ku MS, Yen YF (2016) Hydrogen cyanamide breaks grapevine bud dormancy in the summer through transient activation of gene expression and accumulation of reactive oxygen and nitrogen species. BMC Plant Biol 16:202

Sumitomo K, Narumi T, Satoh S, Hisamatsu T (2008) Involvement of the ethylene response pathway in dormancy induction in chrysanthemum. J Exp Bot 59:4075–4082

Sun TP (2011) The molecular mechanism and evolution of the GA-GID1-DELLA signaling module in plants. Curr Biol 21:R338–R345

Suttle JC (2004) Physiological regulation of potato tuber dormancy. Am J Potato Res 81:253–262

Tarkowski LP, Van den Ende W (2015) Cold tolerance triggered by soluble sugars: a multifaceted countermeasure. Front Plant Sci 6:203

Torres Acosta JA, Fowke LC, Wang H (2011) Analyses of phylogeny, evolution, conserved sequences and genome-wide expression of the ICK/KRP family of plant CDK inhibitors. Ann Bot 107:1141–1157

Tuan PA, Bai S, Saito T, Ito A, Moriguchi T (2017) Dormancy-Associated MADS-box (DAM) and the abscisic acid pathway regulate pear endodormancy through a feedback mechanism. Plant Cell Physiol 58:1378–1390

Tylewicz S, Petterle A, Marttila S, Miskolczi P, Azeez A, Singh RK, Immanen J, Mahler N, Hvidsten TR, Eklund DM, Bowman JL, Helariutta Y, Bhalerao RP (2018) Photoperiodic control of seasonal growth is mediated by ABA acting on cell-cell communication. Science 360:212–215

Uemura M, Steponkus PL (1999) Cold acclimation in plants: Relationship between the lipid composition and the cryostability of the plasma membrane. J Plant Res 112:245–254

Unterberger C, Brunner L, Nabernegg S, Steininger KW, Steiner AK, Stabentheiner E, Monschein S, Truhetz H (2018) Spring frost risk for regional apple production under a warmer climate. PLoS ONE 13:e0200201–e0200201

Velasco R, Zharkikh A, Affourtit J, Dhingra A, Cestaro A, Kalyanaraman A, Fontana P, Bhatnagar SK, Troggio M, Pruss D, Salvi S, Pindo M, Baldi P, Castelletti S, Cavaiuolo M, Coppola G, Costa F, Cova V, Ri AD, Goremykin V, Komjanc M, Longhi S, Magnago P, Malacarne G, Malnoy M, Micheletti D, Moretto M, Perazzolli M, Si-Ammour A, Vezzulli S, Zini E, Eldredge G, Fitzgerald LM, Gutin N, Lanchbury G, Macalma T, Mitchell JT, Reid J, Wardell B, Kodira C, Chen Z, Desany B, Niazi F, Palmer M, Koepke T, Jiwan D, Schaeffer S, Krishnan V, Wu C, Chu VT, King ST, Vick J, Tao Q, Mraz A, Stormo A, Stormo K, Bogden R, Ederle D, Stella A, Vecchietti A, Kater MM, Masiero S, Lasserre P, Lespinasse Y, Allan AC, Bus V, Chagné D, Crowhurst RN, Gleave AP, Lavezzo E, Fawcett JA, Proost S, Rouzé P, Sterck L, Toppo S, Lazzari B, Hellens RP, Durel C-E, Gutin A, Bumgarner R, Gardiner SE, Skolnick M, Egholm M, Van de Peer Y, Salamini F, Viola R (2010) The genome of the domesticated apple (*Malus* × *domestica* Borkh.). Nat Genet 42:833–841

Vergara R, Noriega X, Aravena K, Prieto H, Perez FJ (2017) ABA represses the expression of cell cycle genes and may modulate the development of endodormancy in grapevine buds. Front Plant Sci 8:812

Verkest A, Manes CL, Vercruysse S, Maes S, Van Der Schueren E, Beeckman T, Genschik P, Kuiper M, Inze D, De Veylder L (2005) The cyclin-dependent kinase inhibitor KRP2 controls the onset of the endoreduplication cycle during Arabidopsis leaf development through inhibition of mitotic CDKA;1 kinase complexes. Plant Cell 17:1723–1736

Vertucci CW, Stushnoff C (1992) The state of water in acclimating vegetative buds from *Malus* and *Amelanchier* and its relationship to winter hardiness. Phys Plant 86:503–511

Vitasse Y, François C, Delpierre N, Dufrêne E, Kremer A, Chuine I, Delzon S (2011) Assessing the effects of climate change on the phenology of European temperate trees. Agric Forest Meteorol 151:969–980

Vitasse Y, Schneider L, Rixen C, Christen D, Rebetez M (2018) Increase in the risk of exposure of forest and fruit trees to spring frosts at higher elevations in Switzerland over the last four decades. Agric Forest Meteorol 248:60–69

Wang SY, Faust M (1990) Seasonal changes of membrane lipids in apple shoots. J Am Soc Hortic Sci 115:462–467

Wang D, Gao Z, Du P, Xiao W, Tan Q, Chen X, Li L, Gao D (2015) Expression of ABA metabolism-related genes suggests similarities and differences between seed dormancy and bud dormancy of peach (*Prunus persica*). Front Plant Sci 6:1248

Welling A, Palva ET (2006) Molecular control of cold acclimation in trees. Physiol Plant 127:167–181

Wen LH, Zhong WJ, Huo XM, Zhuang WB, Ni ZJ, Gao ZH (2016) Expression analysis of ABA- and GA-related genes during four stages of bud dormancy in Japanese apricot (*Prunus mume* Sieb. et Zucc). J Hortic Sci Biotech 91:362–369

Wenden B, Mariadassou M, Chmielewski FM, Vitasse Y (2020) Shifts in the temperature-sensitive periods for spring phenology in European beech and pedunculate oak clones across latitudes and over recent decades. Glob Chang Biol 26:1808–1819. https://doi.org/10.1111/gcb.14918

Wiegand KM (1906) The occurrence of ice in plant tissue. Plant World 9:25–39

Wisniewski M, Bassett C, Gusta LV (2003) An overview of cold hardiness in woody plants: seeing the forest through the trees. HortScience 38:952–959

Wisniewski M, Norelli J, Bassett C, Artlip T, Macarisin D (2011) Ectopic expression of a novel peach (*Prunus persica*) CBF transcription factor in apple (*Malus* × *domestica*) results in short-day induced dormancy and increased cold hardiness. Planta 233:971–983

Wisniewski M, Nassuth A, Teulieres C, Marque C, Rowland J, Cao PB, Brown A (2014) Genomics of cold hardiness in woody plants. Crit Rev Plant Sci 33:92–124

Wisniewski M, Norelli J, Artlip T (2015) Overexpression of a peach CBF gene in apple: a model for understanding the integration of growth dormancy and cold hardiness in woody plants. Front Plant Sci 6:85

Wisniewski M, Nassuth A, Arora R (2018) Cold hardiness in trees: a mini-review. Front Plant Sci 9:1394

Wisniewski M, Artlip T, Liu J, Ma J, Burchard E, Norelli J, Dardick C (2020) FOX hunting in wild apples: searching for novel genes in *Malus sieversii*. J Int Mol Sci 21:9516. https://doi.org/10.3390/ijms21249516

Wolbang CM, Ross JJ (2001) Auxin promotes gibberellin biosynthesis in decapitated tobacco plants. Planta 214:153–157

Wolbang CM, Chandler PM, Smith JJ, Ross JJ (2004) Auxin from the developing inflorescence is required for the biosynthesis of active gibberellins in barley stems. Plant Physiol 134:769–776

Wu R, Tomes S, Karunairetnam S, Tustin SD, Hellens RP, Allan AC, Macknight RC, Varkonyi-Gasic E (2017) *SVP*-like MADS box genes control dormancy and budbreak in apple. Front Plant Sci 8:477

Wu SW, Kumar R, Iswanto ABB, Kim JY (2018) Callose balancing at plasmodesmata. J Exp Bot 69:5325–5339

Xie Y, Chen P, Yan Y, Bao C, Li X, Wang L, Shen X, Li H, Liy X, Niu C, Zhu C, Fang N, Shao Y, Zhao T, Yu J, Zhu J, Xu L, van Nocker S, Ma F, Guan Q (2018) An atypical R2R3 MYB transcriptions factor increases cold hardiness by CBF-dependent and CBF-independent pathways in apple. New Phytol 218:201–218

Yamaguchi M, Fabian T, Sauter M, Bhalerao RP, Schrader J, Sandberg G, Umeda M, Uchimiya H (2000) Activation of CDK-activating kinase is dependent on interaction with H-type cyclins in plants. Plant J 24:11–20

Yamane H, Wada M, Honda C, Matsuura T, Ikeda Y, Hirayama T, Osako Y, Gao-Takai M, Kojima M, Sakakibara H, Tao R (2019) Overexpression of *Prunus DAM6* inhibits growth, represses bud break competency of dormant buds and delays bud outgrowth in apple plants. PLoS ONE 14:e0214788

Yang W, Liu X-D, Chi X-J, Wu C-A, Li Y-Z, Song L-L, Liu X-M, Wang Y-F, Wang F-W, Zhang C, Liu Y, Zong J-M, Li H-Y (2010) Dwarf apple MbDREB1 enhances plant tolerance to low temperature, drought, and salt stress via both ABA-dependent and ABA-independent pathways. Planta 233:219–229

Yang X, Brown SK, Davies PJ (2013) The content and in vivo metabolism of gibberellin in apple vegetative tissues. J Am Soc Hortic Sci 138:173–183

Yordanov S, Ma C, Strauss SH, Busov VB (2014) *EARLY BUD-BREAK 1* (*EBB1*) is a regulator of release from seasonal dormancy in poplar trees. Proc Natl Acad Sci USA 111:10001–10006

Yue C, Cao H, Hao X, Zeng J, Qian W, Guo Y, Ye N, Yang Y, Wang X (2018) Differential expression of gibberellin- and abscisic acid-related genes implies their roles in the bud activity-dormancy transition of tea plants. Plant Cell Rep 37:425–441

Zawaski C, Busov VB (2014) Roles of gibberellin catabolism and signaling in growth and physiological response to drought and short-day photoperiods in *Populus* trees. PLoS ONE 9:

Zhang L, Wang Y, Zhang X, Zhang M, Han D, Qiu C, Han Z (2012a) Dynamics of phytohormone and DNA methylation patterns changes during dormancy induction in strawberry (*Fragaria* x *ananassa* Duch.). Plant Cell Rep 31:155–165

Zhang H, Bishop B, Ringenberg W, Muir WM, Ogas J (2012b) The CHD3 remodeler PICKLE associates with genes enriched for trimethylation of histone H3 lysine 27. Plant Physiol 159:418–432

Zhang R-F, Guo Y, Li Y-Y, Zhou L-J, Hao Y-J, You C-X (2016) Functional identification of MdSIZ1 as a SUMO E3 ligase in apple. J Plant Physiol 198:69–80

Zhang Z, Zhuo X, Zhao K, Zheng T, Han Y, Yuan C, Zhang Q (2018) Transcriptome profiles reveal the crucial roles of hormone and sugar in the bud dormancy of *Prunus mume*. Sci Rep 8:5090

Zhao T, Liang D, Wang P, Liu J, Ma F (2012) Genome-wide analysis and expression profiling of the DREB transcription factor gene family in *Malus* under abiotic stress. Mol Genet Genomics 287:423–436

Zhao C, Zhang Z, Xie S, Si T, Li Y, Zhu JK (2016) Mutational evidence for the critical role of CBF transcription factors in cold acclimation in Arabidopsis. Plant Physiol 171:2744–2759

Zhao K, Zhou Y, Ahmad S, Yong X, Xie X, Han Y, Li Y, Sun L, Zhang Q (2018) *PmCBFs* synthetically affect *PmDAM6* by alternative promoter binding and protein complexes towards the dormancy of bud for *Prunus mume*. Sci Rep 8:4527

Zheng C, Halaly T, Acheampong AK, Takebayashi Y, Jikumaru Y, Kamiya Y, Or E (2015) Abscisic acid (ABA) regulates grape bud dormancy, and dormancy release stimuli may act through modification of ABA metabolism. J Exp Bot 66:1527–1542

Zheng C, Kwame Acheampong A, Shi Z, Halaly T, Kamiya Y, Ophir R, Galbraith DW, Or E (2018)

Distinct gibberellin functions during and after grapevine bud dormancy release. J Exp Bot 69:1635–1648

Zhong W, Gao Z, Zhuang W, Shi T, Zhang Z, Ni Z (2013) Genome-wide expression profiles of seasonal bud dormancy at four critical stages in Japanese apricot. Plant Mol Biol 83:247–264

Zhu JK (2016) Abiotic stress signaling and responses in plants. Cell 167:313–324

Zhu Y, Li Y, Xin D, Chen W, Shao X, Wang Y, Guo W (2015) RNA-Seq-based transcriptome analysis of dormant flower buds of Chinese cherry (*Prunus pseudocerasus*). Gene 555:362–376

Zhuang W, Gao Z, Wen L, Huo X, Cai B, Zhang Z (2015) Metabolic changes upon flower bud break in Japanese apricot are enhanced by exogenous GA4. Hortic Res 2:15046

Genetics and Genomics of Fruit Color Development in Apple

13

Nan Wang and Xuesen Chen

Abstract

Apple (*Malus* × *domestica* Borkh.) is one of the most widely produced and economically important fruits in temperate regions. Fruit color development in apples is a major focus for both breeders and researchers as consumers associate brightly colored red apples with ripeness and good flavor. In this chapter, we will discuss research efforts on the coloration of the apple fruit, including the development of important genomic databases to identify important genomic regions and genes, genetic and transcriptional factors that regulate pigment accumulation, environmental factors influencing anthocyanin synthesis, and current goals for breeding for red-skinned and red-fleshed apples. We will also cover key transcription factors, such as MYB, bHLH, and WD40 that are involved in the regulation of anthocyanin synthesis and of fruit color development in apples. We will also discuss the regulation of apple color development by external environmental factors such as light, temperature, and water. Furthermore, we will offer insights into the molecular mechanisms underlying anthocyanin biosynthesis in apples. This knowledge will provide valuable guidance for the breeding of high-quality red-skinned and red-fleshed apple cultivars.

N. Wang · X. Chen (✉)
State Key Laboratory of Crop Biology, College of Horticulture Science and Engineering, Shangdong Agricultural University, Shangdong, China
e-mail: chenxs@sdau.edu.cn

13.1 Introduction

Apple (*Malus* × *domestica* Borkh.) is one of the most widely grown and economically important fruits in temperate regions (Velasco et al. 2010). It is popular with growers and consumers alike due to its strong ecological adaptability and high nutritional value (Eberhardt et al. 2000; Boyer and Liu 2004; Hyson 2011). Although the apple has been widely distributed in Europe, Asia, and North America, following nearly 2000 years of domestication and cultivar breeding efforts, apple cultivars are now grown in all five continents around the world. According to statistical data in the World Apple Review (2017), apples account for 12.26% of the world's fruit production from 2012 to 2014, second only to bananas and citrus (Belrose, Inc. 2018; see Chap. 1 in this volume). According to the Food and Agriculture Organization of the United Nations (FAO), global apple production in 2015 has exceeded 80 million tons (FAO 2015). The major apple cultivars can be roughly divided into either red-colored bearing fruit, such as 'Red Delicious', 'Fuji', and 'Royal Gala', among others, and yellow/green-colored bearing fruits, such as 'Golden Delicious', 'Granny Smith', and 'Orin', among others. Consumers associate red-colored apples with

S. S. Korban (ed.), *The Apple Genome*, Compendium of Plant Genomes,
https://doi.org/10.1007/978-3-030-74682-7_13

ripeness and good flavor. Consequently, red apple cultivars are often highly marketable along with higher economic value than their counterparts (King and Cliff 2002; Ban et al. 2007). Therefore, fruit color development is of major interest for both apple breeders and researchers.

Fruit color is mainly determined by anthocyanins (Allan et al. 2008), an important class of secondary metabolites synthesized in higher plants. Anthocyanins contribute to different color development, depending on the pH of cells, thus conferring leaves, flowers, fruits, and seeds of plants with different colors, such as red, purple, or blue (Ogata et al. 2005; Jaakola 2013).

Biochemically, anthocyanins are glycosylated derivatives of anthocyanidins which are generated via the formation of glycosidic bonds between anthocyanidin and various monosaccharides, including glucose, rhamnose, galactose, xylose, or arabinose. The basic structure of an anthocyanidin is C6-C3-C6; i.e., a C15 skeleton comprising two benzene rings and a single oxygen-containing heterocyclic ring (Fig. 13.1). Based on functional groups at the R1 and R2 sites, anthocyanidins in plants can be subdivided into the following six types: pelargonidin, cyanidin, delphinidin, peonidin, petunidin, and malvidin. In apple, cyanidin is the main type of anthocyanidin, and it mainly combines with galactose to form cyanidin-3-galactoside, which accounts for about 80% of the total anthocyanins (Treutter 2001).

The biosynthesis pathway of anthocyanins in plants has been thoroughly investigated. It is reported that the metabolic pathway is highly

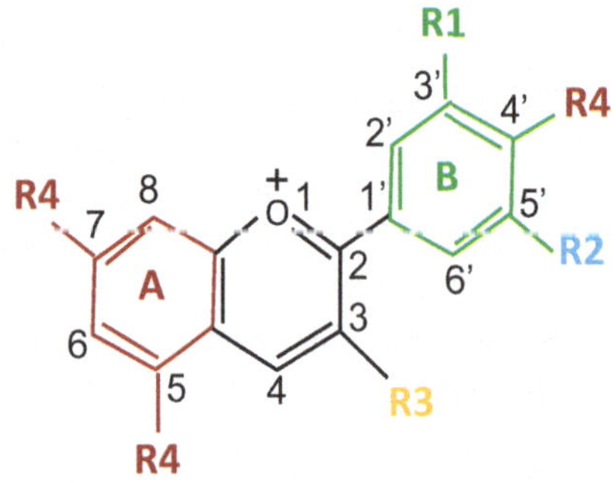

Fig. 13.1 The basic structure of an anthocyanidin

conserved in various species; moreover, genes encoding key enzymes have been cloned and characterized (Winkel-Shirley 2001; Honda et al. 2011). In particular, anthocyanins are synthesized through the phenylalanine metabolic pathway (Fig. 13.2). In the early stages of the anthocyanin biosynthetic process, phenylalanine ammonia lyase (PAL) converts phenylalanine into cinnamic acid, which in turn is converted into 4-coumaroyl-CoA via activities of both cinnamate-4-hydroxylase (C4H) and 4-coumarate-CoA ligase (4CL). Then, chalcone is produced from the substrate 4-coumaroyl-CoA and malonyl-CoA by chalcone synthase (CHS). Subsequently, chalcone isomerase (CHI) catalyzes the formation of naringenin, which is then converted into dihydroflavonol by flavonoid 3-hydroxylase (F3H). In turn, dihydroflavonol-4-reductase (DFR) converts dihydroflavonol into leucocyanidin, which is then converted to cyanidin by anthocyanidin synthase (ANS). Finally, unstable anthocyanidins are glycosylated by UDP-Glc:flavonoid-3-O-glucosyltransferase (UFGT) to form stable anthocyanins (Hugo and Timothy 1991; Holton and Cornish 1995; Wang et al. 2018a, b, c, d, e).

In addition to structural genes involved in this biosynthetic pathway, transcription factors (TFs) that regulate anthocyanin synthesis have also attracted interest. Many TF families related to anthocyanin synthesis including MYB (myeloblastosis), bHLH (basic helix-loop-helix), WD40 (a short structural motif of ~ 40 amino acids, often terminating in a tryptophan-aspartic acid (W-D) dipeptide), WRKY (a domain of 60–70 amino acids, DNA binding domain, characterized by a highly conserved core WRKYGQK motif and a zinc-finger region), and zinc-finger proteins have been identified in and isolated from a range of plants (Walker et al. 1999; Borevitz et al. 2000a, b; Johnson et al. 2002; Liu et al. 2019). A tertiary protein complex comprising MYB, bHLH, and WD40 proteins (MBW) is deemed as an important element in the regulation of anthocyanin synthesis, and *MYBs* would serve as main genes (Gonzalez et al. 2008).

The first *MYB* genes to be identified and isolated from apple are *MdMYB1* (*MdMYBA*) and

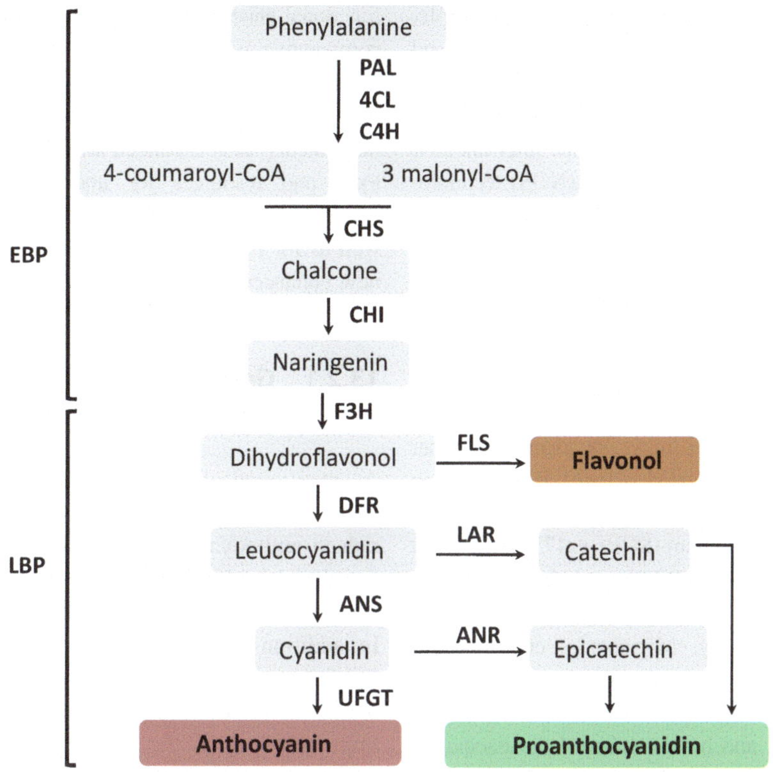

Fig. 13.2 A simplified schematic diagram of the biosynthetic pathway of anthocyanins. PAL, phenylalanine ammonia lyase; C4H, cinnamate-4-hydroxylase; 4CL, 4-coumaroyl-CoA synthase; CHS, chalcone synthase; CHI, chalcone isomerase; F3H, flavonoid 3-hydroxylase; F3'H, flavonoid 3'-hydroxylase; F3'5'H, flavonoid-3',5'-hydroxylase; FLS, flavonol synthase; DFR, dihydroflavonol-4-reductase; ANS, anthocyanidin synthase; LAR, leucoanthocyanidin reductase; ANR, anthocyanidin reductase; UFGT, UDP-Glc:flavonoid-3-O-glucosyltransferase

MdMYB10, encoding TFs regulating anthocyanin synthesis in red-skinned apple and red-fleshed apple, respectively (Takos et al. 2005; Ban et al. 2007; Espley et al. 2007). Subsequent studies have identified additional MYB TFs involved in promoting anthocyanin synthesis in apples, including MdMYB3, MdMYB9, MdMYB11, MdMYB24, MdMYBPA1, and MdMYB110a (Vimolmangkang et al. 2013; Chagné et al. 2013; An et al. 2015; Wang et al. 2018a, b, c, d, e; Wang et al. 2019a, b). Other studies have identified MYB TFs that negatively regulate anthocyanin synthesis in apples, such as MdMYB16, MdMYB17, MdMYB111, and MdMYBL2 (Linwang et al. 2011; Xu et al. 2017a, b; Wang et al. 2018a, b, c, d, e). In addition to MYB TFs, MdbHLH3 and MdbHLH33 of the bHLH TF family are involved in apple fruit coloration (Xie et al. 2012; Xu et al. 2017a, b). Most of the WD40 proteins play important roles in enhancing the stability of MBW complexes, but they do not have catalytic activities (Baudry et al. 2004; Hichri et al. 2011).

The whole-genome sequence of apple, ~742.3 Mb, has been first reported by Velasco et al. (2010). Of this first draft genome sequence, 603.9 Mb has been assembled, and 593.3 Mb has been mapped to 17 chromosomes of the haploid genome of apple. Genomic analyses have revealed that ~500.7 Mb (67%) of the apple genome comprises repetitive sequences. In total, 57,386 genes (including 4021 TFs, 178 miRNAs, 992 resistance genes, and 1246 biosynthetic genes) have been annotated in this first draft of the apple genome (Velasco et al. 2010). This information is essential for pursuing systematic quantitative genetic analyses in apple, such as the construction of fine-linkage maps and mapping of quantitative trait loci (QTLs) (Velasco et al. 2010). Since then, high-quality apple genome sequences have been assembled, one after another, using next-generation sequencing (NGS) technologies or homozygous apple lines (Li et al. 2016; Daccord et al. 2017; Zhang et al. 2019a, b). As the genome has been completely sequenced, the apple has become a model perennial fruit tree species, and this has contributed to rapid progress in research on the development of apple fruit color, biotic and abiotic resistance, fruit flavor, and fruit storage attributes, among other traits. For example, Kumar et al. (2012) have located a fruit color locus on linkage group (LG) 9 by identifying a single nucleotide polymorphism (SNP), ss475879555, that is associated with *MdMYB10*, and this is found to be consistent with earlier findings of Espley et al. (2007). Subsequently, Migicovsky et al. (2016) have conducted a genome-wide association study (GWAS) using 8657 SNP markers and 36 phenotypes of apple, and have confirmed an association between fruit color and the *MdMYB1* locus (Migicovsky et al. 2016).

With the continuous improvement and refinement of the apple genome sequence along with the development of genetic maps, research on apple fruit color development has entered a new era, as this has offered new opportunities for investigating the genetic basis of red-skin color and red-flesh color traits. Moreover, with the development of biotechnological/molecular tools, molecular-assisted selection (MAS) and molecular-assisted breeding (MAB) are now gradually being pursued in apple breeding. These methods can greatly shorten the duration of the breeding period, as well as improve the breeding efficiency for developing red-skinned and red-fleshed-colored apple cultivars.

13.2 Advances in Genomics of Apple Fruit Color Development

With the development of de novo genome assembly, increased numbers of robust molecular markers, such as SNPs, as well as an expanded knowledge of functional genes, these tools can be used to conduct comparative genomics studies. Such analyses can reveal structural variations in chromosomes, such as replicated regions, deletions, inversions, or rearrangements. This expanded knowledge will provide tools and resources for undertaking population genetics and functional gene mapping analyses, as well as for pursuing MAS/MAB efforts to develop new enhanced cultivars.

13.2.1 Genomic Basis of Fruit Color Development in Apple

As mentioned above, red-colored fruits are not only attractive to consumers, but they are also rich in anthocyanins, known to have high nutritional and health benefit values. Therefore, color is an important agronomic trait in apple research. In addition to the common red-skinned apple fruit, the red-fleshed apple has attracted increasing attention in recent years. With continuous improvements of the apple genome sequence along with the development of genetic maps, research on apple fruit color development has entered a new era with new opportunities to analyze red-skinned and red-fleshed traits.

Mapping of quantitative trait loci (QTLs) is the first step in understanding complex genetic traits, wherein the mapping of a large number of QTLs allows for the construction of more detailed genetic maps. Earlier studies have mapped many QTLs related to flavonoids, anthocyanins, and other polyphenols in the apple genome. A total of 488 fruit skin and 181 flesh metabolite QTLs (mQTLs) have been identified and located using the mapping population of 'Prima' × 'Fiesta' (Khan et al. 2012). Interestingly, four LGs (LG1, LG8, LG13, and LG16) are found to contain mQTL hotspots, harboring QTLs related to the regulation of the phenylpropanoid metabolic pathway. In particular, the most important mQTL hotspot is located on LG16 as this included 33 QTLs for skin-related phenolic compounds and 17 QTLs for flesh-related phenolic compounds. Moreover, a structural gene encoding the enzyme leucoanthocyanidin reductase (LAR1), which is involved in flavonoid synthesis, is detected at this locus followed by the analysis of the apple genome sequence (Khan et al. 2012). Later, Verdu et al. (2014) have detected 69 and 72 phenolic-related QTLs in the female parent 'X5210' and the male parent 'X8402'. These QTLs are located on 14 LGs in the female parent and on 11 LGs in the male parent (Verdu et al. 2014). Furthermore, Kumar et al. (2012) have detected a fruit color locus on LG9 associated with the SNP molecular marker ss475879555, and this has allowed for identifying *MdMYB10*, which is consistent with earlier findings of Espley et al. (2007).

Coincidentally, Chagné et al. (2012) used a segregating population of 'Royal Gala' × 'Braeburn' to locate QTLs related to chlorogenic acid, quinine, anthocyanin, catechin, epicatechin, quercetin, and phloridzin on LG1, LG6, LG7, LG9, LG13, LG14, LG15, LG16, and LG17, with limits of detection (LODs) ranging from 3.03 to 41.28. These QTLs could explain 8.4–71.8% of the phenotypic variation observed in their corresponding traits. A QTL related to anthocyanins (cyanidin 3-O-galactoside and cyanidin 3-O-arabinoside) is also detected on LG9 (Chagné et al. 2012). Concomitantly, two homologous genes to *MdMYB10*, *MdMYB110a* and *MdMYB110b*, were found to co-locate with a QTL for the type II red-flesh trait on LG17. These genes were found to control a similar trait for red coloration of apple fruit flesh (Chagné et al. 2013).

As GWAS investigations are based on the linkage disequilibrium (LD) of a population's genome, they do not depend on genetic mapping. Thus, high-density molecular markers, selected from the whole genome, are used to scan a population, and correlations between molecular marker data and phenotypic traits are analyzed. In other words, LD within the whole genome is used to identify those genes associated with phenotypic or quantitative traits (Hirschhorn and Daly 2005). Furthermore, genetic mapping of complex traits can be achieved by breaking associations of polymorphic loci in LD or chromosome regions by using recombination events that must have occurred over generations during a population's evolution (Flintgarcia 2013).

Efforts to conduct GWAS in apples have been launched over the past decade, and these studies focused mainly on fruit color and texture. Kumar et al. (2012) assessed the accuracy of genomic selection (GS) in a population of 1120 seedlings genotyped using an Illumina Infinium chip comprising 8,000 SNPs. They located a fruit color locus on LG9 associated with the molecular marker ss475879555 and the gene *MdMYB10*, and this was consistent with the findings of Espley et al. (2007). Recently, McClure et al. (2019) carried out GWAS using ~100,000 SNP markers and polyphenol content data for two diverse apple populations. A single strong association signal was identified on chromosome 16 for concentrations of catechin, epicatechin, and procyanidins B1, B2, and C1. This genomic region contained a candidate gene encoding leucoanthocyanidin reductase (LAR1). Moreover, a strong GWAS peak for cyanidin-3-galactoside was identified on chromosome 9 (chr9:33717323). Interestingly, the most strongly associated SNP at this locus was the same as that demonstrating the strongest association with total anthocyanins (Mcclure et al. 2019). This locus, located 666 kb downstream from the SNP ss475879531 (chr9:33001375), was successfully used to predict apple skin color (Chagné et al. 2016).

Migicovsky et al. (2016) carried out GWAS using 8657 SNP markers and 36 apple phenotypes and confirmed the existence of an association between fruit color and the *MdMYB1* locus. Recently, Zhang et al. (2019a, b) conducted a comparative genomics analysis of 148 natural apple populations and a single segregating population and demonstrated that a gypsy-like retrotransposon insertion, 1 kb upstream of *MYB1* (chr9:35,541,127–35,541,721), functions as an enhancer for controlling apple coloration. It is reported that weak coloration of some cultivars is due to the absence of this enhancer, termed redTE, which prevents anthocyanin synthesis (Zhang et al. 2019a, b).

Therefore, it is important to summarize that various genomic studies have focused on *MdMYB1* and *MdMYB10* as the two core genes regulating apple skin and flesh color development. However, with continuous progress in the availability of new molecular biology tools, additional regulatory factors have been isolated and identified, and their functions in the anthocyanin synthesis pathway have been widely demonstrated. This has laid out the foundation for efforts to enhance the anthocyanin metabolic network, and for manipulating/modulating apple fruit color development.

13.3 Transcriptional Regulation of Fruit Color in Apple

Anthocyanin biosynthesis is catalyzed by a series of structural enzymes, and the expression of encoding genes is regulated by various TFs, as well as transcription complexes. Transcriptional regulation of anthocyanin synthesis has been first investigated in model plants such as maize, *Arabidopsis*, and petunia, over 30 years ago, but recent research has focused on fruit trees. It is known that TFs regulate the expression of most genes by binding to *cis*-elements in promoter regions of these genes. Moreover, some TF genes possess their own *cis*-elements that can be regulated by other TFs (Fickett and Wasserman 2000; Hobert 2018). Here, we will cover those TF families that regulate anthocyanin synthesis. We will introduce the basic characteristics of those widely studied MBW ternary complex proteins, including MYB, bHLH, and WD40, as well as other TF families such as WRKY, NAC, and ERF, and we will discuss their roles in anthocyanin synthesis.

13.3.1 The MYB TF Family

In plants, the structure of MYB TFs is characterized by a highly conserved MYB domain that binds to DNA. The MYB domain usually contains one to four amino acid sequence repeats (R motifs). Based on the number of R motifs, MYBs can be subdivided into four categories, including R1R2R3-MYB, R2R3-MYB, R1-MYB, and R4-MYB, among which R2R3 MYBs with two motifs are the most abundant (Dubos et al. 2010). It is commonly known that MYB TFs recognize various DNA sequences (Dubos et al. 2010).

Structural analyses of R2R3-MYBs have shown that both R2 and R3 are necessary for the recognition of DNA sequences, and that the specificity of R3 is higher than that of R2 (Ogata et al. 1996). Most of the R2R3-MYB proteins can recognize the TAACTAAC (MBSII) sequence, and some MYB proteins can recognize T/CAACG/TGA/C/TA/C/T (MBSI) sequences (Hartmann et al. 2005).

It is reported that the first MYB TF identified to regulate anthocyanin biosynthesis in plants is *ZmC1,* encoding chalcone synthase, in maize (Paz-Ares et al. 1987). Subsequently, *PhAn2* and *PhAn4,* R2R3-type MYB factors, in *Petunia hybrida* are found to play important roles in the regulation of anthocyanin biosynthesis in petals and anthers, respectively (Quattrocchio et al. 1993). Later, Arabidopsis *production of anthocyanin pigment1* (*AtPAP1*) and *AtPAP2,* whose products are involved in regulating anthocyanin biosynthesis in *Arabidopsis thaliana*, have also been isolated and characterized (Borevitz et al. 2000a, b).

The first MYB TFs isolated from apple are MdMYB1 and MdMYBA, involved in regulating anthocyanin synthesis in apple peel (Takos et al. 2005). Expression levels of their two encoding genes are reported to significantly increase under strong light, and they are positively correlated with anthocyanin accumulation (Takos et al. 2005; Ban et al. 2007). Subsequently, MdMYB10 is identified as an allele of both MdMYB1 and MdMYBA (Espley et al. 2007), and it has been demonstrated that MdMYB10 promotes anthocyanin synthesis not only in apple peel and leaves but also in fruit flesh of apples, resulting in a red-flesh phenotype (Espley et al. 2013). Interestingly, it has been found that five repeated 23-bp motifs in the promoter region render MdMYB10 to be self-activating, and that this is the main driving force for a strong MdMYB10 transcription and anthocyanin synthesis (Espley et al. 2009).

Following completion of the first draft of the apple genome sequence (Velasco et al. 2010), additional MYB TFs have been cloned and characterized. It has been reported that anthocyanin synthesis in the pericarp of apple fruit is also regulated by MdMYB10, a homolog of MdMYB110a (Chagné et al. 2013). Coincidently, these two homologs have conserved functions in particular apple cultivars; however, their responses during fruit ripening along with their expression patterns are different (Chagné et al. 2013).

Studies have revealed that MdMYB3 promotes apple skin coloration by binding to promoters of *MdDFR* (encoding dihydroflavonol-4-reductase) and *MdUFGT* (encoding UDP-glucose: flavonoid 3-*O*-glucosyltransferase) (Vimolmangkang et al. 2013), while MdMYBPA1 in red-fleshed apple can switch proanthocyanidin (PA) synthesis to anthocyanin synthesis in response to low temperature (Wang et al. 2018a, b, c, d, e). For other TFs, MdMYB88 and MdMYB124 have been found to promote anthocyanin accumulation, and positively regulate cold hardiness in apples (Xie et al. 2018). In addition, some MYB TFs can negatively regulate anthocyanin synthesis in apples. Using transient expression analysis in tobacco leaves, it is found that MdMYB16, MdMYB17, and MdMYB111 can inhibit the promoter activity of *MdDFR* (Linwang et al. 2011). Specifically, these TFs block the activation of the *MdDFR* promoter via the MYB10-bHLH3 complex, thus inhibiting anthocyanin synthesis (Linwang et al. 2011). Furthermore, the inhibitory effect of MdMYB16 on anthocyanin synthesis is related to an ethylene-responsive element-binding factor-associated amphiphilic repression (EAR) domain inhibitory sequence at its C-terminal. Accordingly, the removal of the EAR sequence from MdMYB16 eliminates its inhibitory effect (Xu et al. 2017a, b). On the other hand, MdMYB6 can bind to promoters of the structural gene *MdANS* (encoding anthocyanin synthase) and the anthocyanin transport-related gene *MdGSTF12* (encoding glutathione S-transferase F12) to inhibit their expression, thus directly inhibiting anthocyanin accumulation (Xu et al. 2020). In addition to regulating anthocyanin synthesis in apples, MYB TFs also regulate the synthesis of other flavonoids. For example, MdMYB9 and MdMYB11 bind to the promoter of *DFR*, thereby influencing the synthesis of proanthocyanidins (PAs) and anthocyanins in the skin of apple fruit (Gesell et al. 2014; An et al. 2015). However, MdMYB12 and MdMYB22 play important roles in regulating the synthesis of PAs and flavonols in red-fleshed apples (Wang et al. 2017). On the other hand, MdMYB23 is involved in cold tolerance and accumulation of PAs in apples (An et al. 2018a, b, c, d). Furthermore, MdMYB8 regulates the synthesis of flavonols in apple skin by regulating the expression of *MdFLS* (encoding flavonol synthase) (Li et al. 2020). A listing of apple MYB TFs involved in the regulation of anthocyanin or flavonoid synthesis that has been cloned and identified to date is presented in Table 13.1.

13.3.2 The BHLH TF Family

The bHLH (basic helix-loop-helix) family is the second-largest TF family in plants. Members of this family contain a bHLH domain comprising about 60 amino acids (Heim et al. 2003; Feller et al. 2011). The N-terminal of the bHLH domain contains 13–17 hydrophilic basic amino acid residues, followed by a helix-loop-helix (HLH) comprising hydrophobic residues (Heim et al. 2003). In the basic amino acid sequence of the bHLH domain, the HER (His5-Glu9-Arg13) motif is highly conserved, and this motif combines with the E-box motif (CANNTG) in the promoter region of its target gene to regulate transcription (Feller et al. 2011). Proteins containing HLH motifs usually form either homologous or heterologous dimers. This is a prerequisite for a protein to recognize and bind to specific DNA sequences, and it is also an important factor in determining the expression pattern of a protein. For example, the bHLH TFs can bind to the [DE]Lx2[RK]x3Lx6Lx3R motif in the R3 domain of MYB TFs to form a protein complex that participates in the regulation of anthocyanin synthesis (Dubos et al. 2010; Feller et al. 2011).

The first bHLH TFs that have been identified are those encoded by *ZmR* and *ZmB*, both of

Table 13.1 MYB transcription factors associated with anthocyanin synthesis

Gene	Gene ID	Regulatory target[a]	Positive/negative effect	Reference(s)
MdMYB1	DQ886414	Anthocyanin synthesis	Positive	Takos et al. (2005)
MdMYB3	AEX08668	Anthocyanin synthesis	Positive	Vimolmangkang et al. (2013)
MdMYB6	DQ074461	Anthocyanin synthesis	Negative	Xu et al. (2020)
MdMYB8	MD06G1217200	Flavonoid biosynthesis	Positive	Li et al. (2020)
MdMYB9	ABB84757	Anthocyanin and PA biosynthesis	Positive	An et al. (2015), Gesell et al. (2014)
MdMYB10	ABB84753	Anthocyanin biosynthesis	Positive	Espley et al. (2007)
MdMYB11	AAZ20431	Anthocyanin and PA biosynthesis	Positive	An et al. (2015)
MdMYB12	XP_008337875	PA biosynthesis	Positive	Wang et al. (2017)
MdMYB16	HM122617	Anthocyanin biosynthesis	Negative	Linwang et al. (2011), Xu et al. (2017a, b)
MdMYB17	HM122618	Anthocyanin biosynthesis	Negative	Linwang et al. (2011)
MdMYB22	AAZ20438	Flavonol biosynthesis	Positive	Wang et al. (2017)
MdMYB23	MDP0000230141	PA biosynthesis	Positive	An et al. (2018a, b, c, d)
MdMYB24L	XM_008343218	Anthocyanin biosynthesis	Positive	Wang et al. (2019a, b)
MdMYB88	KY569647	Anthocyanin biosynthesis	Positive	Xie et al. (2018)
MdMYB110a	EB710109	Anthocyanin biosynthesis	Positive	Chagné et al. (2013)
MdMYB111	EB137767	Anthocyanin biosynthesis	Negative	Linwang et al. (2011)
MdMYB124	KY569648	Anthocyanin biosynthesis	Positive	Xie et al. (2018)
MdMYB308L	MDP0000950559	Anthocyanin biosynthesis	Positive	An et al. (2020a, b)
MdMYBL2	NP_001281006.1	Anthocyanin biosynthesis	Negative	Wang et al. (2019a, b)
MdMYBPA1	AIF70441	Anthocyanin and PA biosynthesis	Positive	Wang et al. (2018a, b, c, d, e)

[a]PA, proanthocyanidin

which are involved in regulating anthocyanin biosynthesis in *Zea mays* (Chandler et al. 1989). Subsequently, it has been demonstrated that the interaction between ZmC1 and ZmR in maize could significantly enhance the promoter activity of *ZmUFGT*, a key gene in anthocyanin biosynthesis (Roth et al. 1991). Further studies have identified PhAN1 which plays an important role in the regulation of anthocyanin biosynthesis in *P. hybrida*. Moreover, it is found that PhAN2 and PhAN4 form transcriptional complexes with PhAN1 and PhAN11, respectively, and participate in the regulation of anthocyanin biosynthesis in either petunia petals or anthers (Quattrocchio et al. 1993, 1999; Spelt et al. 2000). In *Arabidopsis*, bHLH TFs AtGL3, AtEGL3, and AtTT8 are also found to be involved in anthocyanin and PA biosynthesis (Nesi et al. 2000; Heim et al. 2003), and that the transcription of *AtTT8* is induced by AtPAP1 (Baudry et al. 2006).

In apples, MdbHLH3 and MdbHLH33 are two of the core bHLH TFs involved in anthocyanin synthesis. Both MdbHLH3 and MdbHLH33 can interact with MdMYB10 to promote anthocyanin synthesis by promoting the activity of the *MdDFR* promoter (Espley et al. 2007). MdbHLH3 can also be induced by low temperature and binds to promoters of *MdDFR, MdUFGT, MdMYB10*, and *MdMYBPA1* to regulate their expression (Xie et al. 2012; Wang et al. 2018a, b, c, d, e). Interestingly, MdbHLH33 can also interact with the inhibitor MdMYB16, thus weakening the inhibitory effect of MdMYB16 on anthocyanin synthesis (Xu et al. 2017a, b). Wang et al. (2017) have found that both MdbHLH3 and MdbHLH33 can interact with MdMYB12 to promote PA synthesis (Wang et al. 2017).

13.3.3 The WD40 Protein Family

Members of the WD40 protein family have four to ten random WD repeat domains, each of which comprises about 40 amino acid sequences ending in tryptophan (W) and aspartic acid (D). The WD40 protein itself has no catalytic function, but it can reversibly interact with a variety of proteins to provide either a platform or a link for the assembly of large protein–protein complexes (Smith et al. 1999; Van Nocker et al. 2003). In *P. hybrida,* the WD40 protein encoded by PhAN11 has a regulatory function, as it can activate the transcription of PhAN2 to promote anthocyanin synthesis (De Vetten et al. 1997). Another WD40 protein that can regulate anthocyanin synthesis is identified in *A. thaliana,* as this WD40 protein encoded by *AtTTG1* can interact with bHLH TFs to enhance anthocyanin synthesis in *Arabidopsis* seeds (Walker et al. 1999).

The first WD40 protein to be isolated and identified from apple is MdTTG1. Similar to AtTTG1 in *Arabidopsis*, MdTTG1 can interact with both MdTT2 and MdTT8 and enhance the promoter activity of a downstream gene, *MdBAN* (Brueggemann et al. 2010). Furthermore, MdTTG1 can also interact with both MdbHLH3 and MdbHLH33, but not with MdMYB1, and it does not affect the promoter activity of the downstream structural genes *MdDFR* and *MdUFGT* (An et al. 2012). This finding differs from the reported interactions of AtTTG1 with MYB and with bHLH in *A. thaliana*, thus highlighting the differences present among various plant species (Baudry et al. 2004, 2006).

Studies have demonstrated that the anthocyanin synthesis pathway is mainly regulated by the ternary MBW (MYB-bHLH-WD40) complex in various plant species (Feller et al. 2011). Other TF families, such as WRKY, ERF, NAC, and bZIP, can also either coordinate or participate in the regulation of anthocyanin synthesis, thereby enriching the anthocyanin regulatory network (Zhou et al. 2015; Zhang et al. 2018; An et al. 2018a, b, c, d; Liu et al. 2019).

13.3.4 The WRKY TF Family

The WRKY family of TFs is only found in higher plant species. A total of 74 WRKY proteins identified in *Arabidopsis* can be subdivided into three groups according to the number of N-terminal conserved domains and the type of C-terminal zinc-finger structure (Rushton et al. 2010). Proteins in group I have two conserved WRKY domains, and they have higher molecular weights than proteins in the other two groups. Although proteins in both groups I and II have the same C2-H2 zinc-finger structure, group II proteins have a single zinc-finger motif, while group I proteins have two zinc-finger motifs. In addition, group II proteins can be further subdivided into five subfamilies based on levels of homology within the WRKY domain and the presence of additional domains. Group III proteins are characterized by having a C2-H-C zinc-finger structure (Journot-Catalino et al. 2006). Interestingly, the WRKY proteins with the conserved amino acid sequence WRKYGQK can specifically bind W-box (TTGAC[C/T]) *cis*-acting elements in promoter regions of target genes to either promote or inhibit the expression of these genes (Eulgem et al. 2000).

Previous studies have confirmed that WRKY TFs play important roles in regulating various processes, including developmental processes and stress responses (Rushton et al. 2010). Moreover, there is increasing evidence that WRKY TFs are also involved in regulating anthocyanin biosynthesis. For instance, overexpression of *BnWRKY41-1* in the *A. thaliana* WRKY41-2 mutant rescues the high-anthocyanin content phenotype in rosette leaves (Duan et al. 2018). Compared to the wild type, the AtWRKY75-silenced strain is more likely to accumulate anthocyanins under conditions of phosphate deficiency (Devaiah et al. 2007). Furthermore, transcriptome analysis of red-fleshed apples has revealed that MdWRKY11 is involved in regulating the biosynthesis of both flavonoids and anthocyanins (Wang et al. 2018a, b, c, d, e). Subsequently, it is confirmed that MdWRKY11 participates in anthocyanin accumulation via interactions with MYB TFs and the photoresponse factor MdHY5 (Liu et al. 2019). It is observed that overexpression of MdMYB111 in red-fleshed apple callus decreases the anthocyanin content, and the color changes from red to yellow. However, overexpression of MdWRKY40 in callus overexpressing MdMYB111 leads to a change in color from yellow to pink (Zhang et al. 2019a, b). It is proposed that MdWRKY40 also interacts with MdMYB1, and enhances binding of MdMYB1 to target genes in response to wounding, thus promoting anthocyanin biosynthesis (An et al. 2019).

Some WRKY TFs promote the synthesis of both flavonoids and PAs. For example, TRANSPARENT TESTA GLABRA2 (TTG2), a trichome and seed coat development gene in *Arabidopsis*, encodes a WRKY TF that promotes PA accumulation (Gonzalez et al. 2016). In fact, it is proposed that TTG2 must interact with TTG1 in order to be functional, thus the WRKY TF is added to the MBW ternary complex to form a new MBWW quaternary complex (Lloyd et al. 2017). Interestingly, PhPH3 and VvWRKY26 are homologs of AtTTG2, and these require TTG1 to induce PA and flavonoid synthesis. These findings provide evidence for the existence of an MBWW quaternary complex (Verweij et al. 2016; Amato et al. 2016).

13.3.5 The NAC TF Family

It is known that NAC TFs are present only in plants, widely identified in various plant species, and are among the largest TF families. The *NAM* gene has been first cloned from *P. hybrida*, and then *ATAF1/2* and *CUC2* have been identified in *A. thaliana*. As all these genes encode proteins with a highly conserved amino sequence at the N-terminal, the conserved sequence is termed the NAC domain (i.e., the first letters of the three genes mentioned above), and the proteins carrying the NAC domain are known as NAC TFs (Souer et al. 1996; Aida et al. 1997). The NAC domain comprises about 150 amino acid residues, and includes five subdomains, A, B, C, D, and E. Subdomain A is related to the formation of a protein dimer, while B and E are related to functional differentiation of NACs, and C and D subdomains are related to DNA binding (Ernst et al. 2004; Jensen and Skriver 2014). The NAC TFs play important roles in plant growth and development, metabolic regulation, as well as hormone and stress response networks. These TFs are of particular interest in studies of plant gene function, as well as in the regulation of gene expression.

Some members of the NAC TF family participate in the control of flavonoid and anthocyanin synthesis. In *A. thaliana*, AtNAC078 is a membrane-bound protein that is released from the membrane following stimulation by either internal or external signals. Following its release, AtNAC078 becomes available for promoting expression of genes related to flavonoid biosynthesis and anthocyanin production under high-light stress (Morishita et al. 2019), while another protein, AtNAC032, inhibits anthocyanin biosynthesis under high-sucrose, oxidative, and abiotic stresses (Mahmood et al. 2016). In Norway spruce, PaNAC03 is induced in response to biotic and abiotic stresses to form a homodimer that subsequently inhibits flavonoid biosynthesis (Dalman et al. 2017). However, LcNAC13 in litchi directly binds to promoters of genes related to anthocyanin biosynthesis and inhibits their transcription. Interestingly, the interaction between LcR1MYB1 and LcNAC13 reverses the negative effect of LcNAC13 (Jiang et al. 2019). In peach, *BLOOD (BL),* encoding a NAC TF, has been identified using gene-linkage analysis of a red-flesh line, and its silencing can reduce anthocyanin accumulation in blood peach (Zhou et al. 2015).

In apple, overexpression of MdNAC029 in callus of 'Orin', a yellow-skinned Japanese cultivar, can significantly promote expression of *MdMYB1,* as well as of downstream genes related to anthocyanin synthesis. A transient expression analysis in tobacco has confirmed that MdNAC029 can directly activate the expression of *MdMYB1,* thereby promoting anthocyanin synthesis (An et al. 2018a, b, c, d). For another related TF member, MdNAC52 can bind to promoters of both *MdMYB9* and *MdMYB11* to promote the biosynthesis of anthocyanins and procyanidins, as well as to regulate PA metabolism by directly regulating *MdLAR,* encoding leucoanthocyanidin reductase (Sun et al. 2019). Recently, it is reported that the abundance of MdNAC42 is positively correlated with anthocyanin content as fruits ripen, and it interacts with MdMYB10 to participate in the regulation of anthocyanin synthesis in the red-fleshed apple 'Redlove', released from a breeding program in Switzerland (Zhang et al. 2020).

13.4 Environmental Regulation of Fruit Color in Apple

Plant secondary metabolism has evolved during interactions between plants and biotic/abiotic environmental factors. Secondary metabolites such as flavonoids, anthocyanins, terpenoids, and alkaloids play important roles in plant protection and competitiveness, and are involved in coordinating the relationships between plants and the environment (Yan et al. 2007). Therefore, in addition to genetic and transcriptional regulation, environmental factors such as light, temperature, water, sugars, and plant hormones have important regulatory effects on anthocyanin biosynthesis and in apple fruit color development (Winkel-Shirley 2002; Takos et al. 2005; Lepiniec et al. 2006).

13.4.1 Light

Light is one of the most important environmental factors influencing color development in apple fruit. Bagging of apple fruits significantly inhibits anthocyanin synthesis in the skin, and removal of bags to expose fruit to light results in anthocyanin accumulation in the skin (Ju et al. 1999). Expression of genes encoding structural genes in the anthocyanin biosynthesis pathway and TFs is promoted by light and inhibited by its absence (Kondo et al. 2002; Jeong et al. 2004). As reported by Takos et al. (2005), light induces the expression of *MdMYB1* and promotes anthocyanin accumulation in apple skin (Takos et al. 2005). However, under dark conditions, an apple ubiquitin E3 ligase CONSTITUTIVE PHOTOMORPHOGENIC1 (MdCOP1) interacts with and ubiquitinates MdMYB1. Light inhibits the transport of COP1 into the nucleus to degrade MYB1, thus stabilizing the function of MYB1. Therefore, light induces anthocyanin synthesis, as well as the development of apple fruit color (Li et al. 2012).

Among all light qualities, ultraviolet (UV) light and blue-violet light have the strongest effects in promoting anthocyanin synthesis, while far-infrared light has the weakest or even inhibitory effects (Mol et al. 1996). Ultraviolet-A (wavelength > 320 nm) can burn apple skin and turns it brown in color, while UV-B (wavelength 280–320 nm) can promote phenylalanine ammonia lyase (PAL) activity and expression of related structural genes, thus promoting anthocyanin synthesis (Song et al. 2009). The photoreceptor of UV-B is UVR8, which normally exists as a homodimer, but UV-B can induce depolymerization of the UVR8 dimer into two monomers (Rizzini et al. 2011; Wu et al. 2012; Christie et al. 2012). It is reported that the TFs COP1 and ELONGATED HYPOCOTYL5 (HY5) play key roles in the UV-B signaling pathway, wherein COP1 has a ubiquitinase activity and inhibits photomorphogenesis in the dark, but promotes photomorphogenesis in response to a UV-B signal (Lau and Deng 2012). Under UV-B conditions, UVR8 can interact with COP1 to form a protein complex (Favory et al. 2014). This UVR8-COP1 complex can bind to SPA (suppressor of phya-105) and promotes the transcriptional activity of the downstream TF HY5, thus regulating the expression of genes related to anthocyanin synthesis (Oravecz et al. 2006). A model summarizing the light regulation of fruit coloration is presented in Fig. 13.3.

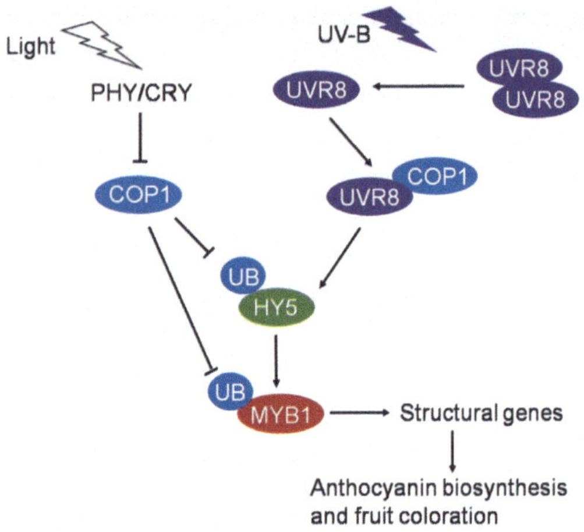

Fig. 13.3 A schematic model of light- and UV-B-regulated anthocyanin synthesis. A constitutive photomorphogenic1 (COP1) and an elongated hypocotyl 5 (HY5) play key roles in the light and the UV-B signaling pathway. HY5 and MYB1 proteins accumulate in light but are degraded by COP1 via a ubiquitin-dependent pathway in the dark. UV-B can induce the depolymerization of the UV Resistance Locus 8 (UVR8) dimer into two monomers. Under UV-B conditions, UVR8 can interact with COP1 to form a protein complex. This UVR8-COP1 complex can promote the transcriptional activity of the downstream TF HY5, thus regulating the expression of genes related to anthocyanin synthesis

13.4.2 Temperature

Temperature is another environmental factor influencing the synthesis of anthocyanins and color development in apple fruit. Various studies have reported that low temperature can significantly promote the synthesis of anthocyanins in apples. For example, Ubi et al. (2006) have demonstrated that low temperature and UV-B radiation can significantly increase expression of several genes involved in the anthocyanin biosynthesis pathway, including those encoding for chalcone synthase (*CHS*), anthocyanin synthase (*ANS*), and UDP-Glc:flavonoid-3-O-glucosyltransferase (*UFGT*), as well as promote accumulation of anthocyanins in apple skin (Ubi et al. 2006). Further studies have shown that MdBBX20 responds to low temperature with the participation of MdbHLH3, and together they regulate the UV-B signaling pathway along with a low-temperature signal to promote anthocyanin synthesis in apple by interacting with MdHY5 (Fang et al. 2019). Tian et al. (2015) have reported that low temperatures can induce expression of *McMYB10, McbHLH3, McbHLH33*, and *McTTG1* in *Malus* crabapples, thus leading to anthocyanin accumulation in leaves. In addition, it is reported that low temperature leads to the phosphorylation of MdbHLH3, thereby enhancing its ability to bind to promoters of the downstream genes *MdDFR* and *MdUFGT*, and promoting anthocyanin synthesis in apple skin (Xie et al. 2012). Recently, Wang et al. (2019a, b) have reported that the promoter of *MdMYBPA1* contains a low temperature-responsive *cis*-element, LTR, that can respond to low temperature. Therefore, binding of MdbHLH33 to this element causes a switch from PA to anthocyanin synthesis (Wang et al. 2019a, b).

On the other hand, high temperature inhibits anthocyanin synthesis and fruit coloration in apples. It has been observed that under high-temperature conditions, the rate of anthocyanin synthesis slows down, while its rate of degradation increases, thus contributing to reduced anthocyanin content (Shaked-Sachray et al. 2010). In particular, it has been observed that high temperature inhibits expression of *MdMYB10*, thereby resulting in reduced anthocyanin biosynthesis and accumulation in apple skin (Linwang et al. 2011).

13.4.3 Water

In general, plants require water as well as appropriate air humidity and soil moisture conditions to maintain normal growth. With global climate change, drought stress has become a critical worldwide problem due to rising temperatures and expanded arid lands. Drought affects growth and productivity of the most important crops (Fedoroff et al. 2010).

It has long been reported that water deficit can significantly promote expression of genes related to anthocyanin synthesis pathways (Castellarin et al. 2007; Ma et al. 2014). In *Arabidopsis*, overexpression of either MYB12/PFG1 (production of flavonol glycosides 1) or MYB75/PAP1 (production of anthocyanin pigment 1) alone, or co-expression of both MYB12 and MYB75 have resulted in overaccumulation of flavonoids, thereby mitigating accumulation of reactive oxygen species (ROS) in vivo under drought stress conditions (Nakabayashi et al. 2014). In another study, overexpression of UGT79B2/B3 significantly increases anthocyanin accumulation and enhances antioxidant activity in response to low temperature, drought, and salt stress (Li et al. 2017).

In a recent study in apple, MdERF38 (ethylene-response factor) is found to promote anthocyanin biosynthesis by enhancing the transcriptional activity of MdMYB1 in response to drought stress (An et al. 2020a, b).

13.4.4 Sugars

Sugars are not only precursors for anthocyanin biosynthesis, but also serve as signaling substances that can induce anthocyanin synthesis (Smeekens 2000). Interestingly, unstable anthocyanidins are glycosylated by UDP-Glc: flavonoid-3-O-glucosyltransferase (UFGT) to form stable anthocyanins (Hugo and Timothy 1991; Holton and Cornish 1995). Furthermore, sugars serve as synthetic substrates for various glycosides of anthocyanins, and such synthesis occurs in the cytoplasm (Vogt and Jones 2000).

It is reported that different types of sugars have different effects on anthocyanin synthesis. For example in *Arabidopsis*, sucrose has the strongest effect in promoting anthocyanin accumulation, as sucrose can regulate expression of both *AtPAP1* and the structural gene *AtDFR* to induce anthocyanin synthesis (Lloyd and Zakhleniuk 2004; Teng et al. 2005; Solfanelli et al. 2006). Moreover, it is proposed that sucrose transporters (SUCs) may serve as signal receptors involved in sucrose-induced anthocyanin synthesis as it is observed that anthocyanin accumulation in a *suc1* mutant is inhibited when this mutant is grown on a medium containing 3% sucrose (Solfanelli et al. 2006; Sivitz et al. 2008).

In apple, UDP-galactose is the main substrate for anthocyanin synthesis, and it plays an important role in fruit coloration (Ban et al. 2009). It has been reported that MdSUT4 transfers sucrose from the vacuole to the cytoplasm through the vacuolar membrane, thereby leading to increased contents of precursors for anthocyanin synthesis, thus promoting anthocyanin synthesis (Xu et al. 2017a, b). In a recent study, it is observed that MdMYB6 can inhibit the synthesis of anthocyanins by regulating the monosaccharide transporter MdTMT1 to reduce the contents of the substrates UDP-glucose and UDP-galactose (Xu et al. 2020). However, glucose can activate the activity of the hexokinase MdHXK1, which phosphorylates the

MdbHLH3 TF to mediate apple fruit coloration (Hu et al. 2016).

13.4.5 Phytohormones

It has long been known that phytohormones participate in the regulation of anthocyanin synthesis and in fruit color development. Moreover, different phytohormones have different effects on anthocyanin synthesis.

For example, ethylene is an important plant hormone that is involved in many aspects of plant growth and development. It promotes leaf senescence and fruit ripening, and it induces root hair and adventitious root formation (Iqbal et al. 2017). In apple orchards, applications of the ethylene inhibitor 1-methylcyclopropene (1-MCP) to trees can significantly inhibit ethylene release and anthocyanin accumulation (Liu et al. 2012). As ERF is a key TF in the ethylene signaling pathway, it is reported that MdERF1B binds to promoters of both *MdMYB9* and *MdMYB11* to regulate both anthocyanin and PA accumulation in apple (Zhang et al. 2018). Furthermore, ethylene treatment markedly induces fruit coloration, as well as expression of *MdMYB1* (An et al. 2018a, b, c, d). It has been observed that MdMYB1 can interact with the promoter of *MdERF3,* thereby providing positive feedback for the regulation of ethylene biosynthesis (An et al. 2018a, b, c, d).

Abscisic acid (ABA) is another important phytohormone that participates in many plant growth and development processes (Mehrotra et al. 2014). In strawberry, ABA plays a key role in fruit color development during ripening by up-regulating ethylene and anthocyanin production, and stimulating PAL activity (Jiang and Joyce 2003). In grape and litchi, ABA promotes anthocyanin biosynthesis at the fruit coloration stage (Ban et al. 2000; Singh et al. 2014). In apple, the Bric-à-brac, tramtrack, and broad complex (BTB) protein MdBT2, which is a MdbZIP44-interacting protein, degrades MdbZIP44 through the ubiquitin-26S proteasome system, thus inhibiting MdbZIP44-modulated anthocyanin biosynthesis (An et al. 2018a, b, c, d).

Jasmonic acid (JA) regulates diverse processes in plants, including defense against biotic and abiotic stresses, organ development, and the biosynthesis of secondary metabolites (Sasaki et al. 2001). It is reported that JA induces degradation of JAZ (JA ZIM DOMAIN) repressor proteins in *Arabidopsis*, resulting in the release of bHLH and MYB TFs, thus promoting anthocyanin accumulation (Qi et al. 2011). In apple, methyl jasmonate can rapidly degrade MdJAZ2 and destroys the competitive binding between MdJAZ2 and MdbHLH3, thus releasing MdbHLH3 and promoting anthocyanin synthesis (An et al. 2015). Furthermore, the JA signaling factors MdJAZ8 and MdJAZ11 form transcription complexes with MdMYB24L, thus affecting anthocyanin synthesis (Wang et al. 2019a, b).

Cytokinin (CK) is yet another phytohormone that induces anthocyanin accumulation (Deikman and Hammer 1995). The ability of CK to induce anthocyanin biosynthesis is dependent on the redox state of the photosynthetic electron transport mediated by sugar-induced Suc signaling (Das et al. 2012). In *Arabidopsis*, it has been observed that increasing the concentration of CK can promote the expression of *MYBD* that is conducive to anthocyanin synthesis (Nguyen et al. 2015). Ji et al. (2015) have also found that CK can induce anthocyanin accumulation in apple calli, and that a high concentration of auxin inhibits anthocyanin synthesis (Ji et al. 2015).

Finally, auxin (Aux) is known to induce rapid cellular responses and gene expression to regulate plant growth, and its function mainly depends on auxin response factor (ARF)-Aux/indole acetic acid (IAA) interactions (Tiwari et al. 2001). Furthermore, a high concentration of auxin can inhibit anthocyanin synthesis. In apple, it has been reported that a high concentration of auxin inhibits expression of *MdMYB10, MdbHLH3,* and structural genes involved in anthocyanin synthesis (Ji et al. 2015). In addition, a high concentration of auxin significantly promotes expression of *MdARF13*, and in turn, MdARF13 interacts with MdMYB10 to form a protein complex that inhibits anthocyanin synthesis (Wang et al. 2018a, b, c, d, e).

13.5 Breeding of Red-Skinned and Red-Fleshed Apples

With the increasing demand for high-quality apples, fruit with a bright color, rich aroma, and excellent flavor has become an important goal for breeders, as consumers associate red-colored apple cultivars with ripeness and good flavor (King and Cliff 2002; Ban et al. 2007). In addition, the red-fleshed apple, as a novel fruit, has attracted more attention in recent years by researchers and breeders (Wang et al. 2018a, b, c, d, e).

13.5.1 Breeding of Red-Skinned Apples

Understanding the genetic basis of red color development is critical for breeding red-skinned apples. Early on, it has been proposed that the red-skin fruit color trait is controlled by a single dominant gene, designated as *Rf* (Klein 1958). Maliepaard et al. (1998) have constructed a genetic map with 17 linkage groups (LGs) using apple cultivars 'Prima', 'Fiesta', and their segregating progeny as test materials, speculating that the 17 LGs correspond to the 17 chromosomes of the apple genome. It is found that the *Rf* gene is located on LG9, close to the sequence-characterized amplified region (SCAR) marker BC226 (Maliepaard et al. 1998). With the development of additional molecular markers, it has become clear that red-skinned apples are either heterozygous or homozygous dominant, while yellow (green)-skinned apples are homozygous recessive (Cheng et al. 1996; Zhao et al. 2006).

Over time, it has been observed that the red-skinned fruit color trait is a quantitative trait controlled by multiple genes. Lespinasse et al. (1985) have reported that the red-skin color trait is controlled by two dominant complementary genes, wherein two dominant genes contribute to a yellow-skinned color fruit; otherwise, the fruit's skin color is red. Subsequently, Takos et al. (2005) have found that *MdMYB1-1*-linked fragments could be amplified from most red-colored fruit cultivars, but not from non-red cultivars. Furthermore, it is confirmed that *MdMYB1-1* is closely related to the red-skin fruit color trait (Takos et al. 2005; Ban et al. 2007; Migicovsky et al. 2016).

It is widely known that red fruit color sports/mutants serve as valuable genetic resources for the red-skinned color fruit trait. In fact, about 30% of current red-skinned apple cultivars are derived from sport selection; moreover, ~50% of all apples grown worldwide are sport mutants, and most are red-colored sports (Yi et al. 2006). A typical case of using red sports to breed red-skinned apples is the series of red sport cultivars selected from the original 'Delicious'. In 1872, 'Delicious' has been first discovered as a chance seedling by Jesse Hiat in an apple orchard on a farm in Peru, Iowa, USA. By 1921, a red sport mutant of 'Delicious' with good red color, 'Starking', is identified and selected in New Jersey, thus 'Starking' has served as a second-generation red sport derived from 'Delicious'. In 1953, a third-generation red sport mutant, 'Starkrimson', is discovered in a population of 'Starking' (Fig. 13.4a). Over the decades, several hundreds of red sports, with varying levels of color pigmentation, have been selected and grown for generations. Many red sport cultivars have also been derived from 'Fuji'. During early breeding efforts, the Japanese 'Red Fuji' apple cultivar 'Changfu 2' has been poorly colored; however, subsequently, breeders in Yantai city, China have selected a better red sport, 'Yanfu 3'. Over time, additional red sport cultivars with enhanced and red-colored fruits have been selected from 'Yanfu 3', including 'Yanfu 8-10' and 'Yuanfuhong' (Fig. 13.4b).

Recently, the mechanism of apple red sport/mutant development has been investigated. It is reported that RNA-directed DNA methylation (RdDM) may play essential roles in apple fruit color development, as it is involved in repressive epigenetic regulation that can initiate transcriptional gene silencing (Jiang et al. 2020).

Fig. 13.4 'Delicious' and 'Fuji' apple cultivars along with their corresponding red sport/mutant cultivars. **a** 'Delicious' apple and its red sports, including 'Starking' (https://www.21food.com/products/starking-apple-547405.html) and 'Starkrimson' (https://www.planfor.co.uk/buy,apple-tree-starkrimson,0113,EN). **b** 'Fuji' apple and its red sports, including 'Yanfu 3' and 'Yanfu 8'

Moreover, it is reported that argonaute 4 (AGO4) is one of the key components of the RdDM pathway. Therefore, it has been observed that AGO4 proteins can recognize the ATATCAGA sequence in the *MdMYB1* promoter, and, in turn, these recruit domains rearranged methyltransferase2 (DRM2s) to catalyze CHH (H = Adenine (A), Thiamine (T), or Cytosine (C) base) methylation of the *MdMYB1* promoter. It is this demethylation of the *MdMYB1* promoter that promotes synthesis of anthocyanin in the skin of apple fruits (Jiang et al. 2020).

13.5.2 Breeding of Red-Fleshed Apples

In recent years, red-fleshed cultivars of sweet orange, peach, kiwi fruit, and other fruits have been bred. Consequently, there is increased interest in breeding for the red-flesh fruit trait in various fruit crops (Pinheiro et al. 2015; Rumainum et al. 2016).

Breeding efforts of red-fleshed apple can be traced back to the 1800s. In 1897, the horticulturist Niels Hansen encountered an unusual apple species in the wild fruit forests of Turkestan. This cold, hardy wild species, *M. pumila* var. *niedzwetzkyana*, a close relative of *M. sieversii*, produces fruit with red-colored skin and flesh, as well as red-colored blossoms and young foliage. Thus, *M. pumila* var. *niedzwetzkyana* has been hybridized with the cultivated apple yielding a series of red-fleshed apple cultivars such as 'Almata'. These cultivars produce fruit with dark-red flesh and skin, as well as dark-red blossoms and juvenile foliage, and these have been designated as Type I red-fleshed apple

(Volz et al. 2009; Deacon 2017). An earlier horticulturist, Albert Etter, has bred 'Pink Pearl' and other red-fleshed apple cultivars with 'Surprise' as a parent. However, only the skin layer of 'Pink Pearl' fruit has pink flesh, and there is no red pigmentation in its leaves, stems, or other vegetative tissues; thus, these cultivars are designated as Type II red-fleshed apples (Greenmantle 2017).

Research efforts to investigate this red-color pigmentation trait have been undertaken using molecular analysis tools. It is found that Type I red-fleshed apple has a minisatellite-like structure comprising six 23-bp repetitive sequences in the promoter of *MdMYB10*. This repeated sequence endows *MdMYB10* with self-activation characteristics, whereas in white-fleshed apple, the same promoter region has only one of these 23-bp motifs (Espley et al. 2009). However, *MdMYB10* is not expressed in Type II red-fleshed apple. Instead, another MYB TF, MdMYB110a, is correlated with the red pigmentation of the flesh of Type II apples (Chagné et al. 2013).

In recent years, breeders in the United Kingdom, Japan, Germany, New Zealand, Switzerland, and China have bred a series of red-flesh apple cultivars such as 'jpp35', 'Weirouge', 'Baya Marisa', 'Redlove', and 'Meihong' (Fig. 13.5). However, breeding objectives have gradually changed over time from earlier ornamental and processing traits to current interests in fresh-eating or functional fruit traits (Brooks and Olmo 1997; Sadilova et al. 2006; Sekido et al. 2010; Wang et al. 2018a, b, c, d, e).

13.6 Conclusions

The complexity of the regulation mechanism of anthocyanin biosynthesis in apple is the embodiment of germplasm diversity. With the rapid development of molecular biology technologies, further research will provide additional details on the regulation of anthocyanin synthesis and fruit color development. Such information will be useful in enhancing the anthocyanin content and the overall quality of fruits. Current research efforts are mainly focused on the regulation of anthocyanin biosynthesis at the transcriptional and translational levels. In addition to genetic factors, many biotic and abiotic factors influence anthocyanin synthesis, and these regulatory relationships may depend on post-translational modifications of proteins. It is clear that the metabolism and regulation mechanisms of anthocyanins are rather complex. In particular, regulatory mechanisms of anthocyanin synthesis are not only complex, but they are highly specific, rendering it difficult to gain a full picture of apple fruit color development.

As it has been found that *MdMYB10* regulates flesh color development in red-fleshed apples, it is also closely linked with malate accumulation and astringency, thereby affecting the eating quality of apple (Hu et al. 2016). Therefore, further breeding efforts of red-fleshed apple cultivars with good fresh eating qualities will require pursuing linkage analysis of genes related to the regulation of sugar content, acidity, and astringency. Moreover, as apples are grown for either fresh eating or for processing, research on the cultivation physiology of red-fleshed apple is limited. At present, there is a lack of supporting cultivation technologies such as suitable tree types, planting areas, and rootstocks.

Against the current background of big data, integrated genomic, transcriptomic, proteomic, and metabolomic analyses will shed new light on unknown segments of anthocyanin regulatory models and metabolic pathways in apple. Moreover, these will further delineate relationships among color and sugar content, acid content, texture, and other flavor attributes. This is an important direction for pursuing studies on apple fruit color development, and for breeding of high-quality red-colored apple cultivars.

Fig. 13.5 Some examples of red-fleshed apple cultivars

References

Allan AC, Hellens RP, Laing WA (2008) MYB transcription factors that colour our fruit. Trends Plant Sci 13(3):99 102

Amato A, Cavallini E, Zenoni S, Finezzo L, Begheldo M, Ruperti B, Tornielli GB (2016) A grapevine TTG2-like WRKY transcription factor is involved in regulating vacuolar transport and flavonoid biosynthesis. Front Plant Sci 7:1976. https://doi.org/10.3389/fpls.2016.01979

An X, Tian Y, Chen K, Wang XF, Hao YJ (2012) The apple WD40 protein MdTTG1 interacts with bHLH but not MYB proteins to regulate anthocyanin accumulation. Plant Physiol 169(7):710–717

An X, Tian Y, Chen K, Liu XJ, Liu DD, Xie XB, Cheng CG, Cong PH, Hao YJ (2015) *MdMYB9* and *MdMYB11* are involved in the regulation of the JA-induced biosynthesis of anthocyanin and proanthocyanidin in apples. Plant Cell Physiol 56(4):650–662

An J, Song L, Zhao L, You CX, Wang XF, Hao YJ (2018a) Overexpression of MdNAC029 promotes anthocyanin accumulation in apple calli. Acta Hortic Sinica 45(5):35–44

An J, Wang X, Li Y, Song LQ, Zhao LL, You CX, Hao YJ (2018b) EIN3-LIKE1, MYB1, and ETHYLENE RESPONSE FACTOR3 act in a regulatory loop that synergistically modulates ethylene biosynthesis and anthocyanin accumulation. Plant Physiol 178:808–823

An J, Li R, Qu F, You CX, Wang XF, Hao YJ (2018c) R2R3-MYB transcription factor MdMYB23 is involved in the cold tolerance and proanthocyanidin accumulation in apple. Plant J 96(3):562–577

An J, Yao J, Xu R, You CX, Wang XF, Hao YJ (2018d) Apple bZIP transcription factor MdbZIP44 regulates abscisic acid-promoted anthocyanin accumulation. Plant Cell Environ 41(11):2678–2692

An J, Zhang X, You C, Bi SQ, Wang XF, Hao YJ (2019) MdWRKY40 promotes wounding-induced anthocyanin biosynthesis in association with MdMYB1 and undergoes MdBT2-mediated degradation. New Phytol 224(1):380–395

An J, Wang X, Zhang X, Xu HF, Bi SQ, You CX, Hao YJ (2020a) An apple MYB transcription factor regulates cold tolerance and anthocyanin accumulation and undergoes MIEL1-mediated degradation. Plant Biotech J 18(2):337–353

An J, Zhang X, Bi S, You CX, Wang XF, Hao YJ (2020b) The ERF transcription factor MdERF38 promotes drought stress induced anthocyanin biosynthesis in apple. Plant J 101(3):573–589

Aida M, Ishida T, Fukaki H, Fujisawa H, Tasaka M (1997) Genes involved in organ separation in Arabidopsis: an analysis of the cup-shaped cotyledon mutant. Plant Cell 9:841–857

Ban T, Shiozaki S, Ogata T, Horiuchi S (2000) Effects of abscisic acid and shading treatments on the levels of anthocyanin and resveratrol in skin of kyoho grape berry. Acta Hortic 514:83–90

Ban Y, Honda C, Hatsuyama Y, Igarashi M, Bessho H, Moriguchi T (2007) Isolation and functional analysis of a MYB transcription factor gene that is a key regulator for the development of red coloration in apple skin. Plant Cell Physiol 48(7):958–970

Ban Y, Kondo S, Ubi BE, Honda C, Bessho H, Takaya M (2009) UDP-sugar biosynthetic pathway: contribution to cyanidin 3-galactoside biosynthesis in apple skin. Planta 230:871–881

Baudry A, Heim MA, Dubreucq B, Caboche M, Weisshaar B, Lepiniec L (2004) TT2, TT8, and TTG1 synergistically specify the expression of *BANYULS* and proanthocyanidin biosynthesis in *Arabidopsis thaliana*. Plant J 39(3):366–380

Baudry A, Caboche M, Lepiniec L (2006) TT8 controls its own expression in a feedback regulation involving TTG1 and homologous MYB and bHLH factors, allowing a strong and cell-specific accumulation of flavonoids in *Arabidopsis thaliana*. Plant J 46(5):768–779

Borevitz JO, Xia Y, Blount J, Dixon RA, Lamb C (2000a) Activation tagging identifies a conserved MYB regulator of phenylpropanoid biosynthesis. Plant Cell 12:2383–2394

Borevitz JO, Xia Y, Blount J, Dixon RA, Lamb C (2000b) Activation tagging identifies a conserved MYB regulator of phenylpropanoid biosynthesis. Plant Cell 12(12):2383–2394

Boyer J, Liu RH (2004) Apple phytochemicals and their health benefits. Nutr J 3:5. http://www.nutritionj.com/content/3/1/5

Brooks RM, Olmo HP (1997) Register of Fruit and Nut Varieties. University of California Press, Berkeley, CA, p 707

Brueggemann J, Weisshaar B, Sagasser M (2010) A WD40-repeat gene from *Malus × domestica* is a functional homologue of *Arabidopsis thaliana* transparent testa glabra1. Plant Cell Rep 29(3):285–294

Castellarin SD, Matthews MA, Gaspero GD, Gambetta GA (2007) Water deficits accelerate ripening and induce changes in gene expression regulating flavonoid biosynthesis in grape berries. Planta 227:101–112

Chagné D, Krieger C, Rassam M, Sullivan M, Fraser J, Andre C, Pindo M, Troggio M, Gardiner SE, Henry RA, Allan AC, McGhie TK, Laing WA (2012) QTL and candidate gene mapping for polyphenolic composition in apple fruit. BMC Plant Biol 12:12. https://doi.org/10.1186/1471-2229-12-12

Chagné D, Linwang K, Espley RV, Volz RK, How NM, Rouse S, Brendolise C, Carlisle CM, Kumar S, De Silva N, Micheletti D, McGhie T, Crowhurst RN, Storey RD, Velasco R, Hellens RP, Gardiner SE, Allan AC (2013) An ancient duplication of apple MYB transcription factors is responsible for novel red fruit-flesh phenotypes. Plant Physiol 161(1):225–239

Chagné D, Kirk C, How N, Whitworth C, Fontic C, Reig G, Sawyer G, Rouse S, Poles L, Gardiner SE, Kumar S, Espley R, Volz RK, Troggio M, Iglesias I (2016) A functional genetic marker for apple red skin coloration across different environments. Tree Genet Genomes 12(4):67

Chandler VL, Radicella JP, Robbins TP, Chen J, Turks D (1989) Two regulatory genes of the maize anthocyanin pathway are homologous: Isolation of B utilizing R genomic sequences. Plant Cell 1(12):1175–1183

Cheng FS, Weeden NF, Brown SK (1996) Identification of co-dominant RAPD markers tightly linked to fruit skin color in apple. Theor Appl Genet 93:222–227

Christie JM, Arvai AS, Baxter KJ, Heilmann M, Pratt AJ, O'Hara A, Kelly SM, Hothorn M, Smith BO, Hitomi K, Jenkins GI, Getzoff ED (2012) Plant UVR8 photoreceptor senses UV-B by tryptophan-mediated disruption of cross-dimer salt bridges. Science 335 (6075):1492–1496

Daccord N, Celton J, Linsmith G, Becker C, Choisne N, Schijlen E, van de Geest H, Bianco L, Micheletti D, Velasco R, Di Pierro EA, Gouzy J, Rees DJG, Guérif P, Muranty H, Durel CE, Laurens F, Lespinasse Y, Gaillard S, Aubourg S, Quesneville H, Weigel D, Van de Weg E, Troggio M, Bucher E (2017) High-quality de novo assembly of the apple genome and methylome dynamics of early fruit development. Nat Genet 49(7):1099–1106

Dalman K, Wind JJ, Nemesio-Gorriz M, Hammerbacher A, Lundén K, Ezcurra I, Elfstrand M (2017)

Overexpression of *PaNAC03*, a stress induced NAC gene family transcription factor in Norway spruce leads to reduced flavonol biosynthesis and aberrant embryo development. BMC Plant Biol 17(1):6. https://doi.org/10.1186/s12870-016-0952-8
Das PK, Shin DH, Choi SB, Yoo SD, Choi G, Park YI (2012) Cytokinins enhance sugar-induced anthocyanin biosynthesis in Arabidopsis. Mol Cells 34:93–101
Deacon N (2017) The diversity of red fleshed apples. http://www.suttonelms.org.uk/apple104.html
Deikman J, Hammer PE (1995) Induction of anthocyanin accumulation by cytokinins in *Arabidopsis thaliana*. Plant Physiol 108:47–57
Devaiah BN, Karthikeyan AS, Raghothama KG (2007) WRKY75 transcription factor is a modulator of phosphate acquisition and root development in Arabidopsis. Plant Physiol 143(4):1789–1801
De Vetten N, Quattrocchio F, Mol J, Koes R (1997) The an11 locus controlling flower pigmentation in petunia encodes a novel WD-repeat protein conserved in yeast, plants, and animals. Genes Dev 11(11):1422–1434
Duan SW, Wang JJ, Gao CH, Jin CY, Li D, Peng DS, Du GM, Li YQ, Chen MX (2018) Functional characterization of a heterologously expressed *Brassica napus* WRKY41-1 transcription factor in regulating anthocyanin biosynthesis in *Arabidopsis thaliana*. Plant Sci 268:47–53
Dubos C, Stracke R, Grotewold E, Weisshaar B, Martin C, Lepiniec L (2010) MYB transcription factors in Arabidopsis. Trends Plant Sci 15(10):573–581
Eberhardt MV, Lee CY, Liu RH (2000) Antioxidant activity of fresh apples. Nature 405(6789):903–904
Ernst HA, Olsen AN, Skriver K, Larsen S, Leggio LL (2004) Structure of the conserved domain of ANAC, a member of the NAC family of transcription factors. EMBO Rep 5(3):297–303
Espley RV, Hellens RP, Putterill J, Stevenson DE, Kutty-Amma S, Allan AC (2007) Red colouration in apple fruit is due to the activity of the MYB transcription factor, MdMYB10. Plant J 49:414–427
Espley RV, Brendolise C, Chagné D, Kutty-Amma S, Green S, Volz R, Putterill J, Schouten HJ, Gardiner SE, Hellens RP, Allan AC (2009) Multiple repeats of a promoter segment causes transcription factor autoregulation in red apples. Plant Cell 21(1):168–183
Espley RV, Bovy A, Bava C, Jaeger SR, Tomes S, Norling C, Crawford J, Rowan D, McGhie TK, Brendolise C, Putterill J, Schouten HJ, Hellens RP, Allan AC (2013) Analysis of genetically modified red-fleshed apples reveals effects on growth and consumer attributes. Plant Biotechnol J 11:408–419
Eulgem T, Rushton PJ, Robatzek S, Somssich IE (2000) The WRKY superfamily of plant transcription factors. Trends Plant Sci 5(5):199–206
Fang H, Dong Y, Yue X, Hu JF, Jiang SH, Xu HF, Wang YC, Su MY, Zhang J, Zhang ZY, Wang N, Chen XS (2019) The B-box zinc finger protein MdBBX20 integrates anthocyanin accumulation in response to ultraviolet radiation and low temperature. Plant Cell Environ 42(7):2090–2104
FAO, IFAD, and WFP, McGuire S (2015) The state of food insecurity in the world 2015. Meeting the 2015 international hunger targets: taking stock of uneven progress. Rome, FAO, p 57
Favory JJ, Stec A, Gruber H, Rizzini L, Oravecz A, Funk M, Albert A, Cloix C, Jenkins G, Oakeley EJ (2014) Interaction of COP1 and UVR8 regulates UV-B-induced photomorphogenesis and stress acclimation in Arabidopsis. Plant Physiol 28(5):591–601
Fedoroff NV, Battisti DS, Beachy RN, Cooper PJM, Fischhoff DA, Hodges CN, Knauf VC, Lobell D, Mazur BJ, Molden D, Reynolds MP, Ronald PC, Rosegrant MW, Sanchez PA, Vonshak A, Zhu JK (2010) Radically rethinking agriculture for the 21st century. Science 327:833–834
Feller A, Machemer K, Braun EL, Grotewold E (2011) Evolutionary and comparative analysis of MYB and bHLH plant transcription factors. Plant J 66(1):94–116
Flintgarcia S (2013) Genetics and consequences of crop domestication. J Agric Food Chem 61(35):8267–8276
Fickett JW, Wasserman WW (2000) Discovery and modeling of transcriptional regulatory regions. Curr Opin Biotech 11(1):19–24
Gesell A, Yoshida K, Lan TT, Constable CP (2014) Characterization of an apple TT2-type R2R3 MYB transcription factor functionally similar to the poplar proanthocyanidin regulator PtMYB134. Planta 240(3):497–511
Gonzalez A, Zhao M, Leavitt JM, Lloyd AM (2008) Regulation of the anthocyanin biosynthetic pathway by the TTG1/bHLH/Myb transcriptional complex in Arabidopsis seedlings. Plant J 53:814–827
Gonzalez A, Brown M, Hatlestad G, Akhavan N, Smith T, Hembd A, Moore J, Montes D, Mosley T, Resendez J, Nguyen H, Wilson L, Campbell A, Sudarshan D, Lloyd A (2016) TTG2 controls the developmental regulation of seed coat tannins in Arabidopsis by regulating vacuolar transport steps in the proanthocyanidin pathway. Dev Biol 419(1):54–63
Greenmantle N (2017) The Rosetta Apples. http://www.greenmantlenursery.com/fruit/rosetta-apples.htm
Hartmann U, Sagasser M, Mehrtens F, Stracke R, Weisshaar B (2005) Differential combinatorial interactions of cis-acting elements recognized by R2R3-MYB, BZIP, and BHLH factors control light-responsive and tissue-specific activation of phenylpropanoid biosynthesis genes. Plant Mol Biol 57(2):155–171
Heim MA, Jakoby M, Werber M, Martin C, Weisshaar B, Bailey PC (2003) The basic Helix-Loop-Helix transcription factor family in plants: a genome-wide study of protein structure and functional diversity. Mol Biol Evol 20(5):735–747
Hichri I, Barrieu F, Bogs J, Kappel C, Delrot S, Lauvergeat V (2011) Recent advances in the transcriptional regulation of the flavonoid biosynthetic pathway. J Exp Bot 62(8):2465–2483

Hirschhorn JN, Daly MJ (2005) Genome-wide association studies for common diseases and complex traits. Nat Rev Genet 6(2):95–108

Hobert O (2018) Gene regulation by transcription factors and microRNAs. Science 319(5871):1785–1786

Honda C, Kotoda N, Wada M, Kondo S, Kobayashi S, Soejima J, Zhang Z, Tsuda T, Moriguchi T (2011) Anthocyanin biosynthetic genes are coordinately expressed during red coloration in apple skin. Plant Physiol Biochem 40(11):955–962

Holton TA, Cornish EC (1995) Genetics and biochemistry of anthocyanin biosynthesis. Plant Cell 7:1071–1083

Hu DG, Sun CH, Zhang QY, An JP, You CX, Hao YJ (2016) Glucose sensor MdHXK1 phosphorylates and stabilizes MdbHLH3 to promote anthocyanin biosynthesis in apple. PLoS Genet 12(8): https://doi.org/10.1371/journal.pgen.1006273

Hugo KD, Timothy PR (1991) Genetic and developmental control of anthocyanin biosynthesis. Annu Rev Genet 25:173–199

Hyson DA (2011) A comprehensive review of apples and apple components and their relationship to human health. Adv Nutr 2:408

Iqbal N, Khan NA, Ferrante A, Trivellini A, Francini A, Khan MIR (2017) Ethylene role in plant growth, development and senescence: interaction with other phytohormones. Front Plant Sci 8:475. https://doi.org/10.3389/fpls.2017.00475

Jaakola L (2013) New insights into the regulation of anthocyanin biosynthesis in fruits. Trends Plant Sci 18 (9):477–483

Jensen MK, Skriver K (2014) NAC transcription factor gene regulatory and protein–protein interaction networks in plant stress responses and senescence. IUBMB Life 66(3):156–166

Jeong ST, Goto-Yamamoto N, Kobayashi S, Esaka M (2004) Effects of plant hormones and shading on the accumulation of anthocyanins and the expression of anthocyanin biosynthetic genes in grape berry skins. Plant Sci 167(2):247–252

Ji XH, Wang YT, Zhang R, Wu SJ, An MM, Li M, Wang CZ, Chen XL, Zhang YM, Chen XS (2015) Effect of auxin, cytokinin and nitrogen on anthocyanin biosynthesis in callus cultures of red-fleshed apple (*Malus sieversii* f. *niedzwetzkyana*). Plant Cell Tiss Org Cult 120(1):325–337

Jiang Y, Joyce DC (2003) ABA effects on ethylene production, PAL activity, anthocyanin and phenolic contents of strawberry fruit. Plant Grow Regul 39:171–174

Jiang GX, Li ZW, Song YB, Zhu H, Lin S, Huang RM, Jiang YM, Duan XW (2019) LcNAC13 physically interacts with LcR1MYB1 to coregulate anthocyanin biosynthesis-related genes during litchi fruit ripening. Biomolecules 9(4):135. https://doi.org/10.3390/biom904013

Jiang S, Wang N, Chen M, Zhang R, Sun QG, Xu HF, Zhang ZY, Wang YC, Sui XQ, Wang SF, Fang HC, Zuo WF, Su MY, Zhang J, Fei ZJ, Chen XS (2020) Methylation of MdMYB1 locus mediated by RdDM pathway regulates anthocyanin biosynthesis in apple. Plant Biotech J 18(8):1736–1748

Journot-Catalino N, Somssich IE, Roby D, Kroj T (2006) The transcription factors WRKY11 and WRKY17 act as negative regulators of basal resistance in *Arabidopsis thaliana*. Plant Cell 18(11):3289–3302

Johnson CS, Kolevski B, Smyth DR (2002) *TRANSPARENT TESTA GLABRA2*, a trichome and seed coat development gene of Arabidopsis, encodes a WRKY transcription factor. Plant Cell 14:1359–1375

Ju Z, Duan Y, Ju Z (1999) Effects of covering the orchard floor with reflecting films on pigment accumulation and fruit coloration in 'Fuji' apples. Sci Hortic 82(1–2):47–56

Kondo S, Hiraoka K, Kobayashi S, Honda C, Terahara N (2002) Changes in the expression of anthocyanin biosynthetic genes during apple development. J Am Soc Hortic Sci 127(6):971–976

Khan SA, Chibon PY, de Vos RCH, Schipper BA, Walraven E, Beekwilder J, van Dijk T, Finkers R, Visser RG, van de Weg EW, Bovy A, Cestaro A, Velasco R, Jacobsen E, Schouten HJ (2012) Genetic analysis of metabolites in apple fruits indicates an mQTL hotspot for phenolic compounds on linkage group 16. J Exp Bot 63(8):2895–2908

Klein LG (1958) The inheritance of certain fruit characters in the apple. Proc Am Soc Hortic Sci 72:1–14

King MC, Cliff MA (2002) Development of a model for prediction of consumer liking from visual attributes of new and established apple cultivars. J Am Pomol Soc 4:223–229

Kumar S, Chagné D, Bink MCAM, Volz RK, Whitworth C, Carlisle C (2012) Genomic selection for fruit quality traits in apple (*Malus* × *domestica* Borkh.). PloS ONE 7(5):e36674. https://doi.org/10.1371/journal.pone.0036674

Lau OS, Deng XW (2012) The photomorphogenic repressors COP1 and DET1: 20 years later. Trends Plant Sci 17(10):584–593

Lepiniec L, Debeaujon I, Routaboul JM, Baudry A, Pourcel L, Nesi N, Caboche M (2006) Genetics and biochemistry of seed flavonoids. Annu Rev Plant Biol 57(1):405–430

Lespinasse Y, Lespinasse JM, Ganne B (1985) Inheritance of two agronomical characters in the apple tree (*Malus pumila* MILL.): compact type habit and fruit colour. Acta Hortic 159:35–48

Li X, Kui L, Zhang J, Xie Y, Wang L, Yan Y, Wang N, Xu J, Li C, Wang W, Van Nocker S, Dong Y, Ma F, Guan, Q (2016) Improved hybrid de novo genome assembly of domesticated apple (Malus x domestica). Gigascience 5: s13742–016

Li Y, Mao K, Zhao C, Zhao XY, Zhang HL, Shu HR, Hao YJ (2012) MdCOP1 ubiquitin E3 ligases interact with MdMYB1 to regulate light induced anthocyanin biosynthesis and red fruit coloration in apple. Plant Physiol 160(2):1011–1022

Li P, Li Y, Zhang F, Zhang GZ, Jiang XY, Yu HM, Hou BK (2017) The Arabidopsis UDP-glycosyltransferases UGT79B2 and UGT79B3, contribute to cold, salt and drought stress tolerance via modulating anthocyanin accumulation. Plant J 89:85–103

Li H, Li Y, Yu J, Wu T, Zhang J, Tian J, Yao YC (2020) MdMYB8 is associated with flavonol biosynthesis via the activation of the *MdFLS* promoter in the fruits of *Malus* crabapple. Hortic Res 7(1):19. https://doi.org/10.1038/s41438-020-0238-z

Linwang K, Micheletti D, Palmer JW, Volz R, Lozano L, Espley R, Hellens RP, Chagne D, Rowan DD, Troggio M, Iglesias I, Allan AC (2011) High temperature reduces apple fruit colour via modulation of the anthocyanin regulatory complex. Plant Cell Environ 34(7):1176–1190

Lloyd JC, Zakhleniuk OV (2004) Responses of primary and secondary metabolism to sugar accumulation revealed by microarray expression analysis of the *Arabidopsis* mutant, *pho3*. J Exp Bot 55(400):1221–1230

Lloyd AM, BrockmanA AL, Campbell A, Bean A, Cantero A, Gonzalez A (2017) Advances in the MYB–bHLH–WD repeat (MBW) pigment regulatory model: Addition of a WRKY factor and co-option of an anthocyanin MYB for betalain regulation. Plant Cell Physiol 58(9):1431–1441

Liu J, Wei JL, Liu MY, Yang S, Feng SQ, Wang CZ, Chen XS (2012) The relationships between the enzyme activity of anthocyanin biosynthesis, ethylene release and anthocyanin accumulation in fruits of precocious apple cultivars. Acta Hortic Sinica 39 (7):1235–1242

Liu WJ, Wang YC, Yu L, Jiang HY, Guo ZW, Xu HF, Jiang SH, Fang HC, Zhang J, Su MY, Zhang ZY, Chen XL, Chen XS, Wang N (2019) MdWRKY11 participates in anthocyanin accumulation in red-fleshed apples by affecting MYB transcription factors and the photoresponse factor MdHY5. J Agric Food Chem 67(32):8783–8793

Ma DY, Sun DX, Wang CY, Li YG, Guo TC (2014) Expression of flavonoid biosynthesis genes and accumulation of flavonoid in wheat leaves in response to drought stress. Plant Physiol Biochem 80:60–66

Mahmood K, Xu Z, El-Kereamy A, Casaretto JA, Rothstein SJ (2016) The *Arabidopsis* transcription factor ANAC032 represses anthocyanin biosynthesis in response to high sucrose and oxidative and abiotic stresses. Front Plant Sci 7:1548. https://doi.org/10.3389/fpls.2016.01548

Maliepaard C, Alston FH, Van Arkel G, Brown LM, Chevreau E, Dunemann F, Evans KM, Gardiner S, Guilford P, van Heusden AW, Janse J, Laurens F, Lynn JR, Manganaris AG, den Nijs APM, Periam N, Rikkerink E, Roche P, Ryder C, Sansavini S, Schmidt H, Tartarini S, Verhaegh JJ, Vrielink-van Ginkel M, King GJ (1998) Aligning male and female linkage maps of apple (*Malus pumila* Mill.) using multi-allelic markers. Theor Appl Genet 97(1):60–73

Mcclure KA, Gong Y, Song J, Vinqvist-Tymchuk M, Palmer LC, Fan LH, Burgher-MacLellan K, Zhang ZQ, Celton JM, Forney CF, Migicovsky Z, Myles S (2019) Genome-wide association studies in apple reveal loci of large effect controlling apple polyphenols. Hortic Res 6(1):107. https://doi.org/10.1038/s41438-019-0190-y

Mehrotra R, Bhalothia P, Bansal P, Basantani MK, Bharti V, Mehrotra S (2014) Abscisic acid and abiotic stress tolerance-different tiers of regulation. J Plant Physiol 171(7):486–496

Migicovsky Z, Gardner KM, Money D, Sawler J, Bloom JS, Moffett P, Chao CT, Schwaninger H, Fazio G, Zhong GY, Myles S (2016) Genome to phenome mapping in apple using historical data. Plant Genome 9(2). https://doi.org/10.3835/plantgenome2015.11.011

Mol J, Jenkins G, Schäfer E, Weiss D, Walbot V (1996) Signal perception, transduction, and gene expression involved in anthocyanin biosynthesis. Crit Rev Plant Sci 15(5–6):525–557

Morishita T, Kojima Y, Maruta T, Nishizawa-Yokoi A, Yabuta Y, Shigeoka S (2019) Arabidopsis NAC transcription factor, ANAC078, regulates flavonoid biosynthesis under high-light. Plant Cell Physiol 50 (12):2210–2222

Nakabayashi R, Yonekura-Sakakibara K, Urano K, Suzuki M, Yamada Y, Nishizawa T, Matsuda F, Kojima M, Sakakibara H, Shinozaki K, Michael AJ, Tohge T, Yamazaki M, Saito K (2014) Enhancement of oxidative and drought tolerance in arabidopsis by overaccumulation of antioxidant flavonoids. Plant J 77:367–379

Nesi N, Debeaujon I, Jond C, Pelletier G, Caboche M, Lepiniec L (2000) The *TT8* gene encodes a basic Helix-Loop-Helix domain protein required for expression of *DFR* and *BAN* genes in Arabidopsis siliques. Plant Cell 12(10):1863–1878

Nguyen NH, Jeong CY, Kang GH, Yoo SD, Hong SW, Lee H (2015) MYBD employed by HY5 increases anthocyanin accumulation via repression of MYBL2 in *Arabidopsis*. Plant J 84(6):1192–1205

Ogata K, Kanei-Ishii C, Sasaki M, Hideki H, Aritaka N, Masato E, Haruki N, Yoshifumi N, Shunsuke I, Akinori S (1996) The cavity in the hydrophobic core of Myb DNA-binding domain is reserved for DNA recognition and trans-activation. Nat Struct Mol Biol 3 (2):178–187

Ogata J, Kanno Y, Itoh Y, Tsugawa H, Suzuki M (2005) Plant biochemistry: anthocyanin biosynthesis in roses. Nature 435(7043):757–758

Oravecz A, Baumann A, Máté Z, Brzezinska A, Molinier J, Oakeley EJ, Adám E, Schäfer E, Nagy F, Ulm R (2006) CONSTITUTIVELY PHOTOMORPHOGENIC1 is required for the UV B response in Arabidopsis. Plant Cell 18(8):1975–1990

Paz-Ares J, Ghosal D, Wienand U, Petersont PA, Saedler H (1987) The regulatory *c1* locus of *Zea mays* encodes a protein with homology to myb proto-oncogene products and with structural similarities to transcriptional activators. EMBO J 6(12):3553–3558

Pinheiro TT, Nishimura DS, Figueira A, Figueira A, Latado RR (2015) Selection of reference genes for expression analyses of red-fleshed sweet orange (*Citrus sinensis*). Genet Mol Res 14(4):18440–18451

Qi T, Song S, Ren Q, Wu D, Huang H, Chen Y, Fan M, Peng W, Ren C, Xie D (2011) The Jasmonate-ZIM-Domain proteins interact with the WD-Repeat/bHLH/MYB complexes to regulate jasmonate-mediated anthocyanin accumulation and trichome initiation in *Arabidopsis thaliana*. Plant Cell 23(5):1795–1814

Quattrocchio F, Wing JF, Leppen H, Mol JNM, Koes RE (1993) Regulatory genes controlling anthocyanin pigmentation are functionally conserved among plant species and have distinct sets of target genes. Plant Cell 5(11):1497–1512

Quattrocchio F, Wing JF, Woude K, Souer E, Vetten N, Mol J, Koes R (1999) Molecular analysis of the *anthocyanin2* gene of petunia and its role in the evolution of flower color. Plant Cell 11(8):1433–1444

Rizzini L, Favory JJ, Cloix C, Faggionato D, O'Hara A, Kaiserli E, Baumeister R, Schäfer E, Nagy F, Jenkins GI, Ulm R (2011) Perception of UV-B by the Arabidopsis UVR8 protein. Science 332(6):103–106

Roth BA, Goff SA, Klein TM, Fromm ME (1991) C1- and R-dependent expression of the maize *Bz1* gene requires sequences with homology to mammalian myb and myc binding sites. Plant Cell 3(3):317–325

Rumainum IM, Worarad K, Yamaki Y, Yamane K (2016) Effects of developmental stages, light, and an auxin polar transport inhibitor on the skin and flesh pigmentation of red-fleshed peach fruit. Hortic J 85 (2):141–147

Rushton PJ, Somssich IE, Ringler P, Shen QJ (2010) WRKY transcription factors. Trends Plant Sci 15 (5):247–258

Sadilova E, Stintzing FC, Carle R (2006) Chemical quality parameters and anthocyanin pattern of red-fleshed Weirouge apples. J Appl Bot Food Qual 80:82–87

Sasaki Y, Asamizu E, Shibata D, Nakamura Y, Kaneko T, Awai K, Amagai M, Kuwata C, Tsugane T, Masuda T, Shimada H, Takamiya K, Ohta H, Tabata S (2001) Monitoring of methyl jasmonate-responsive genes in Arabidopsis by Cdna macroarray: self-activation of jasmonic acid biosynthesis and crosstalk with other phytohormone signaling pathways. DNA Res 8(4):153–161

Schnable PS, Ware D, Fulton RS, Stein JC, Wei FS, Pasternak S, Liang CZ, Zhang JW, Fulton L, Graves TA, Minx P, Reily AD, Courtney L, Kruchowski SS, Tomlinson C, Strong C, Delehaunty K, Fronick C, Courtney B, Rock SM, Belter E, Du FY, Kim K, Abbott RM, Cotton M, Levy A, Marchetto P, Ochoa K, Jackson SM, Gillam B, Chen WZ, Yan L, Higginbotham J, Cardenas M, Waligorski J, Applebaum E, Phelps L, Falcone J, Kanchi K, Thane T, Scimone A, Thane N, Henke J, Wang T, Ruppert J, Shah N, Rotter K, Hodges J, Ingenthron E, Cordes M, Kohlberg S, Sgro J, Delgado B, Mead K, Chinwalla A, Leonard S, Crouse K, Collura K, Kudrna D, Currie J, He RF, Angelova A, Rajasekar S, Mueller T, Lomeli R, Scara G, Ko A, Delaney K, Wissotski M, Lopez G, Campos D, Braidotti M, Ashley E, Golser W, Kim H, Lee S, Lin J, Dujmic Z, Kim W, Talag J, Zuccolo A, Fan CZ, Sebastian A, Kramer M, Spiegel L, Nascimento L, Zutavern T, Miller B, Ambroise C, Muller S, Spooner W, Narechania A, Ren LY, Wei S, Kumari S, Faga B, Levy M, McMahan L, Buren PV, Vaughn MW, Ying K, Yeh CT, Emrich SJ, Jia Y, Kalyanaraman A, Hsia AP, Barbazuk WB, Baucom RS, Brutnell TP, Carpita NC, Chaparro C, Chia JM, Deragon JM, Estill JC, Fu Y, Jeddeloh JA, Han YJ, Lee H, Li PH, Lisch DR, Liu SZ, Liu ZJ, Nagel DH, McCann MC, Miguel PS, Myers AM, Nettleton D, Nguyen J, Penning BW, Ponnala L, Schneider KL, Schwartz DC, Sharma A, Soderlund C, Springer NM, Sun Q, Wang H, Waterman M, Westerman R, Wolfgruber TK, Yang LX, Yu Y, Zhang LF, Zhou SG, Zhu QH, Bennetzen JL, Dawe RK, Jiang JM, Jiang N, Presting GG, Wessler SR, Aluru S, Martienssen RA, Clifton SW, McCombie WR, Wing RA, Wilson RK (2009) The B73 maize genome: complexity, diversity, and dynamics. Science 326(5956):1112–1115

Sekido K, Hayashi Y, Yamada K, Shiratake K, Matsumoto S, Maejima T, Komatsu H (2010) Efficient breeding system for red-fleshed apple based on linkage with S3-RNase allele in 'Pink Pearl'. Infec Immun 45(4):534–537

Shaked-Sachray L, Weiss D, Reuveni M, Nissim-Levi A, Oren-Shamir M (2010) Increased anthocyanin accumulation in aster flowers at elevated temperatures due to magnesium treatment. Physiol Plant 114(4):559–565

Singh SP, Saini MK, Singh J, Pongener A, Sidhu GS (2014) Preharvest application of abscisic acid promotes anthocyanins accumulation in pericarp of litchi fruit without adversely affecting postharvest quality. Postharv Biol Technol 96:14–22

Sivitz AB, Reinders A, Ward JM (2008) Arabidopsis sucrose transporter AtSUC1 is important for pollen germination and sucrose-induced anthocyanin accumulation. Plant Physiol 147:92–100

Smeekens S (2000) Sugar-induced signal transduction in plants. Plant Mol Biol 51(51):49–81

Smith TF, Gaitatzes C, Saxena K et al (1999) The WD repeat: a common architecture for diverse functions. Trends Biochem Sci 24(5):181–185

Solfanelli C, Poggi A, Loreti E, Alpi A, Perata P (2006) Sucrose-specific induction of the anthocyanin biosynthetic pathway in Arabidopsis. Plant Physiol 140:637–646

Song Z, Li TZ, Xu GX, Gu DJ, He ML, Zhang CB (2009) Effect of different light spectra on the surface coloration of 'Red Fuji' apple. Acta Ecol Sin 11 (8):234–241

Souer E, Houwelingen A, Kloos D, Mol J, Koes R (1996) The no apical meristem gene of Petunia is required for pattern formation in embryos and flowers and is expressed at meristem and primordia boundaries. Cell 85(2):159–170

Spelt C, Quattrocchio F, Mol JNM, Koes R (2000) *Anthocyanin1* of petunia encodes a basic helix-loop-helix protein that directly activates transcription of structural anthocyanin genes. Plant Cell 12(9):1619–1631

Sun Q, Jiang S, Zhang T, Xu H, Fang H, Zhang J, Su M, Wang Y, Zhang Z, Wang N, Chen X (2019) Apple NAC transcription factor MdNAC52 regulates biosynthesis of anthocyanin and proanthocyanidin through MdMYB9 and MdMYB11. Plant Sci 289:

Takos AM, Jaffe FW, Jacob SR, Bogs J, Robinson SP, Walker AR (2005) Light-induced expression of a *MYB* gene regulates anthocyanin biosynthesis in red apples. Plant Physiol 142:1216–1232

Teng S, Keurentjes J, Bentsink L, Koornneef M, Smeekens S (2005) Sucrose-specific induction of anthocyanin biosynthesis in Arabidopsis requires the *MYB75/PAP1* gene. Plant Physiol 139:1840–1852

Tian J, Han ZY, Zhang LR, Song TT, Zhang J, Li JY, Yao YC (2015) Induction of anthocyanin accumulation in crabapple (*Malus* cv.) leaves by low temperatures. HortScience 50(5):640–649

Tiwari SB, Wang XJ, Hagen G, Guilfoyle TJ (2001) AUX/IAA proteins are active repressors, and their stability and activity are modulated by auxin. Plant Cell 13:2809–2822

Treutter D (2001) Biosynthesis of phenolic compounds and its regulation in apple. Plant Grow Regul 34:71–89

Ubi B, Honda C, Bessho H, Kondo S, Wada M, Kobayashi S, Moriguchi T (2006) Expression analysis of anthocyanin biosynthetic genes in apple skin: effect of UV-B and temperature. Plant Sci 170(3):571–578

Van Nocker S, Ludwig P (2003) The WD-repeat protein superfamily in Arabidopsis: conservation and divergence in structure and function. BMC Genomics 4:50

Velasco R, Zharkikh A, Affourtit JP, Dhingra A, Cestaro A, Kalyanaraman A, Fontana P, Bhatnagar SK, Troggio M, Pruss D, Dhingra A, Cestaro A, Kalyanaraman A, Fontana P, Bhatnagar SK, Troggio M, Pruss D, Salvi S, Pindo M, Baldi P, Castelletti S, Cavaiuolo M, Coppola G, Costa F, Cova V, Ri AD, Goremykin V, Komjanc M, Longhi S, Magnago P, Malacarne G, Malnoy M, Micheletti D, Moretto M, Perazzolli M, Si-Ammour A, Vezzulli S, Zini E, Eldredge G, Fitzgerald LM, Gutin N, Lanchbury G, Macalma T, Mitchell JT, Reid J, Wardell B, Kodira C, Chen Z, Desany B, Niazi F, Palmer M, Koepke T, Jiwan D, Schaeffer S, Krishnan V, Wu C, Chu VT, King ST, Vick J, Tao Q, Mraz A, Stormo A, Stormo K, Bogden R, Ederle D, Stella A, Vecchietti A, Kater MM, Masiero S, Lasserre P, Lespinasse Y, Allan AC, Bus V, Chagné D, Crowhurst RN, Gleave AP, Lavezzo E, Fawcett JA, Proost S, Rouzé P, Sterck L, Toppo S, Lazzari B, Hellens RP, Durel C-E, Gutin A, Bumgarner R, Gardiner SE, Skolnick M, Egholm M, Van de Peer Y, Salamini F, Viola R (2010) The genome of the domesticated apple (*Malus* × *domestica* Borkh.). Nat Genet 42(10):833–839

Verdu CF, Guyot F, Childebrand N, Bahut M, Celton JM, Gaillard S, Lasserre-Zuber P, Troggio M, Guilet D, Laurens F (2014) QTL analysis and candidate gene mapping for the polyphenol content in cider apple. PLoS ONE 9(10):

Verweij CW, Spelt CE, Bliek M, de Vries M, Wit N, Faraco M, Koes R, Quattrocchio FM (2016) Functionally similar WRKY proteins regulate vacuolar acidification in petunia and hair development in Arabidopsis. Plant Cell 28(3):786–803

Vimolmangkang S, Han Y, Wei G, Korban SS (2013) An apple MYB transcription factor, *MdMYB3*, is involved in regulation of anthocyanin biosynthesis and flower development. BMC Plant Biol 13(1):176. https://doi.org/10.1186/1471-2229-13-176

Vogt T, Jones P (2000) Glycosyltransferases in plant natural product synthesis: characterization of a supergene family. Trends Plant Sci 5(9):380–386

Volz RK, Oraguzie NC, Whitworth CJ, How N, Chagné D, Carlisle CM, Gardiner SE, Rikkerink EHA, Lawrence T (2009) Breeding for red flesh colour in apple: progress and challenges. Acta Hortic 814:337–342

Walker AR, Davison PA, Bolognesi-Winfield AC, James CM, Srinivasan N, Blundell TL, Esch JJ, Marks MD, Gray JC (1999) The TRANSPARENT TESTA GLABRA1 locus, which regulates trichome differentiation and anthocyanin biosynthesis in Arabidopsis, encodes a WD40 repeat protein. Plant Cell 11:1337–1350. https://doi.org/10.1105/tpc.11.7.1337

Wang N, Xu H, Jiang S, Zhang Z, Lu N, Qiu H, Qu C, Wang Y, Wu S, Chen X (2017) MYB12 and MYB22 play essential roles in proanthocyanidin and flavonol synthesis in red-fleshed apple (*Malus sieversii* f. *niedzwetzkyana*). Plant J 90(2):276–292

Wang N, Liu W, Zhang T, Jiang S, Xu H, Wang Y, Zhang Z, Wang C, Chen X (2018a) Transcriptomic analysis of red-fleshed apples reveals the novel role of MdWRKY11 in flavonoid and anthocyanin biosynthesis. J Agric Food Chem 66(27):7076–7086

Wang N, Qu C, Jiang S, Chen Z, Xu H, Fang H, Su M, Zhang J, Wang Y, Liu W, Zhang Z, Lu N, Chen X (2018b) The proanthocyanidin-specific transcription factor MdMYBPA1 initiates anthocyanin synthesis under low-temperature conditions in red-fleshed apples. Plant J 96(1):39–55

Wang N, Jiang S, Zhang Z, Fang H, Xu H, Wang Y, Chen X (2018c) *Malus sieversii*: the origin, flavonoid synthesis mechanism, and breeding of red-skinned and red-fleshed apples. Hortic Res 5(1):70. https://doi.org/10.1038/s41438-018-0084-4

Wang N, Liu W, Wang Y, Xu H, Fang H, Yang G, Wu S, Zhang Z, Chen X (2018d) A new high-flavonoid apple cultivar 'meihong'. Acta Hortic Sin 45(S2):2705–2706

Wang Y, Wang N, Xu H, Jiang S, Fang H, Su M, Zhang Z, Zhang T, Chen X (2018e) Auxin regulates anthocyanin biosynthesis through the Aux/IAA–ARF signaling pathway in apple. Hortic Res 5(1):59. https://doi.org/10.1038/s41438-018-0068-4

Wang Y, Liu W, Jiang H, Mao Z, Wang N, Jiang S, Xu H, Yang G, Zhang Z, Chen X (2019a) The R2R3-MYB transcription factor MdMYB24-like is involved in methyl jasmonate-induced anthocyanin biosynthesis in apple. Plant Physiol Biochem 139:273–282

Wang Y, Sun J, Wang N, Xu H, Qu C, Jiang S, Fang H, Su M, Zhang Z, Chen X (2019b) MdMYBL2 helps regulate cytokinin-induced anthocyanin biosynthesis in red-fleshed apple (*Malus sieversii* f. *niedzwetzkyana*) callus. Funct Plant Biol 46(2):187–196

Winkel-shirley B (2001) Flavonoid biosynthesis. a colorful model for genetics, biochemistry, cell biology, and biotechnology. Plant Physiol 126(2):485–493

Winkel-Shirley B (2002) Biosynthesis of flavonoids and effects of stress. Curr Opin Plant Biol 5(3):218–223

World Apple Review (2018) World Fruit Market Analysis Belrose, Inc., p 171

Wu D, Hu Q, Yan Z, Chen W, Yan CY, Huang X, Zhang J, Yang PY, Deng HT, Wang JW, Deng XW, Shi YG (2012) Structural basis of ultraviolet-B perception by UVR8. Nature 484(7393):214–219

Xie X, Li S, Zhang R, Zhao J, Chen Y, Zhao Q, Yao Y, You C, Zhang X, Hao Y (2012) The bHLH transcription factor MdbHLH3 promotes anthocyanin accumulation and fruit colouration in response to low temperature in apples. Plant Cell Environ 35 (11):1884–1897

Xie Y, Chen P, Yan Y, Bao C, Li X, Wang L, Shen X, Li H, Liu X, Niu C, Zhu C, Fang N, Shao Y, Zhao T, Yu JT, Zhu JH, Xu LF, van Nocker S, Ma FW, Guan QM (2018) An atypical R2R3 MYB transcription factor increases cold hardiness by CBF-dependent and CBF-independent pathways in apple. New Phytol 218(1):201–218

Xu H, Qu C, Liu J, Wang Y, Wang D, Zuo W, Jiang S, Wang N, Zhang Z, Chen X (2017a) Expression analysis and functional identification of a vacuolar sucrose transporter gene *MdSUT4* in apple. Acta Hortic Sin 44(7):1235–1243

Xu H, Wang N, Liu J, Qu C, Wang Y, Jiang S, Lu N, Wang D, Zhang Z, Chen X (2017b) The molecular mechanism underlying anthocyanin metabolism in apple using the *MdMYB16* and *MdbHLH33* genes. Plant Mol Biol 94(94):149–165

Xu H, Zou Q, Yang G, Jiang S, Fang H, Wang Y, Zhang J, Zhang Z, Wang N, Chen X (2020) MdMYB6 regulates anthocyanin formation in apple both through direct inhibition of the biosynthesis pathway and through substrate removal. Hortic Res 7:72. https://doi.org/10.1038/s41438-020-0294-4

Yan XF, Wang Y, Li YM (2007) Plant secondary metabolism and its response to environment. Acta Ecol Sin 27(6):2554–2562

Yi K, Yan ZY, Liu Z, Wang DM, Yang F, Zhang J (2006) Review on identification and utilization of apple sport selection. J Fruit Sci 23(5):745–749

Zhang J, Xu H, Wang N, Jiang S, Fang H, Zhang Z, Yang G, Wang Y, Su M, Xu L, Chen X (2018) The ethylene response factor MdERF1B regulates anthocyanin and proanthocyanidin biosynthesis in apple. Plant Mol Biol 98:205–218

Zhang L, Hu J, Han X, Li J, Gao Y, Richards CM, Zhang C, Tian Y, Liu G, Gul H, Wang D, Tian Y, Yang C, Meng M, Yuan G, Kang G, Wu Y, Wang K, Zhang H, Wang D, Cong P (2019a) A high-quality apple genome assembly reveals the association of a retrotransposon and red fruit colour. Nat Commun 10 (1):1494

Zhang T, Xu H, Yang G, Wang N, Zhang J, Wang Y, Jiang S, Fang H, Zhang Z, Chen X (2019b) Molecular mechanism of MYB111 and WRKY40 involved in anthocyanin biosynthesis in red-fleshed apple callus. Plant Cell Tiss Org Cult (PCTOC) 139(3):467–478

Zhang S, Chen Y, Zhao L, Li C, Yu J, Li T, Yang W, Zhang S, Su H, Wang L (2020) A novel NAC transcription factor, MdNAC42, regulates anthocyanin accumulation in red-fleshed apple by interacting with MdMYB10. Tree Physiol 40(3):413–423

Zhao J, Tian Y, Wang C, Dai H, Wang D (2006) Identification of RAPD marker linked to the red skin traits of apples. J Fruit Sci 23:165–168

Zhou H, Lin-Wang K, Wang H, Gu C, Dare AP, Espley RV, He H, Allan AC, Han Y (2015) Molecular genetics of blood-fleshed peach reveals activation of anthocyanin biosynthesis by NAC transcription factors. Plant J 82(1):105–121

14 Genomics of Fruit Acidity and Sugar Content in Apple

Bo Zhang and Yuepeng Han

Abstract

Soluble sugars and organic acids play important roles in determining fruit taste. For apple, fructose is the most abundant soluble sugar, while the predominant organic acid is malic acid (Ma) that accounts for up to 90% of the total organic acids. The earliest genetic studies of inheritance have uncovered that apple fruit acidity is controlled by a single major locus on linkage group (LG) 16, designated *Ma*. Later, another important quantitative trait locus (QTL) for fruit acidity has been identified on LG 8. Besides these two QTLs, five other QTLs for apple fruit acidity are also found on LGs 2, 10, 13, 15, and 17. Sugar content is a heritable quantitative trait, and two QTLs for sugar content have been identified on LGs 1 and 3. Several strong candidate genes for fruit acidity, such as *Ma1* and *MdPP2CH* encoding a protein phosphatase2C that inactivates H^+-ATPases and Ma1 via dephosphorylation, have been reported. However, few candidates for sugar content have been identified. This review will cover these findings as well as assess their impact on pursuing genetic enhancement efforts to manipulate/modify these important components of fruit taste in apples.

B. Zhang · Y. Han (✉)
CAS Key Laboratory of Plant Germplasm Enhancement and Specialty Agriculture, Wuhan Botanical Garden, the Innovative Academy of Seed Design, Chinese Academy of Sciences, Wuhan 430074, China
e-mail: yphan@wbgcas.cn

Y. Han
Center of Economic Botany, Core Botanical Gardens, Chinese Academy of Sciences, Wuhan 430074, China

Y. Han
Sino-African Joint Research Center, Chinese Academy of Sciences, Wuhan 430074, China

14.1 Introduction

Key components of fruit taste consist of soluble sugars and organic acids, which, along with aromas, have a strong influence on the overall flavor of fruits (Borsani et al. 2009). In fruits, primary soluble sugars consist of sucrose, fructose, and glucose; whereas, organic acids are mainly composed of malic, citric, and tartaric acids. However, it is important to point out that fructose, glucose, and sucrose are not equally sweet, and malic, citric, and tartaric acids differ in acidity (Doty 1976). Therefore, organoleptic fruit quality (or fruit taste) is highly associated with the composition and content of soluble sugars and organic acids (Pangborn 1963). Moreover, the sugar–acid ratio is also important in determining the taste and flavor of fruits. In addition to their impacts on the flavor of fresh fruit, soluble sugars and organic acids also highly influence the flavor of processed products, such as juices and ciders. Hence, it is clear that the

S. S. Korban (ed.), *The Apple Genome*, Compendium of Plant Genomes,
https://doi.org/10.1007/978-3-030-74682-7_14

metabolism and accumulation of soluble sugars and organic acids in fruits are very critical in fruit tree breeding programs.

Over the past two decades, reports on the mechanisms of soluble sugar and organic acid accumulation in fleshy fruits have increased significantly (Borsani et al. 2009; Sweetman et al. 2009). Among the key steps for soluble sugar and organic acid accumulation in fleshy fruits include the following: (1) phloem unloading of sucrose and/or sorbitol into fruit cells; (2) sugar and organic acid metabolism; and (3) solute accumulation in vacuoles. The network of sugar and organic acid metabolism in mesocarp cells consists of multiple pathways, and this is related to the cytosol, as well as to several organelles, such as mitochondria, chloroplasts, and glyoxysomes. Similarly, vacuolar storage of both soluble sugars and organic acids is involved in different types of transport, including simple diffusion, facilitated diffusion, primary active transport, secondary active transport, and electrochemical potential gradient (Etienne et al. 2013). Therefore, those mechanisms underlying the metabolism of sugars and organic acids in mesocarp cells are very complex.

The complex network of soluble sugar and organic acid metabolism, as well as accumulation involves numerous genes, and each gene is present in multiple copies, as most fruit crops are of polyploid origin. Theoretically, it is difficult to identify major genes responsible for variations in soluble sugar and organic acid in fruits using reverse genetic approaches. Genetic map-based approaches such as linkage mapping are practical tools for identifying major genes controlling complex traits, such as those for sweetness and acidity in fruit tree crops.

Linkage mapping is to identify molecular markers that are tightly linked to a gene of interest. To identify candidate genes, it is important to develop a physical map that covers the candidate gene region. Finally, the functionality of a candidate gene(s) must be validated through genetic complementation and/or gene knockout studies. Due to the availability of whole-genome sequences for numerous plant species, map-based cloning of genes of interest has been greatly facilitated. Whole-genome sequences are very helpful for the development of DNA markers, such as Indels, simple sequence repeats (SSRs), and single nucleotide polymorphisms (SNPs), that can be used to saturate a genetic map spanning the target gene region. This fine mapping delimits candidate genes to a narrow target region, wherein candidate genes can be predicted based on reference genome sequences.

The apple belongs to the Rosaceae family. The domesticated apple (*Malus* × *domestica* Borkh.) is one of the most popular and important fruit tree crops worldwide. Despite an allopolyploid origin, the apple is a diploid (2n = 34) with a relatively small genome size of 750 Mb/haploid (Tatum et al. 2005; Korban et al. 2009). The apple is highly heterozygous due to self-incompatibility and has a juvenile period of 6–10 years or more, which makes it difficult to conduct genetic studies. The apple genome is known to be a rich source of genes involved in pest and disease resistance, with major effects that are simply inherited (Gardiner et al. 2007; Korban and Tartarini 2009). Moreover, phenotypes for resistance to diseases and insects are relatively easily identifiable, and phenotypic data are usually reliable. Therefore, the identification of resistance genes is the predominant goal of most initial genetic studies through the development of molecular tools and resources for linkage mapping. With the release of whole-genome sequences of apple (Velasco et al. 2010; Daccord et al. 2017; Zhang et al. 2019), along with the rapid development of high-throughput 'next-generation sequencing' (NGS) technology, genotyping-by-sequencing (GBS), and SNP-arrays, these have provided good opportunities for pursuing forward genetics studies to delineate mechanisms controlling complex traits, such as those of soluble sugar and organic acid accumulation in fruits of apple (Chagné et al. 2007; Khan and Korban 2012; Kumar et al. 2013). In this chapter, we will provide an overall description of genomics studies for apple fruit acidity and sugar content.

14.2 Composition and Content of Organic Acids and Soluble Sugars in Apple Fruits

There are two distinct characteristics related to sugar and organic acid accumulation in apple fruits. One is that apple fruits are rich in fructose, accounting for 44–75% of the total soluble sugars, while sucrose and glucose account for 11–40% and 9–36%, respectively (Hecke et al. 2006; Wu et al. 2007). Moreover, there are distinct differences between cultivated and wild apple fruits in their contents of major components of soluble sugars (Ma et al. 2015a). In cultivated apple fruits, average concentrations of fructose and sucrose are 2- to 3-folds higher than those of glucose; whereas, in wild apple fruits, average contents of fructose and glucose are over 3-folds higher than those of sucrose. Overall, fructose and sucrose are the major components of soluble sugars in cultivated apples, while fructose and glucose are the major components of soluble sugars in wild apples. However, both cultivated and wild apple fruits contain high amounts of fructose. Another is the predominance of malic acid in apple fruits, which accounts for up to 90% of the total organic acids (Wu et al. 2007; Zhang et al. 2010).

Overall comparisons of sugar and organic acid contents between wild and cultivated apples have revealed that wild apple fruits are significantly more acidic compared with cultivated apple fruits, with an average malic acid content of over 2.2-fold higher than that of cultivated apple fruits (Ma et al. 2015a). The malic acid content ranges from 0.8 to 22.7 mg/g in wild apples and from 0.5 to 18.9 mg/g in cultivated apples (Fig. 14.1). Thus, variations in fruit acidity are greater in wild genotypes than in cultivated genotypes. By contrast, average concentrations of total sugars are slightly higher in fruits of cultivated apple than in wild apple, but this difference is not markedly significant. Moreover, wild genotypes have higher variations in total sugar concentrations than those of cultivated genotypes.

As mentioned above, the sugar–acid ratio plays an important role in determining the organoleptic fruit quality. In apple fruits, there is a positive correlation between the malic acid content and glucose content, but the malic acid content is negatively correlated with sucrose content (Ma et al. 2015a). This finding is similar to those correlations found between organic acid and sucrose accumulation detected in peach fruits (Dirlewanger et al. 1999). Thus, metabolic networks of soluble sugars and organic acids are likely to be communicating with each other. For example, it has been reported that genetic modifications of malate metabolism result in dramatic effects on the soluble solid accumulation in tomato fruits (Centeno et al. 2011). Therefore, it is worth to investigate those mechanisms underlying these observed interactions between soluble sugars and organic acid metabolism in fruits.

14.3 Genetic Mapping of QTLs for Fruit Acidity and Sweetness in Apple

QTL mapping for sugar content and acidity in apple fruits has been widely reported. For example, Liebhard et al. (2003) reported on QTL mapping for sugar content in apple fruits using a cross between ‘Fiesta’ and ‘Discovery’. A total of five QTLs for soluble solids were identified, and these were located on linkage groups (LGs) 3, 6, 8, 9, and 14 (Table 14.1). Of these five QTLs, two were detected on LG6 and LG8 of both ‘Fiesta’ and ‘Discovery’, accounting for 10–17% of the observed variance. Subsequently, Kenis et al. (2008) assessed the inheritance of fruit sugar content using progeny derived from ‘Telamon’ x ‘Braeburn’. A major QTL for Brix (sweetness) was identified on LG10 in both parents over 2 successive years and accounting for up to 30% of the observed phenotypic variation. Furthermore, a minor QTL was also detected on LG2 in both parents over 2 successive years, and this accounted for about 8% of the observed phenotypic variation. These findings suggested that the genetic control for fruit sweetness was polygenic or quantitatively controlled (Kenis et al. 2008).

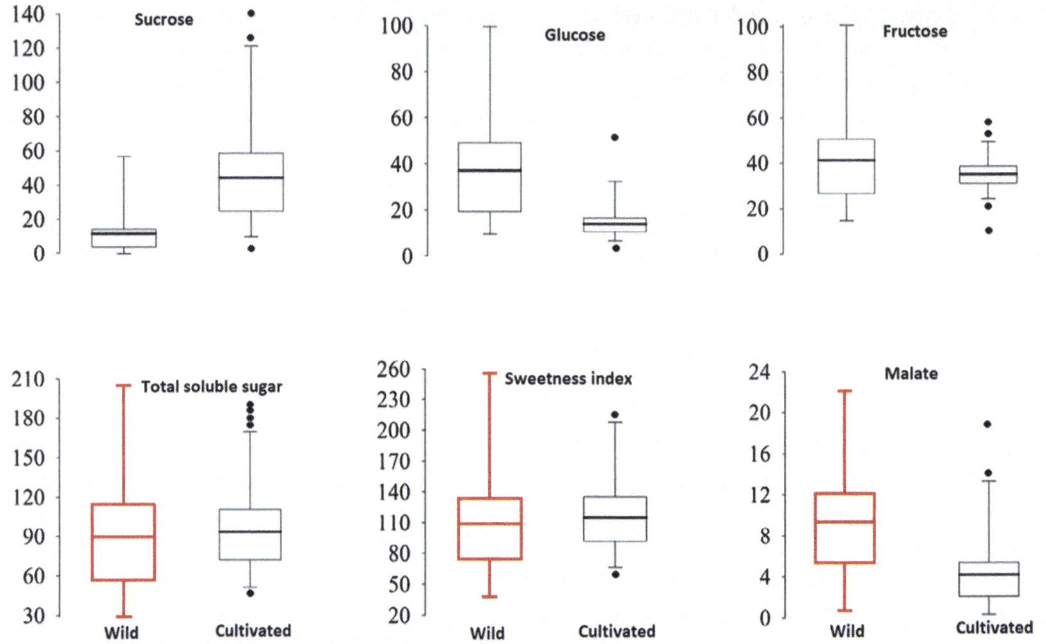

Fig. 14.1 Range and distribution of malic acid and soluble sugar contents (mg/g FW), along with the sweetness index in both cultivated and wild fruits of apple. Horizontal lines within boxes correspond to mean values. A box corresponds to the distribution for 50% of the data. Approximately 99% of the data fall within whiskers. Data outside these whiskers are denoted by solid dark circles

Recently, QTL mapping for contents of individual sugars has also been reported. Guan et al. (2015) identified QTLs for individual sugars, including those for fructose, glucose, sucrose, and sorbitol using multiple apple breeding populations. A single QTL region on LG1 was consistently identified for both fructose and sucrose contents in ripe fruits in 2 successive years, accounting for 34–67 and 13–41% of the total observed phenotypic variation, respectively. However, two QTLs for soluble solids content (SSC) in ripe fruits were located on LG3 and LG2, accounting for 22% and 3% of total phenotypic variations, respectively. These reported discrepancies between QTLs for SSCs and individual sugars indicated that map cloning of some of these major QTLs responsible for fruit sweetness would be rather difficult. Subsequently, Ma et al. (2016) conducted QTL mapping for individual sugars using a cross between 'Jiguan' and 'Wangshanhong'. It was found that four QTLs for glucose, sucrose, fructose, and sorbitol contents were clustered within a single region on LG3, accounting for 11.7–28.0% of the total observed phenotypic variation. Moreover, an additional QTL for glucose content was detected on LG4, accounting for 16.7% of the total phenotypic variation. These two clustered QTLs on LG1 and LG3 controlling individual sugar contents might serve as promising genomic regions for marker-assisted selection (MAS) for sweetness in apple breeding programs.

On the other hand, QTL mapping for fruit acidity has identified a major QTL for malic acid, *Ma* gene, initially mapped onto LG16 (Table 14.2), and explaining ~30% of the observed variance (Maliepaard et al. 1998; Liebhard et al. 2003; Kenis et al. 2008). Later on, Xu et al. (2012) have identified several SSR markers within the *Ma* region using the whole-genome sequence of 'Golden Delicious'. These newly developed SSR markers have been used to screen two half-sib populations, and have allowed for delimiting the *Ma* locus to a physical region no longer than 150 kb on chromosome 16 of the 'Golden Delicious' genome.

Table 14.1 Linkage mapping for soluble sugar content in apple fruit

Trait[a]	Cross	Molecular markers[a]			References
		Name	Distance (cM)	Linkage group	
Sugar content	'Fiesta' × 'Discovery'	D02-700	3.2	3	Liebhard et al. (2003)
		E38M35-0298	0.7	6	
		E35M32-200	0.3	8	
		E32M38-0188	1.6	9	
		CH05g07z	1.1	14	
Brix	'Telamon' × 'Braeburn'	EAATMCCT108	<16.0	2	Kenis et al. (2008)
		CH03d11	<4.0	10	
Fructose content	Multiple populations	ss475883868	<25.9	1	Guan et al. (2015)
Glucose content		ss475876871	<24.6	1	
Sucrose content		ss475876912	<21.7	1	
Sorbitol content		ss475878145	<7.6	5	
		ss475882286	<22.8	1	
SSC		ss475875869	<8.6	2	
		ss475877706	<23.2	3	
Glucose content	'Jiguan' × 'Wangshanhong'	MdSNPui01586	<8.7	4	Ma et al. (2016)
		WBGCAS142	<8.1	3	
Sucrose content		MdSNPui08437	<8.1	3	
Fructose content		WBGCAS134	<17.4	3	
Sorbitol content		WBGCAS130	<22.0	3	

[a]Of those markers identified to date, only those most closely linked to fruit sugar-related traits are listed. All sugar-related traits are derived from ripe fruits. SSC, soluble solids content

Besides the *Ma* gene, other QTLs for apple fruit acidity have also been reported on LGs 2, 8, 10, 13, 15, and 17 (Liebhard et al. 2003; Kenis et al. 2008). Interestingly, the QTL on LG8 accounts for up to 46% of the observed variance in an F_1 population derived from a cross between 'Fiesta' and 'Discovery' (Liebhard et al. 2003). This QTL has been confirmed to be linked SSR marker CH05a02y on LG8 of apple cv. 'Jonathan' (Zhang et al. 2012).

As malic acid is the major organic acid in apple fruit, QTL mapping for malate content has also been conducted using a cross of 'Jiguan' × 'Wangshanhong' (Ma et al. 2016). Genotyping of this segregating apple population using a 1536 EST-derived SNP GoldenGate genotyping platform has allowed for the construction of a high-density linkage map spanning 1368.4 cM and with an average of 2.28 cM per marker. Interestingly, two QTLs for malic acid content have been identified on LG8 and LG16, while no QTL is detected for citric acid content. Interestingly, the QTL detected on LG8 is consistent with the findings reported in previous studies (Liebhard et al. 2003; Zhang et al. 2012). Taken together, these results strongly indicate that the genetic control for fruit acidity in apples is complex, and it is likely that major gene(s),

Table 14.2 Linkage mapping for fruit acidity in apple

Trait	Cross	Molecular markers[a]			References
		Name	Distance (cM)	Linkage group	
Ma	'Prima' × 'Fiesta'	OPT-16-1000	0.0	16	Maliepaard et al. (1998)
Fruit acidity	'Fiesta' × 'Discovery'	E31M38-0193	0.2	8	Liebhard et al. (2003)
		CH05b06x-162	0.5	16	
Fruit acidity	'Telamon' × 'Braeburn'	EAATMCAT85	<1.3	2	Kenis et al. (2008)
		EAATMCAC314	<2.2	8	
		ECAAMCGA385	<2.7	10	
		CH05c04	<5.1	13	
		ECAAMCGA48	<3.0	15	
		CH05c06	<8.8	16	
		CH05g03	<7.0	17	
Malate content	'Jonathan' × 'Golden Delicious'	CH05a02y	<16.7	8	Zhang et al. (2012)
pH/TA	'Royal Gala' × *Malus sieversii*	Hi02h08/Hi22f06	<4.7	16	Xu et al. (2012)
Malate content	'Jiguan' × 'Wangshanhong'	MdSNPui10630	<9.7	8	Ma et al. (2016)
		MdSNPui08981	<8.8	16	

[a]Of those markers identified to date, only those most closely linked to fruit sugar-related traits are listed. TA, titratable acidity

other than the *Ma* locus, are also present that play roles in the regulation of fruit acidity.

As mentioned above, mature fruits of domesticated apple cultivars have lower contents of malic acid than those of wild species. This suggests that apple domestication is likely accompanied by directional selection for fruit acidity, resulting in the loss of functionality of gene(s) responsible for fruit acidity during the period of apple domestication. It has been reported that the *D* locus in peach controls fruit acidity and that a low level of acidity is determined by a dominant allele (Boudehri et al. 2009); whereas, high acidity is controlled by a dominant allele in apple. This observed difference suggests that the *Ma* gene in apple must be different from that of the *D* gene in peach.

As fruit trees are known to have long juvenility periods, an F_1 progeny resulting from full-sib crosses between two outbred parents is often used for pursuing QTL mapping in fruit crops. Thus, the incidence of a recombination event in an F_1 cross corresponds to a single meiotic event, thereby resulting in low resolution of QTLs. Moreover, QTL mapping is both time-consuming and laborious in long-lived forest trees and in fruit crops, particularly when a trait is controlled by several genes of small effects. Hence, QTL mapping is not a promising approach to identify genes for complex traits in fruit trees (Khan and Korban 2012). Unlike linkage mapping, association mapping reveals recombination events present within the gene pool of a natural population and does not require the often time-consuming development of segregating progenies. It is known that apple trees are usually self-incompatible, resulting in random mating, unstructured populations, and rapid decay of linkage disequilibrium (LD). Thus, association mapping can overcome the limitations of pedigree-based mapping in apple and has recently emerged as a highly powerful approach for discovering links between markers or genes and complex traits of interest (Peace et al. 2019).

In apple, genome-wide association mapping studies (GWAS) have been undertaken in various studies (Farneti et al. 2017; Moriya et al. 2017; Urrestarazu et al. 2017; McClure et al. 2018, 2019). Larsen et al. (2018) have conducted a GWAS by combining GBS-generated SNPs with sugar-related traits, including fructose as a percentage of total sugar content ('fructose%'), sucrose as a percentage of total sugar content ('sucrose%'), and sucrose content. SNPs associated with the 'fructose%' trait are located on Chr1, while SNPs associated with both sucrose content and the 'sucrose%' trait are located on Chr7 and Chr11. Lee et al. (2017) have conducted GWAS for soluble solids using 237 apple cultivars genotyped using the GBS strategy. SNPs with significant associations with SSC have been detected on chromosomes 2, 3, and 15, while SNPs with significant associations with fruit acidity are located on chromosomes 1, 3, 8, 9, and 16. However, these fruit acidity-associated SNPs on both Chr8 and Chr16 could not match up to previously reported QTLs. In another study, Amyotte et al. (2017) have detected a significant locus for SSC on Chr8, thus confirming a previously reported QTL on Chr8 that accounted for 11% of the total phenotypic variation observed in 'Fiesta' × 'Discovery' (Liebhard et al. 2003). However, this locus has no significant association with the sensory *sweet* taste. Furthermore, no significant locus is detected for either titratable acidity (TA) or for the sensory acid taste.

In brief, the above GWAS studies have confirmed the presence of QTLs for fruit sugar-related traits on both Chr1 and Chr3, and that SSC is weakly associated with *sweet* taste.

14.4 Identification and Functional Analysis of Candidate Genes for Fruit Acidity

In an effort to identify the candidate gene *Ma*, Bai et al. (2012) have pursued a fine-mapping strategy to narrow down the region of the *Ma* locus to 65–82 kb. It is found that this region contains 12–19 predicted genes, depending on the haplotypes, and that two aluminum-activated malate transporter-like (ALMT) genes, designated *Ma1* and *Ma2*, are deemed to be strong candidates of *Ma*. An analysis of expression profiling has further indicated that *Ma1* is a major determinant at the *Ma* locus controlling fruit acidity, as *Ma1* expression is found to be highly correlated with fruit TA. Furthermore, a low-acidity allele of *Ma1* is primarily characterized by the presence of a mutation at base 1455 in the open reading frame. Later, the candidate *Ma1* gene has been further confirmed to co-segregate with malic acid content in yet another study (Khan et al. 2013). It is found that the dominant allele of the *Ma1* gene has an increased level of expression, resulting in an observed 3-fold increase of malate concentration, and a pH reduction from 4 to 3 in mature apple fruits (Khan et al. 2013). By contrast, the recessive allele of the *Ma1* gene (*ma1*) is lowly expressed in low-acidic apple cultivars. The presence of a premature stop codon mutation at base 1455 in the open reading frame of *Ma1* leads to a truncated protein with a deletion of 84 amino acids at its C-terminal. As the C-terminal domain is conserved in all tonoplast-localized ALMTs, a deletion of 84 amino acids has a significantly negative impact on malate transport activity (Li et al. 2020). Moreover, the mutant *ma1* allele has a reduced capacity to mediate malate influx into the vacuole, resulting in a low-acid trait in apple fruit. Furthermore, it is observed that frequencies of the *Ma1Ma1* genotype are low in apple cultivars, but are high in wild relatives; thus, suggesting that domestication of the apple is likely accompanied by selection for the *Ma1* gene (Ma et al. 2015b).

It is noteworthy that high levels of expression of the dominant *Ma1* allele are not always consistent with increased malate accumulation (Ma et al. 2015b). Moreover, malic acid content variations in mature fruits have also been observed among accessions of the same genotype in the *Ma* locus (Ma et al. 2015b). Therefore, this suggests that the *Ma1* gene is not necessarily the single genetic determinant of fruit acidity in apples. In fact, there is evidence demonstrating that complex traits are usually controlled by genetic regulatory module networks, wherein major genes interact with regulatory genes to form functional units (Bar-Joseph et al. 2003). This seems to be also true for the regulation of fruit acidity in apples. A fine-mapping study has revealed the presence of a candidate gene in the QTL controlling fruit acidity on Chr8 that encodes protein phosphatase 2C (PP2C) (Jia et al. 2018). As PP2C is well known to be involved in dephosphorylation (Spartz et al. 2017), the candidate *MdPP2C* gene is deemed to regulate malate accumulation in apple fruit via dephosphorylation of the Ma1 protein (Jia et al. 2018). Moreover, transcription of the *Ma1* gene can be regulated by *MdMYB1* that plays an important role in the anthocyanin coloration of apple fruit (Hu et al. 2016). These findings are consistent with previous reports proposing that genes encoding tonoplast transport proteins are also regulated at both transcriptional and post-transcriptional levels (Hong-Hermensdorf et al. 2006; Wingenter et al. 2011; Delhaize et al. 2012). Thus, those genetic regulatory networks related to the *Ma1* gene will facilitate marker-assisted breeding (MAB) for the genetic improvement of fruit acidity in apple or in related fruit tree crops.

In addition to forward genetics, reverse genetic approaches have also aided in our understanding of some of the mechanisms controlling organic acid accumulation in fruits (Famiani et al. 2005; Martınez-Esteso et al. 2011). In apple, some structural genes involved in metabolic networks have been deemed responsible for variations in organic acid contents in fruits (Yao et al. 2011b). However, there is increasing evidence demonstrating that organic acid accumulation in fruits is regulated by vacuolar storage (Yao et al. 2009; Bai et al. 2012). It is reported that transport of organic acids into a vacuole requires the presence of a proton gradient across the vacuolar membrane (tonoplast) that is generated by a variety of tonoplast proton pumps, such as V-ATPase, V-PPase, and P-ATPase (Faraco et al. 2014). Interestingly, several studies have demonstrated that *CitPH1* and *CitPH5* genes encoding a vacuolar proton-pumping P-ATPase complex are responsible for the hyperacidification of *Citrus* fruits by creating a steep proton gradient across the tonoplast to drive massive transport of citrate into the vacuole (Butelli et al. 2019; Strazzer et al. 2019).

Recently, Ma et al. (2019) have conducted a comparative transcriptome analysis between two apple cultivars with significant differences in fruit acidity, although they carry non-functional homozygous alleles at the *Ma1* locus controlling apple fruit acidity. As a result, a candidate gene, designated *Ma10*, for fruit acidity is identified. This *Ma10* gene codes for a P-type proton pump, P3A-ATPase, that is targeted to the tonoplast. High levels of expression of the *Ma10* gene can contribute to the incidence of an electrochemical gradient across the vacuole membrane, thereby inducing transport of malate into the vacuole through this *Ma1* channel (Fig. 14.2). Furthermore, this *Ma10* gene has been found to be highly associated with fruit malate content and accounting for ~7.5% of the phenotypic variation in the apple germplasm. Therefore, the *Ma10* gene plays an important role in fruit vacuolar acidification in apples. In addition, an apple transcription factor, *MdMYB73,* is also found to regulate malate accumulation and vacuolar pH in apple fruit (Hu et al. 2017). This *MdMYB73* gene can activate transcription of several genes involved in vacuolar transportation of organic acids, such as *MdALMT9, MdVHA-A,* encoding a vacuolar ATPase subunit A, and *MdVHP1,* encoding a vacuolar pyrophosphatase 1, by directly binding to their promoters. Taken together, all the above findings suggest that there is a complex mechanism for fruit acidity regulation in apples.

14.5 Identification and Functional Analysis of Candidate Genes for Fruit Sugar Accumulation

As mentioned above, several QTLs for fruit sugar accumulation have been reported in apple. However, few candidate genes controlling apple fruit sugar accumulation have been thus far identified. A recent GWAS analysis has revealed that *MDP0000149570* on Chr1 is a good candidate gene for the QTL controlling fructose content in apple fruit (Larsen et al. 2019). This gene

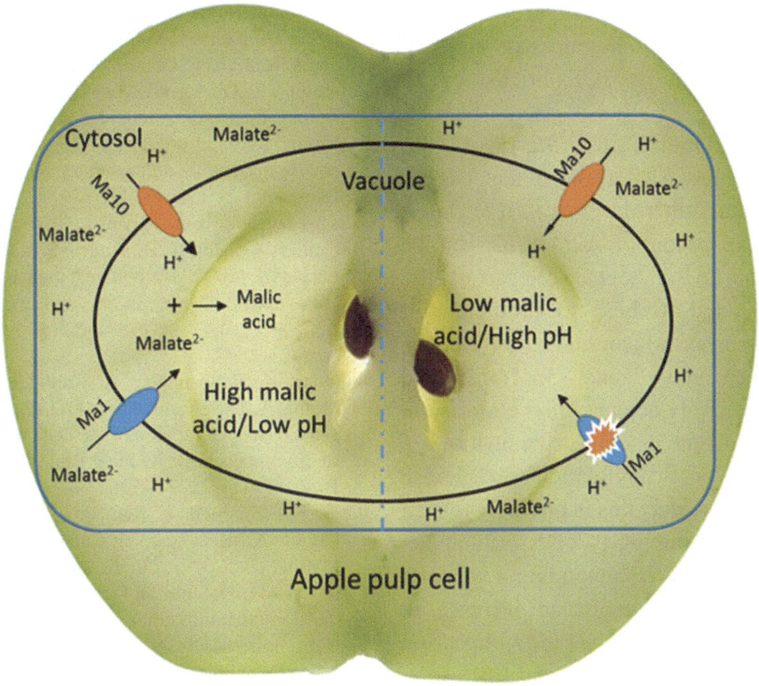

Fig. 14.2 A proposed working model for malic acid accumulation in apple fruit (Ma et al. 2019). The transporter Ma10 pumps protons from the cytosol into the vacuole, while Ma1 mediates the transport of malate from the cytosol to the vacuole. High levels of expression of the *Ma10* allele, wide arrows, are prone to cause an electrochemical gradient across the vacuole membrane, which induces transport of malate into the vacuole through the Ma1 channel. Low levels of expression of the *Ma10* allele, narrow arrows, are insufficient to activate the Ma1 channel, thus resulting in low levels of malate accumulation

codes for vacuolar invertase, and it is the same *VIN1* gene that has been reported earlier by Hyun et al. (2011).

Other than QTL mapping, a reverse genetics approach has also been carried out to investigate the mechanism underlying sugar accumulation in apple fruits. Several studies have reported that structural genes of the soluble sugar metabolic pathway are important for sugar accumulation in apple fruit (Teo et al. 2006; Li et al. 2012; Zhu et al. 2013). However, an increasing number of studies indicate that genes encoding the tonoplast monosaccharide transporter may play important roles in controlling sugar accumulation in fruits (Jung et al. 2015; Cheng et al. 2018; Peng et al. 2020). Moreover, Wang et al. (2016) have conducted a comparative evolutionary study, wherein evolutionary rates, gene duplication, and selective patterns of genes involved in sugar metabolism and transport between grasses and eudicots have been assessed. Overall, it is observed that divergent evolutionary patterns are detected in sugar transporter genes, while similar evolutionary patterns are detected in sugar metabolism genes between monocots and eudicots. This suggests that sugar transporter genes are likely to have undergone positive selection in eudicots, but not in grasses (Wang et al. 2016). These findings further confirm that it is sugar transporter genes rather than sugar metabolism genes that play important roles in sugar accumulation in plants. In apples, *MdTMT1,* coding for a tonoplast monosaccharide transporter, is reported to play an important role in fruit sugar accumulation, as silencing of this gene results in a sharp drop in glucose content (Ma et al. 2017). In addition to tonoplast sugar transporter encoding genes, *TSTs*, other sugar transporter genes, such as *SWEETs* may be also involved in fruit sugar accumulation in apples (Zhen et al. 2018).

It has been reported that sugar transporter genes are regulated at both transcriptional and post-transcriptional levels (Lecourieux et al. 2010). Recently, an abscisic acid (ABA)-responsive transcription factor, *MdAREB2* has been reported to play a vital role in sugar accumulation in apple fruit (Ma et al. 2017). Interestingly, the expression of *MdAREB2* can be induced by an ABA signal. Subsequently, *MdAREB2* is found to bind to promoters of several sugar transporter genes, such as an apple sugar (SUT) encoding gene, *MdSUT2,* and *MdTMT1,* to positively activate their expression, thus leading to increased sugar accumulation.

As the vacuolar accumulation of sucrose requires an energy source, this is probably related to vacuolar proton-pumping P-ATPase. In apple, overexpression of *MdVHP1* encoding a vacuolar H^+-pyrophosphatase results in an increase in sucrose accumulation, but it has little impact on the accumulation of both glucose and fructose in transgenic apple calli (Yao et al. 2011a, b). However, there is no correlation between expression levels of *MdVHP1* and *MdSUT1* suggesting that *MdVHP1* cannot preferentially energize transport of soluble sugars. By contrast, overexpression of *MdVHA-A* induces a significant increase in the accumulation of both sucrose and hexoses, such as fructose and glucose. Moreover, expression levels of *MdVHA-A* are well correlated with expression levels of *MdSUT1*. These findings indicate that *MdVHA-A*, instead of *MdVHP1*, plays an important role in sugar accumulation in apple fruit.

Acknowledgements This work was supported by grants from the National Key Research and Development Program (2019YFD1000200) and National Natural Science Foundation of China (Grant Number 31420103914).

References

Amyotte B, Bowen AJ, Banks T, Rajcan I, Somers DJ (2017) Mapping the sensory perception of apple using descriptive sensory evaluation in a genome wide association study. PLoS ONE 12:2. https://doi.org/10.1371/journal.pone.0171710

Bai Y, Dougherty L, Li MJ, Fazio G, Cheng LL, Xu KN (2012) A natural mutation-led truncation in one of the two aluminum-activated malate transporter-like genes at the *Ma* locus is associated with low fruit acidity in apple. Mol Genet Genomics 287:663–678

Bar-Joseph Z, Gerber GK, Lee TI, Rinaldi NJ, Yoo JY, Robert F, Gordon DB, Fraenkel E, Jaakkola TS, Young RA, Gifford DK (2003) Computational discovery of gene modules and regulatory networks. Nat Biotechnol 21:1337–1342

Borsani J, Budde CO, Porrini L, Lauxmann MA, Lombardo VA, Murray R, Andreo CS, Drincovich MF, Lara MV (2009) Carbon metabolism of peach fruit after harvest: changes in enzymes involved in organic acid and sugar level modifications. J Exp Bot 60:1823–1837

Boudehri K, Bendahmane A, Cardinet G, Troadec C, Moing A, Dirlewanger E (2009) Phenotypic and fine genetic characterization of the D locus controlling fruit acidity in peach. BMC Plant Biol 9:59. https://doi.org/10.1186/1471-2229-9-59

Butelli E, Licciardello C, Ramadugu C, Durand-Hulak M, Celant A, Reforgiato Recupero G, Froelicher Y, Martin C (2019) Noemi controls production of flavonoid pigments and fruit acidity and illustrates the domestication routes of modern citrus varieties. Curr Biol 29:158–164

Centeno DC, Osorio S, Nunes-Nesi A, Bertolo AL, Carneiro RT, Araújo WL, Steinhauser MC, Michalska J, Rohrmann J, Geigenberger P, Oliver SN, Stitt M, Carrari F, Rose JK, Fernie AR (2011) Malate plays a crucial role in starch metabolism, ripening, and soluble solid content of tomato fruit and affects postharvest softening. Plant Cell 23:162–184

Chagné D, Carlisle CM, Blond C, Volz RK, Whitworth CJ, Oraguzie NC, Crowhurst RN, Allan AC, Espley RV, Hellens RP, Gardiner SE (2007) Mapping a candidate gene (*MdMYB10*) for red flesh and foliage colour in apple. BMC Genom 8:212e. https://doi.org/10.1186/1471-2164-8-212

Cheng JT, Wen SY, Xiao S, Lu BY, Ma MR, Bie ZL (2018) Overexpression of the tonoplast sugar transporter CmTST2 in melon fruit increases sugar accumulation. J Exp Bot 69:511–523

Daccord N, Celton J-M, Linsmith G, Becker C, Choisne N, Schijlen E, van de Geest H, Bianco L, Micheletti D, Velasco R, Di Pierro EA, Gouzy J, Rees DJG, Guérif P, Muranty H, Durel CE, Laurens F, Lespinasse Y, Gaillard S, Aubourg S, Quesneville H, Weigel D, Van de Weg E, Troggio M, Bucher E (2017) High-quality de novo assembly of the apple genome and methylome dynamics of early fruit development. Nat Genet 49:1099–1106. https://doi.org/10.1038/ng.3886

Delhaize E, Ma JF, Ryan PR (2012) Transcriptional regulation of aluminium tolerance genes. Trends Plant Sci 17:341–348

Dirlewanger E, Moing A, Rothan C, Svanella L, Pronier V, Guye A, Plomion C, Monet R (1999) Mapping QTLs controlling fruit quality in peach (*Prunus persica* (L.) Batsch). Theor Appl Genet 98 (1):18–31

Doty TE (1976) Fructose sweetness: a new dimension. Cereal Foods World 21:62–63

Etienne A, Génard M, Lobit P, Mbeguié-A-Mbéguié D, Bugaud C (2013) What controls fleshy fruit acidity? A review of malate and citrate accumulation in fruit cells. J Exp Bot 64:1451–1469

Famiani F, Cultrera NGM, Battistelli A, Casulli V, Proietti P, Standardi A, Chen ZH, Leegood RC, Walker RP (2005) Phosphoenolpyruvate carboxykinase and its potential role in the catabolism of organic acids in the flesh of soft fruit during ripening. J Exp Bot 56:2959–2969

Faraco M, Spelt C, Bliek M, Verweij W, Hoshino A, Espen L, Prinsi B, Jaarsma R, Tarhan E, de Boer AH, Di Sansebastiano GP, Koes R, Quattrocchio FM (2014) Hyperacidification of vacuoles by the combined action of two different P-ATPases in the tonoplast determines flower color. Cell Rep 6:32–43

Farneti B, Guardo MD, Khomenko I, Cappellin L, Biasioli F, Velasco R, Costa F (2017) Genome-wide association study unravels the genetic control of the apple volatilome and its interplay with fruit texture. J Exp Bot 68:1467–1478

Guan Y, Peace C, Rudell D, Verma S, Evan K (2015) QTLs detected for individual sugars and soluble solids content in apple. Mol Breed 35:135. https://doi.org/10.1007/s11032-015-0334-1

Gardiner SE, Bus VGM, Rusholme RL, Chagné D, Rikkerink EHA (2007) Apple. In: Kole C (ed) Genome mapping and molecular breeding in plants. Fruits and nuts, vol 4. Springer Publ, NY, pp 1–62

Hecke K, Herbinger K, Veberic R, Trobec M, Toplak H, Stampar F, Keppel H, Grill D (2006) Sugar-, acid- and phenol contents in apple cultivars from organic and integrated fruit cultivation. Eur J Clin Nutr 60:1136–1140

Hong-Hermensdorf A, Brux A, Gruber A, Schumacher K (2006) A WNK kinase binds and phosphorylates V-ATPase subunit C. FEBS Lett 580:932–939

Hu DG, Sun CH, Ma QJ, You CX, Cheng L, Hao YJ (2016) *MdMYB1* regulates anthocyanin and malate accumulation by directly facilitating their transport into vacuoles in apples. Plant Physiol 170:1315–1330

Hu DG, Li YY, Zhang QY, Li M, Sun CH, Yu JQ, Hao YJ (2017) The R2-R3-MYB transcription factor *MdMYB73* is involved in malate accumulation and vacuolar acidification in apple. Plant J 19:443–454

Hyun TK, Eom SH, Kim JS (2011) Genomic analysis and gene structure of the two invertase families in the domesticated apple (*Malus* × *domestica* Borkh.). Plant Omics 4:391–399

Jia D, Shen F, Wang Y, Wu T, Xu X, Zhang X, Han Z (2018) Apple fruit acidity is genetically diversified by natural variations in three hierarchical epistatic genes: *MdSAUR37*, *MdPP2CH* and *MdALMTII*. Plant J 95:427–443

Jung B, Ludewig F, Schulz A, Meißner G, Wöstefeld N, Flügge U, Pommerrenig B, Wirsching P, Sauer N, Koch W, Sommer F, Muhlhaus T, Schroda M, Cuin TA, Graus D, Marten I, Rainer H, Neuhaus HE (2015) Identification of the transporter responsible for sucrose accumulation in sugar beet taproots. Nat Plants 1:14001. https://doi.org/10.1038/nplants.2014.1

Kenis K, Keulemans J, Davey MW (2008) Identification and stability of QTLs for fruit quality traits in apple. Tree Genet Genomes 4:647–661

Khan MA, Korban SS (2012) Association mapping in forest trees and fruit crops. J Exp Bot 63:4045–4060

Khan SA, Beekwilder J, Schaart JG, Mumm R, Soriano JM, Jacobsen E, Schouten HJ (2013) Differences in acidity of apples are probably mainly caused by a malic acid transporter gene on LG16. Tree Genet Genomes 9:475–487

Korban SS, Tartarini S (2009) Apple structural genomics. In: Folta K, Gardiner S (eds) Genetics and genomics of rosaceae. Springer-Science, NY, pp 85–119

Korban SS, Wannarat W, Rayburn CM, Tatum T, Rayburn AL (2009) Genome size and nucleotypic variation in *Malus* germplasm. Genome 52:148–155

Kumar S, Garrick DJ, Bink MC, Whitworth C, Chagné D, Volz RK (2013) Novel genomic approaches unravel genetic architecture of complex traits in apple. BMC Genomics 14:393. https://doi.org/10.1186/1471-2164-14-393

Larsen B, Migicovsky Z, Jeppesen AA, Gardner KM, Toldam-Andersen TB, Myles S, Ørgaard M, Petersen MA, Pedersen C (2018) Genome-wide association studies in apple reveal loci for aroma volatiles, sugar composition, and harvest date. Plant Genome 12: https://doi.org/10.3835/plantgenome2018.12.0104

Larsen B, Migicovsky Z, Jeppesen AA, Gardner KM, Toldam-Andersen TB, Myles S, Ørgaard M, Petersen MA, Pedersen C (2019) Genome-wide association studies in apple reveal loci for aroma volatiles, sugar composition, and harvest date. Plant Genome 12:180104

Lecourieux F, Lecourieux D, Vignault C, Delrot S (2010) A sugar-inducible protein kinase, VvSK1, regulates hexose transport and sugar accumulation in grapevine cells. Plant Physiol 52:1096–1106

Lee SJ, Ban H, Kim GH, Kwon SI, Kim JH, Choi C (2017) Identification of potential gene associated major traits using GBS-GWAS for Korean apple germplasm collections. Plant Breed 136:977–986

Li M, Feng F, Cheng L (2012) Expression patterns of genes involved in sugar metabolism and accumulation during apple fruit development. PLoS ONE 7: https://doi.org/10.1371/journal.pone.0033055

Li C, Dougherty L, Coluccio A, Meng D, El-Sharkawy I, Borejsza-Wysocka E, Liang D, Piñeros M, Xu K, Cheng L (2020) Apple ALMT9 requires a conserved c-terminal domain for malate transport underlying fruit acidity. Plant Physiol 82:992–1006

Liebhard R, Kellerhals M, Pfammatter W, Jertmini M, Gessler C (2003) Mapping quantitative physiological traits in apple (*Malus* × *domestica* Borkh.). Plant Mol Biol 52:511–526

Ma B, Chen J, Zheng H, Fang T, Ogutu C, Li S, Han YP, Wu B (2015a) Comparative assessment of sugar and

malic acid composition in cultivated and wild apples. Food Chem 172:86–91
Ma B, Liao L, Zheng H, Chen J, Wu B, Ogutu C, Li S, Korban SS, Han Y (2015b) Genes encoding aluminum-activated malate transporter II and their association with fruit acidity in apple. Plant Genome 8 (3). https://doi.org/10.3835/plantgenome2015.03.0016
Ma B, Zhao S, Wu B, Wang D, Peng Q, Owiti A, Fang T, Liao L, Ogutu C, Korban SS, Li S, Han Y (2016) Construction of a high density linkage map and its application in the identification of QTLs for soluble sugar and organic acid components in apple. Tree Genet Genomes 12:1. https://doi.org/10.1007/s11295-015-0959-6
Ma QJ, Sun MH, Lu J, Liu YJ, Hu DG, Hao YJ (2017) Transcription factor AREB2 is involved in soluble sugar accumulation by activating sugar transporter and amylase genes. Plant Physiol 174:2348–2362
Ma B, Liao L, Fang T, Peng Q, Ogutu C, Zhou H, Ma F, Han YP (2019) A *Ma10* gene encoding P-type ATPase is involved in fruit organic acid accumulation in apple. Plant Biotech J 17:674–686
Maliepaard C, Alston FH, Van Arkel G, Brown LM, Chevreau E, Dunemann F, Evans KM, Gardiner S, Guilford P, van Heusden AW, Janse J, Laurens F, Lynn JR, Manganaris AG, Den Nijs APM, Periam N, Rikkerink E, Roche P, Ryder C, Sansavini S, Schmidt H, Tartarini S, Verhaegh JJ, Vrielink-Van Ginkel M, King GJ (1998) Aligning male and female linkage maps of apple (*Malus pumila* Mill.) using multi-allelic markers. Theor Appl Genet 97:60–73
Martınez-Esteso MJ, Selles-Marchart S, Lijavetzky D, Angeles Pedren M, Bru-Martınez R (2011) A DIGE-based quantitative proteomic analysis of grape berry flesh development and ripening reveals key events in sugar and organic acid metabolism. J Exp Bot 62:2521–2569
McClure KA, Gardner KM, Douglas GM, Song J, Forney CF, DeLong J, Fan L, Du L, Toivonen PMA, Somers DJ, Rajcan I, Myles S (2018) A genome-wide association study of apple quality and scab resistance. Plant Genome 11: https://doi.org/10.3835/plantgenome2017.08.0075
McClure KA, Gong Y, Song J, Vinqvist-Tymchuk M, Palmer LC, Fan L, Burgher-MacLellan K, Zhang Z, Celton JM, Forney CF, Migicovsky Z, Myles S (2019) Genome-wide association studies in apple reveal loci of large effect controlling apple polyphenols. Hortic Res 6(1):107. https://doi.org/10.1038/s41438-019-0190-y
Moriya S, Kunihisa M, Okada K, Iwanami H, Iwata H, Minamikawa MF, Katayose Y, Matsumoto T, Mori S, Sasaki H, Matsumoto T, Nishitani C, Terakami S, Yamamoto T, Abe K (2017) Identification of QTLs for flesh mealiness in apple (*Malus* × *domestica* Borkh.). Hortic J 86:159–170
Pangborn RM (1963) Relative taste intensities of selected sugars and organic acids. J Food Sci 28:726–733
Peng Q, Wang L, Ogutu C, Liu J, Liu L, Mollah MDA, Han YP (2020) Functional analysis reveals the regulatory role of *pptst1* encoding tonoplast sugar transporter in sugar accumulation of peach fruit. Int J Mol Sci 21:1112. https://doi.org/10.3390/ijms2103111
Peace CP, Bianco L, Troggio M, van de Weg E, Howard NP, Cornille A, Durel CE, Myles S, Migicovsky Z, Schaffer RJ, Costes E, Fazio G, Yamane H, van Nocker S, Gottschalk C, Costa F, Chagné D, Zhang X, Patocchi A, Gardiner SE, Hardner C, Kumar S, Laurens F, Bucher E, Main D, Jung S, Vanderzande S (2019) Apple whole genome sequences: recent advances and new prospects. Hortic Res 6:59. https://doi.org/10.1038/s41438-019-0141-7
Spartz AK, Lor VS, Ren H, Olszewski NE, Miller ND, Wu GS, Spalding EP, Gray WM (2017) Constitutive expression of *Arabidopsis SMALL AUXIN UP RNA19* (*SAUR19*) in tomato confers auxin-independent hypocotyl elongation. Plant Physiol 173:1453–1462
Strazzer P, Spelt CE, Li S, Bliek M, Federici CT, Roose ML, Koes R, Quattrocchio FM (2019) Hyperacidification of citrus fruits by a vacuolar proton-pumping P-ATPase complex. Nat Commun 10:744. https://doi.org/10.1038/s41467-019-08516
Sweetman C, Deluc LG, Cramer GR, Ford CM, Soole KL (2009) Regulation of malate metabolism in grape berry and other developing fruits. Phytochemistry 70:1329–1344
Tatum TC, Stepanovic S, Biradar DP, Rayburn AL, Korban SS (2005) Variation in nuclear DNA content in Malus species and cultivated apples. Genome 48:924–930
Teo G, Suzuki Y, Uratsu SL, Lampinen B, Ormonde N, Hu WK, DeJong TM, Dandekar AM (2006) Silencing leaf sorbitol synthesis alters long-distance partitioning and apple fruit quality. Proc Natl Acad Sci USA 103:18842–18847
Urrestarazu J, Muranty H, Denancé C, Leforestier D, Ravon E, Guyader A, Guisnel R, Feugey L, Aubourg S, Celton JM, Daccord N, Dondini L, Gregori R, Lateur M, Houben P, Ordidge M, Paprstein F, Sedlak J, Nybom H, Garkava-Gustavsson L, Troggio M, Bianco L, Velasco R, Poncet C, Théron A, Moriya S, Bink MCAM, Laurens F, Tartarini S, Durel CE (2017) Genome-wide association mapping of flowering and ripening periods in apple. Front Plant Sci 8:1923. https://doi.org/10.3389/fpls.2017.01923
Velasco R, Zharkikh A, Affourtit J, Dhingra A, Cestaro A, Kalyanaraman A, Fontana P, Bhatnagar SK, Troggio M, Pruss D, Salvi S, Pindo M, Baldi P, Castelletti S, Cavaiuolo M, Coppola G, Costa F, Cova V, Dal Ri A, Goremykin V, Komjanc M, Longhi S, Magnago P, Malacarne G, Malnoy M, Micheletti D, Moretto M, Perazzolli M, Si-Ammour A, Vezzulli S, Zini E, Eldredge G, Fitzgerald LM, Gutin N, Lanchbury J, Macalma T, Mitchell JT, Reid J, Wardell B, Kodira C, Chen ZT, Desany B, Niazi F, Palmer M, Koepke T, Jiwan D, Schaeffer S, Krishnan V, Wu CJ, Chu VT, King ST, Vick J, Tao QZ, Mraz A, Stormo A, Stormo K, Bogden R, Ederle D, Stella A,

Vecchietti A, Kater MM, Masiero S, Lasserre P, Lespinasse Y, Allan AC, Bus V, Chagné D, Crowhurst RN, Gleave AP, Lavezzo E, Fawcett JA, Proost S, Rouzé P, Sterck L, Toppo S, Lazzari B, Hellens RP, Durel CE, Gutin A, Bumgarner RE, Gardiner SE, Skolnick M, Egholm M, Van de Peer Y, Salamini F, Viola R (2010) The genome of the domesticated apple (*Malus* × *domestica* Borkh.). Nat Genet 42:833–839

Wang W, Zhou H, Ma B, Owiti A, Korban SS, Han YP (2016) Divergent evolutionary pattern of sugar transporter genes is associated with the difference in sugar accumulation between grasses and eudicots. Sci Rep 6:29153. https://doi.org/10.1038/srep29153

Wingenter K, Trentmann O, Winschuh I, Hörmiller II, Heyer AG, Reinders J, Schulz A, Geiger D, Hedrich R, Neuhaus HE (2011) A member of the mitogen-activated protein 3-kinase family is involved in the regulation of plant vacuolar glucose uptake. Plant J 68:890–900

Wu JH, Gao HY, Zhao L, Liao XJ, Chen F, Wang ZF, Hu XS (2007) Chemical compositional characterization of some apple cultivars. Food Chem 103(1):88–93

Xu K, Wang A, Brown S (2012) Genetic characterization of the Ma locus with pH and titratable acidity in apple. Mol Breed 30:899–912

Yao YX, Li M, Liu Z, You CX, Wang DM, Zhai H, Hao YJ (2009) Molecular cloning of three malic acid related genes *MdPEPC*, *MdVHA-A*, *MdcyME* and their expression analysis in apple fruits. Sci Hortic 122:404–408

Yao YX, Dong QL, You CX, Zhai H, Hao YJ (2011a) Expression analysis and functional characterization of apple *MdVHP1* gene reveals its involvement in Na+, malate and soluble sugar accumulation. Plant Physiol Biochem 49(10):1201–1208

Yao YX, Li M, Zhai H, You CX, Hao YJ (2011b) Isolation and characterization of an apple cytosolic malate dehydrogenase gene reveal its function in malate synthesis. J Plant Physiol 168:474–480

Zhang Y, Li P, Cheng L (2010) Developmental changes of carbohydrates, organic acids, amino acids, and phenolic compounds in 'Honeycrisp' apple flesh. Food Chem 123(4):1013–1018

Zhang Q, Ma B, Li H, Chang Y, Han Y, Li J, Wei G, Zhao S, Khan MA, Zhou Y, Gu C, Zhang X, Han Z, Korban SS, Li S, Han Y (2012) Identification, characterization, and utilization of genome-wide simple sequence repeats to identify a QTL for acidity in apple. BMC Genomics 13:537. https://doi.org/10.1186/1471-2164-13-537

Zhang L, Hu J, Han X, Li J, Gao Y, Richards CM, Zhang C, Tian Y, Liu G, Gul H, Wang D, Tian Y, Yang C, Meng M, Yuan G, Kang G, Wu Y, Wang K, Zhang H, Wang D, Cong P (2019) A high-quality apple genome assembly reveals the association of a retrotransposon and red fruit colour. Nat Commun 10:1494. https://doi.org/10.1038/s41467-019-09518-x

Zhen Q, Fang T, Peng Q, Liao L, Zhao L, Owiti A, Han Y (2018) Developing gene-tagged molecular markers for evaluation of genetic association of apple *SWEET* genes with fruit sugar accumulation. Hortic Res 5:14. https://doi.org/10.1038/s41438-018-0024-3

Zhu Z, Liu R, Li B, Tian S (2013) Characterization of genes encoding key enzymes involved in sugar metabolism of apple fruit in controlled atmosphere storage. Food Chem 141:3323–3328

Metabolomic Approaches for Apple Fruit Quality Improvement

15

Stefano Brizzolara, Leonardo Tenori, and Schuyler S. Korban

Abstract

Knowledge of plant metabolism is fundamental for understanding mechanisms involved in plant growth and development, as well as those related to plant responses to various abiotic and biotic stresses. Metabolomics is a scientific field that provides valuable information and contributes to our understanding of the biology of any organism, such as plants, by unravelling the metabolic profile of an organism. Essentially, metabolomics is a key analytical tool useful to study plant systems and it has already contributed to our understanding of a number of different biological processes. This valuable tool has become even more important and useful with the development of high-throughput metabolomic technologies. These tools have allowed for our global understanding of functional outcomes of metabolites in apples, including those associated with fruit quality traits. Therefore, as the amount of metabolomics data has increased, public databases have evolved in order to collect and catalog these data to aid in exploiting the fundamental knowledge of these metabolite datasets in order to unravel metabolic pathways. In this chapter, we will explore the global value of metabolomics, provide an overview of analytical tools that are commonly used in metabolic profiling, as well as assess their values in developing a better understanding of apple fruit quality traits.

15.1 Introduction

Genetic enhancement of crops is nowadays highly dependent on knowledge of the molecular mechanisms underlying plant development using various comprehensive methodologies, such as systems biology approaches, including various 'omics' approaches such as those of genomics, transcriptomics, proteomics, and metabolomics, among others. Although major efforts have been devoted to genomics, transcriptomics, and proteomics leading to significant advances in these 'omics' approaches, more recently major efforts have been directed toward metabolomics. The detection of metabolic changes in plant tissues and organs at different developmental stages has

S. Brizzolara (✉)
Institute of Life Sciences, Scuola Superiore Sant'Anna, Piazza Martiri della Libertà 33, 56127 Pisa, Italy
e-mail: stefano.brizzolara@santannapisa.it

L. Tenori
Interuniversity Consortium for Magnetic Resonance of Metalloproteins (CIRMMP), via Luigi Sacconi 6, 50019 Sesto Fiorentino, Florence, Italy

S. S. Korban
Department of Natural Resources and Environmental Sciences, University of Illinois at Urbana-Champaign, Urbana, IL 61801, USA

S. S. Korban (ed.), *The Apple Genome*, Compendium of Plant Genomes,
https://doi.org/10.1007/978-3-030-74682-7_15

contributed to new knowledge of metabolites and metabolic markers. In fact, plant metabolomics aids plant breeders in identifying biomarker metabolites that integrate the genetic background of a plant when subjected to various stress conditions, thereby rendering a selected biomarker as a valuable diagnostic metabolite for plant stress (Gauthier et al. 2015).

15.1.1 The Role of Metabolomics in Systems Biology

The comprehensive study of plant metabolism, or metabolomics, is highly valuable in identifying phenotypic alterations during plant growth and development, as well as in plant response to abiotic and biotic stresses (Brunetti et al. 2013). This will further expand our knowledge of the systems biology of plants and unravel those metabolic mechanisms involved (Bais et al. 2010). Metabolomics measures pool sizes of metabolites (small molecules of Mr $\leq$ 1,000 D) that collectively compose the metabolome of a biological sample (Hall et al. 2002). Essentially, metabolomics provides a comprehensive overview of cellular metabolites that participate in various cellular events, thus elucidating the absolute physiological state of a cell. It is estimated that there are over 200,000 different metabolites within the plant kingdom (Goodacre et al. 2004). Thus, generating the entire metabolome of a plant is not simple. However, with the rapid technological advances in recent decades, it has become possible to generate well-defined profiles of diverse plant systems and investigate alterations induced by biological responses during plant growth and development, as well as respond to varying environmental conditions. As plants grow, develop, and respond to various biotic and abiotic pressures, they undergo various changes in their metabolic profiles. As an example, plant growing conditions constantly vary, such as changes in nutrient and water availability, temperature, irradiance, and humidity, and in turn these specific conditions will influence the composition and the content of metabolites present in different plant tissues as plants adapt their metabolic profiles in response to these changes. In addition to natural heterogeneity, evolutionary influences on the environment such as atmospheric CO_2 enrichment and nitrogen deposition contribute to additional environmental variation (Stevens et al. 2004).

Phenotypic plasticity enables plants to withstand biological and environmental changes, within both short and long time scales, as these are governed by genes that not only determine the character of an organism but also levels of the response of that character to biological and environmental stimuli (Brunetti et al. 2013; Bradshaw 2006). This phenotypic plasticity is measured by a plant's ability to modify its growth (morphological traits) and function (metabolic traits) (Brunetti et al. 2013). There is a wide variation in the extent to which these metabolic traits respond to environmental pressures, ranging from slight to major shifts in metabolite turnover, such as those observed in response to changes in C/N/P/K within an environment (Rivas-Ubach et al. 2012; Brunetti et al. 2013), characterized by the activation or upregulation of highly specialized metabolic pathways (Kusano et al. 2011). Over time, plant traits have been selected through evolutionary processes of natural selection under various conditions prevalent within habitats and distribution of a given species. Thus, understanding of resultant variations in metabolic traits is critical to clarify how environmental factors influence plant growth, and how these responses are ecologically manifested (Brunetti et al. 2013). As the amount of metabolomics data has exploded, public databases have come along to collect and catalog these data to allow for exploiting knowledge of metabolite data generation, annotation, and integration along with metabolic pathway data (Bais et al. 2010; Hall et al. 2002).

15.1.2 Methodologies for Metabolome Analysis

Metabolomics depends largely on methodologies and instrumentations for collective identification, quantification, and localization of metabolites.

The two primary methodologies are dependent on the non-destructive nuclear magnetic resonance (NMR) spectroscopy and mass spectroscopy (MS) (Fig. 15.1). Large-scale analysis of highly complex mixtures can be conducted using a variety of integrated technologies and methodologies based on MS and NMR spectroscopy, including gas chromatography-mass spectroscopy (GC-MS), liquid chromatography-mass spectroscopy (LC-MS), and capillary electrophoresis-mass spectroscopy (CE-MS), as well as Fourier transform (FT) ion cyclotron resonance (ICR) MS, FT-ICR-MS, and field asymmetric waveform ion mobility spectrometry (FAIMS) (Hong et al. 2016; Kumar et al. 2017). NMR is extensively used to identify metabolites of small molecular weight (< 50 kDa), including those that have physicochemical properties of ligands, binding sites on proteins, uncovering structures of protein–ligand complexes, and direct binding of target proteins; however, this technique is of low sensitivity due to its limited coverage of low-abundance biomarkers. On the other hand, MS has a higher sensitivity than NMR, thus providing wide coverage of metabolome data, and identification of novel metabolic biomarkers. Recently, MS has achieved higher accuracy due to advances in ionization methods, such as atmospheric pressure chemical ionization (APCI), electrospray ionization (ESI), and matrix-assisted laser desorption/ionization (MALDI)-time of flight (TOF) (Kumar et al. 2017). Therefore, to enhance throughput analysis, MS is usually combined with chromatography techniques, such as in the case of GC-MS, LC-MS, CE-MS, FT-ICR, and FAIMS, as mentioned above.

For the purpose of the topic of interest in this chapter, we will focus mainly on GC-MS, LC-MS, and NMR as these are the most appropriate and commonly used methods.

The above-mentioned methodologies are accompanied by other technologies for sampling to allow for analysis of metabolites at the subcellular level, as well as those of single cells (Kueger et al. 2012; Misra et al. 2014). As each analytical platform has an intrinsic limitation, it is essential that combined approaches are used in metabolomics analysis (Hong et al. 2016). This is attributed to a plant's high chemical diversity, broad dynamic range of concentrations, and specific cellular compartmentalization of plant metabolites, thus no single analytical platform can cover the whole plant metabolome. As a consequence, different extraction techniques and a combination of complementary analytical technologies must be used to pursue a comprehensive analysis of a plant metabolome (Alseekh and Fernie 2018; Rodrigues et al. 2019). As metabolomics is downstream of other functional

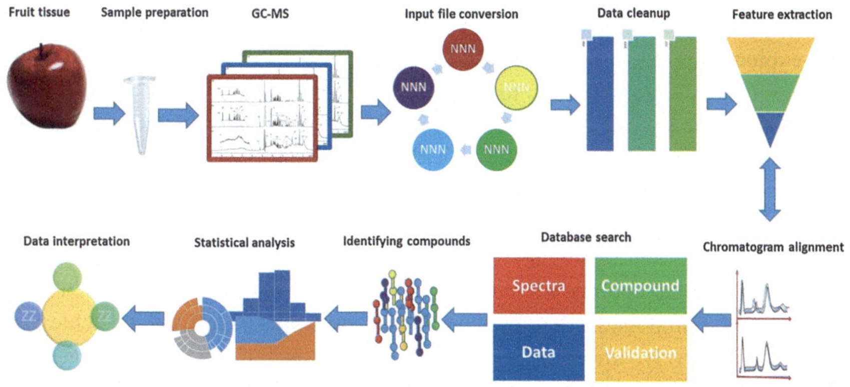

Fig. 15.1 A schematic diagram of a high-throughput data analysis of a metabolome. Modified from Kumar et al. (2017)

genomics tools, such as transcriptomics and proteomics, the metabolome of a plant species, unlike its transcriptome or proteome, cannot be inferred via known genomic information, as it is genome-independent (Hong et al. 2016; Alseekh and Fernie 2018).

It is known that metabolomics allows for collecting large amounts of information to pursue the discovery of genes and associated metabolic pathways through high-throughput corollary peak annotation (Tohge et al. 2014). Given the complicated regulatory network among small molecules in plants, including interactions among these metabolites, only metabolomics analysis can contribute to our understanding of relationships between a genotype and its metabolic outputs by unraveling such key network components (Toubiana et al. 2013). There are examples of metabolomic analysis, integrated with transcriptomic analysis, that have been successfully exploited to investigate the coordinated patterns of metabolic fluxes and metabolite concentrations in plants (Toubiana et al. 2013; Kleessen and Nikoloski 2012). Recently, high-throughput and low-cost approaches have yielded significant omics outputs within short periods of time to reconstruct metabolic models in microbial organisms (Hong et al. 2016). However, such integration of sequential multiple omics data remains challenging in plant systems, as the relationships between each of the omics tools are complex and not always linear. Nevertheless, plant metabolomics has become a powerful tool to explore various aspects of plant physiology and biology, contributing to our expanded knowledge of metabolic and molecular regulatory mechanisms regulating plant growth, development and stress responses, and genetic improvement of crop productivity and quality.

Overall, NMR-based technologies have been outperformed by MS-based approaches due to the lower number of compounds that can be resolved given its relatively lower sensitivity (Rodrigues et al. 2019). While NMR is mostly used to elucidate structures of natural compounds, quality control of medicinal plants, and pharmaceutical drug analysis, MS technologies have been used to characterize metabolomes to discern and develop networks of metabolic pathways (Gomes et al. 2018; Rodrigues et al. 2019). Although advances in MS technology are ongoing, it remains challenging to cover the plant metabolome. Currently, only a few thousand metabolites (>14,000) can be measured, when an estimation of 200,000 up to one million metabolites are expected to be present in the plant kingdom, and associated analyses are concentration-dependent (Alseekh and Fernie 2018). These analytical approaches have been widely used in plant metabolomics studies of various plant species, including tomato, rice, wheat, and maize, as well as in fruit crops for various purposes (Hong et al. 2016; Kumar et al. 2017).

15.1.3 Computational Resources for Metabolomics

A critical component of metabolomics is the availability of computational resources to build large-scale libraries of metabolites to identify and characterize structures of metabolites, as this is accomplished by matching mass spectral features of measured metabolites to spectra of reference standards (Bais et al. 2010; Frainay et al., 2018; Rodrigues et al. 2019). These MS-based plant metabolomics databases have been extensively reviewed by de Souza et al. (2017). Nevertheless, despite the availability of bioinformatic tools for metabolite annotation, there are many unknown molecules, thereby curation of any putative metabolomics information is critical to secure uniformity and quality of data available in databases (de Souza et al. 2017; Töpfer et al. 2018; Rodrigues et al. 2019). As the access to large-scale libraries of 'omics' data has expanded, the challenge of integrating genomics, transcriptomics, and proteomics information into metabolic networks is critical to achieving a better understanding of the structure and regulation of metabolic pathways (Fernie and Stitt, 2012; Tohge et al. 2015; Frainay et al., 2018; Töpfer et al. 2018). Furthermore, integration of metabolomics and physiological data to assess growth, yield, and various agronomic traits offers yet another critical role for crop improvement (Tohge et al. 2015; Hong et al. 2016; Rodrigues et al. 2019).

As metabolomics relies heavily on large datasets, it is imperative that good data management practices are applied and implemented, as this will allow for these data to be exploited for knowledge discovery and innovation, and for subsequent data and knowledge integration, as well as for reuse by the research community following data publication (Wilkinson et al. 2016). Therefore, it is critical to enhance reusability of data. As a result, a set of guiding principles, designated as the FAIR principles and consisting of **f**indability, **a**ccessibility, **i**nteroperability, and **r**eusability, have been established and endorsed by members of academia, industry, funding agencies, and scholarly publishers (Wilkinson et al. 2016). For reusability purposes, data must comply with the FAIR data principles. It is critical that deposition of metabolomics data in dedicated electronic repositories is identified and presented in a standardized format to allow for integration with data or tools obtained from non-cooperating resources; i.e., interoperable, as this would allow for discovery, sharing, and reusability of these data (Rodrigues et al. 2019; Stall et al. 2019).

Datasets generated from high-throughput analysis tools are processed using a variety of data processing platforms such as MET-COFEA, Met-Align, ChromaTOF, and MET-XAlign, among others (Kessler et al. 2014; Zhang et al. 2014, 2015; Ma et al. 2016; Misra and van der Hooft 2016). This processing includes baseline correction, alignment, separation of co-eluting peaks (deconvolution), and normalization prior to identification of compounds (Fig. 15.1). Metabolome databases, such as METLIN, NIST, GOLM, and PM, among others, can be used for identification of metabolites (Bais et al. 2010, 2011; Johnson and Lange 2015; Kumar et al. 2017). Moreover, identified metabolites data are subject to statistical analysis using correlations, principal component analysis (PCA), partial least squares (PLS), K-means clustering, boxplot, heatmap, and reconstruction of metabolic pathways, by using various web tools and softwares, such as MetaboAnalyst, Cytoscape, and Statistical analysis, to monitor and identify metabolic markers associated with agronomic traits of interest (Tsugawa et al. 2015; Xie et al. 2015; Kumar et al. 2017).

There are several web-based available repositories for depositing metabolomics data and relevant information to allow access of all these data by the research community. These include MetaboLights (Haug et al. 2013), hosted by the European Bioinformatics Institute (EMBL-EBI), and the U.S. National Institutes of Health (NIH) repository Metabolomics Workbench (Sud et al. 2016) housing datasets from various organisms. The Metabolomic Repository Bordeaux (MeRy-B) for ^{1}H-NMR metabolite profiles provides experimental and analytical metadata for plant datasets (Ferry-Dumazet et al. 2011). However, the Metabolic Phenotype Database (MetaPhen) hosts a majority of plant datasets, and it is part of the MetabolomeExpress, a web-server for GC-MS metabolomics datasets (Carroll et al. 2010). In addition to these databases, the MetabolomeXchange service, a database/management system that allows for metadata management to aggregate data from several data providers, including MetaboLights, Metabolomics Workbench, MeRy-B, and Metabolonote (Ara et al. 2015). Yet, another comprehensive and dedicated source for these data is hosted by the BioSharing Initiative (McQuilton et al. 2016).

15.1.4 Trends and Advantages of Metabolomics in Plant Improvement Efforts

The fields of genomics, transcriptomics, and proteomics have contributed to our expanded knowledge of plant growth and development, as well as their responses to biotic and abiotic stresses. As metabolites correspond to end-products of cellular regulatory processes involved in plant growth and development and in response to various stresses, they serve as phenotypic signatures of a plant system and its variants (Rodrigues et al. 2019). As numerous plant species have been sequenced due to advances in sequencing technologies, the pursuit of functional genomics studies of plant development has been made possible due to advances in metabolomics tools along with quantitative trait loci (QTL) analysis, genome-wide association studies (GWAS), and gene knock-out or knock-down technologies (Rodrigues et al. 2019). As QTLs are distributed along various regions of a chromosome, and large numbers of alleles have accumulated during the process of plant domestication, this contributes to difficulties in mapping of complex QTLs due to limitations in obtaining accurate phenotypic data. Furthermore, as a plant phenotype depends on synthesis and accumulation of a series of metabolites in specific organs, at specific developmental stages, and in response to various environmental signals, different metabolites have organ and tissue-specific characteristics (Hong et al. 2016). Interestingly, in metabolome-based QTL (mQTL) studies, primary metabolites often have high heritability, while secondary metabolite loci have higher r^2 for a metabolic phenotype (Soltis and Kliebenstein 2015). Furthermore, metabolome-based GWAS (mGWAS) has been used to understand genetic mechanisms underlying metabolic diversity, as well as their associations with complex traits in plants (Hong et al. 2016).

As a plant phenotype depends on synthesis and accumulation of a series of metabolites in specific tissues and organs, at specific developmental stages and under random environmental signals, various metabolites exhibit tissue- and organ-specific characteristics (Hong et al. 2016). As numerous biochemical components vary at different cell levels or even at subcellular levels within a plant, presence of 40 different cell types in plants renders regulation of metabolic processes via asymmetric distribution of regulatory elements highly complex and requiring high-resolution of spatially resolved plant metabolomic technology for accurate tracing of these metabolites and demonstrating their regulation (Bartels and Svatoš 2015; Etalo et al. 2015; Sumner et al. 2015). As plants are exposed to various biotic and abiotic stresses, they undergo adaption at both transcriptional and post-transcriptional levels, thereby resulting in a reconfiguration of regulatory networks to maintain homeostasis (Hong et al. 2016). Once plant receptors are induced via stress signals, plants adapt to such stress signals by inducing stress response genes, which in turn result in synthesis of specialized metabolites, particularly of secondary metabolites. Rapid qualitative and quantitative analyses of metabolic responses allow for identification of phenotypic responses to abiotic and biotic stresses, and for screening and selection of stress-tolerant plant genotypes.

15.2 Metabolomics in Fruit Crops

The tools of metabolomics have been used in various fruit crops to dissect maturity and ripening events, as well as to elucidate biochemical pathways and networks, particularly those of various valuable horticultural traits, such as fruit biology, including fruit quality (volatile compounds, antioxidants, carbohydrates, and flavonoids, among other metabolites) and fruit ripening (Kumar et al. 2017; Brizzolara et al. 2020; Moing et al. 2020).

As fruit shelf-life is highly critical for commercial viability of these high-value crops, it is a common practice that fruit crops are subjected to various environmental conditions and stresses to prolong shelf-life. Therefore, metabolome profiles of fruits in response to various cultural practice manipulations and environmental stresses, such as postharvest treatments and storage, including storage at low temperatures alone or in combination with modified atmospheres (MAs) and controlled atmospheres (CAs) have been investigated in various fruit crops, such as apple, peach, grape, and strawberry, among other tropical and subtropical fruits (Brizzolara et al. 2017, 2018, 2020; Duan et al. 2019; Vallarino et al. 2018).

15.3 Apple Metabolomics

Metabolomics has been used to assess genetic diversity for apple fruit quality traits, particularly for carbohydrates, organic acids, antioxidants, as well as for flavonoid and phenolic compounds, among others (Cuthbertson et al. 2012; De Paepe et al. 2015; Farneti et al. 2015; Eisenmann et al. 2016). As an example, metabolic profiles of 'Golden Delicious' apples have revealed that this fruit has high myo-inositol, sugars, and succinic acid contents, while 'Red Delicious' and 'Fuji' fruits have relatively higher levels of triterpene/sterols, flavonoids, phenolic acids, stearic acid, anthocyanins and carbohydrates (Cuthbertson et al. 2012; Kumar et al. 2017). Moreover, Farneti et al. (2015) have analyzed metabolic profiles of 190 apple accessions to characterize aroma composition of this large collection of apples. Other studies focused on analysis of different apple fruit tissues. As an example, it has been reported that peel tissues of 'Fuji' have higher levels of carbohydrates, including glucose and sorbitol, while peel tissues of 'Red Delicious' and 'Granny Smith' have higher levels of unsaturated fatty acids (oleic and linoleic acid) than those in other cultivars (Cuthbertson et al. 2012). Furthermore, metabolomic analysis has been conducted to assess metabolome profiles within single apple fruits (of sunlit vs. shaded areas of a specific fruit) across its width versus length for phenolic compounds, sugars, and organic acids (Cebulj et al. 2017).

Several studies have been conducted to study metabolomic profiles and modulations of apple fruits during postharvest, as well as during storage (Hatoum et al. 2014, 2016; Brizzolara et al. 2017, 2018). For instance, it has been observed that increased levels of mannose and xylose during postharvest correspond to break down of cell wall hemicellulose, thereby promoting fruit senescence (Hatoum et al. 2014).

15.3.1 The Role of Metabolomics in Postharvest Storage of Apple

During the past decade, many efforts have been undertaken for use of omics tools in fruit science, particularly for apples. This is mainly due to worldwide economic importance of apple fruit, and to its long-term storability. In fact, it is possible to store apple fruit for up to 12 months under low-temperature conditions coupled with controlled atmosphere within storage chambers.

Metabolomics can be exploited via either targeted or untargeted approaches based on the desired final outcome. In the first case, a specific class of compounds, or even a single specific molecule is targeted in order to gain specific information pertaining to such compounds/molecules. As for the second case, a whole group of untargeted compounds will be analyzed in order to obtain an overall metabolic profile of a fruit.

Apple fruit storage is an essential practice for sustaining and/or prolonging shelf-life of apple fruits by maintaining high-quality standards. There are various available storage technologies used in the commercial apple industry to prolong shelf-life of fruits during storage, and to either alleviate and/or avoid fruit storage disorders that may impact fruit quality and salability. In fact, these practices undoubtedly have major impacts on primary metabolism with effects on composition and overall flavor of apple fruit. For instance, controlled atmosphere (CA) storage is applied mainly by lowering the level of oxygen (O_2), as low as possible, while increasing the concentration, up to a certain level, of carbon dioxide (CO_2). CA storage works synergistically with refrigeration (low temperature) to maintain high-quality fruit throughout long storage periods. This affords apple producers to stay in the market for very long periods of time, and likewise allowing consumers to benefit from the availability of apples all year long.

Since early days of application of CA technology in apple storage, there has been a common trend of using a steady lowering of O_2 levels along with increasing of CO_2 concentrations within cold storage chambers. Often, the standard CA technology for storage of some horticultural crops is based on O_2 levels of about 2–3 kPa; however, over the past 15 years, dramatic changes and innovations in CA-based methods have been made, based on significant reductions in O_2 levels. However, it is important to point out that while reducing O_2 concentration and increasing CO_2 levels are effective in slowing down the overall general metabolism of stored fruits, these environmental conditions exert high levels of stress on these fruits. Therefore, it is important to keep in mind that beyond certain limits of low O_2/high CO_2 concentrations and storage duration, negative effects are incurred, and symptoms of physiological disorders can be observed. In particular, when there is a shift from aerobic to anaerobic metabolism, the concentration of oxygen should not be lower than that corresponding to the anaerobic compensation point (ACP). This dual positive versus negative effect, depending on storage time by stress level interaction, must always be taken into consideration when CA protocols are applied. This is especially critical when extreme atmospheric conditions (very low O_2 concentrations and/or high CO_2 levels) are used.

With our ever-expanding knowledge of stress physiology and stress monitoring tools, metabolic responses are accompanying technical advances, as well as applications of such advanced CA technologies. Examples of such advances include the ultra-low oxygen (ULO) technology, wherein O_2 is maintained near 1 kPa, and initial low O_2 stress (ILOS), wherein O_2 levels are maintained as low as 0.25–0.7 kPa for short time periods following fruit harvest. Furthermore, due to advances in technologies that allow for sensing fruit responses to hypoxic stress conditions, dynamic CA (DCA) represents yet an innovative step in most recent advanced CA-based storage protocols. With this new technology (currently used only on a few crops, and especially on apples), fruit is stored at oxygen atmospheres of about 0.4–0.5 kPa for as long as possible until the earliest stress symptoms/conditions are observed. Only at that point, O_2 concentration is promptly adjusted to a higher level, deemed as 'safe'. As DCA protocols dictate that reduction of O_2 levels should either reach or are very close to the lowest level tolerated by the fruit (ACP), accompanied with high risks of severe quality losses due to an anaerobic metabolism, evaluations of physiological and metabolic conditions of the fruit are critical. Therefore, the following major parameters are used to monitor metabolic responses and to identify early symptoms of stress conditions, including measurements of fruit ethanol release/production, chlorophyll fluorescence (CF), and/or respiratory quotient (RQ). If appropriately managed, DCA maintains fruit quality in terms of aroma, firmness, acidity, and also prevents the onset of the most important disorders better than other static CA protocols.

As for CO_2 levels in CA technology, most protocols suggest that these should be set around 1.5–3.0 kPa, depending on the specific commodity and O_2 levels. Higher levels are detrimental to the quality of the fruit due to cellular toxicities of the CO_2 gas. However, specific crops such as cherries, blueberries, raspberries, and strawberries respond positively to higher CO_2 levels and storage conditions based on high (>10 kPa) CO_2 atmospheres along with more limited reductions of O_2 levels are applied; whereas, high CO_2 levels should be avoided for apples.

The above-described technical advances coupled with the use of appropriate monitoring systems have contributed to better and prolonged maintenance of apple fruit quality parameters, as well as to longer marketable shelf-life. Moreover, further reductions of O_2 concentrations, characterized by DCA in comparison with traditional CA protocols, have generally contributed to higher scores for fruit crispness, acidity, firmness, and flavor, as observed in several apple cultivars.

15.4 Apple Metabolomics Using GC-MS

15.4.1 Features of the GC-MS Platform

Among target metabolites, one of the classes that have been heavily investigated due to its importance in both environmental and physiological studies is that of volatile molecules. As the plant volatilome contains all volatile organic molecules present in plant tissues, it can also include molecules that are exogenously generated as these can be adsorbed/incorporated by plant organs (e.g., pollutants and pesticides, among others). Over the last 50 years, these exogenous compounds have been widely investigated. Moreover, there are many ongoing efforts due to continuous emissions of new molecules used in agricultural practices and by various industries, as these molecules can interact with biological systems, such as those of plants, animals, and human beings. Furthermore, organoleptic profiles, including aroma and flavor, of agricultural produce are gaining more and more importance due to increased consumers' interest and awareness of their nutraceutical values.

Gas chromatography coupled with mass spectrometry, GC-MS, is the most important and powerful analytical technique used to study the volatilome. This technology allows for the separation, via gas chromatographic equipment, of gaseous mixtures of compounds that can be identified and quantified by mass spectrometry. With increased use of GC-MS, reliable and vast mass spectral and retention index libraries have been generated, and these are useful in identifying compounds without *a priori* knowledge of the molecules present in tested samples. The most commonly used MS equipment coupled with a GC component remains a simple quadrupole device. This is primarily due to the availability of a large set of compound spectra produced using this MS system that has been included in libraries, as these can be used in an untargeted manner for analyzing biological samples. The fact that GC-MS libraries have been widely shared among laboratories, together with relatively cheap and simple maintenance protocols of this equipment, has contributed to increased use of this analytical approach as it is deemed both robust and reliable.

GC-MS analysis relies on the volatility of analytes, that also have to be sufficiently thermostable so that these can be analyzed in their original states. Volatile and semi-volatile fractions of a sample analyzed using GC-MS equipment will yield a typical spectrum, also referred to as a 'signature' or a 'fingerprint'. The information present in a signature can then be used in several different ways, such as for classification of samples for food authentication purposes (Cubero-Leon et al. 2014). Volatilomes have also been used to study pathogen profiles, in particular for isolating specific compounds emitted by microorganisms of interest that can be used as marker molecules to monitor their presence in fruit and vegetables, among others. As an example, GC-MS has been used for early detection of onion postharvest diseases by volatile organic compound (VOC) profiling, and as a result 16 volatiles have been assigned as biomarkers for *Botrytis allii* and *Burkholderia cepacian* infection (Li et al. 2011). In general, GC-MS fingerprinting is more effective than other profiling techniques for detection of pathogenic bacteria present in food (Cevallos-Cevallos et al. 2011). During recent decades, more sophisticated GC-MS protocols, such as GCxGC-TOF-MS (multidimensional GC coupled with high-resolution spectra produced by a Time of Flight MS detector), have been developed and used for similar purposes (Bean et al. 2012).

As aroma and flavor are quality parameters of critical importance for food quality and consumer acceptance, GC-MS analysis has been often used for aroma profiling. Moreover, additional innovative and powerful analytical protocols have been used, such as solid phase micro extraction (SPME) fibers due to their reliability and relatively easy protocols (Merkle et al. 2015; Zhang et al. 2020). In recent years, new sampling and desorbing techniques for VOCs analysis have been developed with enhanced analytical capacities (Arbulu et al. 2013; Maragò et al. 2016). As a result, higher numbers of compound classes can be detected, along with higher sensitivity, with the desorption procedure for VOC sampling traps. Furthermore, to increase analytical resolution, multidimensional gas chromatography, either 2D-GC or GCxGC, has been explored and successfully implemented for compounds that can be isolated and identified by GC-MS. Specifically, this allows for improved instrument resolution wherein for the same run two separate columns with different stationary phases can be used in order to increase compound separation along two different dimensions. The retention times of both first and second dimensions are recorded by the detector, and the results can be plotted onto a chromatographic plane. This approach is highly useful for quali-quantification of volatile and semi-volatile compounds for a wide range of fruits, vegetables, and foods (Cordero et al. 2010; Humston et al. 2010; Risticevic et al. 2012).

As long as non-volatile compounds are turned into volatiles using different protocols, the most important of which is derivatization, GC-MS can also be used for detection and quantification of molecules that are normally non-volatile at the cellular level. However, it is important to be aware that specific reaction protocols must be carefully optimized for each metabolite in order to avoid producing artifacts. Whereas, for non-volatile molecules, a typical GC-MS metabolomic workflow for investigating polar metabolites involves the following phases: sample collection and preprocessing (e.g., cutting, seed removal, and separation of pulp and peel, among others); quenching in liquid nitrogen and crushing (samples can then be stored at −80 prior to analysis); extraction (combination of organic solvents); drying; derivatization (multistep procedure); transfer to tubes for GC-MS analysis; chromatographic separation; metabolite quali-quantification; and data analysis (Fig. 15.2).

As mentioned above, optimization of each phase of this workflow is of critical importance and must be undertaken based on the specific compound(s) of interest. For example, the extraction phase is required to maximize the yield and range of a metabolite's chemical characteristics while reducing as much as possible its enzymatic, chemical, and physical degradation (Villas-Bôas et al. 2005). Moreover, the specific solvent (or a mixture of solvents) that will be used must also be optimized for such a specific compound(s). However, as repeated extraction(s) tends to increase yield, adding of such steps to an already complex workflow will also dramatically impact reproducibility.

GC-MS metabolomics has been employed for various different investigations of biological samples, including fruits and other foodstuffs (Gorzolka et al. 2012; Beale et al. 2018; Wu et al. 2019; Adebo et al. 2020). One of the most widely used applications for which GC-MS equipment is employed is for screening of food contaminants (McGhie and Rowan 2012; Pinu 2016; López-Ruiz et al. 2019; Carraturo et al. 2020). In fact, depending on specific contaminants of interest, GC-MS is deemed as the reference technique for most environmental pollutants, such as PCBs, PAHs, and PBDEs. Typically, MS/MS system is used for such investigations. In recent years, MS/MS has highly improved efficacy of metabolomics tools, as the application of tandem mass spectrometry and the use of ion traps have enhanced the power of identifications, as well as of sensitivity and speed of analysis (Fernandes et al. 2011; Farcuh et al. 2018; Xu et al. 2018; Maoz et al. 2019; Schulz et al. 2020). The development of these new mass spectrometry tools has contributed to rapid GC separation via the use of fast-GC instruments, using narrower and shorter columns that exert optimal separation with shorter temperature ramps (Du et al. 2012; Tranchida and Mondello 2012). Given the vast heterogeneity of food pollutants, the previously discussed bi-dimensional GC (GCxGC) is particularly interesting as it allows for better separation with an improved peak capacity of sample mixtures containing compounds that are chemically very different from each other (Tranchida et al. 2011; Kalachova et al. 2012).

GC-MS analysis of food derivatized samples has also been used to study fingerprints of

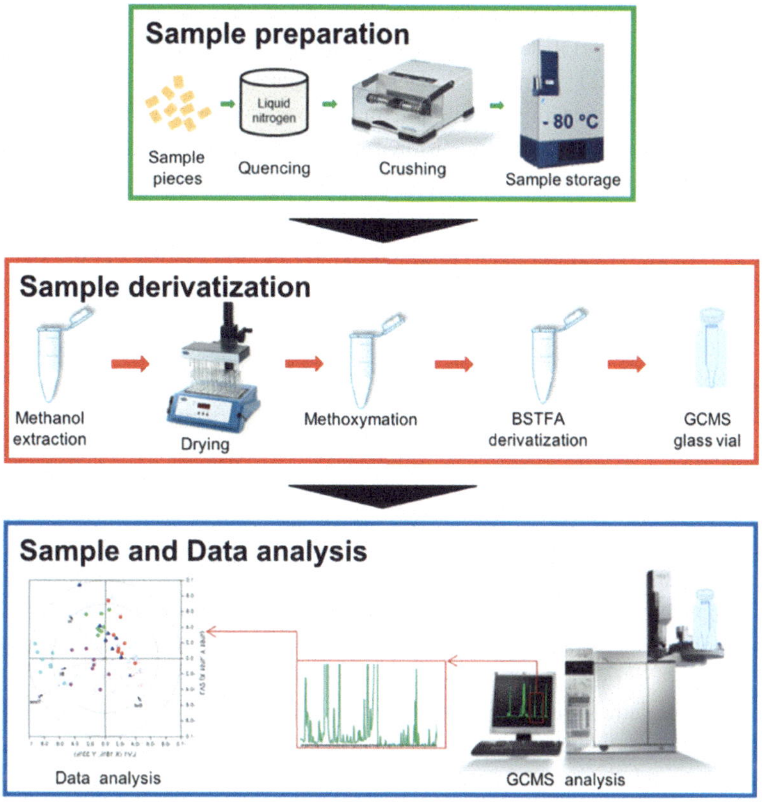

Fig. 15.2 A typical GC-MS metabolomics workflow

foodstuffs in terms of primary non-volatile metabolites, as well as (and more specifically for fruits and vegetables) for physiological studies related to several different aspects, such as metabolite (e.g., nutraceuticals) production/degradation under different growing or storage environments (Farag et al. 2014), metabolic reset of plant cells following exposure to different biotic or abiotic stresses, as well as for related physiological disorders (Brizzolara et al. 2018). Considering these latter types of physiological studies, many efforts have specifically focused on apple fruits.

15.4.2 Apple Metabolomics Using GC-MS

The apple fruit metabolome has been quite extensively investigated using GC-MS for quality attributes, fingerprinting, screening of contaminants, response to postharvest treatments, metabolic reset under different storage and growing conditions, development of storage disorders, and pathogen-related damage.

The production of VOCs by apple fruit has been widely investigated, and hundreds of compounds have been identified (Espino-Díaz et al. 2016). Although VOCs of many apple cultivars are yet to be characterized, aroma profiles of the most important commercial cultivars, such as 'Granny Smith', 'Red Delicious', 'Golden Delicious', 'Royal Gala', 'Fuji', 'Pink Lady', and 'Braeburn' have been identified (Dunemann et al. 2009; Aprea et al. 2012; Both et al. 2014; Brizzolara et al. 2017; Giannetti et al. 2017; Chitarrini et al. 2020). In addition to cultivar characterization, apple fruit VOC profiles have been also used to discriminate among samples collected from different growing conditions and geographic origins (Masi et al. 2017; Perestrelo et al. 2019). Moreover, studies have reported on apple VOCs fingerprinting to detect pathogens responsible for fruit spoilage (Cellini et al. 2016).

The production of high-resolution mass spectra, such as TOF-MS, have also been used (Aprea et al. 2011; Song et al. 1997). In other studies, a combined high-resolution mass spectrometry along with multidimensional gas chromatography has been used, whereby there is a coupling of advantages of high-resolution mass spectra to a better chromatographic separation of compounds of interest (Risticevic et al. 2012). Lastly, thermal desorption gas chromatography has also been successfully employed. As an example, Roberts and Spadafora (2020) have reported on the identification of several volatile molecules never detected before in apple fruit using this recent analytical tool for sampling and desorption of volatiles. Other authors have also reported similar results (Maragò et al. 2016; Lucht et al. 2020).

Apple volatiles have also been evaluated in parallel using both GC-MS and olfactometry techniques in order to identify the most intense odorant volatiles, and to associate these to specific sensory descriptors. Examples of such volatiles include hexanal and 2-hexenal (abundant compounds often emitted by fruits, and at high levels in apple) that have been referred to as 'green apple-like' odors (Mehinagic et al. 2006; Vanzo et al. 2013; Fuhrmann and Grosch 2002). Furthermore, also electronic nose and tongue have been coupled with GC-MS equipment to specifically evaluate the capabilities of these two innovative approaches to discriminate among flavor differences (Zhu et al. 2020). In this latter study, differences related to sourness, saltiness, and umami have been identified using E-tongue and E-nose analyses on five different apple cultivars, and specific signals from these E-tools have been associated to related volatiles coupling these results with the output of an SPME-GC-MS analysis.

Considering apple quality assessment, non-volatile compounds such as antioxidant compounds and, more in general, nutraceuticals have been used as quality markers for apple fruit. For example, apple metabolomics with an emphasis on antioxidants, such as phenolics and in particular anthocyanins, has been employed in order to isolate marker compounds, including cyanidin-3-O-galactoside, (−)-epicatechin, and (+)-catechin, among others, that are useful for selection in apple breeding programs in order to either maintain or enhance fruit quality traits (Cuthbertson et al. 2012).

Another important quality-related aspect that has been widely investigated in apple fruit is the identification and quantification of contaminants. Among these efforts, apple fruits have been screened for pesticides used in apple growing protocols, such as PCBs, PAHs, and PBDEs (Abdulra'uf and Tan 2015; Branković et al. 2019; El Hawari et al. 2019). Moreover, efforts have been undertaken to monitor compounds widely used as postharvest treatments, such as diphenylamine (DPA) that has been recently banned in several countries, in fruit tissues to prevent commercial fraud (Song et al. 2014). Furthermore, efforts have been employed to detect presence of mycotoxins, such as patulin (Cunha et al. 2009), as these are also of primary importance in the apple industry. In these efforts, GC-MS has been widely used for detection of such contaminants, as well as for assessing apple fruit safety. Furthermore, volatile markers for bacterial spoilage have also been detected using GC-MS, such as acetone, ethyl acetate, ethyl alcohol, and acetaldehyde, that in turn would allow for early detection of the presence of *Staphylococcus*, *Salmonella,* and *Shigella* bacterial infections of apple fruit (Ezhilan et al. 2018). However, screening for contaminants is often exerted via an integrated analytical approach using GC-MS complemented by LC-MS equipment in order to expand the range of detectable pesticides, as well as for more precise detection of non-volatile pesticide residues (Kim et al. 2016).

In addition, GC-MS metabolic profiling has focused on apple fruit physiology, particularly relative to non-volatile metabolites. Many efforts have been undertaken to study apple fruit development using isotopically-labeled substrates, as well as in adopting GC-MS-based metabolomics approaches to gain information on carbon allocation in sink fruit during fruit development. For example, Beshir et al. (2017)

have used such an approach to investigate apple fruit carbon intake and re-allocation by tracking labeled metabolites, including glucose, fructose, sucrose, sorbitol, pyruvate, citrate, succinate, fumarate, malate, alanine, aspartate, GABA, serine, valine, proline, epicatechin, catechin, among others, throughout different metabolic pathways. In this latter work, *in vivo* ^{13}C labeling experiments have been carried out using apple pulp tissue discs. In this study, imported molecules have been demonstrated to be involved not only in carbohydrate metabolism, but also in the production of amino acids and related proteins, as well as in the biosynthesis of secondary metabolites, such as polyphenols, during fruit development up to 149 days after full bloom (Beshir et al. 2017).

15.4.3 GS-MS Metabolomics in Apple Postharvest

In recent years, GC-MS analysis has been employed in apple fruit postharvest management. Many different treatments used to extend apple postharvest life, such as ultraviolet irradiation and antioxidant treatments, have been evaluated (Rudell and Mattheis 2009; Lee et al. 2011). Moreover, various studies have been dedicated to apple fruit storage technologies, as understanding of apple stress physiology is critical for optimizing fruit storage technologies. In addition to investigating the influence of low temperature, the main pillar of apple storage, various studies have been specifically focused on CA (low O_2 and relatively high CO_2 levels) storage, which is widely employed in the apple industry.

Studies of apple fruit metabolic responses to storage stress conditions have accompanied technical advances in CA, including those of advanced CA protocols, in particular of DCA. As mentioned above, it is critical that fruit responses to hypoxic stress conditions are constantly monitored in DCA, as fruits are stored at approximately 0.4–0.5 kPa O_2 until the earliest stress symptoms appear. When the monitored stress level exceeds a certain threshold, oxygen levels are adjusted to higher concentrations, deemed 'safe' for apple fruit.

The influence of different CA technologies on the apple fruit metabolome and volatilome has been widely investigated. In fact, the influence of these CA technologies on VOC profiles has been of particular interest to assess and better understand observed differences in flavor and in consumer acceptance. It is reported that depending on CA conditions, apple cultivar, and cultivar-specific ACP, production of volatiles can be enhanced, deemed either as a positive or a negative outcome (Thewes et al. 2017; Brizzolara et al. 2017).

On the other hand, non-volatile metabolites have also been targeted to better understand the metabolic reset of apple fruit coping with low oxygen stress. Moreover, the influence of different CA protocols has been evaluated to unravel those specific mechanisms involved in oxygen sensing, as well as those subsequent cascading events coupled with these mechanisms (Bekele et al. 2016; Cukrov et al. 2016). It is reported that specific amino and organic acids, such as alanine, gamma-aminobutyric acid (GABA), aspartate, pyruvate, lactate, succinate, as well as proline, glycine, and serine have been identified in these mechanisms, thus partly confirming a common re-arrangement of energy metabolism in apple fruit. Furthermore, a metabolism shift to alcoholic and lactic fermentation, with the accumulation of fermentation products, such as ethyl alcohol and acetaldehyde, is likely accompanied by a metabolic re-arrangement involving molecules that allow for partial storage of carbon and nitrogen reserves at the cellular level. These findings are also in agreement with other studies addressing the effects of a partial re-oxygenation phase, often used in DCA protocols when apple fruits display high stress levels (Brizzolara et al. 2019). Following such partial re-oxygenation, a quick down-regulation of fermentation metabolism and a GABA shunt are observed, thereby counter-balancing activation of these two metabolic routes, often deemed as responses to low oxygen stress. Interestingly, it is reported that during long-term CA storage, metabolic profiles of apple fruit samples are observed to undergo specific changes, wherein alanine, galactose, mannitol, sorbitol, and xylose levels are found to increase over time, while, in contrast, malate and sucrose levels are found to decrease (Hatoum et al. 2014).

Interestingly, CA technology is widely used as it positively influences fruit quality traits, such as acidity and color, while reducing the incidence of chilling injuries (e.g., superficial scald). However, given the fact that in DCA protocols oxygen levels either reach or are slightly higher than the lowest oxygen levels tolerated by fruit (ACP) the risk of severe fruit quality losses due to an anaerobic metabolism or to the development of storage disorders remains high.

Following new advances in CA technology, research efforts have focused on the metabolic causes of storage-related disorders. As knowledge of these metabolic causes would further enhance apple fruit postharvest life, as well as fruit storage protocols. For example, an integrated metabolomics approach employing both GC-MS and LC-MS technologies have allowed for the determination of specific compounds, such as GABA and several antioxidant molecules, that are likely altered in fruit tissues exhibiting soggy breakdown (soft, brown, and spongy tissue within the fruit cortex) symptoms (Leisso et al. 2015). Moreover, metabolic alterations preceding soft and superficial scald appearance have also been appropriately delineated (Rudell et al. 2009; Leisso et al. 2016). In fact, metabolic profiling has revealed specific alterations in metabolite content involving different pathways, both before and after the onset of apple soft scald, including elevated GABA, 1-hexanol, acylated steryl glycosides, and free p-coumaryl acyl esters. On the other hand, elevated levels of methanol and methyl ester have been associated with fruit peel exhibiting scald symptoms (Gapper et al. 2017). Major efforts have also been undertaken to study fruit browning symptoms and those metabolic alterations induced by browning, a commercially important disorder that is sometimes observed during CA storage (Hatoum et al. 2014, 2016). It is reported that levels of several compounds are likely to increase in injured brown fruit tissues, such as alanine, galactose, mannitol, sorbitol, xylose, while other compounds, such as malate and sucrose, are likely to decrease. Moreover, the radial distribution of different metabolites has also been observed, with levels of some compounds, such as galactose and mannitol, decreasing, while levels of other compounds, such as mannose and sucrose, increasing, moving from the outer to the inner apple cortex (Hatoum et al. 2014).

In order to prolong postharvest life and avoid several storage-related fruit disorders, apple fruits are often treated with the ethylene inhibitor 1-methylcyclopropene (1-MCP) prior to storage. The metabolic effects of this chemical compound have been widely investigated. It is reported that 1-MCP influences both volatile and non-volatile compounds in apple fruits (Bekele et al. 2015; Lee et al. 2012a, b, 2017). 1-MCP reduces levels of 'off-flavor' volatile compounds, such as 1-butanol and butyl butanoate, while it promotes accumulation of non-volatile metabolites, including alanine and succinate in apple fruits (Bekele et al. 2015; Lee et al. 2012a, b, 2017).

Interestingly, GC-MS metabolomics has been used to characterize apple cider and juice aroma in order to develop fingerprints useful for identifying the geographical origin(s) of the processed apple fruit (Maragò et al. 2016; Medina et al. 2019; Sousa et al. 2020). Moreover, GC-MS profiling has also been used to optimize apple processing protocols. For example, effects of the cultivar, fruit ripening stage at harvest, and yeast strain used in cider processing have been investigated on apple cider composition and VOCs profiles (Nicolini et al. 2018; Rosend et al. 2019). In addition, GC-MS analysis is used to assess the effects of storage on alterations of apple juice quality (Kebede et al. 2020). Thus, identification of shelf-life markers is of critical importance for processed apple products too. In fact, specific volatile compounds have been identified as potential shelf-life markers for apple juice storage, including hexanal, trans-2-hexenal, ethyl acetate, and 1-pentanol, and these compounds are useful for early detection of product quality decay (Kebede et al. 2020). The above studies are critical in maintaining high quality of processed apple, as these products are highly valued in commercial markets.

15.5 Metabolomics Using LC-MS

15.5.1 Features of the LC-MS Platform

As mentioned earlier, LC-MS is often used in plant metabolic studies, and it is commonly used as a complementary analytical platform to the above described GC-MS equipment. In LC-MS, metabolites are separated based on their varying chemical properties, thereby requiring selection of an appropriate column for the separation of these metabolites. These columns could include reversed-phase (RP), porous graphitic carbon (PGC), anion exchange (AEC), and hydrophilic interaction chromatography (HILIC) for target analysis of polar metabolites (Jorge et al. 2016a, b). Nevertheless, most LC-MS applications capitalize on RP stationary phases, such as C_{18} functional groups covalently bonded to the surface of a silica particle, and mobile phases composed of aqueous/organic solvent mixtures, such as water/acetonitrile or water/methanol. Although polar analytes, including most primary metabolites, such as carbohydrates, glycolytic intermediates, and sugar phosphates, have minor interactions with the apolar C_{18} stationary phase, they elute close to the void volume with no chromatographic retention (Rodrigues et al. 2019). Therefore, alternative columns, including HILIC, PGC, and AEC are then used for analysis of a wide variety of polar metabolites commonly detected in plant metabolomes (Jorge et al. 2016a,b; Rodrigues et al. 2019).

In LC-MS, the most commonly employed ionization method is that of electrospray ionization (ESI), a soft-ionization technique that introduces little internal energy to result in little fragmentation, thereby yielding low structural information (Jorge et al. 2016b). However, this can be overcome by undertaking collision-induced dissociation experiments, usually by coupling ESI with tandem MS instruments (ESI-MS/MS) to allow two (MS/MS) or multiple (MSn) sequential stages of mass spectrometric analysis (Jorge et al. 2016b; Rodrigues et al. 2019). There are two types of tandem MSn instruments used in routine LC-MS metabolite analysis. These are tandem-in-time and tandem-in-space (Rodrigues et al. 2019). The most common tandem-in-time instruments used are ion-trap mass spectrometers, such as the linear ion trap (LTQ)-MS, orbitrap, and Fourier transform ion cyclotron resonance (FT-ICR)-MS, while tandem-in-space instruments include triple quadrupoles (QqQ)-MS and quadrupole time-of-flight (qTOF)-MS (Jorge et al. 2016b; Rodrigues et al. 2019). As ion-trap MS instruments can perform MSn sequential stages, they are deemed well-suited for elucidating structures of metabolites, particularly of unknown plant metabolites. Whereas, tandem-in-space MS instruments consist of two mass analyzers separated by a collision cell; thus, they are used in an MS/MS tandem mode through spatial separation of the precursor and its products (Jorge et al. 2016b; Rodrigues et al. 2019).

Often, LC-QqQ-MS/MS methods are used in quantitative targeted approaches for absolute quantification of low-abundance metabolites, as QqQ-MS technology allows for highly sensitive and selective detection of these metabolites in selected reaction monitoring (SRM) experiments. This will allow for monitoring of specific precursor-to-product ion transitions for each metabolite. In this technology, the first quadrupole (Q1) specifically selects the mass of a molecule, m/z value, of a target precursor ion, while in the second quadrupole (Q2), or collision cell, product ions are generated as a result of a collision of the precursor ion with an inert collision gas (such as Ar, N_2, or He), and those resultant product ions are then transferred into a third quadrupole (Q3) whereby ions of m/z values corresponding to one or more structure-characteristic product ions are filtered (Jorge et al. 2016b; Rodrigues et al. 2019). As this technology reduces background chemical noise, it significantly increases selectivity and limits detection and quantification, and it has been often used to quantify phytohormones and other important metabolites in plant tissues (Rodrigues et al. 2019).

15.5.2 LC-MS Analysis of Various Apple Fruit Traits

As the apple fruit is a rich source of phytochemicals, including phenolic compounds, there has been a great interest in investigating the metabolic profiles of these phenolic compounds. Some of the major phenolic compounds isolated and identified from apples include procyanidins, anthocyanins, chlorogenic acid, hydroxycinnamic acid, flavan-3-ols, such as (–)-epicatechin, (+)-catechin, gallaocatechin, phloridzin, and quercetin glycosides (Khan et al. 2012). Phenolic contents of apples are indeed highly variable among cultivars (Francini and Sebastiani 2013), but they are also highly variable within tissues of a single fruit (peel vs. flesh) (Kim et al. 2019a, b, 2020). For example, total phenolic contents have been found to vary widely among 12 apple cultivars in both peel and flesh tissues—ranging between 1157 and 5119 μg/g in peel and 423 and 1534 μg/g in flesh (Kim et al. 2019a, b). Furthermore, while quercetin-3-*O*-galactoside, quercetin-3-*O*-glucoside, catechin, phloridzin, and cyanidin-3-*O*-galactoside are the predominant phenolics in apple peel (Tsao et al. 2003), it is chlorogenic acid, epicatechin, phloridzin, and protocatechuic acid that are the predominant phenolics in apple flesh tissues (Veberic et al. 2005). Furthermore, these observed differences in chemical compositions in apple fruits are undoubtedly influenced by genetic effects, but these chemical compounds are also influenced by environmental factors during the growing season, along with various agricultural practices and postharvest practices, such as the application of 1-MCP, an ethylene antagonist, and CA storage conditions (Kim et al. 2020).

Beshir et al. (2019) using a non-aqueous fractionation (NAF) method, a promising approach that allows for studying a broader range of metabolite pool sizes at the subcellular level by combining LC-MS and GC-MS-based metabolomics, have investigated subcellular compartmentalization of metabolites during apple fruit (cv. 'Braeburn') development, particularly in three main cellular compartments, namely cytosol, plastids, and vacuoles. It is observed that most of the sugars and organic acids are predominantly located in the vacuole, while some of the amino acids are distributed between the cytosol and the vacuole.

In another study, Zupan et al. (2016) have used LC-MS to determine phenolic content, as well as sugar levels in the fruit of three apple cultivars, including 'Delicious', 'Gloster', and 'Fuji', that are susceptible to the development of the physiological disorder of watercore, wherein the flesh has a water-soaked appearance due to development of fluid-filled intercellular spaces. It is observed that total phenolic content is higher in flesh of fruit without watercore, attributed mainly due to higher individual and total flavanol contents. Whereas, total and individual dihydrochalcone contents are higher (2.4-fold) in watercore flesh than that without watercore. Furthermore, it is observed that peroxidase activity is higher in watercore flesh due to oxidative stress. On the other hand, sugar content analysis has revealed that sorbitol content is higher (4-fold) and fructose content is lower (0.7- to 0.85-fold) in flesh with watercore than in flesh without this disorder. Furthermore, glucose content is found to be significantly higher in flesh without watercore in cvs. 'Delicious' and 'Fuji', while sucrose content is not different between these two tissues; whereas, 'Gloster' apple has a higher sucrose content.

Seo et al. (2015) used an isotope dilution (ID)-LC/MS/MS as a higher-order reference method for accurate value-assignment of certified reference materials (CRMs) of mycotoxins, such as patulin produced by mold in apple fruits, among other fruits, which significantly impact quality and storability, as well as salability of these fruits.

15.5.3 LC-MS Analysis of Apple Fruit Response to Postharvest Storage

It is well known that apple fruit storage technologies can extend commercial life for several months while having major impacts on primary metabolism resulting in changes in the composition of these primary metabolites, thereby influencing the overall flavor of apple fruit, as well as marketability and consumer acceptance. By reducing the general metabolism, cold storage protocols are applied to delay fruit ripening and senescence processes by controlling the respiration rate, as low temperature reduces catalytic activities of different enzymes, including those involved in various steps of respiration, and measured using a Q10 temperature coefficient corresponding to the rate of a reaction due to a 10 °C temperature increase (Brizzolara et al. 2020). In general, Q10 values associated with respiration range between 2.0 and 3.0; therefore, when lowering storage temperature from 20 to 10 °C, respiration is decreased by a factor of 2–3, thereby prolonging fruit shelf- or commercial-life.

Among physiological disorders of apple fruit that are impacted by storage conditions, bitter pit is characterized by early symptoms that appear as water-soaked spots caused by plasma membrane breakdown, but subsequently, these turn brown in color, and over time these become desiccated. This disorder has been attributed to calcium (Ca^{2+}) deficiency in apple fruit and it is initiated during preharvest, while symptoms normally develop further along during storage (Zupan et al. 2013). Pits are mostly located in the outer cortex of the fruit and underneath the peel. Pitting often results in small depressions in the peel that can extend deeper into the flesh. Bitter pit is one of the most important physiological disorders as affected fruits are highly undesirable, often culled and diverted for processing. Although bitter pit incidence can be reduced by calcium sprays, it remains a problem. Early studies have reported that resistance to bitter pit is a genetic trait and controlled by two major genes (Korban and Swiader 1984). Later studies have indicated that there are significant genotype × environment interactions for some genetic populations (Volz et al. 2006). Moreover, fruit loss during storage (within one or two months) due to bitter pit can be high, up to almost 50% losses, in apples with no calcium treatments against bitter pit. Zupan et al. (2013) have used LC-MS to assess phenolic content in healthy and bitter pit spots of both peel and flesh tissues of three apple cultivars, including 'Golden Delicious', 'Jonagored', and 'Pinova' stored at temperatures of 0.5 °C (for 'Golden Delicious') or 1.5 °C to 2 °C (for 'Jonagored' and 'Pinova') for a period of three months. They have observed that of 15 phenolic compounds analyzed, flesh tissues with bitter pit have higher levels than those of healthy flesh tissues. Furthermore, it is reported that chlorogenic acid and catechin are five-fold higher in bitter pit flesh tissues. However, hydroxycinnamic acids and flavanols in the peel of bitter pit tissues are higher, while flavonols and anthocyanins are higher in healthy peel tissues. Moreover, anthocyanins in a healthy peel of 'Jonagored' are ten-fold higher than in bitter pit peel tissues.

15.6 Metabolomics Using Nuclear Magnetic Resonance (NMR) Spectroscopy

15.6.1 Features of NMR Spectroscopy

NMR spectroscopy is an analytical technique that exploits magnetic properties of certain atomic nuclei. When placed in a magnetic field, nuclei with non-zero nuclear spin (such as ^{1}H, ^{13}C, and ^{31}P) absorbs electromagnetic radiation at characteristic frequencies (resonance frequencies) that can be revealed as signals in NMR spectra. The resonance frequency and the corresponding intensity of each signal are dependent, respectively, on the chemical environment where that particular nucleus is located (i.e., molecular structure) and on the concentration of that molecule (Calabrò et al. 2014).

NMR spectroscopy is a highly reproducible spectroscopic tool capable of providing detailed information on the molecular structure of both pure compounds and complex mixtures (Claridge 2008). In a typical biological fluid, all hydrogen-containing molecules will yield a ^{1}H-NMR spectrum as long as they are present at concentrations above the detection limit. The NMR spectrum of a biological fluid is therefore the superposition of spectra of hundreds of different small molecules present in a sample at concentrations >1 μM (Wishart et al. 2007). One of the advantages of NMR is that the biological fluid requires only minimal preparation prior to analysis. Further, the development of high-resolution (HR) ^{1}H magic angle spinning (MAS) spectra has allowed for viable acquisition of data using small-sized tissues without any pretreatment (Cheng et al. 1998; Tomlins et al. 1998; Garrod et al. 1999; Cacciatore et al. 2013).

One-dimensional ^{1}H NMR spectra are particularly useful for high-throughput metabolomic studies (Emwas et al. 2019; Vignoli et al. 2019). This is due to the fact that the 1D ^{1}H NMR technique is highly automatable, very reliable, and very fast (Takis et al. 2019). Indeed, 1D ^{1}H NMR acquisition times for a single spectrum are often in the order of a few minutes. Furthermore, the chemical information contained in a single 1D ^{1}H NMR spectrum of a biological mixture or tissue extract is often sufficient to identify and quantify 50–100 metabolites at a time (Vignoli et al. 2019). This identification process is highly supported by the fact that many reference ^{1}H NMR spectra from hundreds of known metabolites have been compiled and stored in a number of public databases (Wishart et al. 2018). However, a major disadvantage of NMR is its relatively low sensitivity, thus limiting (for ^{1}H NMR spectra acquired at 600 MHz) detection to molecules in the order of the micromolar range (Pan and Raftery 2007). Another disadvantage of NMR is the difficulty in identifying all metabolites in samples, as ^{1}H-NMR spectra of biological fluids are very complex, thus likely requiring additional experiments to assign previously unidentified molecules in complex mixtures.

As agricultural practices, postharvest treatments, and food processing incur significant effects on the metabolic composition of resulting foods (Santucci et al. 2015a, b; Tenori et al. 2018), NMR-based metabolomics has been successfully employed in monitoring the effects of different manufacturing procedures on food quality, safety, and traceability (Meoni et al. 2020).

15.6.2 Apple Metabolomics Using NMR

Eisenmann et al. (2016) investigated the metabolic diversity of apple cultivars, apple pulp, and peel extracts of 14 different popular and new apple cultivars, using an untargeted NMR metabolomic approach. Multivariate analyses performed on NMR spectra of pulp and peel extracts were able to unequivocally identify all cultivars, with peel extracts showing a higher discriminative power. Furthermore, it was observed that while sugar levels among cultivars varied within a limited range, levels of both polyphenols and acids were widely variable among these cultivars (Eisenmann et al. 2016).

As mentioned above (Sect. 15.6.1), NMR coupled with HR-MAS probe has allowed for analysis of semisolid materials. This approach provides direct access to the chemical composition of food matrices and without the necessity for sample preparation by extraction protocols, as these protocols could potentially modify or change the metabolic composition. Vermathen et al. (2011) have utilized this approach to identify and characterize metabolic profiles of different apple cultivars.

One of the important benefits of metabolite profiling of various apple cultivars is the ability to better assess and fine-tune flavors in apple-processed products. For example, apple cider is a major beverage. Cider is usually prepared from a blend of different apple cultivars possessing various characteristics, including higher acidity and phenolic content, compared to fresh-eating apple, thus contributing peculiar flavoring to the final product (Del Campo et al. 2005). The proportion of each cultivar in a blend is selected to achieve balanced ratios of phenolic compounds, sugars, and acids required for a particular taste of the product. Therefore, to produce a suitable cider blend of homogeneous and consistent quality, it is important to know of and characterize those physicochemical features of the contributing different apple cultivars using advanced molecular techniques (Del Campo et al. 2005).

For the above reasons, Del Campo et al. (2006) used NMR-based metabolomics for characterization and chemometric differentiation of 52 apple juices derived from six apple cultivars. A discriminant analysis was conducted for 40 samples in one year, resulting in 100% accuracy of classification of these apple juice samples. The most discriminant features corresponded to polyphenols, epicatechin, phloridzin-phloretin, p-coumaric, chlorogenic, and malic acids. More strikingly, this classification model was used for 12 samples of apples harvested in two subsequent years, and the prediction ability resulted in 91.7% accuracy, thus demonstrating the robustness of this approach, independently of the year of apple harvest (Del Campo et al. 2006).

A more integrated approach was attempted in a similar study, wherein 35 cider apple juices obtained from five different apple cultivars were analyzed (Del Campo et al. 2005). A combination of techniques, including NMR, HPLC, and enzymatic assays, was used to identify discriminant features, including such physicochemical parameters as sugars, acids, and phenolic compounds. It was found that titratable acidity, malic acid, pH, (−)-epicatechin, and glucose, as well as (−)-epicatechin/chlorogenic acid and glucose/fructose ratios were critical parameters, and these had the highest discriminating power in identifying these cider apple juices (Del Campo et al. 2005).

The metabolic composition of any agricultural product is influenced by several cultural and environmental factors, including growing conditions such as the chemical composition of soil, irrigation, temperature, and altitude, among others. All these factors contribute to the phenotype of a cultivar, and in turn, this contributes to a specific metabolic fingerprint, which indirectly encodes geographical information about the cultivation site. This property is widely employed to assess the origin of such processed products as wine (Pereira et al. 2005; Caruso et al. 2012) and olive oil (Sacchi et al. 1998; Mannina et al. 2009; Rongal et al. 2017; Girelli et al. 2018). The same approach, *mutatis mutandis,* has been applied to apple fruit. For example, ^{1}H NMR-based metabolic profiling has been used to characterize five apple cultivars grown either in Japan or New Zealand (Tomita et al. 2015). Using multivariate statistical analysis, it is observed that there is a distinct separation of spectra according to the two geographical apple production regions. The major contributions to this discrimination are attributed to sugar signals, including those of glucose, fructose, and sucrose. However, minor metabolites, such as 2-methylmalate, aspartic acid, and L-rhamnitol (the latter identified integrating data from LC/MS and 2D NMR analyses) have also contributed to the geographical phenotype (Tomita et al. 2015).

The study of plant physiology and diseases is yet another important field of application of NMR metabolomics. For example, NMR metabolomics has been used to analyze for metabolite characters of apple fruits with browning disorders (Vandendriessche et al. 2013).

15.6.3 NMR Analysis of Apple Fruit Response to Postharvest Storage

One of the important uses of NMR analysis is that of apple fruit response to postharvest storage conditions. In particular, sample form along with its storage and preparation methods serve as critical steps in obtaining robust and reproducible data of metabolite status in apple fruits.

It is well known that different procedures for collection, handling, and storage result in different molecular profiles of samples; consequently, non-optimized procedures could severely bias analytical data (Bernini et al. 2011; Ghini et al. 2019). In particular, for complex matrices such as fruit and vegetable extracts, it is very important to account for effects induced by preanalytical manipulations. For example, Santucci et al. (2015a, b) have evaluated the quality of NMR spectra of apple pulp extracts (juice), obtained from fresh and frozen biological materials, to develop a robust protocol for analysis of apple fruit samples (cortex tissue). This study has been undertaken to assess the effects of specific postharvest treatments on apple fruit, with an effort to avoid likely bias and any confounding source of variance (Santucci et al. 2015a, b). Thus, the following two protocols for postharvest treatment have been evaluated: (1) frozen pulp tissue is thawed, homogenized into a puree, placed in a 0.2 μm nitrocellulose filter, centrifuged to separate debris from the liquid phase, and 700 μl of this liquid phase is collected in 1.5 ml Eppendorf tubes; and (2) frozen juice is thawed, and 700 μl are collected in 1.5 ml Eppendorf tubes. Both sets of samples are then subjected to NMR analysis. Following analysis, multivariate and univariate techniques are employed to evaluate differences in NMR spectra resulting from the two sets of samples. Although both protocols have resulted in high-quality NMR spectra, these spectra have allowed for distinguishing samples obtained from these two preparations. Specifically, some organic molecules, particularly organic acids, are detected only in fresh apple pulp, and these samples are also characterized by low inter-sample variance. Thus, it is suggested that use of samples from fresh fruit tissues is best for monitoring metabolic changes due to specific growth or storage conditions (Santucci et al. 2015a, b).

Using an integrated approach to investigate responses of apple fruit to hypoxic storage conditions, Cukrov et al. (2016) compared NMR and RNA-seq profiles of 'Granny Smith', harvested at commercial ripening time, and kept at 1 °C under either normoxic, used as control, or two hypoxic, 0.4 and 0.8 kPa oxygen, treatment conditions for up to 60 days. It was observed that fruit maintained under extremely low oxygen conditions underwent marked changes in overall metabolism, with some metabolic processes initiated at the earliest stages of exposure to hypoxia. In particular, apple fruit cortex tissues were found to be highly sensitive even to subtle differences in oxygen concentrations close to ACP. This was also reflected in ethanol concentrations, expression of hypoxia marker genes, such as those encoding alanine aminotransferase (AlaAT), β-amylase (β-amy), sucrose synthase (SuSy), and phosphofructokinase (PFK), along with detection of more than 1000 differentially expressed genes that were highly responsive to small changes in oxygen concentrations (Cukrov et al. 2016). Interestingly, it was observed that fruit maintained at the lowest oxygen concentration (0.4 kPa) induced higher expression of transcription factors (TFs) belonging to auxin/indole-3-acetic acid (AUX/IAA), WRKY, homeobox (HB), zinc-finger families, while fruit maintained under 0.8 kPa oxygen induced higher levels of expression of MADS box family genes. Moreover, it is suggested that ethylene-responsive factors (ERFs) are also involved in modulating hypoxia-dependent gene expression.

Although the above findings are informative, they cannot be extended to other apple cultivars, as fruit responses to storage conditions under low oxygen conditions depend on the genetic background of these cultivars. Therefore, Brizzolara et al. (2017) have investigated metabolic responses of fruits of cvs. 'Granny Smith' and 'Red Delicious' to hypoxia. Fruits have been stored under two different low -oxygen protocols (ULO, at 0.9 kPa O_2 and DCA based on chlorophyll fluorescence [CF], DCA-CF, between 0.2 and 0.55 kPa O_2) for up to 214 days of storage. Subsequently, using an integrated metabolomics approach, consisting of ^{1}H NMR, GC–MS, and headspace (HS)-SPME-GC–MS analyses, a total of 130 metabolites have been quantified. It is observed that most of these metabolites (117) are shared by both cultivars, while a total of 95 metabolites are found to be significantly different between these two cultivars; however, 13 volatile compounds, identified via HS-SPME-GC–MS, are found to be specific for either 'Granny Smith' or 'Red Delicious'. After four months of storage, it is observed that 'Red Delicious' must have undergone higher marked metabolic changes than 'Granny Smith', and this is likely to be attributed to differences in ripening periods (Brizzolara et al. 2017). Based on analytical data related to pyruvate-derived metabolites, including alanine, acetaldehyde, ethanol, and lactate, it is suggested that there are different major metabolic reconfiguration strategies in each of 'Granny Smith' and 'Red Delicious' in addressing energy deficiencies due to hypoxia. While 'Granny Smith' has demonstrated more distinct responses in nitrogen metabolism along with limited induction of ethanol fermentation, 'Red Delicious' has exhibited higher induction of ethanol fermentation (Brizzolara et al. 2017).

Thus, NMR-based metabolomics is an efficient and increasingly utilized tool to characterize apple fruit for various traits of interest, as well as for a wide variety of breeding and/or production purposes. However, the full potential of this tool is yet to be fully exploited.

15.7 Conclusions and Future Efforts

It is clear that the field of metabolomics has allowed for pursuing studies of changes in the chemical composition of apple cultivars, mutants, or genetic populations, as well as in responses of apple fruit to various treatments, including CA storage conditions and their impact on metabolomics changes in apple fruit tissues. Those major high-throughput metabolomics technology platforms, including GS-MS, LC-MS, LC-MS/MS, and ^{1}H NMR have proven to yield valuable information and knowledge of various apple fruit traits of interest. In particular, metabolomics has been critical in identifying genomic regions associated with metabolite traits of valuable fruit quality traits, including those of VOCs. Moreover, metabolomics has been demonstrated to be a highly powerful tool in characterizing the chemical behavior of fruit under storage conditions, in particular to modern CA fruit storage conditions. This has significantly contributed to our expanded knowledge of the mechanisms involved in responses of fruit tissues subjected to the stress of storage conditions, including those of low temperatures and low oxygen levels, among others. Moreover, metabolomics can be exploited via either targeted or untargeted approaches based on the desired final outcome.

High-throughput metabolomics is a critical tool in pursuing biochemical phenotyping of genetic resources for collection, maintenance, or breeding programs. Our understanding of fruit growth, ripening, and postharvest is particularly critical in pursuing fast-breeding approaches in apple breeding programs. Furthermore, metabolomics is highly useful in expanding our knowledge of apple fruit responses to biotic and abiotic stresses. As apple fruit, compared with other fruits, are complex organs consisting of different tissues, thus the distribution of metabolites and bioactive compounds throughout these different fruit tissues, such as pericarp, cortex, and seeds, can influence fruit quality, and its sensory, nutritional, and associated health qualities.

References

Abdulra'uf LB, Tan GH (2015) Chemometric approach to the optimization of HS-SPME/GC–MS for the determination of multiclass pesticide residues in fruits and vegetables. Food Chem 177:267–273

Adebo OA, Oyeyinka SA, Adebiyi JA, Feng X, Wilkin JD, Kewuyemi YO, Abrahams AM, Tugizimana F (2020) Application of gas chromatography–mass spectrometry (GC-MS)-based metabolomics for the study of fermented cereal and legume foods: A review. Int J Food Sci Technol. https://doi.org/10.1111/ijfs.14794

Alseekh S, Fernie AR (2018) Metabolomics 20 years on: what have we learned and what hurdles remain? Plant J 94:933–942

Aprea E, Gika H, Carlin S, Theodoridis G, Vrhovsek U, Mattivi F (2011) Metabolite profiling on apple volatile content based on solid phase microextraction and gas-chromatography time of flight mass spectrometry. J Chromatogr A 1218(28):4517–4524

Aprea E, Corollaro ML, Betta E, Endrizzi I, Demattè ML, Biasioli F, Gasperi F (2012) Sensory and instrumental profiling of 18 apple cultivars to investigate the relation between perceived quality and odour and flavour. Food Res Int 49(2):677–686

Ara T, Enomoto M, Arita M, Ikeda C, Kera K, Yamada M, Nishioka T, Ikeda T, Nihei Y, Shibata D, Kanaya S, Sakurai N (2015) Metabolonote: a wiki-based database for managing hierarchical metadata of metabolome analyses. Front Bioeng Biotechnol 3:38. https://doi.org/10.3389/fbioe.2015.00038

Arbulu M, Sampedro MC, Sanchez-Ortega A, Gómez-Caballero A, Unceta N, Goicolea MA, Barrio RJ (2013) Characterisation of the flavor profile from Graciano *Vitis vinifera* wine variety by a novel dual stir bar sorptive extraction methodology coupled to thermal desorption and gas chromatography–mass spectrometry. Anal Chim Acta 777:41–48

Bais P, Moon-Quanbeck, SM, Nikolau BJ, Dickerson JA (2011) Plantmetabolomics.org: mass spectrometry-based Arabidopsis metabolomics database and tools update. Nucleic Acids Res 40:D1216–D1220

Bais P, Moon SM, He K, Leitao R, Dreher K, Walk T, Sucaet Y, Barkan L, Wohlgemuth G, Roth MR, Wurtele ES, Dixon P, Fiehn O, Lange BM, Shulaev V, Sumner LW, Welti R, Nikolau BJ, Rhee SY, Dickerson JA (2010) PlantMetabolomics.org: a web portal for plant metabolomics experiments. Plant Physiol 152:1807–1816

Bartels B, Svatoš A (2015) Spatially resolved in vivoplant metabolomics by laser ablation-based mass spectrometry imaging (MSI) techniques: LDI-MSI and LAESI. Front Plant Sci 6:471. https://doi.org/10.3389/fpls.2015.00471

Beale DJ, Pinu FR, Kouremenos KA, Poojary MM, Narayana VK, Boughton BA, Kanojia K, Dayalan S, Jones OAH, Dias DA (2018) Review of recent developments in GC–MS approaches to metabolomics-based research. Metabolomics 14 (11):152. https://doi.org/10.1007/s11306-018-1449-2

Bean HD, Dimandja JMD, Hill JE (2012) Bacterial volatile discovery using solid phase microextraction and comprehensive two-dimensional gas chromatography-time-of-flight mass spectrometry. J Chromatogr B 901:41–46

Bekele EA, Beshir WF, Hertog M, Nicolai BM, Geeraerd AH (2015) Metabolic profiling reveals ethylene mediated metabolic changes and a coordinated adaptive mechanism of 'Jonagold' apple to low oxygen stress. Physiol Plant 155:232–247. https://doi.org/10.1111/ppl.12351

Bekele EA, Ampofo-Asiama J, Alis RR, Hertog M, Nicolai BM, Geeraerd AH (2016) Dynamics of metabolic adaptation during initiation of controlled atmosphere storage of 'Jonagold' apple: effects of storage gas concentrations and conditioning. Postharv Biol Technol 117:9–20. https://doi.org/10.1016/j.postharvbio.2016.02.003

Bernini P, Bertini I, Luchinat C, Nincheri P, Staderini S, Turano P (2011) Standard operating procedures for pre-analytical handling of blood and urine for metabolomic studies and biobanks. J Biomol NMR 49(3–4):231–243. https://doi.org/10.1007/s10858-011-9489-1

Beshir WF, Mbong V, Hertog ML, Geeraerd AH, Van den Ende W, Nicolaï BM (2017) Dynamic labeling reveals temporal changes in carbon re-allocation within the central metabolism of developing apple fruit. Front Plant Sci 8:1785. https://doi.org/10.3389/fpls.2017.01785

Beshir WF, Tohge T, Watanabe M, Hertog MLATM, Hoefgen R, Fernie AR, Nicolaï BM (2019) Non-aqueous fractionation revealed changing subcellular metabolite distribution during apple fruit development. Hortic Res 6:98. https://doi.org/10.1038/s41438-019-0178-7

Both V, Brackmann A, Thewes FR, de Freitas Ferreira D, Wagner R (2014) Effect of storage under extremely low oxygen on the volatile composition of 'Royal Gala' apples. Food Chem 156:50–57

Bradshaw A (2006) Unravelling phenotypic plasticity – why should we bother? New Phytol 170: 644-648

Branković M, Anđelković D, Kocić H, Kocić G (2019) Assessment of GC–MS response of selected pesticides in apple matrices related to matrix concentration. J Environ Sci Health B 54(5):376–386

Brizzolara S, Santucci C, Tenori L, Hertog M, Nicolai B, Stürz S, Zanella A, Tonutti P (2017) A metabolomics approach to elucidate apple fruit responses to static and dynamic controlled atmosphere storage. Postharv Biol Technol 127:76–87

Brizzolara S, Hertog M, Tosetti R, Nicolai B, Tonutti P (2018) Metabolic responses to low temperature of three peach fruit cultivars differently sensitive to cold storage. Front Plant Sci 9:706. https://doi.org/10.3389/fpls.2018.00706

Brizzolara S, Cukrov D, Mercadini M, Martinelli F, Ruperti B, Tonutti P (2019) Short-term responses of apple fruit to partial reoxygenation during extreme hypoxic storage conditions. J Agric Food Chem 67:4754–4763. https://doi.org/10.1021/acs.jafc.9b00036

Brizzolara S, Manganaris GA, Fotopoulos V, Watkins CB, Tonutti P (2020) Primary metabolism in fresh fruits during storage. Front Plant Sci 11:80. https://doi.org/10.3389/fpls.2020.00080

Brunetti C, George RM, Tattini M, Field K, Davey MP (2013) Metabolomics in plant environmental physiology. J Exp Bot 64:4011–4020. https://doi.org/10.1093/jxb/ert244

Cacciatore S, Hu X, Viertler C, Kap M, Bernhardt GA, Mischinger H-J, Riegman P, Zatloukal K, Luchinat C, Turano P (2013) Effects of intra- and post-operative ischemia on the metabolic profile of clinical liver tissue specimens monitored by NMR. J Proteome Res 12(12):5723–5729. https://doi.org/10.1021/pr400702d

Calabrò A, Gralka E, Luchinat C, Saccenti E, Tenori L (2014) A metabolomic perspective on coeliac disease. Autoimmune Dis 2014:1–13. https://doi.org/10.1155/2014/756138

Carraturo F, Libralato G, Esposito R, Galdiero E, Aliberti F, Amoresano A, Fontanarosa C, Trifuoggi M, Guida M (2020) Metabolomic profiling of food matrices: Preliminary identification of potential markers of microbial contamination. J Food Sci 85 (10):3467–3477

Caruso M, Galgano F, Castiglione Morelli MA, Viggiani L, Lencioni L, Giussani B, Favati F (2012) Chemical profile of white wines produced from 'Greco Bianco' grape variety in different Italian areas by Nuclear Magnetic Resonance (NMR) and conventional physicochemical analyses. J Agric Food Chem 60(1):7–15. https://doi.org/10.1021/jf204289u

Carroll AJ, Badger MR, Harvey Millar A (2010) The MetabolomeExpress Project: enabling web-based processing, analysis and transparent dissemination of GC/MS metabolomics datasets. BMC Bioinform 11:376. https://doi.org/10.1186/1471-2105-11-376

Cebulj A, Cunja V, Mikulic-Petkovsek M, Veberic R (2017) Importance of metabolite distribution in apple fruit. Sci Hortic 214:214–220. https://doi.org/10.1016/j.scienta.2016.11.048

Cellini A, Biondi E, Blasioli S, Rocchi L, Farneti B, Braschi I, Savioli S, Rodriguez-Estrada MT, Biasioli F, Spinelli F (2016) Early detection of bacterial diseases in apple plants by analysis of volatile organic compounds profiles and use of electronic nose. Ann Appl Biol 168(3):409–420

Cevallos-Cevallos JM, Danyluk MD, Reyes-De-Corcuera JI (2011) GC-MS based metabolomics for rapid simultaneous detection of *Escherichia coli* O157:H7, *Salmonella typhimurium*, *Salmonella muenchen*, and *Salmonella hartford* in ground beef and chicken. J Food Sci 76:M238–M246

Cheng LL, Chang IW, Louis DN, Gonzalez RG (1998) Correlation of high-resolution magic angle spinning proton magnetic resonance spectroscopy with histopathology of intact human brain tumor specimens. Cancer Res 58:1825–1832

Chitarrini G, Dordevic N, Guerra W, Robatscher P, Lozano L (2020) Aroma investigation of new and standard apple varieties grown at two altitudes using gas chromatography-mass spectrometry combined with sensory analysis. Molecules 25(13):3007. https://doi.org/10.3390/molecules25133007

Claridge TDW (2008) High-resolution NMR techniques in organic chemistry. Elsevier Publications, The Netherlands

Cordero C, Liberto E, Bicchi C, Rubiolo P, Schieberle P, Reichenbach SE, Tao Q (2010) Profiling food volatiles by comprehensive two-dimensional gas chromatography coupled with mass spectrometry: Advanced fingerprinting approaches for comparative analysis of the volatile fraction of roasted hazelnuts (*Corylus avellana* L.) from different origins. J Chromatogr A 1217:5848–5858

Cubero-Leon E, Peñalver R, Maquet A (2014) Review on metabolomics for food authentication. Food Res Int 60:95–107

Cukrov D, Zermiani M, Brizzolara S, Cestaro A, Licausi F, Luchinat C, Santucci C, Tenori L, Van Veen H, Zuccolo A, Ruperti B, Tonutti P (2016) Extreme hypoxic conditions induce selective molecular responses and metabolic reset in detached apple fruit. Front Plant Sci 7:146. https://doi.org/10.3389/fpls.2016.00146

Cunha SC, Faria MA, Fernandes JO (2009) Determination of patulin in apple and quince products by GC–MS using 13C5–7 patulin as internal standard. Food Chem 115(1):352–359

Cuthbertson D, Andrews PK, Reganold JP, Davies NM, Lange BM (2012) Utility of metabolomics toward assessing the metabolic basis of quality traits in apple fruit with an emphasis on antioxidants. J Agric Food Chem 60:8552–8560. https://doi.org/10.1021/jf3031088

De Paepe D, Valkenborg D, Noten B, Servaes K, Diels L, de Loose M, Van Droogenbroeck B, Voorspoels S (2015) Variability of the phenolic profiles in the fruits from old, recent and new apple cultivars cultivated in Belgium. Metabolomics 11:739–752. https://doi.org/10.1007/s11306-014-0730-2

de Souza LP, Naake T, Tohge T, Fernie AR (2017) From chromatogram to analyte to metabolite. How to pick horses for courses from the massive web resources for mass spectral plant metabolomics. Giga Sci 6:1–20. https://doi.org/10.1093/gigascience/gix037

Del Campo G, Santos JI, Berregi I, Munduate A (2005) Differentiation of Basque cider apple juices from different cultivars by means of chemometric techniques. Food Control 16(6):549–555. https://doi.org/10.1016/j.foodcont.2004.05.013

Del Campo G, Santos JI, Iturriza N, Berregi I, Munduate A (2006) Use of the ^{1}H Nuclear Magnetic Resonance spectra signals from polyphenols and acids for chemometric characterization of cider apple juices. J Agric Food Chem 54(8):3095–3100. https://doi.org/10.1021/jf051818c

Du G, Xiao Y, Yang HR, Wang L, Song YL, Wang YT (2012) Rapid determination of pesticide residues in herbs using selective pressurized liquid extraction and fast gas chromatography coupled with mass spectrometry. J Sep Sci 35:1922–1932

Duan S, Wu Y, Fu R, Wang L, Chen Y, Xu W, Zhang C, Ma C, Shi J, Wang S (2019) Comparative metabolic profiling of grape skin tissue along grapevine berry developmental stages reveals systematic influences of root restriction on skin metabolome. Int J Mol Sci 20(3):534. https://doi.org/10.3390/ijms20030534

Dunemann F, Ulrich D, Boudichevskaia A, Grafe C, Weber WE (2009) QTL mapping of aroma compounds analysed by headspace solid-phase microextraction gas chromatography in the apple progeny 'Discovery' × 'Prima'. Mol Breed 23(3):501–521

El Hawari K, Mokh S, Al Iskandarani M, Halloum W, Jaber F (2019) Pesticide residues in Lebanese apples and health risk assessment. Food Addit Contamin B 12(2):81–89

Eisenmann P, Ehlers M, Weinert CH, Tzvetkova P, Silber M, Rist MJ, Luy B, Muhle-Goll C (2016) Untargeted NMR spectroscopic analysis of the metabolic variety of new apple cultivars. Metabolites 6(3):29. https://doi.org/10.3390/metabo6030029

Emwas A-H, Roy R, McKay RT, Tenori L, Saccenti E, Nagana Gowda GA, Raftery D, Alahmari F, Jaremko L, Jaremko M, Wishart DS (2019) NMR spectroscopy for metabolomics research. Metabolites 9(7):123. https://doi.org/10.3390/metabo9070123

Espino-Díaz M, Sepúlveda DR, González-Aguilar G, Olivas GI (2016) Biochemistry of apple aroma: a review. Food Technol Biotech 54(4):375–394

Etalo DW, de Vos RC, Joosten MH, Hall RD (2015) Spatially resolved plant metabolomics: some potentials and limitations of laser-ablation electrospray ionization mass spectrometry metabolite imaging. Plant Physiol 169:1424–1435

Ezhilan M, Nesakumar N, Babu KJ, Srinandan CS, Rayappan JBB (2018) An electronic nose for royal delicious apple quality assessment–a tri-layer approach. Food Res Int 109:44–51

Farag MA, Mohsen M, Heinke R, Wessjohann LA (2014) Metabolomic fingerprints of 21 date palm fruit varieties from Egypt using UPLC/PDA/ESI–qTOF-MS and GC–MS analyzed by chemometrics. Food Res Int 64:218–226

Farcuh M, Rivero RM, Sadka A, Blumwald E (2018) Ethylene regulation of sugar metabolism in climacteric and non-climacteric plums. Postharv Biol Technol 139:20–30

Farneti B, Khomenko I, Cappellin L, Ting V, Romano A, Biasioli F, Costa G, Costa F (2015) Comprehensive VOC profiling of an apple germplasm collection by PTR-ToF-MS. Metabolomics 11:838–850. https://doi.org/10.1007/s11306-014-0744-9

Fernandes AR, Rose M, Mortimer D, Carr M, Panton S, Smith F (2011) Mixed brominated/chlorinated dibenzo-p-dioxins, dibenzofurans and biphenyls: Simultaneous congener-selective determination in food. J Chromatogr A 185(1218):9279–9287

Fernie AR, Stitt M (2012) On the discordance of metabolomics with proteomics and transcriptomics: coping with increasing complexity in logic, chemistry, and network interactions. Plant Physiol 158:1139–1145

Ferry-Dumazet H, Gil L, Deborde C, Moing A, Bernillon S, Rolin D, Nikolski M, de Daruvar A, Jacob D (2011) MeRy-B: a web knowledgebase for the storage, visualization, analysis and annotation of plant NMR metabolomic profiles. BMC Plant Biol 11:104. https://doi.org/10.1186/1471-2229-11-104

Frainay C, Schymanski E, Neumann S, Merlet B, Salek R, Jourdan F, Yanes O (2018) Mind the gap: mapping mass spectral databases in genome-scale metabolic networks reveals poorly covered areas. Metabolites 8:51. https://doi.org/10.3390/metabo8030051

Francini A, Sebastiani L (2013) Phenolic compounds in apple (*Malus* x *domestica* Borkh.): compounds characterization and stability during postharvest and after processing. Antioxidants 2:81–193. https://doi.org/10.3390/antiox2030181

Fuhrmann E, Grosch W (2002) Character impact odorants of the apple cultivars Elstar and Cox Orange. Nahrung/Food 46(3):187–193

Gapper NE, Hertog MLATM, Lee J, Buchanan DA, Leisso RS, Fei Z, Qu G, Giovannoni JJ, Johnston JW, Schaffer RJ, Nicolaï BM, Mattheis JP, Watkins CB, Rudell DR (2017) Delayed response to cold stress is characterized by successive metabolic shifts culminating in apple fruit peel necrosis. BMC Plant Biol 17 (1):77. https://doi.org/10.1186/s12870-017-1030-6

Garrod S, Humpfer E, Spraul M, Connor SC, Polley S, Connelly J, Lindon JC, Nicholson JK, Holmes E (1999) High-resolution magic angle spinning ^{1}H NMR spectroscopic studies on intact rat renal cortex and medulla. Magn Reson Med 41(6):1108–1118. https://doi.org/10.1002/(SICI)1522-2594(199906)41:6

Gauthier L, Atanasova-Penichon V, Chéreau S, Richard-Forget F (2015) Metabolomics to decipher the chemical defense of cereals against *Fusarium graminearum* and deoxynivalenol accumulation. Int J Mol Sci 16:24839–24872

Ghini V, Quaglio D, Luchinat C, Turano P (2019) NMR for sample quality assessment in metabolomics. N Biotechnol 52:25–34. https://doi.org/10.1016/j.nbt.2019.04.004

Giannetti V, Mariani MB, Mannino P, Marini F (2017) Volatile fraction analysis by HS-SPME/GC-MS and chemometric modeling for traceability of apples cultivated in the Northeast Italy. Food Control 78:215–221

Girelli CR, Del Coco L, Zelasco S, Salimonti A, Conforti FL, Biagianti A, Barbini D, Fanizzi FP (2018) Traceability of "Tuscan PGI" extra virgin olive oils by ^{1}H NMR metabolic profiles collection and analysis. Metabolites 8(4):60. https://doi.org/10.3390/metabo8040060

Gomes NGM, Pereira DM, Valentão P, Andrade PB (2018) Hybrid MS/NMR methods on the prioritization of natural products: applications in drug discovery. J Pharm Biomed Anal 147:234–249

Goodacre R, Vaidyanathan S, Dunn WB, Harrigan GG, Kell DB (2004) Metabolomics by numbers: acquiring and understanding global metabolite data. Trends Biotechnol 22:245–252

Gorzolka K, Lissel M, Kessler N, Loch-Ahring S, Niehaus K (2012) Metabolite fingerprinting of barley whole seeds, endosperms, and embryos during industrial malting. J Biotechnol 159:177–187

Hall R, Beale M, Fiehn O, Hardy N, Sumner L, Bino R (2002) Plant metabolomics: the missing link in functional genomics strategies. Plant Cell 14:1437–1440

Hatoum D, Annaratone C, Hertog MLATM, Geeraerd AH, Nicolai BM (2014) Targeted metabolomics study of 'Braeburn' apples during long-term storage. Postharv Biol Technol 96:33–41. https://doi.org/10.1016/j.postharvbio.2014.05.004

Hatoum D, Hertog MLATM, Geeraerd AH, Nicolai BM (2016) Effect of browning related pre- and postharvest factors on the 'Braeburn' apple metabolome during CA storage. Postharv Biol Technol 111:106–116. https://doi.org/10.1016/j.postharvbio.2015.08.004

Haug K, Salek RM, Conesa P, Hastings J, de Matos P, Rijnbeek M, Mahendraker T, Williams M, Neumann S, Rocca-Serra P, Maguire E, González-Beltrán A, Sansone SA, Griffin JL, Steinbeck C (2013) MetaboLights — an open-access general-purpose repository for metabolomics studies and associated meta-data. Nucl Acids Res 41:D781–D786

Hong J, Yang L, Zhang D, Shi J (2016) Plant metabolomics: An indispensable system biology tool for plant science. Int J Mol Sci 17:767. https://doi.org/10.3390/ijms1706076

Humston EM, Knowles JD, McShea A, Synovec RE (2010) Quantitative assessment of moisture damage for cacao bean quality using two- dimensional gas chromatography combined with time-of-flight mass spectrometry and chemometrics. J Chromatogr A 1217:1963–1970

Johnson SR, Lange BM (2015) Open-access metabolomics databases for natural product research: present capabilities and future potential. Front Bioeng Biotechnol 3:22. https://doi.org/10.3389/fbioe.2015.00022

Jorge TF, Mata AT, António C (2016a) Mass spectrometry as a quantitative tool in plant metabolomics. Philos Trans R Soc A 374:20150370. https://doi.org/10.1098/rsta.2015.0370

Jorge TF, Rodrigues JA, Caldana C, Schmidt R, van Dongen JT, Thomas-Oates J, António C (2016b) Mass spectrometry-based plant metabolomics: metabolite responses to abiotic stress. Mass Spectrom Rev 35:620–649. https://doi.org/10.1002/mas.21449

Kalachova K, Pulkrabova J, Cajka T, Drabova L, Hajslova J (2012) Implementation of comprehensive two-dimensional GC-time-of-flight-MS for the simultaneous determination of halogenated contaminants and polycyclic aromatic hydrocarbons in fish. Anal Bioanal Chem 403:2813–2824

Kebede B, Ting V, Eyres G, Oey I (2020) Volatile changes during storage of shelf stable apple juice: Integrating GC-MS fingerprinting and chemometrics. Foods 9:165. https://doi.org/10.3390/foods9020165

Kessler N, Walter F, Persicke M, Albaum SP, Kalinowski J, Goesmann A, Niehaus K, Nattkemper TW (2014) Allocator: an interactive web platform for the analysis of metabolomic LC-ESI-MS datasets, enabling semi-automated, user-revised compound annotation and mass isotopomer ratio analysis. PLoS ONE 9: https://doi.org/10.1371/journal.pone.0113909

Khan SA, Chibon PY, de Vos RC, Schipper BA, Walraven E, Beekwilder J, van Dijk T, Finkers R, Visser RG, van de Weg EW, Bovy A, Cestaro A, Velasco R, Jacobsen E, Schouten HJ (2012) Genetic analysis of metabolites in apple fruits indicates an mQTL hotspot for phenolic compounds on linkage group 16. J Exp Bot 63(8):2895–2908. https://doi.org/10.1093/jxb/err464

Kim JH, Kim YJ, Kwon YS, Seo JS (2016) Development of multi-residue analysis of 320 pesticides in apple and rice using LC-MS/MS and GC-MS/MS. Kor J Pest Sci 20(2):104–127

Kim I, Ku KH, Jeong MC, Kim SS, Mitchell AE, Lee J (2019a) A comparison of the chemical composition and antioxidant activity of several new early- to mid-season apple cultivars for a warmer climate with traditional cultivars. J Sci Food Agric 99:4712–4724

Kim I, Ku KH, Jeong MC, Kim SS, Mitchell AE, Lee J (2019b) A comparison of the chemical composition and antioxidant activity of several new early- to mid-season apple cultivars for a warmer climate with traditional cultivars. J Sci Food Agric 99:4712–4724

Kim I, Ku K-H, Jeong M-C, Kwon S-I, Lee J (2020) Metabolite profiling and antioxidant activity of 10 new early- to mid-season apple cultivars and 14 traditional cultivars. Antioxidants (Basel) 9:443. https://doi.org/10.3390/antiox9050443

Kleessen S, Nikoloski Z (2012) Dynamic regulatory on/off minimization for biological systems under internal temporal perturbations. BMC Syst Biol 6:16

Korban SS, Swiader JM (1984) Genetic and nutritional status in bitter pit-resistant and -susceptible apple seedlings. J Am Soc Hortic Sci 109:428–432

Kueger S, Steinhauser D, Willmitzer L, Giavalisco P (2012) High-resolution plant metabolomics: from mass spectral features to metabolites and from whole-cell analysis to subcellular metabolite distributions. Plant J 70:39–50

Kumar R, Bohra A, Pandey AK, Pandey MK, Kumar A (2017) Metabolomics for plant improvement: status and prospects. Front Plant Sci 8:1302. https://doi.org/10.3389/fpls.2017.01302

Kusano M, Tohge T, Fukushima A, Kobayashi M, Hayashi N, Otsuki H, Kondou Y, Goto H, Kawashima M, Matsuda F, Niida R, Matsui M, Saito K, Fernie AR (2011) *Metabolomics* reveals comprehensive reprogramming involving two independent metabolic responses of *Arabidopsis* to UV-B light. Plant J 67:354–369

Lee J, Rudell DR, Davies PJ, Watkins CB (2011) Metabolic changes in 1-methylcyclopropene (1-MCP)-treated "Empire" apple fruit during storage. Metabolomics 8:742–753

Lee J, Rudell DR, Davies PJ, Watkins CB (2012a) Metabolic changes in 1-methylcyclopropene (1-MCP)-treated 'Empire' apple fruit during storage. Metabolomics 8(4):742–753. https://doi.org/10.1007/s11306-011-0373-5

Lee J, Mattheis JP, Rudell DR (2012b) Antioxidant treatment alters metabolism associated with internal browning in 'Braeburn' apples during controlled atmosphere storage. Postharv Biol Technol 68:32–42

Lee YY, Jeong MC, Jang HW (2017) Determination of volatile compounds by headspace-solid phase microextraction-gas chromatography/mass spectrometry: quality evaluation of Fuji apple. Anal Sci Technol 30(2):68–74

Leisso RS, Buchanan DA, Lee J, Mattheis JP, Sater C, Hanrahan I, Watkins CB, Gapper N, Johnston JW, Schaffer RJ, Hertog M, Nicolai BM, Rudell DR (2015) Chilling-related cell damage of apple (*Malus* × *domestica* Borkh.) fruit cortical tissue impacts antioxidant, lipid and phenolic metabolism. Physiol Plant 153(2):204–220. https://doi.org/10.1111/ppl.12244

Leisso RS, Gapper NE, Mattheis JP, Sullivan NL, Watkins CB, Giovannoni JJ, Schaffer RJ, Johnston JW, Hanrahan I, Hertog MLATM, Nicolaï BM, Rudell DR (2016) Gene expression and metabolism preceding soft scald, a chilling injury of 'Honeycrisp' apple fruit. BMC Genomics 17(1):798. https://doi.org/10.1186/s12864-016-3019-1

Li C, Schmidt NE, Gitaitis R (2011) Detection of onion postharvest diseases by analyses of headspace volatiles using a gas sensor array and GC-MS. LWT-Food Sci Technol 44:1019–1025

López-Ruiz R, Romero-González R, Frenich AG (2019) Metabolomics approaches for the determination of multiple contaminants in food. Curr Opin Food Sci 28:49–57

Lucht F, Rinken M, Haberlandt S (2020) Differentiation of volatile aroma components in organically and conventionally produced apples by stir bar sorptive extraction combined with gas chromatography and EI/CI TOF mass spectrometry. Int J Food Sci Technol. https://doi.org/10.1111/ijfs.14738

Ma P, Zhang Z, Zhou X, Yun Y, Liang Y, Lu H (2016) Feature extraction from resolution perspective for gas chromatography-mass spectrometry datasets. RSC Adv 6:113997–114004. https://doi.org/10.1039/C6RA17864B

Mannina L, D'Imperio M, Capitani D, Rezzi S, Guillou C, Mavromoustakos T, Vilchez MDM, Fernández AH, Thomas F, Aparicio R (2009) ^{1}H NMR-based protocol for the detection of adulterations of refined olive oil with refined hazelnut oil. J Agric Food Chem 57 (24):11550–11556. https://doi.org/10.1021/jf902426b

Maoz I, De Rosso M, Kaplunov T, Vedova AD, Sela N, Flamini R, Lewinsohn E, Lichter A (2019) Metabolomic and transcriptomic changes underlying cold and anaerobic stresses after storage of table grapes. Sci Rep 9:2917. https://doi.org/10.1038/s41598-019-39253-8

Maragò E, Michelozzi M, Calamai L, Camangi F, Sebastiani L (2016) Antioxidant properties, sensory characteristics and volatile compounds profile of apple juices from ancient Tuscany (Italy) apple varieties. Eur J Hortic Sci 81:255–263

Masi E, Taiti C, Vignolini P, Petrucci AW, Giordani E, Heimler D, Romani A, Mancuso S (2017) Polyphenols and aromatic volatile compounds in biodynamic and conventional 'Golden Delicious' apples (*Malus domestica* Bork.). Euro Food Res Technol 243 (9):1519–1531

McGhie TK, Rowan DD (2012) Metabolomics for measuring phytochemicals, and assessing human and animal responses to phytochemicals in food science. Mol Nutr Food Res 56:147–158

McQuilton P, Gonzalez-Beltran A, Rocca-Serra P, Thurston M, Lister A, Maguire E, Sansone SA (2016) BioSharing: curated and crowd-sourced metadata standards, databases and data policies in the life sciences. Database (Oxford) 2016: baw075. https://doi.org/10.1093/database/baw075

Medina S, Perestrelo R, Santos R, Pereira R, Câmara JS (2019) Differential volatile organic compounds signatures of apple juices from Madeira Island according to variety and geographical origin. Microchem J 150: https://doi.org/10.1016/J.MICROC.2019.104094

Mehinagic E, Royer G, Symoneaux R, Jourjon F, Prost C (2006) Characterization of odor-active volatiles in apples: influence of cultivars and maturity stage. J Agric Food Chem 54(7):2678–2687

Meoni G, Tenori L, Luchinat C (2020) Nuclear magnetic resonance-based metabolomic comparison of breast milk and organic and traditional formula milk brands for infants and toddlers. OMICS: J Integr Biol 24 (7):424–436. https://doi.org/10.1089/omi.2019.0125

Merkle S, Kleeberg KK, Fritsche J (2015) Recent developments and applications of solid phase microextraction (SPME) in food and environmental analysis—a review. Chromatograph 2:293–381

Misra BB, van der Hooft JJ (2016) Updates in metabolomics tools and resources: 2014–2015. Electrophoresis 37:86–110. https://doi.org/10.1002/elps.201500417

Misra VV, Assmann SM, Chen S (2014) Plant single-cell and single-cell type metabolomics. Trends Plant Sci 19:637–646

Moing A, Pétriacq P, Osorio S (2020) Fruit metabolism and metabolomics, vol 10, p 230. https://doi.org/10.3390/metabo10060230

Nicolini G, Roman T, Carlin S, Malacarne M, Nardin T, Bertoldi D, Larcher R (2018) Characterisation of single-variety still ciders produced with dessert apples in the Italian Alps. J Inst Brew 124:457–466

Pan Z, Raftery D (2007) Comparing and combining NMR spectroscopy and mass spectrometry in metabolomics. Anal Bioanal Chem 387(2):525–527. https://doi.org/10.1007/s00216-006-0687-8

Pereira GE, Gaudillere J-P, Van Leeuwen C, Hilbert G, Lavialle O, Maucourt M, Deborde C, Moing A, Rolin D (2005) ^{1}H NMR and chemometrics to characterize mature grape berries in four wine-growing areas in Bordeaux, France. J Agric Food Chem 53(16):6382–6389. https://doi.org/10.1021/jf058058q

Perestrelo R, Silva CL, Silva P, Medina S, Pereira R, Camara JS (2019) Untargeted fingerprinting of cider volatiles from different geographical regions by HS-SPME/GC-MS. Microchem J 148:643–651

Pinu FR (2016) Early detection of food pathogens and food spoilage microorganisms: application of metabolomics. Trends Food Sci Technol 54:213–215

Risticevic S, DeEll JR, Pawliszyn J (2012) Solid phase microextraction coupled with comprehensive two-dimensional gas chromatography-time-of-flight mass spectrometry for high-resolution metabolite profiling in apples: implementation of structured separations for optimization of sample preparation procedure in complex samples. J Chromatogr A 1251:208–218

Rivas-Ubach A, Sardans J, Perez-Trujillo M, Estiarte M, Penuelas J (2012) Strong relationship between elemental stoichiometry and metabolome in plants. Proc Natl Acad Sci USA 109:4181–4186

Roberts G, Spadafora ND (2020) Analysis of apple flavours: the use of volatile organic compounds to address cultivar differences and the correlation between consumer appreciation and aroma profiling. J Food Qual 2020. Article ID 8497259. https://doi.org/10.1155/2020/8497259

Rodrigues AM, Miguel C, Chaves I, António C (2019) Mass spectrometry-based forest tree metabolomics. Mass Spectrom Rev 40(2):126–157. https://doi.org/10.1002/mas.21603

Rongai D, Sabatini N, Del Coco L, Perri E, Del Re P, Simone N, Marchegiani D, Fanizzi FP (2017) ^{1}H NMR and multivariate analysis for geographic characterization of commercial extra virgin olive oil: a possible correlation with climate data. Foods 6(11):96. https://doi.org/10.3390/foods6110096

Rosend J, Kuldjärv R, Rosenvald S, Paalme T (2019) The effects of apple variety, ripening stage, and yeast strain on the volatile composition of apple cider. Heliyon 5: https://doi.org/10.1016/j.heliyon.2019.e01953

Rudell DR, Mattheis JP (2009) Superficial scald development and related metabolism is modified by postharvest light irradiation. Postharv Biol Technol 51:174–182

Rudell DR, Mattheis JP, Hertog MLATM (2009) Metabolomic change precedes apple superficial scald symptoms. J Agric Food Chem 57:8459–8466

Sacchi R, Mannina L, Fiordiponti P, Barone P, Paolillo L, Patumi M, Segre A (1998) Characterization of Italian extra virgin olive oils using ^{1}H-NMR spectroscopy. J Agric Food Chem 46(10):3947–3951. https://doi.org/10.1021/jf970666l

Santucci C, Brizzolara S, Tenori L (2015a) Comparison of frozen and fresh apple pulp for NMR-based metabolomic analysis. Food Anal Methods 8:2135–2140. https://doi.org/10.1007/s12161-015-0107-9

Santucci C, Tenori L, Luchinat C (2015b) NMR fingerprinting as a tool to evaluate post-harvest time-related changes of peaches, tomatoes and plums. Food Res Int 75:106–114. https://doi.org/10.1016/j.foodres.2015.05.046

Schulz M, Seraglio SKT, Della Betta F, Nehring P, Valese AC, Daguer H, Gonzaga LV, Oliveira Costa AC, Fett R (2020) Determination of phenolic compounds in three edible ripening stages of yellow guava (*Psidium cattleianum* Sabine) after acidic hydrolysis by LC-MS/MS. Plant Foods Human Nutr 75(1):110–115

Seo M, Kim B, Baek S-Y (2015) An optimized method for the accurate determination of patulin in apple products by isotope dilution-liquid chromatography/mass spectrometry. Anal Bioanal Chem 18:5433–5442. https://doi.org/10.1007/s00216-015-8705-3

Soltis NE, Kliebenstein DJ (2015) Natural variation of plant metabolism: genetic mechanisms, interpretive caveats, and evolutionary and mechanistic insights. Plant Physiol 169:1456–1468

Song J, Gardner BD, Holland JF, Beaudry RM (1997) Rapid analysis of volatile flavor compounds in apple fruit using SPME and GC/time-of-flight mass spectrometry. J Agric Food Chem 45(5):1801–1807

Song J, Forney CF, Jordan MA (2014) A method to detect diphenylamine contamination of apple fruit and storages using headspace solid phase micro-extraction and gas chromatography/mass spectroscopy. Food Chem 160:255–259

Sousa A, Vareda J, Pereira R, Silva C, Câmara JS, Perestrelo R (2020) Geographical differentiation of apple ciders based on volatile fingerprint. Food Res Int 137: https://doi.org/10.1016/j.foodres.2020.109550

Stall S, Yarmey L, Cutcher-Gershenfeld J, Hanson B, Lehnert K, Nosek B, Parsons M, Robinson E, Wyborn L (2019) Make scientific data FAIR. Nature 570:27–29

Stevens C, Dise N, Mountford J, Gowing D (2004) Impact of nitrogen deposition on the species richness of grasslands. Science 303:1876–1879

Sud M, Fahy E, Cotter D, Azam K, Vadivelu I, Burant C, Edison A, Fiehn O, Higashi R, Nair KS, Sumner S, Subramaniam S (2016) Metabolomics Workbench: an international repository for metabolomics data and metadata, metabolite standards, protocols, tutorials and training, and analysis tools. Nucl Acids Res 44: D463–D470

Sumner LW, Lei Z, Nikolau BJ, Saito K (2015) Modern plant metabolomics: Advanced natural product gene discoveries, improved technologies, and future prospects. Nat Prod Rep 32:212–229

Takis PG, Ghini V, Tenori L, Turano P, Luchinat C (2019) Uniqueness of the NMR approach to metabolomics. Trac-Trends Anal Chem 120: https://doi.org/10.1016/j.trac.2018.10.036

Tenori L, Santucci C, Meoni G, Morrocchi V, Matteucci G, Luchinat C (2018) NMR Metabolomic fingerprinting distinguishes milk from different farms. Food Res Int 113:131–139. https://doi.org/10.1016/j.foodres.2018.06.066

Thewes FR, Brackmann A, de Oliveira Anese R, Bronzatto ES, Schultz EE, Wagner R (2017) Dynamic controlled atmosphere storage suppresses metabolism and enhances volatile concentrations of 'Galaxy' apple harvested at three maturity stages. Postharv Biol Technol 127:1–13

Tohge T, de Souza LP, Fernie AR (2014) Genome-enabled plant metabolomics. J Chromatogr 966:7–20

Tohge T, Scossa F, Fernie AR (2015) Integrative approaches to enhance understanding of plant metabolic pathway structure and regulation. Plant Physiol 169:1499–1511

Töpfer N, Seaver SMD, Aharoni A (2018) Integration of plant metabolomics data with metabolic networks: progresses and challenges. In: António C (ed) Plant metabolomics: methods in molecular biology 1778. Humana Press, New York, pp 297–310

Tomita S, Nemoto T, Matsuo Y, Shoji T, Tanaka F, Nakagawa H, Ono H, Kikuchi J, Ohnishi-Kameyama M, Sekiyama Y (2015) A NMR-based, non-targeted multistep metabolic profiling revealed l-rhamnitol as a metabolite that characterised apples from different geographic origins. Food Chem 174:163–172. https://doi.org/10.1016/j.foodchem.2014.11.028

Tomlins A, Foxall PJD, Lindon JC, Lynch MJ, Spraul M, Everett JR (1998) High resolution magic angle spinning ^{1}H nuclear magnetic resonance analysis of intact prostatic hyperplastic and tumour tissues. Anal Commun 35(3):113–115

Toubiana D, Fernie AR, Nikoloski Z, Fait A (2013) Network analysis: Tackling complex data to study plant metabolism. Trends Biotechnol 31:29–36

Tranchida PQ, Mondello L (2012) Current-day employment of the micro-bore open-tubular capillary column in the gas chromatography field. J Chromatogr A 1261:23–36
Tranchida PQ, Purcaro G, Dugo P, Mondello L (2011) Modulators for comprehensive two-dimensional gas chromatography. Trends Anal Chem 30:1437–1461
Tsao R, Yang R, Young JC, Zhu H (2003) Polyphenolic profiles in eight apple cultivars using high-performance liquid chromatography (HPLC). J Agric Food Chem 51:6347–6353
Tsugawa H, Cajka T, Kind T, Ma Y, Higgins B, Ikeda K, Kanazawa M, Jean VanderGheynst J, Fiehn O, Arita M (2015) MS-DIAL: Data independent MS/MS deconvolution for comprehensive metabolome analysis. Nat Methods 12(6):523–526. https://doi.org/10.1038/nmeth.3393
Vallarino JG, de Abreu e Lima F, Soria C, Tong H, Pott DM, Willmitzer L, Fernie AR, Nikoloski Z, Osorio S (2018) Genetic diversity of strawberry germplasm using metabolomic biomarkers. Sci Rep 8:14386. https://doi.org/10.1038/s41598-018-32212-9
Vandendriessche T, Schäfer H, Verlinden BE, Humpfer E, Hertog MLATM, Nicolaï BM (2013) High-throughput NMR based metabolic profiling of Braeburn apple in relation to internal browning. Postharv Biol Technol 80:18–24. https://doi.org/10.1016/j.postharvbio.2013.01.008
Vanzo A, Jenko M, Vrhovsek U, Stopar M (2013) Metabolomic profiling and sensorial quality of 'Golden Delicious', 'Liberty', 'Santana', and 'Topaz' apples grown using organic and integrated production systems. J Agric Food Chem 61:6580–6587
Veberic R, Trobec M, Herbinger K, Hofer M, Grill D, Stampar F (2005) Phenolic compounds in some apple (*Malus domestica* Borkh.) cultivars of organic and integrated production. J Sci Food Agric 85:1687–1694. https://doi.org/10.1002/jsfa.2113
Vermathen M, Marzorati M, Baumgartner D, Good C, Vermathen P (2011) Investigation of different apple cultivars by high resolution magic angle spinning NMR. A Feasibility Study. J Agric Food Chem 59 (24):12784–12793. https://doi.org/10.1021/jf203733u
Villas-Bôas SG, Mas S, Akesson M, Smedsgaard J, Nielsen J (2005) Mass Spectro Rev 24:613–646
Vignoli A, Ghini V, Meoni G, Licari C, Takis PG, Tenori L, Turano P, Luchinat C (2019) High-throughput metabolomics by 1D NMR. Angew Chem 58(4):968–994. https://doi.org/10.1002/anie.201804736
Volz RK, Alspach PA, Fletcher David J, Ferguson IB (2006) Genetic variation in bitter pit and fruit calcium concentrations within a diverse apple germplasm collection. Euphytica 149:1–10. https://doi.org/10.1007/s10681-005-9000-8
Wilkinson MD, Dumontier M, Aalbersberg IJ, Appleton G, Axton M, Baak A, Blomberg N, Boiten JW, da Silva Santos LB, Bourne PE, Bouwman J, Brookes AJ, Clark T, Crosas M, Dillo I, Dumon O, Edmunds S, Evelo CT, Finkers R, Gonzalez-Beltran A, Gray AJ, Groth P, Goble C, Grethe JS, Heringa J, 't Hoen PA, Hooft R, Kuhn T, Kok R, Kok J, Lusher SJ, Martone ME, Mons A, Packer AL, Persson B, Rocca-Serra P, Roos M, van Schaik R, Sansone SA, Schultes E, Sengstag T, Slater T, Strawn G, Swertz MA, Thompson M, van der Lei J, van Mulligen E, Velterop J, Waagmeester A, Wittenburg P, Wolstencroft K, Zhao J, Mons B (2016) The FAIR guiding principles for scientific data management and stewardship. Sci Data 3:160018. https://doi.org/10.1038/sdata.2016.18
Wishart DS, Tzur, D, Knox C, Eisner R, Guo AC, Young N, Cheng D, Jewell K, Arndt D, Sawhney S, Fung C, Nikolai L, Lewis M, Coutouly MA, Forsythe I, Tang P, Shrivastava S, Jeroncic K, Stothard P, Amegbey G, Block D, Hau DD, Wagner J, Miniaci J, Clements M, Gebremedhin M, Guo N, Zhang Y, Duggan GE, Macinnis GD, Weljie AM, Dowlatabadi R, Bamforth F, Clive D, Greiner R, Li L, Marrie T, Sykes BD, Vogel H J, Querengesser L (2007) HMDB: the human metabolome database. Nucl Acids Res 35(1362–4962 (Electronic)):D521–D526
Wishart DS, Feunang YD, Marcu A, Guo AC, Liang K, Vázquez-Fresno R, Sajed T, Johnson D, Li C, Karu N, Sayeeda Z, Lo E, Assempour N, Berjanskii M, Singhal S, Arndt D, Liang Y, Badran H, Grant J, Serra-Cayuela A, Liu Y, Mandal R, Neveu V, Pon A, Knox C, Wilson M, Manach C, Scalbert A (2018) HMDB 4.0: The human metabolome database for 2018. Nucl Acids Res 46(D1):D608–D617. https://doi.org/10.1093/nar/gkx1089
Wu H, Huang W, Chen Z, Chen Z, Shi J, Kong Q, Sun S, Jiang X, Chen D, Yan S (2019) GC–MS-based metabolomic study reveals dynamic changes of chemical compositions during black tea processing. Food Res Int 120:330–338
Xu J, Zhang Y, Qi D, Huo H, Dong X, Tian L, Zhang X, Liu C, Cao Y (2018) Postharvest metabolomic changes in *Pyrus ussuriensis* Maxim. wild accession 'Zaoshu Shanli'. J Sep Sci 41:4001–4013
Zhang W, Chang J, Lei Z, Huhman D, Sumner LW, Zhao PX (2014) MET-COFEA: a liquid chromatography/mass spectrometry data processing platform for metabolite compound feature extraction and annotation. Anal Chem 86:6245–6253. https://doi.org/10.1021/ac501162k
Zhang W, Lei Z, Huhman D, Sumner LW, Zhao PX (2015) MET-XAlign: a metabolite cross alignment tool for LC/MS-based comparative metabolomics. Anal Chem 87:9114–9119. https://doi.org/10.1021/acs.analchem.5b01324
Zhang H, Chen H, Wang W, Jiao W, Chen W, Zhong Q, Yun Y-H, Chen W (2020) Characterization of volatile

profiles and marker substances by HS-SPME/GC-MS during the concentration of coconut jam. Foods 9:347. https://doi.org/10.3390/foods9030347

Zhu D, Ren X, Wei L, Cao X, Ge Y, Liu H, Li J (2020) Collaborative analysis on difference of apple fruits flavor using electronic nose and electronic tongue. Sci Hortic 260:108879

Zupan A, Mkulic-Petkovesk M, Cunja V, Stampar F Veberic R (2013) Comparison of phenolic composition of healthy apple tissues and tissues affected by bitter pit. J Agric Food Chem 61:12066–12071

Zupan A, Mikulic-Petkovsek M, Stampar F, Veberic R (2016) Sugar and phenol content in apple with or without watercore. J Sci Food Agric 96:2845–2850. https://doi.org/10.1002/jsfa.7453

The Apple Microbiome: Structure, Function, and Manipulation for Improved Plant Health

16

Susan R. Whitehead, Michael E. Wisniewski, Samir Droby, Ahmed Abdelfattah, Shiri Freilich, and Mark Mazzola

Abstract

Apple trees host diverse communities of bacteria, fungi, viruses, and other microbes that occupy all plant surfaces and internal tissues. In the past two decades, our understanding of these communities has burgeoned due to new technologies that allow culture-independent characterisation of microbial communities. In this chapter, we provide a comprehensive review of our current understanding of the composition and assembly of the apple tree microbiome. We discuss factors that shape variation in the microbiome, including host genotype, domestication, plant chemistry, and agricultural practices. Next, we discuss the consequences of the microbiome for plant health and the opportunities and challenges associated with attempts to manipulate the microbiome for improved sustainability of orchard ecosystems. Finally, we suggest future research priorities that promise to facilitate a paradigm shift in orchard management from viewing trees as individual organisms to viewing trees as diverse ecosystems that must be managed holistically to maximise the environmental and economic sustainability of production systems.

The original version of this chapter was revised (Fig. 16.3 was updated). The correction to this chapter can be found at https://doi.org/10.1007/978-3-030-74682-7_19

S. R. Whitehead (✉) · M. E. Wisniewski
Department of Biological Sciences, Virginia Tech, Blacksburg, VA 24061, USA
e-mail: swhitehead@vt.edu

S. Droby
Department of Postharvest Science, Agricultural Research Organization (ARO), The Volcani Center, 7505101 Rishon LeZion, Israel

A. Abdelfattah
Department of Ecology, Environment and Plant Sciences, Stockholm University, Stockholm, Sweden

Institute of Environmental Biotechnology, Graz University of Technology, Graz, Austria

S. Freilich
Institute of Plant Sciences, Newe Ya'ar Research Center, the Agricultural Research Organization, 30095 Ramat Yishay, Israel

M. Mazzola
Department of Plant Pathology, Stellenbosch University, Matieland 7600, South Africa

16.1 Introduction

Plants have been evolving for hundreds of millions of years in the midst of intimate associations with microbes. Microbes occupy all plant surfaces and internal tissues, including roots, leaves, stems, flowers, and fruits (Vorholt 2012; Philippot et al. 2013; Aleklett et al. 2014; Droby and Wisniewski 2018; Compant et al. 2019). Although there is a long history of research on plant–microbe interactions, recent advances in high-throughput sequencing and bioinformatics

S. S. Korban (ed.), *The Apple Genome*, Compendium of Plant Genomes,
https://doi.org/10.1007/978-3-030-74682-7_16

have revolutionised our ability to characterise the complex microbial communities associated with plants and their environments (Di Bella et al. 2013; Knight et al. 2018).

As our understanding of microbiome diversity and function has grown, there has been increasing recognition of the fundamental roles that microbes play in plant physiology and ecology (Berg et al. 2016; Cordovez et al. 2019). Microbes can directly affect growth, developmental processes, nutrient uptake, stress tolerance, and pest resistance (Berg et al. 2016; Cordovez et al. 2019). Often, microbial colonisation induces rapid physiological changes in plants, such as the upregulation of phytochemicals, leading to complex feedback loops among plants, microbes, and the environment (Chisholm et al. 2006; Kiers et al. 2011; Pieterse et al. 2013). In fact, plant phenotype may be best understood as a manifestation of the combined genomes of plants and their complex communities of associated bacteria, fungi, viruses, and other symbionts (Zilber-Rosenberg and Rosenberg 2008; Vandenkoornhuyse et al. 2015).

New technologies allow for near complete characterization of organisms associated with plants, as well as a breadth of plant responses (e.g., metabolomic analyses of phytochemicals; Hall 2006; Wang et al. 2016b). This provides novel and exciting opportunities to understand how microbiome interactions may alter plants in profound ways that impact productivity, resistance to pests and pathogens, nutritional content, and phytochemical profiles of plant tissues. Clearly, plant–microbe interactions have immense implications for improving crop production, and we are poised to make significant advances that will improve agricultural sustainability (APS 2016; Busby et al. 2017; Toju et al. 2018). However, the composition and function of plant microbiomes are highly dependent on the plant genome and the environment in which they reside (Turner et al. 2013; Cordovez et al. 2019). Thus, translating basic knowledge into agricultural practices will require more comprehensive studies of microbial communities and an interdisciplinary systems-level approach to understand how these communities interact with other components of the agroecosystem.

A large number of studies have been published on the apple microbiome in recent years, and this chapter is intended to provide an overview of what is known and an outlook for applications in sustainable agriculture. First, we provide a systematic review of studies reporting on the composition and assembly of microbial communities associated with different microhabitats of the apple tree. Second, we discuss major factors that can shape the microbiome, including apple genotypes, domestication, phytochemistry, and orchard management practices. Third, we discuss how the apple microbiome can be manipulated for improved plant health, focusing particularly on some of the major diseases, such as fire blight and apple replant disease, affecting commercial orchards. Lastly, we provide a summary and perspectives on promising new directions for future research.

16.2 Composition and Assembly Across Apple Tree Microhabitats

From a microbe's perspective, an apple tree represents a complex ecosystem with numerous distinct microhabitats (Fig. 16.1). These include the rhizosphere (soil immediately surrounding the roots), root endosphere (internal root tissues), phyllosphere (used here to strictly refer to leaves, as it was originally defined; Last 1955; Ruinen 1956; Leveau 2015; Shade et al. 2017), caulosphere (stems), anthosphere (flowers), and carposphere (fruits). Research in apple and other systems has shown that these different plant microhabitats support distinct microbial assemblages (Shade et al. 2013b; Ottesen et al. 2013; Berg et al. 2016; Cregger et al. 2018). We review key ecological factors that may shape microbiome composition and assembly in each microhabitat, and provide a brief historical context of their study in apple. Next, we provide a comprehensive review of studies that have used culture-independent methods to characterize apple microbiome diversity and composition. Our review is based on a systematic literature search that resulted in 69 relevant studies

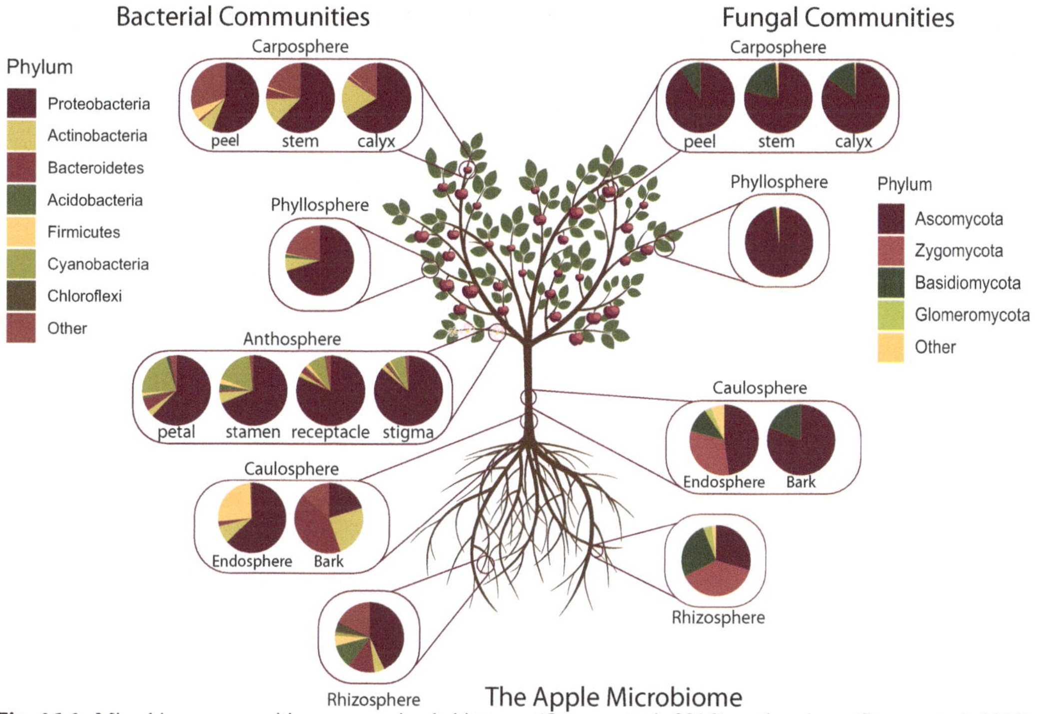

Fig. 16.1 Microbiome composition across microhabitats of the apple tree. Charts show relative abundances of microbial communities as reported in the literature for the carposphere (Abdelfattah et al. 2020), phyllosphere (Ottesen et al. 2016), anthosphere (Steven et al. 2018), caulosphere (Liu et al. 2018; Arrigoni et al. 2018), and rhizosphere (Wang and Mazzola 2019a). Taxa grouped as "other" include unidentified and identified minor phyla

published between 2000 and 2020 (Table 16.1). These include early studies on the rhizosphere and phyllosphere using either cloned libraries or molecular fingerprinting technologies, such as terminal restriction fragment length polymorphism (T-RFLP) or denaturing gradient gel electrophoresis (DGGE), as well as more recent work using either amplicon sequencing or whole-genome shotgun sequencing to understand how microbial composition is affected by genotype, habitat, disease state, or other factors (Table 16.1).

16.2.1 The Root Microbiome

Plant root systems provide a unique and resource-rich habitat for microbes. The root microbiome includes microbes living within root tissues (endosphere), on root surfaces (rhizoplane), and in soils immediately surrounding roots (rhizosphere). The rhizosphere is the most diverse microbial habitat on earth—a small sample of soil can host thousands of unique microbial taxa that are an order of magnitude more diverse than those on plant surfaces or tissues (Thompson et al. 2017). Although the rhizosphere microbiome is largely derived from surrounding soils, the factors shaping its composition are distinct from other soil habitats (Bakker et al. 2013; Philippot et al. 2013). Microbe communities in bulk soil are shaped largely by soil pH and other edaphic factors. The rhizosphere microbiome is shaped and supported by nutrients, exudates, secondary metabolites, mucilage, and border cells secreted by roots (i.e., rhizodeposits). Rhizodeposits can differ among plant species and genotypes, leading to host-specific rhizosphere communities of substantially different compositions from those of bulk soils

Table 16.1 Results from a systematic literature review of studies using culture-independent methods to evaluate the apple microbiome[a]

Microbial habitat	Technology[b]	Factors evaluated[c]	References
Bulk orchard soil, rhizosphere, rhizoplane, and/or root endosphere	Fingerprinting (DGGE, T-RFLP, or ARISA) and/or culture-independent clone libraries	Rootstock genotype; disease status (ARD, phytoplasma, apple scab, violet root rot); ARD mitigation treatments (amendments, fumigation, and row position, among others); fertilization; ground management; time since planting	Rumberger et al. (2004, 2007); Yao et al. (2005, 2006); St. Laurent et al. (2008, 2010); Shishido et al. (2008); Shanmugam et al. (2011); Kelderer et al. (2012); Bulgari et al. (2012); Yim et al. (2013, 2016); Chen et al. (2014); Manici et al. (2015); Caputo et al. (2015); Franke-Whittle et al. (2015); Hewavitharana and Mazzola (2016); Peruzzi et al. (2017); Lucas et al. (2018); Wang and Mazzola (2019a, b); Mazzola et al. (2020)
	Amplicon sequencing	Rootstock genotype; microhabitat (e.g., rhizosphere vs. rhizoplane); disease status (ARD, rapid apple decline); ARD mitigation treatments; planting history; soil amendments (compost, biochar, manure); fertilisation; ground management; streptomycin treatment; time since planting; spatial structuring; analytical methods	Shade et al. (2012a, 2013a); Zhang et al. (2013); Sun et al. (2014, 2017); Walsh et al. (2014); Mazzola et al. (2015); Franke-Whittle et al. (2015); Hewavitharana and Mazzola (2016); Wang et al. (2016a, 2017); Abujabhah et al. (2016); Qin et al. (2016); Jiang et al. (2017); Yim et al. (2017); Peruzzi et al. (2017); Tilston et al. (2018); Zheng et al. (2018); Deakin et al. (2018, 2019); Liang et al. (2018); Chai et al. (2019, 2020); Wang and Mazzola (2019a); Singh et al. (2019)
	Shotgun sequencing	Disease status (ARD); microhabitat (rhizosphere vs. bulk soil); ground management; fertilisation	Zheng et al. (2019); Radl et al. (2019)
Phyllosphere	Fingerprinting (T-RFLP) and/or culture-independent clone libraries	Scion genotype; disease status (*Alternaria* leaf spot); management system (conventional vs. organic); streptomycin use; time	Ottesen et al. (2009); Yashiro et al. (2011); Yashiro and McManus (2012); Hirakue and Sugiyama (2018)
	Amplicon sequencing	Management system (conventional vs. organic)	Glenn et al. (2015); Ottesen et al. (2016)
Caulosphere	Amplicon sequencing	Scion genotype; rootstock genotype; microhabitat (young vs old bark)	Liu et al. (2018); Arrigoni et al. (2018)
Anthosphere	Amplicon sequencing	Scion genotype: microhabitat (different floral parts); *Erwinia amylovora* inoculation; temporal succession; spatial variation	Shade et al. (2013b); Steven et al. (2018); Cui et al. (2020)

(continued)

Table 16.1 (continued)

Microbial habitat	Technology[b]	Factors evaluated[c]	References
Carposphere	Fingerprinting (DGGE)	Scion genotype, fruit vs. cider microbiota	Alonso et al. (2015)
	Amplicon sequencing	Microhabitat (fruit part); management system (conventional vs. organic); storage time; post-harvest treatment; site; analytical methods	Leff and Fierer (2013); Abdelfattah et al. (2016, 2020, 2021); Vepštaitė-Monstavičė et al. (2018); Shen et al. (2018); Wassermann et al. (2019a, b); Meakem (2020); Sare et al. (2020)
	Shotgun sequencing	n/a	Angeli et al. (2019)

[a]A search was conducted on July 2, 2020 of all databases in Web of Science for studies between 2000 and 2020 with the following keywords: Malus AND microbiome OR rhizosphere OR phyllosphere OR caulosphere OR anthosphere OR carposphere. This search produced 295 articles, and an additional 42 sources were identified through Google Scholar searches with the same terms. Abstracts and full-texts of these articles were screened, and we excluded any studies that did not report results from culture-independent studies of the microbial communities associated with apple trees or harvested fruit

[b]ARISA, automated method of ribosomal intergenic spacer analysis; DGGE, denaturing gradient gel electrophoresis; T-RFLP, terminal restriction fragment length polymorphism

[c]ARD, Apple replant disease

(Mazzola 1999; Bais et al. 2006; Bakker et al. 2013; Philippot et al. 2013; Leisso et al. 2017). An even smaller and more specialised subset of soil microbes is able to penetrate and establish themselves in host tissues, forming a root endophyte community (Gaiero et al. 2013; Liu et al. 2018; Van Horn and Mazzola 2019).

Microbiota of the apple rhizosphere have been studied for decades (e.g., Covey et al. 1981; Čatská et al. 1982; Mazzola 1999; Rumberger et al. 2007). In particular, the role of microbial ecology in soil-borne disease has been widely investigated. Among these soil-borne diseases, the most economically important is apple replant disease, caused by a complex of pathogens, including fungi (*Ilyonectria* spp. and *Rhizoctonia solani*), oomycetes (*Phytophthora* spp. and *Pythium* spp.), and nematodes (Turechek 2004; Mazzola and Manici 2012). Disease symptoms include stunted, uneven growth, shortened internodes, root-tip necrosis, and reduced root biomass, which can ultimately result in either tree death or reduced productivity that is maintained over the lifetime of an orchard (Mazzola and Manici 2012). Thus, the need for effective management of this disease in commercial orchards has led to extensive studies of how disease-causing agents persist in the soil and interact with other members of the microbial community (see Sect. 16.5 below on 'Manipulating the Microbiome'). In addition to this focus on disease-causing agents, there is a long history of studies of beneficial microbes in the apple rhizosphere, including arbuscular mycorrhizal fungi and plant-growth-promoting bacteria (reviewed in Mosa et al. 2016).

With advances in technology that allow culture-independent characterization of microbial communities, new studies have greatly increased our understanding of the diversity of organisms present in the apple root microbiome (Table 16.1). Most of these studies have continued to focus largely on the rhizosphere community and its role in the development and control of apple replant disease. The first culture-independent studies applied fingerprinting technologies to illustrate the complexity of microbial communities associated with apple replant disease and the clear differences in composition in soils subjected to different management practices (Rumberger et al. 2004, 2007; Yao et al. 2005, 2006). Amplicon sequencing has further elucidated the diversity and taxonomic composition of rhizosphere communities, as well as differences between the rhizosphere and root endosphere (Shade et al. 2013a; Mazzola et al. 2015; Lucas et al. 2018; Wang and Mazzola 2019a; Van Horn and Mazzola 2019). For example, Wang and Mazzola (2019a) estimated that fungal communities can contain over 400 operational taxonomic units (OTUs) at 97% sequence similarity. These fungal taxa belonged to seven phyla, but were dominated by Ascomycota, Zygomycota, and Basidiomycota. Bacterial communities were estimated to contain over 1100 OTUs from 20 phyla, and were dominated by Proteobacteria, accounting for over 40% of sequences (Wang and Mazzola 2019a). The actual level of diversity is likely to be even higher, and communities may include many rare taxa present at very low abundance. Shade et al. (2012a, 2013a) used culture-based approaches in conjunction with amplicon sequencing to explore rare biota of the apple rhizosphere and found that more than 60% (out of >1,000 total OTUs) of bacterial taxa detected using culture-dependent approaches were not detected with amplicon sequencing alone. These included sequences from many genera with known functional significance in the rhizosphere, such as *Psuedomonas* and *Bacillus* (Shade et al. 2012a).

Current knowledge of microbial community composition, and particularly of rare taxa, will also likely increase as costs of whole-genome shotgun sequencing continue to decrease (Caporaso et al. 2012; but see Tessler et al. 2017). Shotgun sequencing provides a random sampling of sequences across genomes of bacteria, fungi, archaea, and viruses. An advantage of this approach for understanding composition is that it provides an estimation of the relative abundance of taxa across kingdoms. Our literature search found only two studies that have applied shotgun sequencing to the apple rhizosphere (Zheng et al. 2019; Radl et al. 2019; Table 16.1). Radl et al. (2019) assessed apple replant disease and control soils and reported that, in all cases, apple rhizosphere communities were highly dominated by bacteria. Of sequences assigned at the kingdom level (63% of all sequences), 96% were bacteria, 3% were eukaryotes, <1% were archaea, and <0.1% were viruses (Radl et al. 2019). Annotation of gene function indicated that microbes from apple replant disease soils had a higher prevalence of genes related to both stress sensing and degradation of aromatic compounds, thereby providing hints to functional mechanisms shaping microbial influence on the development of apple replant disease (Radl et al. 2019). Continued application of such approaches will provide additional insights into both taxonomic and functional compositions of communities and will offer novel opportunities for improving plant health (see Sect. 16.6 below on ‘Future Research Priorities’).

16.2.2 Phyllosphere

The phyllosphere refers to the microbial habitat associated with leaves (Last 1955; Ruinen 1956; Leveau 2015; Shade et al. 2017) and can be further subdivided into leaf surfaces (phylloplane) and internal leaf tissues (leaf endosphere). Leaf surfaces can be considered extreme and unstable habitats for microorganisms. In contrast to the rhizosphere, where large amounts of photoassimilates are exuded into the surrounding soil, leaf surfaces are nutrient-limited and exhibit highly fluctuating physicochemical constraints, including high levels of light, ultraviolet radiation, temperature, and desiccation (Vorholt 2012; Bulgarelli et al. 2013; Griffin and Carson 2015). In contrast, leaf and stem endospheres may provide a more stable habitat, but access to this niche requires special adaptations to penetrate leaf surfaces, and the endosphere typically harbours many fewer taxa (Vorholt 2012; Thompson et al. 2017). Nevertheless, fungal and bacterial endophytes are ubiquitous among plants, and the close symbiosis between plants and endophytes suggests that these organisms are particularly important for plant health and physiology (Arnold and Lutzoni 2007; Griffin and Carson 2015).

In either annual or deciduous plants, the phyllosphere is an ephemeral environment in which microbial assembly is linked to the annual process of leaf emergence (Vorholt 2012). The assembly of both epiphytic and endophytic microbial communities of leaves is non-random and shaped by host genotype, abiotic conditions, and microbe–microbe interactions, among other factors (Vorholt 2012). Leaf surfaces may be colonized from either ad hoc sources, such as air, dust particles, and rain droplets, or local sources, including soils and other plant tissues (Vorholt 2012). Leaf communities do exhibit some level of stability from year to year, thus suggesting important contributions from local pools, but rules governing assembly of microbial communities in above-ground tissues remain poorly understood (Vorholt 2012; Bulgarelli et al. 2013; Griffin and Carson 2015).

Most early studies conducted on microbial communities of the apple phyllosphere focused on causal agents of specific apple diseases or on potential biocontrol agents. For example, researchers used culture-based techniques to understand the ecology of bacterial communities in relation to biological control of the fire blight organism, *Erwinia amylovora*, and its resistance to streptomycin and oxytetracycline (Sobiczewski et al. 1991; Norelli et al. 1991; Burr et al. 1993; Chiou and Jones 1993; Huang and Burr 1999; Schnabel and Jones 1999; Johnson and Stockwell 2000; Lindow and Brandl 2003). Other studies explored microbial communities of apple leaves to search for fungal and bacterial antagonists that could protect apple against post-harvest pathogens (Falconi and Mendgen 1994) and *Venturia inaequalis*, the causal agent of apple scab (Andrews et al. 1983; Heye and Andrews 1983; Burr et al. 1996; Kucheryava et al. 1999). At least one culture-based study focused specifically on yeast diversity and documented hundreds of strains from the phylloplane belonging to at least 13 species, the most common of which were *Aureobasidium pullulans, Cryptococcus laurentii,* and *Metschnikowia pulcherrima* (Slávikovà et al. 2009). A number of studies have investigated the immigration dynamics of leaf epiphytic communities in the field (Kinkel et al. 1989a, b; Andrews et al. 2002; McGrath and Andrews 2006, 2007; Woody et al. 2007). Kinkel et al. (1989a, b) provided valuable insights into community assembly by demonstrating rapid immigration to disinfested leaves, with up to 29,200 individuals colonising a leaf within a 12 h period during mid-summer. Immigration rates varied widely over time (correlating largely with air spora) and depended on leaf area, but, in general, communities assembled in a relatively predictable manner with the regular arrival of common genera, such as *Aureobasidium, Alternaria, Cladosporium,* and *Microsphaeropsi* (Kinkel et al. 1989a).

More recently, culture-independent studies have broadened our understanding of the diversity of organisms that occupy the apple phyllosphere. Ottesen et al. (2009) used 16S clone libraries to assess bacterial epiphytes collected from surfaces of pooled leaf and fruit samples and identified 136 OTUs from eight bacterial phyla. Communities were found to be largely dominated by Proteobacteria, including Alpha-, Beta-, and Gammaproteobacteria, collectively accounting for approximately 80% of observed sequences (Ottesen et al. 2009). Yashiro and MacManus (2012) and Yashiro et al. (2011) observed similar patterns of composition using 16S clone libraries to study leaf epiphytes across eight orchards in Wisconsin. Yashiro et al. (2011) also directly compared culture-independent clone libraries and culture-based approaches using the same samples, and, similar to patterns reported from rhizosphere communities (Shade et al. 2012a), they found that many phyllosphere taxa captured by culturing were not detected using culture-independent methods and vice versa. Notably, Actinobacteria were only detected using culture-based techniques, but several orders in the phyla Bacteroidetes and Proteobacteria were only detected using clone libraries (Yashiro et al. 2011).

More recently, amplicon-based studies have further documented a large diversity of bacteria and fungi associated with above-ground tissues, including leaves (Glenn et al. 2015; Ottesen et al. 2016). In both of these studies, samples were obtained from washes of pooled leaf and fruit samples, thus leaving uncertainty about the specific composition of leaves versus other microhabitats. However, these findings have informed our understanding of the impacts of pest management (see Sect. 16.3 below on 'Factors Affecting Community Structure'). Undoubtedly, there is much to learn about the composition of the apple phyllosphere and the rules governing its structure and assembly through leaf phenology.

16.2.3 Caulosphere

The caulosphere includes epiphytic and endophytic microbiomes associated with stem and bark tissues of plants. This habitat can also be further subdivided to account for bark surfaces (sometimes referred to as the dermosphere or cauloplane) and internal tissues (endosphere). The caulosphere is perhaps the least-studied plant microhabitat, although it hosts diverse communities that may be critical for plant health and physiology. Bark and stem tissues can, at least in some cases, be highly diverse relative to other above-ground plant parts, and they have unique compositions that include many pathogenic as well as mutualistic organisms (Martins et al. 2013; Leff et al. 2015; Cregger et al. 2018; Vitulo et al. 2019). The bark surface is an extreme habitat for microbial life, characterized by dry, nutrient-poor conditions, along with a recalcitrant substrate composed largely of lignin, cellulose, and hemicellulose (Valentín et al. 2010; Arrigoni et al. 2018). Still, within the complex surface structure of bark, there are many microsites, such as lenticels, that can provide a more favourable habitat and support diverse microbial communities (Sivak and Person 1973; Buck et al. 1998). Furthermore, both bark and stems serve as stable above-ground habitats for microbes, as compared to ephemeral tissues, such as deciduous leaves, flowers, and fruits. In temperate regions, many pre- and post-harvest pathogens, such as *E. amylovora* (causal bacterium of fire blight), *Peltaster fructicola, Geastrumia polystigmatus, Leptodontium elatus,* and *Zygophiala jamaicensis* (causal fungi of sooty blotch), *Neofabraea* species (causal fungi of bull's eye rot), are known to overwinter in bark or twigs (Turechek 2004). Many commensal or mutualistic microbes are likely to do the same. These factors likely contribute to the high microbial diversity in the caulosphere and make it a key microbial niche that influences plant health and shapes the microbial assembly of other above-ground microhabitats.

The caulosphere of apples has been investigated for many years in the context of canker and rot diseases, especially *E. amylovora*, that can infect both stem and bark tissues (Schroth et al. 1974; Johnson and Stockwell 1998; Wang et al. 2011; Weber 2014). However, we found only two studies that have investigated the broader community composition of these tissues using culture-independent approaches (Table 16.1). Liu et al. (2018) investigated bacterial and fungal communities associated with the stem endosphere of different apple genotypes by surface-sterilizing shoots and removing all bark tissues. They detected 513 bacterial OTUs from nine phyla, primarily belonging to Proteobacteria or Firmicutes. Fungal richness was more than double that of bacteria, with 1246 fungal OTUs from six phyla. These were mostly from the phyla Ascomycota and Zygomycota, including several interesting genera, such as *Zoophthora*, which includes well-known insect pathogenic species (Liu et al. 2018). Arrigoni et al. (2018) also studied the apple caulosphere, but focused on fungal and bacterial communities in bark tissues. Fungal communities were largely dominated by Ascomycota, particularly *Aureobasidium*, which accounted for 58–83% of sequences across young and old bark tissues (Arrigoni et al. 2018). Dominant bacteria included Bacteroidetes (especially *Hymenobacter*), Actinobacteria, Proteobacteria, and *Deinococcus-Thermus*. Additional studies on the caulosphere will likely reveal many new taxa and downstream consequences for both apple growth and disease resistance.

16.2.4 Anthosphere

Flowers are reproductive plant organs that function primarily to attract insects and other pollinators. Although studies of anthosphere communities have lagged behind those of the phyllosphere, it has long been recognized that flowers harbour microbiota, including pathogens, beneficial mutualists, and commensal symbionts (Aleklett et al. 2014; Rebolleda-Gómez et al. 2019). Flowers are highly ephemeral, but they provide a more protected and nutrient-rich habitat for microbes than leaves. In some cases, anthosphere diversity can exceed that of phyllosphere diversity (Ottesen et al. 2013; Wei and Ashman 2018), though in other cases, the opposite may be true (Junker et al. 2011). Furthermore, flowers are complex structures, thereby creating diverse niches for microbes. Microbial communities can vary substantially among different floral microhabitats, such as petals, stigma, pollen, or nectar (Pozo et al. 2012; Aleklett et al. 2014; Steven et al. 2018; Rebolleda–Gómez et al. 2019). Similar to leaves, flower communities are assembled as floral organs emerge, and microbes may originate from either ad hoc or local reservoirs in soils or other plant tissues. Additionally, flowers are regularly visited by insects that transport pollen among different flowers and/or individual plants, and there is good evidence that both insects and pollen can serve as reservoirs for introducing microbes to flowers (reviewed in Aleklett et al. 2014). As flowers have discrete lifespans, and offer relatively pristine environments at initiation, they provide rare opportunities for studying assembly of ecological communities "from scratch" along with ongoing changes over time in response to changing abiotic and biotic conditions, a topic of broad conceptual interest in ecology (Fierer et al. 2010; Shade et al. 2013b).

Although research is still limited, the apple has served as an important model system for studies on the floral microbiome. In particular, much work has been undertaken to understand the biology of *E. amylovora* along with the ecology of its interactions with its hosts and with other microbes. Much is known about both the specific abiotic conditions that facilitate *E. amylovora* establishment and the potential for other co-occurring microbes to act as antagonists and biocontrol agents (Steiner 1990; Johnson and Stockwell 1998; Bubán and Orosz-Kovács 2003; Pusey and Curry 2004; Pusey et al. 2009). Culture-based studies have suggested floral bacterial communities are dominated by Proteobacteria, especially *Pseudomonas*, and Actinobacteria (Pusey et al. 2009), while fungal communities are dominated by yeasts and yeast-like organisms, including *Aeurobasidium* and *Metschnikowia* (Ascomycota), as well as *Cryptococcus* (Basidiomycota) (Pusey et al. 2009; Vadkertiová et al. 2012).

The first culture-independent study of the apple flower microbiome was conducted by Shade et al. (2013a), who investigated temporal patterns of microbial succession under a management programme of treating flowers with streptomycin to control fire blight. They detected a total of 1,677 OTUs (97% similarity) from 13 phyla, and found that members of the *Saccharibacteria,* formerly designated as phylum TM7, and *Deinococcus-Thermus* were predominant. Furthermore, they identified distinct successional communities that develop in a predictable manner as flowers age (Shade et al. 2013a). Additional studies by Steven et al. (2018) and most recently by Cui et al. (2020) reported that communities were dominated by Proteobacteria, especially *Pseudomonas*, and several genera of Enterobacteriaceae (Table 16.1). Steven et al. (2018) compared communities across different floral tissues and found the highest taxonomic richness in petals (1274 OTUs at 97% similarity), followed by receptacles (1023 OTUs), stamens (790 OTUs), and stigmas (434 OTUs). To date, the most extensive sampling of the apple floral microbiome was reported by Cui et al. (2020), who sequenced 140 samples that assessed temporal succession, spatial variation, and the effects of inoculation by *E. amylovora.* Although this study focused only on stigmas, it detected a total of 46,809 amplicon sequence variants (ASVs) from 24 phyla (averaging 222 per sample), suggesting that the floral microbiome is highly diverse and variable in space and time.

16.2.5 Carposphere

In most plant systems, studies of the fruit microbiome have lagged behind those focused on leaves and flowers. However, the carposphere represents an exciting frontier for improving our basic understanding of plant–microbe interactions and developing innovative agricultural practices (Shade et al. 2017; Nelson 2018; Droby and Wisniewski 2018). Natural selection has produced fleshy fruits, such as apples, as a nutritional reward to attract seed-dispersing animals, but these same rewards also attract and support diverse microbial communities (Howe and Smallwood 1982; Cipollini and Stiles 1992; Tewksbury 2002; Whitehead et al. 2016). Further selection during crop domestication has produced fruits that are larger in size, more nutrient-rich, and lower in chemical defences against pests and pathogens (Meyer et al. 2012; Whitehead et al. 2017; Whitehead and Poveda 2019). As a result, domesticated fruits represent a resource-rich habitat for microbes. Furthermore, as fruits are harvested organs of many crops, fruit-microbe interactions directly affect fruit production economics and human health. Similar to flowers, fruits are complex organs composed of distinct tissues that provide unique habitats for microbes, including the seed (embryo, endosperm, and seed coat) and pericarp (endocarp, mesocarp, and exocarp). Microbes may either colonize any of these tissues as endophytes, or they may colonize fruit surfaces as epiphytes. Of the distinct tissues in fruits, the seed microbiome has received the most interest in plants due to its relevance for plant fitness and its potential importance for vertically transmitted microbes (Shade et al. 2017; Nelson 2018). The seed community and several other fruit tissues have been shown to share many taxa with floral communities (Compant et al. 2011; Mitter et al. 2017), suggesting that the floral microbiome provides an initial source of inoculum for developing fruits. However, fruits are much longer-lived than flowers, and many other microbes may colonize fruits during development. Little is known about processes that shape community assembly and succession during fruit development.

As with other plant organs, studies of fruit-associated microorganisms in apple have been largely focused on pathogenic microbes of economic importance and on natural antagonists that could be used as biological control agents (Turechek 2004; Droby et al. 2016). Many pre-harvest pathogens colonize both external surfaces and internal tissues of fruits, resulting in significant economic losses (Grove et al. 2003; Turechek 2004). Classic microbiological studies of these diseases have provided an important foundation for understanding the broader fruit microbiome. For example, sooty-blotch flyspeck is a common disease that affects the waxy layer of fruit surfaces. Previously, it was thought that this disease was caused by two co-occurring pathogens, *Gloeodes pomigena* (sooty blotch) and *Schizothyrium pomi* (flyspeck). However, culture-based studies have identified more than 60 fungal species associated with this disease, requiring a paradigm shift in our understanding of its epidemiology and control (Gleason et al. 2011). Several of the fungi detected along the fruit surface or found within the fruit are also responsible for causing significant economic losses after harvest. Major post-harvest diseases caused by fruit pathogens include blue mould (*Penicillium* spp.), grey mould (*Botrytis cinerea*), bitter rot (*Colletotrichum gloeosporioides*), and bull's eye rot (*Neofabrea* spp.) (Jijakli and Lepoivre 2004; Henriquez et al. 2004). Other microbes resident on the fruit surface, both pre- and post-harvest, have been isolated and studied for their potential beneficial role(s) in reducing disease severity by directly interacting with pathogens or inducing host resistance (Droby et al. 2016; Wisniewski et al. 2016). As we continue to learn more about the diversity and composition of the apple microbiome, it is likely that many other taxa will be discovered that either decrease (pathogen antagonism) or increase (pathogen facilitation) severity of fruit diseases (Busby et al. 2017).

A growing number of studies have used high-throughput amplicon sequencing to study the apple fruit microbiome (Table 16.1). The first was Leff and Fierer (2013), who evaluated bacterial communities associated with apples and other fresh produce in supermarkets and detected high levels of taxonomic richness with communities dominated by Proteobacteria. This pattern has been confirmed in subsequent studies (Vepštaitė-Monstavičė et al. 2018; Wassermann et al. 2019b; Abdelfattah et al. 2020, 2021). Fungal communities associated with apple fruit were first reported by Abdelfattah et al. (2016) who also focused on fruit at the consumer point-of-purchase. They demonstrated high levels of fungal diversity, including 1,591 OTUs from four phyla, dominated by Ascomycota. Furthermore, they also detected major differences in fungal communities across different fruit microhabitats, including the calyx-end, stem-end, and equatorial zone. *Penicillium* was the most abundant group in the fruit equatorial region, while *Alternaria* was more abundant in calyx- and stem-end samples (Abdelfattah et al. 2016). Recently, it was further confirmed that fruit microhabitat may be one of the most important drivers of composition for both bacteria and fungi (Wassermann et al. 2019b; Abdelfattah et al. 2020, 2021; Meakem et al. 2020), along with the specific orchard or site of fruit harvest (Vepštaitė-Monstavičė et al. 2018; Shen et al. 2018; Abdelfattah et al. 2021). In addition to these differences in composition, tissues might also differ in the total abundance of microbes harboured. Wasserman et al. (2019b) provided quantitative assessments of bacterial abundance across tissues, and estimated that the apple stem and seeds carried the highest numbers of bacterial cells (Wassermann et al. 2019b), but these tissues have been under-investigated in apple relative to the fruit surface.

Most recently, a global analysis of the microbiome of 'Royal Gala' apples at harvest maturity was conducted to assess whether apple fruits possess a core microbiome shared across regions, climates, and management regimes (Abdelfattah et al. 2021). This effort involved the participation of plant pathologists in various apple production regions, including the United States (West Virginia, New York, and Washington State), Canada (Ontario and New Brunswick), Uruguay, Israel, Turkey, Italy, Spain, and Switzerland. Altogether, 504 samples were analyzed from 21 sites with eight replicate trees per site and three fruit microhabitats (stem-end, calyx-end, and equatorial region). Taxa present in at least 75% of samples were considered part of the core microbiome, consisting of two bacterial genera (*Sphingomonas* and *Methylobacterium*) and six fungal genera (*Aureobasidium*, *Cladosporium*, *Alternaria*, *Filobasidium*, *Vishniacozyma*, and *Sporobolomyces*). Two fungal genera, *Aureobasidium* and *Cladosporium,* were the most prevalent, occurring in up to 96% of samples. These core taxa represented only a small subset of the microbial diversity detected globally in apple fruit, which included a total of 558 bacterial and 822 fungal genera (Abdelfattah et al. 2021). Although many fewer taxa were identified as part of a core microbiome in apples compared to other plants (Xu et al. 2018; Abdelfattah et al. 2021), this information provides a critical starting point for the development of biological strategies for disease management that are consistently effective across sites and regions (Toju et al. 2018).

As technologies have advanced, studies of the fruit microbiome have also moved beyond composition and diversity to further explore the metagenome and describe the microbiome function. Recently, Angeli et al. (2019) used a shotgun metagenomic sequencing approach to taxonomically and functionally characterize bacterial and fungal communities present on surfaces of apple cv. 'Pinova', collected from an organic orchard in Belgium, with a particular interest in biocontrol agents. Taxonomic assignment with MG-RAST (Metagenomics Rapid Annotation using Subsystem Technology; Meyer et al. 2008) confirmed that fungal communities were dominated by Ascomycota, and bacterial communities were dominated by Bacteroidetes, followed by Firmicutes, and then Proteobacteria (Angeli et al. 2019). Moreover, gene function was annotated using both MG-RAST (Meyer et al. 2008) and Integrated Microbial Genomes with Microbiome Samples (IMG/M) systems (Chen et al. 2017), but most genes (>70%) could not be annotated. Nevertheless, 65 genes related to biocontrol were identified, though many were assigned to known pathogenic species (Angeli et al. 2019). The power of this approach will likely increase dramatically with increasing numbers of annotated sequences deposited in databases.

16.3 Factors Influencing Spatial and Temporal Variation in Community Structure

A complex suite of factors drives variation in plant microbiome diversity and composition. These include host factors such as genotype, phytochemical composition, and other functional traits (Köberl et al. 2013; Kembel et al. 2014; van Dam and Bouwmeester 2016; Wagner et al. 2016), microbial community processes, such as competition and facilitation (Agler et al. 2016; Hassani et al. 2018), as well as abiotic factors, such as climate, season, soil type, nutrient availability, and pest management practices (Schreiter et al. 2014; Haesler et al. 2014; Sanchez–Barrios et al. 2017; Fitzpatrick et al. 2018; Timm et al. 2018). Communities may vary in time during plant or plant organ development (Shade et al. 2013b; Chaparro et al. 2014; Wagner et al. 2016; Arrigoni et al. 2018), or even throughout the day based on the circadian clock of the plant (Hubbard et al. 2018). Furthermore, these many factors may interact in complex ways to shape communities. For example, it has been demonstrated that host genotype and abiotic factors shape *Arabidopsis* microbial communities by interacting with "hub" taxa—strongly interconnected taxa that have cascading effects on other members of the microbial community (Agler et al. 2016). As of now, we still have a limited understanding of the major drivers of community structure in the apple microbiome. However, as sequencing costs continue to drop and new bioinformatics tools become available, more broad-scale sampling efforts are allowing us to assess variation in the apple microbiome at multiple spatial and temporal scales. In this section, we discuss four interrelated factors that are likely to be critical in shaping the apple microbiome structure. These include host genotype, domestication, phytochemical composition, and pest management practices.

16.3.1 Genotype Effects

The extent to which plant microbiome composition is, in part, determined by plant genotype is of considerable interest to both evolutionary biologists and plant breeders, as this determines the potential for selection to act on the combined plant and microbial genome, that is, the holobiont (Zilber-Rosenberg and Rosenberg 2008; Vandenkoornhuyse et al. 2015; Douglas and Werren 2016; Rosenberg and Zilber-Rosenberg 2018). Strong host genotype effects on the microbiome have been reported in apple (Liu et al. 2018; Arrigoni et al. 2018; Van Horn and Mazzola 2019), poplar (Bálint et al. 2013), wild mustard (Wagner et al. 2016), sunflower (Leff et al. 2017), and pine (Pérez–Izquierdo et al. 2017; Gallart et al. 2018). In other cases, host genotype may have only weak effects (Weinert et al. 2010). Host genotype effects on the microbiome may be driven either by particular suites of host traits that act as filters of environmental communities or by vertical transmission of microbes during sexual (parent to seed) or asexual (vegetative) reproduction.

In domesticated apples, trees are vegetatively propagated by grafting/budding, and the microbiome may be shaped by both the rootstock and scion genotypes. The rootstock genotype has repeatedly been shown to shape recruitment and ultimate composition of the rhizosphere and the root endophytic microbiome (St. Laurent et al. 2010; Liu et al. 2018; Wang and Mazzola 2019a; Deakin et al. 2019). For instance, in an orchard field trial, the rhizosphere bacterial community structure differed between 'Gala'/Geneva (G.) 41 and 'Gala'/Malling (M.) 26 trees (Wang and Mazzola 2019a). In addition to rhizosphere communities, the root endophytic microbiome can differ among apple rootstock genotypes when planted in the same orchard soil (Van Horn and Mazzola 2019). Specifically, Geneva series rootstocks (G.41 and G.890) were found to differ from one another and from those of Malling series (M.9 and M.26), but Malling series rootstocks were similar to one another (Van Horn and Mazzola 2019).

In some, but not all, cases, microbiomes of above-ground tissues have also been shown to differ based on both scion and rootstock genotypes. Arrigoni et al. (2018) found differences in community composition between 'Golden Delicious' and 'Gala' cultivars for both bacteria and fungi. However, Steven et al. (2018) found no differences in floral bacterial communities among 'Braeburn', 'McIntosh', and 'Sunrise' apples. Thus, the strength of host genotypic effects may vary across different tissues or different microbial taxa. Recently, Liu et al. (2018) compared bacterial and fungal stem endophyte communities among three scion cultivars ('Royal Gala', 'Golden Delicious', and 'Honeycrisp') grafted onto two rootstocks (M.9 and M.111), as well as of shoots of ungrafted rootstocks. Although no clear host genotype effects were detected for the bacterial community, their findings suggest that both rootstock and scion genotypes can influence fungal community composition, with the highest differences attributed to scion genotypes (Liu et al. 2018; Fig. 16.2a). Interestingly, the composition of fungal microbiota of 'Golden Delicious' and 'Royal Gala' was similar, but the community associated with 'Honeycrisp' was distinctly different from the other two cultivars. These findings may be explained by the relative genotypic relatedness of 'Royal Gala' and 'Golden Delicious' compared to that of 'Honeycrisp' (Fig. 16.2b). Based on these preliminary findings, a more comprehensive analysis of the relationship between apple genotypes/pedigrees and endophytic fungal microbiomes is currently in progress, wherein endophytic microbes are being characterised from several apple progenitor species and apple evolutionary clusters.

Host plant genotypes can interact in complex ways with the environment to shape both host plant phenotypes and microbiome compositions (Leon et al. 2016; Wagner et al. 2016; Hewavitharana and Mazzola 2020). For instance, Wang and Mazzola (2019b) found that apple rootstock genotypes vary in their abilities to recruit beneficial microbial constituents to the rhizosphere from a soil microbiome specifically modified via applications of bioactive soil amendments. However, there are limited studies on how genotype × environment interactions shape the apple microbiome. This is likely due, in part, to sequencing costs as these studies require large enough sample sizes to effectively disentangle complex interactions. With advances in technology, these large-scale studies are now becoming more feasible. Our ongoing research aims to assess how genotypes interact with pest management practices to shape the microbiome using a large experimental orchard with replicated mixed-cultivar blocks. Additional work in this area will improve our fundamental understanding of host-microbiome interactions while helping to optimize the design of management practices to account for host genotype effects.

16.3.2 Domestication

Throughout the history of agriculture, humans have, mostly inadvertently, altered the plant microbiome. In natural ecosystems, selection favours plants that support a beneficial microbiome; however, during domestication, many of these selective pressures are removed (Vandenkoornhuyse et al. 2015; Pérez-Jaramillo et al. 2016; Wallenstein 2017; Hassani et al. 2018). Apples were originally domesticated from wild populations of *Malus sieversii* (Lbd.) Roem native to the Tian Shan mountains of Central Asia. Cultivated apples travelled west along the Silk Road trading routes and came into contact with several other wild species, including *M. sylvestris, M. orientalis,* and *M. baccata* that made secondary contributions to the gene pool of modern apples (Cornille et al. 2014). Domestication has greatly altered fruit quality traits, including sugar and acid content (Khan et al. 2014; Ma et al. 2015), and has reduced both concentration and diversity of phenolic metabolites (Whitehead and Poveda 2019). However,

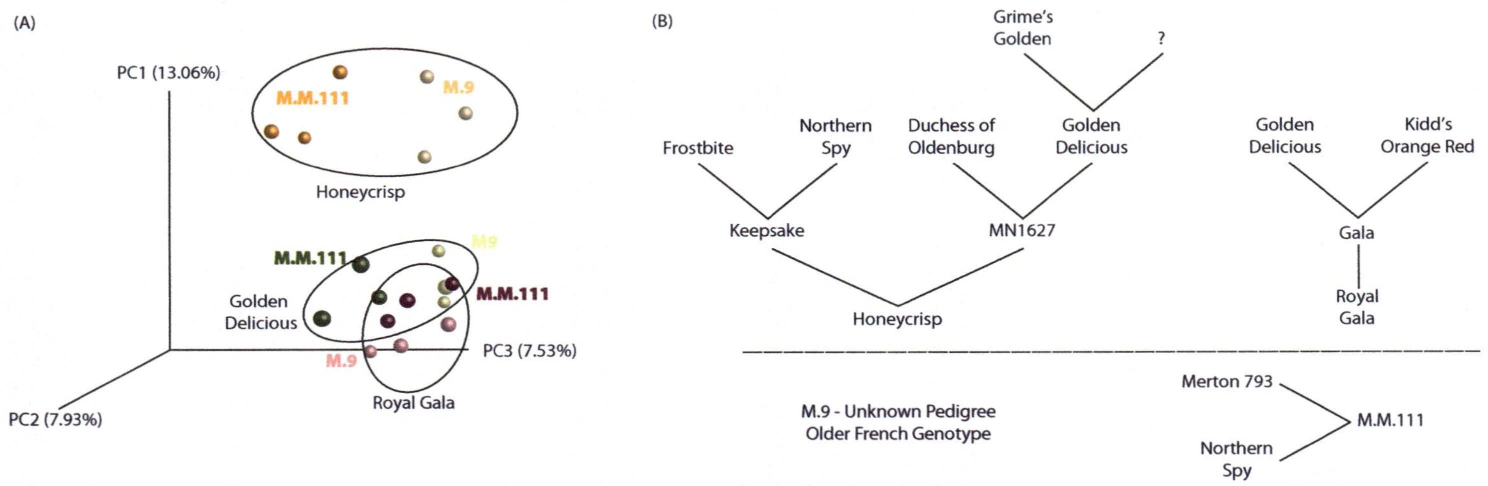

Fig. 16.2 Principle Coordinate Analysis (PCoA) with data from Liu et al. (2018) showing differences in the composition of caulosphere fungal communities across scion and rootstock genotypes (**a**) and known pedigrees of these genotypes (**b**). The PCoA is based on Bray–Curtis dissimilarity, revealing that scion cultivars 'Golden Delicious' and 'Royal Gala' are more similar than either cultivar is to 'Honeycrisp', which has a less similar, but more complex pedigree. Figures are adapted from Liu et al. (2018)

these traits were historically highly variable among apple cultivars. Although most crops experience extreme genetic bottlenecks during domestication, a relatively high genetic diversity was initially maintained in the cultivated apple, perhaps due to its mating system, which requires outcrossing and regional variations in tastes and uses (Cornille et al. 2012). However, genetic diversity has markedly declined in recent decades due to the low numbers of cultivars selected for modern global markets based on productivity, size, texture, disease resistance, and storability (Khan et al. 2014; Cornille et al. 2014).

To our knowledge, there are no studies that have specifically assessed the role of domestication in shaping the apple microbiome. However, we do know that domestication greatly alters plant phenotypes, such as chemical defences, that shape plant-biotic interactions (Meyer et al. 2012; Whitehead et al. 2017; Whitehead and Poveda 2019). Moreover, there is accumulating evidence in other crops that domestication and modern breeding have contributed to complex shifts in plant microbiome structure (Bulgarelli et al. 2015; Leff et al. 2017; Pérez-Jaramillo et al. 2017). A meta-analysis of the effects of domestication on bacterial communities revealed a broad pattern wherein wild plants are more enriched in Actinobacteria and Proteobacteria, while domesticated crops are more enriched in Bacteroidetes (Pérez-Jaramillo et al. 2018). Considering the long evolutionary history of plant–microbe interactions, wild ancestors of crops may serve as important resources for the discovery of beneficial microbes that can help improve the performance and stress tolerance of modern cultivars (Berg and Raaijmakers 2018). Thus, future studies should seek to characterize the diversity associated with wild apples and understand how the process of domestication and modern breeding have shaped the structure and function of the apple microbiome.

16.3.3 Host Plant Chemistry

The phytochemical profile of plants is undoubtedly a key factor shaping microbiome composition, and may provide much of the mechanistic underpinning behind demonstrated genotype and domestication effects. Phytochemicals are key in shaping biotic interactions, as they are present in all plant organs/tissues, and they can function to attract, reward, deter, kill, or modulate consumer physiology (Dicke and Takken 2006). Moreover, phytochemicals are responsible for conferring characteristic odours, flavours, and colours for different plant species and cultivars and have important health consequences for humans. In fact, many of the health benefits of a plant-rich diet are attributed to antioxidant and free-radical scavenging activities of phytochemicals (Dillard and German 2000; Wang et al. 2014). Studies of the rhizosphere have provided strong evidence that the phytochemical composition of root exudates is one of the most important factors shaping the composition of rhizosphere communities (Bais et al. 2006; Sasse et al. 2018). Root exudates include small molecules, such as sugars, amino acids, organic acids, phenolics, and other secondary metabolites, as well as high-molecular-weight compounds such as polysaccharides (mucilage) and proteins (Bais et al. 2006). These molecules provide a carbon source that fuels microbial activity and can also specifically function as signalling (Venturi and Keel 2016) and/or defence compounds (Baetz and Martinoia 2014). Above-ground, it is well-known that phytochemistry influences patterns of pathogen incidence and increasing evidence suggests it may also influence the broader microbial community (Pusztahelyi et al. 2015; Aizenberg-Gershtein et al. 2015).

Apples are rich in several groups of phytochemicals, including phenolics, a large class of compounds with diverse consequences for ecology and human health, as well as volatile organic compounds that are essential determinants of flavour and also have numerous ecological roles (Lee et al. 2003; Natale et al. 2003; Piskorski and Dorn 2010; Zhang et al. 2010a; Cuthbertson et al. 2012). The composition of these compounds in various apple tissues and exudates are known to depend highly on both scion and rootstock genotypes (Holderbaum et al. 2014; Kviklys et al. 2015; Leisso et al. 2017, 2018; Whitehead and Poveda 2019) and may also change substantially in response to insect and microbial interactions (Dangl and Jones 2001; Kessler and Baldwin 2002; Chisholm et al. 2006; Whitehead and Poveda 2011). In the apple rhizosphere, roots exude a diverse group of phenolics, especially the dihydrochalcones phloretin and phloridzin (Hofmann et al. 2009; Leisso et al. 2017, 2018). These compounds are more abundant in trees affected by apple replant disease, particularly during the early onset of disease symptoms (Hofmann et al. 2009). Furthermore, several bacteria and fungi are known to degrade these and related compounds (Jayasankar et al. 1969; Zhang et al. 2010b; Huang et al. 2013), suggesting that phenolic root exudates may play a key role in shaping both trophic structure and community composition of the apple rhizosphere. Aboveground, phenolic diversity in apple fruits is correlated with fungal endophyte diversity (Meakem et al. 2020), and further efforts are underway to explore the complex relationships between the apple metabolome and microbiome. The complex feedback loops through which chemistry may affect the microbial composition, and vice versa, render the causal nature of these relationships difficult to decipher. However, technologies such as axenic plant propagation (Leisso et al. 2018) or the production of transgenic trees with altered metabolism (Li et al. 2018) provide opportunities to study the role of host plant chemistry as a key mechanistic link shaping plant–microbe interactions.

16.3.4 Agricultural Practices

Plant cultivation reduces competition, stabilizes water regimes, and decouples the process of decomposition from plant growth by removing organic matter and detritus from fields. Since the development of the Haber–Bosch process, we have dramatically altered the way in which cultivated plants access nitrogen, often a major limiting factor for growth. Cultivated plants are also treated with pesticides, biocontrol agents, and various amendments to reduce damage from insect and microbial pests. These practices likely have major consequences for plant–microbe interactions (Yeoh et al. 2016; Turcotte et al. 2017; Wallenstein 2017; Hartman et al. 2018).

Apples are attacked by numerous insects and pathogens, and they require intensive management to produce marketable fruit (Beers et al. 2003; Grove et al. 2003; Turechek 2004). Although pest management practices often target specific microbial pathogens that are causal agents of disease, these practices are also likely to affect the broader microbial community, including many commensal or mutualistic organisms (Walter et al. 2007). As many non-pathogenic microbes may either interact with pathogens or serve as biocontrol agents (Droby and Wisniewski 2018), understanding the impacts of pest management practices on the microbiome is critical for improving the sustainability of apple production.

Several studies have suggested that there are large differences in microbial communities between apple trees managed with organic versus conventional systems (Purin et al. 2006; Ottesen et al. 2009; Leff and Fierer 2013; Abdelfattah et al. 2016; Wassermann et al. 2019b), though this might not be a universal pattern (Ottesen et al. 2016). The most extensive sampling was conducted by Ottesen et al. (2009), who collected samples at six-time points from an experimental orchard with replicated organic and conventionally managed blocks. Epiphytic bacterial communities were washed from pooled leaf and fruit samples, and community composition was assessed using clone libraries of 16S rRNA sequences. They found differences in community composition between conventional and organic blocks at four of the six-time points (Ottesen et al. 2009). Notably, samples from the same experimental site that were collected in subsequent years and analyzed using amplicon sequencing did not reveal any consistent differences between conventional and organic treatments (Ottesen et al. 2016). This suggested the relative importance of management systems versus other factors in shaping the apple microbiome might be context-dependent and variable over time. Other studies have focused on differences between organic and conventional fruits at the consumer point-of-purchase and have found differences in community composition of both bacterial (Leff and Fierer 2013) and fungal communities (Abdelfattah et al. 2016). These studies demonstrate the potential effects of different production and processing systems on the microbial community. However, the mechanisms behind these effects remain unclear due to wide variations in specific management practices both within and across organic and conventional systems. Furthermore, post-harvest variables might contribute to differences in the fruit microbiome at the consumer point-of-purchase, including post-harvest sanitation practices, storage, supply chain, or handling practices at stores. Thus, there is much more to learn about the key drivers of microbial community composition in orchard environments and harvested fruits.

Several studies have investigated whether or not the antibiotic streptomycin, commonly used to control *E. amylovora*, can alter the apple microbiome. Unexpectedly, these studies have not detected any effect of streptomycin on bacterial communities associated with leaves, flowers, or soil (Yashiro and McManus 2012; Shade et al. 2013a, b; Walsh et al. 2014). In a study employing both culture-dependent and culture-independent techniques, Yashiro and McManus (2012) reported that streptomycin applications in apple orchards do not have long-term effects on the diversity or phylogenetic composition of phyllosphere bacterial communities. Surprisingly, the proportion of cultured bacteria resistant to streptomycin was lower in orchards with a history of streptomycin exposure than in non-exposed orchards. It is suggested that this unexpected finding may be attributed to the higher abundance of both *Sphingomonas* and *Pseudomonas* in non-exposed orchards, as these genera are known to have high levels of stable, intrinsic, or acquired resistance to streptomycin regardless of exposure (Yashiro and McManus 2012).

In another study, Shade et al. (2013b) compared bacterial community composition and succession on apple flowers that were either treated with streptomycin or controls. Although flowers treated with streptomycin had slightly lower phylogenetic diversity, they did not differ in overall community structure or succession compared to untreated flowers. Most recently, Walsh et al. (2014) reported that there were no differences in rhizosphere bacterial communities as a result of streptomycin applications across three experimental orchards in Switzerland. Furthermore, Duffy et al. (2014) provided quantitative analysis of the presence of antibiotic resistance genes in flower, leaf, and soil microbial communities from the same orchards and detected no effects of streptomycin use on the evolution of antibiotic resistance.

Collectively, the above-mentioned studies suggest that some major components of the apple microbiome may be surprisingly stable to disturbances caused by antibiotic use, though there is much more to learn about the context-dependency of these effects as well as the extent to which stability is due to resistance (a lack of sensitivity to disturbance) versus resilience (a quick recovery following a disturbance) (Shade et al. 2012b).

In addition to the effects of pre-harvest management practices on the microbiome, post-harvest treatments of fruits can substantially alter the fruit microbiome. Abdelfattah et al. (2020) evaluated the effects of apple fruit washing, waxing, and cold-temperature storage on bacterial and fungal compositions associated with different fruit tissues (stem-end, calyx-end, and equatorial region). They found a decrease in diversity along with marked differences in community composition following either washing or washing-waxing of fruits. The composition was also highly affected by storage time over a period of six months, and there were highly significant two- and three-way interactions among tissue, treatment, and time. These findings suggested that post-harvest practices had major effects on the microbiome associated with fruits, and therefore they might influence the effectiveness of any post-harvest chemical or biological control of pathogens.

Applications of commercial fruit waxes on apple have also been shown to affect the survival of food-borne pathogens infecting humans in various ways, including increased survival of *Listeria* (Macarisin et al. 2019), but with decreased survival of both *Escherichia coli* and *Salmonella* (Kenney and Beuchat 2002). Thus, an improved understanding of the functional consequences of post-harvest treatments for fruit microbiome structure will provide new opportunities to both improve food safety and control major post-harvest diseases associated with economic losses.

16.4 Consequences of the Microbiome for Plant Health

The significance of the plant and soil microbiome as determinants of agricultural ecosystems' overall health and productivity is well recognized (Berendsen et al. 2012; Stone et al. 2018). Belowground, the rhizosphere and root endosphere contribute significantly to plant health status by directly influencing growth or developmental processes (Bonkowski and Brandt 2002; Molina-Favero et al. 2008; Panke-Buisse et al. 2015; Santoyo et al. 2016), enhancing nutrient uptake or availability (Oldroyd 2013; Alegria Terrazas et al. 2016), and reducing abiotic or biotic stress (Tanaka et al. 2005; Zamioudis and Pieterse 2011; Raaijmakers and Mazzola 2016; Lata et al. 2018; Caddell et al. 2019). The historical study of disease suppressive soils has also provided clear insights into the role of the rhizosphere soil microbiome in limiting the development and severity of soil-borne plant diseases (Raaijmakers and Mazzola 2016). Aboveground, there is increasing evidence that the microbiome associated with both leaves and other above-ground organs can also modulate plant growth and nutrient acquisition and play a key role in plant defence against major pests and pathogens (Vorholt 2012; Compant et al. 2019). Overall, the microbiome has major consequences for crop yield and quality of harvested organs (Saminathan et al. 2018; Rho et al. 2020), the risks of food-borne pathogens (Berg et al. 2014; Macarisin et al. 2019), and the likelihood of post-harvest food loss (Buchholz et al. 2018; Droby and Wisniewski 2018).

Investigation of the apple rhizosphere and root endosphere has served as an important model for understanding the role of the microbiome in plant health, particularly with regard to disease suppression. Evidence suggests that the apple rhizosphere microbiome functions as an initial barrier to apple root infection by soil-borne pathogens, such as those associated with apple replant disease, through either direct pathogen suppression (Weerakoon et al. 2012) or induction of host defence responses (Zhu et al. 2016). In addition, the rhizosphere microbiome includes mycorrhizal fungi and growth-promoting bacteria that enhance uptake of nutrients and influence root development (Covey et al. 1981; Resendes et al. 2008; Dohroo et al. 2014). For instance, plant growth stimulation by *Azospirillum brasilense*, a diazotroph commonly recovered from the apple rhizosphere soil and associated microflora (Cohen et al. 2004), has been attributed to both promotion of lateral root development and enhanced levels of plant-available nitrogen (Dobbelaere et al. 2003). Those microbes that can colonize root tissues as endophytes may have even broader consequences for tree physiology. Rho et al. (2020) found that inoculation of apple trees with select endophytes resulted in decreased stomatal density, delayed leaf senescence, and increased lateral root biomass for trees growing under poor nutrient conditions. Notably, they also observed significant increases in both fruit glucose content and biomass, emphasizing the systemic effects that endophytes may have on developmental processes across plant organs.

Although most studies documenting consequences of the microbiome for apple tree health have focused on below-ground communities, several studies have investigated links between the microbiome of above-ground organs and plant physiology or disease status. For example, several epiphytic yeasts from fruit surfaces that have been evaluated for use as biocontrol agents, including *Rhodotorula glutinis* and *Aureobasidium pullulans*, are known to cause apple russet when applied on fruits (Matteson Heidenreich et al. 1997; Gildemacher et al. 2006). The capability of yeast isolates to persist on fruit surfaces and cause russet may be determined by cutinase activity, which could disrupt the fruit cuticle and facilitate both epiphytic survival and phytotoxin production responsible for epidermal cell death (Matteson et al. 1995; Matteson Heidenreich et al. 1997). Another recent study on the apple flower microbiome investigated links between the floral microbiome and *E. amylovora* infection (Cui et al. 2020). They observed significant shifts in the community composition of flowers in response to *E. amylovora* inoculations, with high abundance of these bacteria developing on stigmas of almost all inoculated flowers. However, only 42% of inoculated flowers developed disease symptoms (Cui et al. 2020). Although they were unable to definitively link variation in disease development to the broader microbiome community structure, the pattern of low disease incidence despite high pathogen abundance suggests that disease development is the result of complex interactions between microbial communities and host-level factors, such as chemical defences and disease tolerance (Pusey et al. 2009; Medzhitov et al. 2012; Cui et al. 2020).

Microbes may also interact with major insect pests of apples to determine plant health. For example, the attractiveness of apple fruits to codling moth, a worldwide pest, depends on fruit colonization by *Metschnikowia* yeasts (Witzgall et al. 2012). Such yeasts are suggested to act as codling moth mutualists by decreasing mould incidence in feeding tunnels of larvae, thereby decreasing larval mortality (Witzgall et al. 2012). Thus, the use of yeasts as biocontrol agents against the disease should be evaluated within the context of potential unintended increases in codling moth susceptibility. In one case, applying a yeast biocontrol agent against powdery mildew resulted in increased codling moth oviposition (Alaphilippe et al. 2008). In general, these intricate feedback loops among the microbiome, host plant, and other community members, such as insect herbivores and pollinators, are likely ubiquitous. However, we still have a limited understanding of the cascading consequences of plant–microbe interactions.

16.5 Manipulating the Microbiome to Improve Sustainability of Orchard Ecosystems

As we expand our knowledge of the structure and function of the microbiome, there are wide opportunities to shape and engineer communities that will increase the sustainability of agricultural systems (Busby et al. 2017; Wallenstein 2017). In fact, extending our management systems to include the microbiome will likely be one of the most transformative agricultural advances of the twenty-first century (APS 2016). To date, however, effective management of the microbiome in a manner that yields consistent and predictable plant responses has largely remained elusive. Different strategies for microbiome management all aim to increase the yield and quality of harvested products and/or increase resistance to pests and pathogens. Generally, these strategies can be divided into at least three broad categories: (1) inoculations with beneficial microbes (Jones et al. 2007; Berg 2009; Trabelsi and Mhamdi 2013; Coleman-Derr and Tringe 2014); (2) environmental/resource manipulation to alter microbiome structure and function (Bakker et al.

2012; Raaijmakers and Mazzola 2016); and (3) breeding or gene editing of plant traits to alter microbiome structure and improve the function of the holobiont (Bakker et al. 2012; Nogales et al. 2016). Thus far, efforts to manipulate the microbiome in apple have focused on the first two of these categories. Here, we briefly review these efforts and the challenges involved.

16.5.1 Microbial Inoculations for Enhanced Plant Growth and Disease Resistance

Perhaps the most obvious strategy to manipulate the microbiome is direct inoculation with microorganisms that can increase the yield or quality of harvested products (i.e., biostimulants; du Jardin 2015). The application of microbial inoculants to improve either growth or manage apple diseases has long been explored, but examples of successful adoption in commercial plant production systems are limited. A long-acknowledged limitation of inoculants is the inconsistent ability of beneficial microbes to establish, persist, or achieve the same functional outcome in complex field environments (Sundin et al. 2009). However, opportunities may be identified that will enable effective use of such inoculants as advances are made in formulating/designing conditions optimal for their activities.

Several studies in apples have demonstrated increased growth following soil inoculations with either specific growth-promoting bacteria or mycorrhizal fungi (Mosa et al. 2016; Rho et al. 2020). Plant growth-promoting bacteria include strains of *Pseudomonas, Aztobacter, Bacillus*, and *Burkholderia* (Aslantaş et al. 2007; Khosravi et al. 2009; Kurek et al. 2013). Many of these organisms likely improve growth by producing phytohormones that directly increase plant growth, such as indole acetic acid (IAA) and cytokinin (Aslantaş et al. 2007; Kumar et al. 2015). Various bacteria and mycorrhizal fungi also improve growth by increasing the availability of soil nutrients or associating with roots to enhance nutrient uptake (Covey et al. 1981; Fortuna et al. 1996; Sharma et al. 2002; Yang et al. 2014). Although soil inoculations can provide clear benefits, their effectiveness can be highly variable and depend on plant genotype (Aslantaş et al. 2007). Inoculations with consortia, rather than individual strains, may be one approach to improve the survival and resilience of the inoculum and increase its effectiveness (Grzyb et al. 2012; Woo and Pepe 2018; Rho et al. 2020). For example, Rho et al. (2020) used a mixture of nine endophytes with known growth-promoting properties from wild cottonwood and willow trees to inoculate 'Honeycrisp' apple trees. These inoculations decreased stomatal density, delayed leaf senescence, increased lateral root biomass, increased trunk width, and increased fruit size and sugar content, thereby highlighting the broad potential downstream consequences of plant–microbe interactions for plant growth and fruit quality.

Microbial inoculations have also been widely investigated in apples as biocontrol agents against below-ground pathogens (Mazzola and Freilich 2017), including crown and root rots and apple replant disease (Turechek 2004; Mazzola and Manici 2012). In general, these efforts have focused on a relatively small set of organisms, including the biocontrol fungi *Trichoderma* and *Gliocladium* (Smith et al. 1990; Roiger and Jeffers 1991; Singh et al. 2017) and bacteria such as *Bacillus, Rhizobium, Enterobacter,* and *Pseudomonas* (Utkhede 1987; Kawaguchi et al. 2012; Ju et al. 2014; Singh et al. 2017; Sharma et al. 2017). Although inoculations with single biocontrol strains often show promise in initial trials, these strategies have often failed under variable field conditions. This may be due largely to the complex aetiology of soil-borne diseases such as apple replant disease, the causal agents of which are often variable from site to site (Mazzola and Manici 2012).

The efficacy of inoculants may improve as we increase our knowledge of how plant genotypes shape associations with potentially beneficial microbial inoculants. For instance, the apple rootstock G.41 supports a significantly greater abundance of the disease suppressive bacterium

Bacillus megaterium in the root endosphere than M.9 rootstock (Zheng and Sinclair 2000; Van Horn and Mazzola 2019). Success may also be improved with the development of consortia of microbial communities that are more multifunctional and resilient or with the combination of inoculations with manipulation of the soil habitat (e.g., anaerobic soil disinfestation or *Brassica* seed meal amendments; see below) to selectively favour persistence of an introduced agent. However, the biocontrol model for the management of soil-borne disease may be generally quite limited. Inoculations are often restricted to the time of planting, and biocontrol agents must be established within complex pre-existing communities. These challenges are reflected in the fact that only a very limited number of biocontrol agents for soil-borne diseases have been brought to market (Pertot et al. 2015).

Aboveground, biocontrol has also been of considerable interest for the control of several key diseases, including fire blight and apple scab (Johnson and Stockwell 2000; Carisse and Dewdney 2002; Bonaterra et al. 2012). The most extensive area of research has been in the development of biocontrol agents for fire blight disease (Johnson and Stockwell 1998; Bonn and van der Zwet 2000). Most strategies have focused on inoculations of flowers with antagonistic bacteria from several genera, including *Psuedomonas, Pantoea, Bacillus,* and the yeast *Aureobasidium*, that limit either establishment or growth of *E. amylovora* (e.g. Ishimaru et al. 1988; Wilson and Lindow 1993; Pusey 1998; Pusey et al. 2009). Although at least six commercially available biological control products have been developed based on these efforts (Mechan Llontop et al. 2020), control has not always been satisfactory (Sundin et al. 2009). As technologies for screening microbial communities have advanced, the search for effective fire blight biocontrol agents has continued, and many potential antagonists have been identified in recent years (Sharifazizi et al. 2017; Ait Bahadou et al. 2018; Mikiciński et al. 2020; Mechan Llontop et al. 2020). Numerous studies have also explored the potential for biocontrol of fire blight using bacteriophages (Svircev et al. 2006; Jones et al. 2007; Nagy et al. 2012). A large number of phages have been isolated from apple soils and plant tissues that are capable of infecting *E. amylovora,* and in some cases, field trials have demonstrated efficacy levels rivalling those of antibiotic use (Gill et al. 2003; Svircev et al. 2006; Nagy et al. 2012; Akremi et al. 2018; Sharma et al. 2019). Resources for the development of phages as biocontrol agents for fire blight and for other bacterial diseases continue to expand as a result of both whole-genome sequences for many potential taxa (Esplin et al. 2017; Sharma et al. 2018) and methods for genetic engineering of phages to improve host colonization (Born et al. 2017).

Post-harvest quality and marketability of fruit are also negatively impacted by many pathogens that cause significant economic losses, and this has been another key area where biocontrol agents have been investigated (Droby et al. 2016). Apples in storage are damaged by organisms such as *Botrytis cinerea*, *Penicillium expansum*, and *Colletotrichum*, which can cause 5–25% fruit loss if not effectively controlled (Jijakli and Lepoivre 2004). Therefore, developing biocontrol agents as alternatives to chemical treatments of these pathogens has been the focus of considerable research efforts over the last 30 years (Droby et al. 2016). Efforts for the discovery of effective agents have covered a wide range of potential microorganisms derived from apple leaves (Janisiewicz 1988) to rainwater (Mechan Llontop et al. 2020) to Antarctic soils (Vero et al. 2013). Several products have been developed based on these efforts for use on apples, such as BioSave, a formulation of *Pseudomonas syringae* (Janisiewicz and Jeffers 1997; Janisiewicz and Peterson 2004), yet most have had limited commercial use. With the availability of next-generation sequencing (NGS) technologies, post-harvest biocontrol efforts are moving toward new approaches that will take into account not only one-to-one antagonisms between a pathogen and a biocontrol agent, but the community and environmental context of these interactions (Droby and Wisniewski 2018).

Overall, there is clear evidence that microbial inoculations have the potential to impact diverse aspects of plant physiology, growth, and biotic resistance, ultimately shaping the yield and quality of harvested fruits along with the sustainability of orchard ecosystems. The main shortcoming that must be overcome to develop a successful biostimulant or biocontrol product is similar—field trials generally show weak and/or variable effectiveness compared to chemical fertilisers and pesticides (Burr et al. 1996; Sundin et al. 2009; Nagy et al. 2012). As knowledge of the broader community structure and functional interactions among natural microbial communities on aboveground tissues increases, our ability to design inoculants will likely dramatically improve. A number of strategies have been proposed to improve the effectiveness of biocontrol agents. These include the following: designing microbial consortia that form resilient networks and function synergistically, developing formulations of inoculants that include nutritional supplements to improve survival, improving biocontrol agents through artificial selection under conditions of osmotic stress to increase physiological adaptation to harsh abiotic conditions in the field, and improving biocontrol agents with genetic engineering (reviewed in Bonaterra et al. 2012). Much additional work is necessary to optimize the design of beneficial inoculants for apples, characterize the mechanisms of their effects, and understand how outcomes will vary depending on the cultivar, rootstock, and agricultural environment.

16.5.2 Manipulation of the Resource Environment to Re-Structure Microbial Communities

Considering the fundamental limitations of single-strain inoculations or simple consortia as a means to direct the structure of complex natural microbial communities (Mazzola and Freilich 2017), achieving desired outcomes may instead require more holistic approaches that take into account complex processes of succession and microbial syntrophy that determine microbiome function (Morris et al. 2013; Hewavitharana et al. 2019). This could include directed manipulation of the functional microbiome indigenous to the orchard ecosystem via alterations to the resource environment (analogous to the concept of prebiotics) in ways that benefit apple production systems.

In apples, a number of management practices have been evaluated for their abilities to support a beneficial below-ground microbiome. These efforts have focused in particular on structuring healthy rhizosphere communities that can suppress apple replant disease. Historically, management of apple replant disease has focused on chemical fumigation during the period of orchard establishment. Although this treatment yields a positive growth response of young trees on replant sites, this effect is generally of limited duration, and it corresponds with a transient reduction in populations of soil-borne pathogens. For example, when evaluated at 12–18 months post-planting, trees cultivated in fumigated soils had either equivalent or greater levels of root infestations from *Pratylenchus penetrans* and *Pythium* spp. compared to those in untreated orchard replant soil (Mazzola et al. 2015; Wang and Mazzola 2019a). Furthermore, the microbiome structure in fumigated soils is indistinguishable from that of trees established in untreated orchard replant soil (Mazzola et al. 2015; Wang and Mazzola 2019a).

One strategy that can selectively modify the microbiome to a structural and functional state that limits the activity and proliferation of soil-borne pathogens is the use of selective orchard soil amendments. Several studies have addressed the usage of soil amendments of non-specific composition, such as compost and biochar, as a means for control of soil-borne diseases in orchard ecosystems, with mixed results (Rumberger et al. 2004; Khorram et al. 2019). In contrast, defined soil amendments that selectively amplify disease suppressive components of the soil/rhizosphere microbiome have consistently provided control of apple root pathogens in field trials (Mazzola et al. 2015; Hewavitharana et al. 2019; Wang and Mazzola 2019a). For example, pre-planting application of a Brassicaceae seed meal formulation in replant orchards induced transformation of the soil and rhizosphere microbiome and conferred disease suppression and tree growth similar to that achieved with fumigation (Mazzola et al. 2015; Wang and Mazzola 2019a). Interestingly, the microbiome composition in seed meal amended soils differed across orchard sites, but there was consistent amplification of certain functional characteristics within the modified community (Somera et al. 2019). Specifically, while the mycoparasitic/antibiotic-producing fungus *Chaetomium globosum* dominated in the apple rhizosphere post-seed meal application in certain orchards, fungi from the *Hypocrea/Trichoderma* complex with the same attributes dominated in others (Somera et al. 2019). This outcome raises the need to focus on functional consortia rather than individual microorganisms when attempting to define elements with the potential to generate a specific desired outcome.

As an alternative to specific amendments, the soil microbiome can be transformed through the implementation of specific processes that increase the production of anti-microbial metabolites. One such process, termed anaerobic soil disinfestation (Blok et al. 2000; Momma et al. 2013), brings about sequential transformations through aerobic, facultatively anaerobic, and anaerobic soil phases, thus shifting the composition of the soil microbiome and leading to effective soil-borne disease control in an orchard environment (Hewavitharana and Mazzola 2016, 2020; Hewavitharana et al. 2019). In particular, changes in the Firmicutes community during the anaerobic phase have been linked to increased specific metabolites with known antimicrobial activity, such as small chain organic acids, methyl sulphide compounds, hydrocarbons, and p-cresol (Hewavitharana et al. 2019). Such identified metabolites, along with microbes producing these metabolites, may serve as useful biomarkers in monitoring the soil environment during anaerobic soil disinfestation to determine the required treatment duration and predict treatment efficacy.

Overall, the success of holistic strategies, such as *Brassica* seed meal amendments and anaerobic soil disinfestation, demonstrate that management practices that change the resource environment of the microbial community may provide more robust and long-term solutions to microbiome management than inoculations. Even simple management changes, such as rotating rootstock genotypes or re-planting trees in orchard grass lanes rather than previously planted rows, can have profound impacts on the apple tree microbiome and reduce populations of plant pathogens (Rumberger et al. 2004, 2007; Kelderer et al. 2012; Deakin et al. 2019). Many other practices have been used for decades by small-scale, holistic, and no-spray growers to support beneficial microbial associations, but most are yet to be scientifically investigated (Jacke and Toensmeier 2005; Phillips 2012). As we develop a more sophisticated understanding of what constitutes a "healthy" apple microbiome, we can also further explore other holistic management practices that can support these beneficial communities. Ultimately, combining practices that support a healthy indigenous microbiome with specific interventions, such as inoculations, may prove the most effective approach to tree and microbiome management.

16.6 Future Research Priorities

16.6.1 Further Defining the Core Microbiome

Despite many factors that may lead to spatial and temporal variation, some microbial taxa are ubiquitous, while others may consistently appear in particular microhabitats. A critical challenge in microbial ecology is to determine whether or not certain environments possess a core microbiome, defined as a suite of microbial taxa that repeatedly appear in a particular habitat regardless of other environmental factors (Hamady and Knight 2009; Shade and Handelsman 2012). Defining a core provides a critical baseline for understanding which components of a community are stable in response to disturbance and, thus, may play critical functional roles.

Recently, a large collaborative effort defined a core microbiome for fruit of 'Royal Gala'

(Abdelfattah et al. 2021). By identifying a number of stable fungal and bacterial genera that occur across the globe (see Sect. 16.2.5 on the Carposphere), this study provides a starting point for prioritizing taxa for further research or for inclusion in culture collections (Busby et al. 2017; Toju et al. 2018). Similar global studies have been conducted on the citrus rhizosphere microbiome (Xu et al. 2018), as well as the carposphere of nine cultivars of wine grapes (Zhang et al. 2019). Continued work identifying a core microbiome associated with additional apple microhabitats (e.g., rhizosphere and phyllosphere) or across microhabitats of healthy trees should be a top research priority in apple. Additional studies should also identify the extent to which core microbiomes vary across apple genotypes, further informing our understanding of the extent to which host genotype shapes microbiome composition. These studies will require coordinated efforts across research groups and standardized methods for sample collection, metadata, molecular analysis, and downstream data processing in order to provide broad-scale comparative data across apple-growing regions.

16.6.2 Functional Analysis of the Microbiome

New sequencing technologies, particularly whole-genome shotgun approaches, will provide data beyond community structure that will also support efforts in exploring functional outcomes of microbiome interactions (Quince et al. 2017). While amplicon sequencing targets only specific biomarker genes, shotgun metagenome sequencing targets total genomic DNA of a sample. Therefore, metagenome analysis can identify all genes present in the plant microbiome, allowing for both gene- and pathway-based functional analysis. This approach has already been exploited to study rhizosphere and phyllosphere communities of several crop plants (Ofek-Lalzar et al. 2014; Xu et al. 2018). In a recent study, Angeli et al. (2019) used a shotgun metagenomics approach to study the apple carposphere microbiota, leading to the characterization of functions associated with plant pathogens and biocontrol agents. Further utility of these types of datasets will be achieved with increasingly advanced analysis approaches. For example, metabolic network analyses are becoming a central framework for translating discrete data from ecological samples into a structured view of biological functions, and also allow for the subsequent pursuit of simulations that explore associations between environments and communities (Ponomarova and Patil 2015; Noecker et al. 2016; Widder et al. 2016). Such simulations facilitate the formulation of predictions for functional roles of beneficial/non-beneficial taxonomic groups and potential modifications that will encourage a desired function (Mazzola and Freilich 2017). The application of such a network-based approach for the analysis of metagenomics data from rhizosphere communities was demonstrated by Ofaim et al. (2017). Root-specific effects linked utilization of specific plant exudates (e.g., flavonoids and organic acids) with specific taxonomic groups, pointing to roles of these compounds as determinants of microbial community structure. Beyond community-level analysis based on metagenomics gene catalogues, the sequencing depth of current genome sequencing projects will allow for the recovery of significant segments of genomes of highly abundant species (Anantharaman et al. 2016). Genome-scale metabolic models (GSMM) and respective simulation algorithms, such as Flux Balance Analysis (FBA), will further enable in silico analysis of microbial functions (Faust 2019). Computational simulations can compare performances of different species under different conditions and aid in prioritising potential amendments to support desired groups and suppress undesired groups (Xu et al. 2019; Faust 2019).

The capacity to explore the wealth of functional attributes within the apple microbiome is ever-increasing, and continued progress will rely on the integration of various currently available approaches. For example, profiling the apple microbiome using amplicon sequencing is now relatively common and low-cost, and can be integrated with field-based approaches with large enough sample sizes to determine the impacts of management practices, environmental conditions, and genotype (scion and rootstock) on consortia known to have beneficial functional roles. However, many orchard ecosystem functions are likely dependent on diverse plant–microbe and microbe–microbe interactions that are yet to be elucidated. Understanding the mechanisms shaping these interactions will require studies that go beyond documentation of microbial assemblages to integrate additional analytical tools, such as metabolomics, to assess how interactions between members of microbial assemblages and their host induce and regulate the production of functional metabolites. Shotgun metagenomics and metatranscriptomics can also be integrated to increase our knowledge of how the microbiome modulates gene expression in apple. This information can be subsequently used to optimize plant–microbe interactions for improved apple production, especially with regards to plant diseases and biotic/abiotic stress tolerance.

16.6.3 Systems Approaches—Integrating Across the Agroecosystem

Transformative progress toward sustainable agricultural solutions will require systems-based approaches that embrace a holistic view of agroecosytems (Schut et al. 2014). Microbiome composition is shaped by numerous interacting factors, many of which can also be shaped by the microbiome through complex ecological and evolutionary feedback loops (e.g., phytochemistry). Moreover, both the composition and diversity of the apple microbiome have numerous downstream consequences for plant growth, pest resistance, harvestable fruit yield, fruit quality, economic crop value, and human health (Fig. 16.3). Although no single study can address all of these factors simultaneously, a more systems-based view can guide research priorities and identify fruitful areas for collaboration and integrative studies.

An important first step for acquiring a more integrative understanding of the apple microbiome is to consider the apple tree as a holistic ecosystem. Most microbiota can occupy multiple niches within the orchard. Thus, studies should move beyond exploring individual sub-divisions (i.e., phyllosphere, rhizosphere, endosphere, and carposphere) to account for multiple sources of specific microbes or microbial consortia (Fig. 16.1). For instance, *Penicillium* and *Alternaria* species are significant post-harvest pathogens of apple, and studies addressing their control have, understandably, focused largely on the carposphere. However, potential inocula for these fungi are diverse and include the orchard soil microbiome. Thus, management of the orchard soil microbiome may facilitate reducing inoculum potential of these post-harvest pathogens. In fact, a specific type of *Brassica* seed meal formulation has consistently resulted in a reduction in the relative abundance of these two genera of fungi over two orchard growing seasons (Wang and Mazzola 2019a). Assessment of microbial resources in an ecosystem-inclusive manner, rather than as individual components, is likely to reveal numerous opportunities that will benefit apple production, health, and fruit quality.

A second priority is to integrate studies of the apple microbiome across apple production systems; that is, from soils and farmers to consumer markets (Fig. 16.3). Harnessing of the microbiome for improved sustainability and ecosystem health will require not only technological advances but also social and institutional innovations. Involving various stakeholders during the early stages of studies will likely lead to faster research advances and improved translation to practices that can be readily adopted by the apple industry.

16.7 Conclusions

Most research on plant microbiomes has been conducted in a few model annual plant species (e.g., corn, soybean, and *Arabidopsis*), and less is known about the microbiota of trees (APS 2016).

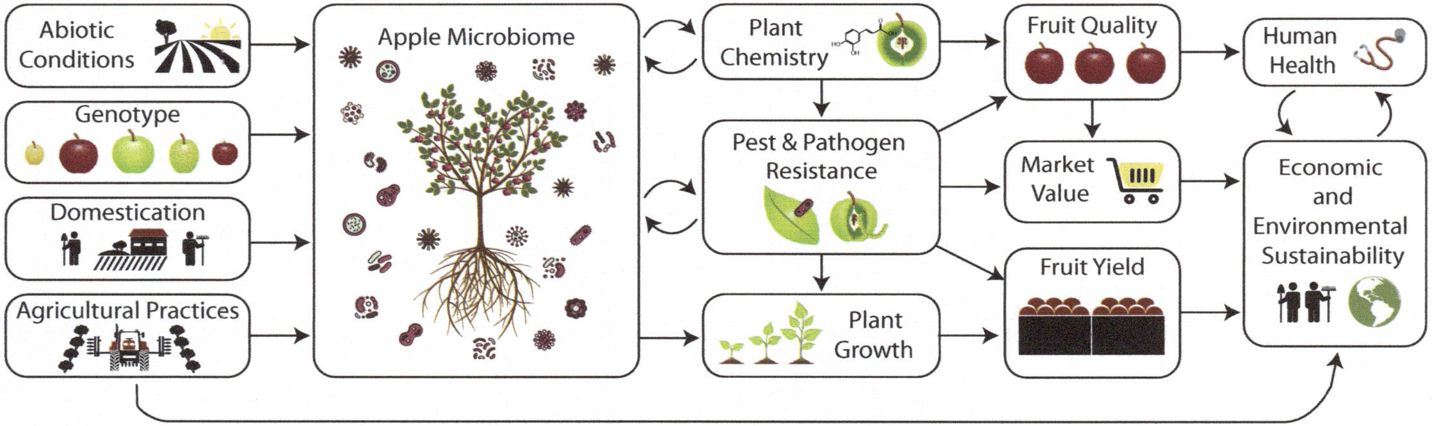

Fig. 16.3 A systems-agriculture approach for understanding the apple microbiome, including factors that influence microbiome composition and downstream consequences of microbes for plant quality, pest/pathogen resistance, yield, human health, and the sustainability of apple production

Yet, there is enormous potential for microbiome management in long-lived perennials to maximise beneficial interactions and minimize the accumulation of damaging pest populations over time. As we begin to harness the power of microbiomes for more sustainable agriculture, a key priority is the development of new model systems for microbiome research (APS 2016). The apple tree is poised to fill this role. Currently, we have a strong foundational knowledge of the apple microbiome composition (Table 16.1 and Fig. 16.1), and we are beginning to move toward an understanding of both factors shaping the microbiome and downstream consequences of the microbiome for plant health. Other chapters in this volume provide details of expansive resources available for understanding the biology of the apple as a host, including whole-genome sequencing (Velasco et al. 2010; Daccord et al. 2017; see Chap. 8 in this volume), as well as an expanded knowledge of apple phytochemistry (Lee et al. 2003; Zhang et al. 2010a; Cuthbertson et al. 2012; see Chap. 15 in this volume).

There is a strong need to pursue new strategies for sustainable apple production, as apples are heavily attacked by insects and diseases. They require intensive management practices to produce marketable fruit (Beers et al. 2003; Grove et al. 2003). Developing strategies to manage the apple microbiome to address these challenges will require a comprehensive understanding of microbial communities occupying various niches in the orchard ecosystem and the functional contributions of individual members of these communities to interactions with their hosts. Coordinated efforts are required among experts in various components of production systems, including the orchard ecosystem (the tree, its associated microbiome, and other abiotic and biotic components of the orchard environment), post-harvest systems of processing, storage, and distribution, and finally, the various consumer markets involved (fresh eating, processing, and specialty use) (Fig. 16.3). We hope this review will inspire more coordinated research efforts among these groups and help harness the power of the microbiome for increased economic and environmental sustainability of apple production.

Acknowledgements This research was funded by Agriculture and Food Research Initiative Competitive Grant no. 2018-07366 from the USDA National Institute of Food and Agriculture to S.R.W., U.S.—Israel Binational Agricultural Fund (IS-5040-17) to S.D. and M.W., U.S.—Israel Binational Agricultural Fund (US-5046-17R) to M. M. and S.F., Agriculture and Food Research Initiative Competitive Grant no. 2016-51181-25406 from the USDA National Institute of Food and Agriculture to M. M., and the European Union's Horizon 2020 under "Nurturing excellence by means of cross-border and cross-sector mobility" programme for MSCA-IF-2018-Individual Fellowships, grant agreement 844114 to A.A.

References

Abdelfattah A, Freilich S, Bartuv R, Zhimo VY Kumar A, Biasi A, Salim S, Feygenberg O, Burchard E, Dardick C, Liu J, Khan A, Ellouze W, Ali S, Spadaro D, Torres R, Teixido N, Ozkaya O, Buehlmann A, Vero S, Mondino P, Berg G, Wisniewski M, Droby S (2021) Global analysis of the apple fruit microbiome: are all apples the same? Environmental Microbiology [early view] https://doi.org/10.1111/1462-2920.15469

Abdelfattah A, Whitehead SR, Macarisin D, Liu L, Burchard E, Freilich S, Dardick C, Droby S, Wisniewski M (2020) Effect of washing, waxing and low-temperature storage on the postharvest microbiome of apple. Microorganisms 8:944. https://doi.org/10.3390/microorganisms8060944

Abdelfattah A, Wisniewski M, Droby S, Schena L (2016) Spatial and compositional variation in the fungal communities of organic and conventionally grown apple fruit at the consumer point-of-purchase. Hortic Res 3:1–12. https://doi.org/10.1038/hortres.2016.47

Abujabhah IS, Bound SA, Doyle R, Bowman JP (2016) Effects of biochar and compost amendments on soil physico-chemical properties and the total community within a temperate agricultural soil. Appl Soil Ecol 98:243–253. https://doi.org/10.1016/j.apsoil.2015.10.021

Agler MT, Ruhe J, Kroll S, Morhenn C, Kim S-T, Weigel D, Kemen EM (2016) Microbial hub taxa link host and abiotic factors to plant microbiome variation. PLoS Biol 14:e1002352. https://doi.org/10.1371/journal.pbio.1002352

Ait Bahadou S, Ouijja A, Karfach A, Tahiri A, Lahlali R (2018) New potential bacterial antagonists for the biocontrol of fire blight disease (*Erwinia amylovora*) in Morocco. Microb Pathog 117:7–15. https://doi.org/10.1016/j.micpath.2018.02.011

Aizenberg-Gershtein Y, Izhaki I, Santhanam R, Kumar P, Baldwin IT, Halpern M (2015) Pyridine-type alkaloid composition affects bacterial community composition of floral nectar. Sci Rep 5:11536. https://doi.org/10.1038/srep11536

Akremi A, Jlidi M, Brahim AH, Ali MB, Daoud L, Hmani H, Bejar S, Feto NA, Ali MB (2018) Isolation of *Erwinia amylovora* bacteriophage from soil and pear trees plant tissues. Proc MOL2NET 2018 Int Conf Multidiscip Sci 4th Ed 1. https://doi.org/10.3390/mol2net-04-05914

Alaphilippe A, Elad Y, David DR, Derridj S, Gessler C (2008) Effects of a biocontrol agent of apple powdery mildew (*Podosphaera leucotricha*) on the host plant and on non-target organisms: an insect pest (*Cydia pomonella*) and a pathogen (*Venturia inaequalis*). Biocontrol Sci Technol 18:121–138. https://doi.org/10.1080/09583150701818964

Alegria Terrazas R, Giles C, Paterson E, Robertson-Albertyn, S Cesco S, Mimmo T, Pii Y Bulgarelli D (2016) Plant–microbiota interactions as a driver of the mineral turnover in the rhizosphere. In: Sariaslani S, Michael Gadd G (eds) Advances in applied microbiology. Academic Press, pp 1–67

Aleklett K, Hart M, Shade A (2014) The microbial ecology of flowers: an emerging frontier in phyllosphere research. Botany 92:253–266. https://doi.org/10.1139/cjb-2013-0166

Alonso S, Laca A, Rendueles M, Mayo B, Díaz M (2015) Cider apple native microbiota characterization by PCR-DGGE. J Inst Brew 121:287–289. https://doi.org/10.1002/jib.220

Anantharaman K, Brown CT, Hug LA, Sharon I, Castelle CJ, Probst AJ, Thomas BC, Singh A, Wilkins MJ, Karaoz U, Brodie EL, Williams KH, Hubbard SS, Banfield JF (2016) Thousands of microbial genomes shed light on interconnected biogeochemical processes in an aquifer system. Nat Commun 7:13219. https://doi.org/10.1038/ncomms13219

Andrews JH, Berbee FM, Nordheim EV (1983) Microbial antagonism to the imperfect stage of the apple scab pathogen, Venturia inaequalis. Phytopathology 73:228–234

Andrews JH, Spear RN, Nordheim EV (2002) Population biology of *Aureobasidium pullulans* on apple leaf surfaces. Can J Microbiol 48:500–513. https://doi.org/10.1139/W02-044

Angeli D, Sare AR, Jijakli MH, Pertot I, Massart S (2019) Insights gained from metagenomic shotgun sequencing of apple fruit epiphytic microbiota. Postharvest Biol Technol 153:96–106. https://doi.org/10.1016/j.postharvbio.2019.03.020

APS (2016) Phytobiomes: a roadmap for research and translation. Am Phytopathol Soc

Arnold AE, Lutzoni F (2007) Diversity and host range of foliar fungal endophytes: are tropical leaves biodiversity hotspots? Ecology 88:541–549. https://doi.org/10.1890/05-1459

Arrigoni E, Antonielli L, Pindo M, Pertot I, Perazzolli M (2018) Tissue age and plant genotype affect the microbiota of apple and pear bark. Microbiol Res 211:57–68. https://doi.org/10.1016/j.micres.2018.04.002

Aslantaş R, Çakmakçi R, Şahin F (2007) Effect of plant growth promoting rhizobacteria on young apple tree growth and fruit yield under orchard conditions. Sci Hortic 111:371–377. https://doi.org/10.1016/j.scienta.2006.12.016

Baetz U, Martinoia E (2014) Root exudates: the hidden part of plant defense. Trends Plant Sci 19:90–98. https://doi.org/10.1016/j.tplants.2013.11.006

Bais HP, Weir TL, Perry LG, Gilroy S, Vivanco JM (2006) The role of root exudates in rhizosphere interactions with plants and other organisms. Annu Rev Plant Biol 57:233–266

Bakker MG, Manter DK, Sheflin AM, Weir TL, Vivanco JM (2012) Harnessing the rhizosphere microbiome through plant breeding and agricultural management. Plant Soil 360:1–13. https://doi.org/10.1007/s11104-012-1361-x

Bakker PAHM, Berendsen RL, Doornbos RF, Wintermans PCA, Pieterse CMJ (2013) The rhizosphere revisited: root microbiomics. Front Plant Sci 4. https://doi.org/10.3389/fpls.2013.00165

Bálint M, Tiffin P, Hallström B, O'Hara RB, Olson MS, Fankhauser JD, Piepenbring M, Schmitt I (2013) Host genotype shapes the foliar fungal microbiome of balsam poplar (*Populus balsamifera*). PLoS One 8. https://doi.org/10.1371/journal.pone.0053987

Beers EH, Suckling DM, Prokopy RJ, Avilla J (2003) Ecology and management of apple arthropod pests. In: Ferree DC, Warrington IJ (eds) Apples: botany, production and uses. CABI, Wallingford, pp 489–519

Berendsen RL, Pieterse CMJ, Bakker PAHM (2012) The rhizosphere microbiome and plant health. Trends Plant Sci 17:478–486. https://doi.org/10.1016/j.tplants.2012.04.001

Berg G (2009) Plant–microbe interactions promoting plant growth and health: perspectives for controlled use of microorganisms in agriculture. Appl Microbiol Biotechnol 84:11–18. https://doi.org/10.1007/s00253-009-2092-7

Berg G, Erlacher A, Smalla K, Krause R (2014) Vegetable microbiomes: is there a connection among opportunistic infections, human health and our 'gut feeling'? Microb Biotechnol 7:487–495. https://doi.org/10.1111/1751-7915.12159

Berg G, Raaijmakers JM (2018) Saving seed microbiomes. ISME J 12:1167–1170. https://doi.org/10.1038/s41396-017-0028-2

Berg G, Rybakova D, Grube M, Köberl M (2016) The plant microbiome explored: implications for experimental botany. J Exp Bot 67:995–1002. https://doi.org/10.1093/jxb/erv466

Blok WJ, Lamers JG, Termorshuizen AJ, Bollen GJ (2000) Control of soilborne plant pathogens by incorporating fresh organic amendments followed by tarping. Phytopathology 90:253–259. https://doi.org/10.1094/PHYTO.2000.90.3.253

Bonaterra A, Badosa E, Cabrefiga J, Francés J, Montesinos E (2012) Prospects and limitations of microbial pesticides for control of bacterial and fungal pomefruit tree diseases. Trees 26:215–226. https://doi.org/10.1007/s00468-011-0626-y

Bonkowski M, Brandt F (2002) Do soil protozoa enhance plant growth by hormonal effects? Soil Biol Biochem 34:1709–1715. https://doi.org/10.1016/S0038-0717(02)00157-8

Bonn WG, van der Zwet T (2000) Distribution and economic importance of fire blight. Fire Blight Dis Its Causative Agent Erwinia Amylovora 37:53

Born Y, Fieseler L, Thöny V, Leimer N, Duffy B, Loessner MJ (2017) Engineering of bacteriophages Y2:*dpoL1-C* and Y2:*luxAB* for efficient control and rapid detection of the fire blight pathogen, *Erwinia amylovora*. Appl Environ Microbiol 83. https://doi.org/10.1128/AEM.00341-17

Bubán T, Orosz-Kovács ZS (2003) The nectary as the primary site of infection by *Erwinia amylovora* (Burr.) Winslow et al.: a mini review. Plant Syst Evol 238:183–194. https://doi.org/10.1007/s00606-002-0266-1

Buchholz F, Kostić T, Sessitsch A, Mitter B (2018) The potential of plant microbiota in reducing postharvest food loss. Microb Biotechnol 11:971–975

Buck JW, Lachance M-A, Traquair JA (1998) Mycoflora of peach bark: population dynamics and composition. Can J Bot 76:345–354. https://doi.org/10.1139/b98-009

Bulgarelli D, Garrido-Oter R, Münch PC, Weiman A, Dröge J, Pan Y, McHardy AC, Schulze-Lefert P (2015) Structure and function of the bacterial root microbiota in wild and domesticated barley. Cell Host Microbe 17:392–403. https://doi.org/10.1016/j.chom.2015.01.011

Bulgarelli D, Schlaeppi K, Spaepen S, van Themaat EVL, Schulze-Lefert P (2013) Structure and functions of the bacterial microbiota of plants. Annu Rev Plant Biol 64:807–838. https://doi.org/10.1146/annurev-arplant-050312-120106

Bulgari D, Bozkurt AI, Casati P, Çalayan K, Quaglino F, Bianco PA (2012) Endophytic bacterial community living in roots of healthy and '*Candidatus* Phytoplasma mali'-infected apple (*Malus domestica*, Borkh.) trees. Antonie Van Leeuwenhoek Dordr 102:677–687. https://doi.org/10.1007/s10482-012-9766-3

Burr T, Norelli J, Reid C, Capron LK, Nelson LS, Aldwinckle HS, Wilcox WF (1993) Streptomycin-resistant bacteria associated with fire blight infections. Plant Dis 77:63–63

Burr TJ, Matteson MC, Smith CA, Corral-Garcia MR, Huang T-C (1996) Effectiveness of bacteria and yeasts from apple orchards as biological control agents of apple scab. Biol Control 6:151–157. https://doi.org/10.1006/bcon.1996.0019

Busby PE, Soman C, Wagner MR, Friesen ML, Kremer J, Bennett A, Morsy M, Eisen JA, Leach JE, Dangl JL (2017) Research priorities for harnessing plant microbiomes in sustainable agriculture. PLoS Biol 15: e2001793. https://doi.org/10.1371/journal.pbio.2001793

Caddell DF, Deng S, Coleman-Derr D (2019) Role of the plant root microbiome in abiotic stress tolerance. In: Seed endophytes. Springer, pp 273–311

Caporaso JG, Lauber CL, Walters WA, Berg-Lyons D, Huntley J, Fierer N, Owens SM, Betley J, Fraser L, Bauer M, Gormley N, Gilbert JA, Smith G, Knight R (2012) Ultra-high-throughput microbial community analysis on the Illumina HiSeq and MiSeq platforms. ISME J 6:1621–1624. https://doi.org/10.1038/ismej.2012.8

Caputo F, Nicoletti F, De Luca PF, Manici LM (2015) Rhizospheric changes of fungal and bacterial communities in relation to soil health of multi-generation apple orchards. Biol Control 88:8–17. https://doi.org/10.1016/j.biocontrol.2015.04.019

Carisse O, Dewdney M (2002) A review of non-fungicidal approaches for the control of apple scab. Phytoprotection 83:1–29. https://doi.org/10.7202/706226ar

Čatská V, Vančura V, Hudská G, Přikryl Z (1982) Rhizosphere micro-organisms in relation to the apple replant problem. Plant Soil 69:187–197. https://doi.org/10.1007/BF02374514

Chai X, Wang L, Yang Y, Xie L, Zhang J, Wu T, Zhang X, Xu X, Wang Y, Han Z (2019) Apple rootstocks of different nitrogen tolerance affect the rhizosphere bacterial community composition. J Appl Microbiol 126:595–607. https://doi.org/10.1111/jam.14121

Chai X, Xie L, Wang X, Wang H, Zhang J, Han Z, Wu T, Zhang X, Xu X, Wang Y (2020) Apple rootstocks with different phosphorus efficiency exhibit alterations in rhizosphere bacterial structure. J Appl Microbiol 128:1460–1471. https://doi.org/10.1111/jam.14547

Chaparro JM, Badri DV, Vivanco JM (2014) Rhizosphere microbiome assemblage is affected by plant development. ISME J 8:790–803. https://doi.org/10.1038/ismej.2013.196

Chen I-MA, Markowitz VM, Chu K, Palaniappan K, Szeto E, Pillay M, Ratner A, Huang J, Andersen E, Huntemann M, Varghese N, Hadjithomas M, Tennessen K, Nielsen T, Ivanova NN, Kyrpides NC (2017) IMG/M: integrated genome and metagenome comparative data analysis system. Nucleic Acids Res 45:D507–D516. https://doi.org/10.1093/nar/gkw929

Chen Y, Wen X, Sun Y, Zhang J, Wu W, Liao Y (2014) Mulching practices altered soil bacterial community structure and improved orchard productivity and apple quality after five growing seasons. Sci Hortic 172:248–257. https://doi.org/10.1016/j.scienta.2014.04.010

Chiou C, Jones A (1993) Nucleotide sequence analysis of a transposon (Tn5393) carrying streptomycin resistance genes in *Erwinia amylovora* and other gram-negative bacteria. J Bacteriol 175:732–740

Chisholm ST, Coaker G, Day B, Staskawicz BJ (2006) Host-microbe interactions: shaping the evolution of the plant immune response. Cell 124:803–814. https://doi.org/10.1016/j.cell.2006.02.008

Cipollini ML, Stiles EW (1992) Relative risks of fungal rot for temperate ericaceous fruits: effects of seasonal variation on selection for chemical defense. Can J Bot 70:1868–1877. https://doi.org/10.1139/b92-231

Cohen MF, Han XY, Mazzola M (2004) Molecular and physiological comparison of *Azospirillum* spp. isolated from *Rhizoctonia solani* mycelia, wheat rhizosphere, and human skin wounds. Can J Microbiol 50:291–297. https://doi.org/10.1139/w04-007

Coleman-Derr D, Tringe SG (2014) Building the crops of tomorrow: advantages of symbiont-based approaches to improving abiotic stress tolerance. Front Microbiol 5. https://doi.org/10.3389/fmicb.2014.00283

Compant S, Mitter B, Colli-Mull JG, Gangl H, Sessitsch A (2011) Endophytes of grapevine flowers, berries, and seeds: identification of cultivable bacteria, comparison with other plant parts, and visualization of niches of colonization. Microb Ecol 62:188–197. https://doi.org/10.1007/s00248-011-9883-y

Compant S, Samad A, Faist H, Sessitsch A (2019) A review on the plant microbiome: ecology, functions, and emerging trends in microbial application. J Adv Res 19:29–37. https://doi.org/10.1016/j.jare.2019.03.004

Cordovez V, Dini-Andreote F, Carrión VJ, Raaijmakers JM (2019) Ecology and evolution of plant microbiomes. Annu Rev Microbiol 73:69–88

Cornille A, Giraud T, Smulders MJM, Roldán-Ruiz I, Gladieux P (2014) The domestication and evolutionary ecology of apples. Trends Genet 30:57–65. https://doi.org/10.1016/j.tig.2013.10.002

Cornille A, Gladieux P, Smulders MJM, Roldán-Ruiz I, Laurens F, Cam BL, Nersesyan A, Clavel J, Olonova M, Feugey L, Gabrielyan I, Zhang X.-G., Tenaillon MI, Giraud T (2012) New insight into the history of domesticated apple: secondary contribution of the European wild apple to the genome of cultivated varieties. PLoS Genet 8:e1002703. https://doi.org/10.1371/journal.pgen.1002703

Covey R, Koch B, Larsen H (1981) Influence of vesicular arbuscular mycorrhizae on the growth of apple and corn in low-phosphorous soil. Phytopathology 71:712–715. https://doi.org/10.1094/phyto-71-712

Cregger MA, Veach AM, Yang ZK, Crouch MJ, Vilgalys R, Tuskan GA, Schadt CW (2018) The *Populus* holobiont: dissecting the effects of plant niches and genotype on the microbiome. Microbiome 6:31. https://doi.org/10.1186/s40168-018-0413-8

Cui Z, Huntley RB, Zeng Q, Steven B (2020) Temporal and spatial dynamics in the apple flower microbiome in the presence of the phytopathogen *Erwinia amylovora*. ISME J 1–12. https://doi.org/10.1038/s41396-020-00784-y

Cuthbertson D, Andrews PK, Reganold JP, Davies NM, Lange BM (2012) Utility of metabolomics toward assessing the metabolic basis of quality traits in apple fruit with an emphasis on antioxidants. J Agric Food Chem 60:8552–8560. https://doi.org/10.1021/jf3031088

Daccord N, Celton J-M, Linsmith G, Becker C, Choisne N, Schijlen E, van de Geest H, Bianco L, Micheletti D, Velasco R, Di Pierro EA, Gouzy J, Rees DJG, Guérif P, Muranty H, Durel CE, Laurens F, Lespinasse Y, Gaillard S, Aubourg S, Quesneville H, Weigel D, Van de Weg E, Troggio M, Bucher E (2017) High-quality de novo assembly of the apple genome and methylome dynamics of early fruit development. Nat Genet 49:1099–1106. https://doi.org/10.1038/ng.3886

Dangl JL, Jones JD (2001) Plant pathogens and integrated defence responses to infection. Nature 411:826–833. https://doi.org/10.1038/35081161

Deakin G, Fernández-Fernández F, Bennett J, Passey T, Harrison N, Tilston EL, Xu X (2019) The effect of rotating apple rootstock genotypes on apple replant disease and rhizosphere microbiome. Phytobiomes J 3:273–285. https://doi.org/10.1094/PBIOMES-03-19-0018-R

Deakin G, Tilston EL, Bennett J, Passey T, Harrison N, Fernández-Fernández F, Xu X (2018) Spatial structuring of soil microbial communities in commercial apple orchards. Appl Soil Ecol 130:1–12. https://doi.org/10.1016/j.apsoil.2018.05.015

Di Bella JM, Bao Y, Gloor GB, Burton JP, Reid G (2013) High throughput sequencing methods and analysis for microbiome research. J Microbiol Methods 95:401–414. https://doi.org/10.1016/j.mimet.2013.08.011

Dicke M, Takken W (2006) Chemical ecology: from gene to ecosystem. Springer Science & Business Media

Dillard CJ, German JB (2000) Phytochemicals: nutraceuticals and human health. J Sci Food Agric 80:1744–1756. https://doi.org/10.1002/1097-0010(20000915)80:12%3c1744::AID-JSFA725%3e3.0.CO;2-W

Dobbelaere S, Vanderleyden J, Okon Y (2003) Plant growth-promoting effects of diazotrophs in the rhizosphere. Crit Rev Plant Sci 22:107–149. https://doi.org/10.1080/713610853

Dohroo A, Sharma DR, Dohroo NP (2014) Occurrence of glomus mycorrhizal fungi in rhizosphere soils of apple orchards in North Western Himalayas. Agric Sci Dig 34:64–66. https://doi.org/10.5958/j.0976-0547.34.1.014

Douglas AE, Werren JH (2016) Holes in the hologenome: why host-microbe symbioses are not holobionts. mBio 7. https://doi.org/10.1128/mBio.02099-15

Droby S, Wisniewski M (2018) The fruit microbiome: a new frontier for postharvest biocontrol and postharvest biology. Postharvest Biol Technol 140:107–112. https://doi.org/10.1016/j.postharvbio.2018.03.004

Droby S, Wisniewski M, Teixidó N, Spadaro D, Jijakli MH (2016) The science, development, and commercialization of postharvest biocontrol products.

Postharvest Biol Technol 122:22–29. https://doi.org/10.1016/j.postharvbio.2016.04.006

du Jardin P (2015) Plant biostimulants: definition, concept, main categories and regulation. Sci Hortic 196:3–14. https://doi.org/10.1016/j.scienta.2015.09.021

Duffy B, Holliger E, Walsh F (2014) Streptomycin use in apple orchards did not increase abundance of mobile resistance genes. FEMS Microbiol Lett 350:180–189. https://doi.org/10.1111/1574-6968.12313

Esplin IND, Berg JA, Sharma R, Allen RC, Arens DK, Ashcroft CR, Bairett SR, Beatty NJ, Bickmore M, Bloomfield TJ, Brady TS, Bybee RN, Carter JL, Choi MC, Duncan S, Fajardo CP, Foy BB, Fuhriman DA, Gibby PD, Grossarth SE, Harbaugh K, Harris N, Hilton JA, Hurst E, Hyde JR, Ingersoll K, Jacobson CM, James BD, Jarvis TM, Jaen-Anieves D, Jensen GL, Knabe BK, Kruger JL, Merrill BD, Pape JA, Anderson AMP, Payne DE, Peck MD, Pollock SV, Putnam MJ, Ransom EK, Ririe DB, Robinson DM, Rogers SL, Russell KA, Schoenhals JE, Shurtleff CA, Simister AR, Smith HG, Stephenson MB, Staley LA, Stettler JM, Stratton ML, Tateoka OB, Tatlow PJ, Taylor AS, Thompson SE, Townsend MH, Thurgood TL, Usher BK, Whitley KV, Ward AT, Ward MEH, Webb CJ, Wienclaw TM, Williamson TL, Wells MJ, Wright CK, Breakwell DP, Hope S, Grose JH (2017) Genome sequences of 19 novel *Erwinia amylovora* bacteriophages. Genome Announc 5:e00931-e1017. https://doi.org/10.1128/genomeA.00931-17

Falconi CJ, Mendgen K (1994) Epiphytic fungi on apple leaves and their value for control of the postharvest pathogens *Botrytis cinerea*, *Monilinia fructigena* and *Penicillium expansum*. J Plant Dis Prot 38–47

Faust K (2019) Microbial consortium design benefits from metabolic modeling. Trends Biotechnol 37:123–125. https://doi.org/10.1016/j.tibtech.2018.11.004

Fierer N, Nemergut D, Knight R, Craine JM (2010) Changes through time: integrating microorganisms into the study of succession. Res Microbiol 161:635–642. https://doi.org/10.1016/j.resmic.2010.06.002

Fitzpatrick CR, Copeland J, Wang PW, Guttman DS, Kotanen PM, Johnson MTJ (2018) Assembly and ecological function of the root microbiome across angiosperm plant species. Proc Natl Acad Sci 115: E1157–E1165. https://doi.org/10.1073/pnas.1717617115

Fortuna P, Citernesi AS, Morini S, Vitagliano C, Giovannetti M (1996) Influence of arbuscular mycorrhizae and phosphate fertilization on shoot apical growth of micropropagated apple and plum rootstocks. Tree Physiol 16:757–763. https://doi.org/10.1093/treephys/16.9.757

Franke-Whittle IH, Manici LM, Insam H, Stres B (2015) Rhizosphere bacteria and fungi associated with plant growth in soils of three replanted apple orchards. Plant Soil 395:317–333. https://doi.org/10.1007/s11104-015-2562-x

Gaiero JR, McCall CA, Thompson KA, Day NJ, Best AS, Dunfield KE (2013) Inside the root microbiome: bacterial root endophytes and plant growth promotion. Am J Bot 100:1738–1750. https://doi.org/10.3732/ajb.1200572

Gallart M, Adair KL, Love J, Meason DF, Clinton PW, Xue J, Turnbull MH (2018) Host genotype and nitrogen form shape the root microbiome of *Pinus radiata*. Microb Ecol 75:419–433. https://doi.org/10.1007/s00248-017-1055-2

Gildemacher P, Heijne B, Silvestri M, Houbraken J, Hoekstra E, Theelen B, Boekhout T (2006) Interactions between yeasts, fungicides and apple fruit russeting. FEMS Yeast Res 6:1149–1156. https://doi.org/10.1111/j.1567-1364.2006.00109.x

Gill JJ, Svircev AM, Smith R, Castle AJ (2003) Bacteriophages of *Erwinia amylovora*. Appl Environ Microbiol 69:2133–2138. https://doi.org/10.1128/AEM.69.4.2133-2138.2003

Gleason ML, Batzer JC, Sun G, Zhang R, Arias MMD, Sutton TB, Crous PW, Ivanović M, McManus PS, Cooley DR, Mayr U, Weber RWS, Yoder KS, Del Ponte EM, Biggs AR, Oertel B (2011) A new view of sooty blotch and flyspeck. Plant Dis 95:368–383. https://doi.org/10.1094/PDIS-08-10-0590

Glenn DM, Bassett C, Dowd SE (2015) Effect of pest management system on 'Empire' apple leaf phyllosphere populations. Sci Hortic 183:58–65. https://doi.org/10.1016/j.scienta.2014.12.009

Griffin EA, Carson WP (2015) The ecology and natural history of foliar bacteria with a focus on tropical forests and agroecosystems. Bot Rev 81:105–149. https://doi.org/10.1007/s12229-015-9151-9

Grove GG, Eastwell KC, Jones AL, Sutton TB (2003) Diseases of apple. In: Apples: botany, production and uses. CABI, Wallingford, pp 459–488

Grzyb ZS, Piotrowski W, Bielicki P, Paszt LS, Malusà E (2012) Effect of different fertilizers and amendments on the growth of apple and sour cherry rootstocks in an organic nursery. J Fruit Ornam Plant Res 20:43–53. https://doi.org/10.2478/v10290-012-0004-x

Haesler F, Hagn A, Engel M, Schloter M (2014) Impact of elevated atmospheric O3 on the actinobacterial community structure and function in the rhizosphere of European beech (*Fagus sylvatica* L.). Front Microbiol 5. https://doi.org/10.3389/fmicb.2014.00036

Hall RD (2006) Plant metabolomics: from holistic hope, to hype, to hot topic. New Phytol 169:453–468. https://doi.org/10.1111/j.1469-8137.2005.01632.x

Hamady M, Knight R (2009) Microbial community profiling for human microbiome projects: tools, techniques, and challenges. Genome Res 19:1141–1152. https://doi.org/10.1101/gr.085464.108

Hartman K, van der Heijden MGA, Wittwer RA, Banerjee S, Walser J-C, Schlaeppi K (2018) Cropping practices manipulate abundance patterns of root and soil microbiome members paving the way to smart farming. Microbiome 6:14. https://doi.org/10.1186/s40168-017-0389-9

Hassani MA, Durán P, Hacquard S (2018) Microbial interactions within the plant holobiont. Microbiome 6:58. https://doi.org/10.1186/s40168-018-0445-0

Henriquez JL, Sugar D, Spotts RA (2004) Etiology of bull's Eye Rot of pear caused by *Neofabraea* spp. in oregon, Washington, and California. Plant Dis 88:1134–1138. https://doi.org/10.1094/PDIS.2004.88.10.1134

Hewavitharana SS, Klarer E, Reed AJ, Leisso R, Poirier B, Honaas L, Rudell DR, Mazzola M (2019) Temporal dynamics of the soil metabolome and microbiome during simulated anaerobic soil disinfestation. Front Microbiol 10:2365. https://doi.org/10.3389/fmicb.2019.02365

Hewavitharana SS, Mazzola M (2020) Influence of rootstock genotype on efficacy of anaerobic soil disinfestation for control of apple nursery replant disease. Eur J Plant Pathol 157:39–57. https://doi.org/10.1007/s10658-020-01977-z

Hewavitharana SS, Mazzola M (2016) Carbon source-dependent effects of anaerobic soil disinfestation on soil microbiome and suppression of *Rhizoctonia solani* AG-5 and *Pratylenchus penetrans*. Phytopathology 106:1015–1028. https://doi.org/10.1094/PHYTO-12-15-0329-R

Heye CC, Andrews JH (1983) Antagonism of *Athelia bombacina* and *Chaetomium globosum* to the apple scab pathogen, *Venturia inaequalis*. Phytopathology 73:650–654

Hirakue A, Sugiyama S (2018) Relationship between foliar endophytes and apple cultivar disease resistance in an organic orchard. Biol Control 127:139–144. https://doi.org/10.1016/j.biocontrol.2018.09.007

Hofmann A, Wittenmayer L, Arnold G, Schieber A, Merbach W (2009) Root exudation of phloridzin by apple seedlings (*Malus* x *domestica* Borkh.) with symptoms of apple replant disease. J Appl Bot Food Qual 82:193–198

Holderbaum DF, Kon T, Guerra MP (2014) Dynamics of total phenolic content in different apple tissues and genotypes: impacts and relevance for breeding programs. Sci Hortic 168:58–63. https://doi.org/10.1016/j.scienta.2014.01.020

Howe HF, Smallwood J (1982) Ecology of seed dispersal. Annu Rev Ecol Syst 13:201–228

Huang T-C, Burr TJ (1999) Characterization of plasmids that encode streptomycin-resistance in bacterial epiphytes of apple. J Appl Microbiol 86:741–751. https://doi.org/10.1046/j.1365-2672.1999.00719.x

Huang Y, Zhou G, Yi G, Gong Y, Zhang W, Liu A, Yuan L (2013) Rhizosphere actinomycete isolation methods and phloridzin degradation activity analysis in plants. J South Agric 44:54–58

Hubbard CJ, Brock MT, van Diepen LT, Maignien L, Ewers BE, Weinig C (2018) The plant circadian clock influences rhizosphere community structure and function. ISME J 12:400–410. https://doi.org/10.1038/ismej.2017.172

Ishimaru CA, Klos EJ, Brubaker RR (1988) Multiple antibiotic production by *Erwinia herbicola*. Phytopathology 78:746–750

Jacke D, Toensmeier E (2005) Edible forest gardens, volume I: ecological vision, theory for temperate climate permaculture. Chelsea Green Publishing

Janisiewicz W (1988) Biocontrol of postharvest diseases of apples with antagonist mixtures. Phytopathology 78:194–198

Janisiewicz WJ, Jeffers SN (1997) Efficacy of commercial formulation of two biofungicides for control of blue mold and gray mold of apples in cold storage. Crop Prot 16:629–633. https://doi.org/10.1016/S0261-2194(97)00048-3

Janisiewicz WJ, Peterson DL (2004) Susceptibility of the stem pull area of mechanically harvested apples to blue mold decay and its control with a biocontrol agent. Plant Dis 88:662–664. https://doi.org/10.1094/PDIS.2004.88.6.662

Jayasankar NP, Bandoni RJ, Towers GHN (1969) Fungal degradation of phloridzin. Phytochemistry 8:379–383. https://doi.org/10.1016/S0031-9422(00)85434-5

Jiang J, Song Z, Yang X, Mao Z, Nie X, Guo H, Peng X (2017) Microbial community analysis of apple rhizosphere around Bohai Gulf. Sci Rep 7:8918. https://doi.org/10.1038/s41598-017-08398-9

Jijakli MH, Lepoivre P (2004) State of the art and challenges of post-harvest disease management in apples. In: Mukerji KG (ed) Fruit and vegetable diseases. Springer, Netherlands, Dordrecht, pp 59–94

Johnson KB, Stockwell VO (2000) Biological control of fire blight. In: Vanneste JL (ed) Fire blight: the disease and its causative agent, *Erwinia amylovora*. CABI, Wallingford, pp 319–337

Johnson KB, Stockwell VO (1998) Management of fire blight: a case study in microbial ecology. Annu Rev Phytopathol 36:227–248. https://doi.org/10.1146/annurev.phyto.36.1.227

Jones JB, Jackson LE, Balogh B, Obradovic A, Iriarte FB, Momol MT (2007) Bacteriophages for plant disease control. Annu Rev Phytopathol 45:245–262. https://doi.org/10.1146/annurev.phyto.45.062806.094411

Ju R, Zhao Y, Li J, Liu P, Yang T, Bao Z, Zhou B, Zhou X, Liu X (2014) Identification and evaluation of a potential biocontrol agent, *Bacillus subtilis*, against *Fusarium* sp. in apple seedlings. Ann Microbiol 64:377–383. https://doi.org/10.1007/s13213-013-0672-3

Junker RR, Loewel C, Gross R, Dötterl S, Keller A, Blüthgen N (2011) Composition of epiphytic bacterial communities differs on petals and leaves. Plant Biol 13:918–924. https://doi.org/10.1111/j.1438-8677.2011.00454.x

Kawaguchi A, Kondo K, Inoue K (2012) Biological control of apple crown gall by nonpathogenic *Rhizobium vitis* strain VAR03-1. J Gen Plant Pathol 78:287–293. https://doi.org/10.1007/s10327-012-0388-4

Kelderer M, Manici LM, Caputo F, Thalheimer M (2012) Planting in the "inter-row" to overcome replant disease in apple orchards: a study on the effectiveness of the practice based on microbial indicators. Plant Soil 357:381–393. https://doi.org/10.1007/s11104-012-1172-0

Kembel SW, O'Connor TK, Arnold HK, Hubbell SP, Wright SJ, Green JL (2014) Relationships between phyllosphere bacterial communities and plant functional traits in a neotropical forest. Proc Natl Acad Sci 111:13715–13720. https://doi.org/10.1073/pnas.1216057111

Kenney SJ, Beuchat LR (2002) Survival of *Escherichia coli* O157:H7 and *Salmonella* muenchen on apples as affected by application of commercial fruit waxes. Int J Food Microbiol 77:223–231. https://doi.org/10.1016/S0168-1605(02)00113-7

Kessler A, Baldwin IT (2002) Plant responses to insect herbivory: the emerging molecular analysis. Annu Rev Plant Biol 53:299–328. https://doi.org/10.1146/annurev.arplant.53.100301.135207

Khan MA, Olsen KM, Sovero V, Kushad MM, Korban SS (2014) Fruit quality traits have played critical roles in domestication of the apple. Plant Genome 7:1–18. https://doi.org/10.3835/plantgenome2014.04.0018

Khorram MS, Zhang G, Fatemi A, Kiefer R, Maddah K, Baqar M, Zakaria MP, Li G (2019) Impact of biochar and compost amendment on soil quality, growth and yield of a replanted apple orchard in a 4-year field study. J Sci Food Agric 99:1862–1869. https://doi.org/10.1002/jsfa.9380

Khosravi H, Samar SM, Fallahi E, Davoodi H, Shahabian M (2009) Inoculation of 'Golden Delicious' apple trees on M9 rootstock with *Azotobacter* improves nutrient uptake and growth indices. J Plant Nutr 32:946–953. https://doi.org/10.1080/01904160902870713

Kiers ET, Duhamel M, Beesetty Y, Mensah JA, Franken O, Verbruggen E, Fellbaum CR, Kowalchuk GA, Hart MM, Bago A, Palmer TM, West SA, Vandenkoornhuyse P, Jansa J, Bücking H (2011) Reciprocal rewards stabilize cooperation in the mycorrhizal symbiosis. Science 333:880–882. https://doi.org/10.1126/science.1208473

Kinkel LL, Andrews JH, Nordheim EV (1989) Fungal immigration dynamics and community development on apple leaves. Microb Ecol 18:45–58. https://doi.org/10.1007/BF02011695

Kinkel LL, Andrews JH, Nordheim EV (1989) Microbial introductions to apple leaves: influences of altered immigration on fungal community dynamics. Microb Ecol 18:161–173. https://doi.org/10.1007/BF02030124

Knight R, Vrbanac A, Taylor BC, Aksenov A, Callewaert C, Debelius J, Gonzalez A, Kosciolek T, McCall L-I, McDonald D, Melnik AV, Morton JT, Navas J, Quinn RA, Sanders JG, Swafford AD, Thompson LR, Tripathi A, Xu ZZ, Zaneveld JR, Zhu Q, Caporaso JG, Dorrestein PC (2018) Best practices for analysing microbiomes. Nat Rev Microbiol 16:410–422. https://doi.org/10.1038/s41579-018-0029-9

Köberl M, Schmidt R, Ramadan EM, Bauer R, Berg G (2013) The microbiome of medicinal plants: diversity and importance for plant growth, quality and health. Front Microbiol 4. https://doi.org/10.3389/fmicb.2013.00400

Kucheryava N, Fiss M, Auling G, Kroppenstedt RM (1999) Isolation and characterization of epiphytic bacteria from the phyllosphere of apple, antagonistic in vitro to *Venturia inaequalis*, the causal agent of apple scab. Syst Appl Microbiol 22:472–478

Kumar S, Agarwal M, Dheeman S, Maheshwari DK (2015) Exploitation of phytohormone-producing PGPR in development of multispecies bioinoculant formulation. In: Bacterial metabolites in sustainable agroecosystem. Springer, pp 297–317

Kurek E, Ozimek E, Sobiczewski P, Słomka A, Jaroszuk-Ściseł J (2013) Effect of *Pseudomonas luteola* on mobilization of phosphorus and growth of young apple trees (Ligol)—pot experiment. Scient Hortic 164:270–276. https://doi.org/10.1016/j.scienta.2013.09.012

Kviklys D, Liaudanskas M, Janulis V, Viškelis P, Rubinskienė M, Lanauskas J, Uselis N (2015) Rootstock genotype determines phenol content in apple fruits. Plant Soil Environ 60:234–240

Last FT (1955) Seasonal incidence of *Sporobolomyces* on cereal leaves. Trans Br Mycol Soc 38:221–239. https://doi.org/10.1016/S0007-1536(55)80069-1

Lata R, Chowdhury S, Gond SK, White JF (2018) Induction of abiotic stress tolerance in plants by endophytic microbes. Lett Appl Microbiol 66:268–276. https://doi.org/10.1111/lam.12855

Lee KW, Kim YJ, Kim DO, Lee HJ, Lee CY (2003) Major phenolics in apple and their contribution to the total antioxidant capacity. J Agric Food Chem 51:6516–6520. https://doi.org/10.1021/jf034475w

Leff JW, Fierer N (2013) Bacterial communities associated with the surfaces of fresh fruits and vegetables. PLoS One 8. https://doi.org/10.1371/journal.pone.0059310

Leff JW, Lynch RC, Kane NC, Fierer N (2017) Plant domestication and the assembly of bacterial and fungal communities associated with strains of the common sunflower, *Helianthus annuus*. New Phytol 214:412–423. https://doi.org/10.1111/nph.14323

Leff JW, Tredici PD, Friedman WE, Fierer N (2015) Spatial structuring of bacterial communities within individual *Ginkgo biloba* trees. Environ Microbiol 17:2352–2361. https://doi.org/10.1111/1462-2920.12695

Leisso R, Rudell D, Mazzola M (2018) Targeted metabolic profiling indicates apple rootstock genotype-specific differences in primary and secondary metabolite production and validate quantitative contribution from vegetative growth. Front Plant Sci 9. https://doi.org/10.3389/fpls.2018.01336

Leisso R, Rudell D, Mazzola M (2017) Metabolic composition of apple rootstock rhizodeposits differs in a genotype-specific manner and affects growth of subsequent plantings. Soil Biol Biochem 113:201–214. https://doi.org/10.1016/j.soilbio.2017.06.011

Leon N de, Jannink J-L, Edwards JW, Kaeppler SM (2016) Introduction to a special issue on genotype by environment interaction. Crop Sci 56:2081–2089. https://doi.org/10.2135/cropsci2016.07.0002in

Leveau JHJ (2015) Life of microbes on aerial plant parts. In: Lugtenberg B (ed) Principles of plant-microbe interactions: microbes for sustainable agriculture. Springer International Publishing, Cham, pp 17–24

Li M, Li P, Ma F, Dandekar AM, Cheng L (2018) Sugar metabolism and accumulation in the fruit of transgenic apple trees with decreased sorbitol synthesis. Hortic Res 5:1–11. https://doi.org/10.1038/s41438-018-0064-8

Liang B, Ma C, Fan L, Wang Y, Yuang Y (2018) Soil amendment alters soil physicochemical properties and bacterial community structure of a replanted apple orchard. Microbiol Res 216:1–11. https://doi.org/10.1016/j.micres.2018.07.010

Lindow SE, Brandl MT (2003) Microbiology of the phyllosphere. Appl Environ Microbiol 69:1875–1883. https://doi.org/10.1128/AEM.69.4.1875-1883.2003

Liu J, Abdelfattah A, Norelli J, Burchard E, Schena L, Droby S, Wisniewski M (2018) Apple endophytic microbiota of different rootstock/scion combinations suggests a genotype-specific influence. Microbiome 6:18. https://doi.org/10.1186/s40168-018-0403-x

Lucas M, Balbín-Suárez A, Smalla K, Vetterlein D (2018) Root growth, function and rhizosphere microbiome analyses show local rather than systemic effects in apple plant response to replant disease soil. PLoS ONE 13(10):e0204922. https://doi.org/10.1371/journal.pone.0204922

Ma B, Chen J, Zheng H, Fang T, Ogutu C, Li S, Han Y, Wu B (2015) Comparative assessment of sugar and malic acid composition in cultivated and wild apples. Food Chem 172:86–91. https://doi.org/10.1016/j.foodchem.2014.09.032

Macarisin D, Sheth I, Hur M, Wooten A, Kwon HJ, Gao Z, De Jesus A, Jurick W, Chen Y (2019) Survival of outbreak, food, and environmental strains of *Listeria monocytogenes* on whole apples as affected by cultivar and wax coating. Sci Rep 9:12170. https://doi.org/10.1038/s41598-019-48597-0

Manici LM, Kelderer M, Caputo F, Nicoletti F, De Luca PF, Topp AR (2015) Impact of cover crop in pre-plant of apple orchards: relationship between crop health, root inhabiting fungi and rhizospheric bacteria. Can J Plant Sci 95:947–958. https://doi.org/10.4141/cjps-2015-013

Martins G, Lauga B, Miot-Sertier C, Mercier A, Lonvaud A, Soulas M.-L., Soulas G, Masneuf-Pomarède I (2013) Characterization of epiphytic bacterial communities from grapes, leaves, bark and soil of grapevine plants grown, and their relations. PLoS One 8:e73013. https://doi.org/10.1371/journal.pone.0073013

Matteson Heidenreich MC, Corral-Garcia MR, Momol EA, Burr TJ (1997) Russet of apple fruit caused by *Aureobasidium pullulans* and *Rhodotorula glutinis*. Plant Dis 81:337–342. https://doi.org/10.1094/PDIS.1997.81.4.337

Matteson M, Burr T, Corral Garcia M, Parker D, Koeller D (1995) Russet of apple fruit induced by isolates of *Rhodotorula glutinis* and *Aureobasidium pullulans* [Abstract]. Phytopathology 85:1558

Mazzola M (1999) Transformation of soil microbial community structure and *Rhizoctonia*-suppressive potential in response to apple roots. Phytopathology 89:920–927. https://doi.org/10.1094/PHYTO.1999.89.10.920

Mazzola M, Freilich S (2017) Prospects for biological soilborne disease control: application of indigenous versus synthetic microbiomes. Phytopathology 107:256–263. https://doi.org/10.1094/PHYTO-09-16-0330-RVW

Mazzola M, Graham D, Wang L, Leisso R, Hewavitharana SS (2020) Application sequence modulates microbiome composition, plant growth and apple replant disease control efficiency upon integration of anaerobic soil disinfestation and mustard seed meal amendment. Crop Prot 132:105125. https://doi.org/10.1016/j.cropro.2020.105125

Mazzola M, Hewavitharana SS, Strauss SL (2015) Brassica seed meal soil amendments transform the rhizosphere microbiome and improve apple production through resistance to pathogen reinfestation. Phytopathology 105:460–469. https://doi.org/10.1094/PHYTO-09-14-0247-R

Mazzola M, Manici LM (2012) Apple replant disease: role of microbial ecology in cause and control. Annu Rev Phytopathol 50:45–65. https://doi.org/10.1146/annurev-phyto-081211-173005

McGrath MJ, Andrews JH (2006) Temporal changes in microscale colonization of the phylloplane by *Aureobasidium pullulans*. Appl Environ Microbiol 72:6234–6241. https://doi.org/10.1128/AEM.00744-06

McGrath MJ, Andrews JH (2007) Role of microbial immigration in the colonization of apple leaves by *Aureobasidium pullulans*. Appl Environ Microbiol 73:1277–1286. https://doi.org/10.1128/AEM.01620-06

Meakem V, Burchard E, Teixeira G, Wisniewski M, Whitehead S (2020) The effect of distinct biotic communities on induced phenolic chemistry of apple fruits. In: Induced defenses in apple fruits: linking fruit chemistry, quality, and plant-insect-microbe interactions (M.S. Thesis). Virginia Tech, pp 4–68

Mechan Llontop ME, Hurley K, Tian L, Bernal Galeano VA, Wildschutte HK, Marine SC, Yoder KS, Vinatzer BA (2020) Exploring rain as source of biological control agents for fire blight on apple. Front Microbiol 11. https://doi.org/10.3389/fmicb.2020.00199

Medzhitov R, Schneider DS, Soares MP (2012) Disease tolerance as a defense strategy. Science 335:936–941. https://doi.org/10.1126/science.1214935

Meyer F, Paarmann D, D'Souza M, Olson R, Glass E, Kubal M, Paczian T, Rodriguez A, Stevens R, Wilke A, Wilkening J, Edwards R (2008) The metagenomics RAST server—a public resource for the automatic phylogenetic and functional analysis of metagenomes. BMC Bioinformatics 9:386. https://doi.org/10.1186/1471-2105-9-386

Meyer RS, DuVal AE, Jensen HR (2012) Patterns and processes in crop domestication: an historical review and quantitative analysis of 203 global food crops. New Phytol 196:29–48. https://doi.org/10.1111/j.1469-8137.2012.04253.x

Mikiciński A, Puławska J, Molzhigitova A, Sobiczewski P (2020) Bacterial species recognized for the first time for its biocontrol activity against fire blight (Erwinia amylovora). Eur J Plant Pathol 156:257–272. https://doi.org/10.1007/s10658-019-01885-x

Mitter B, Pfaffenbichler N, Flavell R, Compant S, Antonielli L, Petric A, Berninger T, Naveed M, Sheibani-Tezerji R, von Maltzahn G, Sessitsch A (2017) A new approach to modify plant microbiomes and traits by introducing beneficial bacteria at flowering into progeny seeds. Front Microbiol 8:11. https://doi.org/10.3389/fmicb.2017.00011

Molina-Favero C, Creus CM, Simontacchi M, Puntarulo S, Lamattina L (2008) Aerobic nitric oxide production by *Azospirillum brasilense* Sp245 and its influence on root architecture in tomato. Mol Plant Microbe Interact 21:1001–1009. https://doi.org/10.1094/MPMI-21-7-1001

Momma N, Kobara Y, Uematsu S, Kita N, Shinmura A (2013) Development of biological soil disinfestations in Japan. Appl Microbiol Biotechnol 97:3801–3809. https://doi.org/10.1007/s00253-013-4826-9

Morris BE, Henneberger R, Huber H, Moissl-Eichinger C (2013) Microbial syntrophy: interaction for the common good. FEMS Microbiol Rev 37:384–406

Mosa W-G, Sas-Paszt L, FRĄC M, TRZCIŃSKI P, (2016) Microbial products and biofertilizers in improving growth and productivity of apple–a review. Pol J Microbiol 65:243–251

Nagy J, Király L, Schwarczinger I (2012) Phage therapy for plant disease control with a focus on fire blight. Open Life Sci 7:1–12. https://doi.org/10.2478/s11535-011-0093-x

Natale D, Mattiacci L, Hern A, Pasqualini E, Dorn S (2003) Response of female *Cydia molesta* (Lepidoptera: Tortricidae) to plant derived volatiles. Bull Entomol Res 93:335–342

Nelson EB (2018) The seed microbiome: origins, interactions, and impacts. Plant Soil 422:7–34. https://doi.org/10.1007/s11104-017-3289-7

Noecker C, Eng A, Srinivasan S, Theriot CM, Young VB, Jansson JK, Fredricks DN, Borenstein E (2016) Metabolic model-based integration of microbiome taxonomic and metabolomic profiles elucidates mechanistic links between ecological and metabolic variation. mSystems 1(1):e00013–15. https://doi.org/10.1128/mSystems.00013-15

Nogales A, Nobre T, Valadas V, Ragonezi C, Döring M, Polidoros A, Arnholdt-Schmitt B (2016) Can functional hologenomics aid tackling current challenges in plant breeding? Brief Funct Genomics 15:288–297. https://doi.org/10.1093/bfgp/elv030

Norelli JL, Burr TJ, Cicero AML, Gilbert MT, Katz BH (1991) Homologous streptomycin resistance gene present among diverse gram-negative bacteria in New York State apple orchards. Appl Environ Microbiol 57:486–491

Ofaim S, Ofek-Lalzar M, Sela N, Jinag J, Kashi Y, Minz D, Freilich S (2017) Analysis of microbial functions in the rhizosphere using a metabolic-network based framework for metagenomics interpretation. Front Microbiol 8:1606. https://doi.org/10.3389/fmicb.2017.01606

Ofek-Lalzar M, Sela N, Goldman-Voronov M, Green SJ, Hadar Y, Minz D (2014) Niche and host-associated functional signatures of the root surface microbiome. Nat Commun 5:4950. https://doi.org/10.1038/ncomms5950

Oldroyd GED (2013) Speak, friend, and enter: signalling systems that promote beneficial symbiotic associations in plants. Nat Rev Microbiol 11:252–263. https://doi.org/10.1038/nrmicro2990

Ottesen A, Skaltsas D, White JR, Gorham S, Ramachandran P, Brown E, Newell M, Walsh C (2016) Using next generation sequencing to describe epiphytic microbiota associated with organic and conventionally managed apples. Int J Environ Agric Res 2:36–42

Ottesen AR, González Peña A, White JR, Pettengill JB, Li C, Allard S, Rideout S, Allard M, Hill T, Evans P (2013) Baseline survey of the anatomical microbial ecology of an important food plant: *Solanum lycopersicum* (tomato). BMC Microbiol 13:114. https://doi.org/10.1186/1471-2180-13-114

Ottesen AR, White JR, Skaltsas DN, Newell MJ, Walsh CS (2009) Impact of organic and conventional management on the phyllosphere microbial ecology of an apple crop. J Food Prot 72:2321–2325. https://doi.org/10.4315/0362-028X-72.11.2321

Panke-Buisse K, Poole AC, Goodrich JK, Ley RE, Kao-Kniffin J (2015) Selection on soil microbiomes reveals reproducible impacts on plant function. ISME J 9:980–989. https://doi.org/10.1038/ismej.2014.196

Pérez-Izquierdo L, Zabal-Aguirre M, Flores-Rentería D, González-Martínez SC, Buée M, Rincón A (2017) Functional outcomes of fungal community shifts driven by tree genotype and spatial-temporal factors in Mediterranean pine forests. Environ Microbiol 19:1639–1652. https://doi.org/10.1111/1462-2920.13690

Pérez-Jaramillo JE, Carrión VJ, Bosse M, Ferrão LFV, de Hollander M, Garcia AAF, Ramírez CA, Mendes R, Raaijmakers JM (2017) Linking rhizosphere microbiome composition of wild and domesticated *Phaseolus vulgaris* to genotypic and root phenotypic traits. ISME J 11:2244–2257. https://doi.org/10.1038/ismej.2017.85

Pérez-Jaramillo JE, Carrión VJ, de Hollander M, Raaijmakers JM (2018) The wild side of plant microbiomes. Microbiome 6:143. https://doi.org/10.1186/s40168-018-0519-z

Pérez-Jaramillo JE, Mendes R, Raaijmakers JM (2016) Impact of plant domestication on rhizosphere

microbiome assembly and functions. Plant Mol Biol 90:635–644. https://doi.org/10.1007/s11103-015-0337-7
Pertot I, Alabouvette C, Esteve EH (2015) Mini-paper—The use of microbial biocontrol agents against soil-borne diseases. EIP-AGRI Focus Group Soil-Borne Dis 11
Peruzzi E, Franke-Whittle IH, Kelderer M, Ciavatta C, Insam H (2017) Microbial indication of soil health in apple orchards affected by replant disease. Appl Soil Ecol 119:115–127. https://doi.org/10.1016/j.apsoil.2017.06.003
Philippot L, Raaijmakers JM, Lemanceau P, van der Putten WH (2013) Going back to the roots: the microbial ecology of the rhizosphere. Nat Rev Microbiol 11:789–799. https://doi.org/10.1038/nrmicro3109
Phillips M (2012) The holistic orchard: tree fruits and berries the biological way, illustrated. Chelsea Green Publishing, White River Junction, Vt, Totnes
Pieterse CMJ, Poelman E, Van Wees SCM, Dicke M (2013) Induced plant responses to microbes and insects. Front Plant Sci 4. https://doi.org/10.3389/fpls.2013.00475
Piskorski R, Dorn S (2010) Early-season headspace volatiles from apple and their effect on the apple blossom weevil *Anthonomus pomorum*. Chem Biodivers 7:2254–2260
Ponomarova O, Patil KR (2015) Metabolic interactions in microbial communities: untangling the Gordian knot. Curr Opin Microbiol 27:37–44. https://doi.org/10.1016/j.mib.2015.06.014
Pozo MI, Lachance M-A, Herrera CM (2012) Nectar yeasts of two southern Spanish plants: the roles of immigration and physiological traits in community assembly. FEMS Microbiol Ecol 80:281–293. https://doi.org/10.1111/j.1574-6941.2011.01286.x
Purin S, Filho OK, Stürmer SL (2006) Mycorrhizae activity and diversity in conventional and organic apple orchards from Brazil. Soil Biol Biochem 38:1831–1839. https://doi.org/10.1016/j.soilbio.2005.12.008
Pusey P (1998) Laboratory and field trials with selected microorganisms as biocontrol agents for fire blight. pp 655–662
Pusey PL, Curry EA (2004) Temperature and pomaceous flower age related to colonization by *Erwinia amylovora* and antagonists. Phytopathology 94:901–911. https://doi.org/10.1094/PHYTO.2004.94.8.901
Pusey PL, Stockwell VO, Mazzola M (2009) Epiphytic bacteria and yeasts on apple blossoms and their potential as antagonists of *Erwinia amylovora*. Phytopathology 99:571–581. https://doi.org/10.1094/PHYTO-99-5-0571
Pusztahelyi T, Holb IJ, Pócsi I (2015) Secondary metabolites in fungus-plant interactions. Front Plant Sci 6:573. https://doi.org/10.3389/fpls.2015.00573
Qin S, Jiao K, He J, Lyu D (2016) Forage crops alter soil bacterial and fungal communities in an apple orchard. Acta Agric Scand Sect B—Soil Plant Sci 66:229–236. https://doi.org/10.1080/09064710.2015.1088569
Quince C, Walker AW, Simpson JT et al (2017) Shotgun metagenomics, from sampling to analysis. Nat Biotechnol 35:833–844. https://doi.org/10.1038/nbt.3935
Raaijmakers JM, Mazzola M (2016) Soil immune responses. Science 352:1392–1393. https://doi.org/10.1126/science.aaf3252
Radl V, Winkler JB, Kublik S, Yang L, Winkelmann T, Vestergaard G, Schröder P, Schloter M (2019) Reduced microbial potential for the degradation of phenolic compounds in the rhizosphere of apple plantlets grown in soils affected by replant disease. Environ Microbiome 14:8. https://doi.org/10.1186/s40793-019-0346-2
Rebolleda-Gómez M, Forrester NJ, Russell AL, Wei N, Fetters AM, Stephens JD, Ashman T-L (2019) Gazing into the anthosphere: considering how microbes influence floral evolution. New Phytol 224:1012–1020. https://doi.org/10.1111/nph.16137
Resendes ML, Bryla DR, Eissenstat DM (2008) Early events in the life of apple roots: variation in root growth rate is linked to mycorrhizal and nonmycorrhizal fungal colonization. Plant Soil 313:175–186. https://doi.org/10.1007/s11104-008-9690-5
Rho H, Van Epps V, Kim S-H, Doty SL (2020) Endophytes increased fruit quality with higher soluble sugar production in "Honeycrisp" apple (*Malus pumila*). Microorganisms 8:699. https://doi.org/10.3390/microorganisms8050699
Roiger DJ, Jeffers SN (1991) Evaluation of *Trichoderma* spp. for biological control of *Phytophthora* crown and root rot of apple seedlings. Phytopathology 81:910–917
Rosenberg E, Zilber-Rosenberg I (2018) The hologenome concept of evolution after 10 years. Microbiome 6:78. https://doi.org/10.1186/s40168-018-0457-9
Ruinen J (1956) Occurrence of *Beijerinckia* species in the 'phyllosphere'. Nature 177:220–221. https://doi.org/10.1038/177220a0
Rumberger A, Merwin IA, Thies JE (2007) Microbial community development in the rhizosphere of apple trees at a replant disease site. Soil Biol Biochem 39:1645–1654. https://doi.org/10.1016/j.soilbio.2007.01.023
Rumberger A, Yao S, Merwin IA, Nelson EB, Thies JE (2004) Rootstock genotype and orchard replant position rather than soil fumigation or compost amendment determine tree growth and rhizosphere bacterial community composition in an apple replant soil. Plant Soil 264:247–260. https://doi.org/10.1023/B.PLSO.0000047760.13004.94
Saminathan T, García M, Ghimire B, Lopez C, Bodunrin A, Nimmakayala P, Abburi VL, Levi A, Balagurusamy N, Reddy UK (2018) Metagenomic and metatranscriptomic analyses of diverse watermelon cultivars reveal the role of fruit associated microbiome in carbohydrate metabolism and ripening of mature

fruits. Front Plant Sci 9:4. https://doi.org/10.3389/fpls.2018.00004

Sanchez-Barrios A, Sahib MR, DeBolt S (2017) "I've got the magic in me": the microbiome of conventional vs organic production systems. In: Plant-mmicrobe Interactions in agro-ecological perspectives. Springer, pp 85–95

Santoyo G, Moreno-Hagelsieb G, del Carmen Orozco-Mosqueda Ma, Glick BR, (2016) Plant growth-promoting bacterial endophytes. Microbiol Res 183:92–99. https://doi.org/10.1016/j.micres.2015.11.008

Sare AR, Stouvenakers G, Eck M, Lampens A, Goormachtig S, Jijakli MH, Massart S (2020) Standardization of plant microbiome studies: which proportion of the microbiota is really harvested? Microorganisms 8:342. https://doi.org/10.3390/microorganisms8030342

Sasse J, Martinoia E, Northen T (2018) Feed your friends: do plant exudates shape the root microbiome? Trends Plant Sci 23:25–41. https://doi.org/10.1016/j.tplants.2017.09.003

Schnabel EL, Jones AL (1999) Distribution of tetracycline resistance genes and transposons among phylloplane bacteria in Michigan apple orchards. Appl Environ Microbiol 65:4898–4907

Schreiter S, Ding G-C, Heuer H, Neumann G, Sandmann M, Grosch R, Kropf S, Smalla K (2014) Effect of the soil type on the microbiome in the rhizosphere of field-grown lettuce. Front Microbiol 5 (Article 144). https://doi.org/10.3389/fmicb.2014.00144

Schroth M, Thomson S, Hildebrand D, Moller W (1974) Epidemiology and control of fire blight. Annu Rev Phytopathol 12:389–412

Schut M, Rodenburg J, Klerkx L, van Ast A, Bastiaans L (2014) Systems approaches to innovation in crop protection. A systematic literature review. Crop Prot 56:98–108. https://doi.org/10.1016/j.cropro.2013.11.017

Shade A, Handelsman J (2012) Beyond the Venn diagram: the hunt for a core microbiome. Environ Microbiol 14:4–12. https://doi.org/10.1111/j.1462-2920.2011.02585.x

Shade A, Hogan CS, Klimowicz AK, Linske M, McManus PS, Handelsman J (2012) Culturing captures members of the soil rare biosphere. Environ Microbiol 14:2247–2252. https://doi.org/10.1111/j.1462-2920.2012.02817.x

Shade A, Jacques M-A, Barret M (2017) Ecological patterns of seed microbiome diversity, transmission, and assembly. Curr Opin Microbiol 37:15–22. https://doi.org/10.1016/j.mib.2017.03.010

Shade A, Klimowicz AK, Spear RN, Linske M, Donato JJ, Hogan CS, McManus PS, Handelsman J (2013) Streptomycin application has no detectable effect on bacterial community structure in apple orchard soil. Appl Environ Microbiol 79:6617–6625. https://doi.org/10.1128/AEM.02017-13

Shade A, McManus PS, Handelsman J (2013b) Unexpected diversity during community succession in the apple flower microbiome. mBio 4(2):e00602–12. https://doi.org/10.1128/mBio.00602-12

Shade A, Peter H, Allison SD, Baho D, Berga M, Bürgmann H, Huber DH, Langenheder S, Lennon JT, Martiny JB, Matulich KL, Schmidt TM, Handelsman J (2012b) Fundamentals of microbial community resistance and resilience. Front Microbiol 3 (Article 417). https://doi.org/10.3389/fmicb.2012.00417

Shanmugam V, Verma R, Rajkumar S, Singh Naruka D (2011) Bacterial diversity and soil enzyme activity in diseased and disease free apple rhizosphere soils. Ann Microbiol 61:765–772. https://doi.org/10.1007/s13213-010-0193-2

Sharifazizi M, Harighi B, Sadeghi A (2017) Evaluation of biological control of *Erwinia amylovora*, causal agent of fire blight disease of pear by antagonistic bacteria. Biol Control 104:28–34. https://doi.org/10.1016/j.biocontrol.2016.10.007

Sharma R, Berg JA, Beatty NJ, Cowger AE, Cozzens BJR, Duncan SG, Fajardo CP, Ferguson HP, Galbraith T, Herring JA, Hoj TR, Durrant JL, Hyde JR, Jensen GL, Ke SY, Killpack S, Kruger JL, Lawrence EEK, Nwosu IO, Tam TC, Thompson DW, Tueller JA, Ward MEH, Webb CJ, Wood ME, Yeates EL, Baltrus DA, Breakwell DP, Hope S, Grose JH (2018) Genome sequences of nine *Erwinia amylovora* bacteriophages. Microbiol Resour Announc 7:e00944–e1018. https://doi.org/10.1128/MRA.00944-18

Sharma R, Pal J, Rana S, Kaur M (2017) Suppression of soil borne fungal pathogens associated with apple replant disease by cyclic application of native strains of *Pseudomonas aeruginosa*. J Appl Nat Sci 9:2105–2109. https://doi.org/10.31018/jans.v9i4.1495

Sharma R, Pielstick BA, Bell KA, Nieman TB, Stubbs OA, Yeates EL, Baltrus DA, Grose JH (2019) A novel, highly related jumbo family of bacteriophages that were isolated against *Erwinia*. Front Microbiol 10:1533. https://doi.org/10.3389/fmicb.2019.01533

Sharma S, Bhutani V, Awasthi R (2002) Effect of vesicular-arbuscular mycorrhizae and phosphorus on leaf and soil mineral nutrient status of apple seedling. Indian J Hortic 59:140–144

Shen Y, Nie J, Li Z, Li H, Wu Y, Dong Y, Zhang J (2018) Differentiated surface fungal communities at point of harvest on apple fruits from rural and peri-urban orchards. Sci Rep 8:2165. https://doi.org/10.1038/s41598-017-17436-5

Shishido M, Sakamoto K, Yokoyama H, Momma N, Miyashita S-I (2008) Changes in microbial communities in an apple orchard and its adjacent bush soil in response to season, land-use, and violet root rot infestation. Soil Biol Biochem 40:1460–1473. https://doi.org/10.1016/j.soilbio.2007.12.024

Singh J, Silva KJP, Fuchs M, Khan A (2019) Potential role of weather, soil and plant microbial communities in rapid decline of apple trees. PLoS One 14:e0213293. https://doi.org/10.1371/journal.pone.0213293

Singh N, Sharma D, Kumar V (2017) Managing apple replant disease: the effect of rootstocks and soil treatments on tree performance and biological activities. J Pharmacogn Phytochem 6:2554–2559

Sivak B, Person CO (1973) The bacterial and fungal flora of the bark, wood, and pith of alder, black cottonwood, maple, and willow. Can J Bot 51:1985–1988. https://doi.org/10.1139/b73-254

Sláviková E, Vadkertiová R, Vránová D (2009) Yeasts colonizing the leaves of fruit trees. Ann Microbiol 59:419–424. https://doi.org/10.1007/BF03175125

Smith V, Wilcox W, Harman G (1990) Potential for biological control of *Phytophthora* root and crown rots of apple by *Trichoderma* and *Gliocladium* spp. Phytopathology 80:880–885

Sobiczewski P, Chiou CS, Jones AL (1991) Streptomycin-resistant epiphytic bacteria with homologous DNA for streptomycin resistance in Michigan apple orchards. Plant Dis 75:1110–1113

Somera T, Freilich S, Mazzola M (2019) Exploring the functional capacity of seed meal-structured microbiomes in the control of apple replant disease. In: Proceedings, soil ecology society 2019. https://soilecologysociety2019.exordo.com/programme/presentation/285

St. Laurent A, Merwin IA, Fazio G, Thies JE, Brown MG, (2010) Rootstock genotype succession influences apple replant disease and root-zone microbial community composition in an orchard soil. Plant Soil 337:259–272. https://doi.org/10.1007/s11104-010-0522-z

St. Laurent A, Merwin IA, Thies JE, (2008) Long-term orchard groundcover management systems affect soil microbial communities and apple replant disease severity. Plant Soil 304:209–225. https://doi.org/10.1007/s11104-008-9541-4

Steiner PW (1990) Predicting apple blossom infections by *Erwinia amylovora* using the MARYBLYT model. Acta Hortic 273:139–148. https://doi.org/10.17660/ActaHortic.1990.273.18

Steven B, Huntley RB, Zeng Q (2018) The influence of flower anatomy and apple cultivar on the apple flower phytobiome. Phytobiomes J 2:171–179. https://doi.org/10.1094/PBIOMES-03-18-0015-R

Stone BWG, Weingarten EA, Jackson CR (2018) The role of the phyllosphere microbiome in plant health and function. In: Annual plant reviews online. American Cancer Society, pp 533–556

Sun J, Zhang Q, Li X, Zhou B, Wei Q (2017) Apple replant disorder of Pingyitiancha rootstock is closely associated with rhizosphere fungal community development. J Phytopathol 165:162–173. https://doi.org/10.1111/jph.12547

Sun J, Zhang Q, Zhou J, Wei Q (2014) Pyrosequencing technology reveals the impact of different manure doses on the bacterial community in apple rhizosphere soil. Appl Soil Ecol 78:28–36. https://doi.org/10.1016/j.apsoil.2014.02.004

Sundin GW, Werner NA, Yoder KS, Aldwinckle HS (2009) Field evaluation of biological control of fire blight in the eastern United States. Plant Dis 93:386–394. https://doi.org/10.1094/PDIS-93-4-0386

Svircev A, Lehman S, Kim W, Barszcz E, Schneider K, Castle A (2006) Control of the fire blight pathogen with bacteriophages. Mitteilungen Aus Biol Bundesanst Für Land-Forstwirtsch, pp 259–261

Tanaka A, Tapper BA, Popay A, Parker EJ, Scott B (2005) A symbiosis expressed non-ribosomal peptide synthetase from a mutualistic fungal endophyte of perennial ryegrass confers protection to the symbiotum from insect herbivory. Mol Microbiol 57:1036–1050

Tessler M, Neumann JS, Afshinnekoo E, Pineda M, Hersch R, Velho LFM, Segovia BT, Lansac-Toha FA, Lemke M, DeSalle R, Mason CE, Brugler MR (2017) Large-scale differences in microbial biodiversity discovery between 16S amplicon and shotgun sequencing. Sci Rep 7:6589. https://doi.org/10.1038/s41598-017-06665-3

Tewksbury JJ (2002) Fruits, frugivores and the evolutionary arms race. New Phytol 156:137–139

Thompson LR, Sanders JG, McDonald D, Amir A, Ladau J, Locey KJ, Prill RJ, Tripathi A, Gibbons SM, Ackermann G, Navas-Molina JA, Janssen S, Kopylova E, Vázquez-Baeza Y, González A, Morton JT, Mirarab S, Zech XuZ, Jiang L, Haroon MF, Kanbar J, Zhu Q, Jin Song S, Kosciolek T, Bokulich NA, Lefler J, Brislawn CJ, Humphrey G, Owens SM, Hampton-Marcell J, Berg-Lyons D, McKenzie V, Fierer N, Fuhrman JA, Clauset A, Stevens RL, Shade A, Pollard KS, Goodwin KD, Jansson JK, Gilbert JA, Knight R (2017) A communal catalogue reveals Earth's multiscale microbial diversity. Nature 551:457–463. https://doi.org/10.1038/nature24621

Tilston EL, Deakin G, Bennett J, Passey T, Harrison N, O'Brien F, Fernández-Fernández F, Xu X (2018) Candidate causal organisms for apple replant disease in the United Kingdom. Phytobiomes J 2:261–274. https://doi.org/10.1094/PBIOMES-11-18-0050-R

Timm CM, Carter KR, Carrell AA, Jun S-R, Jawdy SS, Vélez JM, Gunter LE, Yang Z, Nookaew I, Engle NL, Lu T-YS, Schadt CW, Tschaplinski TJ, Doktycz MJ, Tuskan GA, Pelletier DA, Weston DJ (2018) Abiotic stresses shift belowground *Populus*-associated bacteria toward a core stress microbiome. mSystems 3. https://doi.org/10.1128/mSystems.00070-17

Toju H, Peay KG, Yamamichi M, Narisawa K, Hiruma K, Naito K, Fukuda S, Ushio M, Nakaoka S, Onoda Y, Yoshida K, Schlaeppi K, Bai Y, Sugiura R, Ichihashi Y, Minamisawa K, Kiers ET (2018) Core microbiomes for sustainable agroecosystems. Nat Plants 4:247–257. https://doi.org/10.1038/s41477-018-0139-4

Trabelsi D, Mhamdi R (2013) Microbial inoculants and their impact on soil microbial communities: a review. BioMed Res Int. https://www.hindawi.com/journals/bmri/2013/863240/. Accessed 23 June 2020

Turcotte MM, Araki H, Karp DS, Poveda K, Whitehead SR (2017) The eco-evolutionary impacts of domestication and agricultural practices on wild species. Philos Trans R Soc B Biol Sci 372:20160033. https://doi.org/10.1098/rstb.2016.0033

Turechek WW (2004) Apple diseases and their management. In: Naqvi SAMH (ed) Diseases of fruits and vegetables, vol I. Diagnosis and management. Springer, Netherlands, Dordrecht, pp 1–108

Turner TR, James EK, Poole PS (2013) The plant microbiome. Genome Biol 14:209. https://doi.org/10.1186/gb-2013-14-6-209

Utkhede RS (1987) Chemical and biological control of crown and root rot of apple caused by *Phytophthora cactorum*. Can J Plant Pathol 9:295–300. https://doi.org/10.1080/07060668709501860

Vadkertiová R, Molnárová J, Vránová D, Sláviková E (2012) Yeasts and yeast-like organisms associated with fruits and blossoms of different fruit trees. Can J Microbiol 58:1344–1352. https://doi.org/10.1139/cjm-2012-0468

Valentín L, Kluczek-Turpeinen B, Willför S, Hemming J, Hatakka A, Steffen K, Tuomela M (2010) Scots pine (*Pinus sylvestris*) bark composition and degradation by fungi: potential substrate for bioremediation. Bioresour Technol 101:2203–2209. https://doi.org/10.1016/j.biortech.2009.11.052

van Dam NM, Bouwmeester HJ (2016) Metabolomics in the rhizosphere: tapping into belowground chemical communication. Trends Plant Sci 21:256–265. https://doi.org/10.1016/j.tplants.2016.01.008

Van Horn C, Mazzola M (2019) Apple rootstock genotype influences endophytic and rhizosphere microbial community (Abstract). Phytopathology 109 (S3):2. https://doi.org/10.1094/PHYTO-109-11-S3.1

Vandenkoornhuyse P, Quaiser A, Duhamel M, Van AL, Dufresne A (2015) The importance of the microbiome of the plant holobiont. New Phytol 206:1196–1206. https://doi.org/10.1111/nph.13312

Velasco R, Zharkikh A, Affourtit J, Dhingra A, Cestaro A, Kalyanaraman A, Fontana P, Bhatnagar SK, Troggio M, Pruss D, Dhingra, A, Cestaro A, Kalyanaraman A, Fontana P, Bhatnagar SK, Troggio M, Pruss D, Salvi S, Pindo M, Baldi P, Castelletti S, Cavaiuolo M, Coppola G, Costa F, Cova V, Ri AD, Goremykin V, Komjanc M, Longhi S, Magnago P, Malacarne G, Malnoy M, Micheletti D, Moretto M, Perazzolli M, Si-Ammour A, Vezzulli S, Zini E, Eldredge G, Fitzgerald LM, Gutin N, Lanchbury G, Macalma T, Mitchell JT, Reid J, Wardell B, Kodira C, Chen Z, Desany B, Niazi F, Palmer M, Koepke T, Jiwan D, Schaeffer S, Krishnan V, Wu C, Chu VT, King ST, Vick J, Tao Q, Mraz A, Stormo A, Stormo K, Bogden R, Ederle D, Stella A, Vecchietti A, Kater MM, Masiero S, Lasserre P, Lespinasse Y, Allan AC, Bus V, Chagné D, Crowhurst RN, Gleave AP, Lavezzo E, Fawcett JA, Proost S, Rouzé P, Sterck L, Toppo S, Lazzari B, Hellens RP, Durel C-E, Gutin A, Bumgarner R, Gardiner SE, Skolnick M, Egholm M, Van de Peer Y, Salamini F, Viola R (2010) The genome of the domesticated apple (*Malus* × *domestica* Borkh.). Nat Genet 42: 833–839. https://doi.org/10.1038/ng.654

Venturi V, Keel C (2016) Signaling in the rhizosphere. Trends Plant Sci 21:187–198. https://doi.org/10.1016/j.tplants.2016.01.005

Vepštaitė-Monstavičė I, Lukša J, Stanevičienė R, Strazdaitė-Žielienė Ž, Yurchenko V, Serva S, Servienė E (2018) Distribution of apple and blackcurrant microbiota in Lithuania and the Czech Republic. Microbiol Res 206:1–8. https://doi.org/10.1016/j.micres.2017.09.004

Vero S, Garmendia G, González MB, Bentancur O, Wisniewski M (2013) Evaluation of yeasts obtained from Antarctic soil samples as biocontrol agents for the management of postharvest diseases of apple (*Malus* × *domestica*). FEMS Yeast Res 13:189–199. https://doi.org/10.1111/1567-1364.12021

Vitulo N, Lemos WJFJ, Calgaro M, Confalone M, Felis GE, Zapparoli G, Nardi T (2019) Bark and grape microbiome of *Vitis vinifera*: influence of geographic patterns and agronomic management on bacterial diversity. Front Microbiol 9:3203. https://doi.org/10.3389/fmicb.2018.03203

Vorholt JA (2012) Microbial life in the phyllosphere. Nat Rev Microbiol 10:828–840. https://doi.org/10.1038/nrmicro2910

Wagner MR, Lundberg DS, del Rio TG, Tringe SG, Dangl JL, Mitchell-Olds T (2016) Host genotype and age shape the leaf and root microbiomes of a wild perennial plant. Nat Commun 7:12151. https://doi.org/10.1038/ncomms12151

Wallenstein MD (2017) Managing and manipulating the rhizosphere microbiome for plant health: a systems approach. Rhizosphere 3:230–232. https://doi.org/10.1016/j.rhisph.2017.04.004

Walsh F, Smith DP, Owens SM, Duffy B, Frey JE (2014) Restricted streptomycin use in apple orchards did not adversely alter the soil bacteria communities. Front Microbiol 4. https://doi.org/10.3389/fmicb.2013.00383

Walter M, Frampton CM, Boyd-Wilson KSH, Harris-Virgin P, Waipara NW (2007) Agrichemical impact on growth and survival of non-target apple phyllosphere microorganisms. Can J Microbiol 53:45–55. https://doi.org/10.1139/W06-093

Wang L, Li J, Yang F, Huang Q, Shen Q (2017) Application of bioorganic fertilizer significantly increased apple yields and shaped bacterial community structure in orchard soil. Microb Ecol 73:404–416. https://doi.org/10.1007/s00248-016-0849-y

Wang L, Mazzola M (2019) Field evaluation of reduced rate Brassicaceae seed meal amendment and rootstock genotype on the microbiome and control of apple replant disease. Phytopathology 109:1378–1391. https://doi.org/10.1094/PHYTO-02-19-0045-R

Wang L, Mazzola M (2019) Interaction of Brassicaceae seed meal soil amendment and apple rootstock genotype on microbiome structure and replant disease suppression. Phytopathology 109:607–614. https://doi.org/10.1094/PHYTO-07-18-0230-R

Wang L, Yang F, E Y, Raza W, Huang Q, Shen Q (2016a) Long-term application of bioorganic fertilizers improved soil biochemical properties and microbial communities of an apple orchard soil. Front Microbiol 7. https://doi.org/10.3389/fmicb.2016.01893

Wang M, Carver JJ, Phelan VV, Sanchez LM, Garg N, Peng Y, Nguyen DD, Watrous J, Kapono CA, Luzzatto-Knaan T, Porto C, Bouslimani A, Melnik AV, Meehan MJ, Liu W-T, Crüsemann M, Boudreau PD, Esquenazi E, Sandoval-Calderón M, Kersten RD, Pace LA, Quinn RA, Duncan KR, Hsu C-C, Floros DJ, Gavilan RG, Kleigrewe K, Northen T, Dutton RJ, Parrot D, Carlson EE, Aigle B, Michelsen CF, Jelsbak L, Sohlenkamp C, Pevzner P, Edlund A, McLean J, Piel J, Murphy BT, Gerwick L, Liaw C-C, Yang Y-L, Humpf H-U, Maansson M, Keyzers RA, Sims AC, Johnson AR, Sidebottom AM, Sedio BE, Klitgaard A, Larson CB, P CAB, Torres-Mendoza D, Gonzalez DJ, Silva DB, Marques LM, Demarque DP, Pociute E, O'Neill EC, Briand E, Helfrich EJN, Granatosky EA, Glukhov E, Ryffel F, Houson H, Mohimani H, Kharbush JJ, Zeng Y, Vorholt JA, Kurita KL, Charusanti P, McPhail KL, Nielsen KF, Vuong L, Elfeki M, Traxler MF, Engene N, Koyama N, Vining OB, Baric R, Silva RR, Mascuch SJ, Tomasi S, Jenkins S, Macherla V, Hoffman T, Agarwal V, Williams PG, Dai J, Neupane R, Gurr J, Rodríguez AMC, Lamsa A, Zhang C, Dorrestein K, Duggan BM, Almaliti J, Allard P-M, Phapale P, Nothias L-F, Alexandrov T, Litaudon M, Wolfender J-L, Kyle JE, Metz TO, Peryea T, Nguyen D-T, VanLeer D, Shinn P, Jadhav A, Müller R, Waters KM, Shi W, Liu X, Zhang L, Knight R, Jensen PR, Palsson BØ, Pogliano K, Linington RG, Gutiérrez M, Lopes NP, Gerwick WH, Moore BS, Dorrestein PC, Bandeira N (2016) Sharing and community curation of mass spectrometry data with Global Natural Products Social Molecular Networking. Nat Biotechnol 34:828–837. https://doi.org/10.1038/nbt.3597

Wang X, Ouyang Y, Liu J, Zhu M, Zhao G, Bao W, Hu FB (2014) Fruit and vegetable consumption and mortality from all causes, cardiovascular disease, and cancer: systematic review and dose-response meta-analysis of prospective cohort studies. BMJ 349: g4490. https://doi.org/10.1136/bmj.g4490

Wang X, Wei J, Huang L, Kang Z (2011) Re-evaluation of pathogens causing Valsa canker on apple in China. Mycologia 103:317–324. https://doi.org/10.3852/09-165

Wassermann B, Kusstatscher P, Berg G (2019) Microbiome response to hot water treatment and potential synergy with biological control on stored apples. Front Microbiol 10:2502. https://doi.org/10.3389/fmicb.2019.02502

Wassermann B, Müller H, Berg G (2019b) An apple a day: which bacteria do we eat with organic and conventional apples? Front Microbiol 10. https://doi.org/10.3389/fmicb.2019.01629

Weber RWS (2014) Biology and control of the apple canker fungus *Neonectria ditissima* (syn. *N. galligena*) from a Northwestern European perspective. Erwerbsobstbau 56:95–107

Weerakoon DMN, Reardon CL, Paulitz TC, Izzo AD, Mazzola M (2012) Long-term suppression of *Pythium abappressorium* induced by *Brassica juncea* seed meal amendment is biologically mediated. Soil Biol Biochem 51:44–52. https://doi.org/10.1016/j.soilbio.2012.03.027

Wei N, Ashman T-L (2018) The effects of host species and sexual dimorphism differ among root, leaf and flower microbiomes of wild strawberries in situ. Sci Rep 8:5195. https://doi.org/10.1038/s41598-018-23518-9

Weinert N, Meincke R, Gottwald C, Heuer H, Schloter M, Berg G, Smalla K (2010) Bacterial diversity on the surface of potato tubers in soil and the influence of the plant genotype. FEMS Microbiol Ecol 74:114–123. https://doi.org/10.1111/j.1574-6941.2010.00936.x

Whitehead SR, Poveda K (2019) Resource allocation trade-offs and the loss of chemical defences during apple domestication. Ann Bot 123:1029–1041. https://doi.org/10.1093/aob/mcz010

Whitehead SR, Poveda K (2011) Herbivore-induced changes in fruit-frugivore interactions. J Ecol 99:964–969. https://doi.org/10.1111/j.1365-2745.2011.01819.x

Whitehead SR, Tiramani J, Bowers MD (2016) Iridoid glycosides from fruits reduce the growth of fungi associated with fruit rot. J Plant Ecol 9:357–366. https://doi.org/10.1093/jpe/rtv063

Whitehead SR, Turcotte MM, Poveda K (2017) Domestication impacts on plant–herbivore interactions: a meta-analysis. Philos Trans R Soc B Biol Sci 372:20160034. https://doi.org/10.1098/rstb.2016.0034

Widder S, Allen RJ, Pfeiffer T, Curtis TP, Wiuf C, Sloan WT, Cordero OX, Brown SP, Momeni B, Shou W, Kettle H, Flint HJ, Haas AF, Laroche B, Kreft J-U, Rainey PB, Freilich S, Schuster S, Milferstedt K, van der Meer JR, Groβkopf T, Huisman J, Free A, Picioreanu C, Quince C, Klapper I, Labarthe S, Smets BF, Wang H, Soyer OS (2016) Challenges in microbial ecology: building predictive understanding of community function and dynamics. ISME J 10:2557–2568. https://doi.org/10.1038/ismej.2016.45

Wilson M, Lindow S (1993) Interactions between the biological control agent *Pseudomonas fluorescens* A506 and *Erwinia amylovora* in pear blossoms. Phytopathology 83.117–123

Wisniewski M, Droby S, Norelli J, Liu J, Schena L (2016) Alternative management technologies for postharvest disease control: the journey from simplicity to complexity. Postharvest Biol Technol 122:3–10. https://doi.org/10.1016/j.postharvbio.2016.05.012

Witzgall P, Proffit M, Rozpedowska E, Becher PG, Andreadis S, Coracini M, Lindblom TUT, Ream LJ, Hagman A, Bengtsson M, Kurtzman CP, Piskur J, Knight A (2012) "This is not an apple"-yeast mutualism in codling moth. J Chem Ecol 38:949–957. https://doi.org/10.1007/s10886-012-0158-y

Woo SL, Pepe O (2018) Microbial consortia: Promising probiotics as plant biostimulants for sustainable agriculture. Front Plant Sci 9:1801. https://doi.org/10.3389/fpls.2018.01801

Woody ST, Ives AR, Nordheim EV, Andrews JH (2007) Dispersal, density dependence, and population dynamics of a fungal microbe on leaf surfaces. Ecology 88:1513–1524. https://doi.org/10.1890/05-2026

Xu J, Zhang Y, Zhang P et al (2018) The structure and function of the global citrus rhizosphere microbiome. Nat Commun 9:4894. https://doi.org/10.1038/s41467-018-07343-2

Xu X, Zarecki R, Medina S et al (2019) Modeling microbial communities from atrazine contaminated soils promotes the development of biostimulation solutions. ISME J 13:494–508. https://doi.org/10.1038/s41396-018-0288-5

Yang S-J, Zhang Z-L, Xue Y-X et al (2014) Arbuscular mycorrhizal fungi increase salt tolerance of apple seedlings. Bot Stud 55:70. https://doi.org/10.1186/s40529-014-0070-6

Yao S, Merwin IA, Abawi GS, Thies JE (2006) Soil fumigation and compost amendment alter soil microbial community composition but do not improve tree growth or yield in an apple replant site. Soil Biol Biochem 38:587–599. https://doi.org/10.1016/j.soilbio.2005.06.026

Yao S, Merwin IA, Bird GW et al (2005) Orchard floor management practices that maintain vegetative or biomass groundcover stimulate soil microbial activity and alter soil microbial community composition. Plant Soil 271:377–389. https://doi.org/10.1007/s11104-004-3610-0

Yashiro E, McManus PS (2012) Effect of streptomycin treatment on bacterial community structure in the apple phyllosphere. PLoS One 7(5):e37131. https://doi.org/10.1371/journal.pone.0037131

Yashiro E, Spear RN, McManus PS (2011) Culture-dependent and culture-independent assessment of bacteria in the apple phyllosphere. J Appl Microbiol 110:1284–1296. https://doi.org/10.1111/j.1365-2672.2011.04975.x

Yeoh YK, Paungfoo-Lonhienne C, Dennis PG, Robinson N, Ragan MA, Schmidt S, Hugenholtz P (2016) The core root microbiome of sugarcanes cultivated under varying nitrogen fertilizer application. Environ Microbiol 18:1338–1351. https://doi.org/10.1111/1462-2920.12925

Yim B, Hanschen FS, Wrede A, Nitt H, Schreiner M, Smalla K, Winkelmann T (2016) Effects of biofumigation using *Brassica juncea* and *Raphanus sativus* in comparison to disinfection using Basamid on apple plant growth and soil microbial communities at three field sites with replant disease. Plant Soil Dordr 406:389–408. https://doi.org/10.1007/s11104-016-2876-3

Yim B, Nitt H, Wrede A, Jacquiod S, Sørensen SJ, Winkelmann T, Smalla K (2017) Effects of soil pre-treatment with Basamid® granules, *Brassica juncea*, *Raphanus sativus*, and *Tagetes patula* on bacterial and fungal communities at two apple replant disease sites. Front Microbiol 8. https://doi.org/10.3389/fmicb.2017.01604

Yim B, Smalla K, Winkelmann T (2013) Evaluation of apple replant problems based on different soil disinfection treatments—links to soil microbial community structure? Plant Soil 366:617–631. https://doi.org/10.1007/s11104-012-1454-6

Zamioudis C, Pieterse CMJ (2011) Modulation of host immunity by beneficial microbes. Mol Plant-Microbe Interact 25:139–150. https://doi.org/10.1094/MPMI-06-11-0179

Zhang Q, Sun J, Liu S, Wei Q (2013) Manure refinement affects apple rhizosphere bacterial community structure: a study in sandy soil. PLoS One 8:e76937. https://doi.org/10.1371/journal.pone.0076937

Zhang Y, Li P, Cheng L (2010) Developmental changes of carbohydrates, organic acids, amino acids, and phenolic compounds in "Honeycrisp" apple flesh. Food Chem 123:1013–1018. https://doi.org/10.1016/j.foodchem.2010.05.053

Zhang S, Chen X, Zhong Q, Zhuang X, Bai Z (2019) Microbial community analyses associated with nine varieties of wine grape carposphere based on high-throughput sequencing. Microorganisms 7(12):668. https://doi.org/10.3390/microorganisms7120668

Zhang Z-Y, Pan L-P, Li H-H (2010) Isolation, identification and characterization of soil microbes which degrade phenolic allelochemicals. J Appl Microbiol 108:1839–1849. https://doi.org/10.1111/j.1365-2672.2009.04589.x

Zheng W, Zhao Z, Gong Q, Zhai B, Li Z (2018) Effects of cover crop in an apple orchard on microbial community composition, networks, and potential genes involved with degradation of crop residues in soil. Biol Fertil Soils Heidelb 54:743–759. https://doi.org/10.1007/s00374-018-1298-1

Zheng W, Zhao Z, Lv F, Wang R, Gong Q, Zhai B, Wang Z, Zhao Z, Li Z (2019) Metagenomic exploration of the interactions between N and P cycling and SOM turnover in an apple orchard with a cover crop fertilized for 9 years. Biol Fertil Soils Heidelb 55:365–381. https://doi.org/10.1007/s00374-019-01356-9

Zheng XY, Sinclair JB (2000) The effects of traits of *Bacillus megaterium* on seed and root colonization and their correlation with the suppression of *Rhizoctonia* root rot of soybean. Biocontrol 45:223–243

Zhu Y, Shin S, Mazzola M (2016) Genotype responses of two apple rootstocks to infection by *Pythium ultimum* causing apple replant disease. Can J Plant Pathol 38:483–491. https://doi.org/10.1080/07060661.2016.1260640

Zilber-Rosenberg I, Rosenberg E (2008) Role of microorganisms in the evolution of animals and plants: the hologenome theory of evolution. FEMS Microbiol Rev 32:723–735. https://doi.org/10.1111/j.1574-6976.2008.00123.x

Origin of the Domesticated Apples

17

Awais Khan, Ben Gutierrez, C. Thomas Chao, and Jugpreet Singh

Abstract

Genomic, genetic, and archaeobotanical findings have confirmed that alongside interspecific hybridizations between *Malus sieversii* from Central Asia and wild species along the Silk Road, segmental duplications, point mutations, and clonal propagation have led to the fixation of traits in cultivated apples, unlike in annual crops. Moreover, there is minimal evidence for long-term intentional and targeted selection for fruit quality and horticultural traits; whereas, self-incompatibility, long juvenile phase, and clonal propagation have maintained genetic diversity in apples. Only modern (commercial) apple cultivars hint at the reduction of diversity and selection for commercially important traits. Furthermore, the wide phenotypic variations present in pre-breeding and advanced breeding material reveal that a great deal of genetic diversity is still maintained in the cultivated gene pool.

17.1 Introduction

The cultivated apple, *Malus* × *domestica*, belongs to Rosaceae, a vast family of flowering plants, and to the sub-tribe Pyreae (containing pome fruits). Rosaceae consists of over 100 genera and 3,000 species, most of which are perennial trees, shrubs, and herbs (Korban and Tartarini 2009). This family includes several important genera that account for most of our important deciduous fruit crops, including apple (*Malus*), pear (*Pyrus*), loquat (*Eriobotrya*), and stone fruits, such as peach, cherry, plum, apricot, and almond (*Prunus*), as well as various ornamental plants, including *Rosa* (rose), *Mespilus* (medlar), and *Crataegus* (hawthorn). Of these genera, *Malus* is commercially the most valuable.

Most cultivated apples are self-incompatible, highly heterozygous, display a juvenile period of 6–10 years or more, and are maintained through vegetative propagation; i.e., by grafting or budding (Janick et al. 1996). The haploid chromosome numbers of most Rosaceae are 7, 8, or 9. In contrast, Pyreae has a distinctive $x = 17$. Molecular phylogenetic and sequence-based studies suggest that Pyreae must have originated through either autopolyploidization or by hybridization between two sister taxa with $x = 9$, followed by diploidization and aneuploidization to $x = 17$ (Shulaev et al. 2008; Korban et al. 2009; Velasco et al. 2010), as has been proposed earlier (Janick et al. 1996). Most apples are diploid ($2n = 34$), but

A. Khan (✉) · J. Singh
Plant Pathology and Plant-Microbe Biology Section, Cornell University, Geneva, NY 14456, USA
e-mail: awais.khan@cornell.edu

B. Gutierrez · C. T. Chao
USDA-ARS Plant Genetic Resources Unit, New York State Agricultural Experiment Station, Geneva, NY, USA

S. S. Korban (ed.), *The Apple Genome*, Compendium of Plant Genomes,
https://doi.org/10.1007/978-3-030-74682-7_17

there are many significant triploid and tetraploid cultivars (Ordidge et al. 2018). The apple has a relatively small genome size of 1.54 pg DNA/2C or 743.2 Mb per haploid genome along with some variations across different *Malus* species (Tatum et al. 2005; Höfer and Meister 2010; Velasco et al. 2010). The genome of the apple cultivar 'Golden Delicious' has been sequenced using a combination of Sanger and next-generation sequencing (NGS) technologies (Velasco et al. 2010). Recently, a doubled-haploid 'Golden Delicious' has also been sequenced, providing a higher quality genome sequence for apples (Daccord et al. 2017). Furthermore, advances in genomics are leading to the development of a pan-genome for apples and their wild relatives (Peace et al. 2019).

Genome sequencing along with integrated linkage and physical maps of the apple genome reveals that there are large numbers of repetitive sequences, chromosomal rearrangements, and segmental duplications present across the apple genome (Velasco et al. 2010; Daccord et al. 2017; Han et al. 2007, 2011).

Throughout the evolutionary history of plants, domestication has contributed to alterations of plant genomes. During domestication, humans have selected wild plants based on their desirable attributes, thus resulting in changes to the genetic compositions of these species. Those beneficial alleles present in genomes of domesticated plants have either existed in wild plants or arose during domestication via spontaneous mutations (Paran and van der Knaap 2007). Plants undergo various changes over the course of domestication, collectively referred to as the "domestication syndrome". These changes include synchronization of flowering time, larger seed size, loss of seed dormancy, and loss of seed dispersal (Miller and Gross 2011; Meyer et al. 2012; Meyer and Purugganan 2013). While these changes occur over generations of selection, instant domestication can occur when traits of an individual are fixed through vegetative propagation. For fruit crops, traits commonly selected through domestication and fixed through vegetative propagation are larger fruit size, fruit flavor and texture, and shortened life cycle (Meyer and Purugganan 2013).

17.2 Apple Domestication—Along the Silk Road to the World

Apple has an intriguing history that has been noted in many tales and books written about its role in commerce, trade, and ancient civilizations. It has been well documented that the wild *M. sieversii* is the main progenitor of modern apples (Cornille et al. 2019; Duan et al. 2017). Furthermore, bears and other animals must have played key roles in the evolution of apple fruit size and sweetness by disseminating seeds from larger and sweeter fruit (Juniper and Mabberley 2006; Spengler 2019). Meanwhile, the collision of tectonic plates in central Asia along with melting glaciers must have yielded in fertile soils in the Tian Shan Mountain region that may have provided a suitable environment for the germination of dispersed seeds. Subsequently, *M. sieversii* has been transported by humans along the Silk Road to Central Asia and to Europe (Fig. 17.1).

Meanwhile, people of the Fergana Valley in Uzbekistan were known to be pioneers in Neolithic agriculture techniques (Juniper and Mabberley 2006). Thus, around the time when grafting technology was invented, apples similar to modern apples appeared in the Near East, approximately 4,000 years ago (Juniper and Mabberley 2006). From the Middle East, apples were passed on to the Greeks and Romans, who in turn, spread fruit cultivation across Europe. At some point, *M. sieversii* hybridized with the acidic, but small-fruited European crabapple *M. sylvestris* (Cornille et al. 2012). From Europe, the cultivated apple was then introduced to the Americas.

Nowadays, apples are grown throughout the world in temperate climates. As asexual clonal propagation (grafting) has been more broadly introduced, further developments have resulted from active selection of new mutations and of chance seedlings (open-pollinated). Controlled breeding of apples has been initiated in the early 1800s (Kellerhals 2009). Compared to annual crops, with much more protracted domestication histories, interspecific hybridization followed by clonal propagation has contributed to rapid domestication of the apple, as much of the

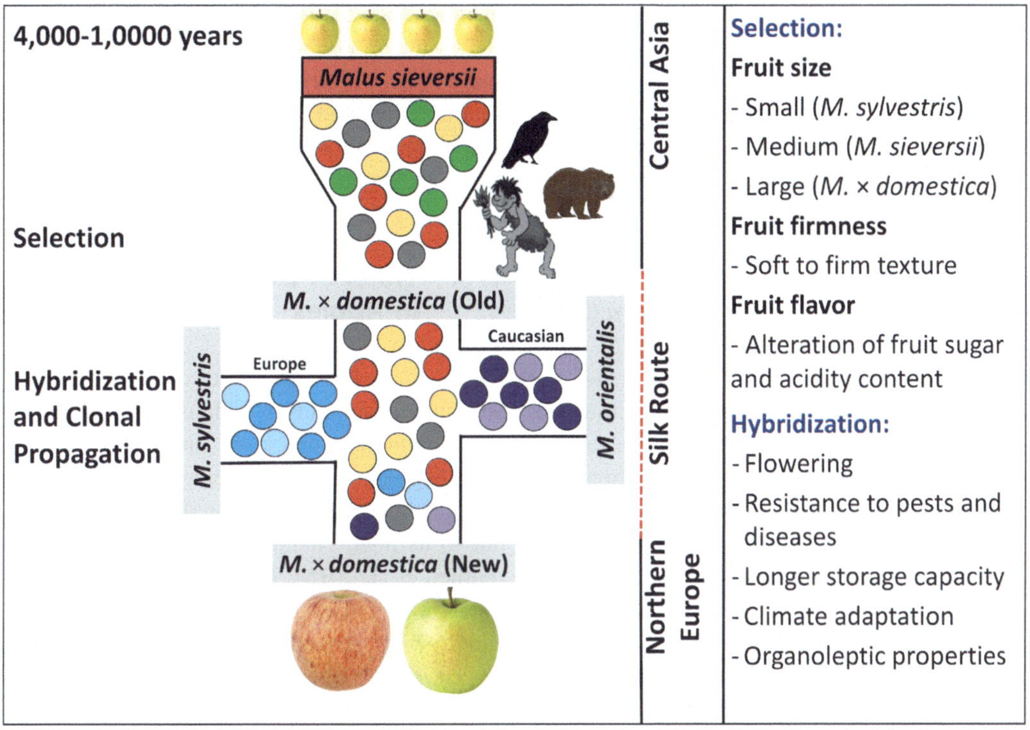

Fig. 17.1 A schematic diagram describing the domestication process in apples. Different *Malus* species contributed alleles (colored circles) to modern domesticated apples through selection and hybridization across the silk route. These processes led to selection of several fruit traits and to introgression of novel traits into the domesticated apple germplasm through hybridization from different wild species in Europe and Caucasian regions

modern diversity in apple has risen fairly recently (Spengler 2019). This would likely explain the absence of a domestication syndrome in apple, as often observed in other crops, thus suggesting that the apple is still in the early stages of domestication.

17.3 Molecular Clues to Domestication

The cultivated apple, *M.* × *domestica,* is a model species for studying domestication in long-lived woody perennial angiosperms. This is due to the availability of an extensive germplasm and genomic resources, including wild apple relatives, landraces, heirloom cultivars, genome references, integrated genetic and physical maps, along with its relatively small genome size. Early on, it has been proposed that morphological comparisons among *Malus* species have proposed that *M. sieversii*, *M. sylvestris*, *M. prunifolia*, and *M. baccata* are likely contributors to the domesticated apple (Kumar et al. 2014). Molecular markers designed for genetic, genomic, and chloroplast DNA have been used to explore the phylogeny and origin of domesticated apples. Molecular studies of diverse *Malus* species have determined that *M. sieversii* is the principal genomic contributor to modern apples (Velasco et al. 2010), with secondary contributions from *M. sylvestris* (Cornille et al. 2012, 2014; Nikiforova et al. 2013). Analysis of chloroplast DNA shows a maternal lineage of *M. sylvestris* in many, but not all modern cultivars (Nikiforova et al. 2013). Subsequently, genome re-sequencing has revealed that 46% of the modern *M.* × *domestica* genome is derived from

M. sieversii in Kazakhstan, 21% from *M. sylvestris*, with the remaining 33% of unclear origin (Duan et al. 2017).

Interestingly, there is minimal evidence for the incidence and presence of a domestication bottleneck in apple. Interspecific hybridization, self-incompatibility, and overlapping generations through clonal propagation have maintained genetic diversity in apple. In one study, genome-wide genetic diversity of *M.* × *domestica* (2.2×10^{-3}) is found to be lower than that of *M. sieversii* in Kazakhstan (2.35×10^{-3}), *M. sylvestris* (2.55×10^{-3}), and other wild *Malus* species (4.26×10^{-3}) (Duan et al. 2017). Moreover, the genetic diversity of *M. sieversii* in Xinjiang, China has the lowest genetic diversity (1.3×10^{-3}), although this is likely influenced by limited sampling. Nevertheless, this study has focused on key founder cultivars of modern breeding programs identified by Noiton and Shelbourne (1992), and therefore this may not be a good representation of apple diversity. In another study, the genetic diversity of apples is evaluated over time from both old and modern cultivars, and it is reported that there is no significant loss of diversity, with the exception of modern cultivars (Gross et al. 2014). Although bottlenecks are presumed to occur early in domestication, new evidence in other crops has suggested otherwise (Allaby et al. 2018). Similar archaeogenomic insights may further elucidate domestication in apples (Cornille et al. 2019).

Resolving relationships between wild and domesticated apples is critical to understanding the history of domestication, but this requires information from as many geographically dispersed *Malus* species and accessions of apples as possible. One challenge is in differentiating representative primary species from hybrid species for pursuing genetic studies. Admixture is common in the *Malus* germplasm (Volk et al. 2015), and domestication of apple is primarily based on interspecific hybridizations. Conversely, wild *Malus* populations are subject to gene flow from *M.* × *domestica,* as observed in *M. sieversii* (Kumar et al. 2014), *M. sylvestris* (Wagner et al. 2014), and *M. coronaria* in North America (Kron and Husband 2009). Further genetic introgression of cultivated apples into wild populations will render future domestication studies even more challenging.

17.4 Traits of Domestication in Apples

Globally, apple is the most economically valuable temperate fruit crop. It is consumed fresh and in multiple processed and cooked forms, including juice, sauce, and cider. The importance of apple in a balanced human diet is well known, due to its fiber, vitamin, and antioxidant content (Farneti et al. 2015). The selection of apples for different end uses and broad consumer preferences across generations and cultures have preserved broad phenotypic diversity in apples. Unlike *M. sieversii*, most wild *Malus* species in both China and North America have not been intensely selected for domesticated traits, except for rootstock and ornamental cultivars. These wild *Malus* species have many different characteristics from those of cultivated apples, including fruit quality, growth habits, and reproduction (apomixis) due to their separate evolutionary histories from apple and its progenitors (Spengler 2019). These highlights of phenotypic variations among apples, progenitors, and distant crabapple relatives from North America are presented in Table 17.1. Commonly, fruits of wild *Malus* species are smaller in size, with thinner pedicels and deciduous calyces, along with more astringent flavors (Gutierrez et al. 2018). Some of these variations in fruit morphology of *Malus* species are presented in Fig. 17.2.

The potential roles of fruit quality traits in the domestication of apples have also been observed using sequence-based single nucleotide polymorphisms (SNPs) to uncover genome-wide linkage disequilibrium (LD) patterns and signatures of selective sweeps in genes and in genomic

Table 17.1 Trait variations and ploidy levels in *Malus* × *domestica* and in wild *Malus* species based on descriptor data available on GRIN-Global (USDA 2014)

Trait	*M.* × *domestica*	*M. sieversii*	*M. sylvestris*	*M. coronaria*	*M. fusca*
Fruit diameter in cm (mean)	3.0 – 12.0 (7.0)	2.2 – 8.5 (4.5)	1.2 – 7.7 (4.0)	2.4 – 5.5 (3.5)	0.6 – 1.4 (1.0)
Total soluble solids in °Brix (mean)	8.0 – 26.3 (13.3)	6.8 – 21.0 (11.6)	5.9 – 17.6 (12.2)	8.5 – 15.9 (12.1)	12.4 – 20.9 (16.7)
Ploidy	2×; but with some important 3× and 4× cultivars	2×; but with some 3×	2×	2×; 3×; 4×	2×
Predominant fruit over color	Yellow and red	Yellow and red	Green; red to brown	Yellow	Brown

Fig. 17.2 Variations in fruits of cultivated (*Malus* × *domestica*) and accessions of wild *Malus* species from the USDA National *Malus* collection at Geneva, NY. Accessions listed left to right. Top row: *M.* × *domestica* 'Honeycrisp' (PI 644,174), 'Jonagold' (PI 588,941), 'Razor' (PI 589,136), 'Red Spitzenburg' (PI 589,100), and 'Gala' (PI 392,303). Middle row: *M. sieversii* from Kazakhstan PI 613,987, PI 613,976, PI 613,997, PI 613,998, PI 657,080. Bottom row (left): *M. sylvestris* PI 589,382, PI 590,061, PI 633,827. Bottom row (right): North American species *M. coronaria* PI 589,344, *M. ioensis* 613,908, *M. coronaria* PI 613,907, and *M. fusca* PI 613,910

regions of a diverse core collection of apples from 30 *Malus* species (Khan et al. 2014). These findings have suggested that sensory attributes, such as fruit color and taste, have been important during the early selection in *M. sieversii* by animals, and later on in *M.* × *domestica,* to some extent.

17.4.1 Fruit Size

Fruit size varies broadly across primary *Malus* species and hybrid species. As mentioned above, animals have contributed to fruit size variation in *Malus*, with larger fruit potentially developing in response to selection by large-animal seed dispersers as opposed to avian-dispersed seeds (Spengler 2019). Descriptor data from the U.S. National *Malus* collection reveal that mean diameters of *M. sieversii* and *M. sylvestris* fruits are about 4.5 and 4.0 cm, respectively, with some medium- to large-fruited outliers. Whereas, mean fruit diameter is about 7.0 cm in *M.* × *domestica* and ranges from 3.0 to 12.0 cm (USDA 2014).

It has been reported that differences in fruit size among *Malus* species are attributed to differences in cell number and/or cell size, which could potentially be revealed through the expression of histone (*MdH1*) and expansin (*MdExp3*) genes (Harada et al. 2005). Both genes are expressed at low levels in small-fruited crab apples, while high levels of expression are detected for *MdH1* at the cell proliferation stage and for *MdExp3* at the cell enlargement stage in large-fruited domesticated apples. Furthermore, large-fruited apples also differ from smaller fruit-sized cultivars both in early cell production (or proliferation) rate and duration of cell production; thus, suggesting that both factors influence final cell number and fruit size in apple (Malladi and Hirst 2010). Expression of cell cycle genes, such as cyclin-dependent kinases (*CDK*s), among others, in 'Gala' and its larger fruit sized mutant 'Grand Gala', has revealed that genes such as *MdCDKA1* are expressed at higher levels in the mutant fruit during the early stages of fruit development (Malladi and Hirst 2010).

Genome resequencing of *M. sieversii,* collected from both Kazakhstan and China, *M. sylvestris*, *M.* × *domestica*, and other wild *Malus* species has revealed that increased fruit size in apple must have involved a two-step process over the course of apple domestication (Duan et al. 2017). Two quantitative trait loci (QTLs) for fruit weight, *fw1* on chromosome 15 and *fw2* on chromosome 8, are part of selective sweeps that harbor 11 and 7 genes, respectively, in regions of *M.* × *domestica* derived from *M. sieversii*; and 8 and 21 genes, respectively, in regions of *M.* × *domestica* derived from *M. sylvestris*. Some of these genes encode regulators of cell division, such as *fw2.2* in tomato and *CNR1* in maize, and the β-galactosidase gene involved in the regulation of fruit weight and size in strawberry. Interestingly, a single cell division regulator gene and two β-galactoside genes are located in the *fw1* region, along with five other cell division regulator genes and a single gene homologous to rice *GS3,* controlling grain size, are found in other selective regions. Moreover, microRNAs genes including *miRNA172p* that regulate apple fruit size by targeting *APETALA 2* (*AP2*) transcription factors (TFs), along with two new *miRNA172* genes, *miRNA172g* and *miRNA172h*, that may also target *AP2* TFs have also been identified within both *M.* × *domestica* and *M. sieversii.*

17.4.2 Fruit Flavor

The organoleptic qualities of apples are influenced by sugars, organic acids, and aromatic composition. Although it is routinely suggested that during domestication, apple has been selected for higher sugar content, it has been reported that total sugar content variation is similar among wild and domesticated apple (Ma et al. 2015). When a collection of apple germplasm is evaluated, it is found that while acid content is higher in wild *Malus* species, total sugar content is not significantly different between domesticated and wild apples (Ma et al. 2015). However, sugar composition is reported to vary, with domesticated apples accumulating fructose at higher concentrations. As acid accumulation is interconnected with sugar metabolism, levels of acidity, not sweetness, are suspected as being targets for selection. Moreover, malic acid is found to be the predominant organic acid impacting acidity in apples. In fact, fruit acidity is regulated by the malate transporter *Ma1,* present in the *Ma* locus on apple linkage group (LG) 16 (Bai et al. 2012). Furthermore, reduced nucleotide diversity in the *Ma1* coding sequence in *M.* × *domestica,* relative to those found in *M. sylvestris* and *M. sieversii,* suggests the incidence of selective pressure in *M.* × *domestica* (Duan et al. 2017). While wild apple fruits also accumulate citric acid, it is present only at low concentrations in mature fruits of *M.* × *domestica, M. sieversii,* and *M. sylvestris* (Ma et al. 2018).

As the apple is known to contain phenolic compounds, it is reported that the phenolic content impacts fruit quality of apples, and it is particularly important for the bitterness and astringency of cider apples. In general, *M. sieversii* and *M. sylvestris* have higher phenolic contents than *M.* × *domestica* (Stushnoff et al. 2002; Volz and McGhie 2011). Interestingly, the phenolic composition of apples has changed over many generations, with phenolic content decreasing over time (De Paepe et al. 2015; Jakobek and Barron 2016), although the genetic potential for high phenolic content is still maintained in breeding materials (Volz and McGhie 2011).

17.4.3 Fruit Color

Fruit color is linked to consumer perception of quality, with many consumers favoring red-colored fruit (Treutter 2010). The anthocyanin organic compound cyanidin-3-O-galactoside confers the red pigment to apple fruit, and it is controlled by *MYB1*, *MYB10*, and *MYBA* TFs involved in regulation of anthocyanin biosynthesis (Espley et al. 2007; El-Sharkawy et al. 2015). Furthermore, *MdMYB10* and *MdMYB110a* are involved in anthocyanin production in red-fleshed apples (Chagné et al. 2013; Espley et al. 2013). Red fruit skin color is commonly found in wild *M. sieversii* fruit, and it has been one of the major traits selected during the domestication process. Recently, a retrotransposon, redTE, upstream of *MdMYB1* has been identified and found to significantly upregulate peel anthocyanin content in red-skinned cultivars (Zhang et al. 2019). The redTE insertion has originated from *M. sieversii* in Xinjiang, China, a population that is reportedly not a direct contributor to apple domestication (Duan et al. 2017).

Apple genes related to secondary metabolites vary greatly in copy number, including those involved in anthocyanin, flavonoids, isoflavones, isoflavonones, and terpene pathways. For example, multiple copies and alleles of genes in the apple genome encode caffeic acid O-methyltransferase (COMT) and are related to such secondary metabolites as lignin, flavonoids, anthocyanins, suberin, and isoflavonoids (Han et al. 2007). Two clusters encoding *COMT* have been identified, one cluster containing three copies and another cluster with four copies. This finding is consistent with results obtained from analyzing a total of over 320,000 apple expressed sequence tags (ESTs), wherein high copy numbers of genes involved in secondary metabolism are found to be frequent in the apple genome (Newcomb et al. 2006; Gasic et al. 2009). It is reported that the clustering of genes involved in secondary metabolism may be the result of segmental duplications. For example, copies of genes involved in either secondary metabolism or abiotic stress in *Arabidopsis* are likely to be retained when derived from segmental duplications but are more rapidly lost following whole-genome duplications (Maere et al. 2005; Newcomb et al. 2006; Han et al. 2011). Furthermore, when comparing 20 *MYB10*-like genes (for pigmentation) from crabapples, domesticated apples, and pears, it is found there are high levels of similarity, except within a region of intron 2 that includes a 2500-bp insertion (Lin-Wang et al. 2010). This insertion could be the result of a local genome rearrangement following the divergence of apple and pear lineages, but before the divergence of apples and crabapples. A few studies highlight the role of genes that may have been involved in the general evolution of *Pyreae*, but no comprehensive study has focused on genes involved in the domestication of apples.

17.4.4 Fruit Texture

Fruit texture is a dynamic and complex trait that affects storability and consumer acceptance. There are several genes and QTLs that have been previously reported for apple flesh texture, wherein many are associated with cell wall composition, organization, and modification through fruit ripening (Di Guardo et al. 2017). Duan et al. (2017) have provided evidence of the presence of selective sweeps for fruit firmness-associated genes across multiple chromosomes in both *M. sieversii,* from Kazakhstan, and *M. sylvestris*. Although there is evidence for selective (either natural or artificial) pressure for fruit texture, broad variation persists in cultivars and in wild *Malus* species, with fruit texture ranging from coarse and dry to crisp and juicy.

17.4.5 Disease Resistance

Disease resistance remains a major goal of apple genetic enhancement efforts. However, unlike fruit quality traits, there is no direct evidence for selection for disease-related traits during apple domestication. Previous studies have reported that wild *Malus* species have mainly contributed towards major resistance against various apple diseases (Gygax et al. 2004; Patocchi et al. 2005; Bus et al. 2005, 2011; Fahrentrapp et al. 2013; Emeriewen et al. 2018). Major resistance genes against apple scab disease, incited by the fungal pathogen *Venturia inaequalis*, have also been identified in *M. sieversii* (Bus et al. 2005, 2011), the direct progenitor of domesticated apples. In contrast, domesticated apples mostly demonstrate susceptible disease responses, and only a few sources of major genetic resistance have been mapped in these genotypes (Crosby et al. 1994; Khan et al. 2006; Bus et al. 2011).

Studies suggest that domestication has indirectly altered the host response against pathogens, but both the extent and magnitude of this effect are not well investigated in apples. There are likely several reasons for observed differences in resistance between wild and domesticated apples. For example, disease resistance traits could have been totally ignored, and their associated genomic regions are, thus, unintentionally filtered out during selection for fruit quality traits, particularly in earlier selection by animal seed dispersers. Alternatively, exposure to new pathogens, as well as severity and frequency of disease incidence due to domestication-related habitat changes may have led to loss of non-specific wild resistance alleles. This is particularly relevant as some of the resistance genes from wild backgrounds have demonstrated gene-for-gene interactions with corresponding pathogens (Singh et al. 2021; Bus et al. 2005; Vogt et al. 2013). Furthermore, a transition from resistant to tolerant disease phenotypes can be expected under these circumstances, whereby domesticated plants can rely on multiple genes with small additive effects to overcome rapid pathogen evolution in monoculture environments. A range of highly susceptible to highly tolerant fire blight disease, incited by the bacterial pathogen *Erwinia amylovora*, phenotypes have been recently identified in apple cultivars (Kostick et al. 2019); however, the genetic mechanism(s) related to these differences is yet to be explored. Some of these differences can be attributed to genetic variations in disease-related susceptibility (*S*) genes and resistance (*R*) genes in domesticated plants (Campa et al. 2019; Pompili et al. 2020). Nonetheless, differences in disease resistance between wild and domesticated apple groups can provide opportunities to explore unique sources of host genetic resistance, as well as their interactions with varying pathogens (Singh et al. 2021).

17.5 Alternative Domestication Events of Apple Landraces

Studies have identified apple landraces that are genetically distinct from cultivated apples. Some are either more closely related to *M. sieversii* or show no signs of genetic introgression of *M. sylvestris.* Presumably, these landraces have their own unique domestication histories, separate from *M.* × *domestica*, and they provide insights into apple domestication and historic uses of apples.

Along the eastern side of the Tian Shan Mountains, *M. sieversii* has been cultivated for fruit production and has undergone selection independently of *M.* × *domestica.* These fruits have distinctive qualities, such as lower acid content due to lack of genetic introgression of *M. sylvestris* (Duan et al. 2017). Interestingly, domestication of wild *M. sieversii* in China has also resulted in the development of the 'Nai' apple (Yao et al. 2010; Ma et al. 2017). 'Nai' apples are among the earliest recorded cultivated apples in China, dating back 2,200 years ago. Until recently, these have been commonly used as either cultivars or rootstocks in Central and Western China, but they are now being replaced by modern apple cultivars and rootstocks.

Similarly, landraces in former Soviet Republics are also more closely related to *M. sieversii* (Gao et al. 2015). Moreover, landraces in Iran and Turkey are not only likely selections of *M. sieversii* but also with secondary contributions from *M. orientalis* instead of *M. sylvestris*. A study of apples from Iran and Turkey, including local cultivars and landraces, wild species, and cultivars, and rootstocks from other countries, have demonstrated that Iranian and Turkish accessions are genetically close to *M. sieversii* from Central Asia and to *M. orientalis* from Turkey and Russia (Gharghani et al. 2009; Burak et al. 2014).

17.6 The Future of the Cultivated Apple and of Wild *Malus* Species

Modern apples will continue with the domestication process. As efforts to search and to utilize novel traits such as disease resistance genes from wild *Malus* species will continue to be of importance for both germplasm collections and to apple breeders (Singh et al. 2021). Although grafting fixes genotypes indefinitely, mutations will continue to arise, and these will significantly impact cultivar phenotypes, as well as serve as new targets for selection. For example, the spontaneous columnar mutation in 'Wijcik McIntosh', with reduced branching and increased spurs, has been a game-changer for apple production in high-density plantings (Morimoto and Banno 2015). Apple sports with mutations for early ripening, delayed ripening, longer tree storage capability, spur bearing, increased attractiveness (red skin, red flesh, and russeting, among others), and uniform ripening have also been selected for and will contribute to the domesticated apple.

References

Allaby RG, Ware RL, Kistler L (2018) A re-evaluation of the domestication bottleneck from archaeogenomic evidence. Evol Appl 12:29–37. https://doi.org/10.1111/eva.12680

Bai Y, Dougherty L, Li M, Fazio G, Cheng L, Xu K (2012) A natural mutation-led truncation in one of the two aluminum-activated malate transporter-like genes at the *Ma* locus is associated with low fruit acidity in apple. Mol Genet Genomics 287:663–678

Burak M, Ergül A, Kazan K, Akçay ME, Yüksel C, Bakir M, Mutaf F, Akpinar AE, Yaşasin AS, Ayanoğlu H (2014) Genetic analysis of Anatolian apples (*Malus* sp.) by simple sequence repeats. J Syst Evol 52:580–588. https://doi.org/10.1111/jse.12099

Bus VGM, Laurens FND, Van de Weg WE, Rusholme RL, Rikkerink EHA, Gardiner SE, Bassett HCM, Plummer KM (2005) The *Vh8* locus of a new gene-for-gene interaction between *Venturia inaequalis* and the wild apple *Malus sieversii* is closely linked to the *Vh2* locus in *Malus pumila* R12740–7A. New Phytol 166:1035–1049

Bus VGM, Rikkerink EHA, Caffier V, Dural C, Plummer KM (2011) Revision of the nomenclature of the differential host-pathogen interactions of *Venturia inaequalis* and *Malus*. Annu Rev Phytopath 49:391–413

Campa M, Piazza S, Righetti L, Oh C, Conterno L, Borejsza-Wysocka E, Nagamangala KC, Beer SV, Aldwinckle HS, Malnoy M (2019) HIPM is a susceptibility gene of *Malus* spp.: Reduced expression reduces susceptibility to *Erwinia amylovora*. Mol Plant Microbe Inter 32:167–175

Chagné D, Lin-Wang K, Espley RV, Volz RK, How NM, Rouse S, Brendolise C, Carlisle CM, Kumar S, Silva ND, Micheletti D, McGhie T, Crowhurst RN, Storey RD, Velasco R, Hellens RP, Gardiner SE, Allan AC (2013) An ancient duplication of apple MYB transcription factors is responsible for novel red fruit-flesh phenotypes. Plant Physiol 161:225–239

Cornille A, Antolín F, Garcia E, Vernesi C, Fietta A, Brinkkemper O, Kirleis W, Schlumbaum A, Roldán-Ruiz I (2019) A multifaceted overview of apple tree domestication. Trends Plant Sci 24:770–782. https://doi.org/10.1016/j.tplants.2019.05.007

Cornille A, Giraud T, Smulders MJ, Roldán-Ruiz I, Gladieux P (2014) The domestication and evolutionary ecology of apples. Trends Genet 30:57–65

Cornille A, Gladieux P, Smulders MJ, Roldán-Ruiz I, Laurens F, Cam BL, Nersesyan A, Clavel J, Olonova M, Feugey L (2012) New insight into the history of domesticated apple: secondary contribution of the European wild apple to the genome of cultivated varieties. PLoS Genet 8:e1002703. https://doi.org/10.1371/journal.pgen.1002703

Crosby JA, Janick J, Pecknold PC, Goffreda JC, Korban SS (1994) 'Gold Rush' apple. HortScience 29:827–828

Daccord N, Celton J-M, Linsmith G, Becker C, Choisne N, Schijlen E, van de Geest H, Bianco L, Micheletti D, Velasco R, Di Pierro EA, Gouzy J, Rees DJG, Guérif P, Muranty H, Durel C-E, Laurens F, Lespinasse Y, Gaillard S, Aubourg S, Quesneville H, Weigel D, van de Weg E, Troggio M, Bucher E (2017) High-quality *de novo* assembly of the apple genome and methylome dynamics of early fruit development. Nat Genet 49:1099–1106. https://doi.org/10.1038/ng.3886

De Paepe D, Valkenborg D, Noten B, Servaes K, Diels L, Loose MD, Van Droogenbroeck B, Voorspoels S (2015) Variability of the phenolic profiles in the fruits from old, recent and new apple cultivars cultivated in Belgium. Metabolomics 11:739–752. https://doi.org/10.1007/s11306-014-0730-2

Di Guardo M, Bink MCAM, Guerra W, Letschka T, Lozano L, Busatto N, Poles L, Tadiello A, Bianco L, Visser RGF, van de Weg E, Costa F (2017) Deciphering the genetic control of fruit texture in apple by multiple family-based analysis and genome-wide association. J Exp Bot 68:1451–1466. https://doi.org/10.1093/jxb/erx017

Duan N, Bai Y, Sun H, Wang N, Ma Y, Li M, Wang X, Jiao C, Legall N, Mao L, Wan S, Wang K, He T, Feng S, Zhang Z, Mao Z, Shen X, Chen X, Jiang Y, Wu S, Yin C, Ge S, Yang L, Jiang S, Xu H, Liu J, Wang D, Qu C, Wang Y, Zuo W, Xiang L, Liu C, Zhang D, Gao Y, Xu Y, Xu K, Chao T, Fazio G, Shu H, Zhong G-Y, Cheng L, Fei Z, Chen X (2017) Genome re-sequencing reveals the history of apple and supports a two-stage model for fruit enlargement. Nat Commun 8:249. https://doi.org/10.1038/s41467-017-00336-7

El-Sharkawy I, Liang D, Xu K (2015) Transcriptome analysis of an apple (*Malus* × *domestica*) yellow fruit somatic mutation identifies a gene network module highly associated with anthocyanin and epigenetic regulation. J Exp Bot 66:7359–7376. https://doi.org/10.1093/jxb/erv433

Emeriewen OF, Richter K, Piazza S, Micheletti D, Broggini GAL, Berner T, Keilwagen J, Hanke M, Malnoy M, Peil A (2018) Towards map-based cloning of *FB_Mfu10*: Identification of a receptor-like kinase candidate gene underlying the *Malus fusca* fire blight resistance locus on linkage group 10. Mol Breed 38:106

Espley RV, Bovy A, Bava C, Jaeger SR, Tomes S, Norling C, Crawford J, Rowan D, McGhie TK, Brendolise C, Putterill J, Schouten HJ, Hellens RP, Allan AC (2013) Analysis of genetically modified red-fleshed apples reveals effects on growth and consumer attributes. Plant Biotechnol J 11:408–419. https://doi.org/10.1111/pbi.12017

Espley RV, Hellens RP, Putterill J, Stevenson DE, Kutty-Amma S, Allan AC (2007) Red colouration in apple fruit is due to the activity of the MYB transcription factor, MdMYB10. Plant J 49:414–427. https://doi.org/10.1111/j.1365-313X.2006.02964.x

Fahrentrapp J, Broggini GAL, Kellerhals M, Peil A, Richter K, Zini E, Gessler C (2013) A candidate gene for fire blight resistance in *Malus* × *robusta* 5 is coding for a CC–NBS–LRR. Tree Genet Genomes 9:237–251

Farneti B, Masuero D, Costa F, Magnago P, Malnoy M, Costa G, Vrhovsek U, Mattivi F (2015) Is there room for improving the nutraceutical composition of apple? J Agric Food Chem 63:2750–2759. https://doi.org/10.1021/acs.jafc.5b00291

Gao Y, Liu F, Wang K, Wang D, Gong X, Liu L, Richards CM, Henk AD, Volk GM (2015) Genetic diversity of *Malus* cultivars and wild relatives in the Chinese National Repository of Apple Germplasm Resources. Tree Genet Genomes 11:106. https://doi.org/10.1007/s11295-015-0913-7

Gasic K, Han Y, Kertbundit S, Shulaev V, Iezzoni AF, Stover EW, Bell RL, Wisniewski ME, Korban SS (2009) Characteristics and transferability of new apple EST-derived SSRs to other Rosaceae species. Mol Breed 23:397–411. https://doi.org/10.1007/s11032-008-9243-x

Gharghani A, Zamani Z, Talaie A, Oraguzie NC, Fatahi R, Hajnajari H, Wiedow C, Gardiner SE (2009) Genetic identity and relationships of Iranian apple (*Malus* × *domestica* Borkh.) cultivars and landraces, wild *Malus* species and representative old apple cultivars based on simple sequence repeat (SSR) marker analysis. Genet Resour Crop Evol 56:829–842. https://doi.org/10.1007/s10722-008-9404-0

Gross BL, Henk AD, Richards CM, Fazio G, Volk GM (2014) Genetic diversity in *Malus* x *domestica* (Rosaceae) through time in response to domestication. Am J Bot 101:1770–1779

Gutierrez BL, Zhong G-Y, Brown SK (2018) Genetic diversity of dihydrochalcone content in *Malus* germplasm. Genet Resour and Crop Evol 65:1485–1502. https://doi.org/10.1007/s10722-018-0632-7

Gygax M, Gianfranceschi L, Liebhard R, Kellerhals M, Gessler C, Patocchi A (2004) Molecular markers linked to the apple scab resistance gene *Vbj* derived from *Malus baccata jackii*. Theor Appl Genet 109:1702–1709

Han Y, Gasic K, Marron B, Beever JE, Korban SS (2007) A BAC-based physical map of the apple genome. Genomics 89:630–637. https://doi.org/10.1016/j.ygeno.2006.12.010

Han Y, Zheng D, Vimolmangkang S, Khan MA, Beever JE, Korban SS (2011) Integration of physical and genetic maps in apple confirms whole-genome and segmental duplications in the apple genome. J Exp Bot 62:5117–5130. https://doi.org/10.1093/jxb/err215

Harada T, Kurahashi W, Yanai M, Wakasa Y, Satoh T (2005) Involvement of cell proliferation and cell enlargement in increasing the fruit size of *Malus* species. Scient Hortic 105:447–456. https://doi.org/10.1016/j.scienta.2005.02.006

Höfer M, Meister A (2010) Genome size variation in *Malus* species. J Bot 2010:e480873. https://doi.org/10.1155/2010/480873
Jakobek L, Barron AR (2016) Ancient apple varieties from Croatia as a source of bioactive polyphenolic compounds. J Food Comp Anal 45:9–15. https://doi.org/10.1016/j.jfca.2015.09.007
Janick J, Cummins J, Brown S, Hemmat M (1996) Apples. In: Janick J, Moore JN (eds) Fruit breeding, vol 1. John Wiley & Sons, Tree and Tropical Fruits, pp 1–77
Juniper BE, Mabberley DJ (2006) The story of the apple. Timber Press, Portland, OR
Kellerhals M (2009) Introduction to apple (*Malus* × *domestica*). In: Folta KM, Gardiner SE (eds) Genetics and genomics of rosaceae. Springer, New York, NY, pp 73–84
Khan MA, Duffy B, Gessler C, Patocchi A (2006) QTL mapping of fire blight resistance in apple. Mol Breed 17:299–306
Khan MA, Olsen KM, Sovero V, Kushad MM, Korban SS (2014) Fruit quality traits have played critical roles in domestication of the apple. Plant Genome 7:1–18. https://doi.org/10.3835/plantgenome2014.04.0018
Korban SS, Tartarini S (2009) Apple structural genomics. In: Folta K, Gardiner S (eds) Genetics and genomics of rosaceae. Springer-Science, NY, pp 85–119
Korban SS, Wannarat W, Rayburn CM, Tatum TC, Rayburn AL (2009) Genome size and nucleotypic variation in *Malus* germplasm. Genome 52:148–155. https://doi.org/10.1139/G08-109
Kostick SA, Norelli JL, Evans KM (2019) Novel metrics to classify fire blight resistance of 94 apple cultivars. Plant Pathol 68:985–996. https://doi.org/10.1111/ppa.13012
Kron P, Husband BC (2009) Hybridization and the reproductive pathways mediating gene flow between native *Malus coronaria* and domestic apple, *M. domestica*. Botany 87:864–874. https://doi.org/10.1139/B09-045
Kumar S, Raulier P, Chagné D, Whitworth C (2014) Molecular-level and trait-level differentiation between the cultivated apple (*Malus* × *domestica* Borkh.) and its main progenitor *Malus sieversii*. Plant Genet Resour 12:330–340
Lin-Wang K, Bolitho K, Grafton K, Kortstee A, Karunairetnam S, McGhie TK, Espley RV, Hellens RP, Allan AC (2010) An R2R3 MYB transcription factor associated with regulation of the anthocyanin biosynthetic pathway in Rosaceae. BMC Plant Biol 10:50. https://doi.org/10.1186/1471-2229-10-50
Ma B, Chen J, Zheng H, Fang T, Ogutu C, Li S, Han Y, Wu B (2015) Comparative assessment of sugar and malic acid composition in cultivated and wild apples. Food Chem 172:86–91. https://doi.org/10.1016/j.foodchem.2014.09.032
Ma B, Liao L, Peng Q, Fang T, Zhou H, Korban SS, Han Y (2017) Reduced representation genome sequencing reveals patterns of genetic diversity and selection in apple. J Integr Plant Biol 59:190–204. https://doi.org/10.1111/jipb.12522
Ma B, Yuan Y, Gao M, Li C, Ogutu C, Li M, Ma F (2018) Determination of predominant organic acid components in *Malus* species: correlation with apple domestication. Metabolites 8:74. https://doi.org/10.3390/metabo8040074
Maere S, Bodt SD, Raes J, Casneuf T, Montagu MV, Kuiper M, de Peer YV (2005) Modeling gene and genome duplications in eukaryotes. Proc Natl Acad Sci USA 102:5454–5459. https://doi.org/10.1073/pnas.0501102102
Malladi A, Hirst PM (2010) Increase in fruit size of a spontaneous mutant of 'Gala' apple (*Malus* ×*domestica* Borkh.) is facilitated by altered cell production and enhanced cell size. J Exp Bot 61:3003–3013. https://doi.org/10.1093/jxb/erq134
Meyer RS, DuVal AE, Jensen HR (2012) Patterns and processes in crop domestication: an historical review and quantitative analysis of 203 global food crops. New Phytol 196:29–48. https://doi.org/10.1111/j.1469-8137.2012.04253.x
Meyer RS, Purugganan MD (2013) Evolution of crop species: genetics of domestication and diversification. Nat Rev Genet 14:840–852. https://doi.org/10.1038/nrg3605
Miller AJ, Gross BL (2011) From forest to field: perennial fruit crop domestication. Am J Bot 98:1389–1414. https://doi.org/10.3732/ajb.1000522
Morimoto T, Banno K (2015) Genetic and physical mapping of *Co*, a gene controlling the columnar trait of apple. Tree Genet Genomes 11:807. https://doi.org/10.1007/s11295-014-0807-0
Newcomb RD, Crowhurst RN, Gleave AP, Rikkerink EHA, Allan AC, Beuning LL, Bowen JH, Gera E, Jamieson KR, Janssen BJ, Laing WA, McArtney S, Nain B, Ross GS, Snowden KC, Souleyre EJF, Walton EF, Yauk Y-K (2006) Analyses of expressed sequence tags from apple. Plant Physiol 141:147–166. https://doi.org/10.1104/pp.105.076208
Nikiforova SV, Cavalieri D, Velasco R, Goremykin V (2013) Phylogenetic analysis of 47 chloroplast genomes clarifies the contribution of wild species to the domesticated apple maternal line. Mol Biol Evol 30:1751–1760. https://doi.org/10.1093/molbev/mst092
Noiton D, Shelbourne C (1992) Quantitative genetics in an apple breeding strategy. Euphytica 60:213–219
Ordidge M, Kirdwichai P, Baksh MF, Venison EP, Gibbings JG, Dunwell JM (2018) Genetic analysis of a major international collection of cultivated apple varieties reveals previously unknown historic heteroploid and inbred relationships. PLoS One 13:e0202405. https://doi.org/10.1371/journal.pone.0202405
Paran I, van der Knaap E (2007) Genetic and molecular regulation of fruit and plant domestication traits in tomato and pepper. J Exp Bot 58:3841–3852. https://doi.org/10.1093/jxb/erm257
Patocchi A, Walser M, Tartarini S, Broggini GA, Gennari F, Sansavini S, Gessler C (2005) Identification by

genome scanning approach (GSA) of a microsatellite tightly associated with the apple scab resistance gene *Vm*. Genome 48:630–636

Peace CP, Bianco L, Troggio M, van de Weg E, Howard NP, Cornille A, Durel C-E, Myles S, Migicovsky Z, Schaffer RJ, Costes E, Fazio G, Yamane H, van Nocker S, Gottschalk C, Costa F, Chagné D, Zhang X, Patocchi A, Gardiner SE, Hardner C, Kumar S, Laurens F, Bucher E, Main D, Jung S, Vanderzande S (2019) Apple whole genome sequences: recent advances and new prospects. Hortic Res 6:1–24. https://doi.org/10.1038/s41438-019-0141-7

Pompili V, Costa LD, Piazza S, Pindo M, Malnoy M (2020) Reduced fire blight susceptibility in apple cultivars using a high-efficiency CRISPR/Cas9-FLP/FRT-based gene editing system. Plant Biotech J 18:845–858

Singh J, Sun M, Cannon S, Wu J, Khan A (2021) An accumulation of genetic variation and selection across the disease-relatedgenes during apple domestication. Tree Geneti Genomes 17:1–11

Shulaev V, Korban SS, Sosinski B, Abbott AG, Aldwinckle HS, Folta KM, Iezzoni A, Main D, Arús P, Dandekar AM, Lewers K, Brown SK, Davis TM, Gardiner SE, Potter D, Veilleux RE (2008) Multiple models for rosaceae genomics. Plant Physiol 147:985–1003. https://doi.org/10.1104/pp.107.115618

Spengler RN (2019) Origins of the apple: the role of megafaunal mutualism in the domestication of *Malus* and rosaceous trees. Front Plant Sci 10:617. https://doi.org/10.3389/fpls.2019.00617

Stushnoff C, McSay AE, Luby J, Forsline PL (2002) Diversity of phenolic antioxidant content and radical scavenging capacity in the USDA apple germplasm core collection. Acta Hortic 305–312. https://doi.org/10.17660/ActaHortic.2003.623.34

Tatum TC, Stepanovic S, Biradar DP, Rayburn AL, Korban SS (2005) Variation in nuclear DNA content in *Malus* species and cultivated apples. Genome 48:924–930

Treutter D (2010) Managing phenol contents in crop plants by phytochemical farming and breeding—visions and constraints. Int J Mol Sci 11:807–857

USDA (2014) National genetic resources program. Germplasm Resources Information Network (GRIN). USDA, ARS Natl. Germplasm Resources Laboratory, Beltsville, MD. http://www.ars-grin.gov/npgs/index.html. Accessed 5 May 2017

Velasco R, Zharkikh A, Affourtit J, Dhingra A, Cestaro A, Kalyanaraman A, Fontana P, Bhatnagar SK, Troggio M, Pruss D, Salvi S, Pindo M, Baldi P, Castelletti S, Cavaiuolo M, Coppola G, Costa F, Cova V, Dal Ri A, Goremykin V, Komjanc M, Longhi S, Magnago P, Malacarne G, Malnoy M, Micheletti D, Moretto M, Perazzolli M, Si-Ammour A, Vezzulli S, Zini E, Eldredge G, Fitzgerald LM, Gutin N, Lanchbury J, Macalma T, Mitchell JT, Reid J, Wardell B, Kodira C, Chen Z, Desany B, Niazi F, Palmer M, Koepke T, Jiwan D, Schaeffer S, Krishnan V, Wu C, Chu VT, King ST, Vick J, Tao Q, Mraz A, Stormo A, Stormo K, Bogden R, Ederle D, Stella A, Vecchietti A, Kater MM, Masiero S, Lasserre P, Lespinasse Y, Allan AC, Bus V, Chagné D, Crowhurst RN, Gleave AP, Lavezzo E, Fawcett JA, Proost S, Rouzé P, Sterck L, Toppo S, Lazzari B, Hellens RP, Durel C-E, Gutin A, Bumgarner RE, Gardiner SE, Skolnick M, Egholm M, Van de Peer Y, Salamini F, Viola R (2010) The genome of the domesticated apple (*Malus* × *domestica* Borkh.). Nat Genet 42:833–839. https://doi.org/10.1038/ng.654

Vogt I, Wohner T, Richter K, Flachowsky H, Sundin GW, Wensing A, Savory EA, Geider K, Day B, Hanke M, Peil A (2013) Gene-for-gene relationship in the host-pathogen system *Malus* × *robusta* 5-*Erwinia amylovora*. New Phytol 197:1262–1275

Volk GM, Henk AD, Baldo A, Fazio G, Chao CT, Richards CM (2015) Chloroplast heterogeneity and historical admixture within the genus *Malus*. Am J Bot 102:1198–1208. https://doi.org/10.3732/ajb.1500095

Volz RK, McGhie TK (2011) Genetic variability in apple fruit polyphenol composition in *Malus* × *domestica* and *Malus sieversii* germplasm grown in New Zealand. J Agric Food Chem 59:11509–11521. https://doi.org/10.1021/jf202680h

Wagner I, Maurer WD, Lemmen P, Schmitt HP, Wagner M, Binder M, Patzak P (2014) Hybridization and genetic diversity in wild apple (*Malus sylvestris* (L.) Mill) from various regions in Germany and from Luxembourg. Silvae Genet 63:81–93. https://doi.org/10.1515/sg-2014-0012

Yao L, Zheng X, Cai D, Gao Y, Wang K, Cao Y, Teng Y (2010) Exploitation of Malus EST-SSRs and the utility in evaluation of genetic diversity in *Malus* and *Pyrus*. Genet Resour Crop Evol 57:841–851. https://doi.org/10.1007/s10722-009-9524-1

Zhang L, Hu J, Han X, Li J, Gao Y, Richards CM, Zhang C, Tian Y, Liu G, Gul H, Wang D, Tian Y, Yang C, Meng M, Yuan G, Kang G, Wu Y, Wang K, Zhang H, Wang D, Cong P (2019) A high-quality apple genome assembly reveals the association of a retrotransposon and red fruit colour. Nat Commun 10:1–13. https://doi.org/10.1038/s41467-019-09518-x

18 Future Prospects of 'Omics' and of Other Technologies for Genetic Improvement of Apple

Schuyler S. Korban

Abstract

Efforts for the genetic improvement of apple have tremendously benefited from the availability of modern tools of genomics, transcriptomics, metabolomics, proteomics, bioinformatics, as well as of genetic engineering, among others, over the past two decades. These tools are continuing to expand, and they have become more critical for use in modern apple breeding programs. All current and future 'omics' technologies should be fully exploited to develop a better understanding of the complex system and biology of apple trees, as well as for pursuing efficient and robust genetic enhancement efforts to develop well-adapted apple cultivars with enhanced resistance to biotic and abiotic stress conditions, as well as those cultivars with desirable targeted nutritional and palatable preferences for consumers and for commercial markets alike. This chapter will cover the prospects of 'omics' technologies that are currently having an impact as well as those with potential impacts on apple biology and genetic enhancement efforts for the near future.

S. S. Korban (✉)
Department of Natural Resources & Environmental Sciences, University of Illinois at Urbana-Champaign, Urbana, IL 61801, USA
e-mail: korban@illinois.edu

18.1 Introduction

Apple is the most important temperate tree fruit crop grown around the world. Apple production requires intensive agricultural practices, including training, pruning, and annual applications of pesticides for control of fungal and bacterial diseases, insects, mites, and weeds; thus, contributing to significant production costs. In comparison to other agricultural crops grown, cultural management practices for apples are labor-intensive and costly. Therefore, it is important to have access to apple cultivars, rootstocks, and grafted trees that are optimal for maintenance in a productive orchard with minimal labor and production costs. Moreover, it is critical that modern apple orchards are growing trees that are tailored for a growing apple industry that is both attentive and catering to discerning consumers, with changing eating habits and preferences, who are highly aware of the nutritional and health benefits of fresh and processed fruits, such as that of apples, among other agricultural food crops.

The importance of apples in a balanced human diet is well known, and this pertains to fiber, vitamins, and antioxidant contents, including various phytochemicals such as quercetin, catechin, phloridzin, chlorogenic acid, and proanthocyanins (PA). Several epidemiological studies have reported that apples are associated with decreased risk of lung cancer in both men

S. S. Korban (ed.), *The Apple Genome*, Compendium of Plant Genomes,
https://doi.org/10.1007/978-3-030-74682-7_18

and women, decreased risk of cardiovascular disease (CVD) in women and thrombotic stroke, decreased risk of asthma and bronchial hypersensitivity, lower risk for Type II diabetes, and reducing total cholesterol, LDL ('bad') cholesterol and triglyceride levels (Boyer and Lui 2004; Hansen et al. 2010; Hodgson et al. 2016; Jedrychowski and Maugeri 2009; Koutsos et al. 2020; Sun and Rui 2008; Woods et al. 2003).

Apples are known to be excellent sources of polyphenols (∼110 mg/100 g) and fiber (∼2–3 g/100 g), and it is these bioactive components that are likely responsible for the observed potential health effects (Koutsos et al. 2015, 2020). Among these polyphenols, flavanols (catechin and PAs) account for the major class (71–90%), followed by hydroxycinnamates (4–18%), flavonols (1–11%), dihydrochalcones (2–6%), and anthocyanins (1–3%) that are present only in red apples (Koutsos et al. 2015). However, it is important to point out that based on their skeleton structure, phenolic compounds can be distinguished into phenolic acids and flavonoids (Cuthbertson et al. 2012). Within phenolic acids, one of the most relevant in apple is chlorogenic acid, while for flavonoids, these include various compounds including flavonols (quercetin), flavan-3-ols or flavanols (catechin and epicatechin), hydroxycinnamates (coumaric acids, 5'-caffeoyl quinic acid), dihydrochalcone (phloridzin), and anthocyanins (Volz and McGhie 2011). The protective role of these compounds on human health is primarily attributed to their redox capacity that enables quenching of singlet oxygen molecules, as well as of scavenging of free radicals and reactive oxygen species (ROS) (Kshonsek et al. 2018; Busatto et al. 2019). It is the consumption of these polyphenols that contributes to either modifications of these molecules into other bioactive compounds or directly leads to interactions with gut microbiota, thus promoting human health (Busatto et al. 2019). Recently, Ichwan et al. (2021) have reported that both flavonoids, in particular quercetin, the most abundant flavanol in apple peel, and 3,5-dihydroxybenzoic acid, unrelated to flavonoids but found in apple flesh, are pro-neurogenic. It is found that quercetin and 3,5-dihydroxybenzoic acid not only activate precursor cell proliferation but also promote cell-cycle exit, cellular survival, and neuronal differentiation in brains of test animals, and therefore they likely promote adult hippocampal neurogenesis, i.e., brain plasticity wherein functional neurons are generated throughout life and integrated into the existing circuitry, thereby mediating particular forms of learning and memory (Ichwan et al. 2021).

It is important to point out that levels of polyphenols along with other compounds in apple fruit vary during ripening, as contents of phenolic acids and flavonoids in epicarp and endocarp tissues change during ripening, and these are likely influenced by growing conditions (Stracke et al. 2009; Alberti et al. 2017). Overall, it is the genetic variability that likely plays a primary role in determining polyphenol levels, as most of these observed variations are attributed to differences among apple cultivars (Drogoudi et al. 2008; Kschonsek et al. 2018; McClure et al. 2019).

In the USA, it is reported that 22% of the polyphenols in the human diet originate from apples, rendering apples as primary dietary sources of these antioxidant compounds (McClure et al. 2019). Furthermore, pectin, the main soluble fiber in apples, is reported to influence transit time, gastric emptying, and nutrient absorption, thereby impacting lipid and glucose metabolism, as well as playing a role in modulating gut microbiota, a key determinant of the bile acid chemical structure, and thus has a signaling potential (Koutsos et al. 2015). Overall, these critical 'health-associated' qualities of apples provide them with advantages that are becoming more readily promoted to expand their marketability.

18.2 The Role of Omics in Understanding the Biology of Traits of Interest in Apple

A major determinant of an apple's nutritional/health-promoting value is likely determined by its genome, and therefore genetic mapping of polyphenols has been pursued in recent years. McClure et al. (2019) have investigated the genetic architecture of polyphenols in apples by combining high performance liquid chromatography (HPLC) data with ~100,000 single nucleotide polymorphisms (SNPs) from two diverse apple populations. They have observed that polyphenol contents can vary by up to two orders of magnitude across cultivars, and that this wide variation is predicted using genetic markers, wherein it is found that this variation is often controlled by a small number of large effect genetic loci. Using genome-wide association study (GWAS), McClure et al. (2019) have identified candidate genes for the production of quercetin, epicatechin, catechin, chlorogenic acid, 4-O-caffeoylquinic acid, and procyanidins B1, B2, and C1. This has demonstrated that often a relatively simple genetic architecture underlies such wide variations in contents of key polyphenols in apples. Therefore, this raises the hope that breeders are likely capable of improving the nutritional value of apples through either marker-assisted breeding (MAB) or gene editing.

Due to recent advances in sequencing technology, this has allowed for taking on wide genetic testing of both populations and individual apple genotypes using both transcriptome and genome sequencing efforts. As a transcriptome sequences the entire protein-coding region of a genome, corresponding to less than 2% of a genome, this region contains approximately 85% of known stress–response trait variants (Horton and Lucassen 2019). On the other hand, genome sequencing encompasses the transcriptome as well as sequences of all non-protein-coding DNA. Therefore, transcriptome sequencing can serve as a diagnostic tool to characterize a trait or traits of interest. Moreover, as sequencing technologies have significantly advanced in both depth and breadth, these have contributed to better and expanded understanding of various traits of interest. It is important to point out that sequencing genomes of various accessions carrying mutant apple genotypes have contributed to rapid identification of driver mutations that allow for uncovering those complex relationships among different variants for traits of interest, such as disease resistance traits, over space and time, demonstrating the wide heterogeneity of reactions to a pathogen (and pathogenic races) infection, as well as of those corresponding difficulties in addressing such disease reactions. As sequencing technologies have advanced to levels wherein individual cells can be sequenced, this allows for uncovering and identification of previously unknown mutational mechanisms, such as those of fruit russeting (Legay et al. 2015; 2017) and resistance/susceptibility to fruit blue mold (Ballester et al. 2017), among others.

Nevertheless, these capabilities of generating genomic data have substantially outstripped our ability to interpret the significance of these resultant data (Horton and Lucassen 2019). While improvements in genomic technologies are in many cases driving improvements in our apple breeding efforts, we will continue to encounter new problems as genomic testing shifts into the diagnostic realm. Moreover, such detailed and targeted approaches for genomic analysis offer opportunities to address questions that go beyond issues of a single trait or a few traits of interest while pursuing incremental improvements in a few selected traits. However, this should allow us to pursue efforts to target and improve multiple traits, such as nutritional fruit quality, keeping quality of fruit, disease and pest resistance, tree architecture and fruiting habit, as well as tolerance to various environmental stresses. So, herein lies in the challenging quandary. How can we make decisions in determining which of the variety of possible outputs obtained from such genomic analyses ought to be considered at any given time? For example, we do recognize that linkages between many genetic variants and disease reactions are often not definitively confirmed or not quite well understood. Therefore, multidisciplinary inputs and collaborations are increasingly key for interpreting the significance of findings of genomic analyses.

Along these lines of thought, it is known that genome sequencing will characterize and identify variants for a trait of interest, this requires detailed and careful filtering to achieve a meaningful output. Thus, this necessitates a significant change in our mindset from an era when most variants have been identified within the context of carefully selected single gene sequencing, and so had a much higher prior probability of being causative. As most traits of interest are complex and polygenic, controlled by quantitative trait loci (QTLs), it is imperative that we remain alert to the importance of how these 'groups of genes' are tandemly arranged, interact, and operate in concert to deliver a favorable or a desirable phenotype.

18.3 The Use of Omics in Discerning and Predicting of Desirable Genetic Variants

In many cases, our understanding of why the same genetic trait may be expressed quite differently among different cultivars is at an early stage at this time; thus, this often renders genetic prediction very challenging. Therefore, functional studies are critically important. For example, the use of a polygenic risk score (PRS) using disease resistance/susceptibility SNPs identified via population GWAS may reveal wide differences in absolute disease incidence among individuals carrying disease variants of higher polygenic risk scores compared to those with lower PRS values.

While it is now easy and inexpensive to sequence genomes of individual cultivars, it is more difficult to turn raw sequence data into actual insights to assess whether a nucleotide misspelling, a C instead of a T, a deletion, or insertion of sequences predicts an increased likelihood of a certain condition. Moreover, linking these data with environmental and cultural management determinants is no small feat. This is attributed to advances in next-generation sequencing (NGS) technologies along with rapid progress ın genomıc data analysis. As these have resulted in yields of high-throughput data for genomes, including SNPs, copy number variants (CNVs), loss of heterozygosity variants, genomic rearrangements, and rare variants; for epigenomes, including DNA methylation, histone modifications, chromatin accessibility, transcription factor (TF) binding; as well as for transcriptomes, including gene expression, alternative splicing, long non-coding RNAs and small RNAs such as microRNAs (Ritchie et al. 2015). Thus, mastering how to utilize these various sources of data in tandem will likely require more researchers with computer science backgrounds, as well as more use of artificial intelligence (AI) technologies.

As most often, apple diseases are rarely attributed to single-gene mutations, they cannot be overcome by replacing a mutated gene with a normal copy, which is the premise for 'gene repair' via gene editing. Gene editing has gradually progressed and likely to be successful in addressing traits controlled by a single gene or a mutation controlled by gene(s) with large effects. Gene repair for single-gene disorders is likely to succeed, but this must be tailored to each individual trait of interest/status.

In recent efforts, new sources of variation for such fruit traits as fruit taste and texture have been identified by analyzing widely diverse apple germplasm using either QTL mapping or association studies (Amyotte et al. 2017). However, an alternative to QTL mapping, wherein designed mapping populations derived from controlled crosses are used, is that of GWAS whereby population(s) of diverse germplasm is/are used instead. For example, Kumar et al. (2013, 2014) have mapped fruit quality traits using a six-parent family of 1200 apple seedlings along with a diverse collection of 115 apple accessions using GWAS. It is observed these two populations (seedlings vs. accessions) differed in terms of their significant associations, wherein titratable acidity (TA) trait is mapped to LG8 in the seedling population while it mapped to LG16 in the diverse accession collection. Therefore, these findings demonstrate that genetic mapping using a diverse germplasm can reveal distinct loci from those found in segregating seedling populations. Moreover, loci detected in both multi-family and diverse germplasm mapping studies have wide practical applications for breeding, as they capture QTLs present in numerous potential breeding parents. Furthermore, GWAS of diverse germplasm allows for a finer resolution of a QTL, as historical recombination events would have reduced the extent of linkage disequilibrium (LD) within a diverse germplasm (Khan and Korban 2012). Thus, significant markers are more likely to be physically close to true genetic regions controlling variation in fruit quality (Khan and Korban 2012).

With increased interest in identifying new sources of variation for fruit quality traits, various phenotyping approaches must be used. For example, genetic components of apple texture have been analyzed via breaking of fruit cell walls by biting and mastication, while earlier texture phenotyping has relied mainly on measuring fruit firmness using the amount of mechanical force (pressure) required to penetrate an apple with a blunt probe, a penetrometer. When components of apple texture are broken down into four acoustic and eight mechanical parameters, significant QTLs have been detected on 13 out of all 17 apple LGs (Longhi et al. 2012). Similarly, when mechanical compression and penetration parameters are used, it is observed that large numbers of texture QTL are detected, and these have been mapped to 14 LGs alongside sensory descriptors of apple texture (Ben Sadok et al. 2015). These studies illustrate the genetic complexity of apple texture and the need for the discovery of new QTLs by mapping fruit quality traits using different phenotyping approaches.

18.4 Challenges of Phenotyping and Phenomics

With all current advances in genomics, metabolomics, transcriptomics, gene editing, and other modern omics and breeding tools, one of the most critical and remaining bottlenecks in pursuing efficient and effective MABs is in the area of pan-phenotyping, i.e., phenomics, which allows for early selection of seedlings carrying multiple desirable traits of interest, including those related to fruiting and fruit quality traits in an apple breeding program.

For example, in a phenotyping approach for fruit texture, trained sensory panels are used to evaluate descriptive sensory parameters, including texture, taste, and flavor, and these collective data are then used to discover new QTLs (Corollaro et al. 2013). These descriptive sensory evaluations for QTL discovery capitalize on incorporating consumer preferences in breeding for new or superior genotypes. Interestingly, although most apple genetic mapping studies for fruit quality traits utilize instrumental measures to phenotype taste, flavor, and texture, such evaluations can widely differ from those obtained from instrumental predictions (Amyotte et al. 2017). Thus, in spite of its potential value, descriptive sensory evaluation for QTL discovery in apple has been rather limited due to its high cost and low throughput as both time and cost are rather prohibitive (Corollaro et al. 2013). Thus, Amyotte et al. (2017) have used a trained sensory panel to evaluate a moderately sized and diverse collection consisting of both heritage and commercial apple cultivars (85 genotypes) over 2 years. However, they have also used instrumental measurements for a few traits, including those for fruit firmness, soluble solids content, and TAs. Furthermore, they have used genotyping by sequencing (GBS) data to conduct a GWAS of apple sensory taste, flavor, and texture. It is reported that using descriptive sensory phenotyping combined with a diverse population and robust GBS data have allowed for the discovery of previously unreported apple fruit quality loci including a novel locus associated with both juiciness and crispness, two critical palatable drivers of consumer preference (Amyotte et al. 2017). Therefore, this study has demonstrated the value of descriptive sensory evaluations in combination with genomic analysis to identify fruit quality loci that likely have direct utility in breeding apple cultivars for desirable consumer preference.

Thus, further advances in apple breeding efforts are dependent on the discovery of new genomic targets for marker-assisted selection (MAS). These advances can be made through the assessment of distinct apple populations as well as via the use of alternative approaches to phenotypic data collection (Amyotte et al. 2017).

As with other desirable target traits, efforts to breed for polyphenolic compounds using MAS are dependent on identifying robust molecular markers associated with such novel and important traits for introgression into elite genetic backgrounds are an essential tool for breeders. This is particularly critical when a donor parent is either of low-quality, heritage, or a wild apple accession (Busatto et al. 2019). To that end, several efforts have been undertaken to characterize various fruit quality traits such as phenolic compound content and composition of apple fruits. Both Khan et al. (2012) and Chagné et al. (2012a) have identified a number of QTLs likely associated with phenolic metabolites, as well as of markers linked to a *leucoanthocyanidin reductase (LAR1)* gene, a major gene controlling flavanol biosynthesis.

In addition, QTLs for metabolite accumulation in both peel and flesh have also been detected (Khan et al. 2012). Specifically, quantitative variations of phenolic compounds in apple fruits have been investigated using a segregating F_1 population of 'Prima' × 'Fiesta' consisting of 113 individuals to map metabolite quantitative trait loci (mQTLs). Untargeted metabolic profiling, using liquid chromatography-mass spectroscopy (LC–MS), of peel and flesh tissues of ripe fruits has been performed on all 113 individuals along with the two parents. In this study, genetic linkage maps of each of 'Prima' and 'Fiesta', representing all 17 linkage groups of apple, have been used, wherein the 'Prima' map consists of 562 markers, while the 'Fiesta' map consists of 452 markers. For untargeted mQTL mapping of metabolites, individual maps of 'Prima' and 'Fiesta' are used; whereas, for targeted mQTL mapping, using MapQTL® 6.0, an integrated map of both parents has been constructed, consisting of 801 markers spanning 1348 cM, and used for the analysis of annotated metabolites. A total of 418 metabolites in peel and 254 metabolites in flesh tissues are detected. As standard QTL mapping software is designed to map individual traits, one at a time, it has not been deemed well suited to map hundreds of metabolites simultaneously; therefore, MetaNetwork has been used instead as it allows for simultaneous genome-wide screening of numerous traits (Fu et al. 2007). Therefore, mQTL mapping using MetaNetwork has resulted in identifying 669 significant mQTLs, wherein 488 are detected in the peel and 181 detected in the flesh. Furthermore, four linkage groups (LGs), including LG1, LG8, LG13, and LG16 are found to contain mQTL hotspots, mainly regulating metabolites belonging to the phenylpropanoid pathway (Fig. 18.1). Therefore, the genetics of annotated metabolites has been investigated further using MapQTL®, and it is found that a number of quercetin conjugates have mQTLs on LG1 or LG13. Moreover, the most important mQTL hotspot with the largest number of metabolites is detected on LG16, wherein mQTLs for 33 peel-related and 17 flesh-related phenolic compounds are located, including the gene encoding LAR1, mentioned above, along with seven transcription factor (TF) genes (Fig. 18.1). Using the apple genome sequence, structural genes involved in the phenylpropanoid biosynthetic pathway have also been located.

Furthermore, QTLs have also been identified for the selection of apple cultivars for cider production (Verdu et al. 2014). To enhance anthocyanin content in apple fruits, molecular markers linked to a *MYB* transcription factor for anthocyanin content have been developed, and these have been used to pre-select red-fleshed apple seedlings (Chagné et al. 2013). As breeding efforts have been designed for selection of red-fleshed apples using wild *Malus* species as sources of anthocyanin and of other phenolic compounds to enhance antioxidant properties in future apple accessions, wherein accumulation of polyphenolic compounds has been associated with expression profiles of candidate genes involved in related biochemical pathways in leaf and fruit of apple (Henry-Kirk et al. 2012).

Interestingly, Busatto et al. (2019) have screened apple fruits for 19 phenolic compounds in both skin and pulp tissues of seven apple accessions across the *Malus* genus collected at two stages, during fruit development and at harvest. They have found that the major difference in phenolic content between wild and domesticated accessions, particularly in the pulp tissue, could be accounted for the higher fruit

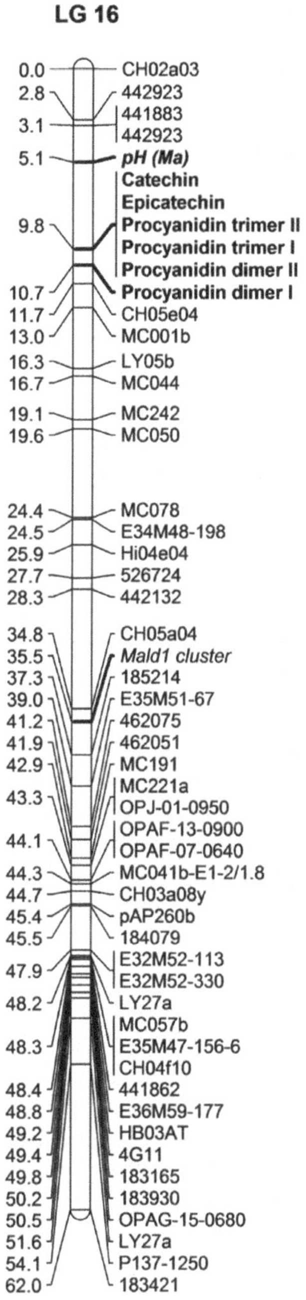

Fig. 18.1 Mapping of (+)-catechin, (–)-epicatechin, several procyanidins, and pH (*Ma*) on linkage group 16. This figure is taken from Khan et al. (2012), courtesy of Dr. Henk Schouten

growth rate of domesticated apples. Furthermore, this proposed dilution effect is also confirmed following observations of increased contents of procyanidin B2 + B4 and phloridzin in russet-skinned apples, known to have higher levels of these phenolic compounds. This metabolite screening is also accompanied by expression analysis of 16 polyphenolic genes revealing that for nine genes, there are higher levels of expression at harvest than during fruit development.

Although there are several available DNA tests, locus-specific, and trait performance-predictive assays for different traits for apple (Evans and Peace 2017), DNA-informed breeding has been rather slow due to lack of predictive DNA tests applicable to breeding germplasm, as well as lack of access to DNA-based diagnostic services (see Chaps. 5 and 6 in this volume). Nevertheless, availability of tools for accelerated trait introgression, termed as 'fast-track' or 'rapid-cycle' breeding, capitalize on developing transgenic intermediate generations overexpressing an early flowering gene, *BpMADS4*, from silver birch to shorten the juvenile phase of apple, is then followed by selection of non-transgenic lines carrying target traits of interest (Flachowsky et al. 2009; Weigl et al. 2015; Schlathölter et al. 2018; also see Chaps. 5 and 11 in this volume). Although the final selection is non-transgenic, it remains to be seen whether or not the apple industry and consumers are willing to accept these new apple cultivars.

As noted above, there has been much progress made in identifying various genomic regions, including markers and LGs, as well as transcriptomic, proteomic, metabolomic datasets critical for various target traits for many desirable traits for apple. Therefore, we should continue to build and expand on this growing knowledge base for desirable phenotypic traits, and undertake serious efforts to develop and establish curated datasets and reliable tools dedicated to pursuing robust, effective, and efficient apple phenomics. By building up pan-apple phenotypic datasets and tools, these will serve as key components for establishing a robust apple phenomics field that will hopefully overcome this critical bottleneck in apple breeding programs (also see Sect. 18.7.1 below).

18.5 Challenges in Genotyping in Apple

It is clear that SNPs are now highly used as genetic markers for pursuing various genetic analyses; moreover, the vast abundance of available SNPs in a genome allows for simultaneous screening of many thousands of polymorphic loci in multiple platforms including SNP arrays, genotyping-by-sequencing (GBS), or resequencing efforts (Vanderzande et al. 2019). These large numbers of generated SNPs require robust curation. SNP arrays are available for apple, including an 8K apple array (Chagné et al. 2012b), a 20K apple array developed by FruitBreedomics (Bianco et al. 2014), both developed on an Illumina Infinium platform, and a 480K apple array by FruitBreedomics on an Affymetrix axiom platform (Bianco et al. 2016). These SNP arrays allow for genotyping individuals via automated scoring of thousands of SNPs. However, with increasing numbers of SNPs on these arrays, the likelihood of potential errors in calling SNPs increases, as it is both time-consuming and less feasible to manually review all automated SNP calls to identify potential errors. Furthermore, incorrect genotype assignments using automated SNP scoring software in SNP arrays are likely to increase when intensity plots deviate from expected patterns.

On the other hand, automated genotyping is based on associations of specific alleles to different fluorescent molecules, detection of such fluorescent molecules, clustering of individual marker data points, based on intensity ratios between different fluorescent dyes across multiple individuals into distinct regions of a genotype-calling space, and leading to final assignments of these clusters to genotypes (Vanderzande et al. 2019). Therefore, such a curation workflow for high-resolution genetic marker data would identify and resolve errors, thereby yielding a robust set of genotypic data. This workflow maximizes genotypic data obtained from high-throughput genome-scanning tools while minimizing the time required to identify and remove errors. Such genotypic data sets can then be used for undertaking multiple efforts for reconstructing pedigrees, determining quantitative genetic relationships, identifying and validating QTLs, and tracing allele sources (Vanderzande et al. 2019).

Recently, Zhang et al. (2019) have assembled a high-quality apple genome, with a N50 contig of 6.99 Mb and a maximum contig length of 18.01 Mb, of an anther-derived homozygous trihaploid 'Hanfu' line, "HFTH1", including 22 telomere sequences, using a combination of PacBio single-molecule real-time (SMRT) sequencing, chromosome conformation capture (Hi-C) sequencing, and optical mapping. It is observed that the assembly size is close to the estimated genome size of the 'Golden Delicious' dihaploid line "GDDH13" (Daccord et al. 2017), but corresponds to 92.99% of the estimated genome size (708.54 Mb) for "HFTH1" by k-mer analysis, and ~97.89% of the Illumina reads of "HFTH1" could be mapped to their assembly. In addition, a 160,068 bp chloroplast genome and 396,939 bp mitochondrial genome have been assembled into two complete contigs. In comparison to the first 'Golden Delicious' reference genome (Velasco et al. 2010), a total of 18,047 deletions, 12,101 insertions, and 14 large inversions have been identified by Zhang et al. (2019). Furthermore, it is proposed that these wide genomic variations among the different genome sequences are largely attributed to regulatory activities of transposable elements (Zhang et al. 2019).

Therefore, significant progress has been made in undertaking robust and reliable genotyping efforts; however, there are some challenges in handling the volume of genotyping datasets that are generated, and this has to be addressed and better integrated with phenomics and other omics tools to contribute to effective genetic enhancement efforts for the apple.

18.6 The Role of Gene Editing in Uncovering and Exploiting Genetic Diversity for Desirable Traits

Genetic diversity is critical for pursuing the improvement of apple, as it is for all other crops. Although there is a wide genetic diversity available in the *Malus* genus, the fact that most traits of interest are controlled by QTLs, and that some of these traits are present in wild *Malus* species, this renders their transfer via sexual hybridization strategies rather highly involved and cumbersome, even if these efforts capitalize on fast-tracking or rapid-cycling approaches. It is now known that much of the genetic changes underlying traits of economic importance are present in *cis*-regulatory regions; i.e., non-coding DNA sequences controlling transcription of genes, as these changes in genes have likely been selected during domestication, thereby resulting in desirable traits caused by altered gene expression (Li et al. 2020; Swinnen et al. 2016).

Therefore, the advent of genome editing or gene editing via clusters of regularly interspaced short palindromic repeats/cas9 associated protein (CRISPR/Cas-9)-based platform offers a powerful tool for engineering *cis*-regulatory regions, *cis*-engineering, to introduce and/or modify genetic traits of interest (Wolter et al. 2019; Li et al. 2020). Despite the importance of regulatory changes in genes, the application of CRISPR/Cas-mediated *cis*-engineering provides a more refined method for modulating gene expression and for creating phenotypic diversity to benefit crop improvement, and therefore it should be assessed and exploited in apple.

CRISPR/Cas-based technologies offer multiple strategies to engineer *cis*-regulatory regions based on prior knowledge of the target region or desired goal. If there is no prior knowledge of the target region, a multiplexed CRISPR/Cas promoter targeting approach can be used to putative 'negative regulators' of the desired traits by generating a collection of reduced function alleles. Furthermore, a well-defined promoter can be exchanged with the promoter of the gene of interest to increase expression level or change the temporal/spatial expression pattern(s) of the target gene. For a given *cis*-regulatory element (CRE) in a target region of interest, this CRE can be either disrupted or deleted based on random indel mutations introduced by a non-homologous end joining (NHEJ) repair pathway (Schmidt et al. 2019). In addition, CRISPR/Cas-mediated point mutations and CRE swaps are also important approaches to manipulate gene expression (Schmidt et al. 2019).

Recently, SNPs that are critically involved in the domestication of crops have been documented as single-nucleotide alterations in regulatory sequences, and they are deemed as contributors to major effects on gene expression (Li et al. 2020). Insertions are sources of genetic diversity that can alter gene expression by either introducing new or disrupting existing CREs. In particular, transposable elements (TEs) play important roles in generating genomic variations by altering gene regulation (Li et al. 2020). For example, the presence of an R6 motif, a binding site of MdMYB10, in the promoter of the apple MdMYB10 results in auto-activation of this gene resulting in higher levels of anthocyanins (Espley et al. 2009); moreover, a 36-bp insertion in the promoter of *MdSAUR37* contributes to high fruit malate content in apple (Jia et al. 2018).

Malnoy et al. (2016) have demonstrated efficient targeted mutagenesis (0.5–6.9%) of the apple gene loci *DIPM-1, 2*, and *4* (DspA/E-interacting proteins, 198 kDa effector proteins homologous to the type III effector AvrE of *Pseudomonas syringae* pv. tomato), using direct delivery of CRISPR ribonucleoproteins (RNPs). Although the mutation efficiency is found to vary with the targeted gene locus and the ratio of Cas9 and sgRNA, mutation patterns and frequency assays have shown that CRISPR RNPs to be effective for targeted mutagenesis of gene loci in apple protoplasts. Subsequently, Pompili et al. (2019) have generated *MdDIPM4* knockouts in apple cvs. 'Gala' and 'Golden Delicious' using the CRISPR/Cas9 system delivered via *Agrobacterium tumefaciens*, using a construct containing a heat-shock inducible flippase (FLP)/

flippase recognition target (FRT) recombination system designed specifically to remove the T-DNA carrying the expression cassette of CRISPR/Cas9, the kanamycin selectable marker gene, and the FLP, and have obtained an editing efficiency of 75%. Among 57 lines screened, they have identified seven edited lines exhibiting loss-of-function mutations. These lines have been inoculated with the bacterial pathogen *Erwinia amylovora*, inciting fire blight disease, and have been found to exhibit highly significant reductions in susceptibility to fire blight compared to control plants.

Other efforts in genome and gene editing in apple are underway in other laboratories (Charrier et al. 2019). Recently, it has been reported that transcription activator-like effector nuclease (TALEN) having a DNA-binding domain that is comprised of a tandem array of 33–34 amino acid-long customizable monomers that likely can be assembled to recognize any genetic sequence following a one-repeat-binds-one-base-pair recognition code (Jain et al. 2021). Therefore, it has been found that TALEN has up to a five-fold increase in editing efficiency compared with that of Cas9 in heterochromatin regions (Jain et al. 2021). Thus, these gene-editing efforts should yield more promising results in apples as more studies are undertaken to capitalize on these new technologies (also see Chap. 10 in this volume).

Thus, the potential and capability of gene editing in apple are promising, yet they are yet to be fully exploited. Efforts should continue to pursue gene editing in apple and to assemble a portfolio of traits of interest that can be readily addressed using this currently available and evolving tool for genetic enhancement of the apple.

18.7 Other Novel Technologies

18.7.1 Advances in Plant Phenomics

As mentioned above, one of the critical bottlenecks of modern apple breeding programs is that of availability of a robust phenotyping system that captures a large number of traits of interest, including those relevant to reproductive growth and fruit quality traits. Currently, phenotyping relies on measurements of selected traits, such as vegetative growth, biotic stress, and perhaps to abiotic stress, using small numbers of plants, and thus offer limited throughput and do not provide a comprehensive analysis of multiple traits within a single plant and across different cultivars (Pieruschka and Schurr 2019; Ubbens and Stavness 2017). This limits a breeding program's capability of determining how expressed phenotypes correlate with underlying genetic factors and with environmental conditions, as well as with ensuing interactions. Collecting multi-dimensional phenotypic data at multiple levels from cell level, organ level, plant level, to population level offer opportunities for pursuing more deliberate and robust plant phenomics (Zhao et al. 2019). This is particularly critical for tree fruit crops, such as apples.

In recent years, a combination of new imaging technologies, robotic and conveyer-belt systems in greenhouses, and ground-based and aerial imaging platforms, including the use of drones, in fields have become more feasible to generate large numbers of photographic images of crops (Fahlgren et al. 2015). In turn, this has allowed for the expansion of high-throughput phenotyping capabilities with increasing amounts of data from high-dimensional imaging sensors along with interests in measuring more complex phenotypic traits (Knecht et al. 2016). As availability and reuse of interoperable data have expanded, this requires standardization from calibration procedures of sensors, via standard operation procedures of experiments and extraction of data from multimodal images, to experimental design and metadata, to data ontology formats and standardized interfaces to generate meaningful data (Pieruschka and Schurr 2019).

As a result, an open-source deep-learning software platform, termed as 'Deep Plant Phenomics' (DPP), has been developed (Ubbens and Stavness 2017). This DPP platform capitalizes on deep convolutional neural networks for plant phenotyping. This easy platform provides pre-trained neural networks for several common plant phenotyping tasks, and a 'proof of concept' of this platform has been used in three complex phenotyping tasks, including leaf counting, mutant classification, and age regression for top-down images of plant rosettes in *Arabidopsis thaliana* (Ubbens and Stavness 2017). This tool is built up to use deep learning in machine learning (ML) for tackling large data analytics (see also Sect. 18.8 below). Deep convolutional neural networks (CNNs) serve as deep learning methods that are particularly well suited to computer vision issues. Thus, a typical setup for a CNN utilizes a raw RGB image as input, deemed as an n $\times$ m $\times$ 3 volume, wherein 'n' is image height, 'm' is image width, and '3' corresponds to the number of color channels in an image, such as red, green, and blue channels. A CNN is based on several different layers of three main types, including convolutional layers, pooling layers, and fully connected layers. As with other supervised methods, CNNs are trained via an iterative optimization procedure to minimize differences between a network's output and a known ground-truth label for each input (Ubbens and Stavness 2017).

Therefore, this is an opportunity to capitalize on this DPP platform, and perhaps others that might become available to pursue efforts in establishing a more robust and high-throughput phenomics system for apple (Pieruschka and Schurr 2019).

18.7.2 Optogenetics

Ochoa-Fernandez et al. (2020) have pursued optogenetics, a genetic approach for controlling cellular processes with light, as this approach provides spatiotemporal, quantitative, and reversible control over biological signaling and metabolic processes, thereby overcoming limitations of chemically inducible systems. However, optogenetics lags in plant studies as ambient light required for growth leads to an undesirable system activation. However, this problem has been overcome by developing plant usable light-switch elements (PULSE), an optogenetic tool for reversibly controlling gene expression in plants under ambient light. The PULSE system combines a blue light-regulated repressor with a red light-inducible switch. This system is comprised of two engineered photoreceptors exerting combinatorial activities over the regulation of transcription initiation. One of the photoreceptors represses gene expression under blue light (B_{Off}), engineered from the EL222 (a 222 amino acid protein isolated from the marine bacterium *Erythrobacter litoralis* HTCC2594) photoreceptor, while the second photoreceptor activates gene expression with red light (R_{On}), based on a phytochrome B (PhyB)–PIF6 optoswitch. Therefore, gene expression is only activated under red light, and it remains inactive under white light or in darkness.

Using a quantitative mathematical model, Ochoa-Fernandez et al. (2020) have characterized PULSE in protoplasts and achieved high induction rates, and they have combined it with CRISPR–Cas9-based technologies to target synthetic signaling and developmental pathways. This approach has been used to control light-inducible immune responses in leaves of *Nicotiana benthamiana* as well as in leaves of transgenic *Arabidopsis* plants. PULSE offers new broad experimental opportunities in plant research and biotechnology. As PULSE provides opportunities for quantitative and spatiotemporal reversible control over gene expression, this can result in high induction rates of gene expression (of up to $\sim$400-fold), while this system is in the 'off' state under white light or in the dark. The developed mathematical model allows for quantitative characterization of the dynamic behavior of the system, and to guide experimental design. Therefore, by combining PULSE with a plant transcription factor (TF) or a CRISPR–Cas9-derived gene activator, this has allowed for demonstrating the functionality of light-controlled activation of both *Arabidopsis* and orthogonal promoters (Ochoa-Fernandez et al. 2020).

Thus, optogenetics offers a possible new opportunity for capitalizing on light activation to modulating gene expression, as well as for targeting synthetic signaling and developmental pathways in apple genetic enhancement efforts.

18.7.3 Nanotechnology

The tools of nanotechnology, including nanomaterials, nanofluidics, microfluidics, and nanosensors offer various opportunities for use in the apple industry whether it is for the purposes of diagnostics or for treatment of biotic or abiotic stresses via delivery of targeted biocontrol agents or other treatments.

It is likely that nanoproteomics could serve as an opportunity for use in apple to enhance probing and evaluation of protein systems. Protein biosensing changes the morphology of a surface from plain to a nanostructured form; moreover, this alters the sensing properties of materials (Kobeissy et al. 2014). This is achieved by increasing available binding sites, as well as by enhancing the accessibility of a target to that of a surface-immobilized probe. This would allow for the detection of specific target(s) with higher sensitivity and faster kinetics (Kobeissy et al. 2014).

In other nanotechnology efforts, Kim et al. (2012) used microencapsulation, uniform core–shell alginate microcapsules of 60–300 μm in diameter, and controlled release of the biocontrol agent *Pantoea agglomerans* strain E325 (E325), an antagonist to the bacterial pathogen *Erwinia amylovora,* causal pathogen of fire blight disease in apple. It was observed that proliferation of the biocontrol E325 agent within these microcapsules was also followed by their subsequent release and colonization activities within apple flowers. This study demonstrated a 'proof of concept' of the use of such a microencapsulation protocol for control of fire blight disease, and these highly promising findings should be pursued further.

18.8 Integrating Omics Datasets

Availability and access to large-scale omics datasets for genomics, transcriptomics, proteomics, metabolomics, metagenomics, and phenomics, among others, require serious efforts to handle these datasets in order to convert this vast volume of data into holistic knowledge and understanding of biological processes (Fig. 18.2; Misra et al. 2019). Currently, these datasets from these different omics are handled separately using various bioinformatics software packages. However, with reduced time and cost to generate these datasets, integration of omics data has become necessary, thereby offering both opportunities and challenges for biologists, computational biologists, and biostatisticians. Data file sizes ranging from tera- to peta-byte sized data files generated on a daily basis, together with differences in nomenclature among these data types, render integration of these multi-dimensional omics data into biologically meaningful context challenging (Misra et al. 2019). These challenges include differences in data cleaning, normalization, biomolecule identification, data dimensionality reduction, biological contextualization, statistical validation, data storage and handling, sharing, and data archiving (Misra et al. 2019). Commonly used approaches are currently limited by the three '**i**'s, consisting of **i**ntegration, **i**nterpretation, and **i**nsights. Post integration, these very large datasets aim to yield unprecedented information of cellular systems at high resolution that will offer transformative insights into processes, activities, and traits of interest through various computational and informatics frameworks (Fig. 18.2).

A trans-omics concept of dynamic networks has been proposed for integrating omics data. This concept relies on the three most commonly used layers of omics datasets, including transcriptomics, proteomics, and metabolomics, as well as of other newer datasets such as protein–protein interactions, DNA–protein interactions, and allosteric regulation, that allow for identifying critical components of dynamic biological networks (Yugi et al. 2016). This concept has been demonstrated by using three case studies in datasets from bacteria and rats, revealing an interplay of omics layers, and introducing phenome-wide association, pathway-wide association, and trans-ome-wide association (Trans-OWAS) studies to connect phenotypes with omics networks that reflect genetic and environmental factors (Misra et al. 2019).

When dealing with individual omics datasets, there are certainly challenges, particularly when

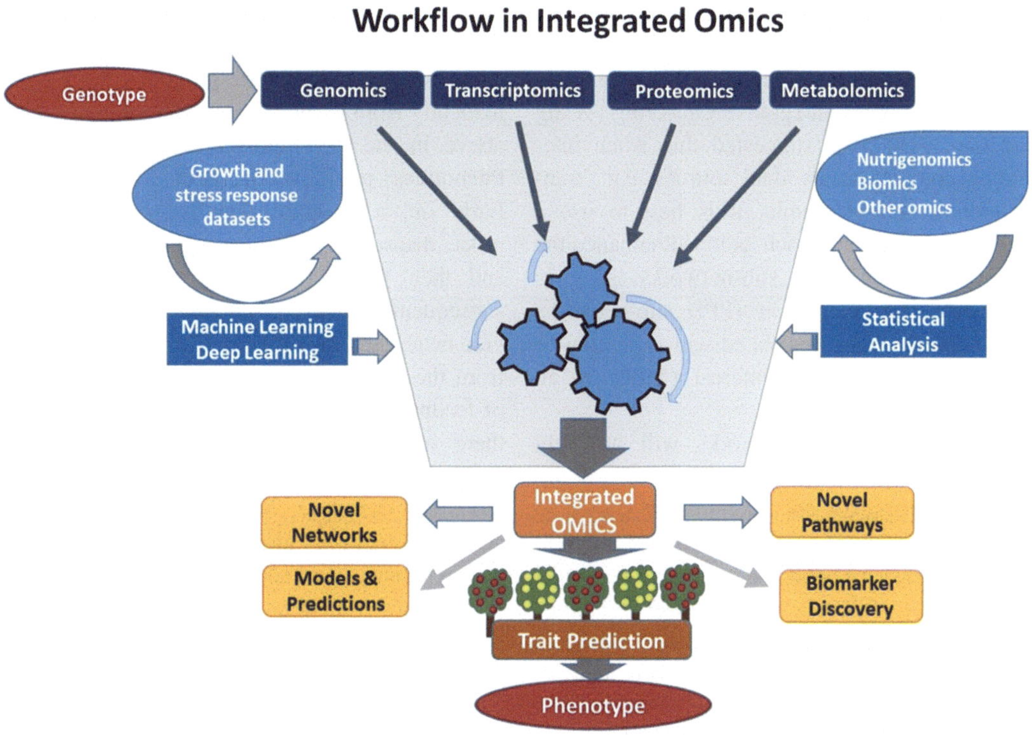

Fig. 18.2 An integrated omics workflow outlining input datasets, output datasets, and outcomes. This capitalizes on omics datasets of genomics, transcriptomics, proteomics, and metabolomics, as well as those of other omics platforms, to integrate these data using statistical and/or advanced machine learning tools. Likely outcomes are either simple pathways or complex networks. This figure is adapted and modified from Misra et al. (2019)

handling large sample numbers, even if they do not have the four 'v's (volume, variety, velocity, and veracity) associated with the integration of 'big data'. Moreover, for high-dimensional datasets of more than 1,000 variables, often referred to as the 'curse of dimensionality', variances among samples are expected to be large and sparse, thus rendering cluster analysis uninformative, and further posing challenges in interpreting integrated omics datasets (Misra et al. 2019; Ronan et al. 2016).

It is known that simplistic, descriptive, and exploratory approaches such as multivariate analysis tools such as principal component analysis (PCA) are often used to reduce data dimensionality, while canonical correlation analysis (CCA) is used to investigate overall correlations between two sets of variables. However, other omics integrative frameworks involve sparse CCA (Parkhomenko et al. 2009), multiple factor analysis (De Tayrac et al. 2009), and multivariate partial least square regression analysis (Palermo et al. 2009). Wanichthanarak et al. (2015) have identified several available tools and packages for integration of genomic, proteomic, and metabolomic datasets using pathway enrichment, biological network, or empirical correlation analysis. Although most of these tools require either standard R-statistical programming, Python, or Galaxy, pursuing this has been deemed rather difficult, particularly when there is high heterogeneity among samples. Therefore, there is a definite need for user-friendly tools (Misra et al. 2019).

It has been proposed that using a fuzzy logic modeling framework will help in efforts to integrate multiple types of omics data with expert-curated biological rules (Pavel et al. 2016; Xu et al. 2008). Furthermore, in order to overcome issues of sample heterogeneity, it has been proposed that the following three clustering categories are taken into consideration, including direct integrative clustering, clustering of clusters, and regulatory integrative clustering (Wang and Gu 2016). It is suggested that when integrating complementary data sources, e.g., transcriptomic and proteomic, it is best to use a 'proteomics-first' approach as it will enhance the discovery of candidate sub-networks, such as protein–protein interaction (PPI) sub-networks that are functionally associated with a phenotype of a biological trait of interest (Nibbe et al. 2010).

Ideally, generating networks will not rely predominantly on known function(s) of a molecule as many genes and proteins are shown to have different activities and functions in different biological systems, and the target system may include key molecules with novel functions and/or novel molecules. Moreover, although weighted gene co-expression network analysis (WGCNA) is heavily used for unbiased integration of genomic and transcriptomic data with quantitative trait data to identify coordinated modules of genes and gene variants associated with changes in phenotypic variation, such an algorithm is yet to be found useful in integrating other omics datasets from diverse analytical platforms, such as proteomics or metabolomics, or other heterogeneous data (Misra et al. 2019). Fortunately, current available tools for integration of omics data include web-based tools that require no computational experience, such as Paintomics, 3Omics, and Galaxy (P, M), as well as more versatile tools for those with computational experience. For those users with expertise in programming and interfacing with computational tools, there are available tools such as IntegrOmics, SteinerNet, Omics Integrator, and MixOmics that allow for customization of various parameters and settings that afford for better control of data analyses (Misra et al. 2019).

Artificial intelligence (AI), specifically machine learning (ML) algorithms, is currently being explored in multi-omics data analysis due to capabilities of making decisive interpretations of large-sized complex data (Biswas and Chakrabarti 2020). Analysis and interpretation of multi-omics data require collaborative efforts of biologists and computer scientists. ML deals with computer programs where programs learn automatically from earlier experiences, as mentioned above in the development of the Deep Plant Phenomics platform (in Sect. 18.7.1 above). Early on, a program initially performs some tasks, measures performance, gains experience, and then learns from these experiences, and subsequently undertakes remaining tasks to provide better performance. As ML algorithms learn from these experiences, depending on the types of feedbacks available from earlier experiences, there are three types of learning, including unsupervised, reinforcement, and supervised learning. Therefore, ML tools offer advanced analytical methods along with predictive capabilities (Biswas and Chakrabati 2020).

Different combinations of auto-encoders have been used to study the most effective approach for multi-omics data integration in human diseases, such as breast cancer (Simidjievski et al. 2019). Furthermore, a multiple-kernel framework is also used to integrate multi-omics datasets and to find closeness between subtypes of disease traits (Simidjievski et al. 2019). These kernels are ML methods wherein a function, termed as kernel function, maps non-linear data sets into a higher dimensional space that will render such data linearly separable, thus allowing for their classification (Hofmann et al. 2008).

18.9 Conclusions

While significant advances have been made in apple genomics, transcriptomics, metabolomics, and proteomics, as well as in ongoing efforts in gene and genome editing, there are yet many other opportunities for fully exploiting the ever-expanding omics technologies, as well as other new technologies for genetic improvement of apple, both scions and rootstocks, as well as that of the overall apple biome, particularly in light of the impact of climate change. Furthermore, these various tools and technologies are critical for monitoring, controlling, and ameliorating biotic and abiotic stress conditions, including postharvest storage issues in the apple industry. Peace et al. (2019) have proposed that studies should expand the concept of a single genome for apple toward that of a pan-genome concept to account for the genomic variation of the entire domesticated apple, wherein this variation is captured, annotated, and cataloged, as well as account for intraspecific sequence differences, and associate such differences with phenotypic observations. Therefore, in light of the current and new technologies described above, it is important to establish an 'Apple Atlas Dataset'. Such an 'Apple Atlas Dataset' will allow us to better understand genetic relationships, construct detailed networks and pathways for biological relationships, identify biomarkers and effectors, identify targets and variants for economic traits of interests, as well as pursue efficient and effective genetic enhancement/breeding efforts.

References

Alberti A, Zielinski AAF, Couto M, Judacewski P, Nogueira A (2017) Distribution of phenolic compounds and antioxidant capacity in apples tissues during ripening. J Food Sci Technol 54:1511–1518

Amyotte B, Bowen AJ, Banks T, Rajcan I, Somers DJ (2017) Mapping the sensory perception of apple using descriptive sensory evaluation in a genome wide association study. PLoS One 12:e0171710. https://doi.org/10.1371/journal.pone.0171710

Ballester AR, Norelli J, Burchard E, Abdelfattah A, Levin E, González-Candelas L, Droby S, Wisniewski M (2017) Transcriptomic response of resistant (PI613981-*Malus sieversii*) and susceptible ('Royal Gala') genotypes of apple to blue mold (*Penicillium expansum*) infection. Front Plant Sci 8:1981. https://doi.org/10.3389/fpls.2017.01981

Ben Sadok I, Tiecher A, Galvez-Lopez D, Lahaye M, Lasserre-Zuber P, Bruneau M, Hanteville S, Robic R, Cournol R, Laurens F (2015) Apple fruit texture QTLs: year and cold storage effects on sensory and instrumental traits. Tree Genet Genomes 11:119. https://doi.org/10.1007/s11295-015-0947-x

Bianco L, Cestaro A, Sargent DJ, Banchi E, Derdak S, Di Guardo M, Salvi S, Jansen J, Viola R, Gut I, Laurens F, Chagné D, Velasco R, van de Weg E, Troggio M (2014) Development and validation of a 20K single nucleotide polymorphism (SNP) whole genome genotyping array for apple (*Malus* × *domestica* Borkh.). PLoS One 9(10):e110377. https://doi.org/10.1371/journal.pone.0110377.

Bianco L, Cestaro A, Linsmith G, Muranty H, Denancé C, Théron A, Poncet C, Micheletti D, Kerschbamer E, Di Pierro EA, Larger S, Pindo M, Van de Weg E, Davassi A, Laurens F, Velasco R, Durel C-E, Troggio M (2016) Development and validation of the Axiom® Apple 480K SNP genotyping array. Plant J 86:62–74. https://doi.org/10.1111/tpj.13145

Biswas N, Chakrabarti S (2020) Artificial Intelligence (AI)-based systems biology approaches in multi-omics data analysis of cancer. Front Oncol 10:588221. https://doi.org/10.3389/fonc.2020.588221

Boyer J, Liu RH (2004) Apple phytochemicals and their health benefits. Nutr J 3:5. https://doi.org/10.1186/1475-2891-3-5

Busatto N, Matsumoto D, Tadiello A, Vrhovsek U, Costa F (2019) Multifaceted analyses disclose the role of fruit size and skin-russeting in the accumulation pattern of phenolic compounds in apple. PLoS One 14(7):e0219354. https://doi.org/10.1371/journal.pone.0219354

Chagné D, Crowhurst RN, Troggio M, Davey MW, Gilmore B, Lawley C, Vanderzande S, Hellens RP, Kumar S, Cestaro A, Velasco R, Main D, Rees JD, Iezzoni A, Mockler T, Wilhelm L, Van de Weg E, Gardiner SE, Bassil N, Peace C (2012) Genome-wide SNP detection, validation, and development of an 8K SNP array for apple. PLoS One 7:e31745

Chagné D, Krieger C, Rassam M, Sullivan M, Fraser J, André C, Pindo M, Troggio M, Gardiner SE, Henry RA, Allan AC, McGhie TK, Laing WA (2012) QTL and candidate gene mapping for polyphenolic composition in apple fruit. BMC Plant Biol 12:12. https://doi.org/10.1186/1471-2229-12-12

Chagné D, Lin-Wang K, Espley RV, Volz RK, How NM, Rouse S, Brendolise C, Carlisle CM, Kumar S, De Silva N, Micheletti D, McGhie T, Crowhurst RN, Storey RD, Velasco R, Hellens RP, Gardiner SE, Allan AC (2013) An ancient duplication of apple MYB transcription factors is responsible for novel red fruit-flesh phenotypes. Plant Physiol 161:225. https://doi.org/10.1104/pp.112.206771

Charrier A, Vergne E, Dousset N, Richer A, Petiteau A Chevreau E (2019) Efficient targeted mutagenesis in apple and first time edition of pear using the CRISPR-Cas9 system. Front Plant Sci 10:40. https://doi.org/10.3389/fpls.2019.00040.

Corollaro ML, Endrizzi I, Bertolini A, Aprea E, Dematte L, Costa F, Biasioli F, Gasperi F (2013) Sensory profiling of apple: methodological aspects, cultivar characterisation and postharvest changes. Postharvest Biol Tec 77:111–120. https://doi.org/10.1016/j.postharvbio.2012.10.010

Cuthbertson D, Andrews PK, Reganold JP, Davies NM, Lange BM (2012) Utility of metabolomics toward assessing the metabolic basis of quality traits in apple fruit with an emphasis on antioxidants. J Agric Food Chem 60:8552–8560

Daccord N, Celton JM, Linsmith G, Becker C, Choisne N, Schijlen E, van de Geest H, Bianco L, Micheletti D, Velasco R, Di Pierro EA, Gouzy J, Rees DJG, Guérif

P, Muranty H, Durel CE, Laurens F, Lespinasse Y, Gaillard S, Aubourg S, Quesneville H, Weigel D, van de Weg E, Troggio M, Bucher E (2017) High-quality de novo assembly of the apple genome and methylome dynamics of early fruit development. Nat Genet 49:1099–1106

De Tayrac M, Lê S, Aubry M, Mosser J, Husson F (2009) Simultaneous analysis of distinct Omics datasets with integration of biological knowledge: multiple Factor Analysis approach. BMC Genomics 10:32. https://doi.org/10.1186/1471-2164-10-32

Drogoudi PD, Michailidis Z, Pantelidis G (2008) Peel and flesh antioxidant content and harvest quality characteristics of seven apple cultivars. Sci Hortic 115:149–153

Espley RV, Brendolise C, Chagne D, Kutty-Amma S, Green S, Volz R, Putterill J, Schouten HJ, Gardiner SE, Hellens RP, Allan AC (2009) Multiple repeats of a promoter segment causes transcription factor autoregulation in red apples. Plant Cell 21:168–183

Evans K, Peace C (2017) Advances in marker-assisted breeding of apples. In: Evans K (ed) Achieving sustainable cultivation of apples. Burleigh Dodds Science Publish, pp 165–191. https://doi.org/10.19103/AS.2016.0017.09

Fahlgren N, Gehan MA, Baxter I (2015) Lights, camera, action: high-throughput plant phenotyping is ready for a close-up. Curr Opin Plant Biol 24:93–99. https://doi.org/10.1016/j.pbi.2015.02.006

Flachowsky H, Hanke M-V, Peil A, Strauss SH, Fladung M (2009) A review of transgenic approaches to accelerate breeding of woody plants. Plant Breed 128:217–226. https://doi.org/10.1111/j.1439-0523.2008.01591.x

Fu J, Swertz MA, Keurentjes JJB, Jansen RC (2007) MetaNetwork: a computational protocol for the genetic study of metabolic networks. Nat Protoc 2:685–694

Hansen L, Dragsted LO, Olsen A, Christensen J, Tjønneland A, Schmidt EB, Overvad K (2010) Fruit and vegetable intake and risk of acute coronary syndrome. Br J Nutr 104:248–255

Henry-Kirk RA, McGhie TK, Andre CM, Hellens RP, Allan AC (2012) Transcriptional analysis of apple fruit proanthocyanidin biosynthesis. J Exp Bot 15:5437–5450

Hodgson JM, Prince RL, Woodman RJ, Bondonno CP, Ivey KL, Bondonno N, Rimm EB, Ward NC, Croft KD, Lewis JR (2016) Apple intake is inversely associated with all-cause and disease-specific mortality in elderly women. Br J Nutr 115:860–867

Hofmann T, Schölkopf B, Smola AJ (2008) Kernel methods in machine learning. Ann Stat 36:1171–1220. https://doi.org/10.1214/009053607000000677

Horton RH, Lucassen AM (2019) Recent developments in genetic/genomic medicine. Clin Sci 133(5):697–708

Ichwan M, Walker TL, Nicola Z, Ludwig-Müller J, Böttcher C, Overall RW, Adusumilli VS, Bulut M, Sykes AM, Hübner N, Ramirez-Rodriguez G, Ortiz-López L, Lugo-Hernández EA, Kempermann G (2021) Apple peel and flesh contain pro-neurogenic compounds. Stem Cell Rep. https://doi.org/10.1016/j.stemcr.2021.01.005

Jain S, Shukla S, Yang C, Zhang M, Fatma Z, Lingamaneni M, Abesteh S, Lane ST, Xiong X, Wang Y, Schroeder CM, Selvin PR (2021) Zhao H (2021) TALEN outperforms Cas9 in editing heterochromatin target sites. Nat Commun 12:606. https://doi.org/10.1038/s41467-020-20672-5

Jedrychowski W, Maugeri U (2009) An apple a day may hold colorectal cancer at bay: recent evidence from a case-control study. Rev Environ Health 24:59–74

Jia D, Shen F, Wang Y, Wu T, Xu X, Zhang X, Han Z (2018) Apple fruit acidity is genetically diversified by natural variations in three hierarchical epistatic genes: *MdSAUR37*, *MdPP2CH* and *MdALMTII*. Plant J 95:427–443. https://doi.org/10.1111/tpj.13957

Knecht AC, Campbell MT, Caprez A, Swanson DR, Walia H (2016) Image Harvest: an open-source platform for high-throughput plant image processing and analysis. J Exp Bot 67:3587–3599. https://doi.org/10.1093/jxb/erw176

Kobeissy FH, Gulbakan B, Alawieh A, Karam P, Zhang Z, Guingab-Cagmat JD, Mondello S, Tan W, Anagli J, Wang K (2014) Post-genomics nanotechnology is gaining momentum: nanoproteomics and applications in life sciences. OMICS 18(2):111–131. https://doi.org/10.1089/omi.2013.0074

Khan MA, Korban SS (2012) Association mapping in forest trees and fruit crops. J Exp Bot 63:4045–4060. https://doi.org/10.1093/jxb/ers105

Khan SA, Chibon PY, de Vos RC, Schipper BA, Walraven E, Beekwilder J, van Dijk T, Finkers R, Visser RG, van de Weg EW, Bovy A, Cestaro A, Velasco R, Jacobsen E, Schouten HJ (2012) Genetic analysis of metabolites in apple fruits indicates an mQTL hotspot for phenolic compounds on linkage group 16. J Exp Bot 63:2895–2908

Kim I-Y, Pusey PL, Zhao Y, Korban SS, Choi H, Kim K (2012) Controlled release of *Pantoea agglomerans* E325 for biocontrol of fire blight disease of apple. J Control Release 161:109–115. https://doi.org/10.1016/j.jconrel.2012.03.028

Koutsos A, Riccadonna S, Ulaszewska MM, Franceschi P, Trošt K, Galvin A, Braune T, Fava F, Perenzoni D, Mattivi F, Tuohy KM, Lovegrove JA (2020) Two apples a day lower serum cholesterol and improve cardiometabolic biomarkers in mildly hypercholesterolemic adults: a randomized, controlled, crossover trial. Am J Clin Nutr 111:307–318. https://doi.org/10.1093/ajcn/nqz282

Koutsos A, Tuohy KM, Lovegrove JA (2015) Apples and cardiovascular health—is the gut microbiota a core consideration? Nutrients 7:3959–3998

Kschonsek J, Wolfram T, Stöckl A, Böhm V (2018) Polyphenolic compounds analysis of old and new apple cultivars and contribution of polyphenolic profile to the in vitro antioxidant capacity. Antioxidants 7:20. https://doi.org/10.3390/antiox7010020

Kumar S, Garrick DG, Bink MC, Whitworth C, Chagne D, Volz RK (2013) Novel genomic approaches unravel genetic architecture of complex traits in apple. BMC Genomics 14:393. https://doi.org/10.1186/1471-2164-14-393

Kumar S, Raulier P, Chagne D, Withworth C (2014) Molecular-level and trait-level differentiation between the cultivated apple (*Malus* x *domestica* Borkh.) and its main progenitor *Malus sieversii*. Plant Genet Resour 12:330–340. https://doi.org/10.1017/S1479262114000136

Legay S, Guerriero G, Deleruelle A, Lateur M, Evers D, André CM, Hausman J-F (2015) Apple russeting as seen through the RNA-seq lens: strong alterations in the exocarp cell wall. Plant Mol Biol 88:21–40. https://doi.org/10.1007/s11103-015-0303-4

Legay S, Cocco E, André CM, Guignard C, Hausman J-F, Guerriero G (2017) Differential lipid composition and gene expression in the semi-russeted 'Cox Orange Pippin' apple variety. Front Plant Sci 8:1656. https://doi.org/10.3389/fpls.2017.01656

Li Q, Sapkota M, van der Knapp E (2020) Perspectives of CRISPR/Cas-mediated *cis*-engineering in horticulture: unlocking the neglected potential for crop improvement. Hortic Res 7:36. https://doi.org/10.1038/s41438-020-0258-8

Longhi S, Moretto M, Viola R, Velasco R, Costa F (2012) Comprehensive QTL mapping survey dissects the complex fruit texture physiology in apple (*Malus* × *domestica* Borkh). J Exp Bot 63:1107–1121. https://doi.org/10.1093/jxb/err326

Malnoy M, Viola R, Jung M-H, Koo O-J, Kim S, Kim J-S, Velasco R, Kanchiswamy CN (2016) DNA-free genetically edited grapevine and apple protoplast using CRISPR/Cas9 ribonucleo proteins. Front Plant Sci 7:1904

McClure KA, Gong Y, Song J, Vinqvist-Tymchuk M, Palmer LC, Fan L, Burgher-MacLellan K, Zhang ZQ, Celton J-M, Forney CF, Migicovsky Z, Myles S (2019) Genome-wide association studies in apple reveal loci of large effect controlling apple polyphenols. Hortic Res 6:107. https://doi.org/10.1038/s41438-019-0190-y

Misra BB, Carl Langefeld C, Olivier M, Cox LA (2019) Integrated omics: tools, advances and future approaches. J Mol Endocrinol 62:R21–R45. https://doi.org/10.1530/JME-18-0055

Nibbe RK, Koyutürk M, Chance MR (2010) An integrative-omics approach to identify functional sub-networks in human colorectal cancer. PLoS Comput Biol 6:e1000639. https://doi.org/10.1371/journal.pcbi.1000639

Ochoa-Fernandez R, Abel NB, Wieland F-G, Schlegel J, Koch L-A, Miller JB, Engesser R, Giuriani G, Brandl SM, Timmer J, Weber W, Ott T, Simon R, Zurbriggen MD (2020) Optogenetic control of gene expression in plants in the presence of ambient white light. Nat Meth. https://doi.org/10.1038/s41592-020-0868-y

Palermo G, Piraino P, Zucht HD (2009) Performance of PLS regression coefficients in selecting variables for each response of a multivariate PLS for omics-type data. Adv Appl Bioinform Chem 2:57. https://doi.org/10.2147/AABC.S3619

Parkhomenko E, Tritchler D, Beyene J (2009) Sparse canonical correlation analysis with application to genomic data integration. Stat Appl Genet Mol Biol 8:Article 1. https://doi.org/10.2202/1544-6115.1406

Pavel AB, Sonkin D, Reddy A (2016) Integrative modeling of multi-omics data to identify cancer drivers and infer patient-specific gene activity. BMC Syst Biol 10:1. https://doi.org/10.1186/s12918-016-0260-9

Peace CP, Bianco L, Troggio, M, van de Weg,E, Howard NP, Cornille A, Durel C-E, Myles S, Migicovsky Z, Schaffer RJ, Costes E, Fazio G, Yamane H, van Nocker S, Gottschalk C, Costa F, Chagné D, Zhang X, Patocchi A, Gardiner SE, Hardner C, Kumar S, Laurens F, Bucher E, Main D, Jung S, Vanderzande S (2019) Apple whole genome sequences: recent advances and new prospects. Hortic Res 6:59 (2019). https://doi.org/10.1038/s41438-019-0141-7

Pieruschka R, Schurr U (2019) Plant phenotyping: past, present, and future. Plant Phenom 2019: Article ID 7507131. https://doi.org/10.34133/2019/7507131

Pompili V, Costa LD, Piazza S, Pindo M, Malnoy M (2019) Reduced fire blight susceptibility in apple cultivars using a high-efficiency CRISPR/Cas9-FLP/FRT-based gene editing system. Plant Biotechnol J 18:845–858

Ritchie MD, Holzinger ER, Li R, Pendergrass SA, Kim D (2015) Methods of integrating data to uncover genotype-phenotype interactions. Nat Rev Genet 16:85–97. https://doi.org/10.1038/nrg3868

Ronan T, Qi Z, Naegle KM (2016) Avoiding common pitfalls when clustering biological data. Sci Signal 9 (432):re6. https://doi.org/10.1126/scisignal.aad1932

Schlathölter I, Jänsch M, Flachowsky H, Broggini GAL, Hanke M-V, Patocchi A (2018) Generation of advanced fire blight-resistant apple (*Malus* × *domestica*) selections of the fifth generation within 7 years of applying the early flowering approach. Planta 247:1475–1488. https://doi.org/10.1007/s00425-018-2876-z

Schmidt C, Pacher M, Puchta H (2019) DNA break repair in plants and its application for genome engineering. Meth Mol Biol 1864:237–266. https://doi.org/10.1007/978-1-4939-8778-8_17

Simidjievski N, Bodnar C, Tariq I, Scherer P, Andres Terre H, Shams Z, Jamnik M, Liò P (2019) Variational autoencoders for cancer data integration: design principles and computational practice. Front Genet 10:1205. https://doi.org/10.3389/fgene.2019.01205

Stracke BA, Rufer CE, Weibel FP, Bub A, Watzl B (2009) Three-year comparison of the polyphenol contents and antioxidant capacities in organically and conventionally produced apples (*Malus domestica*

Borkh. cultivar ‘Golden Delicious’). J Agric Food Chem 57:4598–4605

Sun J, Rui HL (2008) Apple phytochemical extracts inhibit proliferation of estrogen-dependent and estrogen-independent human breast cancer cells through cell cycle modulation. J Agric Food Chem 56:11661–11667

Swinnen G, Goossens A, Pauwels L (2016) Lessons from domestication: targeting *cis*-regulatory elements for crop improvement. Trends Plant Sci 21:506–515

Ubbens JR, Stavness I (2017) Deep Plant Phenomics: A deep learning platform for complex plant phenotyping tasks. Front Plant Sci 8:1190. https://doi.org/10.3389/fpls.2017.01190

Vanderzande S, Howard NP, Cai L, Da Silva LC, Antanaviciute L, Bink MCAM, Kruisselbrink JW, Bassil N, Gasic K, Iezzoni A, Van de Weg E, Peace C (2019) High-quality, genome-wide SNP genotypic data for pedigreed germplasm of the diploid outbreeding species apple, peach, and sweet cherry through a common workflow. PLoS One 14:e0210928. https://doi.org/10.1371/journal.pone.0210928

Velasco R, Zharkikh A, Affourtit J, Dhingra A, Cestaro A, Kalyanaraman A, Fontana P, Bhatnagar SK, Troggio M, Pruss D, Dhingra A, Cestaro A, Kalyanaraman A, Fontana P, Bhatnagar SK, Troggio M, Pruss D, Salvi S, Pindo M, Baldi P, Castelletti S, Cavaiuolo M, Coppola G, Costa F, Cova V, Ri AD, Goremykin V, Komjanc M, Longhi S, Magnago P, Malacarne G, Malnoy M, Micheletti D, Moretto M, Perazzolli M, Si-Ammour A, Vezzulli S, Zini E, Eldredge G, Fitzgerald LM, Gutin N, Lanchbury G, Macalma T, Mitchell JT, Reid J, Wardell B, Kodira C, Chen Z, Desany B, Niazi F, Palmer M, Koepke T, Jiwan D, Schaeffer S, Krishnan V, Wu C, Chu VT, King ST, Vick J, Tao Q, Mraz A, Stormo A, Stormo K, Bogden R, Ederle D, Stella A, Vecchietti A, Kater MM, Masiero S, Lasserre P, Lespinasse Y, Allan AC, Bus V, Chagné D, Crowhurst RN, Gleave AP, Lavezzo E, Fawcett JA, Proost S, Rouzé P, Sterck L, Toppo S, Lazzari B, Hellens RP, Durel C-E, Gutin A, Bumgarner R, Gardiner SE, Skolnick M, Egholm M, Van de Peer Y, Salamini F, Viola R (2010) The genome of the domesticated apple (*Malus* × *domestica* Borkh.). Nat Genet 42:833–839

Verdu CF, Guyot S, Childebrand N, Bahut M, Celton J-M, Gaillard S, Lasserre-Zuber P, Troggio M, Guilet D, Laurens F (2014) QTL analysis and candidate gene mapping for the polyphenol content in cider apple. PLoS One 9:e107103. https://doi.org/10.1371/journal.pone.0107103

Volz RK, McGhie TK (2011) Genetic variability in apple fruit polyphenol composition in *Malus* x *domestica* and *Malus sieversii* germplasm grown in New Zealand. J Agric Food Chem 59:11509–11521

Wang D, Gu J (2016) Integrative clustering methods of multi-omics data for molecule-based cancer classifications. Quant Biol 4:58–67. https://doi.org/10.1007/s40484-016-0063-4

Wanichthanarak K, Fahrmann JF, Grapov D (2015) Genomic, proteomic, and metabolomic data integration strategies. Biomark Insigh 10:1–6. https://doi.org/10.4137/BMI.S29511

Weigl K, Wenzel S, Flachowsky H, Peil A, Hanke M-V (2015) Integration of *BPMADS4* on various linkage groups improves the utilization of the rapid cycle breeding system in apple. Plant Biotechnol J 13:246–258. https://doi.org/10.1111/pbi.12267

Wolter F, Schindele P, Puchta H (2019) Plant breeding at the speed of light: the power of CRISPR/Cas to generate directed genetic diversity at multiple sites. BMC Plant Biol 19:176. https://doi.org/10.1186/s12870-019-1775-1

Woods R, Walter H, Raven J, Wolfe R, Ireland P, Thien F, Abramson M (2003) Food and nutrient intakes and asthma risk in young adults. Am J Clin Nutr 78:414–421

Xu D, Keller JM, Popescu M, Bondugula R (2008) Applications of Fuzzy Logic in Bioinformatics, vol 9, World Scient

Yugi K, Kubota H, Hatano A, Kuroda S (2016) Trans-omics: how to reconstruct biochemical networks across multiple 'Omic' layers. Trends Biotechnol 34:276–290. https://doi.org/10.1016/j.tibtech.2015.12.013

Zhang L, Hu J, Han X, Li J, Gao Y, Richards CM, Zhang C, Tian Y, Liu G, Gul H, Wang D, Tian Y, Yang C, Meng M, Yuan G, Kang G, Wu Y, Wang K, Zhang H, Wang D, Con P (2019) A high-quality apple genome assembly reveals the association of a retrotransposon and red fruit colour. Nat Commun 10:1494. https://doi.org/10.1038/s41467-019-09518-x

Zhao C, Zhang Y, Du J, Guo X, Wen W, Gu S, Wang J, Fan J (2019) Crop phenomics: current status and perspectives. Front Plant Sci 10:714. https://doi.org/10.3389/fpls.2019.00714

Correction to: The Apple Microbiome: Structure, Function, and Manipulation for Improved Plant Health

Susan R. Whitehead, Michael E. Wisniewski,
Samir Droby, Ahmed Abdelfattah,
Shiri Freilich, and Mark Mazzola

Correction to:
Chapter 16 in: S. S. Korban (ed.), *The Apple Genome*, Compendium of Plant Genomes, https://doi.org/10.1007/978-3-030-74682-7_16

The original version of the book was inadvertently published with an incomplete figure (Fig. 16.3) in Chapter 16. The erratum chapter has been updated with the changes and the correct presentation is given here:

The updated version of this chapter can be found at https://doi.org/10.1007/978-3-030-74682-7_16

S. S. Korban (ed.), *The Apple Genome*, Compendium of Plant Genomes,
https://doi.org/10.1007/978-3-030-74682-7_19

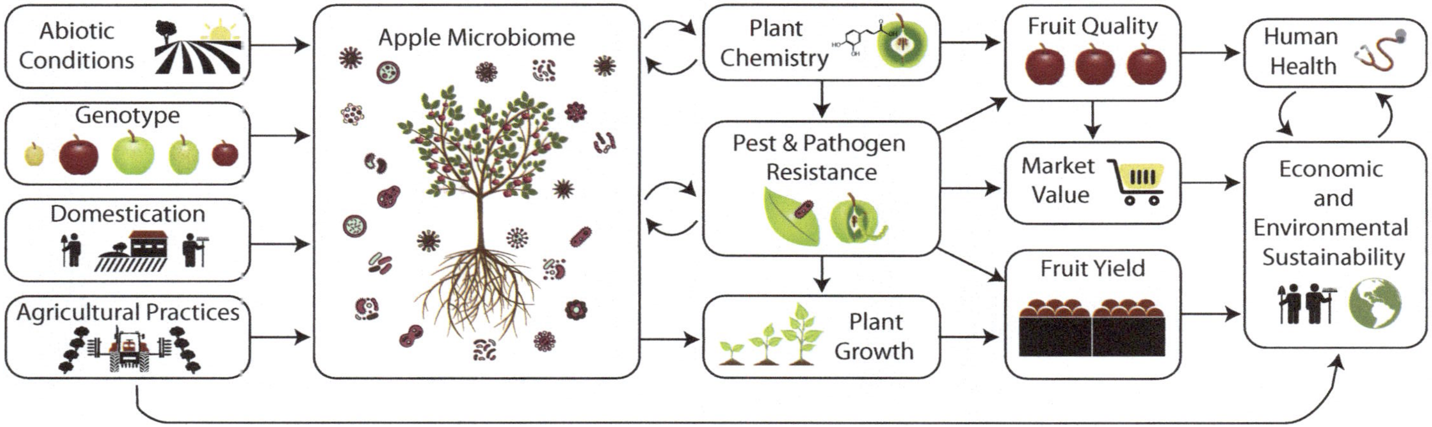

Fig. 16.3 A systems-agriculture approach for understanding the apple microbiome, including factors that influence microbiome composition and downstream consequences of microbes for plant quality, pest/pathogen resistance, yield, human health, and the sustainability of apple production

The manufacturer's authorised representative in the EU is Springer Nature Customer Service Centre GmbH, Europaplatz 3, 69115 Heidelberg, Germany. If you have any concerns regarding our products, please contact ProductSafety@springernature.com

Printed and bound by CPI Group (UK) Ltd, Croydon, CR0 4YY
30/09/2026
02225691-0019